D0744494

REFERENCE

CALIFORNIA PLACE NAMES

The Counties of Present-Day California

California Place Names

The Origin and Etymology of Current Geographical Names

ERWIN G. GUDDE

FOURTH EDITION, REVISED AND ENLARGED

by WILLIAM BRIGHT

UNIVERSITY OF CALIFORNIA PRESS

Berkeley Los Angeles London

University of California Press
Berkeley and Los Angeles, California

University of California Press, Ltd.
London, England

Printed in the United States of America
9 8 7 6 5 4 3 2 1

The paper used in this publication meets the minimum requirements of American National Standards for Information Sciences—Permanence of Paper for Printed Library Materials, ANSI Z 39.48–1984.

Library of Congress Cataloging-in-Publication Data

Gudde, Erwin Gustav, 1889–1969.
California place names : the origin and etymology of current geographical names / Erwin G. Gudde. — 4th ed., rev. and enl. / by William Bright.
p. cm.
Includes bibliographical references (p.).
ISBN 0-520-21316-5 (cloth : alk. paper)
1. Names, Geographical—California.
2. California—History, Local. I. Bright, William, 1928– .
II. Title.
F859.G79 1998
917.94'001'4—dc21 97-43168

This revised fourth edition is dedicated
by the editor to the memory
of Madison S. Beeler

CONTENTS

EDITOR'S PREFACE

Erwin G. Gudde (1889–1969) was a faculty member in the German Department at the University of California, Berkeley, when I was a student there in the 1940s and 1950s. He was a founder of the American Name Society and the founding editor of the society's journal, *Names* (1953–56). He described his own career in the autobiographical memoir "Vita Nostra Brevis Est," published in *Names* 7:1–11 (1959).

I never knew Gudde, but I became acquainted with the first edition of his *California Place Names* shortly after it was published in 1949; I was then a graduate student in linguistics at Berkeley, specializing in California Indian languages. My interest in California place names, particularly those of Native American origin, had been stimulated by Madison S. Beeler (1910–89), to whom this revised edition is dedicated. Beeler was a colleague of Gudde's in the German Department, a specialist in Germanic and Indo-European historical linguistics, and like Gudde a scholar of onomastics; he became Gudde's successor as the editor of *Names*. In the middle of his career, Beeler undertook important philological research and fieldwork on California Indian languages, especially Chumash, and later moved from the German Department to the Linguistics Department. In 1945, he had taught me freshman German; in the 1950s, he taught me Gothic and introduced me to onomastics. We became good friends as well as fellow researchers in California Indian linguistics. After Gudde's death in 1969, Beeler and I talked about working together on a revised edition of *California Place Names*. That plan was never realized; but as I now complete my work on this revised fourth edition of Gudde's book, I am more than ever aware of how much I learned from Beeler and how much inspiration I owe to him.

To make clear the general motivation and organization of Gudde's work and to preserve the acknowledgments that he made to his sources of information, I will quote at length from the introductory material that appears in the the first three editions of his work. First, in this Preface, I give extended excerpts from the prefatory notes that Gudde wrote for his second edition (1960) and third edition (1969). After that, in the Introduction, I give extended excerpts from the introductory remarks that Gudde wrote for his first edition (1949; reprinted in the later editions). Ellipsis (. . .) indicates deletion of passages that have become irrelevant in the context of the present revision. At some points, I have inserted bracketed comments to provide updated information.

In the prefatory note to the second edition, dated by Gudde on April 6, 1958, he wrote the following:

Some friendly critics, including the late Joseph Henry Jackson, literary editor of the San Francisco *Chronicle*, have called this book the definitive work on California geographical nomenclature. Actually, the word "definitive" cannot properly be applied to any historical account; it is especially unsuited when used for the history of the geographical nomenclature of a commonwealth which is still growing at a rapid pace.

An ideal etymological and geographical dictionary would include all place names of the State, either in individual entries or in summarizations of folk, group, or cluster names. Such an ideal work would necessarily have to be based on a geographical gazetteer of the entire State. About ten years ago, Olaf Jenkins, then chief of the California Division of Mines, and I did some preliminary work toward creating such a gazetteer, using the atlas sheets of the United States Geological Survey as a basis. We soon came to the realization that only a large subsidy would make such an undertaking feasible. [There is still no such gazetteer. Approximations are available, however, in the *Geographical Names Information System* (GNIS), available on CD-ROM from the Geographical Names Branch of the U.S. Geological Survey; and in the published *Omni Gazetteer of the United States of America* (Detroit: Omnigraphics, 1991), which incorporates material from the GNIS. These resources have been consulted in the preparation of the present revised edition.]

Until such a geographical gazetteer is published, no "definitive" dictionary of California geographical names can be produced. However, I believe that there is some justification in calling this book "basic." It contains most names in current use that seem to me to be of any importance or interest, and it gives complete or at least partial information about them. It is up to our onomatologists, historians, geographers, and philologists to fill the gaps gradually. The new edition is a step in this direction. . . .

The present edition of this dictionary cannot be definitive for another reason. There are still sources which have not been tapped—manuscripts, maps, newspaper files, forgotten books. *California Place Names* probably provides all fundamental data, but additional research will add substance to many of the items and further round out the picture. To engage in such research would be a rewarding task for county and other local historical societies. Gertrude Steger's pioneer work on Shasta County and the more recent monograph by T. S. Palmer on Death Valley and that by Paul Schulz on Lassen National Park are a good beginning. Further field work and the searching of county records, great registers, newspaper files, and so forth may uncover new facts concerning the origin and meaning of the names of many settlements and topographical features. [The present revision draws on numerous books and other publications that have appeared since 1969; these are listed in the Glossary and Bibliography at the end of this book.]

Although the population of the State of California has about doubled since World War II, the number of new and original names that have been given to places in that period is insignificant. Most of the developers of new residential districts lack the imagination and the courage to bestow upon their subdivisions an unusual name. They simply add "Estates" or "Glorietta" or "Del Rey" or "Woodlands" to an already existing name; they append "town" or "burg" or "vale" to their personal name; or they call their developments "Forest Hills," "Glenhaven," or "Brookdale," or give them some fanciful Spanish name. When a number of towns amalgamated to establish a new and sizable city in 1956, they could find no better name than Fremont, a name borne by dozens of places in our Western states. It is fortunate that the Post Office Department has laid down certain restrictions with respect to the naming of post offices and thus sometimes forces those who would give a place a trite name to find a new or distinctive one.

However, names of communities make up only a fraction of the new names in our growing state. Many of our topographic features are as yet unnamed; a hill which now seems to be too low and insignificant to have a name will be given one when economic development and denser population make it more important. The peculiar climatic conditions of our state make the creation of new reservoirs and artificial lakes necessary; these must have names. The U.S. Board on Geographical Names has begun to give individual names to hundreds of "forks," such as the West Fork of the South Fork of the North Fork of the San Joaquin River. Needless to say, the name-giving authorities show little imagination or originality. A California board on geographic names could be of great service in making future geographic nomenclature more varied and interesting by reviving Indian names, honoring deserving pioneers, and coining and creating entirely new names. [Such a state board now exists and makes recommendations to the federal authority.]

A few of our important names are still subjects of controversy—Marin and Tehama, Mount Saint Helena, and even California, our most important and most beautiful name. Many Southwestern place names—those of pre-Spanish Mexican origin, which form an extremely interesting linguistic group—are still awaiting a specialist who will establish their etymology. [Reference is particularly to names of Aztec (Nahuatl) origin, introduced from Mexico; special attention has been given to these in the present revision.] Native Indian names present a special problem. . . . In the original as well as in this revised edition of *California Place Names* I have included many . . . questionable interpretations; I leave the responsibility for them to students of California native languages. [Again, the revised fourth edition gives special attention to names originating in California Indian languages, drawing on research of linguists and anthropologists from 1950 onward.]

The main criticisms of the original edition of this work have come from dictionary etymologists. It would be a hopeless task to try to refute them. If a Spanish dictionary gives *coche* as meaning 'carriage' or *monte* as meaning 'mountain', these people will continue to insist that *Cañada de los Coches* means 'glen of the carriages', not 'hog valley', and that Monte Diablo means 'devil's mountain' and not 'devil's woods'. And nothing will convince them that neither the spelling nor the evolution of the name Putah Creek offers the slightest evidence that the term means 'whore creek'. In the well-known periodical *Atenea* (March 1951) a Spanish scholar gives ten pages to a review of my book from the point of view of a dictionary etymologist. Cameron Creek was not named for a family named Cameron, he believes; since the country was once Spanish, the name is more likely a misspelling of *camarón*, 'shrimp'—even though there was no Spanish settlement for miles round. The name Aromas cannot refer to the odor of sulphur water, because "los olores sulfuroso no son 'aromas' en español." The reviewer graciously excuses me for not connecting Putah with 'whore': "El profesor Erwin Gudde para quien la razón histórica vale mas que la razón de amor."

The comments of other critics have more justification. A British scholar speaks sarcastically of my statement that California is especially fortunate in having a rich and diversified nomenclature. Provincialisms of this kind will slip in when an author writes about a country dear to him. The nomenclature of California is actually no richer or more diversified than that of Scotland or Switzerland, or of Florida or Wisconsin. One might even be inclined to agree with Mark Twain and Walt Whitman that too many saints' names in California were applied in a purely mechanical manner and not because of any association with the natural, social, or political history of the places named. American men of letters of the nineteenth century were much more interested in applying geographical names in the West than is known. In 1850, when Frémont was still an influential person in California, Longfellow wrote to him, advising him concerning geographical names for the new country. It is unfortunate that this letter is lost.

The most common objection to the first edition was that it contained no obsolete names—a "geographical dictionary" should include vanished names as well as living ones. This criticism came from geographers and historians and even more often from natural scientists. To follow the well-meant advice to list obsolete names in the text or incorporate them in an appendix to the book was not practicable—there are too many of them. But the demand has been so insistent that I have finally decided to compile a gazetteer of vanished place names, which could serve as a supplement to this dictionary of living names. [This listing was published in the third edition. In the present revision, however, the obsolete names have been incorporated into the main listing, with cross-references to the modern names that have replaced them.]

. . . My purpose in writing the book was not only to present the etymology and meaning of the place names but to bring out in the stories of these names the whole range of California history. I was gratified to learn that this was generally recognized and appreciated by readers of the original edition.

Another great satisfaction was the fact that the thousands of readers discovered only four errors of any consequence. Marysville was placed in the wrong county; the name of a mighty peak, Mount Humphreys, was lost somewhere between my desk and the linotype machine; my friend George Schrader of the Forest Service and I fell for the story of an "old-timer" concerning the name Peanut; and one important Mission source, Father Arroyo de la Cuesta's manuscripts of the grammar and vocabularies of a number of Indian dialects, was omitted from the Bibliography. It stands to reason that the revised edition, as well as the original, contains numerous minor errors. No book can be free of them, especially one which contains thousands of names and dates, and which involves many fields of study and dozens of languages and dialects. Fortunately, the majority of the readers of the first edition realized that only positive criticism is of any value; their corrections and suggestions were gratefully received and used for this new edition. If this voluntary coöperation continues, future editions will contain fewer errors and eventually, it is hoped, all will be eliminated.

Manuscript materials which have been examined since the publication of the first edition of *California Place Names* include the George Davidson papers, which Elisabeth K. Gudde sorted in the meantime; the papers of the Whitney Survey which Francis P. Farquhar presented to the Bancroft Library; and the beautiful collection of Western maps which Carl Wheat gave to that institution. Some additional Spanish manuscripts and land-grant papers were also examined; but Dorothy Huggins had done so thorough a job for the original edition that only a few additional items were brought to light. A few new books on California history in general and a number of articles on California names have been published; the names of those from which new information was drawn for this revised edition have been added to the Bibliography.

Of great importance for the study of California geographical names was the publication of Walter N. Frickstad's *A Century of California Post Offices*. When a post office is established and given the name of the community it serves, no new light is thrown on the origin and meaning of the name, but often a name which might otherwise have disappeared is thus perpetuated. If the name of the community is the same as or similar to that of an already existing post office, or if the post office is to serve several settlements and the local pride of any of them prevents the application of the name of one of the others, a new name must be found. Readers seeking additional information about current or obsolete post-office names are referred to Frickstad's excellent compilation.

Two other publications deserve special mention—Carl I. Wheat's monumental work on Western American cartography, the first volume of which appeared just as the revision was finished, and Robert F. Heizer's editions of "California Indian Linguistic Records" in *Anthropological Records*, published by the University of California Press. [Information from vols. 2–6 of Wheat's work was incorporated in Gudde's third edition. A more recent reference book of special value is James D. Hart's *A Companion to California* (revised and expanded; Berkeley and Los Angeles: University of California Press, 1987).]

Many of the old friends of *California Place Names* have liberally contributed to the revision: J. N. Bowman, Thomas P. Brown, Emanuel Fritz, Alfred L. Kroeber, B. H. Lehman, Yakov Malkiel, S. Griswold Morley, Fred B. Rogers, George R. Stewart, Elmer G. Still, Arturo Torres-Rioseco. In the U.S. Forest Service, the regional forester and his staff as well as its supervisors were as generous in supplying information for this as for the first edition. Again, many postmasters supplied important information; the staff of the Bancroft Library gave as valuable assistance for this as for the original edition; and Dorothy H. Huggins edited this revision with great care and eliminated many of the errors which my manuscript contained. August Frugé, Lucie Dobbie, John Goetz, Philip Lilienthal, Harold Small, and other members of the staff of the University of California Press were most helpful in giving the book its final shape.

For regional information the help of two men was especially valuable. Henry Mauldin of Lakeport supplied all the added information about Lake County names; his projected place-name book and his county history, models of painstaking local research, will be published in the near future. [Mauldin died before his place-name book could be published, but his archives are in the Lake County Historical Society.] In J. D. Howard of Klamath Falls, Oregon, I had a unique informant who gave me valuable information about the names of Lava Beds National Monument which he himself had bestowed in his many years of exploring this interesting region.

There were others who supplied new information for this edition besides those given credit in the text: James de T. Abajian and Edwin H. Carpenter of the California Historical Society; Frank Asbill, a grandson of the first settler in western Mendocino County; William E. Ashton; P. A. Bailey; Robert Becker; Stanley T. Borden; Sylvia Broadbent; Lelia Crouch; Ralph H. Cross; Donald C. Cutter; Sydney K. Gally; Carroll D. Hall, Assistant Historian of the California Park Commission; R. E. Hallawell, General Manager of the Southern Pacific; W. E. Hotelling; John K. Howard; Rockwell D. Hunt; Arthur E. Hutson; Olaf Jenkins, former Chief of the Division of Mines; George James; the late Joseph LeConte; John Lyman; Loye H. Miller; E. S. Morby; David F. Myrick; Paul Paine; A. L. Paulson; Howard Randolph; Paul Schulz; A. D. Shamel; Edwin F. Smith; Kenneth Smith; Irwin Solloway, cartographer of the California State Automobile Association; Justin G. Turner; Church Willburn. Professor Fritz Kramer drew the original sketch for the historical chart.

Elisabeth K. Gudde is the co-author of this revised edition, as she was of the first.

In the prefatory note to the third edition, which Erwin Gudde and Elisabeth K. Gudde co-signed on January 20, 1969, they wrote the following:

The third edition of *California Place Names* is considerably enlarged and improved over the third printing of the second edition. Fundamentally, however, the book remains unchanged, and the important items for which special research had been undertaken are essentially the same.

The rapid growth of the population of our commonwealth has naturally been paralleled by an increased number of names in the geographical alignment of the state—towns and subdivisions, post offices, state parks and beaches, redwood memorial groves, historic landmarks, dams and reservoirs. While their number is considerable, the etymological or historical significance and importance of these new names, again, is not great. Most of these are simply taken from geographic features with established names which are already recorded in the book. Another large section of new names consists of commemorative names. Although all the men and women whose names have been put on the map of California doubtless deserve their place, this book does not include all of them. *California Place Names* is not a gazetteer which records all geographical names, regardless of their significance and importance. . . .

Previous editions of *California Place Names* have been criticized by linguists because the pronunciation was not indicated by the International Phonetic Alphabet. We have nevertheless continued to use the symbols of Webster's dictionary wherever it is necessary to indicate the pronunciation. The International Phonetic Alphabet is of great importance for the science of linguistics. It does not belong, however, in a geographical, historical, etymological dictionary. [The present revision follows Gudde's practice in using the kind of phonetic transcription found in most English dictionaries. A simplified system of this type, as used by the Random House dictionaries, is employed; see the explanation in the Key to Pronunciation.]

. . . In spite of our efforts the new edition will again include many errors and omissions. Many minor interesting names will be lacking because we could not discover their origin and meaning. On the other hand, numerous unimportant or even obsolete names of places, like railroad stations, may still be included.

In addition to the long lists of contributors and helpers given in the first and second editions, we wish to express our appreciation to many new persons and agencies who have given voluntary, or solicited, information. Among these are Robert H. Becker, L. Burr Belden, Oscar Berland, Frances E. Bishop, L. J. Bowler, W. Harlan Boyd, Allen K. Brown, Earl E. Buie, Rupert Costo, Carl S. Dentzel, John B. DeWitt, Newton B. Drury, Arda Haenszel, Brother Henry, Robert T. Ives, J. R. R. Kantor, Addison S. Keeler, Fritz Kraemer, Oscar Lewis, Zelma B. Locker, Dale L. Morgan, Harry L. Nickerson, K. H. Patterson, Rodman W. Paul, Edward T. Planer, Sascha Schmidt, Genny Schumacher, Albert Shumate, Peter Tamony, J. B. Tompkins, Grant V. Wallace, Allen W. Welts, Walt Wheelock, and others mentioned in the text. Some government bureaus, like the U.S. Forest Service, the California State Mining Bureau, the California departments of Parks and of Water Resources, as well as many individual postmasters and other government officers, supplied us with information; likewise the historical societies, the Save-the-Redwoods League, the California State Automobile Association, the Pacific Gas and Electric Company. The members of the staff of the University of California Press were again patient and efficient in producing the new edition, and the librarians at the Bancroft Library were as always helpful. We wish also to acknowledge the continued assistance of Fred D. Rogers, J. N. Bowman, and three members of the original advisory committee of *California Place Names*—George R. Stewart, Francis P. Farquhar, and S. Griswold Morley. George Stewart inspired the first edition of this book and has maintained an active interest through successive editions. We shall always remember those members of the advisory committee who are no longer with us, Aubrey Drury, Samuel T. Farquhar, and Edward W. Gifford. Sam Farquhar, who was manager of the University Press, took a special interest in the book and we are grateful for his generous assistance. The invaluable contributions of Dorothy Huggins Harding and Katherine Karpenstein, as well as the services of Henrietta Girgich, we also remember with gratitude. To Doris Sandford, our efficient secretary, we extend our thanks for her excellent work in the preparation of this revised edition.

A few points should be emphasized with regard to this revised fourth edition. First, I have tried to maintain the basic principles laid down by Gudde—as well as most of the information he provided, the value of which is undiminished. I hope, too, that his distinctive "voice" has been preserved.

Second, an effort has been made to provide expanded information on the local pronunciations of place names, employing a somewhat more user-friendly system of phonetic transcription. The whole question of providing pronunciations is controversial among the compilers of place-name dictionaries, partly because it is often difficult to fix on a single "correct" pronunciation. In California and the Southwest, this problem arises especially with regard to names of Spanish origin. The word *cañada* 'valley' is pronounced in Spanish approximately as (kän yä′ thä), but such an utterance is rarely heard among anglophone Californians. What one hears, of course, is a range of pronunciations including (kən yä′ də), (kən yad′ ə), (kə nä′ də), and (kan′ ə də); and it is the researcher's responsibility to report all of these, as has been done in this revision. None are incorrect, although they are used by people with different social backgrounds and attitudes—conservative, innovative, Anglo-oriented, or Hispano-oriented. By contrast, a conceivable pronunciation like (kə nā′ də) simply does not occur among local residents; it could only be heard from the lips of the greenest newcomer.

Third, because of my academic specialization in American Indian linguistics, I have made a special effort to provide information on names originating in Native American languages. For this purpose I have been able to draw on my own research in a few languages, but much more on the work of valued friends and colleagues—many of whom, like me, began their research careers as graduate students in the Linguistics Department at Berkeley. The transcription systems are based on those of my sources; in some cases they are practical orthographies, designed for present-day literacy programs, but in other cases they are technical transcriptions, normalized slightly here for typographical simplicity. (Long vowels have, in general, been written with double letters.) Special thanks are due for the conscientiousness and helpfulness of Catherine Callaghan, Catherine Fowler, Victor Golla, Margaret Langdon, John McLaughlin, Pamela Munro, Robert Oswalt, Alice Shepherd, and William Shipley. Nevertheless it should be noted that some Native American etymologies—especially in the long-missionized southern and central areas of California—are likely to remain forever in obscurity. The original languages are no longer spoken; the cultures have vanished; only the names remain.

Finally, I want to express my warmest gratitude to all the scholars, authors, government employees, teachers, librarians, local historians, and knowledgeable California residents who have provided new information for this revised edition. They are acknowledged by name both in the text and in the Glossary and Bibliography at the end of this book.

William Bright
Boulder, Colorado, November 1996

INTRODUCTION

As indicated in the Preface, I have chosen here to quote at length from Gudde's 1949 Introduction, since the general orientation that he provided for his readers seems to me still of the greatest value. As in the Preface, ellipsis (. . .) indicates points where I have deleted some material that is no longer relevant, and brackets mark my interpolations.

THE PROBLEM. Names belong to the oldest elements of human speech. According to some authorities, they even antedate the verbs for eat, drink, sleep, or the nouns for hand, night, or child. Even in most primitive known societies people bore names—usually descriptive of their looks or their characteristics: "the redhead," "the bear killer," "the hunchback." Primitive people also gave names to places. Identifying locations by description was equally as important when humans were merely foragers and hunters as later when they settled down to agricultural pursuits. "Where the strawberries grow," "where the river can be crossed," "where the waters meet"—such primitive place naming we can easily associate with the earliest stages of culture. Descriptive names probably still outnumber all others in geographical nomenclature, although a diversification has been brought about by names arising from incidents, superstitious beliefs, the forming of landed estates, the desire to honor persons, and other causes.

In European countries the investigation into the origin and meaning of geographical names forms a recognized and assiduously studied branch of etymology. In this as in other fields of philology the Germans have been the pioneers and the teachers, although the classical treatise on geographical names, *Words and Places*, was written by an English scholar, Isaac Taylor. The study of place names has proved of greatest importance for the investigation of the primeval periods, which are thoughtlessly called "prehistoric" because their history cannot be studied on the basis of written documents. In 1821 Wilhelm von Humboldt, brother of the better-known Alexander (for whom our Humboldt Bay is named), succeeded in outlining the demarcations, which otherwise have completely disappeared, of an early stage of the settlement of Spain and Portugal solely on the basis of the place names. In many European geographical names we find apparent remnants of languages and dialects otherwise unrecorded, living witnesses of the onetime presence of people of whose language or culture no trace remains. Teutonic and Romanic languages have long since replaced most of the Celtic dialects once spoken in all western Europe, but the names of the Rhine and the Garonne and the Thames still demonstrate what languages were once spoken on their shores. The name Olympus designated a number of high peaks in Greece and Asia Minor, and simply meant "mountain" in whatever language was spoken in that part of the world before the invading Greeks swept down from the north.

Place-name study in the United States presents a slightly different aspect. The vast majority of our names have been bestowed upon the land within the last few hundred years by explorers, settlers, surveyors, officers of the government, railroad officials. Many of them are descriptive, many are transfer names, or honor our great, or preserve the names of the early settlers; still others owe their origin to an incident or to a misunderstanding, to their pleasant sound, or to artificial coining. With the exception of the names of Indian origin they generally offer few problems to the philologist. The origin, meaning, and application of most of them can be traced. This does not mean that they are less interesting. Our names are an essential part of our living presence, and in their stories are reflected all phases of the nature of the country and of the history of the people. This has been graphically shown in George Stewart's account, *Names on the Land*.

California is especially fortunate in having a rich and diversified nomenclature. Indians who lived here before the coming of the whites, Spanish navigators from aboard their ships, European cosmographers from the narrow confines of their studies, uncouth soldiers and preaching missionaries, Russians and Chinese, French Canadians and Pennsylvania Germans, bawdy miners and hard-working surveyors, postmasters and location engineers, settlers from all states of the Union and from every European country—all have contributed to the names on the California map.

NAMING THE LAND. To include in this Dictionary a full historical account of place naming in California was deemed impractical, but the following sketch will tell briefly how and when our map became dotted with names.

. . . A great many Indian place names, although actually Indian, were not bestowed upon features by the natives but by the white conquerors. Many others arose doubtless when Spanish or Americans misinterpreted Indian words as names of places, and these were then often accepted by the Indians. The first period of European place naming extends from 1542, when Cabrillo sailed as the first navigator along the coast of what is now our state, until 1769, when the Spanish finally settled the country. In this period some of our coastal features were named by explorers and by European map makers, who used the vagueness of California as a geographic conception to indulge in fanciful nomenclature.

In 1769 the land route from Sonora and Lower California was opened up, and the Portolá and Anza expeditions named numerous places along their routes. The padres who accompanied these expeditions had a rather tedious way of applying the name of the saint on or near whose day a certain place was reached. Fortunately the soldiers often applied a name of their own, and many of these names, usually prompted by some incident, were preserved: El Trabuco because they lost a *trabuco* (blunderbuss); Carpinteria because they saw the natives build a boat; El Oso Flaco because they killed a 'lean bear.' In the colonization period which followed, most of the important physical features in Spanish-occupied territory were given names. The names of many of our cities and towns go back to missions, pueblos, and other localities named by civil or ecclesiastical authorities. The monotonous application of saints' names continued, but at the same time many of the Indian names, especially those of rancherias, were preserved. Of the greatest importance in this period were the Spanish and Mexican private land grants through whose *expedientes* and *diseños* hundreds of old names have survived to our day.

Spanish expeditions by sea after 1769, especially those of Bodega and Hezeta, also added to our geographical nomenclature, but these explorations practically ceased with the end of the eighteenth century. A number of exploring expeditions into the interior valley bestowed such important names as Sacramento and San Joaquin, Merced and Mariposa. Of the European and American expeditions, by land and sea, those of Vancouver (1792–94), Beechey (1826–27), and Frémont (1843–47) were of considerable importance for place names. The reports and maps of La Pérouse, Humboldt, the two Kotzebue expeditions, Belcher, Jedediah Smith, Wilkes, and Duflot de Mofras were instrumental in fixing and transmitting many names.

After the American occupation no attempt was made to change the Spanish and Indian names. Some were translated, some were garbled, but on the whole the U.S. Coast Survey, the U.S. Land Office, and other mapping agencies conscientiously kept the names which were current or which were found on maps and in documents. In the meantime the discovery of gold and the rapid infiltration of prospectors into regions of the state outside the old Spanish domain resulted in an entirely new class of names. Many of the Eastern states' names appeared on the map: Mississippi Bar, Virginia Creek, Michigan Bluff; the cosmopolitan composition was attested by the Yankee Hills, the Dutch Flats, the Chinese Camps, the Negro Bars. Colorful, unique, often bawdy names abounded: Henpeck City, Louseville, Petticoat Slide, Pinchemtight, Bloody Gulch, Raggedass Creek. As the settlement of the regions not frozen by the land grants progressed, there appeared the usual array of personal names for first settlers, descriptive names, and transfer names, especially those dear to the American heart like Washington, Lexington, Bunker Hill, Mohawk Valley, Arlington.

Greatly enriched was California geographical nomenclature by the Pacific Railroad Survey (1853–54), the State Geological Survey (1861–73), and the U.S. Geographical Explorations and Surveys West of the 100th Meridian (1875–79). Many of the names of physical features owe their origin or their fixation to the men of these surveys.

Place naming in the subsequent decades followed the common pattern: new communities and subdivisions were named for a promoter, after a town "back East," after a physical feature if it seemed to have some publicity value, or were given a Spanish or supposedly Spanish name, or a manufactured name without any meaning but with a pleasant sound. The coming of the railroad brought new names into existence, often fantastic and farfetched names because the location engineers were not bound by any tradition when they established and named construction camps. The post office usually left the choice with the petitioners, provided they could find a name that would not be confused with one already in use. Our government agencies—Geological Survey, Park Service, Forest Service—in mapping the state depended (and still depend) mainly on local information, and in this manner placed many interesting names on the map; often they applied a descriptive name or created a cluster name by naming various features after one original place name.

To show when, how, and by whom these names were applied, to tell their meaning, their origin and their evolution, their connection with our national history, their relation to the California landscape and the California people—this is the purpose of the book.

SCOPE AND COMPASS. This book is not a gazetteer. It does not contain all the names in the state, nor does it give the latitude and longitude, or the statistics of the places. The book is also no Baedeker which gives information to the tourist concerning our towns and physical features, our historical monuments and our spots of interest. It is restricted to information pertaining to the names and their application.

The term "place name" should be construed to mean the name of a geographic entity: a city or a railroad station, a lake or a river, a mountain range or a hill, a cape or an island. It has been used in this sense by most writers on geographical nomenclature and may be considered as definitely established. "Place name" is more convenient than "geographical name."

The names in this Dictionary have been drawn chiefly from a composite map of the State comprised of the following: the sheets of the topographical atlas of the U.S. Geological Survey as far as they were published and available on January 1, 1947; the atlas sheets made by the Corps of Engineers, U.S. Army, of quadrangles not covered by the Geological Survey, the charts of the Coast Survey, the maps of the Forest Service, and the best available county maps for other areas. [The present revised edition adds material from the *Geographical Names Index System* of the U.S. Geological Survey and from the excellent local maps published by the California State Automobile Association and the Automobile Club of Southern California.]

This composite map is the best available and the most official, but it does not present a perfect geographical picture of the state. A large number of names on the atlas sheets are "map names" created by the whim of surveyors or local enthusiasts. The oilmen of the Kettleman Hills would stare in amazement if they were asked what and where is El Hocico or La Cañada Simada or El Arroyo Degollado. A large number of obsolete names are carried on these atlas sheets: the widely known Orinda crossroads in Contra Costa County are designated on the Concord atlas sheet, issued in 1942, as Bryant—a name that disappeared from common speech a generation ago. On the other hand, many locally current names were not recorded by the mapping agencies and many features were left nameless because no generally accepted local name was obtainable at the time the map was made. In spite of my attempt to make this a book of names which are alive and current, it will contain a number of names which are perhaps known only to the man who put them on a map or into a gazetteer. It will also contain a number of names which cannot be found on any map or which differ in location or in spelling. It was not always possible to be entirely consistent in deciding for one spelling or another. In general I followed the principle that misspellings on maps which have become time-honored and locally accepted, like Coleman instead of Kolmer Valley, should not be corrected.

On the other hand, there was no reason to perpetuate obvious mistakes made by editors of topographical atlas sheets, like Hennerville instead of Hunewill. Wherever reliable local information was available, local usage took precedence over "official standing." Here, too, regional research could do much for California place names.

There are probably more than 150,000 place names in California, not counting street names in our cities. From these the most interesting and important names have been selected for individual treatment, and most of the frequently recurring names have been treated in a summary way. Since there is no absolute standard by which to judge the importance or interest of a name, the decision had to be left to the author and his advisers. As "important" are considered the names of all territorial units, such as counties and national domains; inhabited places of a sizable population, especially when connected with a post office or railroad station; the high peaks of our mountain ranges; other elevations which are landmarks or otherwise prominent; all rivers and the larger bays, creeks, and lakes; generally known valleys, canyons, islands, meadows, and gulches; prominent capes and promontories along the coast. In addition to these the book contains the names of numerous minor places and features interesting from the historical or philological point of view, or because their names hide a story of human interest. As important and interesting among folk names have been considered those of frequent recurrence, like Bear, Red, French, or names typically Californian, like Chinese, Fork, Bally. Here too a selection had to be made. Some common descriptive terms, like 'deep' or 'dark', and some derivatives from fauna and flora, like 'frog' or 'poppy', were not treated because no special significance is attached to them.

Since the book is primarily a dictionary of living names, it could not be burdened with the names of thousands of vanished mining camps, names of peaks and rivers and capes once well known, now lost or supplanted. A study of the obsolete and vanished names of the state would be extremely fascinating, but will have to be left for separate publication. Names no longer in use have been listed only if they are of historical, geographical, political, or anthropological interest, or if they honor a pioneer otherwise forgotten. The former name of many a place is mentioned under the entry of the present name.

Excluded are also the names of all streets and features within the areas of our cities, unless they are of more than local importance; also the names of railroads and highways, schools and churches, seamounts and submarine basins, except for a few which have some special geographical significance or could be listed in connection with a more important feature of the same name. A few new subdivisions are listed because they have unique names.

Included, on the other hand, are the names of Spanish and Mexican land grants. Strictly speaking, these are not "place names," but they have played so important a role in our political, legal, and economic history and have so much enriched the nomenclature of California that their inclusion seems justified. The majority of land-grant names will be found under group or folk names, or under the modern name preserving the land-grant name; a smaller number, unique or rare names, have been listed individually; some have been omitted because they are purely descriptive. There is considerable confusion in these land-grant names because they passed through a number of legal processes. They are given in this book as they are listed in Bowman's Index, that is, as he copied them from the *expedientes;* these sometimes differ from the names as they appear in the cases before the United States courts. Names of unclaimed and unconfirmed grants have been omitted.

STYLE AND METHOD OF PRESENTATION. The order of the entries is alphabetical as determined by the initial letters of the specific part: Pit River, see under Pit, but Point Arena, see Arena, Point; Mount Hilgard, see Hilgàrd, Mount; The Nipple, see Nipple, The. Descriptive adjectives are naturally considered a part of the specific term: North Bloomfield, see under N. For some places, especially towns, the generic term has become part of the specific name: Fort Bragg, see under F. In Spanish names, the preceding generic term, article, or other modifying particle has been treated as a part of the name: La Panza is listed under L, Santa Barbara under S. For frequently occurring Spanish names, however, all items are grouped together under the specific name: Los Coyotes will be found under C, Arroyo las Pozas under P. In names of land grants which are listed individually the Spanish definite article is entirely omitted, and the name is listed under the initial letter of the first noun, ordinarily the specific term. Cross-references have been given when necessary.

For practical purposes several types of entry have been used besides the simple entry with one name only. The individual names of a "cluster name" are listed in chronological order, as far as is determinable, at the beginning of the entry. A cluster name signifies a number of names which were applied to various features in the same locality after one original name (*see* Kaiser). Where a number of features were named for the same reason or after one original feature, but at different times and in different districts, the story of each name is told separately but all are grouped together, again chronologically, in one paragraph. These may be called "group names" (*see* Golden Gate). As for common names of frequent occurrence, the meaning, significance, and frequency of the name are given and added as separate entries, but in the same paragraph are those of special interest (*see* Corral). These are called "folk names" in this book—a term used by George R. Stewart.

The length of the entry naturally does not always correspond to the importance of the place or feature. A small hill with an etymologically interesting name may require many times the space given to the name of a majestic peak. Names applied in the broad daylight of history require less space than names whose origin is lost in the dusk before the coming of the white man.

For names which date back to former centuries, or went through a process of evolution, or were repeatedly changed, I have endeavored to give the first recording or mentioning of the name, variants important to the final version, and former names which lasted for some time or were officially applied.

Every attempt was made to treat uniformly all areas of the state, but a certain unevenness could not be avoided. Regions in which the names have already received special attention and territorial units whose officials supported the undertaking will naturally come closer to adequate presentation than others. A similar unevenness will be noticeable in the treatment of less important individual names. Big names for which special research was undertaken will be found treated in fairly even proportions. Many a small name, however, was included because information about it was available, whereas an equally important (or unimportant) name was omitted because the obscurity of its origin could not be penetrated.

Each entry was written as an entity. Failure to mention other interpretations or explanations of a name does not mean that I was unaware of them. Where different versions of the meaning and origin of a name seem plausible and logical or are supported by some evidence, they are naturally given, even if they do not sound entirely convincing. But notions of the believers in Indian princesses or products of dictionary etymologists are properly kept out.

The pronunciation given is that of the present day. Our place names are a part of our language, and the spoken idiom is a living thing which undergoes slight variations from generation to generation and changes perceptibly from century to century. Pronunciations, when given, represent actual usage of the present time. In other words, this dictionary records what people say, and does not presume to tell them what they should say. In frontier times the Grizzly Peaks were probably Grizzler Peaks since "grizzler" was the common designation for the animal; the name of the state must often have been pronounced Californy; but such pronunciations are not used today. The name Los Angeles had a fully Spanish pronunciation in Spanish and Mexican times, but it has long since become Americanized. Del Norte is a Spanish name, to be sure, but the people of Del Norte County say del nôrt′. There is often a marked difference in the pronunciation of a name by the older and by the younger generation—Tulare: to͞o lâr′ ē, to͞o lâr′. Names which are found in various sections of the state, like Cañada or Coyote, differ naturally according to locality. Usually, names of Spanish origin will be found less Americanized in the southern counties than elsewhere. Sometimes two pronunciations are given, the first being the one most likely to win out. Where no pronunciation is indicated, it may usually be assumed that the common American way prevails. For some names, especially of out-of-the-way places, no pronunciation could be given because no information was obtainable, or because the information from different sources was too contradictory to permit a conclusion. Extensive field work will be required to establish the definite pronunciation of many California place names. The pronunciation of the names of land grants which have not otherwise survived as place names is not indicated.

In general, the location given in brackets after the name is the county, or a well-defined area such as San Francisco Bay [or a National Park or Monument]. . . .

The spellings of Indian names are given as they are found in documents and on maps. Since the natives had no alphabet and could not spell out their names, the Spaniards and later the Americans wrote each name as it sounded to them. Spelling variants, of which there were often as many as a dozen, are given only where they have some bearing on the final evolution of the name, or where they are of etymological interest.

Older versions of Spanish names are likewise quoted as they appear in sources. Their spelling does not always correspond with present-day standards. Initial *y* was almost invariably used for *I*; *s* was usually used instead of *z*; *v* and *b*, and often *ll* and *y*, were used interchangeably. The accent mark was used sporadically in documents, never on maps; on Costansó's *Carta Reducida*, it was inserted by a later hand. [In this revised edition, accents are written in accordance with modern Spanish orthography, as a guide to modern readers, even though the accents may have been absent in the original documents.]

SUMMARY OF SOURCES. Rich manuscript material was at my disposal and was gratefully used: the compilations of the Northern California Writers' Project of the Work Projects Administration; the place-name files of the Atchison, Topeka and Santa Fe Railway System and of the Death Valley National Monument; the files of C. Hart Merriam, of Thomas B. Doyle, of Linnie Marsh Wolfe, of the State Geographic Board; the field copies of Farquhar's and Kroeber's monographs and of the first two editions of Davidson's Coast Pilot; the Bowman Index and the *diseños* of the land grants in the U.S. District Court; the George Davidson papers and correspondence; some of Charles Hoffmann's field notes; the originals or transcripts of numerous manuscripts in the Bancroft Library. These sources are described in the Glossary and Bibliography. At my disposal also were more than one thousand questionnaires containing local information gathered from rangers, surveyors, postmasters, librarians, teachers, county officials, and local historians.

In comparison with the abundance of unpublished material, the number of printed works dealing with place names is meager. The names in the larger part of the Sierra Nevada have been treated by Farquhar, the etymology of many of our Indian names is given by Kroeber, and Wagner's *Cartography* contains a wealth of information on the existing or obsolete names bestowed upon the coast by explorers before 1800. Nellie van de Grift Sanchez's *Spanish and Indian Place Names* is well known.

The only attempt to cover the entire state in a scholarly manner is Phil Townsend Hanna's *The Dictionary of California Land Names*, a broadly conceived work with much interesting information. There are other books, less comprehensive than Mr. Hanna's, which list the interesting and romantic names of the entire state: Thomas Brown, *Colorful California Names;* Laura McNary, *California Spanish and Indian Place Names;* Harry L. Wells, *California Names*, and others.

It seems strange that very few attempts have been made to investigate the names within the limits of a region. Gertrude A. Steger, however, has done an excellent piece of work with the geographical names of Shasta County, and Roscoe D. Wyatt has rendered a similar service to San Mateo County. William M. Maule has done first-class work on the names of the Carson, Walker, and Mono basins, Terry E. Stephenson on the names of Orange County, Elmer G. Still on those of the Livermore district, and Paul Parker on those of the Monterey district. [Regional studies published since 1969 and used as sources for the fourth edition include works on the Eastern Sierra, the Lake Tahoe region, Yosemite National Park, and the Sierra Nevada (from Alpine County to Walker Pass), as well as place-name books for the following counties: Alameda, Humboldt, Kern, Marin, Monterey, Riverside, San Diego, San Francisco, San Luis Obispo, San Mateo, Santa Cruz, Siskiyou, Trinity, and Ventura. Bibliographical references are given in the Glossary and Bibliography at the end of this book.]

Among printed books not directly concerned with place names the most valuable as source material are the reports and diaries of the early explorers, and after the American occupation the reports of the various government agencies. The accounts and journals of travelers and a number of the county histories, particularly the older ones, yielded good material. However, for most of our entries, the information had to be pieced together from manuscripts and maps, local information and county histories, post-office directories and railroad reports, dictionaries and gazetteers.

Maps deserve a special note because they are a most valuable source for historical geography. They do not tell, at least not directly, the meaning and origin of a geographical name, but they are all-important for the application, evolution, and perpetuation of a name. In modern times, to be sure, the names of post offices, railroad stations, and incorporated towns are official and no map maker would undertake to change them. With the publication of the charts of the Coast Survey and the sheets of the topographical atlas by the Geological Survey the name of physical features, too, become fixed, and official if confirmed by decision of the Geographic Board. But as long as the geography of the country was, so to speak, in a liquid state, the fate of a name was more or less determined by the cartographers. No matter how solemnly and ceremoniously a mountain or a cape was christened, the name would not last unless an influential map maker recorded it.

For the earlier periods those maps were quoted which because of their influence and popularity were responsible for the application, fixation, and changing of names along the coast, especially those reproduced or described by Kohl, Nordenskiöld, and Wagner. The many local *planos*, mostly unpublished, which originated after 1769 in California directly, and most of the *diseños* of the land grants, were likewise examined. For the period of transition the most important general maps were those of European or American explorers: Vancouver, La Pérouse, Humboldt, Beechey, Wilkes, Duflot de Mofras, Frémont. After the American occupation the most influential maps in fixing the nomenclature of the state were the charts of the Coast Survey, the maps of the Pacific Railroad Survey, Eddy's official map of the state, the Land Office maps, some private maps like those of Trask and Gibbes, and later those of Goddard and Bancroft. In the 1870s were published the important though uncompleted maps of the State Geological (Whitney) Survey, and later those of the U.S. Survey West of the 100th Meridian (Wheeler). In modern times it is the mapping of the U.S. Geological Survey, the Forest Service, and the War Department, together with the continued activity of the Coast Survey, which is mainly responsible for our geographical nomenclature—if not for its origin, at least for its application.

Californians can take pride in the first three editions of Gudde's *California Place Names*, as one of the most detailed and accurate of the state place-name dictionaries existing in the United States and as a priceless resource for our geographical, historical, and linguistic knowledge of the state's toponymy. I have endeavored to follow in Gudde's footsteps and to make this revised edition a worthy successor to those that preceded it.

William Bright
Boulder, Colorado, November 1996

KEY TO PRONUNCIATION

The English pronunciation of many California place names is obvious from their spelling. In other cases, however, this book indicates pronunciations phonetically within parentheses. The phonetic symbols used are those of the *Random House Dictionary of the English Language*, but similar systems are used in other familiar English dictionaries. The values of symbols are as follows:

a	as in	*cap, bad, act*
ā		*cape, save, day*
â		*care, fair, carrot*
ä		*far, father, spa*
ch		*child, church*
e		*set, red, left*
ē		*eve, east, seed*
ə		the unaccented *a* of *sofa, appear, alone*
g		*go, give, gag*
i		*it, bid, ink*
ī		*ice, side, fine*
ng		*sing, long*
o		*cot, fond, collie:* most Americans pronounce this the same as the *a* in *father*
ō		*oak, over, own*
ô		*law, dawn, caught:* many Americans pronounce this the same as the *o* in *cot* and the *a* in *father*
oi		*boil, boy, coin*
o͝o		*book, look, put*
o͞o		*boot, soon, coop*
ou		*out, loud, down*
th		*thin, thick, think*
t̶h̶		*then, bathe, rather*
u		*up, sun, love*
û		*urge, bird, term*
zh		*measure, azure*

The following additional phonetic symbols occur in languages other than English:

ʼ	represents a glottal stop, i.e., an interruption of the breath in the throat, as in *oh-oh!* or in a careful pronunciation of *the (ʼ) ice*
č	like *ch* in *church*
ḍ	like *d* with the tip of the tongue pulled back (somewhat as in *hard*)
ʰ	a puff of breath after a consonant
ï	like *oo* in *boot*, but with the lips flat instead of rounded
î	vowel with falling pitch
ł	like English *l*, but without vibration of the vocal cords; the same as Welsh *ll*
ñ	like *ny* in *canyon*
ṣ	like *s* with the tip of the tongue pulled back
š	like *sh* in *ship*
ṭ	like *t* with the tip of the tongue pulled back (somewhat as in *heart*)
θ	like *th* in *thin*

The syllable with principal accent has a "prime" mark after it; that is, a name like Mono (mō′ nō) is pronounced "MO-no."

ABBREVIATIONS

BGN	Board on Geographic Names
Co.	County
d.	died
NM	National Monument
NP	National Park
p.c.	personal communication
USFS	United States Forest Service
USGS	United States Geological Survey

CALIFORNIA PLACE NAMES

A

Abadi (ə badʹ ē, ə bäʹ dē) **Creek** [Santa Barbara, Ventura Cos.] was apparently named for the family of William Abadie, of France, who was murdered in Ventura in 1868 (Ricard).

Abalone (ab ə lōʹ nē) is the name of a large mollusk (genus *Haliotis*), valued for its meat and for its shell lined with mother-of-pearl; the term is reflected in the names of several points and coves along the California coast. It comes from the Rumsen language (Costanoan family), in which *awlun* means 'red abalone' (Harrington 1944:37). The Indian word became in Spanish *aulón, avalone,* and both versions entered English. Bayard Taylor, in 1849, gives a detailed description of the "avalones" near Point Pinos; an article in the *Sacramento Union*, Nov. 21, 1856, is written as if "Aulone Shellfish" were the generally accepted name. The Coast Survey map of 1852 has a third version in Point Alones.

Abbot, Mount [Fresno Co.], was named by the Whitney Survey in honor of Henry Abbot (1831–1927), distinguished soldier and engineer, in the 1850s a member of the Pacific Railroad Survey (Farquhar). Sometimes erroneously spelled Abbott.

Abbott Lakes [Monterey Co.]. Ai Hale Abbott settled here around 1880 (Clark 1991).

Abbotts Lagoon [Marin Co.]. Perhaps for the family of John Abbott, who was a justice of the peace for Tomales Township in 1859–60 (Stang).

Abel, Mount [Kern Co.] was probably named for Stanley Abel, a Taft businessman and Kern Co. supervisor in the 1930s (Darling 1988).

Aberdeen [Inyo Co.] was probably named after the city in Scotland (Sowaal 1985).

Abert, Mount. *See* Kaweah Peaks.

Absco [Ventura Co.], a Southern Pacific station, was coined in the 1920s from the name of the American Beet Sugar Company, which had a factory in Oxnard.

Academy [Fresno Co.] was the name given to the post office because the Methodist Episcopal Church South erected an "academy" or secondary school here in 1874. The place is shown on the Land Office map of 1879. The name has outlived the school; the academy building was razed years ago.

Acalanes (ä kə läʹ nēz) [Contra Costa Co.] takes its name from an Indian tribe of the Miwokan family, dwelling south of San Pablo and Suisun Bays; they were mentioned as Sacalanes on June 14, 1797 (PSP 14:13), and with various spellings in the baptismal records of Mission Dolores between 1794 and 1821 and in the reports of the first Kotzebue expedition, 1816. The tribe is now usually referred to as Saklan. The term *Los Sacalanes* was reinterpreted as Los Acalanes in the Acalanes land grant of Aug. 1, 1834. According to the testimony in land-grant case 276 N.D., the rancheria Sacalanes was about a mile and a half northwest of modern Lafayette. In 1854–55 a short-lived post office was spelled Acelanus.

Acampo (ə kamʹ pō) [San Joaquin Co.] was the name applied to the Southern Pacific station when the Sacramento-Stockton section was built in the 1870s. *Acampo* is Spanish for a part of the commons used for pasture, but here may reflect Eng. "camp," which is often *campo* in Californian Spanish (*see* Campo).

"A" Canyon [Death Valley NP] is a canyon that splits in a manner suggesting the letter *A*. Nearby a small canyon is called "B" Canyon.

Acelanus. *See* Acalanes.

Achedomas. *See* Algodones.

Acker Peak [Yosemite NP] was named by the USGS in 1898 for William B. Acker, who for many years was in charge of matters concerning national parks in the Dept. of the Interior (Farquhar).

Ackerson: Meadow, Creek, Mountain [Tuolumne Co.]. For James F. Ackerson, a forty-niner (Browning 1986).

Acolita (ä kō lēʹ tə) [Imperial Co.]. A railroad point apparently named with the Spanish term for 'acolyte', for unknown reasons.

Acorn [Humboldt Co.]. Not named for the fruit of the oak tree, but for Alonzo Acorn; the post office here was named for him in 1891 and was discontinued in 1904 (Turner 1993).

Acrodectes (ak rō dek´ tēz) **Peak** [Kings Canyon NP] was approved by the BGN in list 6903. Although the word resembles an Ancient Greek name, it does not exist in the ancient language. Proposed by Dr. Andrew Smatko of Santa Monica, the name refers to *Acrodectes philopagus*, a rare species of cricket found only in the High Sierra. The zoological name is coined from Greek *akros* 'peak' and *dektēs* 'biter'. (Thanks to Roger Payne of the BGN for this information.)

Acton [Los Angeles Co.] was the name applied to a station of the Saugus-Mojave section of the Southern Pacific, built 1873–76, doubtless after one of the fifteen Actons then in existence in the East.

Adair Lake [Madera Co.]. For Charles F. Adair (1874–1936), a Yosemite Park ranger from 1914. He introduced golden trout into this lake, which gave him the right to name it after himself (Browning 1986).

Adams Canyon [Ventura Co.]. William G. Adams sank an oil well here in 1872 (Ricard).

Adams Springs [Lake Co.] was named for Charles Adams, who settled here in 1869.

Adelaida (ad ə lā´ də) [San Luis Obispo Co.]. A post office and former school district, perhaps named for Adelaida Corelle, a neighbor of the first postmaster in 1977 (Hall-Patton). Locally referred to as "Adelaide."

Adelanto (ad ə lan´ tō) [San Bernardino Co.] was established as a post office in 1917 and named with the Spanish word for 'progress, advance'.

Adele [Humboldt Co.]. *See* Fields Landing.

Adin (ā´ din) [Modoc Co.] was named by residents in 1870 for Adin McDowell, a native of Kentucky, who had settled in Big Valley in 1869.

Adobe (ə dō´ bē), from Spanish, refers to a claylike soil, suitable for making bricks; or to such bricks, or a building constructed with them. The name is given to more than twenty-five features in the state. The cluster name west of Mono Lake originated with Adobe Meadows, recorded on the von Leicht–Craven map of 1874. The name and the method of making sun-dried bricks, introduced into Spain by the Arabs, became common in the southwestern United States, where soil and climate are adapted to the adobe type of structure.

Aeolian (ē ō´ lē ən) **Buttes** [Mono Co.]. Named in the early 1880s by Israel C. Russell, who wrote, "For convenience of description we may name them Aeolian Buttes" (Browning 1986).

Aeroplane Canyon [Lake Co.] was named about 1944 after an army airplane was lost south of Big Valley and the aviator bailed out in the canyon (Mauldin).

Aetna (et´ nə) **Springs** [Napa Co.] was probably named in the 1880s, when John Lawley, owner of the Aetna Mine, discovered the hot mineral spring while mining quicksilver (Len Owens). Mount Aetna in Sicily doubtless suggested the original name; *see also* Etna.

Afton [San Bernardino Co.] was named as a Union Pacific station when the railroad was constructed in 1904–5, probably after one of the many other Aftons "back East." There is another Afton in Glenn Co.

Agalia. *See* Eagle: Eagle Canyon.

Agassiz (ag´ ə sē), **Mount** [Kings Canyon NP], bears the name of Louis Agassiz (1807–73), famous Swiss-American scientist. The name Agassiz Needle was applied by L. A. Winchell in 1879, probably to what is now Mount Winchell. The peak now bearing the name Agassiz is higher than Mount Winchell, but it is not a "needle."

Agate Bay [Placer Co.], on the north shore of Lake Tahoe, was named by the Whitney Survey because of the presence of the variegated waxy quartz. The name appears on the von Leicht–Hoffmann Tahoe map of 1874.

Agate Creek [San Diego Co.] was named by Blake: "I have marked the name Agate Creek . . . as the abundance of this mineral in its bed was its distinguishing characteristic" (p. 67).

Agenda [Monterey Co.]. Perhaps transferred from Kansas, where a town is said to have been so named because "the town meeting was held and someone asked what was on the agenda" (Kelsie Harder, *Names* 40.71–76 [1992]).

Ager (ā´ gər) [Siskiyou Co.]. A stage stop around 1876; named for the founder of the town, J. B. Ager. There was a post office from 1888 to 1940 (Luecke).

Agnew [Santa Clara Co.] was named for Abram Agnew and his family, who settled in the Santa Clara Valley in 1873. The original form,

Agnew's, which referred to their seed farm, is retained in the name of Agnews State Hospital, but the shorter form was adopted for the railroad station and the post office.

Agnew: Meadow, Pass, Lake [Madera Co.]. For Theodore C. Agnew, a miner, who settled here in 1877 and acted as a guide to troops on duty in Yosemite Park (Farquhar).

Agoura (ə go͞or′ ə) [Los Angeles Co.] was first known as Picture City. When the post office was established in 1927 and the Post Office Dept. requested a single-word name, Agoura was chosen because the place was on the Agoura ranch (E. W. Hutton). The name reflects that of Pierre Agoure, a Basque rancher of the 1890s (Atkinson).

Agra (ag′ rə) [San Diego Co.] is a Santa Fe railroad point, originally a railroad cattle loading station on the Santa Margarita Rancho, perhaps named after the city in India (Stein).

Agua, Spanish for 'water', was a common generic geographical term for surface waters, mainly springs, and has survived in numerous place names. The word is now usually pronounced (ä′ gwə), but (ä′ go͞o ə) and (ä′ gwä) are also heard. This combination appeared in eleven land grants as the primary or secondary name and is preserved in some twenty places. **Agua Caliente** (kal ē en′ tē, kä lē en′ tē) [Alameda Co.], meaning 'hot water' (i.e. hot springs), appears on a *diseño* of 1835, and a land grant of that name is dated Oct. 13, 1836. An Arroyo del Agua Caliente appears on Hoffmann's map of the Bay region (1873). Recent maps show Agua Caliente Creek, but a place called Warm Springs is on Agua Fria ('cold water') Creek to the south. **Agua Caliente** [Sonoma Co.] was the name of a land grant dated May 7, 1836. The hot springs that gave the name to the grant are shown on the *diseño* of the re-grant, dated July 13, 1840. **Agua Caliente Creek** and **Cañada Caliente** [San Diego Co.] get their name from an Indian rancheria called El Agua Caliente by the priests of San Diego Mission. On June 4, 1840, the name appears in the title of the Valle de San José y Agua Caliente land grant, and it is shown on Gibbes's map of 1852. (*For* Agua Caliente in Napa Co., *see* Calistoga; in Riverside Co., *see* Palm: Palm Springs.) Less frequent is the combination with *fría* 'cold': **Agua Fria Creek** [Mariposa Co.] is named for the Agua Fria Mine, "from the circumstance of a stream of water gushing from the mountainside, which is so placed and sheltered as never to feel the heat of the sun, consequently the water is unusually cold as it springs from the rock" (Henry Huntly, 1850, quoted in *CHSQ* 12:75). The name of the town that developed around this mine was misspelled Agua Frio. In a number of other California place names, the generic term *agua* has been combined with *buena* 'good', *tibia* 'warm', *mansa* 'quiet', *mala* 'bad', *amargosa* 'bitter', and *escondida* 'hidden'; Mexican land-grant names included Agua Hedionda 'stinking water' [San Diego Co.], Aug. 10, 1842 (Stein, p. 162); Agua Puerca 'dirty water' [Santa Cruz Co.], Oct. 31, 1843 (Clark 1988); Aguas Frias [Butte and Glenn Cos.], Nov. 9, 1844; and Aguas Nieves 'snow waters' [Butte Co.], Dec. 22, 1844. The names Agua Grande ('great') Canyon [Monterey Co.] and Agua Alta ('high') Canyon [Riverside Co.] sound like American applications; the latter first appeared on a 1904 map (Gunther). *For* Agua Dulce [San Diego Co.], *see* Sweetwater River. On several occasions Latin *aqua* (*q.v.*) has been used in place of the Spanish *agua*, probably by error.

Aguaje del Centinela [Los Angeles Co.], a land grant dated Sept. 14, 1844, means 'spring of the sentinel' (*see* Centinela Creek).

Aguajito (ä gwä hē′ tō) **Canyon** [Monterey Co.] is named with the diminutive of *aguaje* 'reservoir, spring, watering place', used for the name of the Aguajito land grant on Aug. 13, 1835. Another Aguajito grant, in Santa Cruz Co., dated Nov. 20, 1837, survives on maps in the misspelled form Aguajita.

Aguanga (ə wäng′ gə): **Valley**, town [Riverside Co.], and **Mountain** [San Diego Co.] reflect a Luiseño village name, *awáanga* 'dog place', from *awáal* 'dog' plus locative *-nga*. The present Cañada Aguanga is shown as Arroyo de Ahanga on a *diseño* of the unconfirmed land grant Camajal y Palomar (1846). About 1850 the name appears in various spellings for an Indian village and a way station.

Aguereberry (ag′ ər bâr ē) **Point** [Death Valley NP] was named for "French Pete" Aguerreberry (also spelled Auguerreberry)—a Basque who, with "Shorty" Harris, opened the Harrisburg mine in 1906 and who built the original road to the point (Sowaal). The current spelling was made official by the BGN in 1966 (list 6604).

Agujas (ä go͞o′ häs), **Cañada de las** [Santa Barbara Co.], Spanish for 'valley of the needles'.

Aha Kwin (əhä′ kwin′) **Park** [Riverside Co.]. Contains Mojave *'ahá* 'water' and *aakwín-* 'to bend' (Munro).

Ahanga, Arroyo de. *See* Aguanga.

Ahart Meadow [Fresno Co.], southeast of Huntington Lake, was patented as a homestead to John Ahart about 1890 (Farquhar).

Ahjumawi (ä jo͞o mä′ wē) **Lava Springs State Park** [Shasta Co.] is named for the Indian group, locally called the Pit River tribe, who lived in the area; the spellings Achumawi and Achomawi also occur. The term is from the native name *ajumaawi* 'river people', from *ajuma* 'river', originally referring to the Fall River band of this tribe.

Ahpah (ä′ pä): **Creek, Ridge** [Humboldt Co.] take their name from Yurok *ohpo*, the name of a large flat near the creek, of unknown origin (Waterman, map 10). Also written Ah Pah.

Ahwahnee (ä wä′ nē) [Yosemite NP] is from Southern Sierra Miwok *awooni* or *owooni* 'Yosemite Valley', from *awwo* or *owwo* 'mouth' (Broadbent; Callaghan). The version Ahwahne was apparently created by Bunnell in 1859 (Hutchings, *California Magazine* 3:503). It was again taken up in the 1890s for advertising purposes, with another *e* added. **Ahwahnee** [Madera Co.] had its name changed by the BGN to Wassamma on June 3, 1908—doubtless on the suggestion of Merriam, who believed that Was-sa′-ma was the original Indian name (*Mewan stock*, p. 346); this corresponds to Southern Sierra Miwok *wasaama*, the location of the Indian roundhouse (Broadbent). However, no attention seems to have been paid to the board's decision.

Ahwiyah (ə wī′ yə) **Point** [Yosemite NP] is from Southern Sierra Miwok *awaaya* 'lake; deep' (Broadbent; Callaghan). Reference may be to Mirror Lake.

Aigare (eg ä′ rē), **Mount** [El Dorado Co.]. Said to be derived from a Basque family name (Witcher).

Airola Peak [Alpine Co.]. John and Emma Airola homesteaded 160 acres at Frogtown in the 1890s. The family has been in the cattle business since 1909 (Browning 1986).

Alabama Hills [Inyo Co.] were named by Southern sympathizers after the Confederate raider *Alabama* sank the Union man-of-war *Hatteras* off the coast of Texas on Jan. 11, 1863 (*see* Kearsarge).

Alabaster Cave [Placer Co.] was discovered in 1860 and was so named because the walls and ceilings were of limestone, a vein of which runs through the Sierra foothills (D. E. Cameron).

Alamar (al ə mär′) **Canyon** [Santa Barbara Co.]. Spanish for 'place of cottonwood trees', from *álamo* 'cottonwood' (*q.v.*).

Alambique (al əm bēk′) **Creek** [San Mateo Co.] reflects the Spanish word for a 'still', for distilling liquor, and is derived from the same Arabic stem as the English word "alembic." Máximo Martínez testified in 1852, in the Pulgas land-grant case, that he knew the creek first about 1839 and that "it took its name from a still placed there" (*WF* 6:374).

Alameda (al ə mē′ də) **Creek**, city, and **County** are named with the Spanish for 'grove of poplar or cottonwood trees' (from *álamo* 'poplar', *q.v.*), but the term also means any group of shade trees. It is common as a place name in Spain and Latin America. The general region of the southern part of the present county is called *la Alameda* in a letter of June 2, 1795 (Prov. Recs. 7:54); it may have been so named earlier. In September of the same year the stream flowing through it was called Río de San Clemente and Río de la Alameda (Arch. MSB 4:192 ff.). On Aug. 8, 1842, the word appears in the name of the land grant Arroyo de la Alameda, and Alameda Creek is mentioned in the *Statutes* of 1850. The name was chosen for the city by popular vote in 1853; the county was created in the same year from parts of Contra Costa and Santa Clara Cos. There is also an Alameda in Kern Co.

Alamilla (ä lä mē′ yə) **Spring** [Amador Co.]. Not from *álamo* 'poplar', but rather named by José María Amador in 1826, when he built his adobe house about a mile west of the spring, *a la milla* 'at the mile' (Browning 1986).

Alamitos (al ə mē′ təs), Spanish for 'little poplars [or cottonwoods]', is contained in the names of three land grants; it has survived in **Alamitos Bay** [Los Angeles Co.] and the names of a number of creeks. **Los Alamitos** [Orange Co.] was named after Rancho Los Alamitos, part of the old Nieto grant, by W. A. and J. R. Clark, who built a beet-sugar factory here in 1896.

Alamo (al′ ə mō) reflects Spanish *álamo* 'poplar, cottonwood tree' (genus *Populus*). The term was frequently used in Spanish names and has survived in a number of places; *see also* Poplar, Cottonwood. The reason for naming was not necessarily an abundance of the trees;

a single specimen sometimes gave rise to the name: "At this place [near the Salinas River] there was a poplar tree [*un álamo*] . . . within our camp, and for this reason the place was called Real del Álamo [Camp of the Poplar]" (Costansó, Sept. 27, 1769). The word Álamo appears in eight land grants or claims. There were two grants named Álamos y Agua Caliente 'poplars and hot spring' in Los Angeles Co., dated Oct. 2, 1843, and May 27, 1846. **Alamo Pintado**, a grant in Santa Barbara Co. dated Aug. 16, 1843, was so called because of a tree painted with Indian symbols. The name was formerly applied to the settlement of Ballard, near Los Olivos. The name of another grant in Santa Barbara Co., dated Mar. 9, 1839, is preserved in the name of **Los Alamos Valley** and the town **Los Alamos**. The latter was laid out and named by J. S. Bell and J. B. Shaw in 1876. Numerous were, and still are, the names in the desert regions where the Fremont cottonwood (*Populus fremontii*) formed landmarks in the treeless plains and promised water to the wanderer: Alamo Solo 'lone cottonwood', Alamo Mocho 'trimmed cottonwood', Alamo Rancho, Alamo River. **Alamo** [Contra Costa Co.] was settled in 1852 and is mentioned in the *Statutes* of 1854.

Alamorio (al ə mə rē´ ō) [Imperial Co.] is on the Alamo River; the name is coined from Spanish *álamo* 'cottonwood' plus *río* 'river'.

Alba (al´ bə) **Creek** [Santa Cruz Co.]. Perhaps for white sands through which the creek flows (Clark 1986).

Albanez (al ban´ əz) **Spring** [San Benito Co.]. For Spanish Albáñez, a family name.

Albanita Meadows [Tulare Co.]. Said to be modifed from Spanish *agua bonita* 'pretty water' (Browning 1986).

Albany [Alameda Co.], in former days jokingly known as O'Shean's View, was named Ocean View at its incorporation in 1908. The next year it was named Albany after the New York birthplace of Frank J. Roberts, the California town's first mayor. *For* Albany Hill, *see* Cerrito.

Alberhill (al bər hil´) [Riverside Co.] was coined from the names of C. H. Albers and James and George Hill, owners of the land on which the town was built about 1890 (Gunther).

Albert Gulch [San Mateo Co.]. From the ranch of the William C. Albrecht family, 1870s to 1954; the gulch was so called by about 1890. "Albert *Canyon* is of course heard in Half Moon Bay" (Alan K. Brown).

Albion (al´ bē ən) **River** and town [Mendocino Co.] bear the ancient name of Britain; the term was applied to the river, as well as to a land grant of Oct. 30, 1844, by William A. Richardson, captain of the port of San Francisco. Richardson, an Englishman, doubtless had in mind New Albion, the name bestowed on northern California by Sir Francis Drake. The town developed around the lumber mill built in 1853 and was named after the river. The post office was established June 29, 1859.

Alcatraces, Isla de. *See* Alcatraz: Alcatraz Island; Yerba Buena.

Alcatraz (al´ kə traz) **Island** [San Francisco Bay]. The name Isla de Alcatraces 'island of pelicans' was given to what is now Yerba Buena Island on Aug. 12, 1775, "on account of the abundance of those birds that were on it" (Eldredge, *Portolá*, p. 58). It thus appears on Ayala's chart of 1775 and on maps of San Francisco Bay in the following decades. Beechey, in 1826, transferred the name—as Alcatrazes Island—to the rock that later became the site of the federal penitentiary. The plural form appeared on most maps until the Coast Survey created the present version in 1851. In Santa Barbara Co., near Gaviota, Alcatraz is the name of a railroad siding and beach.

Alcove Canyon [Lake Co.] was first known as Elk Cove, because elk horns were numerous there. After the elk horns disappeared, the name evolved into Alcove. In the early 1890s, this name was applied to the school district and thus became official (Mauldin).

Alden [Alameda Co.]. Named in 1899 after the founder, pioneer farmer S. E. Alden (Mosier and Mosier).

Alder trees (genus *Alnus*) are widespread in the state (*see* Aliso). The name of the tree has been bestowed on several communities, e.g. Alderpoint [Humboldt Co.], Alder Springs [Glenn Co.], and Alder [Sacramento Co.].

Alessandro (al ə zän´ drō) [Riverside Co.] was platted in 1887 (Brigandi), named after the Indian hero in Helen Hunt Jackson's sentimental novel, *Ramona*, then at the height of its popularity. Jackson perhaps confused Alessandro, the Italian equivalent of Alexander, with the Spanish Alejandro. The site is now part of March Air Force Base.

Alexander Grove [Humboldt Co.] was created as a memorial grove in Humboldt Redwoods

State Park in 1929 and was named for Charles B. Alexander (1848–1927)—lawyer, banker, and businessman of New York and California, whose wife (the daughter of Charles Crocker, one of the Central Pacific "Big Four") had given half the property to the state.

Alexander Valley [Sonoma Co.] was named for Cyrus Alexander, a native of Pennsylvania, who came to California in 1832–33, and in 1847 acquired two leagues of the Sotoyome land grant from Henry D. Fitch.

Alger Creek and **Lakes** [Mono Co.] were named in 1909 by R. B. Marshall for John Alger, a packer for the USGS (Farquhar).

Algodones (al gə dō´ nəs) [Imperial Co.] reflects the name of a Yuman tribe that once dwelt on both sides of the Colorado; the term is not from Spanish *algodón* 'cotton'. The name was spelled Achedomas by Venegas (1759:2.185), and Jalchedunes by Garcés (1774, p. 383). The present spelling was not used until 1841 (Wilkes's map). The form Halchidhoma is used in the anthropological literature, reflecting Mojave *halchidóom* (Munro). The corresponding form in the related Quechan (Yuma) language is *xaalychadhuum*, perhaps analyzable as 'those who turned or faced far off in a different direction' (A. M. Halpern, quoted in *HNAI* 10:84).

Algoma (al gō´ mə) [Siskiyou Co.]. Also spelled Algomah. There were a post office and sawmill here in the early 1900s, but the origin of the name has not been traced (Hendryx).

Algootoon [Riverside Co.] was a name suggested by George Wharton James, in 1901, as a new term for Lakeview, a small community between San Jacinto and Perris; he derived the word from "Algoot," hero of an Indian myth, but it never became current (Brigandi). The origin is probably Luiseño *álwut* 'crow'.

Algoso (al gō´ sō) [Kern Co.] was once known as Weed Patch because weeds grew there in profusion. When the Santa Fe branch line was built in 1922, the Spanish word *algoso* 'weedy' was chosen as the name for the station, to distinguish it from another Weed Patch farther south.

Alhambra (al ham´ brə) [Los Angeles Co.] was laid out in 1874 by George Hansen for the owners, "Don Benito" Wilson and his son-in-law, J. D. Shorb. The name of the Moorish fortress in Spain had been made popular by Washington Irving's book *The Alhambra*. Since the syllable *Al-* is the definite article in Arabic, a name like "The El Alhambra Motel" is historically equivalent to "The the the red (*ḥamrâ*) motel."

Alhambra Valley and **Creek** [Contra Costa Co.] take their name from Cañada del Hambre 'valley of hunger'—no doubt applied because of some incident, although none of the current stories seem to be supported by documentary evidence. The name appears repeatedly on *diseños*, sometimes as Paraje ('stopping place') de Ambre and Arroyo del Hambre; on May 8, 1842, it was applied to a land grant, Cañada del Hambre y las Bolsas del Hambre. Cronise mentions "The Hambre or Hungry Valley" in 1868, and Hoffmann's map of the Bay region records Arroyo del Hambre in 1873. When Mrs. John Strentzel, John Muir's mother-in-law, settled there in the 1880s, she renamed the valley because she disliked the old name.

Alianza (al ē an´ zə) [Santa Cruz Co.]. Spanish for 'alliance' (Clark 1986).

Alice, Mount [Inyo Co.] was the original name of Temple Crag (*q.v.*). Now the name is applied to a lesser peak, 2 miles to the northeast.

Alila. *See* Earlimart.

Alisal. *See* Aliso.

Aliso is Spanish for the alder tree (genus *Alnus*). Since the tree is native only where the source of water supply is permanent, its presence was particularly noted by early travelers. In the southern part of the state, the Spanish form of the name for creeks and canyons still outnumbers the English form. Some of the names referred not to the alder but to the sycamore (*Platanus racemosa*), which the Spanish seem to have likewise called *aliso*. **Aliso** (ə lē´ sō) [Orange Co.] was the name of a station on the land grant Cañada de los Alisos, dated June 18, 1841; it was renamed El Toro before 1900, but an Aliso Creek remains (Brigandi). The word *alisal* 'alder or sycamore grove' was also repeatedly used in place names in Spanish times and is preserved in two streams called **Alisal** (al´ i sal) **Creek** [Santa Barbara Co.; Monterey Co.]. The latter runs through two land grants called Alisal, dated Dec. 19, 1833, and June 26, 1834. The name of another grant, Rincón del Alisal 'corner of the alder grove' [Santa Clara Co.], dated Dec. 28, 1844, has not been preserved in California toponymy; but a little stream south of Mission San Jose [Alameda Co.] is still called Cañada del Aliso. It was so called because of two very large sycamores, later cut down by a squatter named Fallon (*Century Magazine* 1890:190).

Alkali is a term often applied to surface waters in which alkaline chemicals are present. The best known are the lakes in Modoc Co. known locally as Surprise Valley Lakes, the lakes in Modoc Co., and a creek in Yosemite National Park.

All American Canal [Imperial Co.], a name applied by the Imperial Irrigation District when the canal was proposed in 1916, was chosen to indicate that the new canal was entirely within the United States. The earlier canal that irrigated the Imperial Valley traversed Mexico for about 60 miles (M. J. Dowd).

Alleghany [Sierra Co.] was named after the Alleghany Mine tunnel, which was begun in Apr. 1853 and tapped the pay streak in Oct. 1855 (Co. Hist.: Plumas 1882:473 ff.). The post office was established and named in 1859. Like many other Alleghanys, the name goes back to the Delaware (Algonquian) name for the Allegheny River of Pennsylvania, probably meaning 'beautiful stream'.

Allendale [Alameda Co.], probably for real estate broker Charles E. Allen, who came here in 1873 (Mosier and Mosier).

Allen Peak [San Mateo Co.] was named by the county fire department because Benjamin Allen gave land here for a lookout tower (Alan K. Brown).

Allens Camp. *See* Caliente.

Allensworth [Tulare Co.] was named by Lt. Col. Allen Allensworth, who founded the community about 1909 as part of a colonization scheme for African Americans. Unfortunately he was killed in an automobile accident in 1914 before his idea was completely carried out (Mitchell).

Alma (al′ mə) [Santa Clara Co.] is shown in 1876 on the turnpike from which a branch road runs to the New Almaden Mine (*Historical atlas map of Santa Clara County*, p. 65). Another story is that the town was named for a local prostitute. The post office, which had been established under the name Lexington on June 6, 1861, was changed to Alma on Dec. 2, 1873.

Alma Creek [San Diego Co.] was probably named for the nearby Alma Mine (Stein). Alma, Latin for 'nourishing', was once a popular name for women.

Almaden (al mə den′) [Santa Clara Co.], a famous mercury deposit, was identified by Andrés Castillero in 1845; the mineral rights were granted him on Dec. 30. The mine was known as the Santa Clara or Chaboya's mine (Bowman Index). After 1848 the name New Almaden was used, after the famous Almadén mercury mine in Spain (*almadén*, in Spanish, is an obsolete word for 'mine' or 'mineral'). The post office New Almaden was established on July 5, 1861, and reestablished on Dec. 23, 1873. It was again established on Jan. 4, 1934, under the name Almaden. Local pride forced the Post Office Dept. to restore the lost "New" on Dec. 1, 1953.

Almanor (al′ mə nôr): **Lake** and post office [Plumas Co.] were named after ALice, MArtha, and EliNORe, the daughters of Guy C. Earl, president of the Great Western Power Company, which created the reservoir for the hydroelectric power project. The dam was completed in 1917, and the name was applied to the reservoir by Julius M. Howells, in charge of construction, in place of the older name, Big Meadows Reservoir.

Almejas, Punta de las. *See* Mussel; Pedro Point.

Almonte (al mon′ tē) [Marin Co.] was formerly called Mill Valley Junction; the name was changed in 1912 by the Southern Pacific on petition of residents, to avoid confusion with Mill Valley. The proximity of Muir Woods may have suggested the name, which is Spanish for 'at the woods'.

Alpaugh (al′ pô) [Tulare Co.] was named in 1905 for John Alpaugh, one of the officers of the Home Extension Colony, which reclaimed the Tulare Lake "island," successively known as Hog Root, Root, and Atwell Island (Mitchell).

Alpha [Nevada Co.] was originally one of two mining camps on the South Fork of the Yuba River, called Hell-Out-for-Noon and Delirium Tremens (Knave, Sept. 15, 1946). In 1852–53, when the country began to become respectable, these names were changed to Alpha and Omega, the first and the last letters of the Greek alphabet; both were listed as post offices in 1858. Alpha is now a ghost town.

Alpine, or **The Alpine** [San Mateo Co.]. This district was called El Pino in the 1850s, from a large pine beside the Pescadero trail at the Eugenio Soto ranch; it is attested as Alpine and Alpine Rancho in 1859–60. **Alpine Creek**

joins La Honda creek to form San Gregorio Creek (Alan K. Brown).

Alpine County was formed and named by act of the legislature on Mar. 6, 1864, from parts of Amador, El Dorado, Calaveras, and Tuolumne Cos. The alpine character of the High Sierra region suggested the name. **Alpine** town, **Heights**, and **Creek** [San Diego Co.]: The town was established in the 1880s by B. R. Arnold, and the name was proposed by an early resident because the district resembled her native Switzerland. Alpine Butte in Los Angeles Co. and several other features bear the name for the same reason.

Alsace Lake [Inyo Co.]. Ralph Beck, of the Dept. of Fish and Game in Bishop, named this lake in 1954. The Dept. of Fish and Game was asked by the USGS whether the lakes in the basin at the head of French Canyon had names. They didn't, but Beck saw his chance. He had been in France during World War II, so he provided names that were French or had a French association for this lake and several others (Browning 1986).

Alta. The Spanish adjective meaning 'high' or 'upper' has always been a favorite in California place naming. It is sometimes used as a true specific geographical term: Piedras Altas [Monterey Co.], Loma Alta [Monterey, Santa Barbara Cos.], Alta Vista [Sonoma Co.], Agua Alta Canyon ('high water canyon') [Riverside Co.], Altacanyada [Los Angeles Co.]. From 1772 to the American conquest Alta California was the name of the Spanish (later Mexican) province that is now California (*q.v.*). **Altaville** [Calaveras Co.]: The old mining camp, variously known as Forks-of-the-Roads, Low Divide, and Cherokee Flat, adopted its present name at a town meeting in 1857. **Alta** [Placer Co.], the name given to a Central Pacific station in 1866, is said to have been selected because the *Alta California*, a San Francisco newspaper, was favorable to the Central Pacific. **Altamont** [Alameda Co.] was named by the Central Pacific in 1869, probably because of its location at the highest point of Livermore Pass. There were several Altamonts in the eastern United States at that time. **Alta Meadow** and **Peak** [Sequoia NP]: In 1876 W. B. Wallace, Tom Witt, and N. B. Witt camped at the meadow and gave it this name because it was higher than any other meadow in the vicinity. The inhabitants of the Three Rivers settlement apparently extended the name to the mountain on which the meadow is situated. It was formerly known as Tharp's Peak (Farquhar). **Altadena** (al tə dē′ nə) [Los Angeles Co.] was coined from *alta* and the last part of Pasadena and applied in 1887 to the community because of its greater altitude. Byron C. Clark first used the name in 1886 for his nursery. **Alta Loma** [San Bernardino Co.] was applied in 1912 to a station on the Pacific Electric line, as the result of a popular vote. The next year the name was also given to the post office of Ioamosa, a name applied by farmers from Iowa. *See* Hermosa Beach. **Alta Vista** 'high view' appears in Inyo Co. and Monterey Co. **Alta Sierra** [Nevada Co.] means 'high mountain range.'

Altadena. *See* Alta.

Al Tahoe [El Dorado Co.]. From the Al Tahoe Hotel, built in 1907 by Almerin R. Sprague and named for himself (Lekisch). *See* Tahoe.

Altamont. *See* Alta.

Altaville. *See* Alta.

Alton (al′ tən) [Humboldt Co.]. The place was named in 1862 by S. R. Perry after his home town in Illinois. In the 1880s the name was applied to the post office and the station of the Eel River and Eureka Railroad.

Alturas (al to͞or′ əs) [Modoc Co.]. The place was first known as Dorris' Bridge, for Presley Dorris, who built a bridge here across the Pit River. On the von Leicht–Craven map of 1874 it appears as Dorrisville. On June 1, 1876, the legislature, upon petition, gave the Spanish name, meaning 'heights.' *See* Dorris. Alturas is also the name of subdivisions and physical features in other parts of the state.

Alumine (al′ ə mīn) **Peak** [Shasta Co.]. Named after an undeveloped mine at which beautiful alum crystals (aluminum potassium sulfate) have been found (Steger).

Alum Rock: Park, Creek [Santa Clara Co.]. Created as a park by act of the legislature in 1872 and called "The City Reservation." It was later called by its present name because of the striking monolith in its center, on the sides and in the crevices of which alum dust is found.

Alvarado (al və rä′ dō) [Alameda Co.] was founded in 1851 by Henry C. Smith, who named it New Haven after his home in Connecticut. In Mar. 1853 it was made the seat of the newly formed Alameda Co. In the meantime, two San Francisco attorneys, Strode and Jones, had laid out a town nearby and named it in honor of Juan B. Alvarado, governor of California from 1836 to 1842: "The new county officials met in the upper story of Smith's store in Block 81 in New Haven; but the first minutes of their meeting are dated April 11, 1853, in Alvarado. By this date, then, New Haven had discarded its old name and taken that of the neighboring town" (*CHSQ* 12:173–74). In the late 1830s, while Alvarado was governor, San Jose was called Pueblo de Alvarado and San José de Alvarado (DSP Mont. 4:97; DSP San Jose 6:12). Several minor places in the state are named Alvarado for some of the numerous persons having that surname.

Alviso (al vē´ sō): town, **Slough, Channel** [Santa Clara Co.]. Named for Ignacio Alviso (1772–1848), who came to San Francisco from Mexico with the Anza expedition in 1776. On Feb. 10, 1838, he received the grant Rincón de los Esteros, on which was the landing place for Santa Clara and San Jose. Here the town was laid out in 1849 by Chester S. Lyman for three indefatigable promoters of the gold era: Peter H. Burnett, J. D. Hoppe, and Charles B. Marvin. The post office is listed in 1862.

Alvord Mountains and **Well** [San Bernardino Co.] were named for Charles Alvord, who went with a party of prospectors in 1860–61 in search of the lost Gunsight silver lode in Death Valley. He discovered a black manganese ledge that contained "wire gold," but later attempts to find the place failed. Alvord was murdered in 1862 (Weight, pp. 48 ff.). *For* Alvord in Inyo Co., *see* Zurich.

Amador (am´ ə dôr) **Valley** [Alameda, Contra Costa Cos.] was named for José María Amador, a native of San Francisco; he was a soldier in the San Francisco company, 1810–27, then majordomo of Mission San Jose. He became the grantee of Rancho San Ramón on Jan. 22, 1834. His name appears on Duflot de Mofras's map of 1844, and Amador Valley is mentioned in the *Statutes* of 1853. **Amador Creek, County**, and **City**: In 1848 Amador, with several Indians, established a mining camp near the site of the present town. The settlement and creek became known by his name. The county was created and named on May 11, 1854. The post office was established under the name Amador City on Aug. 19, 1863.

Amargo. *See* Boron.

Amargosa (am ər gō´ sə) **River, Desert**, and **Range** [Death Valley NP] contain Spanish *amargoso* (an alternant of *amargo*) 'bitter'. The name is recorded by Frémont in April 1844: "It [the stream] is called by the Spaniards *Amargosa*—the bitter-water of the desert" (Frémont, *Expl. Exp.* p. 383). The Boundary Commission in 1861 records both the river and the range. Formerly, Death Valley was often called the Amargosa Desert because the Amargosa River flows through it as far as Badwater after making a U-turn north of the Avawatz Mountains. What is now called the Amargosa Desert is mainly in Nevada. **Amargosa Creek** [Los Angeles Co.] is doubtless also named for the bitterness of its water.

Amaya (ə mä´ yə) **Creek** [Santa Cruz Co.]. Owned by two Californio brothers, Casimero and Dario Amaya or Amayo, ca. 1860 (Clark 1986).

Amboy [San Bernardino Co.]. Named by the Atlantic and Pacific Railroad (now Santa Fe) in 1883, apparently after one of the Amboys "back East." It was probably Lewis Kingman, a locating engineer of the railroad, who named the stations of the new line between this point and the Arizona border in alphabetical order: Amboy, Bristol, Cadiz, Danby, Edson, Fenner, and Goffs. All these names have counterparts elsewhere in the United States. It appears that another railway official later tried to continue the alphabet: in 1896 the station name Ibex (later Ibis) appears—and in 1904 a Homer, west of Ibex, and a Klinefelter, east of Ibex.

Ambre. *See* Alhambra Valley.

Amedee (am´ ə dē): **Canyon, Hot Spring** [Lassen Co.]. Modified from the name of Amadee Moran, who worked for the Nevada, California and Oregon Railroad in 1890 (Morrill).

Amel (ā´ məl) **Lake** [Lake Co.]. For Amel La Laine or Lalaine, a settler from France (Mauldin).

Amelia Earhart Peak [Yosemite NP]. The name was proposed by the Rocketdyne Mountaineering Club, in honor of the aviator Amelia Earhart Putnam (1897–1937), and was approved by the BGN in 1967 (Browning 1986).

American: River, Basin, Canyon, Canyon Creek, Flat, Hill [Placer, El Dorado, Sacramento Cos.]. The name for the river appears on Wilkes's map of 1841 and was doubtless supplied to Ringgold by Sutter: "I gave the name American River to the stream that now bears it from the fact that about 3 miles above the Fort was a pass [ford], where the Canadian trappers, who were called Americanos by the Spanish speaking Indians, crossed the stream. This place was called 'El Paso de los Americanos.' So I called this stream the American River" (pp. 61–62). **Río de los Americanos** is shown on the *diseño* of the Arroyo Chico grant of 1843; and on Oct. 1, 1844, the name was applied to the Leidesdorff grant. In the *Statutes* of 1850 the name American Fork, used by Sutter previously, is made official, but in the early 1850s the present (and original) name came into current use. Jedediah Smith had given it the name Wild River in 1828, and in 1833 the river and a land grant were called Rio Ojotska. With a few exceptions in other western states, the use of the adjective "American" in place naming is apparently restricted to California, where it is found in several counties, mainly from the many gold camps called American during the Gold Rush. **American Valley** [Plumas Co.] was named after Bradley's American Ranch, which was established in the 1850s. **American House** is in Plumas Co.; **American Canyon** is in Napa and Solano Cos. *For* American Camp, *see* Jamestown.

Americano Creek. *See* Creek.

Amesti (ə mes´ tē) [Santa Cruz Co.]. For José Amesti, a Spanish Basque, who settled here ca. 1822 (Clark 1986).

Amphitheater is a favorite term for features that resemble the Roman open-air theater. The best known are the lake in Sequoia NP, named in 1899 by W. F. Dean; the lake in Kings Canyon NP, named in 1902 by J. N. LeConte; and the canyon near Saratoga Springs in San Bernardino Co., named in 1922 by the USGS.

Anacapa (an ə kap´ ə): **Islands, Passage** [Ventura Co.]. The Portolá expedition in Jan. 1770 called the westernmost island Falsa Vela 'false sail,' because it looked like a ship, and called the other two Las Mesitas 'the little table hills' (Costansó, p. 152). In Mar. 1774 Juan Pérez named the group Islotes de Santo Tomás, a name that was sometimes used on European maps (Humboldt's map, 1811; Duflot de Mofras's map, 1844). Costansó reported (p. 148) that Anajup was the native name for the island now called Santa Cruz. The name was probably later transferred to the westernmost of the Anacapa Islands. Vancouver called all three islands Eneeapah in 1792 but spelled the name Enecapa on his maps. Gibbes has Encapa in 1852, and the Coast Survey has Anacape in the same year. The Parke-Custer map gives the present spelling in 1854. The name is derived from Chumashan *anyapax* 'mirage, illusion' (Applegate 1975, p. 27, after Harrington).

Anada (ə nā´ də) **Creek** [Trinity Co.]. The post office was named in 1898 by John Jeans, combining the names of Ana and Ada, two girls whom he had known in Missouri (Alice Goen Jones, p. 365).

Anaheim (an´ ə hīm): city, **Creek, Bay, Canal** [Orange Co.]. Coined from Santa Ana, the name of the river, plus German *Heim* 'home'. The place was laid out by the Los Angeles Vineyard Society in 1857 and named by the stockholders on Jan. 15, 1858, on the suggestion of T. E. Schmidt. This German colony was the first large-scale agricultural settlement in the Los Angeles area, and the present city still calls itself "the Mother Colony." The early local name was Campo Alemán 'German field'. The area once called Anaheim Landing is now part of Seal Beach; Anaheim Hills is a new name for the old Bixby Hills, along the Santa Ana Canyon (Brigandi).

Anahuac (ä´ nə hwäk) **Spring** [San Diego Co.]. From the Diegueño place name Iñaja (Stein), but confused with Anahuac (ä nä´ wäk), a name which the Aztecs gave to their Mexican homeland. *See* Iñaja Indian Reservation.

Analy (an´ə lē) **Valley** [Sonoma Co.]. Named after Annaly, Ireland, the home of the O'Farrell family. Jasper O'Farrell came to California in 1843, bought the Rancho Estero Americano, and named the valley Annaly. When the township was formed in 1851, one *n* was omitted on the records and has never been restored. The name was applied to a post office on Aug. 7, 1860.

Anapamu (an ə pə mo͞o´) [Santa Barbara Co.]. Now the name of a street in the city of Santa Barbara; once the name of a hill, from Barbareño Chumash *anapamu*' 'ascending place' (Applegate 1975:26, after Harrington).

Anastasia Canyon [Monterey Co.]. Perhaps for Anastasio García, a well-known bandit of the 1850s and 1860s (Clark 1991).

Anaverde. *See* Verde.

Anderson [Shasta Co.]. When the California

and Oregon Railroad (now the Southern Pacific) reached the place in 1872, the station was named for Elias Anderson, owner of the American Ranch, who granted the right of way to the railroad.

Andersonia [Mendocino Co.]. The place was named by and for Jeff Anderson, who built a sawmill there in 1903 (Borden). A post office was established on July 26, 1904.

Anderson Peak [Monterey Co.]. For either James or Peter Anderson, both early settlers in the area (Clark 1991).

Anderson Ridge [Lake Co.] was named for Bob Anderson, a stockman in the region in the 1880s and 1890s (Mauldin).

Anderson Springs [Lake Co.]. Named for Dr. A. Anderson, one of the discoverers of the springs in 1873.

Anderson Valley [Mendocino Co.]. Named for Walter Anderson, who settled here in 1851.

Andesite (an′ də zīt): **Peak, Ridge** [Nevada Co.]. Named by the USGS because the chief component is the volcanic rock known as andesite (BGN, Mar. 1949).

Andover [Placer Co.]. Named by a division superintendent of the Southern Pacific after his home town in Massachusetts (*Grizzly Bear*, Mar. 1935).

Andrade (an drā′ dē, an drä′ dē) [Imperial Co.]. Named by the Imperial Land Company for the Mexican general, Guillermo Andrade, who in 1900 sold the land to the California Development Company for colonization (Tout). The post office was established and named in 1912. The name of the railroad station is Cantu (*q.v.*).

Andreas (an drā′ əs) **Canyon** [Riverside Co.]. Named for old "Captain Andreas," an elder of the Cahuilla tribe. Andreas is a common reinterpretation of Spanish Andrés in Californian names; *see* San Andrés.

Andrew Molera State Park [Monterey Co.]. The land at one time belonged to Andrew Molera (1877–1931), who grew artichokes in the Castroville area. Known locally as "Molera" (Clark 1991).

Andrews Creek [Shasta Co.]. Named for Alexander R. Andrews, a member of the 2d California Constitutional Convention and an assemblyman from Shasta Co. at the 7th, 18th, and 19th sessions of the legislature; according to Steger, he had mining claims on this creek.

Andrews Peak [Monterey Co.]. For the family of George Leslie Andrews, an early settler in the area (Clark 1991).

Andrus Island [Sacramento Co.]. Named for George Andrus, a settler of 1852.

Angel, in the Spanish form *ángel*, has been used in several Californian place names; *see* Angel Island; Los Angeles. *For* Punta del Angel Custodio, *see* Pedro Point.

Angela, Lake [Nevada Co.]. Named in Aug. 1865 for Angela King, sister of Thomas Starr King, by a party of locating engineers for the Union Pacific Railroad (Maule).

Angeles (an′ jə ləs) **National Forest** was created by combining the San Gabriel Forest Reserve and the old San Bernardino National Forest. It was given the present name by proclamation of Pres. Theodore Roosevelt in 1908, because the larger part of the forest is within Los Angeles Co.

Angel Island [San Francisco Bay]. Named Isla de los Ángeles by Ayala when the *San Carlos* anchored there in Aug. 1775. Beechey translated it as "Angel Island" in 1826. Ringgold tried to restore the Spanish version in 1851, but the charts of the Coast Survey kept the American form: 1850, Angel Island; 1851, Los Angeles Island; 1852, Angel Island. The Spanish version was sometimes used on later maps.

Angel Mountain [San Diego Co.]. For a homesteading family named Angel (Stein).

Angelo Creek [San Mateo Co.]. For the roadhouse opened at Belmont by Charles Aubrey Angelo in 1850 (Alan K. Brown).

Angels: Camp, Creek [Calaveras Co.]. George or Henry Angel, said to have been a member of Stevenson's Volunteers, started mining at the creek in June 1848. The place was described by Audubon and other diarists in the early 1850s. The post office was established on May 27, 1853, and the place became well known as one of the richest gold diggings of the southern mines. It was incorporated as Angels in 1912, but popular usage prefers the name Angels Camp, which is still used by the post office.

Angelus Oaks [San Bernardino Co.]. The term "Angelus" refers to a prayer said three times daily in the Catholic Church to commemorate the Annunciation of the Angel Gabriel to the Virgin Mary (beginning *Angelus domini nuntiavit Mariae* 'The Angel of the Lord announced to Mary').

Angiola (an jē ō′ lə) [Tulare Co.]. The railroad point was named for the owner of the site, Angie Bacigalupi (Terstegge).

Angora: Creek, Lakes, Peak [El Dorado Co.]. From a herd of Angora goats that Nathan Gilmore pastured in this region in the 19th century (Lekisch). **Angora Mountain** and **Creek** [Tulare Co.]: The name Sheep Mountain appears on maps through 1931. This was changed by a 1928 BGN decision and appeared as Angora Mountain on a 1939 map (Browning 1986).

Angwin [Napa Co.]. Named about 1874 for Edwin Angwin, who operated a summer resort on his property, which was a part of Rancho La Jota (C. D. Utt).

Animas in place names represents Spanish *ánimas* 'souls' or Día de las Ánimas 'All Souls' Day'. Animas land grant [Santa Clara Co.] was dated Nov. 28, 1808, and Aug. 7, 1835; it was also known as Carnadero and La Brea. *For* Rio de las Animas, *see* Mojave River.

Anklin Meadows [Lassen NP]. "These meadows were in the homestead entry of an early settler, named Anklin" (BGN, 6th Report).

Anlauf (an′ lôf) **Canyon** [Ventura Co.]. Named for the family of Herman Anlauf, who lived in the area in the 1890s (Ricard).

Anna, Lake [Trinity Co.]. For Anna Weber, who with her husband established the Trinity Alps Resort, at the Old Goetze Stock Ranch, in 1922 (Stang). *For* Anna Lake [Mono Co.], *see* Emma Lake.

Anna Herman Island. *See* Paoha Island.

Annaly Valley. *See* Analy Valley.

Anna Mills, Mount [Tulare Co.]. Named for Anna Mills Johnston, who in 1878 was one of the first four women to climb Mount Whitney (Browning 1986).

Annapolis [Sonoma Co.]. The post office, established in the early 1900s, was not named directly after one of the many places of that name in the East, but after the Annapolis Orchards, established by Wetmore Brothers in the 1880s (Leland).

Annette [Kern Co.]. Probably named for James L. Annette, who became head miller of the Kern River Mills in Dec. 1906.

Ano Nuevo (an′ ō n͞oo͞ ā′ vō, an′ yō nwā′ vō) **Point; Ano Nuevo: Bay, Creek, Island** [San Mateo Co.]. The name Punta de Año Nuevo was given to the cape by Vizcaíno on Jan. 3, 1603, because it was the first promontory sighted in the new year. It is one of the few names applied by early navigators that have lasted throughout the centuries for a point in the same general latitude. On May 27, 1842, the name was used for the land grant Punta del Año Nuevo. On some American maps of the 1850s the translation New Years Point can be found. The first Coast Survey charts confused the genders: 1850, Pt. Anno Nueva; 1851, Pta. Año Nueva. Camacho's map of 1785 has two points called Punta Falsa de Año Nuevo (Wagner, p. 427), probably modern Pigeon and Franklin Points. Some modern maps preserve the spelling Año.

Ansares, Llano de los. *See* Goose Lake.

Ansel Adams, Mount [Yosemite NP] bears the name of the preeminent American landscape photographer of the 20th century (1902–84). Adams first saw and photographed this peak in Sept. 1921; in 1934, a Sierra Club group climbed it and named it unofficially for him. Since BGN regulations do not permit naming a geographic feature for a living person, it was not approved by the BGN until 1985 (Browning 1986). **Ansel Adams Wilderness** [Mono Co.] is also named for the photographer.

Ansel Lake [Sequoia NP] was named for Ansel Franklin Hall, a ranger in Sequoia NP, 1916–17 (Browning 1986).

Antelope. About seventy-five places record the presence of the graceful pronghorn antelope, *Antilocapra americana*, which was once abundant in various parts of the state. **Antelope** [Sacramento Co.] was applied to the settlement at Cross's brick warehouse by the Antelope Business Association in 1877. The post office was established on June 9, 1877. The Spanish word for 'antelope', *berrendo*, was repeatedly used for place names in early times but seems to have survived only in the Southern Pacific stations Berendo [Los Angeles Co.] and Berenda [Madera Co.]. **Antelope Creek** [Tehama Co.] is a translation of Arroyo de los Berrendos (Bidwell's map of 1844). **Antelope Bridge** [Lava Beds NM] was named in Jan. 1917 by J. D. Howard, because at its eastern entrance he found the pictograph of an antelope entering an enclosure.

Anthony: Ridge, Peak [Mendocino Co.]. Named for the three Anthony brothers—James, Jesse, and George—who ran sheep on this mountain in the early 1890s (USFS). The oldest brother, James, was mentioned as a farmer in Round Valley as early as 1874.

Anthony Creek [Monterey Co.]. The name probably reflects the proximity of the San Antonio River (Clark 1991).

Anticline (an′ ti klīn) **Ridge** [Fresno Co.]. The

name is a geological term that designates the fold or arch of rock strata in which the layers dip in opposite directions from the crest.

Antimony (an′ ti mō nē) **Peak** [Kern Co.]. In 1854 W. P. Blake of the Pacific Railroad Survey (Pac. R.R. *Reports* 5:2.291 ff.) identified the ore, which prospectors had hoped was silver, as "sulphuret of antimony, commonly known as Grey Antimony or Antimony-Glance," and traced the vein from the canyon high up the side of the mountain. **Antimony Spring, Ridge**, and **Canyon** [Death Valley NP]: Dr. S. G. George and his party were seeking the "lost" Gunsight Mine when they came upon an antimony deposit on Dec. 25, 1860, which they first named "Christmas Gift." Wheeler atlas sheet 65-D records the name Antimony for a mine in the area. Deposits are found elsewhere, including San Benito and Merced Cos., where another peak is named Antimony.

Antioch (an′ tē ok) [Contra Costa Co.]. In 1849 the place was known as Smiths Landing for the first settlers, the twin brothers J. H. and W. W. Smith. In the following year the latter, a minister of the gospel, invited a group of New Englanders to settle on his property. At a picnic on July 4, 1851, the citizens chose the new name, Antioch—the name of a Biblical city in Syria—in preference to Minton and Paradise, other names that had been suggested.

Antoine (an′ tōn) **Cañon** [Placer Co.]. Named for a mixed-blood called Antoine, who was a member of the Bronson prospecting party of 1850 (Co. Hist.: Placer 1882:373).

Antone Meadows [Placer Co.]. Antone Russi, of Swiss extraction, bought land here in the late 1880s (Lekisch).

Antonio [Santa Barbara Co.]. A railroad point, named for adjacent San Antonio Creek and Valley.

Anvil Spring and **Canyon** [Death Valley NP] were so named because Sergeant Neal, of the Bendire expedition of 1867, found at the spring an anvil, wagon rims, and a quantity of old iron. It is possible but not certain that these were the remnants of the blacksmith outfit which Asahel Bennett had brought into the valley in 1849. *See* Bennetts Well.

Anza (an′ zə) [Riverside Co.] commemorates Juan Bautista de Anza (1735–88), leader of the famous expedition that traversed this area in 1774. The name of the old post office Cahuilla was changed to Anza in 1926. **Anza-Borrego Desert State Park** [San Diego Co.] was established in 1933 as Anza Desert State Park and renamed in 1957; *see also* Borrego. **Lake Anza** [Alameda Co.] was named in 1939 by the East Bay Regional Park Board in honor of the same explorer. **Anza Valley** [Riverside Co.] was officially so named by the BGN in list 8104, superseding the name Terwilliger Valley.

Anzar (an zär′) **Lake** [San Benito Co.]. The name commemorates Juan Anzar, grantee of the land grant Aromitas y Agua Caliente on Oct. 12, 1835.

Apache (ə pach′ ē). The name of the Indian tribe that has left an indelible mark on the geographical nomenclature of Arizona is represented in California by Apache Canyon and Potrero, in Ventura Co. The Spanish name Apache was borrowed in the 16th century from Zuni *aapachu* 'Navajos'; at that time, no distinction was made between Navajos and Apaches (*HNAI* 10:385).

Apex. *See* Escondido.

Apperson Creek [Alameda Co.]. For Elbert Apperson, brother of Phoebe Apperson Hearst; he settled here in the 1890s (Mosier and Mosier).

Applegate [Placer Co.]. The place was settled by Lisbon Applegate and was known as Bear River House. The post office was established in the 1870s, with George Applegate as postmaster.

Applesauce Creek [Siskiyou Co.]. The name became known when, in the 1890s, mammoth quartz ledges were discovered in the vicinity. It is derived from the favorite exclamation of a Mr. Sullivan who had a cabin near the mouth of the creek. When he was lucky at card playing, he was wont to exclaim, "That's the applesauce!" (Luddy).

Apple Valley [San Bernardino Co.]. The post office established on Apr. 16, 1949, at the resort city developed by Newt Bass bears the name applied at the turn of the century by Mrs. Ursula M. Poates, a long-time resident of the Mojave Desert. To convince buyers that fruit could be grown in the desert, Mrs. Poates planted three apple trees in her greasewood-covered yard (Corinne K. Flemings).

Aptos (ap′ tōs, äp′ tōs): town, **Creek** [Santa Cruz Co.]. A Costanoan Indian village, *aptoṣ* (*HNAI* 8:485), was mentioned in 1791. There is a local tradition that the original meaning was "the meeting of two streams" (Clark 1986). A Rancho de Aptos, a sheep ranch of Mission Santa Cruz, is mentioned on July 5, 1807 (Arch. Arz. SF 2:61). The name was applied to a provisional land grant on Sept. 4, 1831.

Aqua. The Latin word for 'water' is occasionally found on maps, e.g. in Aqua Buena Spring [San Benito Co.]. The word is an error for Spanish *agua*.

Arana (ə ran′ ə) **Gulch** [Santa Cruz Co.]. The name commemorates José Arana, grantee of the land grant Potrero y Rincón de San Pedro Regalado on Aug. 15, 1842.

Ararat, Mount. The mountain in Armenia on which Noah's ark landed has been a favorite name for peaks in various parts of the globe. California has Mount Ararats in Calaveras, El Dorado, Merced, Plumas, and San Diego Cos.

Arastra Creek [Siskiyou Co.]; **Arastradero Creek** [Santa Clara Co.]; **Arastraville** [Tuolumne Co.]. *See* Arrastre.

Arboga (är bō′ gə) [Yuba Co.]. Named in 1911 by the Rev. N. M. Nielsen, pastor of the Swedish Mission Church, after his former home in Sweden (Paul Erickson).

Arbolado. *See* Sur.

Arbuckle [Colusa Co.]. Named by surveyors of the Central Pacific Railroad when station and town were established in 1875 on the ranch of T. R. Arbuckle, who had settled there in 1866.

Arbuckle Mountain [Shasta Co.]. Named for A. Arbuckle, a prospector of the early 1850s (Steger).

Arcade [Sacramento Co.]. The term may have been originally applied to the tree-lined avenue of horse barns on the Rancho Del Paso, where James Ben Ali Haggin raised thoroughbreds in the late 19th century (Ellis).

Arcadia [Los Angeles Co.]. The city was platted and named by Herman A. Unruh, of the San Gabriel Valley Railroad, about 1888. Arcadia, the name of a district in Greece that became a symbol of rural simplicity in pastoral poetry, is a favorite place name in the United States. However, the famous Arcadia Block in Los Angeles, built by Abel Stearns in 1858, was named for his wife, Arcadia.

Arcane Meadow [Death Valley NP]. The high flat between Manly and Rogers Peaks (formerly Baldy and Sugarloaf) was named by the Park Service in memory of J. B. Arcan, a member of the Bennett-Manly party of 1849 (Death Valley Survey).

Arcata (är kā′ tə) [Humboldt Co.], an offspring of the Humboldt boom of 1850, was founded and named Union Town sometime before Apr. 17 of that year by the Union Company. It appears on most maps as Union. To avoid confusion with Uniontown [El Dorado Co.], the name was later changed to Arcata: "the town of Union . . . is hereby changed to the name of the town of Arcata" (*Statutes* 1860:109). Although the place is in the territory of the Wiyot tribe, Whites took the name from Yurok *oket'oh* 'where there is a lagoon' (referring to Humboldt Bay), from *o-* 'locative' plus *ket'oh* 'to be a lagoon'; the name is also applied by the Yurok to Big Lagoon (Carranco and Genzoli, 374–76; cf. Robins).

Arch. The word is often found along the coast for archlike formations. There are five Arch Rocks in the state, as well as an Arched Rock north of Tomales Bay. Orange Co. has an Arch Beach and a Three Arch Bay south of Laguna Beach.

Arcilla (är sē′ yə) [Riverside Co.]. Spanish for 'clay', from the nearby claypits.

Ardeth Lake [Yosemite NP]. Otto M. Brown, a park ranger, 1927–46, named the lake after his wife (Browning 1986).

Arena (ə rē′ nə), **Point; Arena: Rock, Cove** [Mendocino Co.]. The cape was first sighted by Ferrer in 1543 and called Cabo de Fortunas. It appears on the maps of the following 250 years under various names. The popular name among sailors in the latter part of the 18th century was apparently the descriptive term Barra de Arena 'sand bar' (*see also* Delgada, Point). Vancouver (1792) misspelled it Barro de Arena, and it appears thus on most American maps until 1851. Disturnell 1847 has Punta Arena; Scholfield 1851, Point Barro de Arena; and Bancroft's maps, 1858, Punta de Arena. The Americanized version, Point Arena, is recorded on the map accompanying the English translation (1849) of Schmölder's *Wegweiser* (Wheat, no. 83). It was used again by the Coast Survey in 1853 and won out when it was adopted for the maps of the Whitney Survey. The name Arena occurs in other parts of the state: in Santa Barbara, Kern, and Merced Cos. and on Santa Cruz Island. Sand Point [Marin Co.] is Punta de Arenas on a plano of 1776 (PSP Ben. Mil., vol. 1) and is mentioned by Font on March 27, 1776. Thomes Creek [Tehama Co.] is labeled Arroyo Arenoso 'sandy creek' on the *diseño* of Los Saucos grant and on Bidwell's map of 1844.

Argos (är′ gəs) [San Bernardino Co.]. The Santa Fe station was probably named after the city in Greece. Other classical names—Troy, Trojan, Hector—were applied to stations on the same part of the line and are shown on the Santa Fe maps after 1905. The name may occasionally be confused with Argus (*q.v*).

Arguello (är gwel′ ō), **Point** [Santa Barbara Co.]. Vancouver named the point in 1792 for José Darío Argüello, at that time *comandante* at Monterey. It is called Pedernales on early Spanish maps. Both names refer obviously to the entire headland. The Coast Survey charts misspelled it Point Arguilla until 1874, though the 1858 edition of the Coast Pilot gave the proper spelling (*see* Pedernales, Point). The station, Arguello, is named after the point; the nearby post office, Arlight, is a contraction of "Arguello lighthouse."

Argus: Peak, Range [Inyo Co.]. Like other characters from classical mythology, the many-eyed Argus has provided a convenient name for ships, periodicals, and places. An Argus mining district is shown on Farley's map of 1861; peak and range appear on sheet 65-D of the Wheeler atlas. The name was probably suggested because the peak seems to watch over the country to east, west, and south.

Arica (ə rē′ kə) **Mountain** [Riverside Co.], perhaps from the Arica mines in Chile, perhaps a misspelling of "Eureka" (Gunther).

Arichi. *See* Avichi.

Arlanza [Riverside Co.]. From Arlington, an adjacent community, plus Anza.

Arlight. *See* Arguello.

Arlington. The name, of English derivation, has long been a favorite place name, increasing in popularity after the establishment of the National Cemeteries at Arlington, Virginia, at the close of the Civil War. In California more than ten places and features are named Arlington. **Arlington** [Riverside Co.] was developed by the Riverside Land and Irrigation Company in 1875 and named by popular vote. **Arlington Heights** [Alameda Co.] was named by John H. Spring in 1911, after Arlington Avenue, Berkeley, which led to the subdivision.

Arlynda Corners [Humboldt Co.]. Founded in 1882; the name is said to be "an Indian word signifying merchandise or property" (Turner 1993).

Armada (är mä′ də) [Riverside Co.]. The word is Spanish for 'fleet' or 'squadron'; but here, as in other places in the United States, it was chosen merely for its euphony, when a post office was established on Sept. 25, 1895.

Armijo (är′ mē ō) [Solano Co.]. The high school district preserves the name of José F. Armijo, in 1842 owner of the Tolenas rancho, northeast of Fairfield (D. A. Weir).

Armona (är mō′ nə) [Kings Co.]. The typical railroad name, possibly coined from Ramona by the construction engineer, was applied to a station of the Goshen-Huron branch line in the 1880s. When the San Joaquin branch of the Southern Pacific joined this branch east of the old station in June 1891, the name was transferred to the new junction.

Armstrong Redwoods State Reserve [Sonoma Co.]. The park was deeded by the county to the state on Jan. 8, 1934, and named in memory of Col. James B. Armstrong, the original owner, whose heirs had contributed toward its purchase by the county in 1917–18.

Armuyosa Desert. *See* Death Valley.

Army Pass [Sequoia NP] was so named because a trail was constructed over the pass by Troop K, 4th Cavalry, in the 1890s (Farquhar).

Army Point [Solano Co.]. Appears on Ringgold's map as Navy Point. It was properly changed by the Coast Survey after the army established a reservation there in 1851.

Arndt Lake [Tuolumne Co.]. Named in 1896 by Lt. Harry C. Benson for Sgt. Alvin Arndt of the 4th Cavalry, who had found a route from Matterhorn Canyon to Hetch Hetchy Valley in Sept. 1893 (Farquhar).

Arno [Sacramento Co.]. The post office was established about 1890 and named after the river in Italy. Julio Valensin, an Italian, and Alice McCauley, daughter of the owner of the land, were married in Florence, which is on the Arno River.

Arnold [Calaveras Co.]. The post office was established in 1934 and named for Bernice Arnold (later Mrs. McCallum), owner of the resort and first postmaster. There is also an Arnold in Mendocino Co.

Arnot (är not′) **Peak** [Alpine Co.]. Named for Nathaniel D. Arnot, superior judge of Alpine Co., 1879–1904 (Maule).

Aromas (ə rō′ məs) [Monterey Co.]. The place was formerly known as Sand Cut because of the tunnel built by the Southern Pacific in the 1870s. The present name, applied to the post office about 1895, is derived from the land grant Aromitas y Agua Caliente 'little odors and hot water', dated Oct. 12, 1835. The *diseño* shows Aromitas as well as Aromas; both designations probably refer to the odors of sulfur water. The rancho is called Las Aromas in a report dated Feb. 12, 1847 (Arch. Mont. 13:1).

Arrastre (ə ras′ trə) represents the Spanish word meaning 'the act of dragging; a kind of ore mill'—a mining term derived from *arrastrar* 'to drag', applied in California gold-mining districts to the primitive milling apparatus used for drawing heavy weights over the ore in a circular pit by means of horse, mule, water, or hand power. The name is found in San Bernardino, Los Angeles, and Siskiyou Cos. There is an Arrastra Flat south of Frazier Mountain [Ventura Co.]. Arastraville [Tuolumne Co.] was once a rich gold-mining camp. Arrastre Springs in the Avawatz Mountains [San Bernardino Co.] was so named because in 1907 prospectors found an arrastre in an old Mexican gold camp. **Arrastradero** (ə ras trə dâr′ ō) [Santa Cruz Co.] means 'a place where dragging is done' (Clark 1986).

Arrow. A number of features are so named either because of their shape or because arrowheads were found there. **Arrow Peak** and **Ridge** [Kings Canyon NP]: The peak was named in 1895 by Bolton C. Brown when he made the first ascent (Farquhar).

Arrowbear Lake [San Bernardino Co.]. From Arrowhead Lake (*q.v.*) and Little Bear Lake.

Arrowhead: Springs, Lake [San Bernardino Co.]. Dr. David N. Smith settled near the hot springs in 1860: "The name given to the place is derived from a peculiar configuration resembling an arrow-head. (The springs are situated directly at the foot of the hill bearing this remarkable landmark" (*Wilmington Journal*, Oct. 27, 1866). The lake was originally called Little Bear Lake; a nearby post office, Arrowbear Lake, combines the old and the new name. There is an Arrowhead Mountain in Sonoma Co., probably so named because Indian arrow points were found there.

Arroyo (ə roi′ ō). The word was used in Spanish times for a watercourse—a creek or its bed. Sometimes the word *seco* 'dry' was added to indicate the lack of water. Arroyo occurs in the principal or secondary names of thirty land grants. It has become a generic term in the southwestern United States and is frequently applied in place of "creek" or "canyon." In 1930 the USGS considerably increased the number of Arroyos in the state by calling every gulch in the Kettleman Hills an arroyo. More than 150 waterways are now called Arroyo, chiefly in the San Francisco Bay region and the southern counties. Some have been hybridized: Big Arroyo [Tulare Co.], once called Jenny Lind Creek; Dry Arroyo [Solano Co.]; Surprise Arroyo [Fresno Co.]. There are also some tautologies, e.g. Arroyo Grande Creek [San Luis Obispo Co.] and Arroyo Seco Creek [Monterey Co.]. **Arroyo Alta** [Alameda Co.] was so named in 1874; it is also called El Alto Creek and Arroyita del Alta [*sic*] on an early map (Mosier and Mosier). **Arroyo de en Medio** [San Mateo Co.] means 'in-beween creek'. **Arroyo Grande** (gran′ dē) **Creek, Valley**, and town [San Luis Obispo Co.]: The name was preserved through the land grant Arroyo Grande or San Ramon, dated Apr. 25, 1842. Parke-Custer record the name in 1855. The name of the town dates from 1867–68, when a blacksmith shop and a schoolhouse were built in the valley. **Arroyo Sanitorium** [Alameda Co.] is named after the Arroyo Valle, on which it is situated (*see* Valle de San Jose). **Arroyo Paredon** (pä rā dōn′) [Santa Barbara Co.] was once known as Arroyo Parida or Parida Creek. In 1961 the BGN restored what was apparently the original name, Spanish *paredón* 'thick wall'. **Arroyo Ancho** [Kern Co.] means 'broad creek'. **Arroyo del Chino** (chē′ nō) [Santa Cruz Co.] may be the creek of the Chinese, or the curly-haired person (*see* Chino). **Arroyo Las Positas** [Alameda Co.] is from Las Pocitas 'the little pools' on an 1839 map (Mosier and Mosier). **Arroyo Salada** [San Diego Co.] is 'salty creek', with gender confusion. **Arroyo Seco** 'dry creek' is found in Los Angeles and San Diego Cos. **Arroyo Tapiado** [San Diego Co.] is 'walled up', from *tapiar* 'to block with a mud wall' (Stein, p. 162). **Arroyo Degollado** [Kings Co.] seems to mean 'beheaded', but Degollado is also a Spanish family name.

Arroz (ə rōz′) [Yolo Co.]. Spanish for 'rice', a major crop of the area. *See also* Riz.

Arsenic Spring [Death Valley NP]. The amount of arsenic in the water—if there is any—is negligible; the effects of imbibing the water are not poisonous, but tonic, according to local Indians (Death Valley Survey).

Artesia (är tē′ zhə) [Los Angeles Co.]. Named by the Artesia Company, which drilled artesian wells and established the town in the 1870s. Artesia is the Latin name for the town of Artois (*q.v.*) in France, where artesian wells exist.

Arthur, Lake [Placer Co.]. This is a reservoir built for electric power in 1909 and named for W. R. Arthur, assistant manager of the water district.

Artist Point [Yosemite NP]. From this place in 1855 the artist Thomas Ayres drew the first picture of Yosemite Falls (Browning 1986).

Artois (är′ tois) [Glenn Co.]. The town (originally in Colusa Co.) was named Germantown by popular vote in 1876, most of the early settlers being German. The post office was established on Aug. 2, 1877. In 1918 the American name was replaced by the French Artois, after the district on the Western Front in World War I (*see* Artesia). The Post Office Dept. accepted the new name on May 22, 1918.

Artray Creek [Plumas, Lassen Cos.]. Coined in 1914 from the first names of Arthur Barrett, a forest ranger, and Raymond Orr, a timber reconnaissance man (Stewart).

Arundell Barranca [Ventura Co.]. For Thomas and William Arundell, early settlers in the area (Ricard).

Arvin [Kern Co.]. Named for Arvin Richardson, the first storekeeper in the colony, which was established in 1910 (Santa Fe).

Asbill Creek [Lake Co.]. The stream, also known as Conns Creek, was named for an old pioneer family. *See* Asbill in the Glossary.

Ash. In California the ash tree (genus *Fraxinus*) is inconspicuous, of little value, and not common; hence the number of places named for it is comparatively small. A few settlements, about ten creeks, and five peaks bear the name, and most of these are in the northern counties, where the name was probably given because the Oregon ash (*F. latifolia*) was found growing there. **Ash Meadows** and **Creek** [Inyo Co.] were named for the leather-leaf ash (*F. velutina*), which was formerly abundant (Palmer). *See* Fresno. **Ashland** [Alameda Co.] was named for the Oregon ash tree found growing there (Mosier and Mosier).

Ash Creek [Siskiyou Co.]. So named because of the presence of volcanic ash. Shasta Co. has an Ashpan Butte.

Ashford Canyon and **Peak** [Death Valley NP] were named after the Ashford Mill, built in 1914. The Ashford brothers, of Shoshone, were prominent in the mining boom of that year.

Ash Hill [San Bernardino Co.]. In 1883 the Southern Pacific Railroad named a siding on the newly constructed line (now Santa Fe) and erected a monument for Ben Ash, a surveyor who died of thirst while surveying there (Santa Fe).

Ashrama (äsh′ rä mə) [Santa Cruz Co.]. The term is Sanskrit for a hermitage or religious retreat; the Hindi form, *ashram*, is also used in English.

Asilomar (ə sil′ ō mär, ə sē′ lō mär) [Monterey Co.]. The artificial name, coined from Spanish *asilo* 'refuge' and *mar* 'sea', was given by the National Board of the YWCA to the meeting and vacation grounds that were established in 1913 for YWCA groups and similar organizations (Mrs. B. B. Head). The post office is listed in 1915.

Aspen. The quaking aspen (*Populus tremuloides*) is found on the Sierra slopes at an elevation of 5,000 to 10,000 feet, from Mount Whitney to Oregon, and has given its name to about ten features, including the well-known Aspen Valley in Yosemite. The mountain in Shasta Co. once called Aspen Butte should be Asperin Butte, according to a decision of the BGN, Sept.–Dec. 1963.

Asperin Butte. *See* Aspen.

Asphalt. California is one of the few regions in which extensive deposits of asphaltum or bitumen are found. Their existence in coastal areas from Santa Clara to Orange Cos., and in Kern Co., has given rise to a number of Asphalt, Tar, and Pitch Creeks, Springs, and Lakes (*see* Brea; Pismo). *For* Asphalto, *see* McKittrick.

Associated [Contra Costa Co.]. The name was applied to the post office in 1913, when the Tidewater Associated Oil Company built the company town. The railroad station has the more melodious name Avon.

Assumpcion, La. *See* Asuncion; Natividad.

Asti (as′ tē) [Sonoma Co.]. Named after the city in Italy, a vineyard center. The place is a railroad station serving the wine-producing area of the Italian Swiss Colony, originally organized in 1881. The post office is listed in 1892.

Asuncion (ə sun′ sē ən) [San Luis Obispo Co.]. The Spanish word, referring to the Assumption of the Virgin Mary—her bodily transportation to heaven—is mentioned as a place name, La Assumpción, by Font on Mar. 4, 1776 (*see also* San Buenaventura; Ventura). Rancho de la Asunción, near Mission San Luis Obispo, is mentioned on Nov. 26, 1827 (Registro, p. 17). On June 18, 1845, the name was applied to a land grant. The von Leicht–Craven map (1874) shows the form Ascension, confusing the Assumption of Mary with the Ascension of Christ. But when the station of the Southern Pacific was named, the original Spanish version was again used.

Asylum. *See* Patton.

Atascadero (ə tas kə dâr′ ō): town, **Lake** [San Luis Obispo Co.]. The name is shown on the von Leicht–Craven map (1874) and was applied to a station when the Southern Pacific was built in 1886. The modern community was developed in 1913 by the Atascadero Estates. The name was derived from that of the provisional land grant, Atascadero, dated Dec. 21, 1839. A number of other geographic features called Atascadero were possibly named by Americans who liked the word and did not know that it means 'a place where one gets stuck in the mud', from Spanish *atascar* 'to get mired down'.

Atascoso (at ə skō′ sō) **Creek** [Santa Barbara Co.]. Spanish for 'muddy'.

Atherton [San Mateo Co.]. Named for Faxon D. Atherton, father-in-law of the novelist Gertrude Atherton. Atherton first visited California in 1836; in 1860 he acquired 500 acres of Rancho Las Pulgas, on which the town was built.

Athlone (ath′ lōn) [Merced Co.]. When the Central Pacific Railroad built through the valley in 1872, this station was named Plainsburg after the old settlement a few miles northeast. It appears that the Plainsburgers failed to make use of their chance to move to the new Plainsburg at the railroad, and the station was renamed, probably after the town in Ireland or the place in Michigan.

Atla. *See* Sultana.

Atlanta [San Joaquin Co.]. The name was applied in 1868 by William Dempsey, the first postmaster, and by Lee Wilson, who came from Atlanta, Georgia (Tinkham).

Atlas Peak [Napa Co.]. The peak was apparently not named after the mountain range in Africa but after the Atlas resort, mentioned by the county history in 1881.

Atmore Meadows [Los Angeles Co.]. Named by the USFS for Ted Atmore, an early settler.

Atolia [San Bernardino Co.]. Applied by the Tungsten Mining Company in the early 1900s for two of its officers, Atkins and DeGolia. According to one source, DeGolia supplied the *lia;* P. J. ("Pete") Osdick, for whom the community of Osdick was named, supplied the *o*.

Atwater [Los Angeles Co.]. For Harriet Atwater Paramor, whose husband subdivided the area (Atkinson).

Atwater [Merced Co.]. The railroad station was built in the 1870s on the property of Marshall D. Atwater, a well-known wheat rancher, and named for him. The town was established by the Merced Land and Fruit Company in 1888 and named after the station.

Atwell Grove [Sequoia NP]. The name commemorates A. J. Atwell of Visalia, sometime owner of the lumber mill that began operating here in 1879. In 1920, D. E. Skinner of Seattle bought the site and gave it to the Dept. of the Interior, thus supplying the nucleus for the present grove. In gratitude for the generous gift the grove was called Skinner Grove (BGN, 6th Report). Since the grove was locally better known as Atwell Grove, this name was recommended by the Park Service, and in Apr. 1946 the board reversed its former decision and applied the new name. *For* Atwell Island, *see* Alpaugh.

Atwood [Orange Co.]. The townsite was laid out in 1888, with the name of Richfield (Brigandi). In 1920, that name was rejected by the Post Office Dept., and the Atwood post office was opened by the Santa Fe, named for W. J. Atwood, purchasing agent for the Chanslor Canfield Midway Oil Company.

Auberry (ô′ bâr ē) [Fresno Co.]. Named for Al Yarborough, one of the four hunters who named Dinkey Creek in Aug. 1863. When the post office was established in the 1880s, the spelling reflected the common pronunciation of Yarborough's name.

Auburn: town, **Ravine** [Placer Co.]. Gold was discovered in the ravine by Claude Chana, a Frenchman, in May 1848. Among the miners who flocked to the place were former members of Stevenson's New York Volunteers. In Aug. 1849 the camp—known first as Rich Dry Diggings, then as North Fork Dry Diggings and Wood's Dry Diggings—was named after Auburn, New York, by Samuel W. Holladay: "By virtue of my august authority as Alcalde, in August, 1849, I named our diggings Auburn" (SCP:P 1941:34). As such the place is recorded on Butler's map, 1851. The post office was established on July 21, 1853.

Audrain (ô drēn′) **Lake** [El Dorado Co.]. Named for Thomas Audrain, whose way station at

Echo Summit was burned in 1865 in retaliation for his rejoicing openly at the news of Lincoln's assassination (Knave, Jan. 20, 1959). The spelling Audrian also occurs.

Audubon Canyon [Marin Co.]. Named for the Audubon Society, which sponsored a wildlife preserve here in 1961 (Teather 1986); the society is named in turn for the naturalist and painter John James Audubon (1785–1851).

Auguerreberry. *See* Aguereberry.

Augustine Indian Reservation [Riverside Co.]. For "Captain" Augustin, an Indian leader (Gunther).

Augustine Pass [Riverside Co.]. For Martin Augustine, a miner in the area in 1917 (Gunther).

Auk Auk (ôk′ ôk) **Ridge** [Glenn Co.]. Said to be an Indian word for 'quail' (Rawlins), but the language has not been identified.

Aukum [El Dorado Co.]. Also known as Mount Aukum; the origin is not known. The name Aukum was officially adopted by the BGN in 1975 (list 7504).

Aulon, Point [Monterey Co.]. *See* Abalone; Love.

Aumentos (ô men′ təs) **Rock** [Monterey Co.]. The name was applied to the rock near Point Pinos when the Coast Survey charted Monterey Bay, 1856–57. Since José María Armenta was grantee of the Punta de Pinos grant on May 13, 1833, "Armenta's" was probably the original spelling.

Auras, Cañada de las. *See* Feliz.

Aurora. When Mono Co. was created in 1861, the mining camp of Aurora was made county seat. It was not until Sept. 16, 1863, that it was discovered Aurora was in Nevada, and the county seat was shifted to Bridgeport (Sowaal 1985).

Ausaymas [Santa Clara, San Benito Cos.]. The name of a land grant (also known as Cañada de los Osos, or Ausaymas y San Felipe), dated Nov. 21, 1833, and Feb. 6, 1836. Ausayma was evidently a Mutsun (Costanoan) village near San Juan Bautista (Kroeber).

Austin, Mount. *See* Mary Austin, Mount.

Austin Creek [Sonoma Co.]. Probably named for Henry Austin, a pioneer settler in the region.

Austrian Gulch [Santa Clara Co.]. Named for a group of Austrians who planted vineyards there in the 1870s. It was a thriving colony until a cloudburst in 1889 undermined the foundations of the winery and swept thousands of gallons of wine into Los Gatos Creek (Hoover, p. 539).

Avalanche Peak and **Creek** [Kings Canyon NP]. The mountain was named by Muir in 1891, doubtless because of the marked course of avalanches. The name was subsequently transferred to Palmer Mountain, but it was restored on the Tehipite atlas sheet in 1939. There is another Avalanche Creek in Yosemite NP.

Avalon (av′ ə lon): town, **Bay** [Santa Catalina Island]. The town was founded by George Shatto on property purchased from the Lick estate in 1887: "Mr. and Mrs. Shatto and myself were looking for a name for the new town, which in its significance should be appropriate. . . . I found the name 'Avalon'" (Mrs. E. J. Whitney, quoted in *HSSC:P* 6:30). Avalon was the legendary elysium of King Arthur—a sort of Celtic paradise. The name has been used repeatedly for places; the best known is the southeastern peninsula of Newfoundland. The former name of Avalon Bay, and perhaps the first name on California soil ever applied by an American, was Roussillon Bay: "As I was the first navigator who had ever visited and surveyed the place, I took the liberty of naming it after my much respected friend, M. De Roussillon" (William Shaler, 1803, in *American Register* 3:147–48). The Coast Survey left the bay nameless, but it became popularly known as Timms Cove, for A. W. Timm. *See* Timms Landing.

Avawatz (av′ ə wäts): **Mountains, Pass, Dry Lake** [San Bernardino Co.]. The name appears on the map of the Merriam expedition (1891) as Ivawatch, a name doubtless supplied by the Indians. The Southern Paiute name is *áviwats* or *ávawats* 'gypsum', from *ávi* 'white clay' (Fowler).

Avellanos de Nuestra Señora del Pilar, Los. *See* Hazel.

Avena (ə vē′ nə) [San Joaquin Co.]. The Spanish word for 'oats' was applied to the railroad station when the line from Stockton to Merced was built in the early 1890s, presumably because fields of wild oats were noticeable nearby. **Avena Loca Mesa** [Riverside Co.], a name attested from 1901, is supposed to refer to 'wild oats' (Gunther).

Avenal (av′ ə nəl): **Creek, Ridge, Gap**, city [Kings Co.]. The word is Spanish for 'oat field'. The creek is shown as Avenal on the Mining Bureau map of 1891; as Avendale on the Land Office map of 1901. By decision of May 15, 1908, the BGN adopted the former version. The community was named after the creek by the Standard Oil Company in 1929.

Avendale. *See* Avenal.

Avenue of the Giants [Humboldt Co.]. A scenic drive named for the giant redwood trees through which it passes.

Avichi, Arroyo [Marin Co.]. The name (also found as Arichi) is mentioned in the records of the Novato grant in the 1850s and is shown on Hoffmann's map of the Bay region (1873). The term may be from Coast Miwok *awwič* 'crow' (Callaghan).

Avi Corotath (ə vē′ kôrətäth′) [San Bernardino Co.]. Contains Mojave *'avíi* 'rock, mountain' and *kwalatáth-* 'to be big and round' (Munro); also called Monument Peak.

Avila (ə vil′ ə, av′ i lə) [San Luis Obispo Co.] was laid out in 1867 and named in memory of Miguel Ávila by his sons. Ávila was a corporal at Mission San Luis Obispo and was the grantee of Rancho San Miguelito on Apr. 8, 1839. By a decision of the BGN (July–Sept. 1966), the place is now officially Avila Beach.

Avinsino (ä vin sē′ nō) **Corner** [El Dorado Co.]. Named after a family that operated a winery here (Witcher).

Avisadero (ə vē zə dâr′ ō), **Point** [San Francisco Co.]. The name for the tip of Hunters Point appears as Pta. Avisadera on Duflot de Mofras's plan 16 (1844), possibly taken from an earlier map of Beechey. The Coast Survey has Avisada in 1850, and Avisadera in and after 1851. The present spelling is shown on Hoffmann's map of the Bay region (1873). The word is not in Spanish dictionaries but may mean 'place of advising or warning', from the verb *avisar* 'to advise, warn'. *See* Hunters Point.

Avocado [Fresno Co.]. The name of the fruit (*Persea americana*) is borrowed from Mexican Spanish *ahuacate*, itself from Aztec *ahuacatl*.

Avon (ā′ von) [Contra Costa Co.]. Named after Shakespeare's Avon by the Southern Pacific when a station for shipping was established there in 1877–78. The name of the post office was Associated.

Avonelle Lake [Yosemite NP]. Otto M. Brown, park ranger from 1927 to 1946, named the lake for his daughter (Browning 1986).

Ayala (ə yä′ lə) **Cove** [Angel Island, Marin Co.]. In 1870 the army constructed two hospital buildings here, in what had been called Morgans Cove. It was renamed in 1969 (BGN, list 6903) for Lt. Juan Manuel de Ayala, who directed the first survey of San Francisco Bay in 1775 (Teather 1986).

Ayers Creek [Ventura Co.]. For Robert Ayers, who purchased the Casitas Ranch in 1887 (Ricard).

Azalea [Siskiyou Co.]. For the western azalea, *Rhododendron occidentale*.

Azul (ə zōōl′). The Spanish word for 'blue' is preserved in the names of several valleys, hills, mountains, and *sierras*, some of which may have been so named because the *Ceanothus* or "wild lilac" was observed blooming in spring. **Azule Mountains** and **Springs** [Santa Clara Co.]: Derived from Sierra Azul, which is recorded on a *diseño* of Cañada de los Capitancillos (1842) and elsewhere.

Azusa (ə zōō′ sə) [Los Angeles Co.]. The town was laid out in 1887, the "boom year," by J. S. Slauson and his associates and named after the Azusa (Dalton) grant, on which it is situated. From Gabrielino *ashúkshanga*, with the location suffix *-nga*, but no further etymology is known (Munro). In the *expedientes* of the Azusa land grants of May 10 and Nov. 8, 1841, the modern spelling is used. Local folklore claims that the town was named because a general store (of later years) sold everything "from A to Z in the USA."

B

Babcock Lake [Yosemite NP] was named by Lt. N. F. McClure in 1895 for John P. Babcock, chief deputy, California State Board of Fish Commissioners (Farquhar).

Babel Slough [Yolo Co.]. Named for Frederick Babel, a German who settled here on May 15, 1849, and developed a stock farm.

Baboon Lakes [Inyo Co.]. Name inspired by a CCC gang seen stripping off their clothes (Browning 1986).

Bache, Mount. *See* Loma: Loma Prieta.

Bachelor Valley [Lake Co.]. Named for four early settlers who were bachelors: Richard Lawrence, Green Catran, Daniel Giles, and Benjamin Moore (Mauldin).

Backbone Creek [Shasta Co.]. A party of Germans discovered a rich gold placer here in 1852: "This creek is about fifteen miles north of the mountain heights known as the Back Bone, and as no name was known for it by the party making the discovery, they called it Back Bone Creek" (*San Francisco Alta California*, June 7). The mountains were originally named Devil's Backbone, because the trail across them was very narrow, steep, and rough (Steger). **Backbone Mountain** and **Creek** [Fresno Co.] were named by the USGS in 1904 because of the shape of the ridge. There is a **Devils Backbone** in each of Sonoma and San Bernardino Cos. **Jacks Backbone** [Shasta Co.] got its name when J. M. Simmons's hired man, Jack, was washing his clothes in the stream; his boss remarked, "When you're humped over like that, you look like yonder mountain" (Steger; *see also* Homers Nose).

Bacon Meadow [Fresno Co.]. Named for Fielding Bacon, a pioneer stockman (Farquhar).

Baden (bā′ dən) [San Mateo Co.]. Charles Lux (of the Miller and Lux Company) had his Twelve Mile Farm here in the 1870s. Lux was born on the Rhine, opposite the duchy of Baden. South San Francisco was called Baden until its incorporation in 1908.

Baden-Powell (bā′ dən pō′ əl), **Mount** [Los Angeles Co.]. The BGN (6th Report) approved the present name for the eastern of the two North Baldys, which are within 2 miles of each other. The name honors Sir Robert S. S. Baden-Powell (1857–1941), founder of the Boy Scouts in 1908.

Badger. Only a few places are named for this fur-bearing mammal (*Taxidea taxus*), which never was numerous in California and was almost exterminated before the American conquest. The best-known features are the mountains in Siskiyou Co. and Lassen NP, the ski area at Badger Pass [Yosemite NP], and the lake in Madera Co. Badger Hill on the south bank of the Middle Yuba was a well-known mining town in the 1850s. The settlement **Badger** [Tulare Co.] had a post office called Camp Badger as early as 1892. Some Badger names may be for a person so named, or for a person from Wisconsin, the "Badger State." *See also* Tejon, the Spanish word for badger.

Badwater [Death Valley NP]. The name originated in a notation on the Furnace Creek atlas sheet, published in 1910, that the water in the pool was not potable. According to H. D. Curry, former park naturalist, the water is full of common salt and Glauber's salt (Death Valley Survey).

Bagby [Mariposa Co.]. Named for the owner of the hotel, B. A. Bagby, when the post office was established on June 30, 1897.

Bagdad [San Bernardino Co.]. Named in 1883 by the Southern Pacific Railroad when the original line (now Santa Fe) was constructed. It was presumably named after Baghdad, Iraq, because it is in the desert. In the 1850s there were two Northern California mining towns called Bagdad, one south of Oroville, the other on the Trinity River.

Bagley: Flat, Mountain [Shasta Co.]. Probably named for Alfred Bagley, who filed a claim for 160 acres on Feb. 24, 1852 (Steger).

Bago (bä′ gō), **Mount** [Kings Canyon NP]. The origin is not known (Browning 1986).

Bahia (bä ē′ ə) [Sonoma Co.]. From Spanish *bahía* 'bay', for the location on Suisun Bay.

Bailarin, El. *See* Rincon: Rincon Point.

Baile de las Indias. *See* Graciosa Ridge.

Baird Caves [Shasta Co.]. The caves were named after the former Baird post office, which had been named in 1879 for Spencer E. Baird, U.S. fish commissioner in 1872 (Steger).

Bair Island [San Mateo Co.]. A man named Bair was in charge of the South Belmont Oysterhouse here in the 1880s–90s (Alan K. Brown).

Bakeoven Meadow [Tulare Co.]. The name originated from an old oven constructed of mud and stone, used by early-day sheepherders for baking bread (J. T. Radel). *See* Hornitos.

Baker [San Bernardino Co.]. The station was first known as Berry, for Joe Berry, a prospector. To avoid confusion with other places of the same name, the Tonopah and Tidewater Railroad chose the present name in 1908 in honor of R. C. Baker, its president. The post office is listed in 1933.

Baker: Creek, Lake [Inyo Co.]. William A. Baker settled in this area in 1850 (Browning 1986).

Baker, Fort [Marin Co.]. Named in 1897 by the War Dept. for Col. Edward D. Baker, who commanded a Union regiment, called at first the California regiment, early in the Civil War. Fort Baker was previously called Lime Point Military Reservation.

Bakers Beach [San Francisco Co.]. Named about 1866 for the Baker family, owners of a dairy farm here (State Library).

Bakersfield [Kern Co.]. Col. Thomas Baker, a civil and hydraulic engineer, was state senator in 1861–62 (Darling 1988). He undertook to develop a navigable waterway from Kern Lake to San Francisco Bay, a project that failed. A parcel of land belonging to him became known as Baker's field; this name was transferred to the city in 1868.

Balaklala (bal ək lä′ lə) **Mine** [Shasta Co.]. From Wintu *balakɫili'* (Harrington), with possible influence from Balaklava, the Russian seaport famous for the "Charge of the Light Brigade" during the Crimean War.

Balance Rock [Tulare Co.]. The name, which refers to the huge rock that balances on another at the entrance to the resort, was applied by a Mrs. Shively in 1900 (Ruth Squire). The post office is listed in 1936.

Balboa (bal bō′ ə) [Orange Co.]. The subdivision was developed by the Newport Bay Investment Company in 1905; it was named for Vasco Núñez de Balboa (1475–1517), discoverer of the Pacific Ocean. **Balboa Island** was reclaimed by W. S. Collins in 1910–12 and was named after the older settlement across the channel. *For* Balboa Palisades, *see* Corona: Corona del Mar.

Balch (bôlch) **Park** [Tulare Co.]. The area known as Summer Home was acquired from the Southern California Edison Company by A. C. Balch of Los Angeles, who gave it to the county as a public park in 1923 (Mitchell).

Balcolm (bal′ kəm) **Canyon** [Ventura Co.]. Miles Balcolm and his family arrived here in 1873 (Ricard).

Bald, Baldy. With the exception of "black" and "red," no other adjective is so frequently applied to hills and mountains. More than a hundred mountains, peaks, buttes, hills, rocks, and tops in all sections of the state are called Bald, and several Baldy, though some are no longer bald. **Bald Jesse** [Humboldt Co.] is the name of an isolated peak. **Bald Mill Creek** [Fresno Co.] was not so named because a mill or creek was "bald," but because the Corlew Mill was on the creek that has its source on Bald Mountain. **Bald Rock** [Butte Co.] is a bare landmark in the canyon of the Middle Fork of the Feather River. *See* Ball; Bally; Pelona.

Baldwin Beach [El Dorado Co.]. E. J. ("Lucky") Baldwin, one of the most spectacular promoters in the latter part of the 19th century, bought a hotel here in 1890 (Lekisch). **Baldwin Lake** [San Bernardino Co.] was also named for the financier, who invested much of his Alaska fortune here (Fisk). **Baldwin Park** [Los Angeles Co.] developed around the Pacific Electric station built in 1912 and was likewise named for "Lucky" Baldwin—the onetime owner of the Rancho Puente de San Gabriel, on which the city is situated.

Baldy. *See* Bald; Bally; Old: Old Baldy; Vollmer Peak.

Bale [Napa Co.] commemorates Dr. Edward F. Bale, a spectacular character in the last decade of Mexican rule. Coming to California from England in 1837, he was probably the first practicing surgeon and druggist in the province; he was jailed for bootlegging in 1840 and for shooting at Salvador Vallejo in 1844. On Mar. 14, 1841, he petitioned for a grant of the place bearing the Indian name Calajomanas,

Caligolman, or Kolijolmanok, which the doctor, with a gruesome sense of humor, twisted (in his petition) into Carne Humana 'human flesh'. The name was officially accepted.

Balenas, Puerto. *See* Ballena.

Ball. Several mountains in the state, two granite pillars in Sierra National Forest, and a dome in Tulare Co. are so named because of their shape; but confusion with "bald" is possible (*see also* Bally).

Ballarat (bal′ ə rat) [Inyo Co.]. Named after the gold mining center in Australia, at the time of the gold boom here in the 1890s. The post office was established on June 21, 1897, and continued until 1917; now the place is a ghost town. The earlier name was Post Office Springs.

Ballard: settlement, **Creek** [Santa Barbara Co.]. The place was named in 1881 by George W. Lewis in memory of his wife's first husband, W. N. Ballard. From 1862 to 1870 the latter had been the stage station agent at this place, then known as Alamo Pintado (Co. Hist.: Santa Barbara 1939:384–85). *See* Alamo.

Ballast Point [San Diego Bay] was probably Cabrillo's first landing place in what is now the state of California, on Sept. 28, 1542. The chroniclers of the Vizcaíno expedition of 1602–3 noted that the stones at this point were suitable for ballast. González (1734:305) calls it Punta de los Guijarros ['cobblestones'] o Lastre ['ballast']. The point was frequently called Punta de Guijarros in the years that followed, and until 1835 the fort at the site of Fort Rosecrans was also called Guijarros. When the Boston ships stopped at this point in the early part of the 19th century to take on stones as ballast, the present name came into existence. In 1851, Ballast Point became the official name when the Coast Survey placed it on its sketch of San Diego Bay. **Ballast Point** [Catalina Island] appears in 1852 on the Coast Survey sketch of Catalina Harbor, no doubt for the same reason as mentioned above.

Ballena. The Spanish word for 'whale' was repeatedly used in Spanish times as a place name along the coast. Indeed, it might have become the name of the state: early maps show the southernmost point of Lower California (now Punta de San Lucas, or Cabo San Lucas) as Puerto (or Playa) Balenas or Punta de la California (Wagner, pp. 410, 430). **Ballena** (bə lē′ nə, bə yä′ nə) **Valley** [San Diego Co.] was named after the Indian Ranchería de la Ballena mentioned by Padre José Sánchez in Sept. 1821 (Arch. MSB 4:211). The original name was suggested by a nearby hill shaped like a whale. Richardsons Bay [Marin Co.] was formerly Whalers Harbor, a translation of Puerto de los Balleneros. *See* Bolinas; Pigeon Point.

Balley. *See* Bally.

Ballico (bä lē′ kō) [Merced Co.]. A station on the Santa Fe Railroad; the origin has not been found.

Ballinger Canyon [Ventura Co.]. Probably for Richard A. Ballinger, secretary of the interior in 1909 (Ricard).

Ballona (bə lō′ nə): **Creek, Lagoon** [Los Angeles Co.]. The name is derived from the Ballona or Paso de las Carretas grant, dated Nov. 27, 1839. According to a tradition of the Talamantes family, co-grantees of the rancho, the place was named after the city of Bayona in northern Spain, the home of one of their ancestors (W. W. Robinson, pp. 107 ff.). The letters *ll* and *y* in Mexican Spanish are both pronounced like the English consonant *y*, and are frequently interchanged in writing.

Balloon Dome [Madera Co.]. "A most remarkable dome, more perfect in form than any before seen in the State. It rises to the height of 1800 feet above the river, and presents exactly the appearance of the upper part of a sphere; or, as Professor Brewer says, 'of the top of a gigantic balloon struggling to get up through the rock'" (Whitney, *Geology*, p. 401). It was formerly known also as the Dome (or Great Dome). **Balloon Hill** [San Mateo Co.] was named in 1898 or 1899, when a captive balloon with several passengers escaped from San Francisco and landed on the top of this hill after drifting three days (Alan K. Brown).

Bally, Bolly, Bully. One of the most interesting names in California, unique because it is the only Indian name that has survived here as a generic geographical term; it is from Wintu *buli* 'mountain' (Shepherd). In English, the pronunciations (bal′ ē, bol′ ē, bo͞ol′ ē) all occur. It is now used principally on the east side of the Trinity Mountains, which was originally territory of the north and central Wintun tribes. There is an isolated **Bally Peak** about 6 miles east of Clear Lake; the *diseño* of the Los Putos grant [Solano, Napa Cos.] shows a mountain labeled Buli; that of the Cañada de Pogolimi grant [Sonoma Co.] has a Karsebalo, apparently a mountain ridge, and a Tecabala. The name survived locally in various spellings and was occasionally found in literature. Co. Hist.: Trinity 1858:35 mentions a "Baldy's or Bawly" north of Weaverville, and on p. 110 a "Jollabollas." Hutchings' *Illustrated California Magazine*, June 1861, refers to a Mount Balley between Shasta and Yreka. The name was rescued not long afterward by the men of the Whitney Survey, who surveyed the region in the fall of 1862. The name is repeatedly discussed in Whitney, *Geology* (e.g. pp. 323 ff.): "High mountains rise immediately north of Weaverville, the nearest considerable elevation being known as Mount Balley. 'Balley' appears to be the Indian term for a bare mountain, and the 'Shasta Balley' is to be distinguished from the 'Trinity Balley,' or the one near Weaverville; the orthography of the word is very doubtful." The fact that the men of the Whitney Survey assumed that *balley* means 'bare mountain' suggests that there may have been some confusion with the term "baldy." Other Baldys in the region—Indian Creek Baldy, Little Baldy, as well as Ball Mountain, Bailey Hill, etc.—may or may not have been derived from the same word by folk etymology. On p. 325 Whitney discusses the name again: "[The peak] is known locally as 'Yalloballey,' an Indian name of which the orthography and meaning are doubtful. It is not unlikely that . . . [it] is the same mountain that is indicated on the maps as 'Mount Linn,' a name which . . . is not known to anyone in this part of the State." The map of the survey (von Leicht–Craven, 1874) records Bullet [!] Chup, and North and South Yallo Balley. Today twelve or more peaks and hills in Shasta, Trinity, Tehama, and Lake Cos. are designated by the name, including two Bully Hills in Shasta Co.; *see* Bully Choop, Hay: Hayfork Bally, Shasta: Shasta Bally Mountain, Shoemaker Bally, Trinity: Trinity Bally, Winnibulli, Yolla Bolly Mountains. The name Bally is used sometimes as a true generic term but sometimes tautologically with Mountain or Hill—a confusion that existed already at the time of the Whitney Survey.

Baltic Peak [El Dorado Co.]. The name of the Baltic Sea, between Scandinavia and Germany, was a favorite name for mines in the Gold Rush. The peak was named by the USFS after the Baltic Mine north of the mountain, which was in operation until 1907.

Baltimore Park [Marin Co.]. Named after the Baltimore and Frederick Trading and Mining Company, which brought a sawmill round Cape Horn in 1849 and set it up at this place (Hoover, p. 194).

Bangor (bang′ ər) [Butte Co.]. Named by the Lumbert brothers after their home town in Maine, when they settled here in 1855.

Bankhead Springs [San Diego Co.]. Named in 1916 for U.S. senator John Hollis Bankhead of Alabama, father of actress Tallulah Bankhead (Stein).

Banner [San Diego Co.]. Supposedly named for a small flag that flew over the first gold strike in the canyon, in 1870 (Brigandi).

Banner Hill [Nevada Co.]. The hill was named after the Banner Mine (a favorite name for mines), which was located in 1860. In 1867 there was a Bannerville near the hill (H. P. Davis).

Banner Peak [Madera Co.]. Named in 1883 by Willard D. Johnson, topographer of the USGS, because he noticed cloud banners streaming from the summit (Farquhar).

Banner Ridge [Mono Co.]. The elevation between Lake Crowley and Benton was named for the Banner Mine, which was situated on its slope (Wheelock).

Banning [Riverside Co.]. The town, founded in 1884, was named by Welwood Murray for Phineas Banning (1830–85), who with his brother Alexander operated the first regular stage line between Los Angeles and San Pedro in the 1850s. Banning is remembered as one of the "grand old men" of the southern metropolis. *See* Wilmington.

Bannock [San Bernardino Co.]. The name appears on the Santa Fe Railroad maps after 1903–4. It was derived from the name of an Indian tribe in Idaho, *pannákwatï* (*HNAI* 11:306). The term was changed by folk etymology to Bannock, after a kind of flapjack much used by early traders and settlers.

Banta [San Joaquin Co.]. For Henry Banta, a pioneer settler; applied to the station when the old Western Pacific Railroad from Oakland to Stockton was built in the 1860s.

Bar. This generic term, referring to a flat area alongside a river or creek, became an important word in California geographical nomenclature during the Gold Rush. Gold-washing at a bar soon developed into one of the principal methods of obtaining the precious metal, and maps of the gold district are dotted with mining camps named after such bars. Nine California communities and post offices are still designated as Bars and are listed under their specific names.

Baranca is an Anglo misspelling of Spanish *barranca* 'ravine'. *For* Baranca Colorado, *see* Red: Red Bank Creek.

Barber Creek [Humboldt Co.]. Named for Joseph Tillinghast Barber, who received a patent for land here in the mid 19th century (Turner 1993).

Barber Springs [Lassen Co.]. For an Indian named Barber who, in 1857, was shot from ambush at the springs when he was returning from the Gold Run Mines (Co. Hist.: Lassen 1916:94).

Barcroft, Mount [Mono Co.]. The elevation south of White Mountain was named for Sir Joseph Barcroft (1872–1947), a British scientist especially interested in the physiological effects of high altitude (BGN, May 1954).

Bard [Imperial Co.]. The post office was established on Sept. 3, 1910, and was named for Thomas R. Bard, U.S. senator, 1901–5, in recognition of his promotion of the 13,000-acre irrigation district. **Bardsdale** [Ventura Co.], with a post office in 1892, was also named for him.

Bares Gulch [San Mateo Co.]. For David Bare, who lived here around 1870 (Alan K. Brown).

Barker Peak [Placer Co.]. William A. Barker grazed livestock here in the 19th century (Lekisch).

Barnabe (bär′ nə bē) **Mountain** [Marin Co.]. For the pet white mule of the family of Samuel P. Taylor, who settled here around 1856 (Teather 1986).

Barnard, Mount [Sequoia NP]. Named for Edward E. Barnard (1857–1923), an astronomer at Lick Observatory, University of California, 1888–95, and then at Yerkes Observatory, University of Chicago, until his death. The mountain was named at the first ascent on Sept. 25, 1892 (Farquhar).

Barnes Mountain [Fresno Co.]. For George W. Barnes, who settled near here in 1890 (Browning 1986).

Barney Lake [Mono Co.]. For Barney Peeler, an old resident of Bridgeport (*SCB* 12:126).

Barnwell. *See* Manvel.

Barona (bə rō′ nə) **Indian Reservation** [San Diego Co.]. In 1846 Juan Bautista López received the land grant Cañada de San Vicente y Mesa del Padre Barona, named after a priest at San Diego Mission, 1798–1810. In 1933, the United States purchased land here as a reservation for Diegueño Indians displaced when El Capitan Dam was built (Stein).

Barranca (bə rang′ kə). This Spanish word for 'ravine, gulch' has become naturalized as a generic term in California. **Barranca Colorada** [Tehama Co.] 'red ravine' was a Mexican land grant dated Dec. 24, 1844. *See* Red.

Barr Canyon [Monterey Co.]. For Harry Barr, from Ireland, who homesteaded in the area around 1888 (Clark 1991).

Barrett, Lake [Mono Co.]. For Lou Barrett, an early forester (Browning 1986).

Barrett Reservoir [San Diego Co.]. In 1879 George W. Barrett and his sister homesteaded the land near the dam site, and his name was applied to the reservoir by the City of San Diego in 1919 (B. B. Moore). The former post office for the settlement on Cottonwood Creek was named Barrett on Feb. 4, 1915.

Barroso (bə rō′ sō) **Mountain** [Kern Co.] may contain a Spanish word for 'pimpled', applied because the alternate patches of barren ground and wooded areas have a spotted appearance when viewed from a distance (Crites, p. 268). However, Spanish *barroso* also means 'muddy'.

Barry, Fort [Marin Co.]. The reservation was established by the War Dept. in 1904, from the part of Fort Baker known as Point Bonita, and named in honor of Gen. William F. Barry (1818–79), a veteran of the Mexican and Civil Wars.

Barstow [San Bernardino Co.]. Named in 1886 by the Santa Fe for William Barstow Strong, president of the railroad at that time. Since his name had already been commemorated in Strong City, his middle name was used for this junction on the line. In 1882, when the railroad was first constructed by the Southern Pacific, the name was Waterman Junction, after Robert W. Waterman, governor of California 1887–91. In the 1860s the place had been known as Fishpond and was a stopping place for travelers.

Barsug [Santa Barbara Co.]. Probably contains scrambled letters from the word "sugar", because of sugar beets grown in the vicinity; *see* Garus.

Bartle [Siskiyou Co.]. This was a weigh station on the first wagon road in the area, in 1856, and had a post office in 1889 (Luecke).

Bartlett: Springs, Creek, Flat, Mountain [Lake

Co.]. The health resort was named for Green Bartlett, a native of Kentucky, who drove cattle across the plains to California in 1856. In 1870 he discovered the spring and was cured by its waters, whereupon he bought the land around it and erected a hotel (*CHSQ* 23:296).

Bartolas [Kern Co.]. The plateau in the High Sierra north of Onyx was named for a Frenchman who ran sheep there in early days (Crites, p. 268).

Bartolo Viejo, Paso de [Los Angeles Co.]. A Mexican land grant dated June 12, 1835. On the *diseño* of the San Antonio grant (1838) Paso de Bartolo—pronounced in Spanish (bär tō′ lō), not (bär′ tō lō) as in Italian—is shown where the Camino Viejo crosses the Río de San Gabriel. Perhaps the name originally meant not 'crossing of Old Bartolo' but 'Bartolo's old crossing'. One Bartolo Tapia is listed as the owner of a rancho in the Los Angeles district in 1816 (Bancroft 2:353).

Barton: Peak, Creek [Kings Canyon NP]. The name of the peak was proposed by the Sierra Club, for James Barton, a local cattleman. An earlier name was Mount Moraine. The creek was named after the peak in 1932.

Barton Mound [Orange Co.]. For Sheriff James R. Barton, of Los Angeles, who was killed here in 1857 by a bandit, Juan Flores, and his band. *See* Flores Peak; Sheriffs Springs.

Basin. This generic term seems to have been used in California more extensively than in other states: note American, Blackcap, Big, Butte, Colusa, Sutter, and many minor Basins. *See* Great: Great Basin.

Basket Dome [Yosemite NP]. In Indian legend, this rock was the basket of a woman who, with her husband, was turned to stone here. She is Half Dome, he is North Dome (Browning 1986).

Basset [Los Angeles Co.]. O. T. Bassett, a banker from Texas, opened a subdivision between La Puente and El Monte in 1895 (Patricia A. Warren).

Bass Lake [Madera Co.]. The impounded reservoir for electric power generation was named after 1900, either for a person named Bass or for the fish. **Bass Lake** [Marin Co.] was named for the fresh-water fish (*Micropterus* sp.) that has been planted in many California lakes, though not native to the state (Stang).

Bass Mountain [Shasta Co.]. The older name, Saddleback Mountain, was changed by the USGS in 1901 to honor John S. P. Bass, assemblyman from Shasta and Trinity Cos. in 1880 (Steger). Bass Station, 17 miles from Shasta, was mentioned in Brewer, Notes, Aug. 20, 1862.

Batavia (bə tā′ vē ə) [Solano Co.]. The name was applied to a station of the California Pacific Railroad about 1870. It is the Latin name of the Netherlands and was first applied to a city in New York by early Dutch settlers.

Batequitos (bä tā kē′ tōs): **Lagoon** [San Diego Co.]. Font, on Jan. 10, 1776, mentions "the place called Los Batequitos, a small watering place somewhat apart from the road on the side away from the sea," two leagues from "the village of Indians and the place called San Alexos" (*Compl. diary*, pp. 190–91). *See* San Elijo Lagoon. The name is spelled Batiquitos on the San Luis Rey atlas sheet and Batiguitos by the Coast Survey. This was probably the place where the Portolá expedition dug a *batequi* on May 27, 1769; the word *batequi* is used in northwestern Mexico to mean 'a hole dug in a dry streambed in order to find water'. The origin is Yaqui *bate'ekim* (Blanco, 172).

Bathtub Lake [Lassen Co.]. Named in 1936 because it is shallow, with a smooth cinder bottom (Stang).

Battle. A favorite name for places where an engagement has been fought; in California the name appears about twelve times. **The Battleground** [San Mateo Co.], a flat along the road to Lobitos, was named for a fistfight between W. Smith Downing and Maj. W. W. McCoy, settlers holding adverse titles, around 1858; it lasted all day, "until Mrs. Downing came up with an ax" (Alan K. Brown). **Battle Mountain** [San Diego Co.] was named for the nearby battlefield of San Pasqual, the scene of the engagement between Kearny's forces and the Californios on Dec. 7, 1846. **Battle Island** [Clear Lake, Lake Co.] is the place where a detachment of soldiers under Captain Lyon (*see* Lyons Valley), in Dec. 1849, fought the Indians who had killed Andrew Kelsey (of Kelseyville) and Charles Stone; Gibbs mentions the name in 1851. **Battle Creek** [Tehama, Shasta Cos.] is apparently the creek called Arroyo de la Campaña on a *diseño* of Rancho Breisgau. It appears on the Frémont-Preuss map of 1848 as Noza Creek, but it became known by the present name after a bloody engagement was fought here between Indians and trappers in 1849 (Steger). Beckwith uses the present name in 1854. **Battle Rock** [Shasta Co.] commemorates a "battle" between Modocs and the miners whose operations interfered with the Indians' fishing. **Battle Creek** and **Mountain** [Tulare Co.] are so named because it was here that George Cahoon's burro "Barney" vanquished a mountain lion in a bloody skirmish (Farquhar).

Baulenes [Marin Co.]. A Mexican land grant dated on Oct. 19, 1841, and Nov. 29, 1845. *See* Bolinas; Duxbury; Tomales.

Bautismos, Cañada de los. *See* Cristianitos Canyon.

Bautista (bô tēs′ tə): **Canyon, Creek** [Riverside Co.]. For Juan Bautista ('John the Baptist'), a Cahuilla Indian leader (Gunther).

Bawly. *See* Bally.

Baxter, Mount; Baxter Creek [Kings Canyon NP]. Probably named for John Baxter, a rancher in Owens Valley (Farquhar).

Bay. A convenient name for a locality near a bay, usually in combination with -shore, -side, -view, etc. Five places are named after San Francisco Bay, two after Humboldt Bay, one each after Morro and Newport Bays, and one after the nameless bay on the east shore of San Miguel Island. *For* Bay City [Orange Co.], *see* Seal: Seal Beach; *for* Bay Island [Orange Co.], *see* Modjeska Island; *for* Bay Point, *see* Chicago.

Bayles [Shasta Co.]. The post office was established on Aug. 10, 1875, at Delta; A. M. Bayles, son of Judge Bayles, a settler of 1855, was postmaster (Steger). On Dec. 4, 1884, it was changed to Bayles, and on May 1, 1948, again to Delta.

Bayou Cita. *See* Vallecito.

"B" Canyon [Death Valley NP] was so named because it was close to "A" Canyon (*q.v.*).

Beakban Island [Lake Co.]. For Thomas W. Beakban, an early settler from England; the name is also found as Beakbane (Mauldin).

Beal's Landing. *See* Westport.

Bealville [Kern Co.]. When the Southern Pacific built this section in the 1870s, it named the station for Gen. Edward F. Beale (1822–93), a veteran of the Mexican and Civil Wars, who was superintendent of Indian affairs in California in the 1850s. Beale owned about 200,000 acres in the county. He was also U.S. surveyor general for California, and Abraham Lincoln is said to have complained that Beale made himself "monarch of all he surveyed" (quoting Cowper's poem about Alexander Selkirk, the prototype of Robinson Crusoe). **Camp Beale Air Force Base** [Yuba Co.] was established in Aug. 1942 and named by the War Dept. in honor of the same officer.

Beal Well [Imperial Co.]. Probably for Willis F. Beal, who came to the county in 1903; he later became a county supervisor, an assemblyman, and a director of the irrigation district.

Bean [Kings Co.], a Santa Fe station, and **Asa Bean Ridge** [Mendocino Co.] were apparently named for settlers. There are several Bean Creeks and Bean Canyons in the state, named either for families bearing such a name, or for the plant.

Bean Hollow [San Mateo Co.] was named after the Cañada del Frijol, a name applied in the 1840s; the English term was in use by 1861. Names like "Arroyo de los Frijoles" are errors made by the Coast Survey in 1854 (Alan K. Brown); *see* Lucerne: Lake Lucerne.

Bean Springs [Kern Co.] is for Charles M. Bean, who owned land here in the late 1880s (Darling).

Bear. Approximately five hundred geographic features in California bear the name of its largest and most notable native animal, either the black bear (which is sometimes "cinnamon," *Ursus americanus*) or the grizzly (*Ursus arctos*), which in some parts of the state was a major competitor with humans for food resources. Some of the Bear names may be translations from Spanish *Oso* (*q.v.*). There are numerous Bear Rivers, Bear Mountains, Bear Canyons, Bear Valleys, Bear Creeks, plus an assortment of Bear Lakes, Meadows, Flats, Basins, Buttes, Gulches, and Sloughs. Added to these are a Bear Dome [Fresno Co.], Bear Haven Creek [Mendocino Co.], some Bear Wallow Creeks [Kern, Mendocino, Los Angeles Cos.], and a number of Beartrap Creeks and Beartrap Canyons. **Bear Pen Creek** [Sonoma Co.] recalls the frontiersmen's method of catching bears alive in an enclosure. **Bear Valley** [Mariposa Co.] was named by Frémont in 1848. **Bearpaw: Cave, Butte** [Modoc Co.] was named by Tom Durham, an early-day bear hunter, who had his camp here; he nailed a paw to a juniper tree whenever he killed a bear (Howard). **Bearpaw Meadow** [Sequoia NP] was also supposedly so named because a bear's paw was found nailed to a tree; **Bearskin Meadow**, east of General Grant Grove, because the shape of the last snow patch to melt under the summer sun resembled a bearskin (Farquhar). Many features were named after bears because some encounter with the animal took place in the vicinity: **Bear River** [Humboldt Co.] was so named because Lewis K. Wood of the Gregg party was badly mangled by a wounded grizzly in Jan. 1850; **Bear Creek Gulch** [San Mateo Co.], because of "Grizzly" Ryder's encounter with a bear, which left him nearly dead (Wyatt). **Bear Lake** [San Bernardino Co.] was named as early as 1845 by "Don Benito" Wilson, whose men killed a number of bears there while on an Indian campaign. The present Big Bear Lake is not Wilson's lake but an artificial one created by the Bear Valley Irrigation Company in 1884. The original Bear Lake is probably what is now called Baldwin Lake. **Bear Creek** [Contra Costa Co.] appears on maps since the 1860s; in 1965 it was dammed for the Briones Reservoir. **Bear Paw** [Del Norte, Siskiyou Cos.], a spur of the Siskiyou Mountains, was so named because it is linked to Bear Mountain like a paw to a bear (BGN, 1965). *For* Bear River House, *see* Applegate.

Beard Creek [Alameda Co.]. The name for the tidal channel commemorates E. L. Beard, who settled near Mission San Jose about 1846 and acquired part of the mission lands. His residence north of the mission, as well as Beard Embarcadero, are shown on Hoffmann's map of the Bay region (1873).

Bear Flag Monument [Sonoma Co.]. Dedicated in 1914 to commemorate the raising of the Bear Flag and the proclamation at Sonoma of the short-lived California Republic on June 14, 1846.

Bearup (bēr′ əp) **Lake** [Yosemite NP]. Named by Lt. N. F. McClure in 1894 for a soldier in his detachment (Farquhar).

Beasore (bā′ sôr): **Meadow, Creek** [Madera Co.]. Named for Tom Beasore, an early settler (Durham).

Beatrice [Humboldt Co.]. The original name, Salmon Creek, was changed to the present name to honor Mrs. Beatrice White Hansen, the first postmaster, when the post office was established in 1884 (Turner 1993).

Beatys Icebox [San Diego Co.]. A. A. "Doc" Beaty (1871–1949) came to Borrego in 1912 (Brigandi). When the Truckhaven Trail to Highway 99 was built through the desert in 1929, there was a base camp at Beaty's place, where a cool cave served as a natural icebox. The cave was washed away, but the name remains for the canyon (Parker, p. 33).

Beaumont (bō′ mont) [Riverside Co.]. In 1887 a group of capitalists bought the townsite and gave it the French name, meaning 'beautiful mountain'. Originally the name had been Edgar Station, for a physician who accompanied a survey party there in the early 1850s; in 1875 it was changed to Summit—and in 1884 to San Gorgonio, after the nearby mountain—before receiving its present name.

Beaver. This mammal (*Castor canadensis*) was once found in parts of northern California. Before the American occupation of California, however, it had been almost exterminated by trappers of the Hudson's Bay Company and others; hence the name is found in only a few places. Beaver River was the trappers' name for Scott River.

Becker Peak [El Dorado Co.]. For John S. Becker, who settled here in 1886 (Lekisch).

Beck Lakes [Madera Co.]. Named about 1882 for John Beck, a prospector of the Minaret district (Farquhar).

Beckwourth (bek′ wûrth): post office, **Pass, Butte** [Lassen, Plumas Cos.]. These names commemorate one of the most colorful of frontiersmen, James Beckwourth, an African American from Virginia, who came to California in 1844, participated in the campaign against Micheltorena, and reappeared in California after the discovery of gold. Since "Old Jim" could not spell, his name is found as both Beckwourth and Beckwith; as a Southerner, he may well have pronounced it (bek′ wəth). In 1851 he discovered the pass named for him, which soon became an important immigrant route, later the route of the Sierra Valley and Mohawk Railroad, and finally that of the Western Pacific. The von Leicht–Craven map (1874) shows the spelling Beckworth for the pass, the peak, and a part of Sierra Valley. The Land Office map of 1879 shows the town with the spelling Beckwith; the post office name was so spelled until 1932, and the USGS retained this spelling for town, butte, and pass in 1942. Gannett erroneously assumed that these places were named for Lt. E. G. Beckwith of the Pacific Railroad Survey, who explored the Honey Lake route in 1854 (*see* Pac. R.R. *Reports* in Glossary). In

June 1950 the BGN ended the confusion by adopting for the pass and the peak the spelling used by the post office, which had been Beckwourth since July 1932.

Bee. The presence of the honey-gathering insect is recorded by Bee Tree Ridge [Sonoma Co.], several Bee Tree Creeks, a Bee Wash [Imperial Co.], and an assortment of Bee Creeks, Canyons, and Valleys. **Beehive** [Santa Cruz Co.] was so named because its outline resembles that of a beehive. **Beegum Peak, Creek**, and settlement [Tehama Co.]: The name was probably given to the peak by a Southerner because of its shape; the word "beegum" was applied in the South originally to a bee colony in a hollow gum tree, but since about 1850 it has referred also to a common beehive made of boards. The name appears on the von Leicht–Craven map as Bee Gum Creek, and on the Mining Bureau map of 1891 as Bee Gum Butte. The peak is of limestone and is honeycombed with holes, many of which are actually inhabited by bees.

Beeler, Point. *See* Bihler Landing.

Beeler Canyon [San Diego Co.]. Named before 1900 for German homesteader Julius Buehler (Stein).

Beer Creek. Several surface waters are named for the effervescent drink, probably because of their bubbling waters. Beer Gulch [Siskiyou Co.] appears on maps close to other mouth-watering names: Spirit Lake, Whiskey Creek, Tea Creek, Applesauce Creek, Sauerkraut Gulch.

Begg Rock [Ventura Co.]. Named by the Coast Survey for the ship *John Begg*, which grounded on a nearby rock on Sept. 20, 1824.

Bel-Air [Los Angeles Co.] was named for its developer, Alphonso Bell, in 1923 (Atkinson), with an echo of French *bel air* 'fresh air'. **Belair Channel** [San Mateo Co.] is named for the Au Bel Air House, opened by Charles Soulé near here in 1851–52 (Alan K. Brown).

Belden [Plumas Co.]. The Western Pacific station and the post office were named in 1909 for Robert Belden, a miner and property owner.

Belknap Creek [Tulare Co.]. Corrington G. Belknap patented 120 acres here in 1891 (Browning 1986).

Bell [Los Angeles Co.]. The name was applied in 1898 by James George Bell and his son Alphonso, founders of the town. The post office was called Obed until the name was changed to Bell on May 3, 1898. Alphonso Bell later developed the Bel-Air district in Los Angeles. **Bell Gardens** [Los Angeles Co.] was subdivided in 1930, and the area was named after the nearby town of Bell; Bell Gardens was incorporated in 1961 (Co. Hist.: Los Angeles 1965:4:65).

Bella Vista [Shasta Co.]. The town was founded by the Shasta Lumber Company; the post office was established in 1893 and named, in Spanish, 'beautiful view'. It was moved to the present location in 1920 (Steger).

Bell Canyon [Orange Co.]. So named because a large granite rock, when pounded, gave forth a peal that filled the canyon. It had been known in Mexican times as Cañada de la Campana 'canyon of the bell' (Stephenson).

Bellflower [Los Angeles Co.]. Established by F. E. Woodruff in 1906 and named Somerset in 1909 when the post office was established. The Post Office Dept. rejected the name because there was a town named Somerset in Colorado. The name was suggested by an orchard of bellflower apples in the area (from French *belle fleur* 'beautiful flower').

Bellota (bə lō′ tə) [San Joaquin Co.]. The place is shown as Donnel on Hoffmann's map of 1873, and as Belota on the von Leicht–Craven map of 1874. The present spelling, Spanish for 'acorn,' was adopted when the post office was established in 1879.

Bellows Butte. *See* Liberty Cap.

Bell Spring Mountain [Mendocino Co.]. In 1860, when the settlers pursued Wailaki Indians who had killed two brothers named Sprowell, the Indians were betrayed by two cowbells they carried. Jim Graham found the bells near the spring from which the mountain later received its name (Asbill, p. 253).

Bell Valley [Yuba Co.]. Named for the notorious highwayman Tom Bell, who used the valley as a hideout.

Bellview [Tuolumne Co.; Madera Co.] is for French *belle vue* 'beautiful view'.

Bellyache Canyon [Mariposa Co.]. According to a decision of the BGN, Sept.–Dec. 1963, the name was reportedly applied to the canyon because of an early resident named Billy Aike, or Ache.

Belmont. The name, a variant of Italian *bel monte* or French *beau mont* 'beautiful mountain', is found in various European and American countries. The oldest and most important Belmont in California, in San Mateo Co., developed around the hotel built in 1850–51; it was named by Steinburger and Beard because of the "symmetrically rounded eminence" nearby (Co. Hist.: San Mateo 1878:29). The town is mentioned in the *Statutes* of 1855; it was the county seat in 1856.

Belota is an Anglo misspelling of Spanish *bellota* 'acorn' (*q.v.*).

Belvedere (bel′ və dēr) [Marin Co.]. An international place name of Italian origin, meaning 'beautiful view'. The town was founded about 1890 by the Belvedere Club of San Francisco on what was formerly known as Peninsular (or De Silva's) Island.

Bena (bē′ nə) [Kern Co.] is a railroad name, perhaps transferred from Bena in Wisconsin or Minnesota; it is Ojibwa for 'partridge' and occurs in Longfellow's *Hiawatha*.

Ben Ali (ben ə lē′) [Sacramento Co.]. Named for James Ben Ali Haggin, a San Francisco lawyer, who was part owner of Rancho del Paso in the 1860s; also written Benali.

Benbow (ben′ bō) [Humboldt Co.]. The Hotel Benbow opened here in 1926, owned by the nine children of Arthur Benbow (Turner).

Bend [Tehama Co.]. In early days the place was known as Horsethief Bend because the bend in the Sacramento River afforded an ideal hideout for outlaws. This was later named Sander's Bend for an early settler and finally simplified to Bend (Dorothy Dorland).

Bendire (ben dēr′) **Canyon** [Inyo Co.]. The canyon in the Argus Mountains was named for Lt. Charles Bendire, native of Germany and veteran of the Civil War, who led a military expedition through Death Valley in 1867. The name also appears as Bendere's Cañon.

Benedict Canyon [Los Angeles Co.]. Named for Edson A. Benedict, a Los Angeles storekeeper, who filed a claim to the land in 1868 (W. W. Robinson, p. 166).

Benedict Meadows [Madera Co.]. For Maurice A. Benedict, supervisor of Sierra National Forest from 1909 to 1944, who died in 1959 (BGN, Jan.–Apr. 1960).

Bengal. *See* Bolo.

Ben Graves Canyon [Monterey Co.]. For Benjamin Graves, who settled here in 1885 (Clark 1991).

Ben Hur [Mariposa Co.]. Named by the community about 1890, for the hero of Lew Wallace's novel *Ben-Hur*, then at the height of its popularity (Grace Quick).

Benicia (bə nē′ shə) [Solano Co.]. On Dec. 23, 1846, Mariano G. Vallejo and Robert Semple signed an agreement for the founding of a city on the northern shore of Carquinez Strait, a logical place for a metropolis of the West. It was to be named Francisca, in honor of Vallejo's wife, and the agreement was recorded and translated in Yerba Buena on Jan. 19, 1847, by the *alcalde*, Washington A. Bartlett. This agreement and the town's name prompted Bartlett to bring about the change of the name of Yerba Buena to San Francisco, a name well known because of the mission and the bay. Nothing was done to found Francisca, and on May 19 Vallejo transferred his interest in the venture to Thomas O. Larkin and Semple. Before the end of May, Jasper O'Farrell began laying out the townsite. Lots in Francisca were advertised until June 5; but one week later, on June 12, Semple referred to the town as Benicia (another of Señora Vallejo's Christian names). On June 19, he made an official and public announcement in the newspapers that the change was made (Bowman). Benicia was also the name originally chosen for the county that was named Solano in 1850.

Ben Lomond (lō′ mənd): town, **Mountain** [Santa Cruz Co.]. John Burns, a Scot who settled here by 1852, named the mountain Ben Lomond for the mountain that overlooks Loch Lomond in Scotland. The post office existed as early as 1872, and the town was so named in 1887. The form Ben Lomond Mountain was used by the USGS when it mapped the region about 1900; its literal meaning is 'Mountain Lomond Mountain', as *ben* is a term used in Scotland for 'mountain'. A Ben Lomond in Butte Co. is shown on the von Leicht–Craven map.

Ben Mar Hills [Los Angeles Co.]. Established in 1920 and named by Ben Mark, promoter of the subdivision (R. F. Flickwir). The *k* in the family name was evidently omitted to make it seem more "Spanish."

Benmore: Canyon, Valley, Creek, Glades [Lake Co.]. The canyon was named for Ben Moore, a stock rustler, about 1851 (Mauldin). In 1960 the BGN decided for the spelling Benmore, which was apparently the original spelling of the name.

Bennett, Point [San Miguel Island]. Named by the Coast Survey before 1900, possibly for W. H. Bennett, who was engaged with George Davidson in the defense of Philadelphia in 1863.

Bennett Creek [Tulare Co.]. Named for William F. Bennett, a stockman of the 1870s (Farquhar).

Bennetts Well [Death Valley NP]. The name commemorates Asahel Bennett, a member of a party that crossed Death Valley in 1849–50. Although it is not certain that the Bennett-Manly party camped here, the name Bennetts Spring was affixed to this spot by the von Schmidt–Gibbs map of 1869. On the Wheeler atlas sheet 65-D of 1877 it appears as Bennett's Wells. According to Wheelock, it may not refer to Asahel Bennett but to a teamster, "Texas" Bennett.

Benson Lake and **Pass** [Yosemite NP] were named in 1895 for Harry C. Benson (1857–1924), acting superintendent of the park from 1905 to 1908 (Farquhar).

Benton: town, **Range, Crossing** [Mono Co.]. The town came into existence during the mining boom of 1865; the post office is listed in 1867. Since it was chiefly a silver-mining district, it may have been named for Sen. Thomas H. Benton of Missouri, Frémont's father-in-law, who was an advocate of metallic currency.

Berdoo (bər do͞o′) **Canyon** [Joshua Tree NM; Riverside Co.]. The popular name for San Bernardino, used by many persons but frowned upon by the Chamber of Commerce, was applied to this canyon through the Little San Bernardino Mountains and was placed on the Pinyon Well atlas sheet by the Corps of Engineers in 1944. The names of the triangulation hills of the district are playful abbreviations of San Bernardino, with or without the descriptive adjectives: Bern, Bernard, Dino, North Dino, Berdoo, Round Berdoo, Little Berdoo.

Berdugos, Sierra de los. *See* Verdugo Mountains.

Berenda. *See* Antelope.

Berg [Sutter Co.]. The name was applied by the Central (now Southern) Pacific Railroad in the 1870s, for the owner of the property—possibly John Berg, pioneer of 1850—on which the station was built.

Berkeley: city, **Hills** [Alameda Co.]. The name for the new college town was adopted on May 24, 1866, after many months of discussion, by the trustees of the College of California (now the University of California). The name was proposed by Frederick Billings, inspired by the verses that George Berkeley, Bishop of Cloyne, wrote when he sailed for Rhode Island in 1728, with the ultimate aim of forming a college in Bermuda and extending its benefits to the Americans: "Westward the course of empire takes its way; / The four first acts already past, / A fifth shall close the drama with the day; / Time's noblest offspring is the last."

Berlin. *See* Genevra.

Bermuda Dunes [Riverside Co.]. Founded in 1953 by Ernie Dunlevie and Ray Ryan; named because Ryan owned property in Bermuda (Gunther).

Bernal (bûr′ nəl) **Heights** [San Francisco Co.]. Probably named for Cornelio de Bernal, who acquired the entire territory of what was to become Bernal Rancho in 1839. His ancestor, Juan Francisco Bernal, was a soldier in Anza's party (Loewenstein).

Bernardo Mountain [San Diego Co.]. The mountain and a small settlement (now usually called Lake Hodges) preserve the name of the San Bernardo land grant, dated Feb. 16, 1842. El Paraje o Cañada de San Bernardo is mentioned in mission archives before 1822 (Arch. MSB 3:261) and repeatedly in later years. The San Dieguito River is shown as Arroyo de S. Bernardo on a *diseño* of 1841, and on the Land Office map of 1859.

Bernasconi (bâr nə skō′ nē): **Hot Springs, Pass, Hills** [Riverside Co.]. Named for Bernardo Bernasconi of Switzerland, who developed the springs in the 1880s.

Bernice Lake [Yosemite NP] was named for the wife of W. B. Lewis, superintendent of the park from 1916 to 1927.

Berrendo. *See* Antelope.

Berros (bâr′ ōs); **Los Berros: Creek, Canyon** [San Luis Obispo Co.]. *Berros* 'watercress' in the district was mentioned by Crespí on Sept. 4, 1769. Los Berros Creek is shown as Arroyo de los Berros on Hutton's map of Rancho Nipomo, 1850. The town was founded and named in 1891.

Berry. *See* Baker.

Berry Creek [Butte Co.]. The creek was named after a rich mining bar of the 1850s. The post office was established on May 10, 1875, at Virginia Mills but was named after the creek. Gold mining was carried on until the 1920s. The creek may have been named for Henry Berry, of Pennsylvania.

Berry Creek [San Mateo Co.]. For the homestead of T. G. Berry, 1878 (Alan K. Brown).

Berryessa (bâr ē yes´ ə): post office, **Creek** [Santa Clara Co.]. Named for an old family that came directly from Spain and settled in the Santa Clara Valley. On May 6, 1834, Nicolás Berreyessa was grantee of the land grant Milpitas, through which the creek flows. José Reyes Berreyesa was grantee of the San Vicente grant on Aug. 1, 1842. The family name is spelled in various ways, including Berrelleza.

Berryessa Valley [Napa Co.]. The valley preserves the names of José Jesús and Sisto Berryessa, soldiers in the San Francisco company and grantees of Rancho Las Putas in 1843. By decision of the BGN in 1957, the former Monticello Reservoir is now known as Berryessa Lake.

Berryvale. *See* Strawberry: Strawberry Valley [Shasta Co.].

Berteleda (bûr ti lē´ də) [Del Norte Co.]. Named by Dr. George J. Brown, a Crescent City physician, for his three daughters, one of whom was called Bertha; now called Douglas Park (Saunier).

Berwick Canyon [Monterey Co.]. For Edward Berwick, who bought land here in 1867 (Clark 1991).

Beryl Lake [Fresno Co.]. Possibly named by Scott M. Soule of the Dept. of Fish and Game in 1946 (Browning 1986). The term refers to a mineral but is also a woman's name.

Bestville [Siskiyou Co.]. The one-time metropolis of Salmon River was named for Captain Best, who in 1850 led the first party of prospectors into this wilderness area (Luddy).

Bethany: town, **Ferry** [San Joaquin Co.]. First called Mohr Station, for John Mohr, who deeded the site to the Southern Pacific. In order to avoid confusion with another station called Moore's Station, the name was changed to Bethany, probably after the place in Palestine.

Bethel Island [Contra Costa Co.]. The term "bethel" refers to a sacred area, from Hebrew *beth el* 'house of God'.

Betteravia (bet ə rā´ vē ə) [Santa Barbara Co.]. From French *betterave* 'sugar beet'. The beet industry was introduced here in 1897 by the Union Sugar Refining Company (AGS: *Santa Barbara*, p. 182.)

Betty Lake [Fresno Co.]. Named in 1935 by Ted Anderson, former packer in the Dinkey Creek area, for his daughter (Browning 1986).

Beulah. *See* Mineral King.

Beverly Hills [Los Angeles Co.]. A newspaper account in 1907, which reported that President Taft was spending a few days at Beverly Farms (in Massachusetts), suggested the name to Burton E. Green, president of the Rodeo Land and Water Company. The post office was established in 1907 as Beverly, and the name was changed to Beverly Hills in 1911.

Beville (be´ vil) [Tulare Co.]. Named for a Visalia family that took mountain outings here (Browning 1986).

Bicycle Lake [San Bernardino Co.]. The intermittent lake has become well known because of the army antiaircraft range here, which is named after the lake. The story that an unfortunate traveler left a bicycle here while attempting to cross the desert was confirmed by Washington W. Cahill, a longtime official of the U.S. Borax Company, who told L. Burr Belden that the teamsters of the company found a rusty bicycle here around 1890. However, according to an old resident of Barstow, the young men of Daggett in its flourishing days used the lake for bicycle races when it was dry (Arda Haenszel). In 1957 an aerial photograph, taken by pilot Don Krogh and widely published, shows a corpse lying beside a bicycle in the Mojave Desert not far from the lake.

Biderman (bī´ dər mən), **Mount** [Mono Co.]. Named for J. W. Biderman, prospector and promoter in the Bodie mining district in the early 1860s. The name also appears as Biedeman.

Bidwell: Bar, Creek, Butte, State Park [Butte Co.], **Lake** [Lassen Co.], **Peak** [Modoc Co.]; **Fort Bidwell**: town, **Indian Reservation** [Modoc Co.]. The bar in Feather River was called Bidwell's Bar after John Bidwell (*see* Glossary) discovered gold here in July 1848. The fort in Modoc Co. was named in 1865 for Bidwell, who was at that time a congressman and a general in the California militia. The post office was established in the 1870s. The state park was presented as a memorial by his widow. Bidwell Point in Mendocino National Forest was named by the USFS in 1934 because the pioneer had spent an anxious night there in 1844 in fear of an attack by Indians.

Bieber (bē´ bər) [Lassen Co.]. The post office was established in 1879 and named for Nathan Bieber, who in 1877 built the first store and house at the crossing of Pit River known as Chalk Ford.

Biedeman, Mount. *See* Biderman, Mount.

Bielawski (bē lav′ skē), **Mount** [Santa Clara Co.]. Named by J. D. Whitney in 1861 for the chief draftsman of the surveyor general's office, Capt. Casimir Bielawski; he came to California in 1853, and was connected with the U.S. Land Office in San Francisco for forty-five years.

Big. The adjective is used in geographical nomenclature for various reasons. First, a physical feature is actually "big": Big Basin [Santa Cruz Co.]; Big Lagoon [Humboldt Co.]; Big Creek [Fresno Co.]. **Big Bar** was a name given to many river bars (*see* Bar) during Gold Rush days; the places so named in Amador and Trinity Cos. are best known. (*For* Big Bar in Butte Co., *see* Pulgas: Pulga; in Calaveras Co., *see* Mokelumne Hill.) Second, a feature is called "big" in comparison with a smaller feature named previously or at the same time. The most notable example of this is the **Big Bear** cluster in San Bernardino Co. (*see* Big Bear Lake *below*). The result of this unimaginative method of naming is confusing because it is often not clear whether the specific or the generic term is modified by "big": thus Big Coyote Knoll [San Mateo Co.], Big Salmon Creek [Mendocino Co.]. Sometimes composite names with a humorous effect are evolved: Butte Co. has a Big Chico Creek and a Little Chico Creek, i.e. a big and a little "little" creek; on the Caliente Road in Sequoia National Forest, two water sources are designated as Big Last Chance Can and Little Last Chance Can. Third, the adjective is applied because it clearly modifies the specific term: thus **Big Bird Lake** [Tulare Co.] was named because the tracks of a large bird were found on its shore (*see also* Dollar Lake); Big Oak Flat [Tuolumne Co.] was named for a large oak; and Big Rock Creek [Los Angeles Co.], because of a huge boulder in the creek. Some names are translations of the Spanish *grande:* Big Sur River [Monterey Co.] was Río Grande del Sur 'big river of the south'; Big River [Mendocino Co.] is shown as Río Grande on a *diseño* of the Albion grant dated Oct. 30, 1844. The adjective also occurs in the names of a number of post offices and communities: Big Bear: Lake, City, Park [San Bernardino Co.]; Big Chief [Placer Co.]; Big Flat [Siskiyou Co.]; Big Gum [Shasta Co.]; Bigpipe [Inyo Co.]; Big Trees [Calaveras, Santa Cruz Cos.]; Big Bend [Shasta Co.]; and others. **Big Oak Flat** [Tuolumne Co.] was named for a giant oak near which a rich mining town started to develop in 1849. The post office was established on Jan. 21, 1852, and the place is shown on the maps of that year. The name is preserved in the Big Oak Flat Road. **Big Basin Redwoods** [Santa Cruz Co.]: The state park was established in 1961 with the assistance of the Save-the-Redwoods League and the Sierra Club. The cluster of names around **Big Bear Lake** [San Bernardino Co.] originated with Bear Lake, so named by "Don Benito" Wilson in 1845. The artificial lake was created in 1884 by building the Old Bear Valley Dam. Big Bear City is a community but not an incorporated city; Big Bear Lake is a city. (*For* Big Bear Park, *see* Sugarloaf.) **Big Sandy Creek** and **Rancheria** [Fresno Co.] are the site of a Native American community, with federal recognition since 1980. **Big Valley Rancheria** [Lake Co.] is also the site of a Native American community. *For* Big Coyote [Marin Co.], *see* Tamalpais: Tamalpais Valley; *for* Big Meadows [Mono Co.], *see* Bridge: Bridgeport; *for* Big Meadows [Plumas Co.], *see* Prattville; *for* Big Meadows Reservoir [Plumas Co.], *see* Almanor, Lake; *for* Big Moody Creek [Santa Clara Co.], *see* Campbell Creek; *for* Big Valley City [Lassen Co.], *see* Nubieber.

Bigelow Peak [Yosemite NP] was named for Maj. John Bigelow Jr., acting superintendent of the National Park in 1904 (Farquhar).

Biggs [Butte Co.]. The station was named in 1870 by the California and Oregon Railroad for Maj. Marion Biggs, who was the first rancher to ship grain from there (Co. Hist.: Butte 1882:246).

Bighorn Lake [Fresno Co.]. The name refers to the mountain sheep, *Ovis canadensis*. The lake was named shortly before 1900 by Lincoln Hutchinson's party: "An exclamation of surprise burst from one of the party, and we found directly before us a band of 'big-horn' sheep. We had supposed the animal long since extinct in the Sierra, and at first we could scarcely believe our eyes . . . Deep down in the amphitheater below us lay an azure lake" (*SCB* 4:202 ff.). In a footnote he adds, "This we named Big Horn Lake." **Bighorn Spring** and **Gorge** [Death Valley NP]: E. W. Nelson of the Merriam expedition camped by the spring in 1891 while hunting sheep. The name was applied to the gorge by the Park Service in 1939.

Bigler, Lake. *See* Tahoe.

Bihler (bē′ lər): **Landing, Point** [Sonoma Co.]. Named for William ("Dutch Bill") Bihler, a German butcher who came round the Horn to California in 1848; he became a large landowner and stockbreeder in Sonoma Co. The landing was called Black Point Harbor on the county map of 1879, but the Coast Survey used the name Beeler in 1876. In 1915 the present name was placed on the Plantation atlas sheet of the tactical map by the Corps of Engineers.

Bijou (bē′ zhōō) [El Dorado Co.]. The French word for 'jewel' was applied to the resort in the 1880s, probably because it was good publicity. In 1843 Frémont camped on Bijou Creek in Colorado; but since none of the detailed maps show the place in California before 1890, it is not likely that it was an early transfer name.

Biledo Meadow [Madera Co.]. A cabin was built here in 1890 by Thomas Biledo, a French Canadian; the name is properly spelled Biledeaux (Browning 1986).

Billibokka. *See* Bollibokka.

Billy Hughes Gulch [San Mateo Co.]. Named for "a character with a small goat ranch from the middle 1850s" (Alan K. Brown).

Billys Peak [Placer Co.]. For William Albert Dutton (1951–89), member of the Tahoe Nordic Search and Rescue Team (BGN, 1992).

Bingaman Lake [Yosemite NP]. John W. Bingaman was a park ranger, 1921–56, who first planted trout here in 1930 (Browning 1986).

Biola (bī ō′ lə) [Fresno Co.]. The name was coined by William Kerchoff of Los Angeles from the initial letters of Bible Institute of Los Angeles, when he established the town in 1912.

Birch. The birch is rare among native California trees, hence only a few places are named for it. The water birch, *Betula occidentalis*, is found in eastern and northern regions of the state. There is a Birch Creek in Deep Springs Valley [Inyo Co.], and there are two in Owens Valley, after one of which the stately **Birch Mountain** was named, a name appoved in 1938 by the BGN. Old-timers are of the opinion that **Birch Creek** in the San Bernardino Mountains was named for a tree mistaken for a birch, as there are no birch trees there (Fisk).

Birchim Lake [Inyo Co.]. Possibly named for James G. Birchim, who came to Round Valley in 1865 (Browning 1986).

Birchville [Nevada Co.], now a ghost town, was named in 1853 "by common consent . . . in honor of L. Burch [*sic*] Adsit, a prominent citizen" (Co. Hist.: Nevada 1880:61).

Bird. More than twenty-five geographic features, mainly islands and rocks, were given this name because of the presence of birds. Alcatraz Island [San Francisco Bay] was once known as Bird Island. **Big Bird Lake** [Tulare Co.] was named in 1902 by James Clay because tracks of a large bird were seen on the shore (Farquhar). Santa Clara Co. has a Blackbird Valley, and Orange Co. a Blue Bird Canyon. **Bird Rock** [San Diego Co.], a subdivision started by M. Hall, took its name from a rocky formation that resembled a bird (Hunzicker). *See* Pajaro.

Birds Crossing [Butte Co.]. Probably named for Ralph Bird, one of the founders of Ophir City (Oroville) in 1855.

Birds Landing [Solano Co.]. Formerly the shipping point of John Bird, who had a storage and commission business. Bird, a native of New York, settled here in 1865 (Hoover).

Bishop: Creek, town [Inyo Co.]; **Pass** [Kings Canyon NP]. Named for Samuel A. Bishop, a well-known cattleman, who came from Virginia to California in 1849, served as an officer in the Mariposa Battalion in 1851, lived in Owens Valley from 1860 to 1863, and became a supervisor of Kern Co. in 1866.

Bishop Rocks [San Diego Co.]. These rocks, part of the Cortes Bank, were discovered in 1855 when the Philadelphia clipper *Bishop* struck one of them. The name was applied by the Coast Survey in 1856.

Biskra Palms [Riverside Co.]. Named for the ancient Algerian town of Biskra, once the site of a Roman fort, and known for centuries as a Sahara Desert oasis in which huge gardens of palm and olive trees grow (Gunther).

Bismarck Knob [Napa, Sonoma Cos.]. The peak is probably the same as Carnero Mountain, mentioned in the *Statutes* of 1850. In later years it became known as Mount Nebo (D. T. Davis). When the Napa quadrangle was surveyed by the USGS in 1899, the present name was applied, probably in memory of the German statesman Otto von Bismarck (1815–98).

Bisnaga (biz nä′ gə) **Alta Wash** [San Diego Co.] refers to "tall barrel cactus" (Stein, p. 162).

The term *biznaga* means 'barrel cactus' in Mexican Spanish, but in Spain refers to a wild carrot; it is from Latin *pastinaca* 'carrot'.

Bitter, Bitterwater. The palatability of water, especially in the arid sections of the state, is of great importance to herdsmen, prospectors, and surveyors, and a creek or lake with bitter-tasting water is almost invariably so designated. There are more than fifty Bitter Creeks and Lakes and Bitterwater Canyons in California. *See* Amargosa.

Bivalve [Marin Co.]. This site on Tomales Bay refers to two-shelled mollusks such as clams and oysters.

Bixby: Landing, Creek [Monterey Co.]. Named for Charles H. Bixby, who came to the area in 1868.

Bixby Slough [Los Angeles Co.]. In Spanish days this was known as Cañada de Palos Verdes. In 1884 it appeared on maps as Lagunita, and by 1900 it had come to be known as Bixby Slough, for Jotham Bixby, pioneer and landowner (*HSSC:Q* 19:9 and 19:19).

Black. Of the adjectives of color used in descriptive geographic names, "black" is by far the most common, followed by "red" and "white." It is chiefly applied to orographic features that appear black because of the geological formation, or the dark green chaparral, or the atmospheric condition. California has around one hundred Black Mountains, some thirty Buttes, about ten each of Peaks, Cones, and Rocks, but only a few Hills. The twenty-five or more Black Canyons were doubtless so named because of their dark and forbidding appearance; most of the Black Lakes, Springs, and Creeks are not black but were named after a Black Mountain nearby. Often the generic term is preceded by a specific term modified by black: Black Crater Mountain, Black Lake Canyon, Black Butte River, Blackcap Mountain, Blackoak Mountain, Blackhawk Creek, Black Rabbit Canyon Creek. Merced Co. has a Black Rascal Creek, but the sinister story behind its naming is not known. Black Rascal Hills are mentioned in a railroad survey of 1873. **Black Lassic Creek** [Trinity Co.] was named after a black peak: neither Lassic (an Indian chief) nor the creek was black (*see* Lassic). **Black Bear** [Siskiyou Co.] was named after the Black Bear Mine, started up in 1860; there was a post office from 1869 to 1941 (Luecke). **Black Butte** [Siskiyou Co.] was so known in the early 1850s because it was "black as the darkest iron ore" (Pac. R.R. *Reports* 2:2.50). The BGN changed the name first to Wintoon Butte, then to Cone Mountain, and finally (June 6, 1934) back to the original name. **Black Star Canyon** [Orange Co.]: The Cañada de los Indios, so named because of the large Indian village there, became known by the present name after Joseph Yoch opened the Black Star Coal Mine in the canyon in the late 1870s (Stephenson 1:49). **Black Rock Pass** [Sequoia NP]: This name for the pass was recorded in 1896 by Lt. M. F. Davis because a band of black rock is in noticeable contrast to the red and white formations nearby (Farquhar). **Black Giant** [Kings Canyon NP] was named in 1904 by J. N. LeConte. When the USGS made the map of the Mount Goddard quadrangle, 1907–9, it placed the name Mount Goode on this peak, apparently unaware of the name given by LeConte. The earlier name was restored in 1926 by decision of the BGN, and the name Goode was transferred to another peak (Farquhar). **Black Divide** [Kings Canyon NP]: The ridge south of Black Giant was named by George R. Davis, topographer, when the Mount Goddard quadrangle was surveyed, 1907–9. **Blackeye: Grade, Canyon** [Lake Co.] were named at the turn of the century for "Blackeye" Herndon, who lived at the top of the grade and was so nicknamed because he had big black eyes, or because one of his eyes was blacked in a fight (Mauldin). **Black Butte Dam** [Glenn Co.]: The 140-foot earth-filled dam on Stony Creek, built to create the Stony Gorge Reservoir, was dedicated in June 1963. **Black Hawk** [Contra Costa Co.] was the name of an American Indian chief (1768–1838) who led native tribes of the Midwest against the whites. *For* Black Peak [Madera Co.], *see* Madera Peak; *for* Black Point Harbor [Sonoma Co.], *see* Bihler Landing.

Blackie Lake [Madera Co.]. For the nickname of Fish and Game warden Herb Black (Browning 1986).

Black Mountain [Marin Co.]. The name commemorates James Black, a Scottish sailor, who came to California in 1832. He was grantee of Cañada de Jonive, Feb. 5, 1846; owner of part of Rancho Nicasio and part of Olompali, 1852; a judge in 1850; and an assessor in 1852. The mountain is shown as Black's Mountain on a county map of 1873. **Black Point** [Marin Co.] was probably named for the same pioneer. A surveyor's plot (1859) of Rancho Novato shows Black's Store in Novato Valley. Hoffmann's map of the Bay region (1873) has a settlement called Black John 2 miles north of the point, and J. Black 4 miles north. The three names are on Rancho San Jose, whose owner, the widow of Ignacio Pacheco, became James Black's wife in 1865. The post office, named after the point, was established on Jan. 11, 1865. In 1905, real estate promoters succeeded in having the name changed to Grandview Terrace; on Apr. 1, 1944, the old pioneer name was restored (H. W. Hobbs). *For* Blacks, *see* Zamora.

Blackwells Corner [Kern Co.]. In 1921 George Blackwell established a gasoline and refreshment stop here, at the intersection of state highways 46 and 33 (Darling).

Blackwood Creek [Placer Co.]. Hampton C. Blackwood settled here around 1866 (Lekisch).

Blair. *See* Whiskey: Whiskeytown.

Blairsden [Plumas Co.]. Named about 1903 after the country home of James A. Blair, prominent in the early financing of the Western Pacific Railway (*Headlight*, July 1942).

Blair Valley [Anza-Borrego Desert State Park] was probably named for David Blair, a prospector who disappeared in 1887 (Brigandi).

Blair Valley [San Diego Co.]. Named before 1893 for settler Samuel "Bronco Sam" Blair (Stein).

Blakes: Sea, Ravines [Riverside, San Diego Cos.]. Parker (*see* Glossary) uses these names in his guidebook. They commemorate William P. Blake, geologist with the Pacific Railroad Survey, 1853–54. Blakes Sea is the ancient lake bed (the middle of which is now covered by Salton Sea) that was explored by Blake and named Lake Cahuilla. The Ravines are strange, washed-out canyons, west of Highway 99 (Parker, pp. 39–40, 46–47). *See* Cahuilla; Coachella; Salton Sea.

Blanc Lake [Inyo Co.]. Named in 1954 by Ralph Beck of the State Dept. of Fish and Game, perhaps for Mont Blanc in the Alps; *see* Alsace Lake (Browning 1986).

Blanco (blang′ kō), **Blanca** (blang′ kə). The Spanish word for 'white' is found in the names of about fifteen orographic features, including Peñon Blanco [Tuolumne, Mariposa Cos.], from *peñón* 'rocky mountain'; Muro Blanco [Fresno Co.]; Pico Blanco [Monterey Co.], Blanco Mountain [Mono Co.]; and Piedra Blanca Creek [Ventura Co.].

Blanco [Monterey Co.]. Tom White, who deserted ship at Monterey in 1840 and became known as Tomás Blanco ('white') to his Spanish-speaking neighbors, received a small grant on the Salinas River on Aug. 27, 1844. His place became known as Blanco Crossing, and in the fall of 1872 the name was given to the Southern Pacific station. It is a coincidence that, on Sept. 28, 1769, the Portolá expedition had given the name Real Blanco 'white camp' to a site about 25 miles southeast of the present settlement, because of the white appearance of the earth in the vicinity (Crespí, p. 199).

Blaney Meadows [Fresno Co.]. The original name, Lost Valley, was changed to the present name for a sheepman who camped there every year (Farquhar).

Blanton Creek [Humboldt Co.]. For John and Elizabeth Blanton, who settled here in 1871 (Turner 1993).

Blind Spring: Hill, Valley [Mono Co.]. An underground spring that was hard to find gave the name to the hill and the valley. In early days a barrel casing made the water available to miners on the hill (Robinson). There are several Blind Canyons in the state—which probably, like "blind" roads, lead nowhere.

Bliss Rubicon Point State Park [El Dorado Co.]. Named in memory of Duane L. Bliss and after Rubicon Point at Lake Tahoe. The park was given to the state in 1929 by William, Walter, and Hope Bliss.

Blithedale Canyon [Marin Co.]. Blithedale Sanitarium was built here in the 1870s, apparently named for Nathaniel Hawthorne's *Blithedale romance* (Teather).

Blocksburg [Humboldt Co.]. Named for Benjamin Blockburger, an immigrant of 1853 and participant in Indian fights; he established a store at this place in 1872. When the post office was first established on Jan. 30, 1877, the name was spelled Blocksburgh; since Apr. 29, 1893, the spelling has been Blocksburg.

Blomquist (blo͞om′ kwist) **Creek** [San Mateo Co.]. The Blomquist family lived here from the late 1860s (Alan K. Brown).

Bloody. The names of about twenty features in the state include the word "bloody" or "blood." Usually the name was applied because blood flowed there for some reason, though sometimes it may have been used in jest, or because of a cursing Englishman, or because the rocks looked blood-red. **Bloody Island** [Tehama Co.] was named in 1844 by

Samuel J. Hensley, of Sutters Fort, because Indians had attacked his party there. It is shown as Isla de la Sangre on a *diseño* of Rancho Buenaventura. **Bloody Island** [Lake Co.] was the scene of the "Bloody Island Massacre" in Dec. 1849 (*see* Battle: Battle Island). **Bloody Point** [Modoc Co.]: On the promontory jutting out into Tule Lake, Modoc Indians ambushed parties coming over the Oregon Trail. In Sept. 1852 a party of sixty-four was massacred here (William S. Brown). **Bloody Rock** [Mendocino Co.]: According to Powers (pp. 136–38), the last of the Chumaia (a part of the Yuki) perished here by leaping from the boulder upon which they had been driven by white settlers. **Bloody Canyon** [Mono Co.]: "The horses were so cut by sharp rocks that they [the early prospectors] named it 'Bloody Canyon,' and it has held the name—and it is appropriate" (Brewer, p. 416). **Bloody Mountain** [Mono Co.]: The slope of this peak was said to have been the scene of a bloody fight between escaped convicts and a posse in September 1871 (*see* Convict). But according to Wheelock, the name probably originated from the "bloody" red rocks of the peak. **Blood Gulch** [Amador Co.] was so named because miners noted blood mingled with water and found a dead man when they went upstream. **Bloody Run** [Nevada Co.] "took its name by virtue of the greater than ordinary number of murders once committed along its course" (Ritchie, pp. 363–64).

Bloomfield [Sonoma Co.]. The place was first settled in 1855. When the post office was established on July 12, 1856, it was named for Dr. Frederick G. Blume, a German surgeon who had been a prominent pioneer in the county since 1847. Blumefield, which was probably the intended form, was easily changed to the popular American place name.

Bloomington [San Bernardino Co.]. This popular American name for a town was applied to the development by the Semi-Tropic Land and Water Company, organized in 1887 (Co. Hist.: San Bernardino 1904:619).

Blossom Lakes [Sequoia NP] was named in 1909 by R. B. Marshall of the USGS, for Charles W. Blossom, park ranger (Farquhar).

Blossom Rock [San Francisco Bay]. A rock slightly below the surface of the water halfway between Alcatraz and Yerba Buena Islands. Discovered and named by F. W. Beechey for his ship, *Blossom*, which anchored nearby in Nov. and Dec. 1826. The danger to shipping was removed when A. W. von Schmidt blew off the peak with twenty-three tons of powder on Apr. 23, 1870.

Blucher [Sonoma, Marin Cos.]. The name of a Mexican land grant dated June 5, 1842, and Oct. 14, 1844, and granted to Juan J. Vioget, a native of Switzerland and pioneer surveyor of San Francisco. Vioget was generally known as "Blucher" because he strongly resembled the Prussian field marshal Gebhard von Blücher (1742–1819), Wellington's comrade-in-arms at Waterloo (Bowman).

Blue. There are more than twenty-five Blue Mountains, Ridges, and Hills throughout the state, apparently so named because of their appearance under certain atmospheric conditions, or because of the presence of bluish rock formations, or because the slopes were covered by *Ceanothus* ("wild lilac") or other blue blossoms. Several Blue Canyons were probably named for the same reason. Since the adjective was used in Spanish times (Sierra Azul on several *diseños*), some names may be translations. The many Blue Creeks and Blue Lakes were apparently all named because of the color of the water. The name of the feature was sometimes transferred to settlements: Blue Lake [Humboldt Co.], Blue Rock [Mendocino Co.], Blue Canyon [Placer Co.]. The last received its name from the blue smoke of the camps when extensive lumbering was done in the 1850s. According to Loye Miller, however, the canyon was originally named for Old Jim Blue, who mined there in the 1850s. *For* Blue Mountain [San Francisco Co.], *see* Davidson, Mount. **Blue Nose** [Siskiyou Co.] derived its name from the Blue Nose gold mine, an enterprise started by a group of Nova Scotians, sometimes called Blue Noses. A post office was established on June 1, 1917 (R. W. Bower). Imperial Co. has a railroad stop called Blue Goose, and San Bernardino Co. has a post office named Blue Jay. Often not the generic but the preceding specific name is modified: Blue Canyon Peak [Fresno Co.], Bluejay Mountain [Shasta Co.], Blue Tent Creek [Tehama Co.]. The Blue Nose and Blue Rock Mountains [Plumas, Trinity, Shasta, Mendocino, Monterey Cos.] were probably all named because of the presence of glaucophane or other bluish rock formations. **Blueschist Narrows** [Mendocino Co.]

is a ravine named for "bright blue metamorphosed rocks commonly called blueschist by geologists" (BGN, list 6804).

Bluff. The name is found as a generic term, especially along the coast, for a steeply rising bank or cliff with a broad front. The city of Red Bluff and a number of Bluff Creeks, Coves, and Canyons were named after nearby bluffs. **Bluff Creek** [Humboldt, Del Norte Cos.]: "We camped [Oct. 9, 1851] opposite the high point which forms a landmark from the Bald Hills, and which gives the name of Bluff creek to a stream" (Gibbs, in Schoolcraft 3:147). *See* Gold: Gold Bluffs; Red: Red Bluff.

Blunts Reef [Humboldt Co.]. The reef off Cape Mendocino was discovered by Vancouver, but left nameless. In 1841 Wilkes named it for Simon F. Blunt, a midshipman on the USS *Porpoise*. The name was placed on the map by the Coast Survey in 1850.

Blythe [Riverside Co.]. Established in 1908 by the Palo Verde Land and Water Company and named for Thomas H. Blythe, a San Francisco capitalist who in 1875 had bought the land in Palo Verde Valley on which the irrigation project was later developed.

Boardman: Canyon, Ridge [Lake Co.]. Named in the 1880s for Oscar Boardman and his brother, who homesteaded in the rugged country north of Lake Pillsbury and ran stock on the ridge there (Mauldin).

Boardman, Mount. This mountain, on the peak where the boundaries of Alameda, Santa Clara, San Joaquin, and Stanislaus Cos. meet, was named in memory of W. F. Boardman, surveyor of Alameda Co., city engineer of Oakland (1864–68), and builder of the dams that created Lake Chabot and Lake Temescal (E. T. Planer).

Bob Rabbit Place [Kern Co.]. For Robert Roberts, known as Bob Rabbit, an Indian who hunted cottontails in the area (BGN, list 7403).

Boca. In Spanish times the term for 'mouth' was often found not only for the mouth of a river but also for the entrance to, or the outlet of, an estuary, valley, or port. The term was contained in the names of three land grants: Boca de Santa Monica [Los Angeles Co.], June 19, 1839; Boca de la Cañada del Pinole [Contra Costa Co.], June 21, 1842; Boca de la Playa [Orange Co.], Apr. 18, 1845, and May 7, 1846. **Boca** (bō′ kə) [Nevada Co.] is a modern application to the station of the Central Pacific Railroad when the Truckee-Verdi section was built in 1867, because it is situated not far from the mouth of Little Truckee River.

Bodega (bō dā′ gə): **Bay, Head, Port, Rock**, town [Sonoma Co.]. The bay was entered by Juan Francisco de la Bodega, captain of the *Sonora*, on Oct. 3, 1775, and named for him. Bodega's map of 1775 (Wagner, pl. 34) shows "Pto y Rio del Capitan Vodega." On later maps the bay is usually designated as Puerto de la Bodega. Kotzebue, in the report (1821) of his first expedition, gives the abbreviated form, Port Bodega (*Rurik*, p. 62). Between 1812 and 1841 the bay was used by the Russians as a harbor and often called Romanzov (variously spelled), in honor of the chancellor of the empire (1809–14) and patron of the *Rurik* expedition. Kotzebue, in 1830, records Hafen Bodega oder Port Romanzov. The modern version, Bodega Bay, was established by the Coast Survey in 1850. The cape was called Pte. Bodega by La Pérouse (1797). Duflot de Mofras (1844) follows Kotzebue: Pte. de la Bodega ou C. Romanzoff. Tyson's map of 1850 has Bodega Point; the Coast Survey of 1851 has Bodega Head. A settlement called Bodega on the north shore of the bay is shown on Tyson's map. The present village was settled by George Robinson in 1853 and is recorded as Bodega Corners on Hoffmann's 1873 map of the Bay region. Two land grants dated July 15, 1841, and Sept. 12, 1844, were called Bodega. Presidio de Bodega was a name sometimes used for Fort Ross (*see* Presidio). *For* the current local name of Bodega Head, *see* Campbells Point.

Bodey. *See* Bodie.

Bodfish Creek [Santa Clara Co.]; **Bodfish**: town, **Peak** [Kern Co.]. The name in Santa Clara Co. is shown on a map of the land grant Las Animas in 1855. The name of one Orlando Bodfish is recorded in both Santa Clara and Kern Cos. during the 1860s. However, the Kern Co. site may commemorate George A. Bodfish, a pioneer who owned a store in the vicinity (Darling 1988).

Bodie (bō′ dē): post office, **Creek, Mountain** [Mono Co.]. Waterman S. Body discovered the ore deposits here in 1859. He lost his life in a snowstorm in May 1860, but the settlement preserved his name. The change in spelling was perhaps caused by the desire to keep the proper pronunciation. The name is sometimes spelled Bodey.

Boga [Butte, Sutter Cos.]. The name of two land grants, dated Feb. 21, 1844. The name is probably the Spanish rendering of the name of an Indian village shown on the *diseños* of the Boga and Fernandez grants. The rancho was also known as Rio de las Plumas, because of its location on the Feather River, and as Rancho de Flugge for the grantee, Charles W. Flügge, native of Germany and pioneer of 1841.

Bogess (bō′ gəs) **Creek** [San Mateo Co.] appears first in an 1857 survey, probably named for the family of J. M. Boggess, living in Woodside in 1857 (Alan K. Brown). *See also* Bogus.

Boggs: Lake, Mountain [Lake Co.]. Named for H. C. Boggs, owner of the sawmill that stood on the shore of the lake in the 1870s.

Bogue [Sutter Co.]. Named for Virgil G. Bogue, chief engineer of the Western Pacific (Drury). *See* Virgilia.

Bogus. Like "humbug," a common term for fake or sham, this word was used by disappointed prospectors, mainly for creeks. Placer Co. also has a Bogus Thunder Bar, so named because a nearby waterfall produces a sound like thunder. **Bogus Country** and **Creek** [Siskiyou Co.]: The district drained by Big Bogus Creek and Little Bogus Creek came by its name from the operations of counterfeiters who had a furnace near here (*WF* 6:376–77). Bogus post office existed from 1876 to 1913 (Luecke).

Bohemian Grove [Sonoma Co.]. The name came into existence when, in 1880, the Bohemian Club of San Francisco held a Midsummer High Jinks in the redwood forests of Russian River. It was applied to the present grove in 1891.

Bohemotash (bō hē′ mə tash) **Mountain** [Shasta Co.]. From Wintu *bohema thoos*, lit. 'big camp' (Shepherd).

Boiling Springs [Lassen NP] steams at about 125° Fahrenheit; it was once called Lake Tartarus (Stang).

Bokman's Prairie. *See* Fieldbrook.

Bolam (bō′ ləm) **Creek** [Siskiyou Co.]. From Wintu *bulim*, a form of *buli* 'mountain' (Shepherd); *see* Bally.

Bolanos Bay. *See* Bolinas Bay.

Bolbones, Sierra de los. *See* Diablo, Mount.

Bolcoff [bōl′ kôf] **Adobe** [Santa Cruz Co.]. Named for José Antonio Bolcoff, California's first permanent Russian settler, who deserted a Russian ship at Monterey in 1815. He built the adobe at Rancho Refugio in 1841 (Clark 1986).

Bolinas (bō lē′ nəs): **Bay, Lagoon, Point, Ridge**, town [Marin Co.]. *La Cañada que llaman los Baulenes* 'the valley that they call the Baulenes' is mentioned by Ignacio Martínez in 1834 as one of the boundaries of the pueblo of San Rafael and was probably the valley through which Olema Creek flows (SP Mis. 10:11). A map of about the same date has the word Baulenes on the peninsula that now includes Bolinas Point, Duxbury Point, and the town of Bolinas. The name Baulenes, possibly from a Coast Miwok word of undetermined meaning, probably referred to the Indians who inhabited the region. The Cañada de Baulines is shown on a *diseño* of Punta de los Reyes, 1835, running into the Estero de Tamales (Tomales Bay) on the north. From 1852 to 1910 the name was spelled Ballenas on the charts of the Coast Survey, obviously through mistaken analogy to the Spanish word for 'whale'. In the 1869 edition of his *Coast Pilot*, George Davidson erroneously assumed that the bay was named for Francisco de Bolaños of Vizcaíno's expedition; this gave rise to the spellings Bolanos and Boliñas. In 1873, Baulinas on Hoffmann's maps approximated the original spelling; but the USGS adopted the spelling Bolinas, which had been used when the post office was established in 1863. *See* Baulenes. *For* the name of the bay and the lagoon on Ringgold, General Chart, *see* Rialto.

Bolinas Creek [Alameda Co.]. Named for Antonio Bolena, a native of the Azores, who owned land here in the 1870s (Mosier and Mosier).

Bolivar (bol′ i vär) **Lookout** [Siskiyou Co.]. The ranger station preserves the old name of what is otherwise called Craggy Peak. The name of the South American revolutionary leader, Simón Bolívar (1783–1830), was often used for place names in the United States. A local pronunciation is (bol′ i zär).

Bollibokka (bol ē bok′ ə): **Mountain, Creek** [Shasta Co.]. Apparently from Wintu *buli* 'mountain' (*see* Bally) and *phaqa* 'manzanita' (Shepherd). The spelling Billibokka also occurs.

Bolling (bō′ ling) **Memorial Grove** [Humboldt Co.]. Dedicated in 1921 and named in honor of Col. Raynal Bolling, an American army officer killed in action in World War I (Drury).

Bollinger (bō´ lin jər) **Canyon** [Contra Costa Co.]. In 1855 Joshua Bollinger settled in the canyon east of the present site of Saint Marys College (Co. Hist.: Contra Costa 1882).

Bolo [San Bernardino Co.]. The original name, Bristol, was changed by the Santa Fe in 1898 to Bombay, then to Bengal, and in 1915 to Bolo (Santa Fe).

Bolsa (bōl´ sə). The Spanish word means 'pocket'; it is used in a geographical sense for a semi-enclosed or shut-in place, often a neck of land surrounded by water or a swamp and accessible from one side only. The term, extremely popular in Spanish times, was the chief or secondary name of no fewer than twenth-four land grants and claims. **Bolsa Chica State Park** [Orange Co.] refers to a 'small pocket' of land. **Rancho Bolsa de las Escarpines** [Monterey Co.] was granted in 1837 to Salvador Espinosa. *Escarpines* has been variously explained in terms of Spanish *escarpe* 'bluff, declivity' or *escorpina* 'a type of fish, grouper, perch' (Clark 1991).

Bolton Brown, Mount [Kings Canyon NP]. Chester Versteeg and Rudolph Berls made the first ascent on Aug. 14, 1922, and named the peak "in honor of Bolton C. Brown, of the Sierra Club, who was the first to explore, map and write of the Upper Basin of the So. Fork of the Kings River" (*SCB* 11:426). Bolton C. Brown was professor of drawing and painting at Stanford University, 1891–1902 (Farquhar).

Bonadelle (bon ə del´ ē) **Ranchos** [Madera Co.]. Named for the developer, John Bonadelle of Fresno (Emmert).

Bonanza (bə nan´ zə). The word, derived from the Spanish word for 'prosperity', is used in the West for anything highly profitable, especially a richly yielding mine. Although commonly used in California, it appears as a place name apparently for only four minor features—in Kern, San Bernardino, Merced, and Lake Cos.

Bonaventura. *See* Buenaventura; Sacramento River.

Bond Pass [Yosemite NP] was named for Frank Bond, member of the Yosemite National Park Boundary Commission in 1904, later chairman of the BGN (Farquhar).

Bonete, Punta. *See* Bonita, Point.

Boneyard Hill [San Mateo Co.]. From the Consolidated Chemical Industries fertilizer plant, established here in the 1880s, producing bonemeal (Alan K. Brown).

Boney Mountain [Ventura Co.]. Earlier called Sierra del Conejo; named because of its sharp crags, imagined to outline the face of an old man, and locally called "Old Boney" (Ricard).

Bonita (bō nē´ tə) [San Diego Co.]. In 1884 Henry E. Cooper named his estate Bonita Ranch, and the name was later applied to the post office. Spanish *bonita*, a diminutive of *buena* 'good', means 'pretty' (feminine).

Bonita, Point; Bonita: Channel, Cove [Marin Co.]. The point was named Punta de Santiago by Ayala in 1775 and was so designated as late as 1825. On other maps the point is shown as Punta Bonete (Wagner, p. 377), probably because there were originally three heads, each of which resembled the bonnet of a clergyman (Davidson). The term Punta de Bonetas used in 1809 (PSP 19:226–27) was probably intended for a plural and was not a folk-etymological rendering (*bonetas* 'pieces of extra canvas'). Beechey spelled the name Punta Boneta in 1826 (2:424), and this version was used by Duflot de Mofras (1844), the Coast Survey (1851), and Gibbes (1852). The Williamson-Abbot map (1855) has Pt. Bonita, and the Bancroft maps of the 1860s fixed this now-accepted version, meaning literally 'pretty point'. Until about 1900 the Coast Survey charts spelled the name Bonito, probably because it was assumed that the point was named for the bonito, a fish that lives in channels where there are currents.

Bonita Lake [Inyo Co.]. Given by the Dept. of Fish and Game, inspired by the scientific name for the golden trout, *Salmo aguabonita* (Browning 1986).

Bonner Grade [Modoc Co.]. For John H. Bonner, who planned and managed the construction of the road over the summit of the Warner Range, in northeastern California, in the 1860s.

Bonny Doon (bon´ ē dōōn´) [Santa Cruz Co.]. The name, attested from 1902, presumably comes from the song by Robert Burns, referring to the Doon River in Scotland (Clark 1986).

Bonpland, Lake. *See* Tahoe, Lake.

Bonsall (bon´ səl) [San Diego Co.]. For James Bonsall, a native of Pennsylvania, who set up a nursery here in the 1870s. When a post office was established in 1889, his name was selected from those submitted by the residents (Susie K. Worris).

Bonta (bon´tə): **Creek, Saddle** [Sierra Co.]. Authorized by the BGN in list 7203. Said to be a family name.

Boomer. According to local tradition, the Boomer Mine in Trinity Co. was so called because it was a big producer. The word is also a mining term denoting the contrivance by which the accumulation of water (in placer mining) is suddenly discharged. The name is found in other mining regions of the state: Boomer Creek [Butte Co.].

Boonville [Mendocino Co.]. The town was founded as Kendalls City in 1864. Soon afterward, W. W. Boone bought the store of Levi and Straus and changed the name of the town to Booneville. The *e* was dropped when the post office was established in the 1880s. The place has become famous for Boontling, a "secret language" invented by the inhabitants.

Boothe Lake [Yosemite NP] was named for Clyde Boothe, park ranger 1915–27 (Browning 1986).

Bootjack [Mariposa Co.]. A bootjack is a V-shaped device used for pulling off boots. There is also a Bootjack in Marin Co. (Teather 1968).

Boot Lake [Modoc Co.]. Named by early residents on account of its shape (William S. Brown).

Borax. About twenty features are so named because of deposits of borax (sodium borate). The oldest is probably Borax Lake in Lake Co., discovered and named by Dr. John A. Veatch in Sept. 1856.

Borden [Madera Co.]. When the Southern Pacific reached the place in the spring of 1872, the station was named for Dr. Joseph Borden, a member of the nearby community popularly called the Alabama settlement.

Boreal (bôr´ ē əl) **Ridge** [Nevada Co.]. The elevation between the old and the new route of Highway 40 was so named because it lies north of Donner Pass and the Southern Pacific Railroad crossing of the Sierra Nevada. The name means 'northern', from Greek *Boreas* 'north wind'.

Borego. *See* Borrego.

Borel (bôr el´) **Creek** [San Mateo Co.]. From the estate of Antoine Borel, who owned land here from ca. 1885 (Alan K. Brown, p. 112). **Borel Canal** [Kern Co.] was probably named for the same man, an early partner of Henry E. Huntington and William G. Kerchoff, founders of the Kern Power Company (Darling 1988).

Boron (bôr´on) [Kern Co.]. From 1912 until 1938 the Santa Fe station was called Amargo (Spanish for 'bitter'), probably from the taste of the borax deposits (*see* Borax), and the name is still in use. After the Pacific Coast Borax Company moved from Death Valley to Amargo, the railway station and the nearby post office took the name Boron (Santa Fe).

Boronda (bə ron´ də) **Creek** [Monterey Co.]. Named for the Boronda family, early settlers of the district. In 1839 José M. Boronda was a co-grantee of Rancho Los Laureles, 8 miles northwest of the creek.

Boron Springs [Inyo Co.] reflects a misunderstanding of the name of a homesteader, John C. Borum (Browning 1986).

Borosolvay (bôr ō sol´ vā) [San Bernardino Co.]. The name is a combination of the names of the two firms that established the community in 1916, the Pacific Coast Borax Company and the Solvay Process Company.

Borrego; Borrega. The word for 'yearling lamb', often extended to 'sheep' in general, was repeatedly used for place names in Spanish times, usually in the feminine form. In mountain and desert areas it may refer not to the domestic sheep but to the bighorn sheep, *Ovis canadensis* (*see* Bighorn). It appears in the names of land grants in Contra Costa, Santa Clara, and Santa Barbara Cos. It is also preserved in **Borego** (bə rā´ gō) **Mountain** [San Diego Co.] and in **Borregas** (bə reg´ əs) **Gulch** [Santa Cruz Co.], from Zanjón de Borregas (Clark 1986). In San Diego Co. it has become an important cluster name, which probably originated because of the presence of the bighorn sheep: **Borrego Desert, Mountain, Palm Canyon, Sink, Spring, Springs**, and **Wells**. The name was applied to the area as early as 1883, and a Borego post office existed from 1928 to 1940 (Brigandi). Borrego Springs post office was established on May 16, 1949. The former Borrego State Park is now part of Anza-Borrego Desert State Park.

Bosquejo [Tehama, Butte Cos.]. The name of two land grants, granted to Peter Lassen on July 26, 1844, and in Dec. 1844. On the *diseño* of the former is the following inscription: *Bosquejo del terreno solicitado por Pedro Lawson* 'sketch of the land solicited by Peter Lassen'. Perhaps someone took the Spanish word for 'sketch' as the name of the rancho.

Bostonia [San Diego Co.]. Named after the Boston Ranch in 1895 by the promoters of the town. The former name was Meridian District. *For* Boston Ranch in Butte Co., *see* Hurleton.

Bothe (bō thā′) **Napa Valley State Park** [Napa Co.]. The park was established in 1960 and bears the name of the former owner of the property, who operated a commercial enterprise under the name Bothe's Paradise.

Bottileas. *See* Jackson.

Bottlerock Mountain [Lake Co.]. The names of mountains and other orographic features consisting chiefly of obsidian often include the descriptive term "glass." In Lake Co., according to Mauldin, obsidian is "bottlerock" in common speech.

Boulder. This is an extremely popular place name in the West. California probably leads the other states with more than a hundred Boulder Creeks, Lakes, etc. **Boulder Creek** [Santa Cruz Co.]: In the 1880s the lumber town Lorenzo took the name of the Boulder Creek post office, which had been established in the 1870s. In Siskiyou Co. the spelling variant Bowlder is used for two creeks and a lake.

Boulevard [San Diego Co.]. Town and post office were so named because they are situated on the "boulevard" (U.S. Highway 80) to the Imperial Valley.

Boundary. Surveyors sometimes used the word to name a feature at a boundary. Boundary Hill [Yosemite NP] was named by Lt. Montgomery M. Macomb of the Wheeler Survey because it was on the line of Yosemite State Park; but the hill is now near the center of the national park. Boundary Peak in the White Mountains suffered a similar misfortune. When A. W. von Schmidt established the California-Nevada boundary in 1872, the line ran through the peak; but a later correction placed the mountain well into the state of Nevada. Madera Co. has a Boundary Creek, and Inyo Co. a Boundary Canyon.

Bouquet (bō kā′, bō ket′) **Canyon** [Los Angeles Co.]. When a French sailor named Chari settled in the canyon, it became known by his Spanish nickname, El Buque 'the ship'. Surveyors in the 1850s, believing the name to be French, recorded it as Bouquet, and their version is still used (*HSSC:P* 14:197).

Bourdieu Valley [Monterey Co.]. For Edward Bourdieu, of French descent, who settled here around 1914 (Clark 1991).

Bourland: Mountain, Creek [Tuolumne Co.]. For John L. Bourland, sheriff of Tuolumne Co., 1865–68 (Browning 1986).

Bourns (bôrnz): **Gulch, Landing, Rock** [Mendocino Co.]. The features were named for the Bourn family, early settlers, whose family cemetery is located near the mouth of the gulch (BGN, 1962).

Bourri. *See* Buriburi Ridge.

Bower Cascade [Mono Co.]. "One of the smallest of the cascades [of Bloody Canyon Creek], which I name the Bower Cascade, is in the lower region of the pass, where the vegetation is showy and luxuriant" (Muir, p. 301). Bower is found elsewhere as a descriptive specific term: Bower Cave [Mariposa Co.].

Bowers Gulch [San Mateo Co.]. So named since 1858, perhaps for Abram S. Bowers, a shingle maker killed in the Deer Gulch landslide of 1861 (Alan K. Brown).

Bowlder. *See* Boulder.

Bowles Prairie [Humboldt Co.]. Named for one of the partners of Bowles, Bowles, and Codington, large property owners and, according to the assessment roll of 1853–54, "heavy merchants of Uniontown."

Bowman [Placer Co.]. Named for Harry H. Bowman, a merchant and agriculturist, and the first commissioner of horticulture of Placer Co. (H. P. Davis).

Bowman Lake [Nevada Co.] was named for James F. Bowman, a native of Scotland, who settled at what became known as Little Bowman Lake in the early 1860s. In 1873 Big and Little Bowman Lakes were consolidated, providing a source of water supply for hydraulic mining. The lake is now the principal reservoir of the county irrigation district (H. P. Davis).

Bow Willow Canyon [San Diego Co.]. The name first appears in print in the 1940s and apparently comes from willow trees used by Indians for making bows; however, another source suggests that the original name was Bull Willow Canyon (Brigandi).

Box Canyon [San Diego Co.] refers to the narrowest point on the old Southern Emigrant trail. From 1847 onward it was known as Cooke's Pass; the name Box Canyon dates from around 1900 (Brigandi).

Box Springs: Canyon, Mountains, town [Riverside Co.]. The two springs which were once "boxed in" to supply water for domestic use doubtless suggested the name. *See* Cajon.

Box S Spring [San Bernardino Co.]. Named after the nearby Box S Ranch, which derived its name from the owner's cattle brand, a square enclosing the letter *S*.

Boyden Cave [Sequoia NP] was named for its discoverer, Pete Boyden.

Boyer (bôi′ ā) **Creek** [Santa Cruz Co.]. For early landowners Oliver and Ramenia Boyer, also spelled Boyea (Clark 1986).

Boyes Hot Springs [Sonoma Co.]. Named for Capt. H. E. Boyes, owner and developer of the springs. The post office is listed in 1912.

Boyle Heights [Los Angeles Co.]. Founded by Andrew Boyle in 1858 (Atkinson).

Bozzo Gulch [San Mateo Co.]. So named since 1872; the Emmanuel Bozzo ranch was here from the late 1860s (Alan K. Brown).

Bra-Cut (brā′ kut) [Humboldt Co.]. Short for Brainard's Cut (referring to a railroad cut), also known as Brainard's Point, named for T. Brainard (Turner).

Bradbury [Los Angeles Co.]. Incorporated in 1957 and named for Lewis Leon Bradbury, owner of the property at the turn of the century (Co. Hist.: Los Angeles 1965:3.126).

Bradley [Monterey Co.]. The Southern Pacific reached the place in Oct. 1886 and named the station for the owner of the land, Bradley V. Sargent, state senator, 1887–89. Earlier, the land was called Rancho La Pestilencia by its owners, for unknown reasons (Clark 1991).

Bradley, Mount [Kings Canyon NP]. Named by Robert M. Price, Mrs. Price, and Joseph C. Shinn—who made the first ascent on July 5, 1898—for Cornelius B. Bradley (1843–1936), professor of rhetoric at the University of California, 1894–1911 (Farquhar). Mount Bradley in Siskiyou Co. was also named for Cornelius Bradley. He spent many summers at the Castle Crags Resort and always dragged his friends to the top of the mountain, which commands a splendid view (Schrader).

Bradley Creek [San Mateo Co.]. Charles Bradley's dairy was here from the 1880s (Alan K. Brown).

Brainards Point [Humboldt Co.]. *See* Bra-Cut.

Brainerd Lake [Inyo Co.]. Lawson Brainerd was a district ranger, 1914–29 (Browning 1986).

Braly Mountain. *See* Brawley Peaks.

Branch. The geographical term for a fork of a river is rarely used in California nomenclature: Spring Branch of Willow Creek [Tehama Co.], South Branch of Pleasant Grove Creek [Placer Co.], West Branch of San Pablo Creek [Contra Costa Co.]. Sometimes the term is used to avoid a repetition of "Fork": South Branch of the Middle Fork of the Feather River; Dusy Branch of the Middle Fork of the Kings River; East Branch of the East Fork of the North Fork of Trinity River. *See* Fork.

Branciforte (bran sə fôr′ tē) **Creek** [Santa Cruz Co.]. The creek takes its name from the Pueblo or Villa de Branciforte, established in 1797 at the site of the present town of Santa Cruz and named in honor of the viceroy of New Spain, Miguel de la Grua Talamanco, Marqués de Branciforte. From Feb. 18 to Apr. 5, 1850, Branciforte was the provisional name of Santa Cruz Co. A current Italianized pronunciation is (bran chə fôr′ tē).

Brand Park [Los Angeles Co.]. Named for L. C. Brand, who gave to the park the great Star Fountain, which had been built by Indians under the direction of the padres.

Brandy City [Sierra Co.]. *See* Whiskey.

Branigan Lake [Yosemite NP]. Named by Lt. N. F. McClure in 1894 for a soldier of his detachment, who was later killed in action in the Philippines (Farquhar).

Brannan Island [Sacramento Co.]. The island opposite Rio Vista is the only place in California to commemorate Samuel Brannan, who came to California in 1846 in charge of the Mormon colonization—and after the discovery of gold, which he publicized far and wide, became the most spectacular financier and speculator in central California.

Branscomb [Mendocino Co.]. The settlement and post office (1895) were named for the first postmaster, probably Benjamin Branscomb, a native of Ohio. The Branscomb who "shot three bear in one tree December, 1900" (Co. Hist.: Mendocino 1914:101) may have been the same man.

Branstetter Canyon [Monterey Co.]. For Henry, Oscar, or William Branstetter, all of whom settled near here around 1890 (Clark 1991).

Bratton Valley [San Diego Co.]. For Napoleon Bratton, early settler (Stein 1975).

Brave Lake [Fresno Co.]. The nearby Lake of the Lone Indian was named in 1902; by 1953 maps showed several new Indian-type names: Brave, Warrior, Papoose, Chief, and Squaw Lakes (Browning 1986). Many of these terms are now considered offensive by Native Americans.

Brawley [Imperial Co.]. Laid out and named by the Imperial Land Company in 1902, on property that had belonged to J. H. Braly of Los Angeles. Although the locality had begun to be called after him, he feared a failure of the project and refused permission to use Braly as a name for the proposed town. A. H. Heber, general manager of the land company, suggested as a compromise that Brawley, the name of one of his Chicago friends, be substituted (Tout).

Brawley Peaks [Mono Co.]. Named for James M. Brawley, a prospector of the 1860s (Maule). Recorded as Braly Mountain on the Whitney maps.

Bray [Siskiyou Co.]. When the post office was established on June 7, 1907, it was named after the Bray Ranch, situated about half a mile from the present settlement (D. Steffenson).

Brazil Beach [Marin Co.]. For Antone Brazil, a local farmer who died in 1936 (BGN, list 6701).

Brazos del Rio. *See* Rio: Rio Vista.

Brea. The beds of tar or asphaltum (Spanish *brea*) in Los Angeles Co. were mentioned by the Portolá expedition and repeatedly by others in later years. In these beds thousands of skeletons of extinct animals have been discovered. As a geographical term, the word *brea* appears in the names of five land grants in Santa Clara, Monterey, Los Angeles, and Orange Cos. It is still preserved in Labrea Creek [Santa Barbara Co.] and some minor features. The term "The La Brea Tar Pits" for a famous site in Los Angeles Co. is of course doubly tautological. **Brea** (brā′ ə) town and **Canyon** [Orange Co.] are named after the grant Rincón de la Brea, dated Feb. 23, 1841. *See* Asphalt; Pismo.

Breakfast Canyon [Death Valley NP]. "Dude wranglers" of Furnace Creek Inn are in the habit of eating breakfast in this colorful gorge after a morning ride; hence the name (Death Valley Survey).

Breakneck Canyon [Inyo Co.]. Named in 1871 by Lt. D. A. Lyle of the Wheeler Survey because of the difficulties encountered in crossing it; it is shown on Wheeler atlas sheet 65. Another very steep canyon by that name is in Butte Co., and there is a Breakneck Creek in San Bernardino Co.

Breckenridge Mountain [Kern Co.]. Named after the Breckenridge sawmill on Lucas Creek, which in earlier days provided much of the lumber for Bakersfield. The Pacific Railroad Survey had named the elevation Cañon Mountain. At the time of the Civil War the new name arose, probably having been given by Southern sympathizers for John C. Breckinridge of Kentucky (1821–75), a Confederate general in the Civil War (W. H. Boyd). The mountain was also known as Cross Mountain, named after the Cross Brothers lumber mill.

Breeze: Creek, Lake [Madera Co.]. Named in 1896 by Lt. H. C. Benson for his brother-in-law, William F. Breeze of San Francisco.

Breisgau (brēs′ gou) [Tehama, Shasta Cos.]. A Mexican land grant dated July 26, 1844, and named by the grantee, William Benitz, after his home district in southwestern Germany.

Brentwood [Contra Costa Co.]. Named after Brentwood in Essex, England, the ancestral home of John Marsh, who had owned Rancho Los Meganos (on which the present town is situated). The town was laid out in 1878 on land donated by the owners of the property.

Brewer, Mount [Kings Canyon NP]. Named by the Whitney Survey for William H. Brewer, after he and Charles F. Hoffmann made the first ascent, on July 2, 1864. Brewer, professor of natural sciences at the College of California (1863–64) and professor of agriculture at Yale (1864–1903), was the principal assistant of Whitney in the California State Geological Survey, 1860–64. **Brewer Creek** [Siskiyou Co.] was apparently also named for Prof. Brewer, who identified the weeping spruce, *Picea breweriana*, in the Mount Shasta region.

Brewer Island [San Mateo Co.]. Leveed in and drained by Frank M. Brewer around 1905 (Alan K. Brown).

Brewery Gulch [Mendocino Co.]. The name survives from a brewery that flourished here in the "early days" (Stewart).

Breyfogle (brā′ fō gəl): **Canyon, Buttes** [Death Valley NP]. The names commemorate Charles C. Breyfogle, one of the fabulous prospectors of Death Valley. The Breyfogles, father and several sons, had mined for gold on the North Yuba River near Downieville in 1850. One of the sons, Charles, resided in Oakland later and was assessor and treasurer of Alameda County (1854–59). In 1864 he claimed to have discovered a rich gold deposit in or near the canyon. He was captured by Indians, and after his release he could not find the place of his discovery. In the jargon of Death Valley's gold seekers, "breyfogling" became the synonym for "searching for lost mines." Numerous articles and monographs about the junior Breyfogle's "lost mine" have been published. In some of these his first name is given as Jacob; in others as Anton, identified as a German from Austin, Nevada.

Briceland [Humboldt Co.]. Named for John C. Briceland, a native of Virginia, who about 1889 bought the ranch on which the present town was later developed (Co. Hist.: Humboldt 1915:1190). The post office is listed in 1892.

Bridalveil Fall and **Creek** [Yosemite NP] were apparently named by Warren Baer, editor of the *Mariposa Democrat*, on Aug. 5, 1856. James M. Hutchings asserted that he suggested the name on his first visit to Yosemite in 1855 (Farquhar). The Indians called it Pohono, and at the time of the Whitney Survey the Indian name was apparently still in common use (Brewer, Notes, June 16, 1863).

Bridge. One of the most popular designations for a place ever since people have built bridges and settled near them. **Bridgeport** town, **Valley, Canyon** and **Creek** [Mono Co.]: The town was settled in the late 1850s and became the seat of Mono County in 1864. The valley was first called Big Meadows because of the large area of grass around the settlement. Another Bridgeport [Nevada Co.] was founded on the Yuba River at a covered bridge. *For* Bridgeport in Solano County, *see* Cordelia. **Bridgeville** [Humboldt Co.] was so named because of its location at the bridge over the Van Duzen River. The word occurs in the names of seven other districts or settlements in the state, including Bridge Creek [Lassen Co.] and Bridgeport Landing [Mendocino Co.].

Brier [Inyo Co.]. Named for the Rev. J. W. Brier and his family, who crossed Death Valley in 1849 and settled in Los Angeles (*Desert Magazine*, Sept. 1938).

Bright Dot Lake [Inyo Co.]. Bill Garner, former owner of the Convict Lake Resort, named the lake in the 1930s because it is a bright blue spot—and after his wife, Dolly "Dot" Bright (Browning 1986).

Brighton [Sacramento Co.]. Laid out in 1849 about a mile north of the present site, near Sutter's gristmill (built in 1847), by a party of Sacramento speculators and named after Brighton, England. The name appears in the *Statutes* of 1851, and on Gibbes's map of 1852. *See* Perkins.

Briones (brē ō′ nēz): **Hills, Valley** [Contra Costa Co.]. Named for the family of Felipe Briones, who were allowed to settle on Ignacio Martínez's rancho in 1831. Briones himself was killed by Indians in 1840.

Brisbane [San Mateo Co.]. The town was established in 1908 under the name of Visitacion City. To avoid confusion with the nearby Visitacion Valley, the name was changed to Brisbane, supposedly for the well-known journalist Arthur Brisbane (1864–1936), when the post office was established in 1931. But according to Stanger (pp. 47 ff.), the name was applied by the promoter Arthur Annis, a native of Brisbane, Australia.

Briscoe Creek [Glenn Co.]. Named for Watt Briscoe, a stockman of the district in the early 1850s. Sometimes spelled Brisco.

Bristol. *See* Bolo.

Brites Valley [Kern Co.]. Said to be named for John M. Brite, who came to the Tehachapi region in 1854 (Co. Hist.: Kern 1914:185). But according to Crites (p. 266), it was named for W. F. Brite, one of the first supervisors of the county.

Brittans Knoll [San Mateo Co.]. The elevation near San Carlos was named after the Nat Brittan Ranch, established in 1856 (Alan K. Brown).

Britton, Lake [Shasta Co.]. The reservoir was formed in 1925 and named in memory of John A. Britton, general manager of the Pacific Gas and Electric Company (Steger).

Brizzolara Creek [San Luis Obispo Co.]. For Bartolo Brizzolara, who owned land in the area and died in 1881 (Hall-Patton). Also spelled Brizzolari.

Brockway [Placer Co.]. Before 1900 Frank Brockway Alverson, who was associated with "Lucky" Baldwin, built a hot spring resort on the north side of Fallen Leaf Lake. It was named for "Uncle Nathaniel" Brockway. The post office was established on Mar. 16, 1901.

Broderick (brod′ ə rik) [Yolo Co.]. The town was laid out in Feb. 1850 by the widow of James McDowell and was named Washington. In 1854 there were two Washingtons, one in Yolo County and one in Nevada Co. Both post offices were discontinued before 1858, but the one in Nevada Co. was reestablished before 1867. About 1890, when the town in Yolo Co. again petitioned for a post office, a new name had to be found. It is not known why the name Broderick was chosen; perhaps it was given in memory of Sen. David Broderick (*see* Broderick, Mount).

Broderick, Mount [Yosemite NP]. David C.

Broderick, U.S. senator from California, 1857–59, was killed (near San Francisco) by Judge David S. Terry in 1859 in the last formal duel fought on California soil. The Whitney Survey originally placed his name on what is now known as Liberty Cap. The latter name was given in 1865, and the name Mount Broderick was later transferred to its present location (Farquhar).

Broken Rib Mountain [Del Norte Co.] suggests that a party of the USGS had tough going when they surveyed the Preston Peak quadrangle in 1915—there is a Wounded Knee Mountain 2 miles west.

Bronco. *See* Floriston.

Brook is uncommon in California place names, being generally replaced by "creek"; but it occurs in some compounds such as **Brookdale** [Santa Cruz Co.].

Brooklyn Basin [Alameda Co.]. The name in Oakland's Inner Harbor preserves the memory of the town of Brooklyn, now part of Oakland. The section of Oakland around Mills College had a post office named Brooklyn from 1855 to 1878. A town was created on Mar. 10, 1856, when the supervisors of Alameda Co. amalgamated the settlements Clinton and San Antonio. According to Analeone Patten, the place was named after the ship *Brooklyn*, which brought Sam Brannan and 238 Mormons to San Francisco in 1846.

Brooks [Yolo Co.]. The name was selected at a public meeting on Mar. 3, 1884, as an appropriate name for the post office, because a brook flows past the site (Earl Smith).

Brooks Canyon [Riverside Co.]. The canyon in Cleveland National Forest was named in 1960 in memory of Andrew C. Brooks, a USFS employee who lost his life in the Decker fire of Aug. 1959.

Brooks Island [San Francisco Bay]. The Spanish name of the island was Isla del Carmen (Cañizares). When the Coast Survey charted the bay in 1850, it was used as a triangulation point and called Rocky Island. On Eddy's map (1854) it is labeled Brooks Island, probably for the owner of the island. It is also popularly known as Sheep Island.

Brooks Lake [San Mateo Co.]. From the Patrick Brooks ranch, here from the late 1850s (Alan K. Brown).

Brother Jonathan Rock [Del Norte Co.]. This is the name of a submerged rock about 4 miles off the shore near Crescent City, upon which the steamer *Brother Jonathan* was wrecked on July 30, 1865. The name was applied by the Coast Survey; it is mentioned as Jonathan Rock in a letter to Davidson, Oct. 10, 1869.

Brothers. It is an old international custom to designate a group of two or more orographic features of similar appearance as brothers or sisters. Two rocks at San Pablo Point [Contra Costa Co.] are called The Brothers; the two rocks on the opposite shore [Marin Co.] are The Sisters. Another group of bare rocks called The Brothers is south of Cape Mendocino. Fresno Co. and Riverside Co. have mountain groups called Three Sisters, and four peaks in Del Norte Co. are named Four Brothers. *See* Three Brothers.

Brown. The brown slopes of the California landscape are so common a sight during the long dry season that this color adjective has seldom been used in place naming. Only a few orographic features, including Brown Cone [Fresno Co.], are so named; most of these derive their names from the color of rock formations, but some may contain the family name Brown.

Brown as a family name is given to several California places. **Brown Canyon** [Monterey Co.] was named for the family of Perry L. Brown, who arrived in the area in 1876 (Clark 1991). **Brown Mountain** [Los Angeles Co.] was named for Jason and Owen Brown, whose ranch was at the foot of the mountain; they were sons of the fiery abolitionist John Brown (1800–59). John Brown's family moved to California after the Civil War (*Shasta Courier*, Apr. 22, 1865). *For* Brown [Kern Co.], *see* Mount: Mount Owen; *for* Brown's Lake [Riverside Co.], *see* Lake: Lakeview; *for* Brownsville [Humboldt Co.], *see* Samoa.

Browns Flat [Mariposa Co.]. "Here the adventurous pioneer David Brown made his headquarters for many years, dividing his time between gold-hunting and bear-hunting" (Muir, p. 35).

Browns Valley [Napa Co.]. Apparently named for John E. Brown, who had purchased a slice of Salvador Vallejo's enormous Napa grant "for a horse and buggy" (Doyle) and was the successful claimant for the section in 1853.

Browns Valley: settlement, **Ridge** [Yuba Co.]. Named for a prospector who is reported to have taken within a few weeks' time, in 1850, $12,000 in gold-bearing quartz from near the base of a huge boulder. The name is mentioned in Trask, *Report* (1854:90).

Brownsville [Yuba Co.]. Named for I. E. Brown, who erected the first sawmill there in 1851 (Co. Hist.: Yuba 1879:91).

Bruce, Mount [Yosemite NP]. Albert O. Bruce patented 160 acres here in 1889 (Browning 1986).

Brunnette Creek [Monterey Co.]. For Rooth Brunnette, who had a cabin here in the early 1930s; she was the wife of Frederick C. Brunnette, a King City merchant (Clark 1991).

Brush Creek [Butte Co.]. "Brush" is a term widely used in California to mean 'bushes, undergrowth, chaparral'.

Bryan Canyon [Monterey Co.]. Probably for Henry Clay Bryan, who took up land here in 1891 (Clark 1991).

Bryant. *See* Orinda.

Bryant Creek [Alpine Co.]. Andrew S. Bryant logged near here in 1865 (Browning 1986).

Bryant Creek [Monterey Co.]. Perhaps for Mary E. Bryant, who acquired land here in 1900 (Clark 1991).

Bryn Mawr (brin môr´) [San Bernardino Co.]. The post office was established about 1895 and was probably named after the town in Pennsylvania, which had been named after a place in Wales; the name is Welsh for 'big hill'.

Bryte [Yolo Co.]. Because of its location on the Sacramento River, the place was at first called Riverbank. When the post office was established in 1914, the present name—for George Bryte, a local dairyman—was chosen in order to avoid confusion with Riverbank in Stanislaus Co. (F. V. Wike).

Bubbs Creek [Kings Canyon NP]. The name was applied by the Whitney Survey for John Bubbs, a prospector, who crossed Kearsarge Pass from Owens Valley in 1864 (Farquhar).

Buchon (bə shon´), **Point** [San Luis Obispo Co.]. The Portolá expedition, on Sept. 4, 1769, found in Price Canyon north of Pismo a settlement of Indians whose chief had a swelling that hung from his neck: "On account of this the soldiers named him El Buchón ['the goiter'], which name he and the village retained. I named the place San Ladislao" (Crespí, p. 183.) The San Luis Range is shown as the Sierra de Buchón on Costansó's map of 1770, and San Luis Obispo Bay is called Bahía de Buchón by Padre Palou in a letter of 1772 (Palou 4:317). On 19th-century maps the range appears as Mount Buchon or Monte del Buchón. The Coast Survey mapped the district during the Civil War and called the nameless cape Point Buchon. For several years it also retained the designation Monte del Buchón; the highest elevation on it was later named Saddle Peak.

Buck. More than one hundred geographic features in California are named Buck—almost as many as are named Deer. The word appears mainly with Creek, Mountain, and Peak. Since the name was often used as a nickname, or to designate an Indian man (in contrast to "squaw"—both terms are now considered derogatory), not all such names refer to a male deer or antelope. There are a number of Buckhorn Creeks and Flats, and Lake Co. has a Bucksnort Creek.

Buckeye. The striking California buckeye tree (*Aesculus californica*), native to the Coast Ranges and the Sierra foothills, has given its name to more than fifty geographic features. Since the tree is found chiefly near water, it is not surprising that about half these features are creeks. It was in Buckeye Ravine [Nevada Co.] that a new method of gold mining, "hydraulicking," originated in 1852 (H. P. Davis). **Buckeye**: town, **Creek** [Shasta Co.]: The town was founded in 1856 and named by settlers from Ohio after their home, the "Buckeye State" (named for a related tree, *Aesculus glabra*). Since *A. californica* is not found east of the Sierra Nevada, the cluster name in Mono Co.—**Buckeye Creek, Ridge, Pass**, and **Hot Springs**—probably also reflects an Ohio background; it was taken from the Buckeye Mill, which was owned and operated in the 1860s by E. Roberts (Maule).

Buckingham: Peak, Bluffs, Peninsula, Point, Park, Island [Lake Co.]. Named for the family connected with the well-known western shoe firm Buckingham and Hecht. The family resided here in the 1880s and 1890s (Mauldin).

Buckman Springs [San Diego Co.]. Named for Amos Buckman, who settled in the valley in 1868 or 1869 and owned the property that includes the springs.

Bucks: Valley, town, **Creek, Mountain, Lake, Reservoir** [Plumas Co.]. Named for Horace (or Francis) Bucklin, popularly known as "Buck," who settled here in 1850.

Bucks Flat [Tehama Co.]. In the 1850s the Indians stole an old ox named "Buck" from a rancher, killed it at this place, and hung the meat up to jerk. This act precipitated one of the last Indian fights of the region (Carl Turner).

Bucksport [Humboldt Co.]. Named for David A. Buck—formerly of New York, a member of the Josiah Gregg party of 1849—who laid out the town in 1851.

Budd Creek and **Lake** [Yosemite NP]. The name was applied by the USGS before 1910, for James H. Budd, governor of California from 1895 to 1899.

Buellton [Santa Barbara Co.]. The post office was named in 1916 by William Budd, the first postmaster, for Rufus T. Buell, a native of Vermont who had settled there in 1874.

Buelna (bwel′ nə) **Point** [Santa Cruz Co.]. For a ranch owner, María Hilario Buelna (Clark 1986).

Buena (bwā′ nə): settlement, **Creek** [San Diego Co.]. For Buena Vista rancho (*q.v.*); the settlement was divided in 1908 into Buena and Vista (Stein, p. 162).

Buena Esperanza, Río Grande de. *See* Colorado River.

Buena Guía, Río de. *See* Colorado River.

Buena Park [Orange Co.]. Founded in 1887 and given its hybrid name (Spanish *buena* 'good'). As the town grew, it reached the railroad station Northam, which the Santa Fe renamed Buena Park in 1929.

Buenaventura. After the Lewis and Clark expedition had definitely dispelled the legend of a great "River of the West," traversing the western part of the continent and entering the ocean somewhere in Oregon Territory, the mythical river was shifted south. In the early part of the 19th century a Buenaventura ('good fortune') River was supposed to flow from the Rocky Mountains via Great Salt Lake, and through a gap in the Sierra Nevada into the ocean somewhere in central California. The name Bonaventura was used for the Sacramento by the Hudson's Bay Company trappers, who usually entered central California along its course, and mapmakers found it convenient to use this name for the new "River of the West." After Jedediah Smith's journeys exploded the myth of a river flowing from Salt Lake to the sea, Burr's map of 1839 (based largely on Smith's discoveries) placed the name Buenaventura River on what is now the Sacramento River, and the name Valley of the Buenaventura on the Sacramento and San Joaquin Valleys. Hood's map of 1838 has Buenaventura R. apparently for what is now the Humboldt River in Nevada; a dotted line from Humboldt Sink to the Sacramento Valley indicates a possible connection with the California river system. Coulter's map of 1835 and Hood's map also label the Salinas River as R. Buenaventura. Wilkes's party in 1841 left the name Buenaventura for the Salinas on the map, preceding it with an S (i.e. San) for good measure (*see* San Buenaventura; Ventura). The report of the expedition refers to the San Joaquin Valley as Buena Ventura Valley. After that the name was used only sporadically, and in the 1850s it disappeared altogether. The Salinas is called San Buenaventura as late as 1857 (Pac. R.R. *Reports* 5:2.139). It is interesting to note that the *Mapa de los Estados Unidos Mejicanos* (Paris, 1837) has a complete Rio S. Buenaventura reaching the ocean north of San Luis Obispo, and that Frémont too remained convinced of the existence of the river almost until he reached Sutter's Fort in Mar. 1844 (*Expl. exp.* 1853:286, 300, 309). **Buenaventura** [Shasta, Tehama Cos.]: The former name of the Sacramento River was given to the Mexican land grant, dated Dec. 4, 1844, granted to Pierson B. Reading.

Buena Vista (bwā′ nə vis′ tə). This phrase, meaning 'good view', is probably the most common place name of Spanish origin in the United States. Many places bear the name in commemoration of Gen. Zachary Taylor's decisive victory over the Mexican general Santa Anna at Buena Vista in northwest Mexico on Feb. 23, 1847. However, no evidence has been found to indicate that any of the Buena Vistas in California were named for this battle. The peaks in Amador, Mariposa, Madera, and Tulare Cos. were probably so named because they command a *buena vista*. Other place names go back to Spanish times, when the term was very popular and was used repeatedly for land grants and geographic features. **Buena Vista Lake, Valley, Creek,** and **Hills** [Kern Co.] bear the oldest Spanish place name in the San Joaquin Valley. An Indian village, Buena Vista, is mentioned by Fages in 1772 (*CHSQ* 10:218–19) and again by Zalvidea in 1806 (Arch. MSB 4:51). The lake is recorded as Laguna de Buenavista on Estudillo's sketch map of 1819, and elsewhere in Spanish times. It is repeatedly mentioned in the Pacific Railroad *Reports*. On Apr. 30, 1855, a Buena Vista Co. was created (*Statutes* 1855: 203), but it did not materialize. The names were officially applied to creek, valley, and hills by the BGN on Apr. 7, 1909 (*see also* Tejon: Tejon Pass; Visalia). **Buena Vista** [Monterey Co.]: A place called Buenavista is mentioned by Font on Apr. 14, 1776; it appears on Narváez's Plano and Duflot de Mofras's map. This name, applied to a land grant before 1795 and to another in 1822, is repeatedly found on maps and in records. **Buena Vista Creeks** and **Hills** [San Diego Co.]: The name was applied to a land grant dated June 17, 1845, doubtless after a rancheria called Buena Vista. A rancheria so named is mentioned by Emory in 1848 (*Mil. rec.*, p. 116). In 1890 the San Diego Central extended its line across the rancho and named a station Buena Vista. When a name was needed for another station in 1908, the old one was simply called Buena, and the new one, Vista. Another creek and a cañada in the county, near Warners, take their name from a spring mentioned as Buena Vista in 1821 (Arch. MSB 4:217).

Buenos Ayres. *See* Corral: Corral Hollow.

Bueyes, Rincón de los [Los Angeles Co.]. The name was applied to the land grant of Dec. 7, 1821. *Bueyes* is Spanish for 'oxen.' *See* Rincon.

Buffalo Hill [El Dorado Co.]. This may be a transfer name (e.g. from Buffalo, N.Y.), since the bison is not native to California.

Buhach (byōō′ hak) [Merced Co.]. The origin of the name has not been discovered.

Buhne (bōō′ nə), **Point** [Humboldt Co.]. Named for Captain H. H. Bühne, who entered Humboldt Bay on Apr. 9, 1850, as second officer of the *Laura Virginia*, and who later became a prominent lumberman.

Buli. *See* Bally.

Bull. The old English custom of using the stem "ox" in place names has never taken root in America; "bull" and to a lesser degree "cow" or "cattle" are used instead. In California more than fifty names contain the word "bull," including Bull Barn Gulch [Sonoma Co.] and Bull Tail Valley [Marin Co.]. The name Bullpen, used for a number of features, is western slang for sleeping quarters in camps, or for "pens" in general. **Bullhead Creek** [Siskiyou Co.] was probably named for the fish. **Bulls Head Point** [Suisun Bay] took its name from "Large House (Bull's Head)" shown on surveyor's plats of Rancho Las Juntas (Apr. 1864), and **Bullwheel Ridge** [Kings Co.] from a bull wheel (a large driving gear) used in drilling. Since the Spanish also frequently used *toro* 'bull' as a name, some of the Bull Creeks in the southern counties were doubtless originally Arroyos del Toro. **Bull Creek** and **Flat** [Humboldt Co.] were so named because, in the early 1850s, Indians stole a bull from a white settler near Briceland, slaughtered it at the creek, and were killed by the settlers in retaliation (*Humboldt Times*, Sept. 15, 1931).

Bullards: Bar, Dam, Power Reservoir [Yuba Co.]. Perpetuates the name of a forty-niner, Dr. Bullard of Brooklyn, New York, for whom the settlement Bullards Bar, now covered by the reservoir, had been named.

Bullet Chup. *See* Bally.

Bullfrog Lake [Kings Canyon NP]. John Muir called it Bryanthus Lake, but local custom prevailed. Since trout have been planted, it is no longer distinguished for its bullfrogs; yet the name remains.

Bullion. The term for gold or silver in bars or bulk was used for several place names in California and has survived in Mount Bullion [Alpine Co.], listed as a mining town in 1864. *See* Mount: Mount Bullion. **Bullion Bend** [El Dorado Co.], in the South Fork of American River, became known by this name because in the 1850s a gang of stage robbers hid their stolen gold in a deep hole in the river, from which it was later retrieved (Knave, Sept. 22, 1946).

Bull Run Creek [San Mateo Co.]. So named from 1862, presumably by a Southern sympathizer after the first Battle of Manassas (Alan K. Brown). *See* Run.

Bullskin Ridge [Shasta Co.]. The ridge was named for "Bullskin Jack," a saloon keeper, who was so nicknamed by Jim Smith, a stage driver, when the latter saw a bull's skin drying on the fence in front of the saloon (Steger).

Bullwinkel. *See* Crannell.

Bully. *See* Bally.

Bully Choop (bŏŏl ē chōōp′): **Mountain, Mountains** [Shasta, Trinity Cos.] represent Wintu *buli č'uup* 'mountain peak', from *buli* 'mountain' and *č'uup* 'sharp point, awl' (Shepherd). The name has sometimes been folk-etymologized as "bullet shoot."

Bumpass (bump′ əs) **Mountain** and **Hell**

[Lassen NP] were named for Kendall V. Bumpass, hunter, guide, and prospector, who was last seen at Morgan Springs in 1870 (Doyle). The editor of the *Red Bluff Independent* wrote in 1865: "We took up the line of march with Mr. K. V. Bumpass as guide. . . . On turning the ridge all the wonders of Hell were suddenly before us. . . . This basin was discovered by our guide last year while hunting" (D. J. Tobin). Bumpass Hell is also called Bumpass Hot Springs.

Bunker Hill. The place where American patriotism received its baptism of fire has always been a favorite place name in the United States. There seems to have been no settlement so named in California, but the name is preserved in three "hills" in Amador, Nevada, and Plumas Cos., the one in Plumas Co. being 7,290 feet higher than Bunker Hill near Boston. However, there is also a Bunker Hill between Colton and San Bernardino, and one within the city limits of Los Angeles, which are about the size of the original hill.

Bunnell: Point, Cascade, and **Cliff** [Yosemite NP]. The point was named for Dr. Lafayette H. Bunnell (1824–1903), a member of the Mariposa Battalion, which was the first party of white men to enter Yosemite Valley (Mar. 25, 1851). Bunnell, who proposed the name Yosemite for the valley, was later an army surgeon in the Civil War (Farquhar).

Bunny Lake [Mono Co.]. Possibly for Ivan Bunny, appointed as forest ranger in 1901 (Browning 1986).

Buque, El. *See* Bouquet Canyon.

Burbank [Los Angeles Co.]. The city was laid out in 1887 on the Providencia rancho and was named for one of the subdividers, Dr. David Burbank, a Los Angeles dentist.

Burbank Memorial Park [Sonoma Co.]. Named in honor of the horticulturalist Luther Burbank (1849–1926). The park was his home and experimental garden. He is buried there under a cedar of Lebanon that he planted. The Luther Burbank Grove in the Prairie Creek Redwoods State Park was established through the efforts of the Save-the-Redwoods League in 1949.

Burcham Flat [Mono Co.]. For James Burcham, reported to have been a Confederate deserter, who grazed cattle here in the 1860s (Maule).

Burdell: Mountain, Island, station [Marin Co.]. Named for Dr. Galen Burdell, who came to California on the *Duxbury* in 1849; he married the daughter of James Black, a pioneer of Marin Co., and settled on Rancho Olompali in 1863.

Burgson Lake [Tuolumne Co.]. Probably for Ed Burgson, a cattleman here in the 1940s (Browning 1986).

Buriburi (bûr ē bûr′ ē, byo͞o rē byo͞o′ rē) **Ridge** [San Mateo Co.]. From a Costanoan name, perhaps related to Rumsen *purris* 'needle' (Harrington; Callaghan). The Register of Mission Dolores lists Urebure as one of the rancherias under its jurisdiction soon after its founding. In a letter of July 24, 1798, the new Rancho de Real Hacienda in *el parage nombrado* ('the place named') Buriburi is mentioned (SP Mis. and C. 1:74). In Nov. 1826 Beechey mentions a farm called Burri Burri, with a small cottage, about 12 miles from San Francisco. On his map (1827–28) a place called Bourri is shown near what is now San Bruno. Duflot de Mofras's plan has Buri in the same location. The name Buri Buri was applied to a provisional land grant dated Dec. 11, 1827. Later the entire region between San Bruno Mountain and San Mateo became known as Buri Buri. The name of **Burra Burra Peak** [Santa Clara Co.] may come from the same stem.

Burkhalter [Nevada Co.]. Probably named for M. E. Burkhalter, president of the Pacific Wood and Lumber Company (Myrick). A post office named Burkhalter was established on June 5, 1891.

Burley Creek [San Mateo Co.]. The Burley ranch was here from the 1850s (Alan K. Brown).

Burlingame [San Mateo Co.]. Named in 1868 by William C. Ralston for his friend Anson Burlingame (1822–70), well-known orator and diplomat, who was U.S. Minister to China, 1861–67, and head of the Burlingame mission of the Chinese government, 1867–70. According to Alan K. Brown, a joking pronunciation Blingum (bling′ əm) was used in the 1890s as a dig at the anglophile mannerisms of the country-club set.

Burlington [Humboldt Co.]. The source of the name is not known. At one time the name of the post office was spelled Burhlington, to avoid confusion with an existing Burlington in the state (Turner 1993).

Burney: Creek, town, **Falls** [Shasta Co.]. When the post office was established in 1872, R. M. Johnson named it Burney Valley in memory of Samuel Burney, a trapper and immigrant guide of Scottish origin, who was killed by Indians in 1856. The Post Office Dept. changed the name to Burney in 1894. *See* McArthur-Burney Falls State Park.

Burneyville. *See* River: Riverbank.

Burnham, Mount [Los Angeles Co.]. Named by the BGN in 1951 in honor of Maj. Frederick R. Burnham, explorer and scout leader.

Burns Valley [Lake Co.]. Named for an early settler of the Lower Lake region (Co. Hist.: Napa 1881:140).

Burnt. Although there are several Burnt Mountains, Hills, and Valleys in the state, the number is small in view of the fact that grass and forest fires are of frequent occurrence. The Coso District was once called Burnt District and may have been called that by the Indians in their own tongue (*see* Coso). There is a unique Burned Mountain in Monterey Co. Often a geographic feature is named Burnt because a man-made structure has burned there: Burnt Mill Creek [San Bernardino Co.], Burnt Camp Creek [Sequoia NP], Burnt Bridge Creek [Yuba Co.], Burnt Corral Creek [Fresno Co.]. **Burnt Ranch** [Trinity Co.], a former important mining center that had a post office as early as 1858, was so named because Canadian miners burned down an Indian rancheria in 1849 (De Massey, *A Frenchman in the Gold Rush*, p. 100; Schoolcraft 3:135). The often-told story of the Indians burning down a settler's ranch there is purely imaginative. *See* Quemado.

Burra Burra. *See* Buriburi.

Burrel (bûr′ əl) [Fresno Co.]. The railroad station was named in 1889, for Cuthbert Burrell (or Burrel), an immigrant of 1846 and one of the pioneers in stock raising in Fresno Co., where he lived from 1860 to 1869. Before the coming of the railroad the place had been known as Elkhorn Station, after Burrel's Elkhorn Ranch. The post office was named in 1912.

Burrell (bûr′əl) **Creek** [Santa Cruz Co.]. For Lyman John Burrell, a native of Massachusetts, who homesteaded here in 1852 (Clark 1986).

Burrill (bûr′ əl) **Mountain** [Humboldt Co.]. Named "in the early days" for Robert Burrill, who had a cattle ranch at the foot of the mountain (Irene Quinn).

Burro. The Spanish name for 'jackass' is repeatedly found in place names, though not nearly as often as the English term. The cluster of names in Monterey Co. is derived from the name of the Los Burros Mining District of the 1870s. It is possible that some places were named for the *buro* 'mule deer', a large herd of which was observed by Font on Apr. 3, 1776, near Antioch (*Compl. diary*, p. 382). West of Santa Barbara there is a state park called Arroyo Burro.

Burson: Valley, Springs, post office [Calaveras Co.]. The places were named for a railroad man, David S. Burson, when the post office was established on Dec. 2, 1884 (Frances Bishop). The name of the railroad station, however, was Helisma.

Burt Canyon [Mono Co.]. Named for C. H. Burt, who herded sheep in this canyon before the establishment of the Mono National Forest (Maule).

Burton Creek [Placer Co.]. Homer D. Burton homesteaded here in 1871 (Lekisch).

Burton Mound [Santa Barbara Co.]. This Indian burial mound was named for Lewis T. Burton, who came to California with the Wolfskill party in 1831, lived in an adobe on the site, and later was a claimant for the Bolsa del Chamizal and Jesús María grants. *See* Mesa.

Buscombe. *See* Mill: Millville.

Butano (byo͞o′ tə nō, bo͞o′ tə nō): **Creek, Ridge, Falls, Forest** [San Mateo Co.]. El Butano was mentioned by Padre Jaime Escudet on July 7, 1816 (Arch. Arz. SF 3:101). Arrollo del Butano is shown on a *diseño* of the San Antonio or El Pescadero grant (1833). A land grant called El Butano was granted in 1838 and in 1844; one of its *diseños* shows Bolsa del Butano. According to Frances M. Molera, *butano* is what early Californians called a drinking cup made from the horn of a bull or other animal. An Indian origin is possible but has not been established.

Butcher Ranch [Placer Co.]. Mentioned as a farming settlement in the *Sacramento Union* on June 16, 1858; it was probably named for the original owner of the land. The name Butchers appears on the Placerville atlas sheet.

Bute Mountains. *See* Sutter: Sutter Buttes.

Butt: Valley, Creek, Mountain, Reservoir [Plumas Co.]. Named for Horace Butts, a successful miner at Dutch Hill, who settled in the lower part of the valley now known as Butt Valley. Butt Valley Reservoir was named in 1921 by the Great Western Power Company (Co. Library).

Butte (byo͞ot). The origin of the word was probably a Germanic stem designating a blunt extension or elevation, found in the English "butt." In French it became a geographical term and stood for a small isolated elevation, a knoll, mound, or hillock. It was introduced into the northwestern states by the French-Canadian trappers of the Hudson's Bay Company. The first Americans to use it as a generic term were probably the members of the Lewis and Clark expedition. It was Frémont, however, who gave the term a sort of official blessing: "The French word *butte,* which so often occurs in this narrative, is retained from the familiar language of the country, and identifies the objects to which it refers. It is naturalized in the region of the Rocky mountains, and, even if desirable to render it in English, I know of no word which would be its precise equivalent. It is applied to the detached hills and ridges which rise rapidly, and reach too high to be called hills or ridges, and not high enough to be called mountains" (*Expl. exp.* 1853:214–15). This distinction soon disappeared in California. High peaks like Shasta and Lassen were called buttes (often spelled bute), and for some time it seemed that the term might replace "mount" and "peak." There are in the state today some five hundred Buttes, many of them more than 5,000, and some more than 10,000, feet high. Most of them are in the north, but some are found in almost every county. A great number of creeks, lakes, meadows, etc., and five settlements are named after nearby buttes. **Butte County, Creek, Basin, Slough, Meadows,** and **Sink**: This cluster name had its origin in the three buttes that now are called the Sutter Buttes or Marysville Buttes, the impressive landmark in the Sacramento Valley. When Butte Co. was created and named on Feb. 18, 1850, the buttes were within its borders; they are now in Sutter Co. **Butte City** [Glenn Co.]: "This is the name given to a new town which we learn has been surveyed on the east bank of the Sacramento river" (*San Francisco Alta California,* Mar. 23, 1850). The city is situated on the Llano Seco grant and was likewise named after the Sutter Buttes. The place is shown as Butte on Eddy's map (1854) but apparently did not develop until the Butte Ferry was established by the Marysville-Shasta stage line about 1875. *For* Butte Mills, *see* Magalia; *for* Butteville, *see* Edgewood.

Butterbredt: Canyon, Peak, Well [Kern Co.]. In the 1860s Frederick Butterbredt, a native of Germany, settled here as a "ranchero," married an Indian woman, and had a large family. When the USGS mapped the Mojave quadrangle in 1912–13, the surveyors misspelled the name as Butterbread, an error that was corrected by the adoption of Butterbredt by the BGN in list 7403. One of the pioneer's grandsons has adopted the spelling Butterbread for his family name (H. C. Topp). It is quite possible that the original name of the settler was really *Butterbrot* 'buttered bread, sandwich.'

Butterfield Valley [Riverside Co.]. For the famous stage line, which began overland service to California in 1858 (BGN, list 7003).

Buttermilk: Hill, Country [Inyo Co.]. The name arose in the early 1870s when "Old Joe" Inman, father of State Senator Joseph Inman, had a dairy there, and the teamsters from Tim Lewis's sawmill on Birch Creek stopped at his place for a drink of buttermilk (Robinson).

Buttle Canyon [Monterey Co.]. Perhaps for William Buttle, who acquired land here in 1898 (Clark 1991).

Buttonwillow: town, **Ridge** [Kern Co.]. The name was applied to the station and post office when the branch line from Bakersfield reached the place in 1895. It goes back to the early days of Miller and Lux, when the cowboys used the lone buttonwillow tree by the slough bank as a landmark (W. D. Tracy). There is also a Buttonwillow Peak in Tulare Co. The name is a California localism for the buttonbush (*Cephalanthus occidentalis*) and arose because buttonbush leaves somewhat resemble those of the willow. Under favorable circumstances the bush will grow into a tree.

Buzzard. The heavy, slow-flying turkey vulture (*Cathartes aura*) is a common sight in many sections of the state and appears in more than twenty-five place names, including two mountains called Buzzard Roost [Tulare, San Joaquin Cos.]. The American "buzzard" should not be confused with European buzzards (genus *Buteo*), which are a type of hawk.

By-Day Creek [Mono Co.]. George Byron ("By") Day was the first white man to winter in Bridgeport Valley, around 1879 (Browning 1986).

Byrnes Ferry [Calaveras Co.]. Named for Patrick O. Byrne, who operated a ferry at the place where the road between Copperopolis and

Mountain Pass crosses the Stanislaus River. The river was first bridged in 1853, and the settlement that sprang up retained the original name. The site is also known as O'Byrne Ferry.

Byron: town, **Hot Springs** [Contra Costa Co.]. The name was applied to the station of the Southern Pacific in 1878, probably after one of the numerous Byrons "back East." The springs (without the name) are indicated on Hoffmann's map of the Bay region (1873) as Sulphur, Salt, and Hot Springs.

C

Caballada, Arroyo de. *See* Cavallo Point.

Caballo, Spanish for 'horse', occurs in several California place names but usually in the non-standard spelling "cavallo" (*q.v.*).

Cabarker. *See* El Centro.

Cabazon (kab′ ə zon): town, **Peak, Indian Reservation** [Riverside Co.]. From Spanish *cabezón* 'big head'. The station was named by the Southern Pacific Railroad in the 1870s, after a nearby Indian rancheria that appears as Cabezone on the Land Office map of 1859. The town was laid out and named after the station in 1884; the peak was named by the USGS in 1901. "These settlements in Coahuilla [i.e. Cahuilla] Valley comprise about 800 . . . Coahuilla Indians. . . . They . . . recognize Cabezon as their head chief and supreme authority. . . . His name 'Cabezon' was given him by the Californian-Mexican population, on account of the unusually large size of his head" (*San Francisco Alta California*, July 17, 1862).

Cabernet (kab ər nā′) [Kern Co.]. Named for Cabernet Sauvignon, a red wine grape originating in France.

Cabeza de Milligan. *See* Mulligan Hill.

Cabeza de Santa Rosa [Sonoma Co.]. The name of a land grant dated Sept. 30, 1841. Cabeza was used in a geographical sense like the corresponding English term "head" (of a canyon, etc.).

Cabezone. *See* Cabazon.

Cabrillo (kə brē′ yō, kə bril′ ō), **Point** [Mendocino Co.]. Named in 1870 by the Coast Survey in honor of Juan Rodríguez Cabrillo (*see* Glossary) at the suggestion of George Davidson. **Cabrillo National Monument** [San Diego Co.], which in all probability includes Cabrillo's first landing place in what is now the state of California, was created and named by proclamation of Pres. Woodrow Wilson in 1913. **Cabrillo Point** [Monterey Co.] was named in 1935 by the BGN because Cabrillo anchored near it on Nov. 16, 1542. The BGN Decisions (Sept.–Dec. 1964) also list a **Cerro Cabrillo**, "mountain peak with an elevation of about 912 feet," southeast of Morro Bay in San Luis Obispo Co.

Cabrini Canyon [Los Angeles Co.]. For "Mother" Saint Frances Xavier Cabrini (1850–1917).

Cachagua (kə chä′ gwə) **Creek** [Monterey Co.]. On a *diseño* of the Los Tularcitos grant this stream, a tributary to the upper Carmel River, is unnamed, but near it appear the words *Cañada* and *ojo de agua*. It is not impossible that a later mapmaker garbled these words into Cachagua. There is also a Mexican Spanish word *cachagua* 'sewer, gutter', which might be the source of the name. According to Clark 1991, Cachagua is probably an Esselen name, variously spelled as Tasshaguan, Jachaguan, and Xasauan.

Cache (kash) **Creek** [Yolo Co.]. The name Rivière la Cache was applied to the stream by Hudson's Bay Company trappers sometime before 1832, because they had a cache or hiding place for their traps on its banks. In Mexican documents of the 1840s the creek is called Río (or Arroyo) de Jesús María, a name that was bestowed on the Sacramento River north of its junction with the Feather by Gabriel Moraga in 1808 (*CFQ* 6:73–74). In the *New Helvetia Diary*, on Jan. 25, 1846, the stream is called Cash Creek, and this phonetic spelling is often found in the early American period. Derby's map of 1849 has the present spelling. Cacheville was the name of Yolo in the 1850s, and an Indian village of Kachitule appears on the map of Kroeber's *Handbook* facing p. 354. In pioneer times the name was quite common in the American West, and the spelling was usually "cash." Kern Co. has a Cache Peak and a Cash Creek.

Cachuma (kə cho͞o′ mə): **Creek, Mountain** [Santa Barbara Co.]. From the Chumash village name, which the Spanish spelled Aquitsumu, representing Barbareño *aqitsu'm* 'sign' (Applegate 1975:27, after Harrington). Cachuma Lake was created by impounding the Santa Ynez River in 1960.

Cactus. The four species of cactus native to California have given the name to a number of places in the southern counties, including settlements in Imperial and Riverside Cos. *See* Chollas.

Cadenasso (kad ə nas′ ō) [Yolo Co.]. Named for Nicolo Cadenasso, who bought the Adobe Ranch, west of Capay, in 1880. In 1885 he gave part of his land for a school, and the near vicinity became known as Cadenasso. The spelling Cadanassa is also found.

Cadiz (kā′ diz): station, **Lake, Valley** [San Bernardino Co.]. The station was named in 1883 by the Atlantic and Pacific Railroad, perhaps after a town of the same name in the East; the original Cádiz is in Spain. *See* Amboy.

Cady (kā′ dē) **Mountains** [San Bernardino Co.]. The name preserves the memory of Camp Cady, a military post active in 1864 to protect travelers along the San Bernardino–Fort Mojave road (Myrick). The camp had been named for Albemarle Cady, a major in the Mexican War and a brigadier general in 1865.

Caguenga. *See* Cahuenga.

Cagüilla. *See* Cahuilla.

Cagwin Lake [El Dorado Co.]. Hamden El Dorado Cagwin settled near here in 1896 (Lekisch).

Cahill (kā′ hil) **Ridge** [San Mateo Co.]. Named for Anthony Cahill, a native of Ireland and a farmer in the district in 1867. The spelling Cahil also occurs.

Cahoon: Meadow, Mountain, Creek [Tulare Co.]. The mountain and meadow were named for George Cahoon, who lived nearby (Farquhar). The name was placed on the map by the USGS in 1902–3.

Cahto (kä′ tō) [Mendocino Co.], **Cahto Creek** [Humboldt Co.]. From Northern Pomo *khaṭo* 'lake', containing *khá* 'water' (Oswalt). John P. Simpson and Robert White came here in 1856 and started the settlement known later as Cahto. A post office was established on July 22, 1863. Cahto Mountain is mentioned by John Rockwell in 1878.

Cahuenga (kə hung′ gə, kə weng′ gə): **Pass, Peak, Park** [Los Angeles Co.]. From Gabrielino *kawé'nga* (Munro); this is possibly cognate with Luiseño *qawíinga* 'at the mountain', from *qawíicha* 'mountain', and Cahuilla *qáwinga'* 'at the rock', from *qáwish* 'rock'. A *sitio de Caguenga* is mentioned on June 27, 1802 (Docs. Hist. Cal. 4:117). The name was used for three Mexican land grants, dated May 5, 1843, Feb. 7, 1845, and July 29, 1846, and is found repeatedly, with many spelling variants, in documents and *expedientes* of the following decades. It gained historical significance when Andrés Pico surrendered to Frémont at the Campo de Cahuenga on Jan. 13, 1847. The site is now within the limits of Universal City.

Cahuia. *See* Kaweah.

Cahuilla (kə wē′ ə): **Indian Reservation, Valley, Mountain**, post office [Riverside Co.]. The Cahuilla Indians are a widespread division of the Uto-Aztecan language family, who occupied the territory on both sides of the San Jacinto Mountains and still live on a number of reservations. The name of the tribe is sometimes thought to be from Cahuilla *qáwiy'a* 'leader, chief'; but in fact it is borrowed from local Spanish *cahuilla* 'unbaptized Indian'; this term, used in Mission days, was itself apparently derived from an extinct language of Baja California (cf. W. Bright, "The origin of the name 'Cahuilla'," *Journal of California Anthropology* 4.116–18 (1977), drawing on J. P. Harrington archives). In the records of the 1820s and 1830s the name was usually spelled Cagüilla; on June 21, 1845, the modern spelling variant Cahuillas was recorded for the first time (DSP 6:43). On Warren's map (1859) the habitat of the tribe is indicated along the mountain ranges between longitude 116° west and 118° west, and the name is spelled Coahuillas. As a geographical term, the name was used on the Land Office map of 1859—for what is now called the Coachella Valley (*q.v.*)—with the spelling Cohuilla. It was used by the Whitney Survey (1873) with the spelling Coahuila, showing confusion with the state in Mexico so named, and with similar spellings on all maps published before 1891. Salton Sea was called Lake Cahuilla by the Pacific Railroad Survey and is still so called by some geologists. In more recent times, the names Coahuila and Cahuilla have not referred to features in Coachella Valley but to those in the valley southwest of the San Jacinto range and south of the San Bernardino National Forest, where another band of the Cahuilla lives. The post office Cahuilla in this valley, established Apr. 12, 1888, west of Anza, was apparently the first geographical name from which the cluster name in this region developed. The name of the post office was changed to Anza on Sept. 16, 1926; and the valley was renamed Anza in 1963 (Brigandi). Various spellings, including Coahuila, have been used; but in 1963 the present form, Cahuilla, was made official by the BGN. The name Cahuilla has no relationship to Kaweah (*q.v.*), the name of a Yokuts tribe, although they are similarly pronounced.

Cahwia. *See* Kaweah.

Cajalco (kə hal′ kō) **Canyon** [Riverside Co.]. According to Gunther, this is a Spanish spelling of an Indian name; early forms include Keihal, Keihalker, Cahalco, Cahales, Cohalca. One could consider a relationship to Luiseño *qaxáal*, Cahuilla *qáxal*, meaning 'quail'.

Cajon. Spanish *cajón* 'box' was used as a geographical term to describe boxlike canyons. It is included in the names of five Mexican land grants or claims, all in southern counties. **Cajon** (kə hōn′): **Pass, Canyon, Creek**, town [San Bernardino Co.]. The original Cajon was *el cajón que llaman Muscupiavit* 'the canyon that they call Muscupiavit', as it was referred to before 1806; Muscupiavit was an Indian rancheria, the name of which has been spelled in various ways (SP Mis. and C. 1:241). On Nov. 24, 1819, Padre Nuez named it solemnly "el Caxón de San Gabriel de Amuscopiabit" (Arch. MSB 4.140). The name appears in the following decades with various spellings. The abbreviated form Cajon Pass is used on Gibbes's new map of 1852. The town was laid out when the California Southern Railroad (Santa Fe) began operations through the pass in 1885, but a settlement called Cajon had already appeared on Williamson's map of 1853. *See* El Cajon.

Cal-, Cali-. The first three or four letters of the name of the state have been used in coining a number of place names. Most of them are border names: Calada, Calneva, Calvada are found along the Nevada line; Calor connects with Oregon, Calzona with Arizona, Calexico with Mexico. Some have been coined from the names of business firms: Calwa [Fresno Co.] from California Wine Association; Calpack [Merced Co.] from California Packing Corporation; Calgro [Tulare Co.] from California Growers Wineries, Caldor [El Dorado Co.] from California Door Company. In others, the letters prefix another name: Calipatria [Imperial Co.] from Latin or Spanish *patria* 'fatherland'; Calimesa [Riverside Co.] from Spanish *mesa* 'flat-topped hill'. Califa [Madera Co.] and Calime [Butte Co.] may also belong here. *See* Calistoga.

Calabazal (kal ə bə zäl′) **Creek** [Santa Barbara Co.]. Apparently Spanish *calabazal* 'pumpkin patch', but probably a popular etymology from Ineseño Chumash *kalawashaq'* 'turtle shell' (Applegate 1975:31, after Harrington). The name also occurs as Calaguasa and Calahuasa; *see* Santa Ynez.

Calabazas, Spanish for 'pumpkins, squashes, gourds', was an important term in the cultural history of the Southwest because the gourd was an essential fruit for the Indians, who used it for food as well as for making drinking vessels and other utensils. The word appears sometimes on *diseños* to indicate a patch of pumpkins or squash, and it has survived in several place names in California. **Calabasas** (kal ə bas′ əs): town, **Peak, Canyon** [Los Angeles Co.]: The name was applied to the site, or to a rancheria at the site, prior to Aug. 18, 1795, when it is recorded that Padre Santa María slept there (Arch. MSB 2:11). A *laguna de las calabasas* is shown on *diseños* of the Corte de Madera and Las Pulgas grants. **Calabazas Creek** [Sonoma Co.]: The word Calabazas is shown on a *diseño* of part of the Agua Caliente grant (1840), a little west of Arroyo de los Guilicos. This stream may be the present Calabazas Creek. The name of the Laguna de las Calabazas grant [Santa Cruz Co.], dated Dec. 17, 1833, is not preserved in current toponymy.

Calaboose (kal′ ə bo͞os) **Creek** [Monterey Co.]. The colloquial western American term for 'jail' (from Spanish *calabozo*) was put on the map by the USGS, for unknown reasons, when the Jamesburg quadrangle was surveyed in 1917. There is a Jail Canyon in Inyo Co.

Calada. *See* Cal-.

Calaguasa; Calahuasa. *See* Calabazal.

Calajomanas. *See* Bale.

Calaveras. The Spanish word for 'skulls' was repeatedly used in early times for places where human skeletons testified to a fight or a famine. **Calaveras** (kal ə vâr′ əs) **River, Creek, County, Reservoir**, and **Big Tree State Park**: According to John Currey ("Incidents in California," 1878, MS, Bancroft Library), John Marsh and his party came upon a place in 1836–37 near the present Calaveras River, where they found a great many skulls and skeletons, and afterward always referred to the place as Calaveras. A Spanish name for the river was apparently Rio San Juan; this appears on Wilkes's map of 1841. Eld's sketch has a San Juan as well as a Calaveras. Frémont and Preuss fixed the latter name on the stream, Río de las Calaveras (map of 1845). The county, one of the original twenty-seven, was created and named by act of the legislature on Feb. 18, 1850. The state park was named because of its location in the county. **Calaveras Creek, Valley**, and **Dam** [Alameda, Santa Clara Cos.]: A *parage de las Calaveras* is mentioned as early as 1809 (Arch. SJ 3:75), and Monument Peak at one time was known as Cerro de las Calaveras. José Romero, who was born in San Jose in 1800, testified in 1860 that it was so called "because when it was first discovered by the first inhabitants it was recognized . . . as a place where the Indians . . . kept their idols. Some remains of idols [probably skulls and bones] were found there" (*WF* 6:374). Beechey (2:426) and Duflot de Mofras (plan 16) applied the name (Calavaros) to Coyote Creek, 7 miles to the west. On the Parke-Custer map of 1855, Coyote Hills are called Calaveras Point; Hoffmann's map (1873) designated the upper part of Arroyo Hondo as Calaveras Creek. A land grant called Calaveras in San Joaquin Co., dated July 11, 1846, was later rejected by the U.S. Supreme Court.

Calaveritas (kal ə və rē´ təs): **Creek**, ghost town [Calaveras Co.]. The name of the creek, a tributary to the South Fork of the Calaveras River, is a diminutive in the sense of 'Little Calaveras River'. Hoffmann (1873) has Calaveritas River.

Calavo (kə lä´ vō) **Gardens** [San Diego Co.]. For the trade name used by the California Avocado Growers Association.

Caldor. *See* Cal-.

Caldwell: Butte, Ice Cave [Lava Beds NM] were named for a "professor from Boston" who settled in the vicinity with his wife and two daughters and started a horse ranch (Howard).

Caldwell's Upper Store. *See* Nevada: Nevada City.

Calera. This Spanish word for 'lime kiln' (from *cal* 'lime') was repeatedly used in early times for place names. **Arroyo Calero** (kə lâr´ ō); **Calero Reservoir** [Santa Clara Co.]: A *paraje de la Calera* is mentioned on Aug. 16, 1803 (Prov. Recs. 11:183); the misspelling is found on many maps. **Calera** (kə lâr´ ə) **Creek** [Santa Clara Co.] was named for a *calera* mentioned on Oct. 24, 1807 (Arch. SJ 4:17). As a geographical name, *el paraje de la Calera*, it is used in 1828 (Registro, p. 19). **Calera Valley** [San Mateo Co.]: Cañada de la Calera is shown on a *diseño* of the San Pedro land grant, 1838. The lime kiln was on a point south of Laguna Salada. **La Calera y las Positas** [Santa Barbara Co.], meaning 'limekiln and water holes', was the name applied to a land grant dated May 8, 1843, and July 1, 1846.

Calexico. *See* Cal-.

Calgro. *See* Cal-.

Calico Hills [San Bernardino Co.], **Calico Peaks** [Death Valley NP]. "The colors . . . are a kaleidoscopic mixture of light and dark—the light shades are yellowish white to buff, and the dark shades dull pinkish to red. Such assemblages of particolored Tertiary volcanic rocks are called calico by the prospectors throughout the desert region, and the hills formed from them are called calico hills, the Calico Peaks being themselves an example of this usage" (Geol. Survey *Bulletin* 52:969). According to Weight (*Desert Magazine*, July 1953), Walter Knott bought the ghost town in 1951 to restore "the greatest southern California silver camp as it was in the heyday of its 1880s boom."

Caliente (kal ē en´ tē, kä lē en´ tē): **Mountain, Range** [San Luis Obispo Co.]. The name, meaning 'hot', does not refer to the temperature of the mountains. It has its origin in the Ojo Caliente, recorded on the Parke-Custer map of 1855 in the Cuyama Valley, south of the mountain, and refers obviously to the hot spring. **Caliente**: town, **Creek, Canyon** [Kern Co.]: A trading post for Indians and cowboys was established here in the 1870s and named Caliente because of the hot springs in the canyon. The name was retained by the Southern Pacific Railroad when it established a grading camp in 1874. The settlement had originally been called Allens Camp, because a sheep owner named Allen had his camp here. According to Darling 1988, it was earlier called Agua Caliente (*q.v.*).

Califa. *See* Cal-.

California. The discovery of the New World by Columbus gave a strong momentum to the age-old search for an earthly paradise with unbounded productiveness without labor, with beautiful women, gold, and pearls. Among a number of such utopias, created by the fertile imaginations of fiction writers and ambitious explorers, was the rich island of California, inhabited by handsome black women like Amazons. The story of this extraordinary realm and its queen Calafia was recounted and published about 1500 in *Las sergas de Esplandián*, by the Spanish writer Garci Ordóñez de Montalvo, as a continuation of the famous romance *Amadís de Gaula*. It was the American writer Edward Everett Hale who, in 1862, pointed out the probable connection between the name of the utopia and the name of the state. The location attributed to this island, "at the right hand of the Indies, . . . very close to that part of the Terrestrial Paradise, which was inhabited by black women," places it definitely in the same group as El Dorado, the Seven Cities of Cíbola, Quivira, and other realms eagerly sought by the Spanish conquerors. The name was probably coined and has no definite meaning. It is, of course, possible that the author may have thought of Spanish *califa* 'caliph, Muslim ruler' (from Arabic *khalīf*, lit. 'successor [of Muhammad]') when he formed the word. In spite of Christians' hatred for Muslims, the Islamic Orient had exercised, since the days of the Crusades, a strong influence on Western European culture. Indeed, a similar name for the Arab domain, *Californe*, is found in the French epic, the *Chanson de Roland*. (For a detailed discussion of the various theories about the origin and etymology of the name, see Ruth Putnam, "California: The name," *UCPH* 4:356–62; Chapman, chap. 6; Stewart, pp. 346–54.) The precise date and circumstances of the application of the name to some part of what is now called Lower California are not known. The historian Antonio de Herrera stated definitely in 1601 that it was Cortés himself who gave it this name. However, it seems more probable that the mutinous pilot of the Becerra expedition, Fortún Jiménez, discovered the peninsula in the winter of 1533, and it is not impossible that he called the newly discovered land California. On hearing of the discovery of the territory and the finding of pearls, Cortés himself set out for the new land in Apr. 1535; but he gave the name Santa Cruz to the harbor where he anchored (Chapman, p. 65), and on his map he shows the peninsula as Santa Cruz (Wagner, p. 16). By 1542, however, navigators were using the name California (report of the Cabrillo-Ferrer voyage in Bolton, *Span. expl.*, pp. 3–39). It is a matter of conjecture whether Jiménez applied the name and sailors kept it alive by word of mouth, or whether another navigator applied the name after 1535, believing that he had come upon the land of the Amazons. It took several decades for it to get on the maps and thus become definitely established. If we disregard its appearance on Castillo's map of 1541, which is considered to be a later interpolation (Wagner, p. 32), the name perhaps first appears on two maps of 1562: Golfo de la California on the map of South America in the Olives atlas in the Vatican (Wagner, no. 60), and C[abo] California on Diego Gutiérrez's map (reproduced by Putnam, "California: The name," *op. cit.*). Mercator himself applied the name in 1569 to the peninsula now called Lower California; his map of the North Polar regions (reproduced in Nordenskiöld, p. 95) shows the peninsula with the legend: *Califormia* [*sic*] *regio sola fama Hispanis nota* 'California region known to the Spanish by hearsay only'. The name remained restricted to the peninsula. Map makers frequently placed the mythical kingdom of Quivira in the general vicinity of what is now the state of California. After Sir Francis Drake (1579) applied the name New Albion to the region north of San Francisco, this name, as well as Quivira, was often used. When, throughout most of the 17th century, cartographers believed that the peninsula was really an island, there was no longer any space left for Quivira and New Albion on the maps, but California remained as the name of the island. However, even before Eusebio Kino reestablished the peninsular character of California in 1705, the French cartographer Guillaume Delisle had extended the name into the area included in the present state of California (Wagner, no. 459). The Spaniards, for the sake of priority of claim, applied the name to the entire coast; but the British, for the same reason, again applied the term New Albion to the region that is now within our state. When the Spaniards renewed their interest in colonizing the coast in 1769, the terms Baja California and Alta (or Nueva) California were applied. In common American usage, however, the term California referred to the region now within the state's borders, even before the American occupation. (For further information on the name, see E. Gudde, "The name California," *Names* 2:121–33, 196 [1954]; George R. Stewart, "More on the name California," *Names* 2: 249–54 [1954], and "More California notes," *Names* 2:275–76 [1954].) **California City** [Kern Co.] was established in 1958, incorporated in 1965 (Darling 1988). **California Valley** [San Luis Obispo Co.] was a real estate development promoted in 1960; a post office existed from 1963 to 1974 (Hall-Patton). *For* California, Punta de la, *see* Ballena; *for* California National Forest, *see* Mendocino National Forest; *for* California Range, *see* Sierra Nevada.

Caligolman. *See* Bale.

Calime. *See* Cal-.

Calimesa. *See* Cal-.

Calipatria. *See* Cal-.

Calistoga (kal i stō′ gə) [Napa Co.]. In 1859 Sam Brannan bought the place called Agua Caliente and developed it as a resort. His biographer, Reva Scott, lets Sam tell the story of its naming: "As I was saying, I named Calistoga in the first place. . . . 'Someday I'll make this place the Saratoga of California,' I started to say, but my tongue slipped and what I said was, 'I'll make this place the Calistoga of Sarafornia'" (Brannan, p. 384).

Callahan [Siskiyou Co.]. Named for M. B. Callahan, who built a cabin at the foot of Mount Bolivar in 1851 and opened a hotel there in 1852 (Co. Hist.: Siskiyou 1881:215).

Callan, Camp [San Diego Co.]. Established by the War Dept. in 1940 as a training center for antiaircraft artillery and named in honor of Gen. Robert E. Callan (1874–1936), a veteran of the Spanish-American War and World War I.

Calleguas (kä yā′ gəs): **Creek** [Ventura Co.]. The origin is Ventureño Chumash *kayïwïsh* 'the head' (Applegate 1975:33, after Harrington). Padre Vicente de Santa María visited an Indian rancheria in Aug. 1795, the name of which he spelled Cayegues (Arch. MSB 2:9–17). The name appears with the present spelling as the name of a land grant dated May 10, 1837, and again in the Land Office reports. The Parke-Custer map of 1855 has Cayegua.

Calneva. *See* Cal-.

Calor. *See* Cal-.

Calpack. *See* Cal-.

Calpella (kal pel′ ə) [Mendocino Co.]. The name of Kalpela, chief of a northern Pomo village (Barrett, *Pomo*, p. 143), was applied to his people, and later to all the Indians of Redwood Valley, by the white settlers (Kroeber). Redick McKee mentions the chief on Aug. 21, 1851, and uses the present spelling (Indian Report). The name is from Northern Pomo *khál phíila* 'carrying mussels down', from *khál* 'mussels' (Oswalt). The town was founded in 1858 by C. H. Veeder.

Calpine [Sierra Co.]. The place developed in 1919 around the mill and yards of the Davies Johnson Lumber Company and was known as McAlpine. After the Post Office Dept. rejected the name for the post office in 1922, the abbreviated form was used (L. C. Hjelte).

Caltech Peak [Sequoia NP]. This name, after the California Institute of Technology in Pasadena, was approved in 1962 (Browning 1986).

Calvada. *See* Cal-.

Calwa. *See* Cal-.

Calzona. *See* Cal-.

Camajal y Palomar [San Diego Co.]. A land grant dated Aug. 1, 1846. Mesa de Camajal and Cañada de Palomar 'valley of the pigeon roost' are shown on a *diseño*. Camajal was the name of two Indian villages in 1852, according to B. D. Wilson. *See* Palomar.

Camanche (kə man′ chē) [Calaveras Co.]; **Comanche: Creek, Point** [Kern Co.]. For the Comanche tribe of the Southern Plains; the name is from Ute *kïmánci, kïmáci* 'enemy, foreigner' (McLaughlin). The town in Calaveras Co. was named in 1849 after the town in Iowa. It is now covered by the Camanche Reservoir, completed in 1963.

Camarillo (kam ə ril′ ō, kam ə rē′ ō) [Ventura Co.]. When the Ventura section of the Southern Pacific Railroad was built in the 1880s, the name was applied to the station in memory of Juan Camarillo, owner of Rancho Calleguas from 1859 until his death in 1880.

Camatta (kə mat′ ə): **Creek, Ranch** [San Luis Obispo Co.]. An Indian village called Camate, in the border area between the Migueleño Salinan and Obispeño Chumash tribes, is listed in San Miguel Mission records. Alternative spellings include Comate, Comata, Commatta (Hall-Patton).

Cambria (kām′ brē ə) [San Luis Obispo Co.]. The place was settled in the 1860s and called

Slabtown. When the town became more self-conscious, the names Santa Rosa and Rosaville were suggested because it was built on Rancho Santa Rosa. A Welshman named Llewellyn wanted the place to be known by the Roman name of his homeland; he hung the sign "Cambria Carpenter Shop" over the door of his place of business and won the day. The new name is mentioned in a letter of Nov. 18, 1870 (Pico 3:394), but the von Leicht–Craven map of 1874 still shows both names: Cambria and Santa Rosa.

Camel. Most of the Camel Hills, Humps, and Mounds of the state were so named because of their shape. Some of the creek names may refer to the camels imported in 1857, when an expensive and abortive attempt was made to introduce the "ship of the desert" as a transportation animal into California; others were so called because the fossils of a prehistoric camel were found there.

Cameron Creek [Tulare Co.]. The name is shown on Williamson's map of 1855 and probably refers to a settler. The notion that the name is derived from the Spanish *camarón* 'shrimp' is hardly admissible, since there were no Spanish settlements in this region.

Cameron Valley [San Diego Co.]. Named for the first settlers here. A Cameron Station on the Yuma stage route is mentioned on Oct. 27, 1870, in the *San Diego Union*.

Camiaca (kä mē ä′ kə) **Peak** [Yosemite NP] is possibly from Southern Sierra Miwok *kamyaka* 'from the yarrow', from *kamya* 'yarrow' (Callaghan).

Camino. The Spanish word meaning 'road', 'trail', or 'route' has been applied to communities in El Dorado and Los Angeles Cos., probably just for its sound. **El Camino Real** (kə mē′ nō rä äl′): In Spanish days *camino real* (originally 'royal road'), *camino nacional*, or *camino principal* designated the public roads and trails between presidios, missions, and settlements (Bowman). In modern times the name has been applied to the highways connecting the missions and has often been erroneously interpreted to mean the 'king's highway'.

Camp. Military reservations are listed in this volume under their specific names. Besides military reservations, the names of many work sites and summer resorts are preceded or followed by the generic Camp. Many creeks and canyons and some settlements were named after some camp now disappeared; Camp Two [Siskiyou Co.], Camp Seven [Mendocino Co.], Camp Nine [Humboldt Co.] survived from numbered railroad or lumber camps. **Camp Bartlett** [Ventura Co.] was founded as a resort by George C. Bartlett in the 1920s (Ricard). **Camp Cawatre** [Monterey Co.], a Girl Scout camp, was named by its founders with the letters *ca* for "camp" (or "cabin"), *wa* for "water," and *tre* for "trees" (Clark 1991). **Camp Meeker** [Sonoma Co.] was known in 1876 as Meeker's, later as Camp Meeker, for Melvin C. Meeker, an early settler and lumberman. **Camp Steffani** [Monterey Co.]: Joseph Steffani, a native of Switzerland, settled here about 1888 and later subdivided part of his 2,000-acre ranch for summer homes. The original name, Camp Carmel, was changed to the present name in 1910. It is misspelled Stephani on the Jamesburg atlas sheet (F. Feliz). **Camp Curry** [Yosemite NP] was established on June 1, 1899, by David A. Curry (1860–1917) and Jennie Curry (1861–1948). **Camp Inis** (ē′ nēz) [Lake Co.]: The name, applied in the 1890s, was coined from the name Campini's Campgrounds (Mauldin). **Camp Creek** [El Dorado Co.] was named by the Mormons on their march from Sutters Fort to Salt Lake City. They camped here on July 16, 1848 (Bigler, p. 114). *For* Camp Baldy, *see* Mount: Mount Baldy; *for* Camp Carmel, *see* Camp: Camp Steffani; *for* Camp Independence, *see* Independence; *for* Camp Taylor, *see* Taylorville. *See also* Camphora.

Campaña, Arroyo de la. *See* Battle: Battle Creek.

Campana, Cañada de. *See* Bell Canyon.

Campana, La. *See* Picacho Peak.

Campbell: Creek, town [Santa Clara Co.]. The creek was named for William Campbell, an immigrant of 1846, who established a sawmill here in 1848 and a stage station in 1852. The stream was also known as Arroyo Quito, Big Moody Creek, and Saratoga Creek. The town was founded by his son, Benjamin Campbell, in 1885. By decision of the BGN (May 1954) the stream now officially bears the name Saratoga Creek.

Campbell, Mount [Fresno Co.]. The mountain was named for William Campbell, who operated a store at Pooles Ferry on the Kings River (Co. Hist.: Fresno 1956).

Campbell Creek [Shasta Co.]. Jeremiah B. Campbell was a settler of 1855 and a prominent ichthyologist (Steger).

Campbells: Point, Cove [Sonoma Co.]. The point, officially known as Bodega Head, near the spot where Bodega brought his sloop to anchor in 1775, is known locally as Campbells Point, for Capt. John Campbell, a pioneer ranch owner.

Camphora (kam fôr′ ə) [Monterey Co.]. Mexican railroad workers referred to Camp Four, a construction camp set up here in 1873, as Camfora. The railroad officials adapted the name for the station (Paul Parker, *CFQ* 1:295 [1942]).

Campini's Campgrounds. *See* Camp: Camp Inis.

Campito (kam pē′ tō): **Peak, Meadow** [Mono Co.]. The coined name, suggesting 'little field' or 'little camp', was probably applied by a surveyor who tried to bring a touch of Spanish into the White Mountains. Blanco Peak and Tres Plumas Flat are close by.

Campo. The Spanish word for 'field' is preserved in the names of a number of physical features and two communities in San Diego and Calaveras Cos. In Mexican Spanish the word also means 'mining camp', and in California it has often been used in the general sense of 'camp'. Hence **Campo Seco** [Calaveras Co.] was probably not a 'dry field' but a 'dry camp'; and the land grant Campo de los Franceses [San Joaquin Co.], dated Jan. 13, 1844, was simply 'Camp of the Frenchmen', so named because French-Canadian trappers made their camp there in the 1830s (*see also* Acampo). *For* Campo Alemán, *see* Anaheim.

Campoodle Creek [Tulare Co.]. Origin unknown. This spelling was established by the BGN, as against Compoodle, in list 7904.

Camptonville [Yuba Co.]. In 1850–51, J. M. and J. Campbell built a hotel, the Nevada House, at this place. In 1854 the town was named for the blacksmith, Robert Campton (Co. Hist.: Yuba 1879:98).

Camuesa (kə mo͞o′ sə): **Canyon, Peak** [Santa Barbara Co.]. The name is probably derived from Spanish *camuza, gamuza* 'chamois', used locally in the meaning 'buckskin'; it was applied because Indian women tanned buckskin here while camped on a hunting expedition (William S. Brown). The nearby Buckhorn Creek and Indian Creek seem to support this statement. *Camuesa* is also a Mexican name of the cactus *Opuntia robusta*, but this particular species is not native to California.

Camulos (kə myo͞o′ ləs) [Ventura Co.]. From Ventureño Chumash *kamulus* 'the juniper' (Applegate 1975:31, after Harrington). A *paraje de Camulo* (*Camulus le llaman los Naturales* 'Camulus the natives call it') is mentioned in a letter of Apr. 27, 1804 (Arch. Arz. SF 2:36). In a report of May 19, 1821, a rancho called Camulus is mentioned (Arch. Arz. SF 4:1.61), and the name Camulos appears repeatedly in documents. A *diseño* of the land grant Camulos or Álamos y Agua Caliente, dated Oct. 2, 1843, shows a Río, a Lomería, and a Cañada de Camulos. The name does not seem to be recorded on early American maps, but in the 1850s the Del Valle family applied it to the remnant of the Rancho San Francisco, which is still called Rancho Camulos.

Cañada. The Spanish word for 'valley' was perhaps the most common generic term used before the American occupation. Unlike the related terms *cañon* and *arroyo*, it did not survive as a generic term. In Ventura and Santa Barbara Cos., to be sure, American surveyors not only kept the existing Cañadas but applied the generic term to many gulches. Neither the spelling nor the pronunciation of the word is uniform. The Corps of Engineers has restored the Spanish spelling, Cañada, on most of its atlas sheets. The following pronunciations are attested: (kan′ ə də, kə nä′ də, kən yad′ ə, kən yä′ də), the last coming closest to Spanish pronunciation. **Cañada de la Dormida** [Santa Clara Co.] means 'valley of the sleeping woman'. **Cañada de la Ordeña** [Monterey Co.] means 'canyon of the dairy', referring to what was once the Meadows' dairy farm (Clark 1991). *Ordeña* is Mexican Spanish for 'dairy', from *ordeñar* 'to milk' (Santamaría). **Cañada Montuosa** [Monterey Co.] means 'canyon full of brush', from *monte* 'brush, bushes' (Clark 1991). **Cañada Taravella** [Alameda Co.] is spelled Tarravella on an 1878 map, and there is also a Tarraville Creek (Mosier and Mosier); this is perhaps from a family name, Taravella. **Cañada Verruga** 'wart valley' [San Diego Co.], according to J. Jasper, "was named by the public from the fact that in 1864 an old Indian squatter settled here. He was more intelligent than the rest, but was known particularly by all the settlers of these parts as Verruga because of a very large and prominent wart on the side of his neck. His ranch was always spoken of as Verruga Ranch and so this canyon came to be known as Canada Verruga." The name appears on maps from around 1900, and there was a Verruga post office from 1917 to 1926 (Brigandi). **La Canada** (lä kən yä´ də) [Los Angeles Co.]: The post office was established about 1890 and named after the land grant La Cañada, dated May 12, 1843. After a merger in 1979 with the settlement of Flintridge, the offical name became La Cañada Flintridge. **Cañada Verde** 'green valley' [San Mateo Co.] was the name given to a land grant dated Mar. 25, 1838. **Cañada Larga o Verde** [Ventura Co.] is the name of a land grant dated Jan. 30, 1841. **Rancho Cañada de la Carpentería** 'canyon of the carpenter shop' [Monterey Co.] was granted to Joaquín Soto in 1831 and 1845 (Clark 1991). **Rancho Cañada de la Segunda** 'canyon of the second one' [Monterey Co.] was granted to Lázaro Soto in 1839; the reason for the numbering is not known (Clark 1991). Other names of land grants beginning with Cañada are listed under their specific names.

Canadian Creek [Trinity Co.]. This and several other features were obviously named for settlers from Canada. The Canadian element in California was much stronger than these few names seem to indicate, but the French Canadians were ordinarily called "French," and British Canadians were not distinguished from other English speakers.

Canby; Fort Canby [Modoc Co.]. The post office was named for Gen. Edward R. S. Canby (1817–73), murdered during the Modoc War.

Candlestick: Point, Cove, Park, Stadium [San Francisco Co.]. In 1894 the Coast Survey established a triangulation station here and named it after Candlestick Rock, an eight-foot pinnacle shown on the map of the Board of Tide Land Commissioners in 1869. The extreme eastern part of the stadium structure marks the approximate location of the pinnacle (Rogers).

Canebrake Creek [Kern Co.]. The name, referring to a place overgrown with cane, was used by the USGS in 1908. **Canebrake Canyon** and community [San Diego Co.]: The name Cane Canyon, referring to the *carrizo* 'reed, *Phragmites communis*', was used from around 1900 and appears in its present form by 1913 (Brigandi). *See also* Kane Spring.

Cane Creek. *See* Carrizo Creek.

Canfield [Kern Co.]. The name commemorates C. A. Canfield, who, with other former gold miners, started prospecting for oil with pick and shovel about 1895 and discovered the rich deposits near Fellows.

Cannery Row [Monterey Co.]. An area along Monterey Bay where there was once a row of sardine canneries. After the area had been made famous by the writings of John Steinbeck, the name was made official in 1953 (Clark 1991).

Cannibal Island [Humboldt Co.]. Named for quarrelsome settlers who were said to fight like a "bunch of cannibals" (Turner).

Canoas (kə nō´ əs). The Spanish word for 'canoes', used in Mexico also for 'troughs', is preserved in the names of two creeks [Fresno, Santa Clara Cos.] and in Cinco ('five') Canoas Canyon [Monterey Co.]. The Spanish word was borrowed from the Arawak Indian language of the Caribbean.

Canoga (kə nō´ gə) **Park** [Los Angeles Co.]. The name Canoga was applied to the station when the Southern Pacific branch from Burbank was built in the 1890s. Park was added when the community developed. It was probably named after Canoga, New York, which had taken its name from a Cayuga (Iroquoian) village. The original name of this section of Los Angeles was Owensmouth, which is still used for the name of a street (Gregory Stein). The name of the post office was also Owensmouth from 1912 to 1931.

Canon (kan´ ən). The names of the creeks in Humboldt and Trinity Cos. are probably American renderings of the Spanish term *cañón*. *For* Cañon Mountain [Kern Co.], *see* Breckenridge Mountain; *see also* Canyon.

Cantara (kan târ´ ə) [Siskiyou Co.]. A loop on the railroad near Dunsmuir; the origin is not known (Hendryx).

Canthook (kant´ ho͞ok): **Creek, Mountain, Prairie** [Del Norte Co.]. The name of this important tool used to turn logs seems very appropriate as a place name in a lumber district.

Cantil (kan til´) [Kern Co.]. The name was given to the station when the Nevada and California Railroad was extended from Owens Lake to Mojave in 1908–9. The Spanish word means 'steep rock', but it may have been applied only because the location engineer was fond of names beginning with *C*. Cambio, Cinco, and Ceneda are other station names in this sector.

Cantu (kän to͞o´) [Imperial Co.]. The name of the Mexican colonel Estéban Cantú was applied to the station when the branch of the Inter California Railroad (Southern Pacific) was built from Calexico to Yuma in 1904–9. The post office name is Andrade.

Cantua (kan tōō′ ə) **Creek** [Fresno Co.], mentioned in the 1850s, is said to have been one of the retreats of Joaquín Murieta, the bandit. It is shown on Goddard's map of 1857 and was named for a member of the Cantúa family, in Mexican times prominent in the Monterey district. Waltham Creek at Coalinga was formerly also known as Cantua Creek (BGN, May 6, 1908).

Canyon. "The Spaniards . . . had a word meaning pipe or cannon, and in Mexico they had come to use it also for a narrow watercourse among mountains. The trappers needed such a word; they took over *cañón*, and spread it across all the West as *canyon*" (Stewart, p. 221). In California, as in other western states, the word has become a true generic term and is used more frequently than older English terms. However, it has assumed the meaning of 'narrow valley, ravine, gulch'; hence the numerous tautological Canyon (sometimes Canon) Creeks. **Canyon** [Contra Costa Co.] is abbreviated from New Redwood Canyon, where the post office was established in the 1910s. In the early 1920s the name was temporarily changed to the more melodious Sequoya. The word is found in the names of other inhabited places: Canyondam [Plumas Co.], Canyon Park [Humboldt Co.], Canyon Tank [Tuolumne Co.], Canyon Country [Los Angeles Co.]. *For* Canyon Lake, *see* Railroad Canyon.

Capay (kə pā′): town, **Valley, Canal** [Yolo Co.]. From Hill Patwin *kapay* 'creek'. A rancheria named Capa in what is now Colusa Co. is mentioned on Oct. 27, 1821 (Arch. MSB 4:169–90). The name appears with the present spelling in the land grant Cañada de Capay, dated Jan. 26, 1846. In 1851 Gibbs applied the name with the spelling Copéh to the Indians living on Putah Creek. The town in Yolo Co. was first called Munchville, then Langville, both for early settlers. The name Capay appears also in two other land grants in Tehama, Glenn, and Butte Cos., but it did not survive there as a place name. *For* Río de Capay [Glenn Co.], *see* Grindstone Creek.

Cape Horn. Many pioneers experienced the "rounding of the Horn" on their journey to California, and Cape Horn became a favorite name for places where the going was tough. The name is preserved in San Diego, Los Angeles, Fresno, Mendocino, Shasta, Siskiyou, and Placer Cos. The Southern Pacific Railroad station near Colfax [Placer Co.] was named by the construction engineers in 1866 because of the difficult curve and grade. Lassens Horn, the old name of Fandango Pass [Modoc Co.], was doubtless named in analogy to Cape Horn. Dana jokingly called Point Conception "the Cape Horn of California, where it begins to blow the first of January, and blows until the last of December."

Capell (kə pel′) **Creek** [Humboldt Co.]. The name, also spelled Cappell, is derived from that of the Yurok village *kep'el* mentioned by Heintzelman in 1858 (Waterman, p. 248; Harrington).

Capell (kə pel′) **Valley, Creek** [Napa Co.]. Named for the first settler in the valley.

Capistrano (kap i strä′ nō) **Beach** [Orange Co.]. The post office was established on Oct. 1, 1925. On Mar. 1, 1931, the name was changed to Doheny Park, for E. L. Doheny, a California oil promoter, who had laid out the townsite in the 1920s. On Jan. 1, 1948, the name was changed back to Capistrano Beach—Capistrano being short for San Juan Capistrano (*q.v.*). The community is now part of the city of Dana Point (Brigandi).

Capitan (kap′ i tan); **Cañada del Capitán; El Capitan Beach** [Santa Barbara Co.]. Arroyo de Capitán is mentioned in 1804 (Prov. Recs. 11:104), and *paraje del Capitán* in 1817 (PSP 20:177). The cañada is shown on a *diseño* (1840) of Rancho Cañada del Corral. The name may commemorate el Capitán José Francisco Ortega, though the valley is several miles east of the boundaries established for the Ortega grant. *See* El Capitan.

Capitancillos, Cañada de los [Santa Clara Co.]. The name was applied to a land grant dated June 16, 1842. Capitancillo is a diminutive of *capitán* 'captain, chief.'

Capitan Grande (kap′ i tan gran′ dē) **Indian Reservation** [San Diego Co.]. The land was set aside for Diegueño Indians and named by executive order of Pres. U. S. Grant on Dec. 27, 1875. The area is now under El Capitan Lake, the Indians having been moved to Barona Reservation (Brigandi). *See* El Capitan.

Capitola (kap i tō′ lə) [Santa Cruz Co.]. The place was developed as a resort by F. A. Hihn in 1876 and called Camp Capitola, apparently a publicity name coined from "capitol." The name may have been given in commemoration of Soquel's effort to become the capital of California shortly before. It should also be noted that two popular novels of the 1850s were *The hidden hand, or Capitola's peril*, and *Capitola the madcap*, by Mrs. Emma Dorothy Eliza Nevitte Southworth (Clark 1986).

Cappell Creek [Humboldt Co.]. *See* Capell.

Captain Jacks Stronghold and **Ice Cave** [Lava Beds NM]. The two names honor "Captain Jack," the heroic Indian leader in the Modoc War of 1872–73.

Carbon in several California place names refers to coal (cf. Spanish *carbón* 'charcoal, coal'). **Carbondale** [Amador Co.], along with nearby Lignite, keeps alive the memory of one of the greatest coal booms in California history. The deposits in Ione Valley were discovered before 1870, and in 1877 the Central Pacific built a branch line from Galt, with Carbondale as the principal station for shipping "black diamonds." Although the coal proved useless for locomotives, these mines were profitably worked for a number of years. Carbon [Mendocino Co.], Carbona [San Joaquin Co.], Carbondale [Orange Co.], and various other combinations with "carbon" or "coal" are reminders of other booms.

Carbonera (kär bə nâr′ ə) **Creek** [Santa Cruz Co.]. Named for Rancho la Carbonera, a land grant of 1838; also recorded as La Carbonero. A *carbonera* is a place where charcoal is made (Clark 1986).

Cardiff-by-the-Sea [San Diego Co.]. The town was laid out in 1911 by J. Frank Cullen and named after the seaport in Wales. The place was formerly called San Elijo after the nearby lagoon.

Cardinal Mountain and **Lake** [Kings Canyon NP]. The mountain was thus named by George R. Davis, of the USGS, because the brilliant coloring of the mountain's summit looked like the red hat of a cardinal (Farquhar).

Careaga (kâr ē ä′ gə): **Canyon**, station [Santa Barbara Co.]. Named for Juan B. Careaga of the firm Careaga and Harris, which cultivated a tract near Los Alamos in the early 1880s (Co. Hist.: Santa Barbara 1883:295).

Careys Mill. *See* Woodfords.

Caribou. This large deer (*Rangifer* spp.), native to Alaska and northern Canada, is represented in a number of place names in Trinity, Siskiyou, Shasta, Lassen, and Plumas Cos. There is even a Caribou Creek in the Mojave Desert. These names arose when California prospectors returned from British Columbia, where the famous Cariboo mines were opened in 1858. Some names were probably taken directly from the mines, and others may have been given for persons nicknamed Caribou (Stewart, "Caribou as a place name in California," *CFQ* 5:393–95).

Carillon (kâr′ i lon), **Mount** [Sequoia NP]. The bell-tower shape of the mountain suggested this name to Chester Versteeg, and it was adopted by the BGN in 1938.

Carisal; Cariso; Carizal. *See* Carrizo Creek.

Carlotta [Humboldt Co.]. When the Northwestern Pacific was built in 1903, John M. Vance, a pioneer of 1865, laid out the town and named it for his youngest daughter, Carlotta, later married to Lester W. Hink of Berkeley.

Carlsbad [San Diego Co.]. Known in 1884 as Frazier's Station, for John A. Frazier, who discovered the spring. In 1886 Gerhard Schutte and his associates tested the waters, found them like those of Karlsbad, Bohemia (now Karlovy Vary, Czech Republic), and transferred the name to the town. Carlsbad Beach State Park was named after the town in 1933.

Carmel (kär mel′): **River, Valley, Bay, Mission, Point; Mount Carmel; Carmel-by-the-Sea** [Monterey Co.]. From Spanish Carmelo, the name of Mount Carmel near Jerusalem; this in turn is from Hebrew *karmel* 'vineyard, orchard'. The river was discovered by Vizcaíno on Jan. 3, 1603, and called Río del Carmelo, probably because three friars of the Carmelite order were members of the expedition (Wagner, p. 379). In 1771 Serra and Crespí removed Mission San Carlos from Monterey to the site by the Río del Carmelo (Engelhardt 2:87), and it soon became known as Mission Carmelo, although officially the old name was retained. Except for the town that developed in modern times, and Mount Carmel, which received its name from the Coast Survey in 1856, all the features were named in Spanish times. Alejandro Malaspina's map of Monterey Bay (1791) records Río del Carmelo, Punta del Carmelo, and Ensenada de Carmelo. The form Carmel appears as early as 1798 for the river (English translation of La Pérouse, *Voyage*, 2:204), and Mount Carmel for the triangulation point of the Coast Survey (1856), but the Americanization of the name was slow. Bancroft established the modern version in his popular maps of the 1860s. Wood's *Gazetteer* (1912) still has Carmelo River. Modern Carmel-by-the-Sea was so named to distinguish it from Carmel Valley, some 10 miles inland. A land grant called Carmel, in Yolo Co., dated May 4, 1846, was rejected by the U.S. Supreme Court. *For* Point Carmel as a name for Point Lobos, *see* Lobo: Lobos Rocks.

Carmen, Isla del. *See* Brooks Island.

Carmenita (kär mə nē′ tə) [Los Angeles Co.]. Laid out in the boom year of 1887 on the land grant Los Coyotes. Someone probably considered this the diminutive of the Spanish woman's name Carmen; however, the usual form is Carmencita.

Carmen Lake. *See* Kirman Lake.

Carmichael [Sacramento Co.]. Named in 1910 by the owner of the land, for himself (G. H. Artz).

Carnadero (kär nə dâr′ ō) **Creek** [Santa Clara Co.]. The word *carnadero* (or *carneadero*), probably meaning 'butchering place', is recorded for a place in the vicinity as early as Jan. 23, 1784 (PSP 5:70). In the following decades the name appears repeatedly in documents and on maps and was used as an alternate name for the Las Animas grant. Carnadero River is mentioned by Trask in 1854 and is shown on Hoffmann's map of the Bay region (1873).

Carnaza Creek [San Luis Obispo Co.]. The Spanish word may mean 'an animal's skin with meat attached'.

Carne Humana [Napa Co.]. A land grant dated Mar. 14, 1841, named 'human flesh', appears in the records with various spellings: Huilic Noma, Caligolman, Colijolmanoc. *See* Bale.

Carnelian Bay [Placer Co.]. The bay that indents the northwest shore of Lake Tahoe was named by the Whitney Survey because of the presence of a variety of chalcedony, known as carnelian or Cambay stone. The name is shown on the von Leicht–Hoffmann Tahoe map of 1874. The post office is listed in 1910.

Carneros (kär nâr′ ōs) **Creek** [Napa Co.]. The name (Spanish for 'sheep', plural) is derived from that of the land grant Los Carneros (part of the Entre Napa grant), dated May 9, 1836. Arroyo de los Carneros is shown on *diseños* of the Huichica grant of 1844 and others of the period. Carnero Creek and Carnero Mountain (probably Bismarck Knob) are mentioned in the *Statutes* of 1850; Carnero Ridge, in Hutchings' *Illustrated California Magazine* (3:355). There is also a Carneros Creek in Santa Barbara Co.: A place Los Carneros is shown on a *diseño* of the land grant Dos Pueblos (1842), and an arroyo is labeled Carnero on a *diseño* of La Goleta (1846). Land grants in Monterey Co., dated May 13, 1834, Aug. 16, 1839, and Oct. 5, 1842, were named Los Carneros, a term from which apparently no present-day name has sprung. Kern Co. has a Carneros Spring and a Carneros Canyon (BGN, Apr. 7, 1909).

Carol Col [Fresno Co.] is a pass named after Carol Kassler Ransford, who led hikers here in 1973 (BGN, list 7801); the term "col" is otherwise very rare in California place names.

Carpenter Valley [Nevada Co.]. The small settlement at the north fork of Prosser Creek was probably named for John S. Carpenter, who was engaged in hauling logs to "Old Hobart Mills" in the 1860s (H. P. Davis).

Carpinteria (kär pin tə rē′ ə): **Creek, Lagoon**, town, **State Park** [Santa Barbara Co.]. The Portolá expedition reached the Indian village at this place on Aug. 17, 1769: "The Indians have many canoes, and at the time were building one, for which reason the soldiers named this town La Carpintería ['the carpenter shop'], while I christened it with the name of San Roque" (Crespí, p. 164). The name La Carpintería is repeatedly mentioned in early manuscripts. The post office was named Carpinteria in 1868. The park was created and named in 1932. A land grant Cañada de la Carpintería, in Monterey Co., was dated Oct. 12, 1831, and Sept. 25, 1835.

Carquinez (kär kē′ nəs): **Strait, Point, Bridge** [Solano, Contra Costa Cos.]. This is originally a Spanish plural, Carquines, of a tribal name, Carquin (Karquin, Karkin), based on a Costanoan word meaning 'barter', as recorded by Padre Arroyo de la Cuesta in 1821. The strait and Suisun Bay were discovered by the Fages expedition in 1772, and in 1776 Font (*Compl. diary*, p. 336) called it Boca del Puerto Dulce 'mouth of the freshwater port'. The Karquin Indians were mentioned in the baptismal records of Mission Dolores between 1795 and 1821. In a geographical sense the name was used by Abella in 1807: Ranchería de los Karquines and Estrecho de los Karquines (Arch. Arz. SF 2:55). Belcher

(1:118) used the same spelling in 1837. The Coast Survey used the spelling Karquines until 1905. The current spelling, Carquinez, imitating Spanish family names like Martinez, has been official since the *Statutes* of 1850 but was not generally adopted until the 20th century.

Carreteros (kär ə târ′ ōs) **Creek** [Alameda Co.]. The Spanish word means 'cart-drivers, teamsters'.

Carriger (kâr′ i gər) **Creek** [San Mateo Co.]. Timber cruisers working for the Santa Cruz Lumber Company around 1946 named this for Edward Carriger, secretary-treasurer of the company (Alan K. Brown).

Carriger Creek [Sonoma Co.]. For Nicholas Carriger, pioneer of 1846, mail carrier from Sonoma to San Rafael during the Mexican War, and farmer in the district after 1850.

Carrillo (kə rē′ yō, kə ril′ ō) **Beach State Park** [Los Angeles Co.]. The name was applied in 1953 for the prominent Southern California family. The actor Leo Carrillo later became a member of the State Park Commission.

Carrisalito (kä ri sä lē′ tō): **Spring, Creek** [Merced Co.]. The name is derived from the name of the land grant Panoche de San Juan y Los Carrisalitos 'raw sugar of St. John and the little patches of reeds', dated Feb. 10, 1844. Standard Spanish spelling would be Carrizalito, diminutive of *carrizal* 'patch of reeds', from *carrizo* 'reed' (*see* Carrizo).

Carrizo is Spanish for the common reed, *Phragmites communis*. **Carrizo** (kə rē′ zō) **Creek, Gorge**, and **Station** [San Diego, Imperial Cos.]: This plant was mentioned by Font when the Anza party camped at the junction of Carrizo and San Felipe Creeks on Dec. 13, 1775 (*Compl. diary*, p. 130). The name is recorded on Wilkes's map (1841) as Carisal, and by Emory as Cariso Creek in 1848; it appears with various spellings in the Pac. R.R. *Reports*. An attempt to Americanize it to Cane Creek was unsuccessful. **Carrizo Creek** [San Diego Co.]: Carizal is shown on a *diseño* of Agua Caliente (1840 or 1844). The name has also survived in San Luis Obispo and Riverside Cos.

Carroll Creek [Inyo Co.]. Named for A. W. de la Cour Carroll, of Lone Pine, a charter member of the Sierra Club (Robinson).

Carrville [Trinity Co.]. Named for the James E. Carr family, on whose land the town was built. The Carr ranch was established in 1852 and was first called Ruch Ranch.

Carson [Los Angeles Co.]. After George Henry Carson, who married a daughter of the Dominguez family in 1857 and managed Rancho San Pedro (Atkinson).

Carson: Creek, Flat, Hill [Calaveras Co.]. The town Carson Hill, the Slumgullion of Bret Harte's story, was named for James H. Carson, a native of Virginia who came to California as a soldier in 1847. In 1848 he discovered gold here, and his diggings became known as Carson's Creek. He was a member-elect of the legislature when he died in 1853. Carson Hill and Creek are shown on Gibbes's map of the southern mines (1852). *See* Melones.

Carson: River, Valley, Pass, Hill, Range [Alpine Co.]. The river, which was first so labeled on the Preuss map of 1848, was named by Frémont for his guide, Christopher (Kit) Carson (1809–68), with whom he crossed the Sierra Nevada in 1844. The stream had been named Pilot River on Aug. 5, 1848, by Bigler and the Mormons on their march to Salt Lake City.

Carsons [Humboldt Co.]. Probably named for William Carson, who came to California in 1850 from New Brunswick and was prominent as a banker and lumberman in Humboldt Co.

Cartago (kär tā′ gō) [Inyo Co.]. The Spanish or Latin name for the ancient city of Carthage in North Africa was applied to the station by the Southern Pacific when the Mojave-Owenyo branch reached the place in 1909, and it was given to the post office in 1919. A nearby settlement and creek bear the English version, Carthage.

Cartridge Creek [Kings Canyon NP] was named in the 1870s by Frank Lewis: "While hunting there with a young friend, Harrison Hill, I wounded a bear and told him to finish it. He became excited and threw all the shells out of his Winchester without firing a shot" (Lewis to Farquhar).

Cary Peak [Alpine Co.]. Named for the Carey (or Cary) brothers, settlers in the vicinity in the early 1850s. William Carey owned a small ranch; John Carey established a sawmill at the place that later became known as Woodfords.

Casa. The Spanish for 'house' is found in a number of place names, apparently all applied in American times. **Casa Diablo** (kä′ sə dē ä′ blō) **Mountain, Lake**, and **Hot Springs** [Mono Co.]: The mountain was named after the Casa Diablo Mine on the west slope, for Casa (del)

Diablo 'house of the devil'; a geyser once existed there (Sowaal 1985). The name was also transported to the springs (and lake) about 15 miles west. **Casa Blanca** (blang′ kə) 'white house' [Riverside Co.]: When the railroad station was built in 1887, a house visible from the right-of-way prompted the Santa Fe to bestow the Spanish name. Both Riverside and Orange Cos. have a name Casa Loma, for *casa (de la) loma* 'house of the hill'. **Casa Conejo** [Ventura Co.] was named for nearby Conejo ('rabbit') Mountain and Pass (*q.v.*). **Casa de Oro** [San Diego Co.] means 'house of gold'.

Casaba (kə sä′ bə) [Imperial Co.]. The name was applied to the Southern Pacific station when the branch from Niland was built in 1916–17. The casaba is a melon that was grown originally in Kasaba, Asia Minor.

Cascadel Point [Madera Co.]. The name may have been coined on "cascade," but it is reminiscent of Spanish *cascabel* 'rattle', *culebra de cascabel* 'rattlesnake'.

Cascade Range. The name applies now to the entire range from British Columbia to the gap south of Lassen Peak (BGN, 5th Report, 1920). The name is derived from that of the Cascades of the Columbia River. Cascade Mountains and Cascade Range of Mountains are mentioned in the travel journal (1823–27) of the Scottish botanist David Douglas (1799–1834), after whom the Douglas fir was named (*Journal kept by David Douglas*, London, 1914). Because of the orographic vagueness of the section extending into California, the name did not appear on the maps of the state until recently (for a full discussion, *see* McArthur). In addition to the Cascade Range, which is really an Oregon name, about fifteen features in California, mostly creeks, are named Cascade. The best known are the Cascades and Cascade Cliffs in Yosemite NP.

Case Mountain [Tulare Co.]. Named for Bill Case, who used to run a team of four different animals: a horse, a mule, a burro, and a steer (Farquhar).

Cash Creek. *See* Cache Creek.

Cashlapooda (kash lə po͞o′ də) **Creek** [Humboldt Co.]. Probably from the Cache la Poudre River in Colorado, French for 'hide the [gun]powder', so named by French trappers who cached their supplies there.

Casitas (kə sē′ təs): **Creek, Springs, Pass, Valley** [Ventura Co.]. The Arroyo de las Casitas ('creek of the little houses') is recorded on a plat of the lands of former Mission San Buenaventura in 1864. Lake Casitas was created in the late 1950s by impounding Coyote Creek.

Casitec. *See* Castac: Castaic.

Caslamayomi [Sonoma Co.]. A land grant of this name, also called Laguna de los Gentiles 'lake of the heathen (Indians)', was granted Mar. 21, 1844. Maps of the grant show a Río de Caslamayom (or possibly Caslamayoni). The element *yomi* means 'place' in Lake Miwok and Bodega Miwok (Callaghan). *See* Collayomi.

Casmalia (kaz māl′ yə): town, **Hills** [Santa Barbara Co.]. The name, from Purisimeño Chumash *kasma'li* 'it is the last' (Applegate 1975:32, after Harrington), appears on Apr. 6, 1837, in the *expediente* of a land grant; on *diseños* of various dates it is sometimes spelled Casmali and Casmaria. Maps made soon after the American occupation do not show the name. The name of the town is shown as Casmale on George H. Goddard's map (1860), and, with the modern spelling, on the von Leicht–Craven map (1874).

Casnau (kaz′ nō) **Creek** [Tuolumne Co.]. Misprinted as Casnan on some maps. Sometimes thought to be named for Gen. Thomas N. Casneau or Casnau, but in fact for Thomas Casenave, a French rancher who received a patent here in 1875 (de Ferrari).

Caspar (kas′ pər): town, **Point, Creek, Anchorage** [Mendocino Co.]. Named for Siegfrid Caspar, who settled there before 1860. In 1861 Kelley and Randall built a sawmill on Caspar Creek. When Jacob Green purchased the property in 1864, he named the community Caspar (Borden). The name for the point appears on the Coast Survey chart of 1872. In 1878 the hill southeast of the point was used as a triangulation point and was given the proud name Great Caspar.

Cassel [Shasta Co.]. The former name was Hat Creek; it was changed by the Post Office Dept. in 1888 at the instigation of a real estate promoter named Myers, whose birthplace was Cassel (Kassel), Germany.

Cassina Springs [Riverside Co.]. For Secondo Cassina, from Italy, who bought this land in 1927 (Gunther).

Castac (kas tāk′): **Lake, Valley** [Kern Co.]; **Castaic: Creek**, town [Los Angeles Co.]. From Ventureño Chumash *kashtiq* 'the eye, the face' (Applegate 1975:32, after Harrington). An Indian rancheria Castec is mentioned in 1791 (AGN 46). It is spelled Casteque in 1806

(Arch. MSB 4:49–68). Still another spelling is recorded on June 27, 1824 (DSP 1:48): *paraje llamado Casitec, y por nosotros S. Pablo* 'place called Casitec, and by us San Pablo'. It appears as Castec for a land grant of Nov. 22, 1843. Early American maps do not show the name, but the *Statutes* of 1851 mention a Rancho Casteque, and Blake refers to Casteca Lake in 1853 (Pac. R.R. *Reports*, 5:2.47). The official spelling (BGN, 5th Report) was established by the von Leicht–Craven map as Castac Ranch and Lake, but post office and local usage prefer Castaic. In 1960 the BGN decided for Castac Lake and Valley in Kern Co., but for Castaic Creek, Valley, and town in Los Angeles Co.

Castec. *See* Castac: Castaic.

Castella (kas tel´ ə) [Shasta Co.]. The castle-like formation of the nearby granite pinnacles known as the Crags doubtless inspired the fanciful name (Latin, 'castles'), which appeared on railway maps from 1900. Between 1900 and 1911 the name of an adjacent station to the north, similarly inspired, fluctuated from Castle Crag (1900–1904) to Castle Rock (1905–9) and back to Castle Crag (1910–11), and in 1913 it disappeared entirely.

Castellammare (kas tel ə mä´ rē) [Los Angeles Co.]. From an Italian place name, meaning 'castle at the sea'.

Casteque. *See* Castac: Castaic.

Castilleja (kas ti lā´ yə) [Sequoia NP] was named in 1896 by Bolton C. Brown, for the genus name of the flower called Indian paintbrush: "This we named Castilleja Lake, the castilleja blossoms [Indian paintbrush] being especially perfect and brilliant upon its shores" (*SCB* 2:21). The term was originally the surname of the Spanish botanist D. Castilleja.

Castillo (kas tē´ yō) **Canyon** [Riverside Co.]. For José Jesús Castillo, who was granted the land in 1890. Later misspelled as Castile Canyon and Castle Canyon (Gunther).

Castillo Point [Santa Barbara Co.] reflects either the Spanish for 'castle' or else the common family name.

Castle. A popular descriptive term for orographic features taken to resemble a castle. In California many rocks, peaks, cliffs, etc., are so named. **Castle Crags State Park** [Shasta Co.] was established in 1934 and named after the granite formation known as Castle Crags. Castle Rock, the most outstanding peak, was known as Devil's Castle until 1852, when the name was changed by the boundary survey (*Statutes* 1852:233). There is a Vulcans Castle in Lassen NP, and Siskiyou Co. has a Kings Castle. *For* Castle Peak [Yosemite NP], *see* Tower Peak; *for* Castle Rock [Shasta Co.], *see* Castella; *see also* Dunderberg Peak.

Castoria. *See* French: French Camp.

Castro. Several place names bear this Spanish family name, which is common in California. **Castroville** (kas´ trō vil) [Monterey Co.] was laid out and named in 1864 by Juan Bautista Castro on the Rancho Bolsa Nueva del Cojo, the first parcel of which had been granted to his father, Simeón Castro, on Feb. 14, 1825. **Castro Creek, Point**, and **Rocks** [San Francisco Bay] preserve the name of Joaquín I. Castro, owner of Rancho San Pablo, first granted provisionally to his father, Francisco, on Apr. 15, 1823. **Castro Valley** [Alameda Co.] commemorates Guillermo Castro, who became grantee of parts of the San Lorenzo and San Leandro lands on Feb. 23, 1841. His rancho is shown on Duflot de Mofras's map of 1844. The name for the valley appears on Hoffmannn's map of the Bay region. **Castro Flats** [Santa Clara Co.] is mostly on Las Ánimas Rancho, which was granted on Aug. 17, 1802, to José Mariano Castro, son of Joaquín Castro who came with the Anza expedition in 1776. The name also appears in other counties: for a peak [Los Angeles Co.], a canyon [Santa Barbara Co.], and a settlement [Santa Clara Co.] founded in 1867.

Caswell Memorial State Park [San Joaquin Co.]. Established in 1952 and named for the Caswell family, several members of whom were donors of the park.

Catacombs Cave [Lava Beds NM]. The peculiarly formed niches in the walls make the cave look like a subterranean cemetery of the early Christians.

Catacula [Napa Co.]. A land grant dated Nov. 9, 1844. The valley now called Chiles Valley is shown as Valle de Catuculu on an early map of the grant. The word is of Patwin origin, but its meaning is not known. *See* Chiles.

Catalina Island. *See* Santa Catalina Island.

Cataract Creek [Fresno Co.]. "Down . . . tumbled the foaming stream, a long line of silver, lost here and there amongst the talus-piles. Cataract Creek, we called it" (J. N. LeConte, "The ascent of the North Palisades," *SCB* 5:10.1–19).

Cathay [Mariposa Co.]. The post office was established on Apr. 6, 1882, and apparently named for James or Nathaniel Cathay, both residents of the township.

Cathedral. A number of peaks are so named because of their resemblance to a cathedral. **Cathedral Canyon** and **City** [Riverside Co.]: When Col. Henry Washington made the first survey of the canyon in 1858, he applied the name because he thought the canyon resembled the interior of a cathedral. The city was mapped in 1925 and named because of its location at the desert fan of the canyon (W. R. Hillery). **Cathedral Peak, Pass**, and **Range** [Yosemite NP]: The peak was named in 1862 by Henry G. Hanks, James Hutchings, and Captain Corcoran, representatives of the San Carlos Mining and Exploration Company, while on a trip to the mines near Independence (Chalfant, *Inyo*, pp. 125–26). The peak had first been designated as Cathedral Spires, but the Whitney Survey changed the name to Cathedral Peak. Cathedral Spires and Rocks are now the names of the rock formations opposite El Capitan in Yosemite Valley.

Cathey: Valley, Mountain [Mariposa Co.]. The places were named for Andrew Cathey, a native of North Carolina, who settled in the valley about 1850. The post office, 1879 to 1881, was called Catheys Valley.

Cat Mountain [San Bernardino Co.]. The mountain between Baker and Yermo is called Cat Mountain because the elements have cut the figure of a sitting cat on its front (Doyle).

Cattle. In comparison with the many places named Bull and Cow, there are only a few Cattle Creeks and Hills in the state, including a Wild Cattle Creek in Sonoma Co.

Caution [Trinity Co.]. The little town in the Trinity Alps near the Mendocino Co. line was founded in, or before, 1901. It was probably so named because it is situated in a rugged district, which could be reached only on foot or on horseback.

Cavallada (kä vä yä′ də) **Creek** [Mariposa Co.]. From Spanish *caballada* 'herd of horses'.

Cavallo (kə vä′lō) **Point** [Marin Co.]. A Punta de los Caballos 'point of the horses' is shown on a *diseño* of the Tamalpais grant of 1845, and a Plaza de los Caballos appears on Duflot de Mofras's plan 16. The present version was used by the Coast Survey on the charts of San Francisco Bay in the early 1850s. The Spanish letters *b* and *v* are often interchanged. It may be assumed from a note of Padre Payeras in 1859 (Docs. Hist. Cal. 4:270) that the name arose because horses were kept here for travel in what is now Marin Co. The bay partly formed by the point is called Horseshoe Bay because of its shape. Only a few names containing the Spanish word for 'horse' occur in California. In Spanish and Mexican times, the horse played an important role but was so common a sight and was held in so little esteem that few places were named for it until the American occupation.

Cave City [Calaveras Co.]. Named after the nearby limestone caves (Doyle). The place is shown on Hoffmann's map (1873).

Cawelo (kə wel′ ō) [Kern Co.]. Named for S. A. "Sol" Camp, Harry D. West, and Lawson Lowe, who were partners of the "Camp, West, Lowe Farm Company" here in 1940 (Darling 1988). Previously called Lerdo (*q.v.*).

Caxa. *See* Cojo.

Cayegua; Cayegues. *See* Calleguas.

Cayetano (kä yə tä′ nō) **Creek** [Alameda, Contra Costa Cos.]. A Cañada de San Cayetán appears on a *diseño*, 1834, of the land grant Las Positas. Napa Co. also has a Cayetano Creek. *See* San Cayetano.

Caymus [Napa Co.]. This land grant, dated Feb. 23, 1836, was named after an Indian village at the site of modern Yountville. The Caymos Indians are mentioned on Mar. 29, 1824 (Arch. Arz. SF 4:2.126).

Cayton: Valley, Creek, town [Shasta Co.]. The valley was settled by William Cayton about 1855 (Steger).

Cayucos (kä yo͞o′ kəs): town, **Creek, Point, Landing** [San Luis Obispo Co.]. The word was used to designate small fishing boats in California and occurs elsewhere in American Spanish. It is a Spanish rendering of the Eskimo *kayak*. A minute description of a *cayuco* is given in Font, *Compl. diary*, on Mar. 27, 1776. A document dated July 16, 1806, mentions *canoas o cayucos de las que usan en Noka* 'canoes or kayaks, the kind they use in Nootka [British Columbia]' (PSP 19:134–35), and the word appears frequently thereafter in Spanish documents to designate the *bidarkas* of the Aleuts who were employed in hunting sea otter along the California coast. The word is found also in the name of the land grant Moro y Cayucos, dated Dec. 28, 1837, and Apr. 27, 1842. The town was laid out and named in 1875 and is shown on the Land Office map of 1879.

Cazadero (kaz ə dâr′ ō) [Sonoma Co.]. The station was named when it became the terminus of the North Pacific Coast Railroad in the late 1880s. The name is Spanish for 'hunting place'.

Cazadores [Sacramento Co.]. The name of an unconfirmed land grant dated July 26, 1844, means literally 'hunters.'

Cebada (si bä′ də) **Canyon** [Santa Barbara Co.]. From Spanish *cebada* 'barley', once a principal crop in the Lompoc district.

Cebolla Canyon [Monterey Co.]. Spanish *cebolla* means 'onion', but this is a variant of Chavoya Canyon (*q.v.*).

Cecilville [Siskiyou Co.]. The place was named for John Baker Sissel, who came to Shasta Valley sometime before 1849 (Mary Bridwell). The name was misspelled on Goddard's map of 1857 and again when the post office was established on June 25, 1879.

Cedar. The name has been applied to more than one hundred places throughout the state, as Cedar Hill, Flat, Gulch, etc. The popularity of the name, which rivals Pine, results from the application of the term to a variety of native coniferous trees, many of which are not cedar but cypress. The name is also used for a number of towns and settlements, some of which may have been named for plantings of cedars or similar trees: Cedar Glen, Cedarpines Park [San Bernardino Co.]; Cedar Ridge [Nevada Co.]; Cedar Crest [Fresno Co.]. **Cedarville** [Modoc Co.] is a transfer name: it was named by J. H. Bonner in 1867 after his home town in Ohio.

Cedric Wright, Mount [Fresno Co.]. Named by the BGN in 1961, in memory of the photographer George Cedric Wright (1899–1959), who made significant contributions to appreciation of the natural scene.

Cemetery Hill. *See* Radio Hill.

Center Peak [Tulare Co.]. The peak was named by Cornelius B. Bradley and his party in July 1898. "A third [mountain], standing more detached, and in the very center of the mighty cirque at the head of the valley, we named Center Peak" (*SCB* 2:272).

Centerville. The name has always found favor in the United States. As early as 1854 Lippincott's gazetteer lists sixty-three occurrences of it, all spelled Centreville. **Centerville** [Alameda Co.], one of the earliest, was settled before 1850; it became part of the city of Fremont in 1956. **Centerville** [Fresno Co.] was named about 1870 by residents because of its central location in the Kings River Valley. For many years the post office name was Kings River. There are other Centervilles in Butte and Shasta Cos. *For* Centerville Butte [Modoc Co.], *see* Opahwah Butte; *see also* Grass Valley.

Centimudi (sen tē mo͞o′ dē) **Bay** [Shasta Co.], near Shasta Dam, may be from Wintu *kenti* 'down, inside' plus *mute* 'hear, listen' (Shepherd).

Centinela (sen ti nel′ ə) **Creek** [Los Angeles Co.]. The creek emptying into Ballona Lagoon derives its name from the grant called Aguaje del Centinela ('spring of the sentinel'), dated Sept. 14, 1844, on the territory on which it rises. The post office was established on July 24, 1889, but was discontinued in 1895. *See also* Santa Nella.

Central Valley. As an alternative to Great Central Valley, the term has long been used to refer to the combination of the Sacramento and San Joaquin Valleys. The term Central Valley was made official by the BGN in list 6602. The town of Central Valley [Shasta Co.] was named when construction began in 1938 on Shasta Dam, the main unit of the Central Valley Project; two boom towns were named Central Valley and Project City.

Centreville. *See* Centerville.

Century City [Los Angeles Co.] was named for the Twentieth Century–Fox film studios, on the site of which it was built, starting in 1961.

Cepsey. *See* Sespe.

Ceres (sēr′ ēz) [Stanislaus Co.]. The railway station of the Southern Pacific was named in 1874 by Elma Carter, daughter of one of the first settlers, in honor of the Roman goddess of growing vegetation.

Cerrito (sə rē′ tō). The word means 'little hill' (diminutive of *cerro* 'hill') and was a favorite geographical term in Spanish California. **Cerritos** [Los Angeles Co.]: Los Cerritos was the name of a land grant dated May 22, 1834. For many years this was also the name of the eminence now known as Signal Hill and of a Pacific Electric Railroad station. **El Cerrito** [Contra Costa Co.] is mentioned in the records since 1820 as Cerrito de San Antonio, a name possibly applied by the padres of Mission Dolores in honor of Anthony of Padua, a patron saint of the Franciscans (Jacob N. Bowman). Cerrito Creek is shown on Hoffmann's map of the Bay region (1873). The hill in question is actually in Albany, and is now known as Albany Hill. The community north of the hill was first known as County Line. In 1909 the post office was established and named Rust, for William R. Rust, a pioneer of 1888 and first postmaster. When the city was incorporated in 1917, the name was changed to El Cerrito. **Potrero de los Cerritos** [Alameda Co.], meaning 'pasture of the little hills', was the name given to the land grant dated Mar. 21, 1844.

Cerro (sâr′ ō). This term for 'mountain' or 'high hill' was commonly used in Spanish times and has survived as the generic term in the names of several mountains, particularly in San Diego and San Luis Obispo Cos. According to tradition, the well-known **Cerro Gordo** (lit. 'fat hill') [Inyo Co.] was named in the 1860s when Pablo Flores and two other Mexicans discovered the rich ore deposits there. The various Cerros followed by Alto ('high'), Ultimo ('last'), Lodoso ('muddy'), etc., on the atlas sheets of the Kettleman oil district are affectations of American surveyors and are not used locally. **Cerro Noroeste** [Kern Co.], meaning 'northwest mountain', was formerly called Mount Abel (BGN, 1990). For Cerro Alto [San Luis Obispo Co.], *see* Hollister Peak. **Cerro Romualdo** [San Luis Obispo Co.] commemorates a Chumash Indian, grantee of the Huerta de Romualdo or Chorro land grant dated 1842 and July 10, 1846.

Cespai River. *See* Sespe Creek.

Chabot (shə bō′), **Lake** [Alameda Co.]. An artificial lake, created in 1868–69 and named for Anthony Chabot (1814–88), a Canadian who played an important role in the Gold Rush. Later he was associated with A. W. von Schmidt in establishing the first water supply system for San Francisco. He was the donor of the Chabot Observatory in Oakland.

Chaffee (chā′ fē) **Island** [Los Angeles Co.]. This island in San Pedro Bay was named for astronaut Roger B. Chaffee (1935–1967); *see* BGN, 7902.

Chagoopa [chə gōō′ pə]: **Plateau, Falls, Creek** [Sequoia NP]. The falls were named in 1881 by W. B. Wallace and his party, for an old Paiute chief (Farquhar). The name is almost certainly a Mono word, according to Kroeber, though its meaning is unknown. The spellings Chagoopah and Shagoopah are also found.

Chalaney (chə lā′ nē) **Creek** [Tulare Co.]. Previously known as Chilean Creek and Chanley Creek; possibly from Sp. *chileno* (*q.v.*) 'Chilean'. Authorized by the BGN in list 6603.

Chalfant Lakes [Inyo Co.]. For Pleasant Arthur Chalfant (1831–1901), proprietor of the *Inyo Independent* and the *Inyo Register* in the 1870s–80s; or his son, William Arthur Chalfant, editor and author of several books on regional lore and history (Browning 1986). The Chalfant post office [Mono Co.] was established in 1913 and named for the younger Chalfant (*see* Glossary).

Chalk. The presence of chalk-like deposits is indicated in the names of about twenty peaks, buttes, and mountains in California. Chalk Mountain [Shasta Co.] is mentioned by Reading as early as 1843 (Steger). **The Chalks** [Santa Cruz Co.] is an area marked by light-colored rock, known locally as "chalkrock"; a Chalk Mountain is also nearby (Clark 1986). **Chalk Bluff Ridge** [Nevada Co.] was named after the mining camp Chalk Bluff, which was later called Red Dog.

Challenge [Yuba Co.]. Named about 1856, after the Challenge Lumber Mill, around which the settlement developed (Co. Hist.: Yuba 1924:204).

Chalone (shə lōn′, chə lōn′): **Mountain, Creek** [San Benito, Monterey Cos.]. From a Costanoan place name *čalon*, of unknown meaning (*HNAI* 8:485). *La lengua Chalona* 'the Chalone language' is mentioned in a letter of Jan. 13, 1816 (Arch. Arz. SF 3:1.6). On a *diseño* of the San Lorenzo grant the mountain is shown as Cierro Chalón. The Parke-Custer map (1855) places a Mount Chelone far-ther south, approximately in latitude 36°20' north. Goddard 1860 likewise misspells the name, but he has it in the right position. Hoffmann established the modern spelling for the name of mountain and creek. There is probably no connection with the rancheria Cholam or Cholan, which was in Salinan territory. *See* Cholame. Chalone Peaks are now called The Pinnacles (*see* Pinnacles; Metz).

Chambers Lodge [Placer Co.]. The post office was established in 1928 and named after the resort, Chambers Lodge (Florence Slade).

Chamise (shə mēs′, chə mēs′); **Chamiso**. Botanists now associate the name "chamise" with the greasewood, *Adenostoma fasciculatum*, and "chamiso" with the saltbush, *Atriplex canescens*. Both are apparently derived from regional Spanish *chamiso*, which designates several different plants (Santamaría). Standard Spanish also has *chamiza* 'brushwood'. The Indians of San Fernando Mission used the Spanish word *chamiso* for islay, *Prunus ilicifolia* ("islai, called chamiso by them,"AAE 8:12–13). The Luiseño Indians farther south called the same shrub *chámish* (AAE 8:232), but it appears to be a native word in their language. **Chamisal** designates a place where *chamiso* grows. The terms became extremely popular in Spanish times and appear in various spellings: *chamish, chamisso, chemise*, etc., but did not necessarily identify either the islay or the greasewood. On *diseños* the words *chamisal, chemisal*, and *chamiso* seem to designate brushwood or chaparral. The name appears in three land grants: Chamisal, Nov. 15, 1835 [Monterey Co.]; Bolsa del Chamisal, May 11, 1837 [San Luis Obispo Co.]; Punta de Lobos or Chamisal de los Lobos, June 25, 1846 [San Francisco Co.]. In American times the word was used as a generic name in early documents and plats. At the time of the Whitney Survey the shrub was in some degree identified with greasewood: "dense chaparral, composed more exclusively of the Adenostoma fasiculata [*sic*], or 'chamiso'. . . . Where the chamiso predominates, the thick under growth is usually designated as 'chamisal'" (Whitney, *Geology*, p. 65). Hoffmann's map of the Bay region designates several large areas as Chamisal. The name is preserved in **Chamisal** (cham′ i säl) **Ridge** [Monterey Co.], also spelled Chemisal; Chemissal and Chemisal Creeks [Colusa Co.], Chemise Ridge [San Benito Co.], Chemise Creek [Mendocino Co.], Chamisa Gap [Lake Co.], and some minor places.

Chanac (shə nak′) **Creek** [Kern Co.]. For François Chanac, a farmer from France and early settler in Cummings Valley (Darling 1988).

Chanchelulla (chän chə lo͞o′ lə) **Mountain, Gulch** [Trinity Co.]. From Wintu *son čuluula*, lit. 'rock black' (Shepherd). The Mining Bureau map of 1891 records the name, and the Land Office map of the same year has Chanche Lulla Mountain. A local pronunciation is (chän sə lo͞o′ lo͞o).

Chandler Grove [Humboldt Co.]. The grove in the Avenue of the Giants was established in 1966 and named in memory of the late Harry Chandler, publisher of the *Los Angeles Times* until 1941.

Chandon [Butte Co.]. Named in 1906 by the Northern Electric Railroad for the owner of the land on which the station was built (Florence Campbell).

Chango (chang′ gō) **Lake** [Mono Co.]. Spanish *chango* is used in California to mean 'monkey'.

Channel Islands. The collective name for the islands that are separated from the mainland by the Santa Barbara Channel. The group was discovered by Cabrillo in Oct. 1542 and named for Saint Luke, whose feast day is Oct. 18 (Wagner, p. 503). In the 18th century the individual islands were named and renamed for various saints. In 1841 the Wilkes expedition affixed to the three largest islands the respective names that they still bear: San Miguel, Santa Rosa, Santa Cruz, an alignment that probably appeared for the first time on a map of about 1794 (Wagner, p. 360). *See* San Miguel; Anacapa. **Channel Islands National Park** now includes all the islands except San Nicolas, Santa Catalina, and San Clemente.

Chanslor [Kern Co.]. The name commemorates J. A. Chanslor, one of Charlie Canfield's partners in the discovery and development of the rich oil fields. *See* Canfield.

Chaparral (shap′ ə ral, shap ə ral′). The Spanish word designated a place where the evergreen scrub oak (*Quercus dumosa*, Spanish *chaparro*) grows. In California the word is used to describe the dense brush covering of the hillsides; since these hills are a common sight, only a few bear the name. The name of Chaparrosa Spring, southeast of Bear Valley [San Bernardino Co.], is probably from the same root. *See* Chamisal, the corresponding term usually used on *diseños* of land grants.

Chappo [San Diego Co.]. This cattle-loading station on the Santa Margarita Ranch perhaps takes its name from Mexican Spanish *chapo* 'a short person', from Aztec *tzapa* 'dwarf'.

Chariot: Canyon, Mountain [San Diego Co.]. The canyon, leading off from Banner Canyon, was named after the Golden Chariot Mine, the ore deposits of which were discovered by George N. King on Feb. 13, 1871 (Hunzicker).

Charity Valley. *See* Hope Valley.

Charles Creek [Shasta Co.]. Named for Charles Keluche, son of an Indian doctor; *see* Keluche.

Charley Haupt Creek. *See* Haupt Creek.

Charleys Butte [Inyo Co.]. The lava butte near the highway, about 18 miles north of Independence, preserves the memory of Charley Tyler, an African American who had been a slave. On Mar. 7, 1863, Indians attacked the party with which he was traveling. Tyler gave up his horse to help the women escape; and failing to catch another mount, he was captured and killed by the Indians (Brierly).

Charlotte: Lake, Creek [Fresno Co.]. A Lake Charlotte is recorded on Hoffmann's map of 1873. It was probably named by a member of the Whitney Survey.

Charlton (shär′ əl tən) **Flat** [Los Angeles Co.]. Applied to an open space near Pine Mountain, for the late R. H. Charlton, supervisor of Angeles National Forest from 1905 to 1925.

Charter Oak [Los Angeles Co.]. The community developed in the late 1890s and was probably so named because a large oak tree reminded someone of the famous Charter Oak in Hartford, Connecticut. The post office is listed in 1904. There is no evidence to support the local tradition that the place was named when a party of Americans achieved a victory near the oak at some time in the war with Mexico. Regarding the Charter Oak in Connecticut, there is a legend that Capt. Joseph Wadsworth hid a royal charter in its trunk, giving the residents a claim to a strip of land extending westward to the Pacific.

Charybdis. *See* Scylla.

Chasta; Chasty. *See* Shasta.

Chatsworth: town, **Peak, Reservoir** [Los Angeles Co.]. Named in 1887 after Chatsworth, the estate of the Duke of Devonshire in England (Atkinson).

Chatterdowen (chat′ ər dou ən) **Creek** [Shasta Co.]. From Wintu *čati tawin*, lit. 'diggerpine-nut flat' (Shepherd).

Chauciles, Rancheria de los; Chausila, Rancheria de. *See* Chowchilla.

Chavoya (chə voi′ ə) **Canyon** [Monterey Co.]. For Alejandro Chavoya or Chaboya, who acquired land here in the late 19th century (Clark 1991).

Chawanakee (chə wä′ nə kē, shə- . . .) [Fresno Co.]. Said to be an Indian word, but the origin is unknown.

Cheapskate Hill [San Luis Obispo Co.] was so named because it was visited by people who wanted to watch races at an adjoining racetrack without payment (Hall-Patton).

Cheenitch Creek [Humboldt Co.]. From the Karuk place name *chíinach*, etymology unknown.

Cheeseville [Siskiyou Co.] grew up around J. D. Shelly's cheese factory, founded in 1889 (Luecke).

Chelame Pass. *See* Cholame.

Chelone, Mount. *See* Chalone Mountain.

Chemawa (chə mä′ wə) [Riverside Co.]. Transferred in 1901 from Chemawa, Marion Co., Oregon, by the Riverside and Arlington Railroad; originally a Willamette or Kalapooya Indian name (Gunther 1984).

Chemehuevi (chem ə wā′ vē): **Valley, Mountains, Indian Reservation** [San Bernardino Co.]. The name of a tribe closely related to the Southern Paiute, repeatedly mentioned by this name in Spanish and early American times under a great variety of spellings. The source is Mojave *'achiimuuév*, said to mean 'those who work with fish', containing *'achíi-m* 'with fish' and *uu-év-* 'plural-work' (Munro).

Chemise, Chemisal. *See* Chamise.

Chemurgic (kə mûr′ jik) [Stanislaus Co.]. The term refers to the industrial application of organic chemistry.

Cherokee (châr′ ə kē). The name of the tribe of the southern Appalachian region is one of our romantic and truly American names and is used as a place name in many states. The tribe's own name is Tsalagi, of unknown origin (Scancarelli). At the time of the Gold Rush several parties of Cherokees came to California; of the five mining towns named for them, only one, in Butte Co., survives. The one in Calaveras Co. is remembered in the name of a creek, and the one in Sierra Co. in the name of a creek and a bridge. There is a Cherokee Station on the Central California Traction Railroad in San Joaquin County. Cherokee Flat of the 1850s is now Altaville [Calaveras Co.], and Cherokee [Nevada Co.] is the site of Paterson on the Smartsville atlas sheet. **Cherokee** [Tuolumne Co.] is the site of a mining camp where gold was discovered in 1853 by the Scott brothers, descendants of Cherokee Indians.

Cherry. Several native shrubs have cherry-like fruit, especially the western chokecherry, *Prunus virginiana*, and the islay or holly-leaved cherry, *Prunus ilicifolia*. (*See* Islay.) Most of the twenty-odd places originally named Cherry were so called because of the presence of these shrubs. Some, however, may be named for orchards or for places "back East," where the name is popular. There are also a few Chokecherry Creeks and Canyons in the southern part of the state.

Chester [Plumas Co.]. Named after Chester, Vermont, by Oscar Martin in the 1900s (G. Stover).

Chianti (kē an′ tē) [Sonoma Co.]. The station on the Northwestern Pacific was named, like nearby Asti, after a wine-producing district in Italy.

Chicago. The metropolis of the Middle West has been used repeatedly for place names in various sections of the state. **Chicago Park** [Nevada Co.] was named by Paul Ullrich and a group of German settlers from Illinois in the 1880s (Elizabeth Beukers). **Port Chicago** [Contra Costa Co.] was founded in 1906 by the C. A. Smith Lumber Company and called Bay Point. The present name was chosen for the post office in 1931, after the name Chicago had been rejected by the Post Office Dept.

Chicarita (chik ə rē′ tə) **Creek** [San Diego Co.] is perhaps an error for Spanish *chiquita* 'little' (Stein).

Chickabally (chik′ ə bä lē) **Mountain** [Shasta Co.]. Probably from Wintu *ƚikup'uri* 'a fight', merged with *buli* 'mountain' (Shepherd). *See* Bally; Clikapudi.

Chickahominy (chik ə hom′ i nē) **Slough** [Yolo Co.]. According to the County History (1940: 240), the slough was named because two ranchers, Joseph Griffin and J. McMahon, had a fight here around the time of the Battle of Chickahominy in Virginia, June 27, 1862. The name is originally that of a Virginia Algonquian tribe.

Chicken Ranch Rancheria [Sacramento Co.]. An Indian community, federally recognized since 1980.

Chickering Grove [Humboldt Co.]. The grove in the Prairie Creek Redwoods State Park was established in 1958 in honor of Allen and Alma S. Chickering. Allen Chickering (1877–1958) had been a member of the Sierra Club since 1896, of the Save-the-Redwoods League since 1919 (one of its councillors for twenty-eight years), and of the California Historical Society since 1923 (a director for twenty-seven years).

Chico (chē′ kō) [Butte Co.]. The origin of the name is found in the creek that suggested the name for the land grant Arroyo Chico ('little stream'), granted to William Dickey on Nov. 7, 1844, and sold in 1849 to Bidwell, both formerly connected with Sutters Fort. The city and Creek are recorded in the *Statutes* of 1850, but the modern city was not laid out until Feb. 1860. The original Arroyo Chico is now called Big Chico Creek (lit. 'big little creek'), and the stream to the south, Little Chico Creek ('little little creek'). Chico is found in the names of twenty or more features in California—including some with mixed genders, such as Bolsa Chico Lake [San Luis Obispo Co.]. In some cases it may be a nickname for the Spanish name Francisco. **Chico Martinez Creek** [San Luis Obispo Co.] was named for Chico Martínez, "king of the mustang runners" (mentioned by Rensch-Hoover, p. 139). *See* Chiquito.

Chicopee Canyon [San Bernardino Co.]. *See* Jacoby Canyon.

Chihuahua (chi wä′ wä): **Valley, Creek** [San Diego Co.]. José Melandras, a goatherd who loved solitude, settled first near Warner Hot Springs. When too many people came to the valley, he moved farther inland to a place near Dead Mans Hole, and finally to the valley that became known as Chihuahua, for his native state in Mexico (Mary Connors).

Chilao (chi lā′ ō) [Los Angeles Co.]. Formerly Chileo or Chilleo, a nickname of José González—a herder and a member of Tiburcio Vásquez's bandit gang—who killed a grizzly bear in the area with only a hunting knife. The nickname may be from *chileno* (*q.v.*) 'Chilean'.

Chilcoot (chil′ ko͞ot) [Plumas Co.]. Named about 1900 after Chilcoot Pass, Alaska, the gateway to the Klondike, at the time of the gold rush to the Yukon (R. F. Ramelli). Said to be from an Indian tribal name, also spelled Chilkat.

Chile, Chili, Chilean. Of all South American countries, Chile sent the largest contingent of miners to California during the Gold Rush. Some of the many camps and physical features named for them are still found on the maps. **Chili Bar** [El Dorado Co.], on the South Fork of the American River, was a gold-mining camp in 1851 and was first called Chillian Bar. **Chili Camp** [Calaveras Co.], east of Lancha Plana, was first mentioned in 1850. Gold mining was carried on there until the 20th century. **Chili Gulch** [Calaveras Co.], south of Mokelumne Hill, was mentioned as Chile or Chilean Gulch as early as 1850. The place was so named when a Dr. Concha imported Chilean workers to the gulch, and their presence led to the so-called Chilean War in 1850. The gold deposits proved to be very rich, and in 1857 a post office was established with the name Chili.

Chileno (chi lē′ nō) **Valley** [Marin Co.]. From the Spanish designation for a native of Chile. The term also occurs in Chileno Creek [Merced Co.] and Chileno Canyon [Los Angeles Co.]. *See* Chile: Chili Gulch.

Chileo; Chilleo. *See* Chilao.

Chiles (chīlz): town, **Valley** [Napa Co.]. Named for Joseph B. Chiles, of Kentucky, who came across the plains in 1841 and was grantee of the Catacula grant in Napa Co. on Nov. 9, 1844.

Chillian Bar. *See* Chile, Chili.

Chilnualna (chil nōō äl′ nə) **Creek, Falls, Lake**, and **Ranger Station** [Yosemite NP]. An Indian name, perhaps containing Southern Sierra Miwok *čïlï* 'mosquito' (Broadbent 1964; Callaghan). It is shown as Chilnoialny Creek on the Hoffmann-Gardiner Yosemite map of 1867.

Chimiles [Napa Co.]. The name, of unknown Indian origin, was given to the land grant dated May 2, 1846.

Chimmekanee (chim ə kə nē′) **Creek** [Humboldt Co.]. From the Karuk village *chamiknîi-nach*, etymology unknown.

Chimney. Some of the thirty-odd Chimney Rocks, Canyons, Gulches, etc., were named because of chimney-like rock formations; others from the presence of "pay chimneys," i.e. lodes of gold-bearing quartz; still others possibly from the vertical cleft in rocks called a "chimney" by mountain climbers. **Chimney Peak** [Imperial Co.]: The famous landmark for early explorers and travelers in the Colorado Desert was so named because of its shape. *See* Picacho. **Chimney Rock** in Modoc Co. was so named because Thomas L. Denson, who settled here in the early 1850s, cut a fireplace and flue out of the rock. In Lava Beds NM the term is repeatedly used with reference to "chimneys" created by the cooling of lava.

China; Chinese. China and Chinese are among the most popular of California place names derived from nationalities. They bear witness to the important part that Chinese labor played in the building of the state. Some of the places date from Gold Rush days; others were named in the 1860s and 1870s when large numbers of Chinese were employed in mining and railroad building. Some of the names along the coast point to trade with China, and others to the presence of Chinese fishermen in the district. **China Basin** [San Francisco Co.] was named for ships of the Pacific Steamship Company, the "China Clippers," which tied up near here in the 1860s (Loewenstein). **China Slide** [Trinity Co.] was named not only because of extensive tunneling by Chinese miners but also because the disastrous slide occurred on Chinese New Year's Day, 1890 (W. E. Hotelling). **China Lake** [Kern Co.]: The post office was established in 1948 and named after the large dry lake nearby. **Chinese Camp** [Tuolumne Co.], locally pronounced "Chinee Camp", was named for Chinese miners during the Gold Rush. One California place name is based on a Chinese curse; *see* Tunemah.

Chino (chē′ nō): city, **Creek, Hills, Valley** [San Bernardino Co.]; **Canyon** [Riverside Co.]. The name is derived from that of the Santa Ana del Chino land grant, dated Mar. 26, 1841. The term *chino* is used in most Spanish-American countries for persons of mixed blood; but paradoxically, it also refers in Mexican Spanish to a person with curly hair. The word was used in a geographical sense before 1830 (Bancroft 2:352, map), probably as the name of an Indian rancheria whose chief was a *chino*. Early American maps and reports have the spelling China, but in the *Statutes* of 1853 the original spelling was restored. The settlement is shown on the Land Office map of 1862. The name appears in several other features in the southern part of the state, and in Chino Creek [Butte Co.]. An unclaimed grant, Ranchería del Chino [Santa Clara Co.], might have something to do with Chino Rodríguez, patentee of the San Gregorio grant [San Mateo Co.].

Chinquapin (ching′ kə pin) [Yosemite NP]. The ranger station was named for the nut-bearing "bush chinquapin" (*Castanopsis sempervirens*). The name is originally a Virginia Algonquian word, introduced by Capt. John Smith as *chechinquamins* in 1612 (Cutler, pp. 19, 249). According to Rowntree, there is a good stand of this shrub at the station. The name seems not to have been recorded until the USGS mapped the valley in 1893–94.

Chintache. *See* Tache; Tulare: Tulare Lake.

Chiquito (chi kē′ tō), **Chiquita** (chi kē′ tə). The word is a diminutive of Spanish *chico* 'small' and is found in the name of Chiquita station north of Healdsburg [Sonoma Co.] on the Northwestern Pacific Railroad, and Chiquito

Creek in Shasta Co. **Chiquito: Creek, Ridge, Pass, Lake** [Madera Co.]. The name goes back to the Chiquito Joaquín 'little [San] Joaquin', probably applied to this tributary of San Joaquin River by Mexican prospectors in the mining days. Hoffmann 1873 also applies the name Chiquito to the highest peak of the Ridge and has Chiquito Meadows for Beasore Meadows.

Chiriaco (shi rā′ kō) **Summit** [Riverside Co.]. Joseph L. Chiriaco established a gas station and store here in 1933 (Gunther 1984).

Chirpchatter Mountain [Shasta Co.]. From Wintu *t'arap č'araw*, lit. 'cottonwood field', from *t'arap* 'cottonwood tree' + *č'araw* 'green place' (Pitkin, p. 649).

Chismahoo (chis′ mə ho͞o): **Creek, Mountain** [Ventura Co.]. Probably from Ventureño Chumash *ts'ismuhu* 'it streams out' (Applegate 1975:44, after Harrington).

Chittenden Peak and **Pass** [Yosemite NP] were named for Capt. Hiram M. Chittenden (1858–1917), who became a brigadier general in the Corps of Engineers and later a boundary commissioner for the national park.

Chivo (chē′ vō) **Canyon** [Ventura Co.]. Spanish *chivo* means 'young goat'.

Chocolate. The word is repeatedly used to indicate dark brown coloring. The best-known features so named are the Chocolate Mountains [Imperial Co.], Chocolate Peak [Inyo Co.], and Chocolate Creek [San Diego Co.]. The word was borrowed from Spanish, which in turn borrowed it from Nahuatl, the language of the Aztecs.

Chokecherry Creek. *See* Cherry.

Cholame (shō lam′, chō lam′) [San Luis Obispo Co.]; **Cholame: Creek, Hills** [Monterey, San Luis Obispo Cos.]. Cholam was a Salinan rancheria, fourteen leagues from Mission San Miguel; Harrington derives it from Migueleño Salinan *č'olám*, said to refer to evil people. It is mentioned on Jan. 29, 1804, as Cholan (PSP Ben. Mil. 34:7). The name appears as Cholame on a *diseño* of the San Miguel grant in 1840, and on a *diseño* of the Cholam land grant of Feb. 5, 1844. The Parke-Custer map (1855) has a Chelame Pass for the valley between Shandon and Cholame. The Land Office map of 1859 has Choloma Creek. The current spelling of the name was used by Hoffmann in a letter to Whitney on Apr. 7, 1866, and is so recorded on the von Leicht–Craven map of 1873. *See* Chalone.

Chollas (chol′ əs, choi′ əs): station, **Creek, Reservoir** [San Diego Co.]. Originally the name of a large Indian village near the present Indian Point, mentioned on Nov. 30, 1775 (PSP Ben. Mil. 1:1), and shown on Pantoja's map of 1782 as Ranchería de las Choyas. A. B. Gray's map of 1849 has Las Choyas; that of the Coast Survey, 1857, has Choyas Valley; and the Coast Survey in 1858 has Choya Point. The *Official map of San Diego*, 1870, has the spelling variant Chollas Valley. In standard Spanish, *cholla* is 'skull' or 'head'; but in Mexican and California Spanish the word, pronounced (chō′ yä), refers to the cholla cactus (genus *Opuntia*). There is a Cholla Creek in San Benito Co.

Choloma Creek. *See* Cholame Creek.

Chopines, Arroyo de. *See* Chupines.

Chorro (chôr′ ō) **Creek** [San Luis Obispo Co.]. *Chorro* is Spanish for 'a jet of water'. The creek is shown on the *diseño* of the "Huerta de Romualdo (o del Chorro)" grant (1842 and 1846) as Arroyo del Chorro and Aguage ('spring') del Chorro. An adjoining grant, Cañada del Chorro, dated Oct. 10, 1845, was doubtless named after the same arroyo.

Choual Mountain. *See* Chual, Mount.

Chowchilla (chow chil′ ə): **River, Mountains**, town, **Canal** [Madera, Mariposa Cos.]. According to Hodge and to Kroeber (*Handbook*, p. 443), there was a Yokuts tribe on the lower course of the river and a Miwok tribe probably on the upper course. The first ascertainable recording is *ranchería de Chausila*, by Sebastián Rodríguez in Apr. 1828. A *ranchería de los Chauciles* is mentioned on Dec. 8, 1834 (Pico Docs., p. 128), and *indios Chauciles y Joyimas* are mentioned by the alcalde of San Jose, Feb. 5, 1835 (Arch. SJ 1:40). Both references (which relate to horse stealing) doubtless are to the Yokuts tribe or village. The Chauciles were aggressive, and the name may mean 'killers.' Kroeber ("Place names") thinks it likely "that the Miwok Chauchilas were so named by Americans, or by English-speaking Indians, after the name of the stream." At any rate, in the various Indian reports of the early 1850s the two tribes are obviously confused, and the name is recorded in various spellings. The stream is shown on most maps of the 1850s. The present version is found on the maps of the Pacific Railroad Survey.

Choyas. *See* Chollas.

Christine [Mendocino Co.]. A number of Swiss

families settled here before 1856, and the place became known as Guntley's, for one of the pioneers. When a post office was established in the 1870s it was named Christine, for a daughter of John Geschwend, another pioneer, who had built the first sawmill in the valley (Co. Hist.: Mendocino 1914:35).

Chrysopylae. *See* Golden Gate.

Chual (chōō′ əl), **Mount** [Santa Clara Co.]. The name originated in Mexican times and is recorded on C. S. Lyman's manuscript "Mapa de Nuevo Almadén" (1848) as Picacho de Chual. Hoffmann's map of the Bay region calls it Choual Mountain. The name refers to an edible plant, *Chenopodium album*, the common pigweed or lamb's quarters; it is Mexican Spanish *chual* or *chuale*, from Aztec *tzohualli, tzoalli* (Santamaría). **Chualar** (chōō′ ə lär) [Monterey Co.]: This Spanish word means 'place where the *chual* grows.' In 1830 the place was one of the landmarks delineating the jurisdiction of Monterey (Legis. Recs. 1:147). The name was preserved through two land grants, dated Sept. 23, 1831, and Jan. 5, 1835. The present town is recorded on the von Leicht–Craven map.

Chubbuck [San Bernardino Co.]. When the Santa Fe line from Cadiz to Blythe was built in 1910, the station was named Kilbeck for an employee of the railroad. When the post office was established in 1937, the name was changed to Chubbuck for the owner and developer of the local lime deposits (Santa Fe).

Chuchita (chōō chē′ tə), **The; Chuchitas, Chuchita Creek** [Santa Cruz Co.]. Attested from 1860 (Clark 1986). California Spanish Chucha or Chuchita can be a nickname for Jesusa, a woman's name, or can represent the feminine of *chucho* 'dog'.

Chuchupate (chōō chōō pat′ ē) [Kern Co.]. The name for the district between Mount Pinos and Frazier Mountain in Los Padres National Forest and for the old Tejon ranger station. The term has been derived from the local Indian name of a yellow-flowered plant, probably the arrowleaf balsamroot, the roots of which were valued by the natives as a food (USFS). However, *chuchupate* has also been identified as an umbelliferous plant of Baja California (*Arracacia brandegesi*, Blanco) and as *Lomatium californicum*, a plant resembling angelica (Alice Anderton, "The Spanish of John P. Harrington's Kitanemuk notes," *Anthropological Linguistics* 33:450 [1991]). The Spanish term is probably from Aztec; Frances Karttunen has pointed out the Aztec plant name *xoxouhca-pahtli*, lit. 'blue-medicine' (*Badianus*, p. 268).

Chuckawalla (chuk′ ə wä lə) **Mountains, Valley** [Riverside Co.]; **Chuckwalla: Canyon** [Death Valley NP], **Mountain** [Kern Co.]. These places were named for the chuckwalla (*Sauromalis obesus*), a species of lizard native to the southeastern part of the state. The name of the lizard is borrowed from California Spanish *chacahuala*, or directly from Cahuilla *cháxwal*, perhaps with influence from Chemehuevi *chagwara* (Munro); this is one of the few animal names that have entered English from California Indian languages. Chuckwalla Wells is mentioned in the 1870s as a stopping place of stagecoaches.

Chuckchansi (chu̇k chan′ sē) **Indian Reservation** [Fresno Co.]. The name of a Yokuts tribe is mentioned with various spellings in the Indian reports after 1851; the Yokuts pronunciation is *č^h^ukč^h^ansi* (Gamble).

Chuck Pass [Fresno Co.]. Apparently named by the USGS in 1907–9 after Woodchuck Creek, a tributary to the North Fork of Kings River.

Chukar (chōō kär′) **Springs** [San Bernardino Co.]. Named for a game bird, originally introduced to the United States from Asia; the term is derived from Hindi.

Chula Vista (chōō′ lə vis′ tə) [San Diego Co.]. *Chula* is Mexican Spanish for 'pretty'. The town was laid out and named by the San Diego Land and Town Company in 1888. Through the efforts of Eldredge and Davidson, the post office name was changed from Chulavista to the present form in 1906.

Chumash (chōō′ mash) **Peak** [San Luis Obispo Co.]. The 1,250-foot elevation northwest of San Luis Obispo was named in 1964 for the Indians of the area, who left a great number of place names in their language on the maps (Hall-Patton). *See* Chumash in Glossary.

Chupadero. The word is Spanish American for a 'brackish pool where animals come to drink', from *chupar* 'to suck'. **Rancho Chupadero** was used as the name of a land grant [Monterey Co.] before 1795 (DSP 13:267), and the word is shown on *diseños* of several other grants. A mountain, Chupedero, is mentioned by Trask (*Report* 1854:91) and is shown on Goddard's map at latitude 36°35′ north.

Chupines (chə pē′ nəs): **Creek** [Monterey Co.]. The name perhaps contains Spanish *chopo* 'black cottonwood, *Populus trichocarpa*'; Chupines given as the name of a tree in a report of 1828 (Clark 1991). An Arroyo de los Chupines (or Chopines) is recorded on a *diseño* of the land grant Los Tularcitos.

Church Creek [Monterey Co.]. The Church family, members of which still reside in the vicinity of Monterey and Carmel, was the first to settle on the creek.

Churchwalla Well [Riverside Co.]. Probably a mapmaker's error for Chuck(a)walla (*q.v.*).

Churn Creek [Shasta Co.]. A hole in the rock bottom of the creek, made by a waterfall and resembling an old-fashioned churn, suggested the name. The creek is mentioned in the *Sacramento Union* of July 9, 1851. Churntown, now vanished, was named after the creek.

Chute Landing [Santa Barbara Co.]. Because of the high rates charged at the wharf of Point Sal, a company of farmers commissioned a French Canadian named St. Ores to build a chute, or slide, down the cliff to a wharf (Co. Hist.: Santa Barbara 1883:300–301).

Ciénega. The Spanish generic term (with the alternative form *ciénaga*) was widely used before the American occupation. It means 'marsh', but in the southwestern states was applied also to a marshy meadow. Although the word appears in the names of about ten land grants or claims, it seems to have survived in only a few place names. **Cienega** (sē en′ ə gə) **Creek** and **Valley** [San Benito Co.]: Here the name was preserved through the land grant Ciénega de los Paicines, dated Oct. 5, 1842. The *ciénega* is shown on a *diseño* of 1834. The term is still used at times in a generic sense: Cienaga Seca [San Bernardino Co.], Cooper Cienaga [San Diego Co.]. There is also Cienaga Canyon in the Angeles National Forest, southwest of Liebre Mountain. **Cieneguitas**, meaning 'little swamps', was the name given to a land grant [Santa Barbara Co.] dated Oct. 10, 1845, which was also known as Paraje ('place') de la Cieneguita, or Suerte ('lot') en las Cieneguitas.

Ciervo (sē âr′ vō): **Hills, Mountain; Arroyo Ciervo** [Fresno Co.]. The word for 'elk' was repeatedly used in Spanish times and has survived in this cluster name as well as in Cañada del Cierbo [Contra Costa Co.].

Cima (sē′ mə) [San Bernardino Co.]. This Spanish word for summit was given to the station in 1907 when the San Pedro–Los Angeles–Salt Lake Railroad reached the place. It is at the top of the pass between Kelso and Ivanpah (Gill). **Cima Mesa** [Los Angeles Co.]: So named because of its location above Antelope Valley.

Cinco Canoas Canyon. *See* Canoas.

Cinder Cone. *See* Cone.

Cinnabar. The vermilion-colored mercury ore is found in various parts of the state and has given its name to several places, of which Cinnabar Camp and Springs [Siskiyou Co.] are the best known.

Cipreses, Punta de. *See* Cypress: Point Cypress.

Circle City. *See* Corona.

Cirque (sûrk). The geographical term for a steep-walled, semicircular depression in a mountain, caused by glacial action, is repeatedly found as a specific name in High Sierra features: **Cirque: Mountain** [Mono Co.], **Crest** [Fresno Co.], **Peak** [Inyo Co.]. Glacier Divide, now part of the boundary of Kings Canyon NP, was named Cirque Ridge on LeConte's map of 1905: "We stood [July 1904] upon the brink of an immense cirque, or amphitheater, filled with snow and ice. . . . This cirque was one of a chain of cirques which formed the northern face of the ridge" (*SCB* 5:158). The same party named Cirque Lake, a name likewise disregarded by the USGS: "This little lake is located in the bottom of the cirque into which we had descended, and for that reason we named it 'Cirque Lake'" (ibid., p. 160).

Cisco [Placer Co.]. Named in 1865 by the Central Pacific Railroad, for John J. Cisco, treasurer of the company (1863–69), at the suggestion of Charles Crocker, one of the organizers of the railroad. It was formerly known as Heaton Station.

Citrus. The citrus fruit industry in the state has given rise to a number of place names: Citro [Tulare Co.], Citrona [Yolo Co.], Citrus [Los Angeles Co.], Citrus Heights [Sacramento Co.]. *See* Lemon; Orange.

City of Commerce [Los Angeles Co.]. Incorporated in the 1960s as a manufacturing center for B. F. Goodrich, UniRoyal Rubber, Chrysler, etc. (Atkinson).

City of Industry [Los Angeles Co.]. Created in 1957 on undeveloped land in La Puente Valley and incorporated for the sole purpose of industrial development. Additional non-residential corridors in other areas were annexed later (Co. Hist.: Los Angeles 1965:2.108 ff.).

City of Six Ridge [Sierra Co.]. The ridge preserves the name of the mining town, City of Six, near Downieville, which was often mentioned in the 1850s as yielding large gold nuggets. *See* Jackass: Jackassville.

City of the Angels. *See* Los Angeles.

City Reservation, The. *See* Alum Rock Park.

Ciudad de los Angelos. *See* Los Angeles.

Claiborne Creek [Shasta Co.]. Named by Sim Southern (*see* Sims), owner of the Eagle Hotel in old Shasta, for G. B. Claiborne, a banker from Stockton.

Clair Engle Lake [Trinity Co.]. The artificial lake created by Trinity Dam in 1965 was named in honor of the late Clair Engle, 1911–64, U.S. Senator from California.

Clairville [Plumas Co.]. The transfer point from rail to teams was named for Claire Bowen, who was Mrs. W. E. Clarke (Myrick). The name was applied to the post office from July 20, 1896, to Oct. 15, 1950. *For* Clairville [Sonoma Co.], *see* Geyser: Geyserville.

Claminitt, Clammitte. *See* Klamath.

Claremont [Los Angeles Co.]. The Pacific Land and Improvement Company, a subsidiary of the Santa Fe Railroad, offered to name the town, which it platted in 1887, for H. A. Palmer, owner of the land. Mr. Palmer declined the honor and suggested a number of Spanish names descriptive of the grand view of the mountains. The directors, from Boston, called for equivalents in their own language, and influenced by one among them who formerly lived in Claremont, New Hampshire, chose the present name (Willis Kerr).

Clarence King, Mount [Kings Canyon NP] was named Mount King in 1864 by the Brewer party of the Whitney Survey, for Clarence King, a member of the party. King (1842–1901) was connected with the State Geological Survey from 1863 to 1866 and later became the first chief of the USGS, 1879–81. The full name is recorded on the 1939 edition of the Mount Whitney atlas sheet, but King Spur, named after the mountain, retains the shorter form.

Clark, Mount [Yosemite NP] was named for Galen Clark, mountaineer and first guardian of Yosemite State Park. The Whitney Survey had named the peak The Obelisk because of its odd shape. Clark's Station, a stage stop and home of Clark in the 1860s and 1870s, is now Wawona.

Clark Dry Lake [San Diego Co.]. Named in the 1890s for the Clark brothers, who developed the lake for watering their stock (J. Jasper).

Clarke City. *See* Green: Greenfield.

Clark Mountain [San Bernardino Co.]. The name was placed on the map by the Death Valley Expedition of 1891. Like Clark Co. in Nevada, it honors Sen. William A. Clark of Montana, famous for his copper mines and builder of the Las Vegas and Tonopah Railroad (E. C. Jaeger, *Desert Magazine*, July 1954, p. 14).

Clarks: Lake, Peak, Spring [Lake Co.]. Named for Peter Clark, who took up a homestead in Big Valley soon after the arrival of the earlier settlers of 1854 (Mauldin).

Clarksburg [Yolo Co.]. Named for Judge Clark, who settled there in 1849 (Co. Hist.: Yolo 1940:216).

Clarks Point [San Francisco Co.]. The point was named when William S. Clark, an emigrant of 1846, acquired the property at Montgomery and Sacramento streets and built a wharf and a warehouse here.

Clark's Station. *See* Clark, Mount.

Claus (klôz) [Stanislaus Co.]. Established in the 1890s by the San Francisco and San Joaquin Valley Railroad (now the Santa Fe) and named for Claus Spreckels, who was instrumental in having the railroad built. Earlier called Clauston.

Claussenius (klô sē′ nē əs) [El Dorado Co.]. Apparently from the family name of an early settler, probably German or Scandinavian.

Clawhammer Bar [Siskiyou Co.]. "In '54 there came to Hardscrabble on the 4th of July a stranger attired in clawhammer coat and plug hat to attend the celebration. He had plenty of dust and of congenial spirit, but alas, coat and hat were destroyed by the wild happy crowd, but the incident added the name Clawhammer Bar to one of the richest spots on the Salmon" (Luddy).

Clay [Sacramento Co.]. The post office was established on July 26, 1878, and was possibly so named because of the composition of the soil in the district.

Clayton [Contra Costa Co.]. Named for Joel Clayton, who settled here in 1857. Hoffmann in his notes of Sept. 23, 1861, refers to the place as Clayton's, and two days later as Claytonville. The town was for many years the trade and social center for the Mount Diablo coal mines.

Clear Creek. There are more than forty Clear

Creeks, mostly in the mountainous sections of the state. Clear Creek [Shasta Co.] is the site where gold was discovered by Pierson B. Reading in 1848. *For* Clear Creek Hot Springs [Kern Co.], *see* Hobo Hot Springs.

Clearinghouse [Mariposa Co.]. This was the name of a gold mine on the Merced River named by Frank X. Egenhoft. During the panic of 1907, when clearinghouse certificates were in circulation, the mine was a clearinghouse for gold bullion. It subsequently failed and closed, but the name continues in the railroad station and the post office across the river (J. W. Warford).

Clear Lake; Clearlake City [Lake Co.]. The lake was called Laguna and Laguna Grande ('large lake') in Spanish times and so appears on a number of early American maps. The present name is recorded in Gibbs's journal (1851). The names of three post offices include the name of the lake: Clearlake Highlands, Clearlake Oaks, and Clearlake Park. *For* Clear Lake Villas, *see* Nice. **Clear Lake Reservoir** and **Hills** [Modoc Co.]: The large body of clear fresh water in Tule Lake Valley gave the name to the reservoir, which was surveyed in 1870 and repeatedly enlarged. The outline of the reservoir resembles Rhett Lake on the Frémont-Preuss map of 1848, though the latter was probably what is now Tule Lake.

Clearwater [Los Angeles Co.]. Founded in 1887 by the California Cooperative Land Colony and named by Col. F. A. Atwater after a nearby lake, dry since 1914. Clearwater Creek in Mono Co. had been named McLean's River by Lt. Tredwell Moore in 1852 for a member of his party. The name is used for several other creeks and lakes in the state.

Cleary Reserve [Napa Co.]. The 400-acre reserve in Pope Valley was established in 1962 by the Biological Field Studies Association and the Nature Conservatory, as an area for the study of wildlife. It was named in honor of the generous owners of the land, Ernest W. and Mary Edna Cleary.

Cleghorn, Mount; Cleghorn: Canyon, Pass [San Bernardino Co.]. For Mathew Cleghorn and his son John, who leased land nearby in the 1870s for a lumber business (Co. Library).

Clements [San Joaquin Co.]. For Thomas Clements, on whose land a settlement and trading center developed in the 1870s.

Cleone (klē ōn′) [Mendocino Co.]. According to Hodge, the name is derived from the name of the Keliopoma, the northernmost branch of the Pomo Indians. However, it may be just a girl's name.

Cleveland National Forest. Created by executive order of Pres. Theodore Roosevelt on July 1, 1908, and named in memory of Pres. Grover Cleveland (1837–1908), who had died on June 24. Trabuco Canyon and San Jacinto National Forests were combined to form the new area.

Clews Ridge [San Bernardino Co.]. Apparently named for Joe Clews, a prospector in the 1860s (BGN, Jan.–Apr. 1961).

Clifton. *See* Del Rey.

Clikapudi (klik′ ə po͞o dē) **Creek** [Shasta Co.]. From Wintu *łikup'uri* 'a fight', from *łik-* 'to fight' (Shepherd); also spelled Clickapudi in English. The name appears on the Redding atlas sheet (1901), but it seems not to be recorded on older maps. *See* Chickabally Mountain.

Clinton. *See* Brooklyn Basin.

Clio (klī′ ō) [Plumas Co.]. In the 1870s the post office was established and named Wash, for a pioneer. In 1905 the name was changed to Clio. According to local tradition, the new name was suggested to the postmaster, Fred King, by the trade name of a heating stove (*Headlight*, Oct. 1942). Thus the name of the muse of epic poetry and history entered California by way of a lowly stove. However, Clio was then already a popular place name: a Clio mine in Tuolumne Co. is listed by Browne in 1868, and the Postal Guide of 1892 lists nine Clios in different states.

Clipper. The name was applied in the 1830s to sailing vessels designed primarily for speed. Before the era of the steamship, American clipper ships had a short but brilliant career, and from this the word was transferred in the sense of speed to the terms "clipper mill," "plow," "sled," etc. In California the geographical terms probably owe their origin to nearby clipper mills. **Clipper Mills** [Butte Co.] was established as a post office on Aug. 13, 1861. Besides Clipper Gap [Placer Co.], applied to the Central Pacific station in the summer of 1865, there is a Clipper Mountain northwest of Danby [San Bernardino Co.], a Clipper Creek east of Applegate [Placer Co.], and Clipper Creek and Clipper Ravine in Nevada Co.

Cloud Creek and **Canyon** [Kings Canyon NP]. "I named it 'the Cloud Mine' because the clouds hung so low overhead. At the same time I named the creek Cloud Creek and put the name in my notebook" (W. B. Wallace, *SCB* 12:47–48). On many maps the canyon is erroneously labeled Deadman Cañon, the name of the canyon immediately to the west (Farquhar).

Clouds Rest [Yosemite NP]. So named "because upon our first visit [1851] the party exploring the 'Little Yosemite' turned back and hastened to camp upon seeing the clouds rapidly settling down to rest upon that mountain, thereby indicating the snow storm that soon followed" (Bunnell, p. 201).

Clough (kluf): **Creek, Gulch** [Shasta Co.]. Named for Noah Clough, who settled at the creek in the 1850s (Steger).

Clough (kluf) **Cave** [Tulare Co.]. "So named in honor of William O. Clough (1851–1917), who discovered this cave in 1885" (BGN, 6th Report).

Clover. A favorite name in regions where this fodder grows. About ten places in the state are so named, including a Red Clover Valley [Plumas Co.] and a settlement, Clover Flat [San Diego Co.]. **Cloverdale** [Sonoma Co.]: The post office was established on Aug. 15, 1857. Previously the settlement had been known as Markleville, for R. B. Markle, former owner of the land.

Clovis [Fresno Co.]. The Southern Pacific named its station for Clovis Cole, owner of the large wheat ranch through which it built a branch line in 1889.

Clyde [Contra Costa Co.]. The name originated during World War I, when the Clyde Shipyard was located here (J. Silvas).

Coachella (kō chel′ ə, kō ə chel′ ə): **Valley**, town [Riverside Co.]. The name of the valley between the San Bernardino Co. line and Salton Sea was known until about 1900 as Cahuilla Valley (*q.v.*), because it was within the habitat of the Cahuilla Indians. In 1901 the section north of Palm Springs appears on the atlas sheet of the San Jacinto quadrangle as Coachella Valley. The post office was established on Nov. 30, 1901. No evidence of the early use of the name could be found, but a number of etymologies have been advanced. Since shells could be found in the valley—obviously remnants of the time when the region of Riverside and Imperial Cos. was under water—Dr. Stephen Bowers called it Conchilla Valley in a lecture before the Ventura Society of Natural History in 1888, after Spanish *conchilla* 'shell'. According to Elmo Proctor in *Desert Magazine*, Nov. 1945, the valley was generally called Salton Sea Sink; and when the region was surveyed by the USGS before 1900, A. G. Tingman, a storekeeper in Indio, proposed the change of the name to Conchilla Valley. This name was accepted by the prospectors and homesteaders, and apparently also by W. C. Mendenhall of the USGS. At any rate, he used the name Conchilla as late as 1909 (WSP, no. 225). But the cartographers apparently misread the name, and it appeared as Coachella Valley on the San Jacinto atlas sheet—a "bastard name without meaning in any language," as Mr. Tingman is reported to have remarked. But other inhabitants of the valley considered the name "unique, distinctive and euphonious," and in 1909 the the BGN made the name official.

Coahuila. *See* Cahuilla.

Coal. The Coal Creeks and Coalmine Canyons in various counties are reminiscent of the many "coal booms" that were widely publicized and heavily financed but always failed to produce valuable coal (*see also* Carbon). The hope of discovering coal in California was foremost in the minds of industrial and mining circles from the American occupation until the large-scale exploitation of oil deposits. A land grant in Santa Cruz Co., dated Feb. 3, 1838, was called La Carbonera, but this name probably referred to a place where charcoal was made, not to a coal mine. There is a Charcoal Ravine in Sierra Co.

Coalinga (kō ling′ gə) [Fresno Co.]. The place was known as Coaling Station after the Southern Pacific built a branch line to the district in 1888, when deposits of lignite were widely publicized as great coal seams. According to local tradition, the sonorous name was created by an official of the Southern Pacific who added an *a* to "coaling" (Laura Lauritzen). Coalmine Canyon northeast of the town also recalls the "Coalinga coal boom," which petered out like the other California coal booms and in later years was replaced in this district by a more substantial oil boom.

Coallomi. *See* Collayomi Valley.

Coarsegold [Madera Co.]. The name was given in 1849 by Texan miners, probably because they found coarse gold nuggets in the placer.

Coast Range. The name was used spontaneously by early American explorers and cartographers as a collective term for the various sierras along the coast, some of which had previously borne the names of saints. The term Coast Ranges is now loosely though officially applied to the entire group of mountains from Mexico to Canada. In the singular it is commonly restricted to the coastal mountains of central California.

Cobb: Valley, Creek, Mountain, town [Lake Co.]. Named for John Cobb, a native of Kentucky, who built a combined sawmill and gristmill in the valley in 1859. Cobb Mountain is on the von Leicht–Craven map of 1874.

Coburn Station. *See* Truckee.

Coches. This Mexican Spanish word for 'pigs' was repeatedly used for geographical terms and was applied to a number of land grants and claims. **Coches** (kō′ chəs): **Creek, Canyon** [San Diego Co.]: Cañada de los Coches ('valley of the pigs') was the name of a land grant dated May 16, 1843. In mission days there was a pig farm in the vicinity, and wild pigs were found there for many years afterward. Other grants were Ojo de Agua de la Coche ('pig spring') [Santa Clara Co.], Aug. 4, 1835; Los Coches [Santa Clara Co.], Oct. 10, 1840, at San Jose; and Los Coches [Monterey Co.], June 14, 1841. Another Mexican Spanish term for 'pig' is *cochino*, and Las Pulgas, a land grant in San Mateo Co., was also known as Cochenitos 'little pigs'; but this may be a folk-etymological rendering of the name of an Indian rancheria, Chachanegtac. Arroyo de los Coches, the creek by Milpitas (shown as Arroyo del Monte de los Coches on a *diseño* of Los Tularcitos), has no connection with the two grants in Santa Clara Co. There is a Coche Canyon in Ventura Co.; Los Coches Mountain and Cañada de los Coches in Santa Barbara Co.; and Coche Point and Coches Prietos ('dark-colored pigs') Anchorage on Santa Cruz Island.

Cockatoo Grove [San Diego Co.]. The name was applied by a former Austrian sailor named Seiss, who had a vineyard and winepress here (Emily Richie). It was probably applied in jest; a cockatoo is a popular bird among sailors, and the German *Kakadu* has a decidedly funny connotation. *Katuktu* was the Indian name for Moro Hill in San Diego Co. (AAE 8:157, 191), but this is in Luiseño territory and probably has no connection with Cockatoo.

Cockscomb Crest [Yosemite NP] was so named in 1919 by François E. Matthes of the USGS, because of its appearance (Farquhar). **Coxcomb Mountains** [Riverside, San Bernardino Cos.] is a proper descriptive term, since from certain angles the granite crags look like cocks' combs. On some maps they have been called the Granite Mountains.

Cocopah (kō′ kō pä): **Mountains, Indian Reservation** [Imperial Co.]. The name of an Indian tribe and language of the Yuman family, whose name for themselves was recorded in 1930 as Kokwapá. They were mentioned as Cocapa by Francisco de Escobar in 1605; the current Spanish rendering is Cucapá. The present English spelling was used in 1852 (*HNAI* 10:111).

Codornices (kō dôr nē′ səs): **Creek, Village** [Alameda Co.]: The creek was named by José Domingo Peralta in 1818, when he and his brother found a nest of quail's eggs on its bank: "We ate the eggs and called the creek Codornices Creek" (*WF* 6:371).

Codorniz (kō dôr nēs′) [Orange Co.]. Spanish for the California quail, *Lophortyx californicus*.

Coe State Park [Santa Clara Co.]. The park was named for Henry W. Coe, who gave the land to the county, which in turn transferred ownership to the state. In 1959 it was made a state park.

Coffee. There are a number of Coffee Creeks in the state, the best known of which is in Trinity Co. The color of the water probably gave rise to these names, although an incident like the loss of a bag of coffee in the stream (*see* Sardine; Sugar) may have provided the reason. According to Jessie Gummer (Knave, Sept. 9, 1951), the creek in Trinity Co. was so named because a pack mule loaded with coffee was swept away by the high water.

Coffin: Canyon, Peak [Death Valley NP]. The canyon has been thus known since the Greenwater copper boom of 1906 and was probably so named in analogy to the other morbid names in Death Valley.

Cohasset (kō has′ ət) [Butte Co.]. When the Post Office Dept. rejected the name North Point in 1887, Miss Welch, the teacher, and Marie Wilson chose the name of the town in Massachusetts for the new post office (*WF* 6:266 ff.). The name is from Algonquian, probably *koowas-et* 'high-place-at, promontory-at' (Stewart 1970).

Cohuila. *See* Cahuilla.

Coit Tower [San Francisco Co.]. Named for Lillie Hitchcock Coit, who left funds to beautify the city of San Francisco in 1929. She was one of the city's great eccentrics, given to chasing fire engines and visiting the Barbary Coast in male garb (Loewenstein).

Cojo (kō′ hō), meaning 'lame (person)', was a favorite sobriquet in Spanish times and was transferred several times to a place. The two land grants in Monterey Co. that include the word seem to have left no trace in California geography, but there is a Coja ('lame woman') Creek in Santa Cruz Co., which is the Arroyo de San Lucas or Las Puentes of the Portolá expedition (Crespí, p. 216). **Cañada del Cojo** [Santa Barbara Co.]: The Portolá expedition reached this place on Aug. 26, 1769, and camped by an Indian village. "Their chief is lame in one leg, for which reason the soldiers called it Ranchería del Cojo, but I christened it Santa Teresa" (Crespí, p. 174). The anchorage at this point, which is the Puerto de Todos Santos of Cabrillo, became known as Ensenada 'bay', Puerto 'port', or Rada 'roadstead' del Cojo. When A. M. Harrison of the Coast Survey made the topographical survey of Point Conception in Aug.–Sept. 1850, he called the canyon Valley of the Coxo. On Goddard's map a place called Caxa appears, but when the Coast Survey established a triangulation point at this place in 1873, the spelling was changed to Cojo (letter to Davidson, Sept. 25, 1873).

Col [kol], meaning a pass, is an unusual word in California toponymy; for an example, *see* Carol Col.

Colby Mountain [Yosemite NP] was named in 1909 by R. B. Marshall in honor of William E. Colby, a well-known San Francisco attorney and mountaineer. **Colby Meadow** [Fresno Co.] was named in 1915 by members of the USFS engaged in building the John Muir Trail. **Colby Pass** and **Lake** [Kings Canyon NP]: The pass was discovered by a Sierra Club party led by William E. Colby on July 13, 1912, and was named by the party for its leader (Farquhar). The name was applied to the nearby lake by another Sierra Club party in 1928.

Cold, Coldwater. The adjective appears in the names of almost a hundred creeks, lakes, springs, and canyons. Some are doubtless translations of the common Spanish designation *agua fría*. **Cold Mountain** [Yosemite NP] is not colder than other peaks but was so named simply because of its location west of Cold Canyon. **Cold Creek** [Amador Co.], however, was originally Cole Creek, named for a settler (Co. Surveyor). **Cold Boiling Lake** [Lassen NP] has cold water that continually bubbles with carbon dioxide gas (Stang).

Cole Creek [Lake Co.]. The original name Cold Creek came to be pronounced Cole Creek (Mauldin).

Colegrove Point [Sutter Co.]. The projection into Sutter Basin was named for G. S. Colegrove, a settler of 1851 (Co. Hist.: Sutter 1879:98).

Coleman [Death Valley NP]. The place was named for William T. Coleman, of San Francisco Vigilance Committee fame, who had extensive borax interests there until 1888. *See* Furnace Creek Ranch.

Coleman Valley [Sonoma Co.]. The original name was Kolmer Valley, for Michael Kolmer, a German immigrant, who came to California in 1846 and in 1848 became the first permanent settler in this vicinity. Tyson, placing the name on the map in 1849, apparently misunderstood the name and therefore misspelled it.

Cole Slough [Tulare Co.]. Named for William T. Cole, who dug the irrigation ditch in 1868 that formed a new channel for the Kings River during a flood (Wallace Smith, p. 192).

Coleville [Mono Co.]. Named for Cornelius Cole, congressman 1863–67, U.S. senator 1867–73 (Maule).

Colfax [Placer Co.]. The naming of the station by the Central Pacific Railroad was one of the many honors bestowed on Schuyler Colfax (1823–85) when he visited California in summer 1865. At that time Colfax was Speaker of the House. In 1868 he was elected vice president and served during Grant's first term.

Colijolmanoc. *See* Carne Humana.

Coliseum Peak. *See* Tenaya Peak.

Collayomi (kol ē ō′ mē) **Valley** [Lake Co.]. Probably from Lake Miwok *koyáa-yomi* 'song place' (Callaghan). Luis Argüello, on Nov. 6, 1821 (MS, Bancroft), mentions a rancheria named Cuaguillomic, and the records of Mission San Francisco Solano include the name Collayomi with various spellings. The land grants Caslamayomi on Mar. 21, 1844, Colloyomi on June 17, 1844, Colijolmanoc (Carne

Humana) on Mar. 14, 1841, and possibly Locoallomi (also spelled Coallomi) in 1845 may or may not preserve the same tribal name. The name of the valley was preserved through the Colloyomi grant.

College. A favorite name for real estate subdivisions in the vicinity of colleges and often applied to physical features near them. **Collegeville** [San Joaquin Co.] was named in 1867 by the founders of Morris College, a Presbyterian institution, which was destroyed by fire in 1874. **College City** [Colusa Co.] grew up around the former Pierce Christian College, founded in the 1870s and endowed by the legacy of Andrew Pierce. **College Terrace** [Santa Clara Co.] was laid out in 1888 on the land purchased from Peter Spacher and F. W. Weisshaar and named Palo Alto. Leland Stanford, who called his ranch Palo Alto, objected and suggested College Place. The property owners, however, adopted the present name, proposed by Edgar C. Humphrey in 1891 (Guy Miller, *WF* 6:78–79). **College Heights** [San Bernardino Co.] was named in 1909–10 by L. W. Campbell because of its proximity to Pomona College.

Collins Valley [San Diego Co.]. The place in Coyote Valley was named for John Collins, who tried to file a homestead there in the 1890s but was opposed by the cattlemen (Parker).

Collinsville [Solano Co.]. Named for C. J. Collins, who settled here in 1859. The name was changed to Newport in 1867, but the old name was restored in 1872.

Collis. *See* Kerman.

Collote, Arrollo de. *See* Coyote.

Colluma. *See* Coloma.

Colma (kōl′ mə) [San Mateo Co.]. On older maps the place was simply designated as Station (of the San Francisco–San Jose Railroad). Because of its location near the schoolhouse, it became known as School House Station, and this was the name given to the post office in 1869. The name Colma is given in the Wells Fargo Express Directory of 1872 and is shown on the county map of 1877. The post office is listed in 1891. The origin of the name is not known. William T. Coleman, the "Lion of the Vigilantes," owned part of the Buri Buri Rancho 2 miles south, and a Thomas Coleman was a registered voter in the district in the 1870s; it is not impossible that Colma was coined from this name. Since most of the settlers of those years were apparently German or Irish, it may be a transfer name from the "old country." Switzerland has a Colma, and Alsace a Colmar. Another possible origin is suggested by Beeler 1954: in the San Francisco dialect of Costanoan, *kolma* means 'moon.' However, no Indian rancheria called Colma is shown on any *diseño* or early map of the vicinity. Finally, there is the possibility of a literary origin: Colma occurs as a personal name in MacPherson's *Songs of Selma*, one of the Ossianic fragments (Arthur Hutson).

Coloma (kə lō′ mə) [El Dorado Co.]. The settlement developed around Sutter's mill after the discovery of gold by James Marshall on Jan. 24, 1848. It was named after a nearby Nisenan (Southern Maidu) village and appears in the *New Helvetia Diary* on Mar. 17, 1848, as Culloma. It is spelled Colluma on Derby's map of 1849. The post office, established Nov. 8, 1849, was Culloma until Jan. 13, 1851. The present spelling was first used on Tyson's map of 1850. The statement made by Bancroft (4:27), that the original name means 'beautiful vale', is a figment of American romanticists.

Colony: Meadow, Peak, Mill [Tulare Co.]. The name goes back to the establishment in 1886 of the Kaweah Co-operative Colony, headed by Burnette G. Haskell, whose purpose was the cutting and marketing of lumber on a socialist basis. Internal dissension, the countermeasures of the lumber barons, and the creation of Sequoia National Park brought about the collapse of the colony in 1891 (cf. Ruth R. Lewis, "Kaweah: An experiment in cooperative colonization," *Pacific Historical Review* 17:429–41 [1948]).

Colorado Reef [San Mateo Co.]. The steamship *Colorado* was wrecked on this ledge in 1868 (Alan K. Brown).

Colorado (kol ə rad′ ō) **River**. The name Colorado ('red') was first applied by the Oñate expedition in Oct. 1604, to what is now the Little Colorado in Arizona, "because the water is nearly red" (Bolton, *Span. expl.*, p. 269). The transfer to the larger river (called Río Grande de Buena Esperanza by Oñate in Jan. 1605) seems to have been made by Kino in 1699 (*Kino* 1:193). In Oct. 1700 Kino mentions the "very large volumed, populous, and fertile Colorado River, which . . . is that which the ancient cosmographers . . . called Río del Norte" (ibid., p. 252). On his map of 1701 it appears as Río Colorado del Norte. Other early names had been Río de Buena Guía 'river of good guidance' given by Alarcón in Aug. 1540 (*CHSQ* 3:385–86), and Río del Tizón ('firebrand river', because the Indians carried firebrands) given by Melchor Díaz in 1541 (Wagner, p. 519). The hybridization, Colorado River, is found, apparently for the first time, on Cary's map of 1806. Attempts were made by American cartographers of the 1840s and 1850s to call it Red River of the West or Red River of California. In 1853 William P. Blake of the Pacific Railroad Survey applied the name Colorado Desert, after the river, to the region now known as the Imperial and Coachella Valleys. Mesa de Colorado [Riverside Co.] and the station Colorado [Imperial Co.] were likewise named after the river. Elsewhere the Spanish adjective is found in place names in Contra Costa, Mariposa, Tehama, Alpine, San Benito, and Monterey Cos. The name of the state of Colorado is also derived from that of the river. *See* Palo: Palo Colorado.

Colton [San Bernardino Co.]. The place was settled by the Slover Mountain Colony in 1873. When the Southern Pacific Railroad reached the settlement two years later, it named the station for David D. Colton, financial director of the Central Pacific; he was David Broderick's second in the duel with David Terry.

Columbia. One of the most popular of our geographical names and in the early years of the republic a serious contender for the name of the United States. California has its share of Columbias, including the tall peak, Columbia Finger, in Mariposa Co. **Columbia State Park** [Tuolumne Co.]. The discovery of gold at "Hildreth's Diggings" in Mar. 1850 precipitated a rush to the place. "The honor of bestowing upon the new camp its present name, Columbia, is due to Majors Farnsworth and Sullivan and Mr. D. G. Alexander, who formally named the place on the 29th of April" (Co. Hist.: Tuolumne 1882:26). The name is prominently displayed on Gibbes's map of the southern mines (1852). **Columbia Hill** [Nevada Co.] was named in 1853 by N. L. Tisdale and other miners for the Columbia Consolidated Mining Company.

Columbia Beach [San Mateo Co.]. The steamship *Columbia* was wrecked here in 1896 (Alan K. Brown).

Columbine Lake [Tulare Co.]. The name was given by Joseph Palmer, a pioneer of the Kaweah region, because of the great quantities of columbines (genus *Aquilegia*) growing on the shores of the lake.

Colusa (kə lo͞o′ sə): **County**, town, **Basin**. The rancheria named Coru is mentioned by Padre Blas Ordaz on Oct. 28, 1821 (Arch. MSB 4:169 ff.), obviously in Patwin territory. The name was applied to two land grants: Coluses or Colussas (July 26, 1844) and Colus (Oct. 2, 1845). Although simpler than other Indian names, it seems to set the record for spelling variants, ranging from Corusies to Colouse. The town was founded in 1850 by Charles D. Semple, claimant of the Colus rancho, and was called first Salmon Bend, then Colusi. The post office is listed in 1851 as Colusi's, but newspaper accounts of the same year have the modern spelling. The county, one of the original twenty-seven, was organized on Feb. 18, 1850, and was likewise called Colusi until 1854.

Comanche. *See* Camanche.

Comata [San Luis Obispo Co.]. *See* Camatta.

Combie (kom′ bē) **Lake** [Placer, Nevada Cos.]. Named after Combie (or Coombe) crossing and ranch, now submerged by the lake. Combie was a Frenchman who reached Bear River in the mining days. He is said to have introduced alfalfa into California (Doyle).

Cometa (kō mā′ tə) [San Joaquin Co.]. Spanish for 'comet'.

Comimi (kō mē′ mē) **Canyon** [Yosemite NP]. This name for the canyon traversed by the upper part of the trail to Eagle Peak, and to the top of Upper Yosemite Fall, has fallen into disuse. It may be from the name of the Indian village at the foot of the trail: Koom-i-ne or Kom-i-ne, as recorded by Merriam in 1917 (Browning 1988:174).

Commata [San Luis Obispo Co.]. *See* Camatta.

Commerce. *See* City of Commerce.

Complexion: Canyon, Spring [Lake Co.]. Because the mineral water is reputedly good for the complexion, an attempt was made to commercialize the spring when a toll road was built between Bear Valley and Bartlett Springs in the 1870s (Mauldin).

Compoodle Creek [Tulare Co.]. *See* Campoodle.

Comptche (komp′ chē) [Mendocino Co.]. Probably an Indian name, possibly from the Pomo village Komacho (Barrett, *Pomo*, p. 178).

Compton [Los Angeles Co.]. The name was given to the station of the Los Angeles and San Pedro Railroad in 1869, for Griffith D.

Compton, founder of a Methodist temperance colony on the Dominguez rancho and in 1880 one of the founders of the University of Southern California.

Conard (kon′ ərd) **Meadows; Mount Conard** [Lassen NP]. Black Butte and Black Butte Meadows were renamed in 1948 at the instigation of the park headquarters, in memory of Arthur L. Conard, of Red Bluff, who had been instrumental in the establishment of the national park (Schulz).

Conception, Point; Concepcion (kən sep′ shən) [Santa Barbara Co.]. The cape was named Cabo de Galera by Cabrillo on Oct. 18, 1542, because it looked to him like a seagoing galley. The new name was apparently applied by Vizcaíno in 1602, for he reached the point about Dec. 8, the day of the *Purísima Concepción* ('Immaculate Conception'). It appears on Palacios' plan as Punta de la Limpia Concepción (Wagner, p. 381). On Aug. 27, 1769, the Portolá expedition reached an Indian village north of the cape, where a native stole the sword of a soldier: "For this reason [the men] called it Ranchería de la Espada, so that with this reminder the soldier would be more careful. I gave it the sweet name of Concepción de María Santísima, in view of the neighborhood of the point which has had this name for so many years" (Crespí, pp. 175–76). On May 10, 1837, the name was applied to a land grant, Punta de la Concepcion, on which the town of Concepcion now stands. The Anglicized name, Point Conception, appears in Vancouver's *Voyage* and again on Gibbes's new map (1852). The Coast Survey adopted this form in 1855, although the *Coast Pilot* and many maps continued for years to use the Spanish spelling. The name of the post office is still spelled Concepcion but is pronounced in the American way.

Concha; Conchilla. The Spanish word for 'shell' and its diminutive are often found in geographical nomenclature, especially along the coast. *See* Abalone; Aulon; Coachella.

Conchoso, Arroyo [Kings Co.]. Perhaps an invented Spanish term, intended to mean 'full of shells'.

Concord [Contra Costa Co.]. In 1862 Salvio Pacheco, owner of the Monte del Diablo rancho, selected the site for a town. He called the new settlement Todos Santos and offered a free lot to anyone who would build. The project succeeded, but the strong New England influence soon changed the name (Doyle). In 1868 Lewis Castro made a survey using the name Todos Santos, but Juan Galindo's ledger shows that the name had been changed to Concord before June 1869 (Robert Becker). The new name appears on Hoffmann's map of the Bay region, 1873.

Concow (kon′ kow, kong′ kow): **Valley, Creek** [Butte Co.]. The Concow Indians, a branch of the Maidu, lived in the valley until they were transferred to the Round Valley Reservation. The name in their language is *koyoom k'awi* 'valley earth', from *koyoo* 'valley, flat place' and *k'aw* 'earth, ground' (Shipley, p.c.).

Condor Peak [Los Angeles Co.]; **Condor Point** [Santa Barbara Co.]. These two features, and perhaps **Vulture Rock** [San Luis Obispo Co.], record the presence of the California condor (*Gymnogyps californianus*), which once ranged as far north as British Columbia but now barely survives in the southern part of the Coast Ranges. *See* Sisquoc.

Cone. The Cones and Cone Peaks in the state are apparently all named for their conical shape. "There are a great number of volcanic cones, more or less regular in outline, scattered over the plain to the northwest of Mount Shasta. . . . One of these conical mountains, which was much higher than any of the others, and which lay close at the western base of Mount Shasta, was so beautifully regular in its outline that we gave it the name of Cone Mountain" (Whitney, *Geology*, p. 345). There are three Cinder Cones on the slopes of Mount Shasta, and others in Lassen NP and in Lava Bed NM; also Ventana, South Ventana, and Ventana Double Cone [Monterey Co.]; Brown Cone and Volcanic Cone [Fresno Co.]; Dardanelles Cone [Alpine Co.]; Red Cinder Cone [Shasta Co.]. *For* Cone Creek [Tulare Co.], *see* Kern: Kern-Kaweah River; *for* Cone Mountain [Siskiyou Co.], *see* Black: Black Butte.

Conejo. This Spanish word for 'rabbit' refers primarily to the cottontail (*Sylvilagus audobonii*), rather than the jackrabbit (Spanish *liebre*). The term *conejo* was often used in early place names and is preserved in San Diego, Ventura, Fresno, Stanislaus, and Monterey Cos. **Conejo** (kə nā′ hō, kə nā′ ō) **Mountain, Valley, Creek,** and **Grade** [Ventura Co.]: A watering place, Los Conejos, is mentioned by Font (*Compl. diary*) on Feb. 23, 1776, and appears repeatedly in the provincial records as the name of an Indian village and as the name of a rancho. It is again recorded in the *expediente* of the land grant El Conejo (or Nuestra Señora de Altagracia 'Our Lady of High Grace'), dated Oct. 10, 1822. The place is recorded on the Narváez Plano of 1830, and again on the Parke-Custer map (1855); Conejo Pass is shown on the von Leicht–Craven map of 1874. *See* Newbury Park.

Confidence Springs, Hills, Peak, and **Wash** [Death Valley NP] were named after the Confidence Mill operated by Mormons in the 1870s. It was hoped the name might inspire confidence in the valley, which had been so full of disappointments. A settlement called Confidence in Tuolumne Co., mentioned as early as 1880, was likewise named after a mine.

Confusion, Lake [Kings Canyon NP] was so named by the USGS because of uncertainty over which way it drained and whether it belonged in the national park (Browning 1986).

Conglomerate Mesa [Inyo Co.]. The prominent mesa, listed in the decisions of the BGN, Sept.–Dec. 1963, was apparently so named on account of the conglomerate soil.

Congress Springs [Santa Clara Co.]. The springs, discovered in the 1850s by Jud Caldwell, were originally named Pacific Congress Springs because the water is similar to that of Congress Springs, Saratoga, New York.

Conness (kə nes′), **Mount; Conness Creek** and **Glacier** [Yosemite NP]. "Named by the Whitney Survey in 1864 for John Conness (1821–1909), a native of Ireland; member of the state legislature, 1853–54 and 1860–61, U.S. senator, 1863–69, who sponsored, in 1860, the bill that established the Geological Survey of California, and, in 1864, the bill by which Yosemite Valley and the Mariposa Grove were granted to the state" (Farquhar).

Connick Grove [Humboldt Co.]. The grove in the Prairie Creek Redwoods State Park was established in 1965 and named in honor of Arthur F. and Florence R. Connick. Arthur Connick was president of the Save-the-Redwoods League from 1951 to 1960.

Conns Creek. *See* Asbill Creek.

Conn Valley [Napa Co.]. Commemorates John Conn, who came to California in 1843, possibly with the Chiles-Walker party, and who in 1853 was claimant of the Locoallomi grant. *See* Locoallomi.

Consoli Springs [Riverside Co.]. Owned and operated by Italian-born Giovanni Consoli, who arrived in 1894 (Gunther 1984).

Consultation Lake [Inyo Co.]. Named in 1904 when the men who laid out the first trail to the summit of Mount Whitney from the east had a consultation here on which direction the trail should take.

Contra Costa (kon′ trə kōs′ tə, kôs′ tə) **County** was one of the original twenty-seven counties, organized and named on Feb. 18, 1850. It took its name from the term used by the Mexicans to designate the 'coast opposite' San Francisco. The term Contra Costa (or Contracosta) is found in Spanish documents as early as 1797. It referred sometimes to the northern shore of the Bay: in Hutchings' *Illustrated California Magazine,* Marin Co. is called "Contracosta" as late as Aug. 1860. In a letter of Jan. 22, 1835, however, the East Bay district is clearly so designated (DSP Mont. 4:91). The name lost its appropriateness when Alameda Co. was formed in 1853 and the larger part of the *contra costa* was included in the new county.

Converse: Basin, Mountain [Fresno Co.]. Named for Charles Converse, who came to California in 1849 and took up timber lands in the valley in the 1870s. He built the first jail in Fresno Co. and was the first person confined in it (Farquhar).

Conversión, Punta de la. *See* Mugu.

Convict: Creek, Lake [Mono Co.]. On the morning of Sept. 24, 1871, an engagement was fought here between convicts escaped from Carson City, Nevada, and a posse led by Robert Morrison. Morrison was killed, but the convicts were rounded up a few days later. The name of the creek had formerly been Monte Diablo Creek (Chalfant, *Inyo,* pp. 214 ff.). *See* Bloody; Morrison.

Conway Lake [Trinity Co.]. Named for Frederick E. Conway (1857–1951), early trapper, hunter, miner, and finally settler in the region (BGN, May 1954).

Cony Crags [Yosemite NP] were so named because of the abundance of rabbit-like "conies," or pikas (*Ochotona princeps*), that inhabit the talus around the crags (BGN, May–Aug. 1963).

Cooke, Camp [Santa Barbara Co.]. Named by the War Dept. in honor of Philip St. George Cooke (1809–95), who led the Mormon Battalion to San Diego in 1847 and became a general in the Union army in the Civil War. *For* Cooke's Pass, *see* Box Canyon.

Cool [El Dorado Co.]. An old placer mining camp of the 1850s. Mining was carried on until the end of the 19th century. A post office was established on Oct. 20, 1885.

Coolbrith. *See* Ina Coolbrith.

Coon. About fifty physical features in the state, mainly creeks, are thus named, mostly because of the presence of raccoons (*Procyon lotor*), although some may have been named for settlers named Coon. Santa Clara Co. has a Coon Hunters Gulch.

Cooper Canyon [Los Angeles Co.]. Named for the brothers Ike and Tom Cooper, hunters and trappers in this area between 1870 and 1890 (USFS).

Coopertown [Stanislaus Co.]. The town developed on the Rock River Ranch of William F. Cooper, who served as an officer in the Mexican War under General Scott (Paden and Schlichtmann). A post office with this name was established on July 27, 1901.

Cooskie Mountain [Humboldt Co.]. Supposedly from the Mattole (Athabaskan) Indian name (Turner 1993); the origin is not known.

Copco [Siskiyou Co.]. Coined from California Oregon Power Company and applied to the post office in 1915 when the company's hydroelectric project was under construction.

Copeh. *See* Capay.

Copernicus Peak [Santa Clara Co.]. Named in 1895 by the staff of the Lick Observatory for the great Polish astronomer Nicolaus Copernicus (1473–1543).

Copper. Only a few features in the state bear this name, given because of the presence of the ore or for the copper color of a mountain. San Bernardino Co. has a Copper Mountain Valley, but apparently no Copper Mountain. **Copperopolis** [Calaveras Co.]: Greek *polis* 'city' was combined with the word "copper" to form the name of the town that grew up around the mines. "The Reed Lode was discovered by Mr. W. K. Reed, July 4th, 1861, and the first house built at what is now the thriving town of Copperopolis, Sept. 5th" (Whitney, *Geology*, p. 255). Copper City [Shasta Co.] and Copper Vale [Lassen Co.] were prosperous mining towns and had post offices.

Coquette Creek [Plumas Co.]. Like many Lady and Squaw Creeks, this is probably a euphemistic term for a less respectable local name. The falls of Bolam Creek on the north slope of Mount Shasta are called Coquette Falls.

Cora Lakes [Madera Co.]. Named by R. B. Marshall for Cora Cressey Crow, probably when the Mount Lyell quadrangle was mapped in 1898–99.

Coralillos Canyon. *See* Corralitos.

Corall. *See* Corral.

Coral Reef [Riverside Co.]. "The rocky wall a few miles south of Indio is an isolated hill, close to the main rise of the mountain, noticeable for the strongly marked beach line, which is seen in a broad band of dark brown that reached 10 to 12 feet above the soil. The so called 'Coral' is really travertine or calcium carbonate" (Doyle). *See* Travertine Rock.

Corcoran [Kings Co.]. The establishment of the settlement was promoted by H. J. Whitley of Los Angeles in 1905, and the town was named for a civil engineer of the Santa Fe.

Corcoran, Mount [Sequoia NP] was named for William W. Corcoran (1798–1888), banker and philanthropist. Alfred Bierstadt painted a mountain in 1878, which he named Mount Corcoran but which was earlier called Sheep Mountain and is now called Mount Langley. In 1891 the BGN gave the name Mount Corcoran to this peak; but in 1905 the USGS gave the name Mount Langley to the same peak. In 1933 the USGS replaced this with the name Corcoran Mountain. In list 6801, the BGN restored the name Mount Langley, and gave the name Mount Corcoran to an adjacent peak (Browning 1986).

Cordelia [Solano Co.]. The place was settled in 1853 by clipper ship captain Robert H. Waterman and was named Bridgeport after the town in Connecticut. When the post office was established about 1869, the Post Office Dept. insisted on a less common name, and Cordelia was chosen in honor of Waterman's wife. The name Cordelia for a section of the town had been mentioned since 1858. Hoffmann's map of the Bay region, 1873, still records the town as Bridgeport, but the post office and the slough as Cordelia. The place received the name definitely when the Wells Fargo agency was established in 1880.

Cordell Bank [Marin Co.]. The shoal west of Point Reyes was explored by Capt. Edward Cordell in June 1869 and named for his fellow countryman J. A. Sutter (both were natives of

Baden, Germany). After Cordell's sudden death on Jan. 26, 1870, the Coast Survey renamed the bank in his memory.

Cordero (kôr dâr′ ō). This place name, found for several minor geographic features, is derived either from a Spanish family name or from a word meaning 'lamb'. *For* Cordero Valley [San Diego Co.], *see* McGonigle Canyon.

Cordilleras (kôr dē yâr′ əz). A geographical term designating the mountain ranges bordering a continent, applied in particular to South America. Emory in 1846 or 1847 (*Notes*, p. 102) called apparently both the Sierra Nevada and the Coast Range the "Cordilleras of California." In the 1850s the Pacific Railroad Survey tried to make this official: "I am informed that it is now proposed to term the entire chain of mountains, extending through to the northern part of Oregon, and running south into lower California, 'Cordilleras of Western America'" (Trask, *Report* 1854:13). In 1870 Whitney proposed the name for the entire orographic system of western North America, and it was actually applied until geographers realized that there is no need for such a collective name. San Mateo Co. has a Cordilleras Creek.

Cordova Village [Sacramento Co.]. *See* Rancho: Rancho Cordova.

Cordua's Rancho. *See* Marysville.

Coreopsis (kôr ē op′ sis) **Hill** [San Luis Obispo Co.]. Named for the spectacular flowering plant, *Coreopsis gigantea*, which grows in the Nipomo Dunes (Hall-Patton).

Corkscrew Peak [Death Valley NM]. The name was suggested in 1936 by Don Curry, naturalist of the monument, because, in a certain light, peculiar folds in the rocks give it the appearance of a corkscrew. Corkscrew Canyon has been so called since 1910 because of its meandering course (Death Valley Survey).

Cornez [kôr nez′] **Lake** [Shasta Co.]. Named for Julius Cornez, pioneer settler of Burney Valley, who built a trading post in 1863 (Steger). This form of the name replaces an earlier version, Cornaz (BGN, list 8602). *See* Freaner Peak.

Corning [Tehama Co.]. The town was laid out in 1882 by the Pacific Improvement Company, a subsidiary of the Central Pacific Railroad, and named in memory of John Corning, general manager of the railroad, who died in 1878. The post office, established on Apr. 5, 1881, was first named Riceville, for a settler of 1872.

Corona. The Latin or Spanish word for 'crown' has always been a favorite place name in the United States. In California there are, or were, at least six communities so named. **Corona** (kə rō′ nə) [Riverside Co.] was laid out by the South Riverside Land and Water Company in 1886 and named South Riverside. When the city was incorporated in 1896, the present name was suggested by Baron Harden-Hickey, owner of the El Cerrito orchards (Janet Gould). It was accepted by popular referendum on June 26. Because of the circular drive around the city, it was sometimes called Circle City, a name popularized when Barney Oldfield and Eddie Rickenbacker gained their laurels in the spectacular auto races, 1913–16. **Corona del Mar** [Orange Co.] was developed and named 'crown of the sea' by George E. Hart in 1904. In 1915 the major interests were acquired by the F. D. Cornell Company, and for advertising purposes the name was changed to Balboa Palisades. Public sentiment brought about the restoration of the old name (Sherman, pp. 154–55).

Coronado (kôr ə nä′ dō) [San Diego Co.]. The townsite was surveyed and named in Oct. 1887 by the Coronado Beach Company. The name was doubtless suggested by the islands, called Los Coronados 'the crowned ones', off the coast of Lower California. These had been named by the friars of the Vizcaíno expedition in Nov. 1602, because Nov. 8 was the day of the Cuatro Coronados, four brothers allegedly put to death for their Christian faith in the time of Diocletian.

Corral (kə ral′). The Spanish word for 'enclosed space' is used in the western United States to designate a pen for livestock or almost any outdoor enclosure. The word appears in the names of a number of Mexican land grants and often has survived as a place name. In California the word was first used in a geographical sense when the soldiers of the Portolá expedition, on Aug. 8, 1769, called a village near Castaic [Los Angeles Co.] Ranchería del Corral because "the inhabitants lived without other protection than a light shelter of branches in the form of an enclosure" (Costansó). **Corral de Piedra** (pē ā′ drə) **Creek** [San Luis Obispo Co.] takes its name from the land grant Corral de Piedra, dated Feb. 11, 1841. A *Corral de Piedra*, possibly a natural stone corral, is shown on the *diseño*. Parke, in the Pac. R.R. *Report* (1854), calls the stream

Corral de Piedras Creek. **El Corral de Tierra** [San Mateo Co.]: According to Alan K. Brown, "The phrase, meaning the Earth Corral, was commonly applied in Spanish times to geographic features that made natural enclosures for stock." A headland at Pillar Point was so called by 1818; in the 1830s, it was used for the surrounding ranch area. Monterey Co. has a Rancho Corral de Tierra (Clark 1991). **Corral Hollow, Canyon, Creek**, and **Pass** [Alameda, San Joaquin Cos.]: The canyon appears as Portezuela ('pass') de Buenos Ayres on the *diseño* of Las Positas in 1834, and as Arroyo Buenos Ayres on Gibbes's map of the southern mines (1852). In the 1850s the name was spelled Corall by Trask and Goddard. The canyon was perhaps so named because at its mouth there was a large corral for catching wild horses. In fact, this may have been the "caral" mentioned in the *California Star* on Mar. 18, 1848: "We are credibly informed . . . that a number of our countrymen with several Californians are actively engaged in building an extensive caral, or enclosure, in the valley of the river San Joaquin, for the purpose of capturing wild horses. The caral . . . will enclose twenty-five acres of land." According to local tradition the place was named for Edward B. Carroll, a settler of the 1850s. However, James Capen Adams, who was there in 1855, spells the name of the man Carroll and (as did Brewer in 1861) the name of the place Corral Hollow. California Kerlinger, who was born there in 1873, believes that the name referred to an old sheep corral that she remembers seeing there when she was a child. **Cañada de Corral** [Santa Barbara Co.]: The name of the gulch near El Capitan Beach preserves the name of a land grant dated Nov. 5, 1841. **Corral Canyon** [Los Angeles Co.] has given its name to the adjoining beach—which is, however, often reinterpreted as "Coral Beach." *For* Corral Valley [Napa Co.], *see* Wooden Valley.

Corralitos (kôr ə lē´ tōs): **Lagoon, Creek**, town [Santa Cruz Co.]. A lake was called Laguna del Corral by Portolá's men in Oct. 1769, because "a piece of fence . . . was constructed between the lake and a low hill in order to keep the animals penned by night with few watchmen" (Costansó, p. 91). The present name, meaning 'little corrals', is mentioned in 1807 (Arch. Arz. SF 2:60) and was given to a land grant dated Apr. 18, 1823. The name apparently did not appear on American maps until 1873 (Hoffmann), but Cronise (p. 129) mentions a town of Corallitas of "nearly one thousand five hundred inhabitants" in 1868. **Corralillos Canyon** [Santa Barbara Co.]: A Corral de Guadalupe is shown on the south side of the Arroyo de Guadalupe (probably Corralillos Canyon) on a *diseño*, 1837, of Rancho Guadalupe. The word is another diminutive of *corral*.

Corriganville [Ventura Co.]. Ray 'Crash' Corrigan developed this location as a site for making western movies (Ricard).

Corte Madera. The term *corte de madera*, meaning 'a place where wood is cut', was often found in Spanish times near pueblos or other settlements. It was applied to several land grants and has survived in a number of place names. **Corte Madera** (kôr´ tē mə dâr´ ə, kôr´ tə . . .) town and **Creek** [Marin Co.]: The Corte de Madera del Presidio was so named because it supplied the San Francisco presidio with lumber. On Oct 2, 1834, this name was applied to the grant of John Reed (or Read, or Reid). The sawmill which he built was the most important source of lumber for the Bay district. The present abbreviated form of the name was used by Theodor Cordua, who leased part of the rancho in 1846. A Corte de Madera Creek in San Mateo Co. runs through the two grants, Cañada del Corte de Madera (May 18, 1833) and El Corte de Madera (May 1, 1844). There is a Corte Madera Mountain and Valley in San Diego Co. *For* Arroyo Corte Madera del Presidio, *see* Widow Reed Creek.

Cortes (kôr tez´) **Bank** [San Diego Co.]. The bank was charted by the Coast Survey in 1856 and named for the steamer *Cortes*, from the bridge of which Captain Cropper had first seen the shoal in Mar. 1853. The name of the bank was at first spelled Cortez, but this spelling was later corrected by Davidson. *See* Lost: Lost Islands.

Cortina (kôr tē´ nə): **Creek, Valley**, station [Colusa Co.]. Probably named for an Indian chief. According to Barrett (*Pomo*, p. 324), Kotina was the name of a captain or headman of a Southern Wintun village. The name of the station is spelled Cortena.

Coru; Corusies. *See* Colusa.

Corvallis. *See* Norwalk.

Cosio (kō sē´ ō) **Knob** [Monterey Co.]. For Francisco (Chico) Cosío, who acquired land in the area in 1892 (Clark 1991).

Cosmenes River. *See* Cosumnes River.

Cosmit (kōz mēt′) **Indian Reservation** [San Diego Co.]. The name is Diegueño, but its meaning is not known (Kroeber).

Coso (kō′ sō): **Range, Peak, Hot Springs**, settlement [Inyo Co.]. Probably from the local Panamint Indian language, in which *kosoowa* means 'be steamy', referring to hot springs in the area (McLaughlin); compare the related Chemehuevi *kosowavi* 'steam' (Munro). The name began to appear on maps when prospectors invaded the region. On Goddard's map (1860) the vast territory between Owens Lake and the Nevada line is designated as Coso Diggings. On the von Leicht–Craven map the mountains are named; and Wheeler atlas sheet 65-D shows Coso Peak, while Coso Valley is shown between Coso Mountains and Argus Range. The USGS in 1913 retained the name but changed Coso Mountains to Coso Range. In 1864 the legislature approved a proposal for a new county, south of Mono Co., to be called Coso, but it did not come into existence, and two years later Inyo Co. was established. Coso post office is listed in 1925.

Cosomes, Rio. *See* Cosumnes River.

Costa Mesa (kōs′ tə mā′ sə) [Orange Co.]. The site was named Harper from the 1880s onward. Subdivision began in 1906, and from 1909 there was a Harper post office; but there was confusion with a Pacific Electric station called Harperville. In 1920 there was a contest to select a new name; Alice Plummer suggested Costa Mesa and won 25 dollars (Brigandi). The two words (meaning 'coast' and 'tableland') are Spanish, but the combination is not. The post office was established and named in 1921.

Cosumnes (kə sum′ nəs) **River** [El Dorado, Sacramento Cos.]. The village of the Indians called Cossomenes is indicated on the Plano de San Jose (about 1824), and the name appears in the following years with various spellings; it is probably from Plains Miwok or Northern Sierra Miwok *kooso* 'holly berries, toyon' (Callaghan). The present spelling was established by Sutter, who mentions the *tribas Cosumnes y Cosolumnes* in 1841 in a letter to Alvarado (DSP 17:90). It was also probably Sutter who suggested the name for the river. This is shown on Wilkes's map of 1841 as Cosmenes R. and is mentioned by Eld as Rio Cosomes. Although Frémont and Preuss used Sutter's spelling on their official maps of 1845 and 1848, the spelling was not uniform until the BGN decided on May 5, 1909: "Cosumnes . . . not Consumne, Cosumne, Cosumni, Mokesumne, nor Mokosumne." The board was less successful in trying to force another decision of the same day on the settlement on State Highway 16: "Cosumne . . . not Bridge House, Cosumnes, Howells, nor McCosumne." The name of the place is still spelled like that of the river. Two land grants in Sacramento Co., dated Dec. 22, 1844, and Nov. 3, 1844, included Cosumnes in their names.

Cotati (kō tä′ tē): town, **Valley** [Sonoma Co.]. The land grant, also spelled Cotate, is dated July 6, 1844. On the *diseño* a range north of the grant is designated as Lomas de Cotate. Barrett (*Pomo*, p. 311) lists a former Coast Miwok village, Kotati, north of the present town; but the meaning is unknown.

Coto de Caza (kō′ tō dē kä′ zə) [Orange Co.]. The Spanish phrase means 'territory for hunting; hunting preserve'.

Cottage Grove [Siskiyou Co.]. The name was transferred from one of the several Cottage Groves in the eastern United States. This mining camp had a post office from 1857 to 1898.

Cottamin. *See* Katimin.

Cottaneva (kot ən ē′ və): **Creek, Cove, Needle, Ridge, Rock, Valley** [Mendocino Co.]. The name of the creek emptying into Rockport Bay is shown on older maps as Cottonwood Creek. John Rockwell of the Coast Survey gave the name as Cottaneva Creek in 1878, and the County History (1880:470–71) gives Cotineva as an alternate name for Rockport. The name may be from Kato (Athabaskan) *kaatəneebi* 'place where the trail goes over the hill' (from *təni* 'trail'; Golla). The current spelling was made official by the BGN in 1967 (list 6704).

Cotter, Mount [Kings Canyon NP]. Named for Richard D. Cotter, a packer of the Whitney Survey, who accompanied Clarence King on his memorable ascent of Mount Tyndall in 1864. The name was applied by David R. Brower in 1934 and was approved by the BGN in 1938. The mountain was called Precipice Peak by Bolton C. Brown on his sketch map (*SCB* 3:136).

Cottonwood. Among deciduous trees, the cottonwood (or poplar, genus *Populus*) and the willow are the favorites for place names in California. About a hundred creeks and as many other geographic features are named Cottonwood. There is hardly a county that has not at least one creek so named. The one between Shasta and Tehama Cos. is shown on the Frémont-Preuss map of 1848. Two settlements [Shasta, Inyo Cos.] retain the name. The reason for its popularity is the widespread occurrence of both the black cottonwood (*Populus trichocarpa*) and the Fremont cottonwood (*P. fremontii*), the former throughout the state, the latter mainly in its southern part. The decision for Cottonwood Creek (not Tia Juana River) in Cleveland National Forest was one of the very few contributions to California nomenclature made by the BGN between Jan. 1945 and July 1947. *For* Cottonwood [Siskiyou Co.], *see* Henley; *for* Cottonwood Creek [Mendocino Co.], *see* Cottaneva. *See also* Alamo.

Cougar Canyon [San Diego Co.]. Originally Krueger Canyon, for George Krueger, secretary of the Brawley Chamber of Commerce (Stein 1975). The word "cougar" is little used in the state for the native cat that is more often called "mountain lion" (or, in northern California, "panther"). However, Siskiyou Co. has a place called Cougar.

Coulterville [Mariposa Co.]. Named for George W. Coulter from Pennsylvania, who opened a store there in 1849 and later became one of the first commissioners of the Yosemite Valley grant. The post office was established in 1852 and was called Maxwell Creek until 1872. The place is shown on Trask's map, 1853, and on all other early maps because it became one of the important trading centers of the southern Mother Lode. *See* Maxwell Creek.

County Line. *See* Cerrito.

Courtland [Sacramento Co.]. Named for Courtland Sims, son of the owner of the land on which a steamboat landing was built in the 1860s (*Grizzly Bear*, Apr. 1922).

Courtright Reservoir [Fresno Co.]. The dam and lake on Helms Creek were completed in 1958 and named in memory of H. H. "Kelly" Courtright (1887–1956), general manager of the Pacific Gas and Electric Company.

Coutolenc (kō′ tə lingks) [Butte Co.]. Originally Coutolanezes, for Eugene Coutolanezes, but abbreviated to the present form when a post office was established in the 1880s (S. A. Vandegrift).

Covell. *See* Easton.

Covelo (kō′ və lō) [Mendocino Co.]. According to Bailey, the name was given in 1870 by Charles H. Eberle, after a fortress in Switzerland. Since no such place is mentioned in the gazetteers of Switzerland, the name may be a misspelling of Covolo, the name of an old Venetian fort in Tirol, not far from the Swiss border. Covelo also exists as a place name in Spain. The town in Mendocino Co. is recorded on the Land Office map of 1879.

Covertsburg. *See* Red: Red Bluff.

Covina (kō vē′ nə) [Los Angeles Co.]. This pleasant-sounding name was applied to a subdivision of La Puente Rancho in the late 1880s. It is shown on the Land Office map of 1891. Locally the name is said to mean 'place of vines'.

Cow. Although not as popular as the bull in geographical nomenclature, the cow is represented in some twenty-five features in California. Of these, the oldest is doubtless **Cow Creek** [Shasta Co.], of which Frémont speaks in his *Memoirs:* "So named as being the range of a small band of cattle, which ran off here from a party on their way to Oregon. They are entirely wild, and are hunted like other game" (p. 477). Bidwell, too, says that the animals of this herd were the wildest he ever saw. Besides simple Cow Creek, a Little, an Old, a North, and a South Cow Creek flow through the same valley. **Cow Mountain** [Mendocino Co.] was so named because it was the range of a herd of cattle, probably the stock of Andrew Kelsey, which ran wild after he was killed by the Indians in 1849. **Cowhead Potrero** [Ventura Co.] and **Cowhead Lake** and **Spring** [Modoc Co.] were probably so named because the skull of a cow was found nailed to a post. Kern Co. has a Cow Heaven Canyon, and Mendocino Co. a Curly Cow Creek. **Cow Creek** [Death Valley NP]: When Phi and Cub Lee moved a herd of cattle through the valley, they lost many animals here. The cows had had no water for twenty-four hours and "riled" the water of the small creek in a stampede (Death Valley Survey). **Cowskin Island** [Tuolumne Co.]: The island on Woods Creek became so known because the Mexicans covered their huts here with hides from the slaughterhouse on the island (Doyle). *For* Cow Wells, *see* Garlock.

Cowell (kou′ əl) [Contra Costa Co.]. Named for Henry Cowell, of the Cowell Lime and Cement Company, which built the cement works there in 1908. **Cowell Redwood State Park** [Santa Cruz Co.] was named in honor of Henry Cowell in 1954, after his son donated the large area of redwood groves to the old Welch County Park.

Cowell [San Joaquin Co.]. *See* Manteca.

Cowier Creek. *See* Kaweah.

Cowles. *See* Santee.

Coxo. *See* Cojo.

Coyote. From Spanish *coyote*, from Aztec *coyotl;* an extremely popular place name. Many geographic features in California were named directly or indirectly after *Canis latrans*, including a pass, a ridge, an Indian reservation, a land grant, and several settlements. The pronunciation of the word is usually (kä yō′ tē) in California, although (kä′ yōt) is used in some other areas. The oldest name is probably **Coyote River** [Santa Clara Co.], mentioned by Anza as Arroyo del Coyote on Mar. 31, 1776, and appearing as Arollo de Collote on Joseph Moraga's map of San Jose, 1781. **Coyote Canyon** [San Diego Co.], where Heintzelman defeated the Indians in 1851, was called in Spanish Cañada de los Coyotes (Brigandi); it was Coyote on the von Leicht–Craven map, and now is Los Coyotes Indian Reservation. **Coyote Wells** [Imperial Co.] were discovered in 1857 by James E. Mason of the "Jackass Mail Line," who saw a coyote scratching in the sand for water (*Desert Magazine*, May 1939). Vanished Coyoteville [El Dorado Co.] and several geographic features were named not for the animals but for the type of mining known as "coyoteing." This method of burrowing into the ground and throwing dirt and gravel about the mouths of the shafts was apparently first used in what is now Nevada Co.: "The resemblance of these holes with their burrows of dirt . . . to those made by the coyote [was so strong that] the system of mining was at once called 'coyoteing' and the hill was named Coyote Hill. This was the first working of the great coyote range that was so famous in early days" (Co. Hist.: Nevada 1880:197).

Crabtree: Creek, Meadow [Sequoia NP] were named for W. N. Crabtree, an early stockman of the region (Farquhar).

Crafton; Craf [San Bernardino Co.]. The post office is listed in 1887 as Crafton Retreat, the name of the resort developed by Myron H. Crafts in the 1870s. The present town was laid out and named by the Crafton Land and Water Company about 1885. In 1908 the Santa Fe named its station for the same pioneer but abbreviated it to Craf to avoid confusion with the Southern Pacific station, Crafton.

Crag. This topographical term, like many other generic names, is of Celtic origin. It is often found for a broken cliff or projecting rock: Chaos Crags [Lassen NP], Eagle Crags [San Bernardino Co.], Temple Crag [Inyo Co.], Devils Crags [Kings Canyon NP]. Tulare Co. has a Crag Peak.

Craggy Peak [Siskiyou Co.]. *See* Bolivar Lookout.

Craig Peak [Yosemite NP] was named in 1898 by R. B. Marshall for John White Craig, at that time a second lieutenant in the 5th Cavalry.

Craigy Peak [Siskiyou Co.]. The peak was given this name, a Scottish variant of "craggy," because of its rugged appearance.

Cramer. *See* Milo.

Cramville. *See* High: Highland.

Crane. Crane Flat in Yosemite NP, as well as the places called Crane in Alameda, Lake, Madera, Napa, Sonoma, and Tuolumne Cos., were named because of the presence of the bird. Flocks of the sandhill crane (*Grus canadensis*) are often observed on their journey between the arctic and the tropics; however, the word "crane" is also used locally for the great blue heron (*Ardea herodias*). The soldiers of the Portolá expedition, on Oct. 7, 1769, gave the name Laguna de las Grullas to a lake near Del Monte Junction because they saw many cranes (*grullas*) there (Crespí, p. 209).

Cranes Creek. *See* Parker: Parker Creek.

Crannell [Humboldt Co.]. The post office was named in 1924 for Levi Crannell, president of the Little River Redwood Company. From 1901 to 1924 the place was known as Bullwinkel (or Bulwinkle) for Conrad Bullwinkel, owner of the site.

Crater. The volcanic nature of the Sierra Nevada and the Cascades has given rise to a large number of Crater Peaks, Mountains, Buttes, or simply Craters. Some are really not extinct craters but only have that appearance. The highest is Crater Mountain [Fresno Co.]. The name is sometimes used as a generic term: Black Crater [Siskiyou Co.], Crescent Crater [Lassen NP]. In Lassen Co. there is a miniature Crater Lake, and in Madera Co. a Crater Creek.

Crazy Lake [Fresno Co.]. Named in 1948 by a Dept. of Fish and Game survey party: the lake is in a bleak area, and "anyone visiting this lake is crazy" (Browning 1986). *See also* Mad.

Credow Mountain. *See* Quedow.

Creek. In England the word designates a tidal channel, but in America it has taken the meaning of a stream smaller than a river (Stewart, pp. 60–61). The word "estuary" is now used instead of "creek" to designate an arm of a larger sheet of water or the tidal channel of a stream, but it has not become widespread in geographical nomenclature. The tidal channels of central California are still called, as in England, "creek," or else "slough," the latter word having meant originally a muddy place. The wide channel between Oakland and Alameda, commonly known as the "Estuary," is officially San Antonio Creek; the wide channel between Pittsburg and Browns Island [Contra Costa Co.] is called New York Slough; the long channel north of Suisun Bay has been designated both as Montezuma Slough and as Montezuma Creek. A good example of the use of the term "creek" in American usage can be found in the watercourse along the boundary of Marin and Sonoma Cos.: the tidal channel has been known since Spanish times as Estero Americano ('American estuary'). In Wood's *Gazetteer* and on Punnett's map of Sonoma Co., both the estuary and the stream entering it are called Estero Americano Creek. On some maps, the ambiguity is avoided by calling the estuary Estero Americano and the stream Americano Creek.

Cremer (krā′ mər) **Creek** [Santa Cruz Co.]. For Thomas Cremer (or Cramer), who built a pioneer hotel near here in 1876 (Clark 1986).

Crescent. The name is found in more than twenty places and features, including Crescent Bay [Orange Co.], Crescent Hill [Plumas Co.], and Crescent Lake [Yosemite NP]. Crescent Crater and Cliff [Lassen NP] are both shaped like a crescent. Of the five towns on the maps of the 1850s, two have survived: **Crescent Mills** [Plumas Co.], a post office since 1870, and **Crescent City** [Del Norte Co.]. The latter was founded by J. F. Wendell in 1853 and was so named because of the crescent-shaped bay, formerly known as Paragon Bay. The name appears in the *Statutes* of 1853 and is recorded on the Coast Survey charts of the same year. Los Angeles Co. has two fanciful creations: La Crescenta and Crescentia. The latter may have been suggested by the botanical name of the tropical calabash tree, *Crescentia cujete*, now grown in Southern California.

Crespi (kres′ pē) **Pond** [Monterey Co.]. For Father Juan Crespí (1721–82), who accompanied Father Serra to California in 1769 (Clark 1991); *see* Glossary.

Cressey. *See* Livingston.

Crest. The word is not only a common generic term but a favorite for naming places at or near a crest, e.g. Crest [Lassen Co.], Crestmore [Riverside Co.], Creston [Napa Co.], and Crestview [Mono Co.]. **Creston** [San Luis Obispo Co.] was founded and named in 1887 by J. V. Webster of Alameda Co., owner of a part of the Huerhuero grant. **Crestline** [San Bernardino Co.]: The name was suggested by Dr. Thompson when the post office was established in 1920 (F. M. Spencer). **Crest Park** [San Bernardino Co.]: When the post office was established on Sept. 16, 1949, it was named after the Crest Park Campgrounds, on which it is situated (Doris Olson). The Spanish equivalent is found in Cresta [Butte Co.] and Cresta Blanca [Alameda Co.].

Crissy Field [San Francisco Co.]. In 1919 Maj. H. H. Arnold, an Air Service officer in San Francisco (later chief of Air Staff), chose the site of the racetrack of the Panama-Pacific International Exposition of 1915 for an airfield. He named it in memory of Maj. Dana H. Crissy, who had been killed in a transcontinental air race in the same year.

Cristianitos (kris tē ä nē′ tōs) **Canyon** [San Diego Co.]. The Portolá expedition camped here on July 22, 1769, and Crespí named it for St. Apollinaris, whose feast day is July 23. However, the soldiers, impressed by the fact that the padres of the expedition baptized two little Indian girls here, called the place Los Cristianos 'the Christians', also Cañada de los Bautismos 'valley of the baptisms' (Crespí, p. 135). Later the diminutive form came to be used.

Crocker [San Luis Obispo Co.]. *See* Templeton.

Crocker, Mount [Fresno, Mono Cos.]. When the USGS mapped the region in 1907–9, its chief geographer, R. B. Marshall, named the four peaks of the divide, surrounding the Pioneer Basin, in memory of the "Big Four" of the Central Pacific Railroad: Charles Crocker, Mark Hopkins, Collis P. Huntington, and Leland Stanford (Farquhar). The four peaks are almost equal in height.

Crocker Ridge [Tuolumne Co.]. Named after Crockers Sierra Resort at the foot of the ridge, established in 1880 by Henry R. Crocker.

Crockett [Contra Costa Co.]. The town was laid out by Thomas Edwards in 1881 on the property of Judge J. B. Crockett, who had received 1800 acres as a fee for settling a land case.

Cromberg [Plumas Co.]. Until 1880 this was known as Twenty-Mile House because it was 20 miles from Quincy on the road to Reno. When the post office was established, Gerhard A. Langhorst, owner of the post, chose the Americanized form of his mother's family name, Krumberg (German for 'curved mountain'), as appropriate for the place on the winding road (Minnie Church).

Cronise (krō′ nēs): **Valley, Mountains, Lake** [San Bernardino Co.]. Sometimes spelled Cronese. Probably named for Titus F. Cronise, a California pioneer and author of *The natural wealth of California* (1868).

Cronkhite, Fort [Marin Co.]. Established by the War Dept. in June 1937 as a training center for Coast Artillery troops and named in memory of Adelbert Cronkhite (1861–1937), a major general in World War I.

Crooked Spring Gulch [Calaveras Co.]. A nearby gulch is called Straight Spring Gulch.

Cross Mountain. *See* Breckenridge Mountain.

Cross Roads [San Bernardino Co.]. The name was applied because a road to the river crossed the highway at this spot. The road was washed out by floods, but the name remained and was given to the post office in 1939 (E. U. Oettle).

Crow. More than twenty features in the state bear this name. Most of them, including Crows Nest in San Luis Obispo Co., were doubtless named for the bird *Corvus brachyrhynchos*. Others were named for men named Crow, or may be abbreviations of "Jim Crow," the once widely current nickname for an African American.

Crow Creek [Shasta Co.]. Named for Isaac Crow, a miner, who had his home at this tributary of Cottonwood Creek in the late 1860s.

Crowles Mountain [San Diego Co.]. Named for George A. Crowles, who owned a ranch on the north slope in 1878. It has also been called Black Mountain (B. B. Moore).

Crowley Lake [Mono Co.]. The reservoir of the Los Angeles water system was named in 1941 for Father John J. Crowley, in recognition of his services to the people of Inyo and Mono Cos.

Crown: Mountain, Ridge, Creek [Kings Canyon NP]. The mountain was named about 1870 by Frank Dusy because of a crownlike cap of rocks (Farquhar). The descriptive name is found elsewhere: Crown Point [Mono Co.].

Crows Landing [Stanislaus Co.]. The original Crows Landing perpetuates the name of Walter Crow and his sons, who settled along Oristimba Creek after 1849. According to the County History (1881:218), the place was named for John Bradford Crow, a native of Kentucky, who acquired 4,000 acres of land along the San Joaquin River in 1867. The post office was established on June 27, 1870.

Crucero (kro͞o sâr′ ō) [San Bernardino Co.]. In 1906 the Tonopah and Tidewater Railroad connected with the Los Angeles–Salt Lake Railroad at Epsom. In 1910 the name of the station was changed to the more melodious Crucero, Spanish for 'crossing'.

Crumbaugh Lake [Lassen NP] was named in memory of Peter C. Crumbaugh, an early sheepman and pioneer of Red Bluff (Steger). Also spelled Crumbo.

Crystal. California has more than thirty geographic features the names of which include this descriptive term. Most of them are lakes and creeks so named because of their clear water; but some may be named for rock crystals, or from Crystal as a family name. **Crystal Springs Lake** [San Mateo Co.], the reservoir of the San Francisco water system, was created in the 1870s and named after Crystal Springs, a community that was mentioned as early as 1856 and was situated a little to the northwest of present Crystal Springs Dam. **Crystal Lake** [Placer Co.] was named by the owner of Crystal Lake Hotel on the old Dutch Flat Road (clipping from *Sacramento Union*). **Crystal Cave** [Lava Beds NM]: The deepest cave in the monument was so named by J. D. Howard because the right side of the wall of the lower level was covered with ice crystals. **Crystal Peak** [Santa Clara Co.] was named for the radio communication facility here, to commemorate crystal radio development (BGN, list 8303). *For* Crystal Lake [Siskiyou Co.], *see* Medicine: Medicine Lake; *for* Crystal Peak [El Dorado Co.], *see* Tallac, Mount; *for* Crystal Springs [Napa Co.], *see* Sanitarium.

Cuaguillomic. *See* Collayomi.

Cuarta, Cañada de la [Santa Barbara Co.]. Mexican Spanish *cuarta* 'a short leather whip for riding' comes into English as "quirt."

Cuaslui (kwäs lī′) **Creek** [Santa Barbara Co.]. The name Guaslay is shown on the southwest part of the *diseño* of the land grant La Zaca (1838). It is from Purisimeño Chumash *awashla'y, washlayïk* 'net sack with wooden mouth-ring' (Applegate 1975:27).

Cuate. This Mexican Spanish word for 'twin', from Aztec *coatl*, also occurs in Mexico as the name of a plant (*Eisenhardtia polystachya*). The term may have been applied to some other plant in California. It was used in the names of three Mexican land grants: Huerta ('garden') de Cuati, Oct. 12, 1838, and Próspero o Cuati, May 16, 1843, both in Los Angeles Co., and Corral del Cuate (also spelled Cuati, Quate, and Quati), Nov. 14, 1845, in Santa Barbara Co. A place called Cuate is mentioned on July 21, 1822, by a padre of Santa Ynez (Guerra y Noriega 5:266). *See* Quatal.

Cuba. This became a very popular name at the time of the Spanish-American War. In 1898 California had a post office [Lassen Co.] and several communities of that name. Only Cuba in Merced Co. is still found on maps and in gazetteers.

Cuca [San Diego Co.]. A land grant dated May 7, 1845. *Los gentiles de Cucam* ('the heathens of Cucam') are mentioned on Feb. 27, 1842 (DSP Ang. 6:97). Cuca is said to be the Indian (Diegueño?) name of a root or fruit.

Cucamonga (ko͞o kə mong′ gə): **Canyon**, town, **Peak, Park** [San Bernardino Co.]. From Gabrielino *kúkamonga*, containing *-nga* 'place'; no further etymology is known. *Un parage llamado* ('a place called') *Cucamonga* is mentioned by Padre Nuez on Nov. 23, 1819, when he was on an expedition commanded by Gabriel Moraga. He named it Nuestra Señora del Pilar de Cucamunga (Arch. MSB 4:140). On Jan. 25, 1839, the name was given to a land grant: an Arroyo de Cucamonga is shown on the *diseño*. The settlement developed around the old winery; it is mentioned in 1854 in the Pac. R.R. *Reports* (3:1.134) and is shown on the von Leicht–Craven map of 1873. The post office, the oldest in the county, listed in 1867, is still on the grant, but the railroad station established in 1879 is several miles southeast.

Cucapá. *See* Cocopah.

Cudahy (kud′ ə hē) [Los Angeles Co.]. For Michael Cudahy, who bought the acreage in 1893, founded the Cudahy Meat Packing Company, and subdivided the community in 1908 (Atkinson 1988).

Cuddeback (kud′ ə bak) **Creek** [Humboldt Co.]. For Henry and Martha Cuddeback, who homesteaded here in 1853 (Turner 1993).

Cuddeback (kud′ ē bak) **Lake** [San Bernardino Co.]. The dry lake took its name from John Cuddeback, who hauled ore from Randsburg to Kramer during the latter part of the 19th century and had a mercury mine at the edge of the lake. The valley is known as Golden, Cuddeback, or Willard Valley (Carma Zimmerman).

Cuesta (kwes′ tə): **Pass** [San Luis Obispo Co.]. The word has the general meaning of 'slope, grade' and was applied in Spanish times. It is shown as Cañada de la Cuesta on a *diseño* of the Potrero de San Luis Obispo (1842). The name Cuesta was applied to a camp and a station north of the pass when the construction engineers of the Southern Pacific accomplished, between 1887 and 1894, the difficult task of crossing the Santa Lucia Mountains. The Pacific Railroad Survey had applied the name San Luis Pass when it proposed the same route in the 1850s. **Cuesta de los Gatos** [Santa Cruz Co.] refers to the ridge between Santa Cruz and the town of Los Gatos; the name is attested from the Frémont expedition in 1846 and was made official for modern use in 1978 (Clark 1986); *see also* Santa Cruz: Mountains.

Cuffey: Cove, Inlet [Mendocino Co.]. The name appears on the charts of the Coast Survey in 1870 as Cuffee's Cove, but it is not known who Cuffee (or Cuffey) was.

Cuidado or **Cuidow Mountain** [Tulare Co.]. *See* Quedow.

Culbert L. Olson Grove [Humboldt Co.]. The grove was dedicated by Matthew Gleason, chairman of the California State Park Commission, on June 28, 1940, and named in honor of Culbert L. Olson, governor of California, 1938–42.

Culbertson Lake [Nevada Co.]. The reservoir was named for J. H. Culbertson, who built the dam in the early 1850s.

Culebrino, Arroyo [Kings Co.]. 'Snaky creek', from Spanish *culebra* 'snake'.

Cullama. *See* Cuyama.

Culloma. *See* Coloma.

Culver City [Los Angeles Co.]. Named for Harry H. Culver, who came from Nebraska in 1914, and acquired and subdivided part of the Ballona land grant.

Cumbre. The Spanish word for 'top' or 'summit' is used in the name of **La Cumbre** (ko͞om′ brə) **Peak** [Santa Barbara Co.], formerly called Tip Top Mountain (*q.v.*). Kings Co. has a hill called La Cumbre.

Cummings [Mendocino Co.]. Named for Jonathan Cummings, a native of Maine who came to California in 1856 and acquired a homestead on the Eel River in the early 1870s (Co. Hist.: Mendocino 1914:1025).

Cummings: Valley, Creek, Mountain [Kern Co.]. Named for George Cummings—an Australian who came to California in 1849, took up a homestead in the valley in the 1870s, and became a successful stock raiser (Darling 1988).

Cunningham [Sonoma Co.]. The railroad station was named in 1904 for the Cunningham family, which had large holdings in the vicinity.

Cupertino (ko͞o pər tē′ nō) [Santa Clara Co.]. An Arroyo de San José Cupertino, named in honor of an Italian saint (1603–63), was mentioned by Anza and Font when the expedition camped at the creek on Mar. 25, 1776. In the late 1840s, when Elisha Stephens (or Stevens) settled there, it became known as Stevens Creek. Hoffmann's map of the Bay region (1873) has "Stephen's or Cupertino Creek." The USGS in 1899 decided for Stevens Creek, but the old name was preserved in the name of the post office, established in 1882. After this post office was discontinued, the name was transferred to the one at West Side in 1895 as a result of a petition.

Curious Butte. *See* Striped Mountain.

Curl Creek [Shasta Co.]. After the Wintu Indian family of William Curl (Shepherd).

Curry, Lake [Napa Co.]. The reservoir of the Valley Water Supply was named about 1925 for Congressman Charles F. Curry (D. T. Davis).

Curry-Bidwell Bar State Park [Butte Co.]. In 1947 T. E. Curry gave the land for the park, which includes the place where John Bidwell discovered gold on July 4, 1848.

Cushenbury (ko͝osh′ ən bâr ē): **Canyon, Springs, Grade** [San Bernardino Co.]. The names go back to a short-lived Cushenbury or Cushionberry City at the east end of Holcomb Valley, named for a miner who operated there during the gold rush of 1860 (*Mining and Scientific Press*, Apr. 27, 1861).

Cutca (kut′ kə) **Valley** [San Diego Co.]. This place north of Palomar has also been called Whitka; the origin is unknown (Stein 1975).

Cutler: town, **Park** [Tulare Co.]. Named by the Santa Fe Railroad in 1903 for John Cutler, a pioneer of the district and for many years a county judge. In 1898 Cutler had given the right-of-way through his land, with the proviso that the station be named for him. The park was given by John Cutler Jr. in 1919, to be known as Cutler Park in memory of his parents.

Cutten [Humboldt Co.]. The post office was established in 1930 and named for the Cutten family, prominent since the early 1850s in the industrial and political history of Humboldt Co.

Cut-throat Gulch [Trinity Co.]. So named because "Blind Lee," a Chinese prospector who was thought to have struck it rich, was found there around 1860 with his throat cut (J. D. Beebe).

Cuyama (kwē yä′ mə): **River, Peak**, settlement [Santa Barbara, San Luis Obispo Cos.]. From Chumash *kuyam* 'clam' (Applegate 1975:34, after Harrington). An arroyo *llamado de Cuyam* is mentioned in 1824 (DSP 1:48). The name Cuyama was given to two land grants, dated Mar. 16, 1843, and June 9, 1846. Cuyama Plain and Mountains are mentioned by Parke in 1854–55 (Pac. R.R. *Reports* 7:1.6). The name was doubtless that of an Indian rancheria shown as Cullama and Cuyama on *diseños* of the land grants. Alternative pronunciations are (ko͞o yä′ mə, kwē yam′ ə).

Cuyamaca (kwē ə mak′ ə, ko͞o yə-, -mä′ kə): **Peak, Valley, Lake, State Park** [San Diego Co.]. An Indian rancheria, Cullamac, is recorded on Aug. 10, 1776 (PSP 1:229), and is mentioned as Cuyamac in early records of Mission San Diego. The present spelling is used in the title of a land grant dated Aug. 11, 1845; the Sierra de Cuyamaca is shown on the *diseño;* and an Indian village named Cuyamaca is mentioned as late as 1852. From Diegueño *'ekwiiyemak* 'behind the clouds', from *'ekwii* 'cloud, rain' plus *'emak* 'behind' (Langdon).

Cuyapipe, Cuyapaipa (ko͞o′ yə pīp, kwē ə pī′ pə) [San Diego Co.]. From Diegueño *'ewiiyaapaayp* 'leaning rock', containing *'ewiiy* 'rock' and *aapaayp* 'lean' (Langdon). Also spelled Guyapipe.

Cuyler (kī′ lər) **Harbor** [San Miguel Island Co.]. Named in Feb. 1852 by James Alden of the Coast Survey schooner *Ewing* for Lt. R. M. Cuyler: "a very good anchorage, which I have named 'Cuyler's Harbor' in honor of one of the officers attached to the party" (Coast Survey *Report* 1852:105).

Cypress. In spite of the existence of five different native species of the tree (genus *Cupressus*) and the striking shape of the Monterey cypress (*C. macrocarpa*), the name is connected with only a few geographic features. This is because most members of the cypress family are commonly called cedars. **Point Cypress** [Monterey Co.] is the only feature that is named after Monterey cypress. Tomás de la Peña (June 15, 1774) calls it, in his diary, *La Punta de cipreses* (*HSSC:P* 2:1.85), and it is so named on several maps. It was used for the name of a land grant on Feb. 29, 1836. The trees that suggested the name are mentioned by Palou (2:285–86): "They [Portolá, Crespí, and Fages] also found a grove of cypresses on a point which is on the little bay [Carmel Bay] to the south of the Point of Pines." In 1852 the Coast Survey translated the name as Point Cypress. **Cypress Mountain** [San Luis Obispo Co.] was doubtless named after the Sargent cypress, found in the Santa Lucia Mountains. **Cypress** [Orange Co.], **Cypress Grove** [Los Angeles and Marin Cos.], and **Cypress Lawn** [San Mateo Co.] were named either for plantings of the tree or after other places so named.

D

Daggett [San Bernardino Co.]. Named in 1882 by the Southern Pacific Railroad (the name was retained in 1884 by the Santa Fe) for John Daggett, lieutenant governor of California (1883–87), who built one of the first houses here and laid out the townsite. Daggett Pass in Alpine Co. was apparently likewise named for him.

Dairy Valley [Los Angeles Co.]. A "rural city" surrounded by a populated area was incorporated in 1956 to protect its dairy and agricultural industry (Co. Hist.: Los Angeles 1965:4.106).

Dalton [Los Angeles Co.]. The Big and the Little Dalton Wash, and the Big and Little Dalton Canyons, commemorate Henry Dalton, an English trader from Lima, grantee of San Francisquito Rancho and claimant of Azusa and Santa Anita Ranchos.

Dalton Mountain [Fresno Co.]. So named because Grattan Dalton of the Midwestern Dalton gang hid out in the area after his escape from the Visalia jail in 1891 (Co. Hist.: Fresno 1956).

Daly City [San Mateo Co.]. The city came into existence when many people found a refuge there after the San Francisco earthquake and fire of 1906. It was incorporated in 1911 and named for John Daly, owner of a dairy farm (Wyatt).

Damiana (dä mē ä′ nə) **Gulch** [San Mateo Co.]. For Damiana Padilla de Martínez, widow of Máximo Martínez, who had the Mexican grant; she farmed here from 1863 (Alan K. Brown). The spelling Damiani also occurs.

Damnation Peak [Shasta Co.]. So named because of the desperate struggle between entrapped Indians and a detachment of the 4th Infantry under Lt. George Crook in July 1859. **Damnation Creek** [Del Norte Co.] was doubtless named for some similar reason.

Dana (dā′ nə) [Shasta Co.]. The town was laid out in 1888 and named for Loren Dana, who had settled at the Big Springs in the late 1860s (Steger). The post office is listed in 1892.

Dana, Mount; Dana Creek [Yosemite NP]. In 1889 J. N. LeConte copied, from a record found on the summit, "State Geological Survey, June 28, 1863. J. D. Whitney, W. H. Brewer, Charles F. Hoffmann, ascended this mountain June 28th and again the 29th. We give the name of Mount Dana to it in honor of J. D. Dana, the most eminent American geologist." James Dwight Dana (1813–95) was professor of geology at Yale at that time (*SCB* 11:247). "Messrs. Dana and Lyell expressed themselves as highly gratified over the honor done them; as well they might, for a thirteen-thousand-foot peak is no mean memorial for any man, while of all human monuments few endure like place names" (Whitney, *Life and letters*, July 10, 1863). *See* Lyell.

Dana: Point, Cove; City of Dana Point (city) [Orange Co.]. Neither point nor bay was prominent enough to have found a place on early maps. The bay was called Bahía (on Humboldt's map: Ensenada) de San Juan Capistrano when this mission used it as an anchorage in the early part of the 19th century. Early American maps failed to show a name for the bay, but in 1884 the Coast Survey gave the name Dana to a triangulation point on the promontory, doubtless in memory of Richard H. Dana (1815–82), author of *Two years before the mast*, who once "had swung down [the cliff] by a pair of halyards to save a few hides"; however, the place so called is not where Dana tossed hides. On the Capistrano atlas sheet of 1902, the USGS applied the name to the point and the cove. The Coast Survey accepted the name, but other maps use the name San Juan Capistrano Point. The post office, Dana Point, was established in 1929 and named after the point. What is now called the City of Dana Point incorporates the old communities of Dana Point and Capistrano Beach (Brigandi).

Danby. *See* Amboy.

Dan Hunt: Mountain, Meadows [Shasta Co.].

Named for Dan Hunt, who settled in the county in 1852 and became one of the early cattle kings (Steger).

Dani (dä′ nē) **Creek** [Monterey Co.]. For Gabriel Dani of Vermont, who homesteaded in the area in 1877 (Clark 1991).

Danish (də nish′, də nēsh′) **Creek** [Monterey Co.]. Perhaps from Spanish *Dionisio* 'Dennis' (Clark 1991).

Dantes View [Death Valley NP]. The name, doubtless inspired by Dante's description of the Inferno, had already been used locally when it was chosen in the early 1930s by the National Park Service for the best-known view point of Death Valley. In 1909 W. C. Mendenhall (WSP, no. 224) mentioned Dante Springs near Soda Lake, about 80 miles southeast.

Danville [Contra Costa Co.]. In an issue of the *Danville Sentinel* of 1898, Daniel Inman, who had settled here in 1858, told the story of the naming of the post office in 1867: "Then the people wanted a post office. Of course it had to have a name, and quite a number were suggested. At first they thought of calling it 'Inmanville,' but my brother Andrew and I objected to that. Finally, 'Grandma' Young, my brother's mother-in-law, said: 'Call it Danville' and as much or more out of respect to her, as she was born and raised near Danville, Kentucky, it took that name" (R. R. Veale). However, a post office Danville was already established on Aug. 31, 1860, and it is shown on a plat of Rancho San Ramon.

Dardanelles (där də nelz′), **The; Dardanelles: Cone, Creek; Dardanelle** post office [Alpine, Tuolumne Cos.]. The peaks were apparently named by the men of the Whitney Survey in the 1860s because they saw a resemblance between the volcanic rock formations and the mountain castles that guarded the entrance to the Sea of Marmara in Turkey. By a strange coincidence, the contour lines on the topographical atlas sheet resemble the outline of Gallipoli Peninsula and the opposite coast of Asia Minor. The watercourse, however, does not bear the classical name Hellespont but is called McCormick Creek. Nearby is another group of peaks called Whittakers Dardanelles, shown on the Stanislaus National Forest map (*WF* 6:79, 178). The name of the straits in Turkey was repeatedly used for mines and mining companies, doubtless because the Crimean War (1854) made the name Dardanelles known all over the world. In a letter of Mar. 10, 1861, Whitney mentions "Dardanelles Diggings, way up in Placer County," and Brewer mentions the name on Nov. 7, 1863 (Notes).

Dark Hole, The. *See* Hole.

Darlingtonia (där ling tō′ nē ə) [Del Norte Co.]. For the carnivorous pitcher plant, *Darlingtonia californica*, which grows in the area. The plant was named for the American botanist William Darlington.

Darrah Creek [Shasta Co.]. The branch of Battle Creek was named for Simon Darrah, a pioneer of 1857, who guided lumber rafts down the Sacramento (Steger).

Darwin, Mount; Darwin: Glacier, Canyon, Creek [Kings Canyon NP]. The highest peak in the Evolution Group was named in 1895 by T. S. Solomons, in honor of Charles Darwin (1809–82). *See* Evolution.

Darwin: Wash, Canyon, town [Inyo Co.]. The wash was named in 1860 by Dr. Darwin French when he led a party into Death Valley to search for the mythical "Gunsight Lode." Darwins Canon is mentioned by the Nevada Boundary Survey on July 31, 1861. Darwin Cañon and the mining town appear on Wheeler atlas sheet 65-D. According to Weight (p. 9), Dr. E. Darwin French had a ranch at Fort Tejon and was with a party of prospectors in this canyon in the fall of 1850.

Daubenbiss (dô′ bən bis) **Dam** [Santa Cruz Co.]. Built in 1847 by John Daubenbiss, a native of Bavaria (Clark 1986).

Daulton [Madera Co.]. Named for Henry C. Daulton, who came to Fresno Co. in 1857 and was made chairman of the commission to organize Madera Co. The name was applied by the Southern Pacific in the late 1860s because Daulton gave the railroad 8 miles of right-of-way through his property.

Davenport [Santa Cruz Co.]. Named about 1868 after the nearby whaling station, Davenport Landing, established by Captain John P. Davenport in the 1850s.

Davidson, Mount [San Francisco Co.]. The highest elevation in San Francisco Co. had been known as Blue Mountain since Beechey placed it on a map in 1827. In 1912 it was named after the American astronomer and geodesist George Davidson (1825–1911). The change was made at the instigation of the Sierra Club, and the ceremony of christening was performed by Davidson's friend and associate Alexander McAdie. On some maps the elevation is labeled San Miguel Hills. **Davidson Seamount** (75 miles west of Point Piedras Blancas) was discovered in 1932 by the Coast Survey ship *Guide* and was named for Davidson in 1938: "The generic term 'seamount' is here used for the first time, and is applied to submarine elevations of mountain form whose character and depth are such that the existing terms bank, shoal, pinnacle, etc., are not appropriate" (BGN, 1938).

Davidson City [Los Angeles Co.]. *See* Dominguez.

Davies Mill. *See* Graeagle.

Davis [Yolo Co.]. Originally called Davisville for Jerome C. Davis, who settled here in the early 1850s. With the coming of the railroad in the 1860s, the town grew and became known by its present name.

Davis, Lake [Plumas Co.]. The artificial lake was formed by the Grizzly Valley Dam in the 1950s and named for Lester T. Davis, conservationist and leader in behalf of the project, who died in 1952 (BGN, 1965).

Davis Creek [Inyo Co.]. Named for David Davis, who came to the region about 1866. He married Lizzie Wiles, an Indian woman, and raised cattle and potatoes for the mining camps (Robinson).

Davis Lake [Kings Canyon NP] was named in 1925 in memory of George R. Davis (1877–1922), topographic engineer with the USGS. Davis mapped and assisted in mapping about thirty topographical quadrangles (Farquhar). There is another Davis Lake nearby in Mono Co.

Davis Mountain [Mono Co.]. Named in 1894 by Lt. N. F. McClure, for Lt. Milton F. Davis, who made the first ascent on Aug. 28, 1891. Davis was chief of staff of the Army Air Service during World War I (Farquhar).

Davy Brown Creek [Santa Barbara Co.]. The name has been current since the 1860s, when it was the favorite back-country haunt of an early settler, Davy Brown (William S. Brown).

Dawson, Mount [Los Angeles Co.]. Named for R. W. Dawson, a miner in San Gabriel Canyon about 1876 and operator in 1890 of the first resort in what is now Angeles National Forest.

Day [Modoc Co.]. The post office is listed in 1892 and was perhaps named for Nathaniel H. Day, a pioneer whose name is recorded in the Great Register of 1879.

Day, Mount [Santa Clara Co.]. The peak, near the Alameda Co. line, is recorded on maps since the 1860s. It was probably named for Sherman Day, state senator representing Alameda and Santa Clara Cos. in 1855–56, and U.S. Surveyor General for California in 1868–71. Sherman Island [Sacramento Co.] was also named for Senator Day.

Daylight: Spring, Pass [Death Valley NP]. The name of the spring was first Delightful, then Delight, and finally Daylight Spring (AGS Death Valley).

Days Needle [Sequoia NP]. This pinnacle was named for William C. Day (1857–95), of Johns Hopkins University, later professor of chemistry at Swarthmore College. He was engaged in solar observations on Mount Whitney in 1881 with the party led by the astronomer Samuel P. Langley (Farquhar).

Dead. The adjective is applied to a number of features, mainly mountains, because of their forbidding appearance. Best known are the Dead Mountains extending from San Bernardino Co. into Nevada. More than twenty-five geographic features, including Dead Horse Glen [Del Norte Co.], were named because a dead horse was found at each of them. There is also an assortment of Dead Indian, Dead Cow, Dead Pine Creeks, Gulches, Mountains, etc. *For* Dead Mule Spring, *see* Mule. *See also* Deadman; Death Valley; Muerto.

Deadman. A convenient and often-used term for places where a corpse was found. More than ten Deadman Creeks and as many Deadman Canyons appear on maps, as well as various Deadman Gulches, Sloughs, Forks, Flats, Lakes, Points, and Islands. **Dead Mans Island** [Los Angeles Co.] was so named because, before 1836, an English sea captain was buried on its highest point. In later years it served as a convenient burial ground for persons who died aboard ships. The name was mentioned by Richard H. Dana in 1835, and again in Robert C. Duvall's log of the *Savannah:* On Nov. 11, 1846, William H. Berry "departed this Life from the effects of a wound received in Battle. Sent his body on 'Dead Mans Island,' so named by us" (*CHSQ* 3:118–19). The island, now only a shoal in San Pedro harbor, was also known as Isla del Muerto. **Deadman's Hole** [San Diego Co.] became thus known after a stagecoach driver found a corpse there. The area was subsequently referred to as Holcomb Village and currently as

Sunshine Summit. **Deadman Canyon** [Kings Canyon NP]: At the lower end of the west branch of Roaring River is the grave of a sheepherder, concerning which there are many legends. The name is incorrectly given to the east branch on some maps (Farquhar). **Deadman Creek, Pass**, and **Summit** [Mono Co.]: The creek became known by this name because the headless body of a man was found there in the 1860s. It was probably the body of Robert Hume, of Carson, who had been killed by a fellow miner.

Deadwood. The word was a typical gold miner's term meaning 'sure thing'. During the Gold Rush it was applied to numerous camps; the name of one, in Trinity Co., has survived. Some of the twenty orographic and hydrographic features in the state may actually have been applied for dead or burned forests.

De Angulo (dē ang′ gə lō) **Trail** [Monterey Co.]. For Jaime de Angulo, Spanish-born anthropologist (1887–1949), affiliated with the University of California at Berkeley, who published numerous works on the languages and cultures of California Indians. He homesteaded in the Big Sur area around 1914 (Clark 1991).

Dearborn Park [San Mateo Co.]. Named for Henry Dearborn, a lumberman of the early logging days (Wyatt).

Death Valley; Death Valley: Buttes, Canyon, post office, **National Park, Junction**. The name was probably used for the valley by gold-seekers of the 1850s because of its forbidding appearance and because of the skeletons of unfortunate wanderers found there. Most maps of the period leave the region blank, but Lapham and Taylor (1856) designate the valley as Armuyosa Desert (for Spanish *amargosa* 'bitter'). The present name was probably applied if not bestowed by the Nevada Boundary Survey. Its earliest known mention is an entry written by a member of the Boundary Commission dated Feb. 24, 1861: "Death Valley, which forms the grand retort for all its [Amargosa's] damned ingredients, and which is . . . a vast and deep pit of many gloomy wonders" (*Sacramento Union*, July 9, 1861). Soon afterward everyone called the region Death Valley as if it were an old name (Wheat, *SCB* 24:108). The first map recording of the name seems to be that of Farley 1861. William L. Manly, a member of the ill-fated Manly-Bennett party of 1849, wrote, forty-five years later: "Just as we were ready to leave and return to camp we took off our hats, and then overlooking the scene of so much trial, suffering and death spoke the thought uppermost saying, 'Goodbye, Death Valley!'" (*Death Valley in '49* [Santa Barbara: Hebberd, 1929], p. 221). No contemporary evidence, however, has been discovered to indicate that the name was used before 1861. The names Lost Valley and Amargosa Desert were also used for the valley or parts of it until the BGN by decision of Feb. 7, 1906, made the present name official. The national monument was created and named in 1933 by proclamation of President Hoover. *See* Amargosa.

Decker Creek [Humboldt Co.]. Authorized by the BGN in 1973 (list 7303), instead of Sit-se-tal-ko Creek.

Declezville (də klez′ vil) [San Bernardino Co.]. When the Southern Pacific built a spur to the large granite quarries, it named the junction Declez and the terminal Declezville, for William Declez, owner of the granite works. Declez is now South Fontana.

Decoto (də kō′ tō) [Alameda Co.]. Named by the Southern Pacific for Ezra Decoto, who came from Canada in 1854, bought a large farm with his brothers in 1860, and sold the right-of-way to the railroad in 1867.

Dedeckera (də dek′ ə rə) **Canyon** [Inyo Co.]. Named for a rare shrub of the buckwheat family, first found here, according to the BGN (list 8303).

Dedrick (ded′ rik) [Trinity Co.]. Named around 1890 for Dan Dedrick, original locator of the Chloride Mine (Alice Goen Jones, p. 194).

Deer. The common species in California is the mule deer (*Odocoileus hemionus*). There are more than 150 Deer Creeks, Canyons, Flats, Parks, etc., in the state. In place names derived from the animal kingdom, only Bear outranks Deer as a favorite. There are a number of Deerhorn Meadows and Creeks, probably so named because horns were found there. Tulare Co. has an Old Deer Creek Channel, and Trinity Co. has a settlement, Deer Lick Springs. Some of the names are translations from the Spanish: Deer Creek [Tehama Co.] is shown on *diseños* as Arroyo de los Venados and Río de los Venados. **Deer Creek** [Nevada Co.] was named by Isaac Wistar and a companion named Hunt in Aug. 1849, when they abandoned a freshly killed deer because they feared hostile Indians. "Next day we reached camp before dark, and described to eager listeners our creek—then and there christened Deer Creek—with the promising appearance of its vicinity" (Wistar 1:125). Hunt returned to the creek, struck one of the richest and most famous gold deposits, and named the place Deer Creek Dry Diggings. This was the nucleus of Nevada City.

Deer Flat [Shasta Co.]. So named because "Deer Flat," a chief of the Hat Creek Indians, lived there (Steger).

Deerhorn Mountain [Kings Canyon NP] was named by Joseph N. LeConte in 1895 because of the resemblance of its double summit to two deerhorns (Farquhar).

Deguynos (də gē´ nōs) **Canyon** [San Diego Co.]. Corruption of "Diegueños," the local Indian tribe; the name is derived in Spanish from that of San Diego Mission (Stein).

De Haven [Mendocino Co.]. Named for John J. DeHaven, a native of Missouri, who came to California in 1849; he was district attorney in 1867–69, assemblyman in 1869–71, and finally state senator (Co. Hist.: Mendocino 1880: 177).

Dehesa (də hē´ sə) [San Diego Co.]. The Spanish word for 'pasture ground' was applied to a post office, June 18, 1888. Shown on the Land Office map of 1891. **Dehesa, Mount** [Lake Co.]: According to Mauldin, the name was applied by Salvador Vallejo. It is commonly pronounced (desh´ ə).

Dekkas (dek´ əs) **Creek** [Shasta Co.]. From Wintu *dekes* '[the act of] climbing', from *dek-* 'to climb' (Shepherd).

Delamar [Shasta Co.]. The mining town was named in 1900 by M. E. Dittmar, for Capt. J. H. Delamar of New York, owner of the Bully Hill mines (Steger).

Delaney Creek [Yosemite NP] was named, apparently by John Muir, for Pat Delaney, the sheepman who accompanied him on his first trip to the Sierra in 1869 (Farquhar).

Delano (də lā´ nō) [Kern Co.]. The Southern Pacific reached the place on July 14, 1873, and named the station for Columbus Delano, at that time (1870–75) secretary of the interior.

Delavan [Colusa Co.]. The Southern Pacific station was named, probably after one of the four Delavans then existing in the Middle West, when the line from Woodland to Willows was built in the 1870s. The name of the post office is spelled Delevan.

De Laveaga (də lä vē ā´ gə) **Park** [Santa Cruz Co.]. For José Vincent De Laveaga, a San Francisco landowner who developed his estate here in the 1870s–80s (Clark 1986).

Delgada (del gä´ də), **Point; Delgada Canyon** [Humboldt Co.]. The descriptive name, meaning 'narrow point', was applied to Point Arena, apparently by Bodega in 1775. In 1792 Vancouver changed this name to Barra de Arena, and cartographers were obliged to find a new point for Delgada. La Pérouse (1797) placed it just south of Cape Mendocino, and apparently all subsequent maps, with the exception of Humboldt's, have a Point Delgada somewhere near latitude 40° north. The cautious Duflot de Mofras placed two such names on his map, one south of Cape Mendocino, one north of Point Arena. Wilkes in 1841 finally applied it to the point above Shelter Cove, although this particular point is not at all narrow. The name for the submarine canyon, 7 miles off the coast, was applied by the Coast Survey and approved by the BGN in 1938.

Delhi. The name of the historic city in India is a favored place name in the United States. In California it has been used at least three times. **Delhi** (del´ hī) [Merced Co.] is shown on the Mining Bureau map of 1891; the modern town was laid out in 1911. **Delhi** [Solano Co.] was listed as a railroad station by McKenney in 1884 and still appears in the gazetteers. **Delhi** [Orange Co.] is mentioned in the county history of 1921.

Delight Spring; Delightful Spring. *See* Daylight Spring.

Delirium Tremens. *See* Alpha.

Delkern (del´ kûrn) [Kern Co.]. Name of the Bakersfield post office branch serving the Greenfield area. The name refers to the delta of the Kern River (Darling 1988).

Delleker (del´ ə kər) [Plumas Co.]. The name of W. H. Delleker, a lumberman, was applied to the station of the Western Pacific in 1908 (*Headlight*, Dec. 1942).

Del Loma [Trinity Co.]. The name, intended for Spanish *de la loma* 'of the hill', was given to the post office in 1927.

Del Mar [San Diego Co.]. Spanish for 'of the sea'. The seaside town was founded in 1884 by Col. Jacob S. Taylor of Oklahoma. The wife of T. M. Loop, one of the promoters, is said to have suggested the name because the site was not far from the setting of Bayard Taylor's poem "The fight of Paso del Mar" (Santa Fe).

Del Monte (del mon´ tē) [Monterey Co.]. The name, meaning 'of (or by) the grove,' was given to the hotel by Charles Crocker in 1886, echoing the *monte* in the name Monterey. It was probably suggested by the beautiful oak groves nearby, rather than by the land grant Rincón de la Punta del Monte, 25 miles to the southeast.

Del Norte (del nôrt´) **County**. The bill proposing the creation of the new county in 1857 proposed the name Buchanan, for the newly elected president of the United States. The legislature, in spite of its Democratic majority, rejected the name of the president and insisted on a Spanish name. Del Norte, meaning 'of the north', was chosen in preference to the other suggested names: Alta, Altissima, Rincon, and Del Merritt(!). The area comprising Del Norte Coast Redwoods is now a state park.

Delonegha (də lon´ ə gə) **Hot Springs** [Kern Co.]. For the Delonagha mining district, formed in 1866. The name comes from Dahlonega, Georgia, a Gold Rush town of the 1830s (Darling 1988), derived in turn from Cherokee *adel dalonige* 'gold', lit. 'money-yellow' (Scancarelli). The current spelling was authorized by the BGN in 1975 (list 7501).

Del Paso (del pas´ ō) [Sacramento Co.]. The post office is listed in 1904 and was named after the Mexican grant, Rancho del Paso, dated Dec. 18, 1844, which in turn had been named after El Paso de los Americanos—the ford in American River, northeast of Brighton. The town was platted in 1911. The post office name is now Del Paso Heights. *See* American River; Paso.

Delpiedra (del pē ā´ drə) [Fresno Co.]. When the Santa Fe spur from Reedley was built in 1910, the terminal was called Piedra ('stone'). In 1923 the post office was established and given the name Del Piedra, a pseudo-Spanish combination. (More correct Spanish would be De la Piedra 'of the stone'.) The railroad station retained the original name.

Del Rey (del rā´) [Fresno Co.]. The town was settled by the Wilkinson brothers in 1884, and the post office was established the following year with the name of Clifton. When the railroad reached the place in 1898, the present name was adopted because the station was on the Río del Rey 'river of the king' ranch (Santa Fe).

Del Rosa [San Bernardino Co.]. The post office and the town bore the name Delrosa (possibly derived from a family name) until 1905, when Zoeth S. Eldredge, a Spanish language enthusiast, petitioned the Post Office Dept. for the change. The result is an incorrect combination of a feminine noun with a masculine article; more correct would be De la Rosa 'of the rose'.

Del Sur [Los Angeles Co.]. Spanish for 'of the south'.

Delta. In geographical terminology the fourth letter of the Greek alphabet is applied to any formation having its shape (Δ). Because of the fame and fertility of the Nile Delta, the word is frequently applied to any alluvial deposits at the mouth of a river, even if no triangular shape is evident as, for example, the Sacramento and San Joaquin Delta. There is a Delta Lake in Modoc Co.; a Santa Fe siding, La Delta, in San Bernardino Co.; a Delta Canyon and a Pacific Electric station named Delta in Los Angeles Co. **Delta** [Shasta Co.] was so named because the level land on top of the hill and the intersection of the Sacramento with Dog Creek form the Greek letter (Steger). A post office named Delta was established on Aug. 10, 1875, and again on May 1, 1948. The name is shown on the Land Office map in 1879. Between 1884 and 1948 the post office name was Bayles, for A. M. Bayles, the postmaster, whose father had been a settler of 1855 (Steger).

De Luz (də lo͞oz´): **Creek**, station, town [San Diego Co.]. Spanish for 'of light'. According to local tradition, a pioneer Englishman named Luce built a large enclosure for his horses, and his Spanish-speaking neighbors called it Corral de Luz. When the post office was established in the 1880s, this name was abbreviated to the present form.

Democrat Mountain [Shasta Co.]; **Democrat Hot Springs** [Kern Co.]. Former generations not only were fond of bestowing the names of political leaders upon places but sometimes commemorated their parties in geographical names. *See* Knownothing.

Denair (də nâr´) [Stanislaus Co.]. Named by John Denair, a former superintendent of the Santa Fe division point at Needles, when he purchased the townsite in 1906. The original settlement was known as Elmdale, then as Elmwood Colony (Santa Fe).

Denlin [Tulare Co.]. Originally named Linden, for John and Emma Linden, who sold the right-of-way to the Santa Fe in 1917. To end confusion with another Linden, the syllables were reversed in 1927 to form the present name (Santa Fe).

Denniston Creek [San Mateo Co.]. Probably named for James G. Denniston, an assemblyman from San Mateo Co. who came to California with Stevenson's New York Volunteers in 1847 (Wyatt).

Denny [Trinity Co.]. The place was settled in 1882 and called New River City. When the Post Office Dept. rejected the name a few years later, the miners chose the present name, for Denny Bar, owner of the store. In 1920 the town was abandoned and the post office moved to Quimby, which soon afterward was renamed Denny (W. Ladd).

Denverton [Solano Co.]. Named in 1858 for James W. Denver, California's secretary of state, 1853–55, and congressman from the district, 1855–57, in recognition of his stand against a bill to confirm all existing grants less than ten leagues in area. Denver later became governor of Kansas Territory, and his name is commemorated in Denver, Colorado. Denverton was formerly known as Nurse's Landing, for Dr. Stephen K. Nurse. *See* Nurse Slough.

Derby, Lake. *See* Honey Lake.

Derby Dike [San Diego Co.]. Named for Lt. George H. Derby, of the U.S. Corps of Topographical Engineers, who in 1853 supervised the construction of a dam across the course of the San Diego River, diverting the water into False Bay (now Mission Bay). The original structure was washed out in 1855 and replaced in 1876. Derby was the early California humorist who wrote under the name John Phoenix. *See* Honey Lake.

Dersch Meadows [Lassen NP]. The name commemorates Fred and George Dersch, prominent pioneers in Shasta Co. in the 1860s and 1870s.

De Sabla (də sab′ lə): **Reservoir**, town [Butte Co.]. Named for Eugene De Sabla, who initiated the construction of the power plant for the Pacific Gas and Electric Company in 1903.

Descanso (dəs kan′ sō): town, **Valley, Creek** [San Diego Co.]. The Spanish word for 'rest' or 'repose' was applied to the post office on Feb. 16, 1877. The place is mentioned in a diary of 1849 with the drawing of a mission-style building (*SCP:Q* 6:96), but this may refer to a place on the Descanso land grant, just south of the international boundary in Lower California. There is also a Descanso Bay on Catalina Island.

Desert. The desert has a strong fascination for those who know it, and the name is used not only for a number of Buttes, Creeks, and Springs, but also for several communities: Desert Center, Desert Hot Springs [Riverside Co.]; Desert [San Bernardino Co.]. *For* Desert Lake [Kern Co.], *see* Koehn Lake; *for* Desert Springs [San Bernardino Co.], *see* Piñon Hills.

De Silvas Island. *See* Belvedere.

Desolation. The name is repeatedly used in California nomenclature to designate isolated or uninviting spots. The name Desolation Lake was applied to the largest of the lakes of Humphreys Basin [Fresno Co.] by J. N. LeConte in 1898 (Farquhar).

Destiladera (di stil ə dâr′ ə), **Cañada de la** [Santa Barbara Co.]. Spanish for 'valley of the distillery'.

Destruction River. *See* Sacramento River.

Detachment Meadow. *See* Soldier.

Devair. *See* Ryan.

Devil. There are in California between 150 and 200 topographic features that are named for the Prince of Darkness. Probably no other state can equal this number. We have not only many forbidding places in mountains and forests where he might abide, but also an assortment of weird formations of basalt, sandstone, and lava, as well as numerous evil-smelling pools and wells, which people like to connect with the devil. Besides the common generic terms, the most popular terms found are Gate, Punchbowl, Den, Kitchen, Gap, and Backbone. There are a number of unusual combinations: Devils Speedway [Death Valley NP], Devils Rock Garden [Shasta Co.], Devils Playground [San Bernardino Co.], Devils Half-Acre [Shasta Co.], Devilwater Creek [Kern Co.], Devils Parade Ground [Tehama Co.], Devils Heart Peak [Ventura Co.], Devils Nose [Calaveras Co.], Devils Potrero [Ventura Co.], Devils Pulpit [Contra Costa Co.], and Devils Head Peak [Napa Co.]. (Note that the absence of the apostrophe in all these names is in compliance with a general rule laid down by the U.S. Board on Geographic Names.) The name is also applied to a town: Devils Den [Kern Co.]. The bend in the Sacramento River, north of Colusa, was once known as Devil's Hackle. The "devilish" names in Geyser Canyon [Sonoma Co.] were already current in 1867—Devils Gristmill, Inkstand, Laboratory, Pulpit, Quartzmill, etc. Well-known features are **Devils Postpile National Monument** [Madera Co.], a strange pile of basalt columns established as a national monument, July 6, 1911; **Devils Crags** [Kings Canyon NP], named by J. N. LeConte in 1906; **Devils Golf Course** [Death Valley NP], a wide expanse of jagged salt hummocks on which only the devil could play golf; and **Devils Garden** [Riverside Co.], an immense thicket of cactus. **Devils Homestead** [Lava Beds NM] is so named because of the weird appearance of the formation along the west boundary of the Monument caused by a recent lava flow (i.e. only a few centuries old). **Devil's Mush Pot Cave** [Lava Beds NM]: J. D. Howard originally applied the name Devil's Mush Pot to the small crater near Indian Well, and the name Pots to the cave south of Hercules Leg because there were "two stonewhirl pools" on the floor of the lower chamber. *For* Devils Castle [Shasta Co.], *see* Castle: Castle Crags State Park; *for* Devils Cellar, *see* Yosemite; *for* Devils Mount, *see* Saint Helena, Mount. *See also* Diablo.

Devines Gulch [El Dorado Co.]. The gulch near Georgetown was named for Caleb Devine, who discovered gold here in 1850. According to an unidentified newspaper clipping, the discoverer was later hanged by a mob.

Devore (də vôr′) [San Bernardino Co.]. The station was named for John Devore, a landowner of the district, when the California Southern Railroad (now the Santa Fe) completed the line through Cajon Pass in 1885. It was formerly called Kenwood for a Chicago clothier who owned a large tract of land here (Santa Fe).

Dewey. *See* Wasco.

Dewing Park. *See* Saranap.

Dexter Peak [San Diego Co.]. Named in memory of John Porter Dexter, an early resident of the area and public-spirited citizen, who died in 1960 (BGN, Sept.–Dec. 1960).

Diablo (dī ab′ lō, dē ab′ lō, dē äb′ lō), **Mount; Diablo: Valley, Creek, Range**, post office [Contra Costa Co.]. The earliest name of the mountain was apparently San Juan Bautista (Dalrymple 1790). In Oct. 1811, Ramón Abella's diary mentions the peak as Cerro Alto de los Bolbones ('high hill of the Bolbones', i.e., the Indians who live at its foot). The ridge was later designated as Sierra de los Bolbones (or Golgones, and other variants). The name Monte del Diablo 'devil's woods' appears on the Plano topográfico de la Misión de San José about 1824, where there was an Indian rancheria perhaps near a thicket at the approximate site of the present town of Concord. On Aug. 24, 1828, the name was applied to the Monte del Diablo land grant for which Salvio Pacheco had petitioned in 1827. Vallejo's Report of 1850 tells the story of a fight between a detachment of soldiers from the San Francisco presidio and the Indians at the foot of the mountain in 1806. The appearance of "an unknown personage, decorated with the most extraordinary plumage," made the soldiers take to their heels, believing the devil had allied himself with the Indians; then and there they applied the present name to the mountain. This is quite certainly fanciful. Marshin 1850 stated that Vallejo was incorrect in placing the engagement near the mountain and that it had occurred in the vicinity of a thicket of willows near the house of Salvio Pacheco (i.e. near present-day Concord) at a later date. The name was transferred to the peak by non-Spanish explorers who associated "monte" with a mountain and applied the Italian form Monte Diavolo or Diabolo. Belcher (1:119, writing in 1837) calls the range Sierras Bolbones and speaks of "the high range of the Montes Diavolo" on his left (!) as he entered the Sacramento River. The Wilkes expedition of 1841 definitely fixed the name on the lofty mountain peak. Duflot de Mofras cautiously gives both names in 1844, Monte del Diablo and Sierra de los Bolbones (plan 16). Finally, Frémont and Preuss, on their map of 1848, give the proper latitude and call it Mount Diabolo. Trask's and other maps of the early 1850s established the modern spelling, although Monte Diablo is found as late as 1873 (Hoffmann). The orographic identity of the range headed by Mount Diablo was recognized by cartographers in Mexican and early American times. The BGN (May 15, 1908) designated the entire range from Carquinez Strait to Antelope Valley in Kern Co. as Diablo Range to distinguish it from the other chains of the Coast Ranges. (In recent years the expression "Mount San Diablo" is sometimes heard.) Diablo post office is listed in 1917. This Spanish word for Satan was used repeatedly for other place names in Spanish times, although not as often as the word "devil" in American times. It occurs in two other land grants: Rincon del Diablo [San Diego], May 18, 1843; and Cañada del Diablo (or San Miguel) [Ventura Co.], May 30, 1845. Point Diablo [Marin Co.] and Diablo Point [Santa Cruz Island] can also be traced to Spanish times. Other Diablos in various counties were applied in American times. *See* Devil.

Diamond. The occasional discovery of small genuine diamonds, and of quartz crystals that look like diamonds, has given rise to some twenty place names in California. The oldest is probably **Diamond Spring** [El Dorado Co.], where the finding of pretty specimens of quartz crystals in 1849 brought about the first large-scale rumor of the discovery of diamond fields. Sometimes the name is used because of the diamond shape of the feature. **Diamond Peak** [Inyo Co.] and **Diamond Mesa** [Sequoia NP] received their names because the contour lines on the topographic map form a more or less perfect diamond. There is a city called **Diamond Bar** [Los Angeles Co.] and a post office called **Diamond Springs** [El Dorado Co.].

Diaz (dē′ əz): **Creek**, station [Inyo Co.]. Named for the brothers Rafael and Eleuterio Díaz, who owned a cattle ranch on the creek in the 1860s (Farquhar).

Dickinson. *See* Le Grand.

Dickson. *See* Dixon.

Dictionary Hill [San Diego Co.]. "On Feb. 21, 1911, the pompously named East San Diego Villa Heights was filed as a subdivision. . . . Persons in the Midwest who purchased certain encyclopedia sets received a lot in the present 'Dictionary Hill' as a premium. Somehow the word 'dictionary' soon became a nickname substitute for 'encyclopedia.' Researcher John Davidson reported that the county surveyor of 1911 opposed this legal filing, for he announced to the public that the area lacked water—and therefore was uninhabitable . . . most of these lots were eventually sold for delinquent taxes" (Stein).

Didallas (dī dal′ əs) **Creek** [Shasta Co.]. From Wintu *didalas* 'daybreak', from *did-* 'light' (Shepherd).

Digger. "Root-Diggers. . . . This name seems to embrace Indian tribes inhabiting a large extent of country west of the Rocky Mountains. . . . With these tribes, roots are, for the great portion of the year, their main subsistence" (Schoolcraft 4:221). Indians of the area also valued as food the green cones and the seeds of the *Pinus sabiniana*, whence the common designation "Digger pine" (however, the term "Digger" as an ethnic label is now considered offensive). In California the name seems to have been used in a geographical sense mainly in Wintu territory. With the exception of a Digger Creek, which empties into the ocean near Beaver Point [Mendocino Co.], the existing Digger names are in Shasta, Tehama, and Trinity Cos. The natives on Kings River were called Root Digger Indians in the Indian Report (pp. 236–76).

Digges [digz] **Canyon** [San Mateo Co.]. R. Montgomery Digges ranched here from 1866 (Alan K. Brown).

Diggings. Since the beginning of the 19th century, this word has been used in the American language to designate pits made by searchers for minerals. With the discovery of gold in California, the term gained new significance and was raised to the dignity of a generic term. A number of places on the map of the state are still called Diggings; in this book they are listed under their specific names.

Di Giorgio (di jôr′ jō) [Kern Co.]. The station was named by the Southern Pacific in 1923 for Joseph Di Giorgio, president of the Earl Fruit Company. The post office was established on May 26, 1944.

Dike. The word, also spelled dyke, designates an igneous rock wall that has resisted erosion and is occasionally used as a generic term (*see* Jackass: Jackass Dyke). **Grand Dike** [Kings Canyon NP]: "Just at the junction of the forks [main forks of Kings River], the end of the divide is crossed by a broad red stripe, bearing about northwest. . . . This, which seemed to be a great dyke of volcanic rock, but which was afterwards found to be a vein of granite, led to giving this divide the name of 'Dyke Ridge'" (Whitney, *Geology*, p. 370). The divide itself is now called Monarch Divide, and Dyke Mountain (Hoffmann's map), 10 miles northwest, is Crown Mountain. Dike Creek on the Mount Lyell atlas sheet was doubtless so named because its source is on the dyke formation of the Minarets.

Diller, Mount [Shasta Co.]. Named for Joseph S. Diller, a geologist for the USGS in the 1880s and the author of *Geology of the Lassen Peak district*, 1889.

Dillon Beach [Marin Co.]. The post office was established in 1923 and named for the Dillon family. George Dillon, a native of Ireland, had settled here before 1867.

Dingleberry Lake [Inyo Co.]. Named by sheepman John Schober for the "dingleberries" (bits of excrement) on the hind ends of his sheep (Browning 1986).

Dinkey: Creek, Lake, Meadow, Mountain; Dinkey Creek (town) [Fresno Co.]. The creek was named by four hunters, whose dog Dinkey was injured there in a fight with a grizzly bear in Aug. 1863. John Muir mentions in 1878 a grove of sequoias called Dinkey Grove, on Dinkey Creek (Farquhar). The other features were named after the creek when the USGS mapped the Kaiser quadrangle in 1901–2.

Dino (dē´ nō) **Hill** [San Bernardino Co.]. An abbreviation of (San) Bernardino; *see* Berdoo Canyon.

Dinuba (dī no͞o´ bə, di no͞o´ bə) [Tulare Co.]. A number of theories concerning the origin of the name have been advanced, but none has been substantiated. It is probably a fanciful name applied by the construction engineer when the branch line was built in 1887–88. Stations between Sanger and Exeter bear these typical railroad names: Fortuna, Smyrna, Dinuba, Taurusa. The original name was Sibleyville, for James Sibley, who had deeded 240 acres to the Pacific Improvement Company (Southern Pacific Company).

Dioleo, Rancho de. *See* Yolo.

Dirty Sock Hot Springs [Inyo Co.]. Hydrogen sulfide in the water might have given rise to the name. However, since this odor is hardly noticeable, the story that the miners washed and hung their socks in the natural washroom probably accounts for the name having been given to the place (Wheelock).

Disappearing Creek [Kings Canyon NP]. Applied by T. S. Solomons in 1895 to the creek in the Enchanted Gorge between Scylla and Charybdis (Farquhar). A number of other watercourses are so named because the surface stream disappears.

Disappointment, Mount [Los Angeles Co.]. In the early 1890s a party of the USGS climbed the mountain to use it as a triangulation point; to their "disappointment" they found that nearby San Gabriel Peak was 100 feet higher (Harry Grace).

Disaster: Peak, Creek [Alpine Co.]. Named by Lt. Montgomery M. Macomb of the Wheeler Survey because on Sept. 6, 1877, his topographer, W. A. Cowles, accidentally dislodged a huge boulder and was severely injured. The mountain was formerly known as Kings Peak (*WF* 6:270).

Disneyland [Orange Co.]. The filmmaker Walt Disney (1909–66) created this extensive amusement park in 1955.

Ditch. Although the building of ditches in mining and in irrigation was of great importance in the development of California, the word has survived in only a few place names as a generic term; the most notable are Big Ditch near Yreka, and Milpitas Ditch near Mission San Antonio in Monterey Co. *See* Zanja.

Divide, The [Placer Co.]. The mining country between the North and Middle Forks of the American River, formerly known as Yankee Jim's Divide, is now called the Forest Hill Divide, abbreviated locally to "the Divide" (State Library). The name is found elsewhere in the state—Divide Peak [Santa Barbara Co.], Divide Ridge [Alameda, Contra Costa Cos.]—and is given to two railroad stations in Sonoma and Santa Barbara Cos. *See* Great: Great Western Divide; LeConte: LeConte Divide.

Dixie. There are a number of Dixies in the state, including two cluster names in Lassen and Plumas Cos.; most of these were probably applied by Southerners at the time of the Civil War, when Union was a place name favored by Northerners. The name Dixieland [Imperial Co.] was chosen by promoters in 1909 in the hope that irrigation would make the district suitable for cotton growing.

Dixon [Solano Co.]. Named in 1870 for Thomas Dickson, who gave ten acres for the townsite. The present spelling was adopted by the Post Office Dept. through an error.

Doane (dōn): **Valley, Creek** [San Diego Co.]. Named for the Doane brothers, who took up homesteads in the valley in the late 1860s (Co. Library).

Dobbins: Creek, town [Yuba Co.]. For William M. and Mark D. Dobbins, brothers who settled at the creek in 1849.

Docas (dō´ kəs) [Monterey Co.]. The name of the Southern Pacific station was coined from San Ardo and San Lucas, between which it is situated (Knave, Mar. 17, 1935).

Dockweiler Beach State Park [Los Angeles Co.]. The Venice-Hyperion beach area was made a park in 1947 and named for Isidore B. Dockweiler, a state park commissioner, 1939–47.

Dockwell [Siskiyou Co.] also appears as Doc Well; it was named for "Doc Skeen," a local veterinary.

Doctor Rock [Del Norte Co.]. So named because it was visited by Indian shamans ("doctors") when they were seeking supernatural power.

Dodge Gulch [Shasta Co.] was named for Wilbur S. Dodge, a prospector of 1850, who later owned the American Ranch in Trinity Co. (Steger).

Doe. The word for female deer appears in the names of a number of physical features, including Little Doe Ridge in Mendocino Co.

Dog. More than twenty-five features, mainly creeks, are named for the faithful canine. The usual reason for naming was doubtless the finding of a stray dog, or some incident concerning a dog. **Dog Lake** [Yosemite NP] was named by R. B. Marshall in 1898 because he found a sheep dog with a litter of puppies there (Farquhar). Kern Co. has a Five Dog Creek, and elsewhere features are named for red, black, and yellow dogs. Doghead Peak [Yosemite NP] and Dogtooth Peak [Fresno Co.] are obviously descriptive. *See* Red Dog.

Dogtown. A popular miners' term for camps with huts or hovels "good enough for dogs to live in." The sites of some of them are still shown on the topographical maps (Jackson, Clements, Mother Lode atlas sheets). A well-known **Dog Town** [Mono Co.] was so named because of the makeshift huts of numerous Chinese placer miners in the late 1850s. The creek there should be called Dog Town Creek instead of Dog Creek. **Dog Town** [Butte Co.]: In 1850 the place was a gold camp named Mountain View. According to the County History (1882:252) the name Dogtown was applied because Mrs. Bassett, the wife of the first permanent settler, bred and sold dogs to the miners. In 1861 the name was changed to Magalia, which was the name of the post office.

Dogwood. A few features are named for this flowering tree, the best-known native species of which are *Cornus nuttallii*, distinguished by its handsome white flowers, and *C. californica*.

Doheny (dō hē′ nē) **Park**. A former name of Capistrano Beach (*q.v.*).

Dollar Lake [San Bernardino Co.]. The lake is like a silver dollar in shape and color. The name Dollar Lake was formerly given to Big Bird Lake [Kings Canyon NP], although it is not perfectly round.

Dolly Varden Creek [Humboldt Co.]. The stream was named for the well-known Dolly Varden trout (*Salvelinus malma spectabilis*), common in the waters of northern California. Dolly Varden was a frivolous character in Dickens's novel *Barnaby Rudge* who wore a bright flower-sprinkled dress. About 1870 her name was applied to many gay and colorful things, including the beautifully marked trout.

Dolomite (dol′ ə mīt) [Inyo Co.]. The name was given to the station of the Carson and Colorado Railroad after 1900 because of the presence there of dolomite, a carbonate of calcium and magnesium. White Mountain Peak [Mono Co.] received its name because its summit, composed of this rock, looks as if it were covered with snow.

Dolores, Mission [San Francisco Co.]. The common designation of the San Francisco mission. The name Los Dolores or Nuestra Señora de los Dolores ('Our Lady of Sorrows') was given to a spring and a small stream by Anza on Mar. 29, 1776, and Misión San Francisco de Asís, founded near it later that same year, afterward came to be popularly known as la Misión de los Dolores.

Dome. In California the word is used, almost exclusively, as a generic term for dome-shaped mountains in the Sierra Nevada between latitude 36° and 38° north, where it occurs about forty times. In other parts of the state the word is sometimes used as a specific term: Dome Mountain [Siskiyou, San Bernardino Cos.]. *For* The Dome, *see* Balloon Dome.

Dominguez (dō ming′ gəz): town, **Hills** [Los Angeles Co.]. The place preserves the family name of the grantees of the vast San Pedro or Domínguez Rancho, granted first to Juan José Dominguez before Nov. 20, 1784. From 1923 to 1937 the station was known as Davidson City, for the Davidson Investment Company; prior to that time as Elftman, for the owner of the land. Laguna Dominguez is the name adopted by the Los Angeles Board of Supervisors on Mar. 30, 1938, for the swampy lake formerly known as Nigger Slough.

Dom Pedro. *See* Don Pedro.

Donahue [Sonoma Co.]. The place, for many years the terminus of the Northwestern Pacific Railroad branch from Petaluma, was named in 1870 for Peter Donahue, a native of Scotland of Irish parentage. He was a well-known San Francisco financier, who built the railroad in 1869–70 that was first called the San Francisco and North Pacific.

Don Castro Lake [Alameda Co.]. A man-made lake created in 1964 and named for Don Guillermo Castro, who owned a portion of Rancho San Lorenzo (Mosier and Mosier). The form of the name would be odd in Spanish, where the title *don* is generally used with the first name (thus don Guillermo) rather than the last.

Doney (dō′ nē) **Creek** [Shasta Co.]. The tributary to the Sacramento was named for William K. Doney, who settled there in 1860 (Steger).

Don Gaspar, Bahía de. *See* Drakes Bay.

Donnel. *See* Bellota.

Donnells Reservoir [Stanislaus Co.]. The name of one of the dams and reservoirs of the Tri-Dam project commemorates one of the partners of Donnell and Parsons, of Columbia, who constructed the first water system between Donnells Flat and Columbia in 1855.

Donner: Lake, station, **Peak, Pass, Monument** [Nevada, Placer Cos.]. The names commemorate one of the worst disasters in the history of the American West. Eighty-one immigrants, led by George and Jacob Donner, were obliged to winter in the vicinity of the lake in 1846–47. Thirty-six died from cold and starvation. The lake had been discovered by the Stephens-Murphy-Townsend party in 1844 and called Truckee Lake. The present names for the lake and peak appear on the von Leicht–Hoffmann map of Lake Tahoe (1874) and were doubtless applied by the Whitney Survey.

Donohue Pass and **Peak** [Yosemite NP] were named by Lt. N. F. McClure in 1895, for a soldier in his detachment (Farquhar).

Don Pedro Reservoir [Tuolumne Co.]. A place on Tuolumne River called Don Pedro's Bar is mentioned by Audubon in 1849 and appears on Gibbes's map of the southern mines (1852). It is sometimes incorrectly spelled Dom Pedro. It was named for Pierre ("Don Pedro") Sainsevain, a French pioneer of 1839, who mined here in Aug. 1848 and was a member of the first California Constitutional Convention. The town was covered by the reservoir, but the San Francisco Light and Power Company preserved the old name.

Doré: Cliffs, Pass [Mono Co.]. Named about 1882 by Israel C. Russell of the USGS in honor of Paul Gustave Doré (1832–83), the celebrated French illustrator.

Dorn Valley [Lake Co.]. The valley was named for Max Dorn, who settled here with his family in 1871 (Mauldin).

Dorothy Lake [Yosemite NP] was named by R. B. Marshall for Dorothy Forsyth, daughter of the acting superintendent of the park, 1909–12 (Farquhar).

Dorris [Siskiyou Co.]; **Dorris Reservoir** [Modoc Co.]. The station was named by the Southern Pacific Railroad about 1907. The name commemorates Presley A. Dorris and his brother, Carlos J. Dorris, stock raisers in Little Shasta in the 1860s. In 1870 they left their home to take up claims near Pit River in what is now Modoc Co. The settlement that developed around the bridge built by Presley Dorris was called Dorris' Bridge until 1876, when the name was changed to Alturas.

Dorst Creek [Sequoia NP] was named for Joseph H. Dorst (1852–1916), a captain in the 4th Cavalry and first acting superintendent of Sequoia and General Grant National Parks, 1891–92 (Farquhar).

Dos Cabezas (dōs kə bā′ səs) **Spring** [San Diego Co.]. "Two heads", for two heaps of boulders above a spring. The name appears from around 1891; there was a railway station here from 1919 to 1958 (Brigandi).

Dos Palmas (dōs pä′ məs) [Riverside Co.]. The watering place close to Mecca was known by this name when the Arizona pioneer Hermann Ehrenberg was killed there by Indians in 1868. At that time there were only two palms at the oasis, as indicated by the Spanish name. *See* Palm.

Dos Palos (dôs pal′ ōs): **Slough**, town [Merced Co.]. The name, meaning 'two trees', was applied to the station when the Southern Pacific Railroad reached the place in 1889. The two trees and the words Dos Palos are shown on the *diseño* of the Sanjón de Santa Rita grant of 1841, a few miles northeast. A Dos Palos Colony was established in 1892. The name of the post office was spelled Dospalos for many years, but in 1905 Eldredge succeeded in having the Spanish version restored. The modern town was founded and named by the Pacific Improvement Company in 1907. *See* Palo.

Dos Piedras. *See* Two Rock.

Dos Pueblos (dôs pweb′ lōs): **Canyon, Creek** [Santa Barbara Co.]. The name is derived from two Indian villages (at the mouth of the canyon), the inhabitants of which differed greatly in appearance and speech; these villages were noted by Cabrillo in 1542. The canyon apparently formed the boundary between two dialectal divisions of the Chumash Indians (Wagner, p. 384). It is doubtless the same place as that recorded by Anza on Apr. 28, 1774: "I came to camp for the night at the place which they call Dos Rancherías." The *dos pueblos* are mentioned in 1795 (PSP 13:29) and thereafter appear frequently in the records. On Apr. 18, 1842, the name Dos Pueblos was given to the land grant, which includes the creek and canyon.

Dos Rios (dôs rē´ ōs) [Mendocino Co.]. The situation of the town at the junction of two branches of the Eel River suggested the name, which is Spanish for 'two rivers'.

Double. This adjective is used much less frequently than "twin" for pairs of geographic features. Only some twenty include the word in their names, among them Double Cone Rock [Mendocino Co.], Doublehead and Doublehead Lake [Modoc Co.], Double Point [Marin Co.], and Double Butte [Riverside Co.]. **Double Gate Ridge** [Trinity Co.] was so named because of a drift fence, separating a sheep range from a cattle range, near which a sheepherder named Newtervin was killed in a feud between cattle owners and sheep owners in 1885. **Double Springs** [Calaveras Co.] was once a prosperous mining town and in 1850 was the county seat. Promoters claimed that there were not only double springs near the town but "six or eight never failing springs of the best water in California" (J. A. Smith, Calaveras).

Dougherty (dō´ ər tē) **Creek** [Kings Canyon NP] was named for Bill and Bob Dougherty, pioneer sheepmen (Farquhar); *see* Simpson Meadow.

Doughertys Station [Alameda Co.]. *See* Dublin.

Douglas City [Trinity Co.]. Settled in the early 1860s and named in memory of Stephen A. Douglas (1813–61), the "Little Giant." Although defeated by Abraham Lincoln, Douglas was admired as an upright and uncompromising political leader. The post office is listed in 1867.

Douglas Flat [Calaveras Co.]. The post office was established in the 1870s and named after an old mining camp. A place, Douglass and Raney, is shown 25 miles east of Stockton on Gibbes' map of 1852. Douglass Flat is shown on Goddard's map of 1860 at the location of the present settlement.

Douglas Park [Del Norte Co.]. *See* Berteleda.

Dove. The mourning dove (*Zenaidura macroura*) is a common wild bird of California, but its name has been given to only a few places, such as Dove Canyon [Orange Co.].

Downey [Los Angeles Co.]. In 1865 John G. Downey, governor of California from 1860 to 1862, subdivided Rancho Santa Gertrudis and later gave his name to the new town. The highest peak of the Santa Ana Range had been named for him by the Whitney Survey in 1861.

Downieville [Sierra Co.]. Named for "Major" William Downie, a native of Scotland, who mined at the forks of the Yuba River in Nov. 1849. The place, at first called merely "the Forks," was named Downieville the next spring by a group of miners, at the suggestion of James Galloway.

Doyle [Lassen Co.]. The name appears on the railroad map of 1900 as a station of the Nevada-California-Oregon Railway. John W. and Stephen A. Doyle had been settlers in Long Valley since the 1860s.

Doyles: Camp, Springs [Tulare Co.]. The name commemorates John J. Doyle, one of the farmers who served a jail sentence for taking part in the Mussel Slough affair in 1880 (Wallace Smith, pp. 287–88). *See* Mussel: Mussel Slough.

Dragon Channel [Del Norte Co.]. On Apr. 23, 1792, the explorer George Vancouver christened the cape north of Crescent City in honor of Saint George and at the same time called the rocks (probably in jesting analogy) Dragon Rocks. This name was used for many years by the Coast Survey, but at present the individual rocks have local names. Vancouver's name, however, is preserved in the name of the channel between Mansfield Break and The Great Break. *See* Saint George.

Dragon Peak and **Lake** [Kings Canyon NP]. The peak was so named because its outline as seen from Rae Lake suggests a dragon (Farquhar). **Dragon Head** [Lava Beds NM] was named by J. D. Howard because of the appearance of the rock formation at the "skylight" of a cave.

Drakes: Bay, Estero, Head [Marin Co.]. The bay was discovered by Sebastián Rodríguez Cermeño and was named Puerto y Bahía de San Francisco on Nov. 7, 1595. Vizcaíno, on Jan. 8,

1603, renamed the bay Bahía de Don Gaspar in honor of Gaspar de Zúñiga y Azevedo, Conde de Monterrey. On his map of 1625, Henry Briggs ignored both names and applied the name Puerto Sir Francisco Draco to a port north of Point Reyes—apparently an attempt to support the English claim to the coast of California. In 1790 Arrowsmith's map applies the name Sir Francis Drake to what is now San Francisco Bay (Wagner, no. 744). The Spanish explorer and cartographer Martínez y Zayas placed Drake's name at the present location in 1793: P[uer]to de Fran[cis]co Drak. Vancouver likewise called it Bay of Sir F. Drake. The Coast Survey uses the name Sir F. Drake's Bay in 1850, and the modern abbreviated form in 1854. The actual anchorage of Drake in June–July 1579 has never been established. *See* Estero: Estero de Limantour.

Drakesbad [Plumas Co.]. Formerly known as Hot Spring Valley, or Drake's Place, or Drake's Hot Springs, for E. R. Drake, who settled there in the 1860s. When the Sifford family bought the place in 1900, it also became known as Sifford's. To avoid confusion, the Sifford family added German *Bad* 'bath, watering place', referring to the hot mineral springs, to the name of the former owner, who had always wished that his name could be retained (R. D. Sifford). Nearby Drake Lake was named by the USGS in 1925.

Drapersville. *See* Kings: Kingsburg.

Drennan. *See* Earp.

Dripping Blood Mountain [Inyo Co.]. The peak received the name because of its sanguine coloring.

Drum Barracks [Los Angeles Co.]. Established 1862, abandoned 1866; named by the War Dept. for Richard Drum, who was a lieutenant colonel in the Civil War and assistant adjutant general of the Dept. of California.

Drum Bridge [Plumas Co.]. Pipes about the diameter of a snare drum were filled with concrete to serve as piers and looked from the ends like drums; hence the name. There are other such bridges in the state (Stewart).

Drum Reservoir [Placer Co.]. Named for Frank G. Drum, president of the Pacific Gas and Electric Company, 1907–20. The Drum Powerhouse on Bear River was the first generating plant constructed after the earthquake of 1906. The dam and reservoir were built in 1924.

Drunken Indian. *See* Indian.

Drury Grove [Calaveras Co.]. The grove in Calaveras Big Trees State Park was established in 1965 in memory of Aubrey Drury (1891–1959), who had been associated with the Save-the-Redwoods League from 1918. He was a well-known California author, the son of the pioneer western journalist Wells Drury, and the brother of Newton Drury, former director of the National Park Service.

Dry. There are more than one hundred Dry Creeks and about twenty-five Dry Lakes in the state, together with an assortment of Dry Sloughs, Gulches, Valleys, Arroyos, and Lagoons. Most of these were named in the dry season; very few are dry all year round. **Dry Lake** [Modoc Co.]: The name of Sorass Lake was changed to Dry Lake in 1873 when, in the Modoc War, soldiers expecting to get water there found it dry. At daylight on May 10, 1873, the lake bed was the scene of a crushing defeat inflicted on Captain Jack and his Indians (William S. Brown). **Drytown** [Amador Co.] was so named by the miners in 1849–50 because the camp's rivulet could not supply enough water for washing gold. **Drylyn** [Imperial Co.] contains the Gaelic generic term *lin* or *lyn* 'spring, pool'. **Dry Lagoon Beach State Park** [Humboldt Co.], created and named in 1931, is an example of unimaginative official place naming. *For* Dry Diggings [El Dorado Co.], *see* Placer: Placerville.

Duarte (dōō är′ tē, dwär′ tē) [Los Angeles Co.]. The settlement developed when Rancho Azusa (which had been granted to Andrés Duarte on May 10, 1841) was subdivided in 1864–65. It became the nucleus of the town promoted in the boom years 1886–87.

Dubakella (dōō′ bə kel ə): **Mountain, Creek** [Trinity Co.]. From Wintu *duubit kiili*, the name of a type of 'wild potato'; *duubit* refers to such edible roots, perhaps of genus *Perideridia* (Shepherd).

Dublin: town, **Canyon** [Alameda Co.]. James W. Dougherty of Tennessee acquired a parcel of Amador's Rancho San Ramón and settled on it after 1852. When the post office was established in the 1860s, it was named Dougherty's Station. The story that Dougherty called the part south of the road Dublin because so many Irishmen lived there (Doyle) has its merits: on Hoffmann's map of the Bay region, the settlement north of the road is labeled Dougherty and the one south of it

Dublin. The name of the post office was not changed to Dublin until the 1890s.

Ducor (dōō′ kôr) [Tulare Co.]. In 1899 the Southern Pacific reached the place known as Dutch Corners because four Germans had adjoining homesteads here. For euphony and convenience the name was shortened to Ducor.

Duffields Valley. *See* Pickel Meadow.

Dulzura (dul zōōr′ ə): town, **Creek** [San Diego Co.]. The Spanish name, meaning 'sweetness', was applied to the place because the honey industry was introduced here in 1869 by John S. Harbison. There is a Honey Spring Ranch to the north, and a Bee Canyon to the south. According to local tradition, the name was suggested by Mrs. Hagenbeck, a lover of wild flowers, when the post office was established in 1887 (Hazel Sheckler).

Dumbarton: station, **Point, Bridge** [Alameda, San Mateo Cos.]. The name of the county in Scotland was given to the station when the Oakland–Santa Cruz line was built in 1876. The highway bridge, the first to cross San Francisco Bay, was constructed and named in 1927.

Dumbbell Lake [Kings Canyon NP]. Named in 1903: "a lonely lake. This, from its shape, we called Dumbbell Lake" (J. N. LeConte, *SCB* 5:7).

Dume (dōōm, dōō mā′): **Point, Cove, Canyon** [Los Angeles Co.]. The name was given to the point by Vancouver in honor of his host, Padre Francisco Dumetz of Mission San Buenaventura. Vancouver, who often erred in the spelling of names, gave the form Dume. The maps of the Pacific Railroad Survey recorded it as Duma; this spelling was generally accepted until 1869, when the Coast Survey changed it to Dume, doubtless after consulting Vancouver's map.

Dumont Meadows [Alpine Co.]. Named for one of the French Canadian woodcutters who were engaged in the slaughter of Alpine's forests in the 1870s for timbering in the Comstock mines (Maule). There are Dumont Sand Dunes and Little Dumont Dunes in Death Valley NP, but probably not named for the same man.

Duncan: Creek, Canyon, Peak [Placer Co.]. The tributary of "the Middle Fork of the Middle Fork of the American" was named for Thomas Duncan, who came overland from Missouri in 1848, according to James Marshall, discoverer of gold at Coloma (Co. Hist.: Placer 1882:381).

Duncan Creek [Shasta Co.]. The tributary of Cottonwood Creek was named for the Duncan brothers, early settlers (Steger).

Duncan Reservoir [Modoc Co.]. The reservoir was constructed by Charles Duncan, a pioneer settler and horse raiser (R. B. Sherman).

Duncans Mills [Sonoma Co.]. Named for S. M. and A. Duncan, who built a mill there in 1860. The post office was established on Dec. 20, 1862.

Dunderberg Peak [Mono Co.]. The peak appears as Castle Peak on Goddard's and Hoffmann's maps. On Sept. 19, 1878, a party of the Wheeler Survey headed by Lt. M. M. Macomb renamed it Dunderberg, after the famous mine on Dog Creek on the north slope of the mountain. (Farquhar). This mine was probably named after the Union man-of-war *Dunderberg*, launched in New York in July 1865. The vessel in turn probably derived its name from Dunderberg mountain in New York state (Dutch for 'thunder mountain'). Thus it may well have been that a mountain in California was named after a mountain in New York via a warship and a mine (*see* Kearsarge). In June 1947 Birge M. and Malcolm Clark of Palo Alto found the following record of the Macomb party in a metal can on the peak: "This peak is called 'Dunderberg' instead of 'Castle Peak' in view of the fact that there is a peak north of Summit Station, Cal., on the C.P. [Central Pacific] R.R. to which the name is more appropriate. Also because the name 'Castle Rocks' has been given to some peaks north of the Relief Trail at the head of the Stanislaus." The present name appeared on the Bridgeport atlas sheet in 1911.

Dunlap [Fresno Co.]. The post office was first established on Nov. 13, 1882, and the place is shown on the Land Office map of 1891. It was named for George Dunlap Moss, the first postmaster and a schoolteacher in the early 1880s (Co. Hist.: Fresno 1956).

Dunnigan: Creek, town [Yolo Co.]. When the Northern Railway reached this point in 1876, the station was named for the owner of the land, A. W. Dunnigan, who had settled here in 1853.

Dunsmuir [Siskiyou Co.]. The Southern Pacific Railroad reached this section in Aug. 1886. When Alexander Dunsmuir, a "coal baron" of

British Columbia and San Francisco, passed through Cedar Flat (a station consisting of a box car), he promised the settlers a fountain if they would name the future town for him. The proposal was accepted, and at the beginning of Jan. 1887 the name and the box car were moved to the present site, which previously had been called Pusher. The fountain was erected at the railroad station, where it still stands (Marcelle Masson).

Durasnitos (do͞o räz nē′ tōs) **Spring** [San Diego Co.]. For Spanish *duraznitos* 'little peaches', plural diminutive of *durazno* 'peach'. There is also a **Durasno Valley** in Riverside Co. (Gunther).

Durham [Butte Co.]. When the railroad reached the place in the 1870s, the station was named for W. W. Durham, a mill owner and assemblyman from Butte Co. (1880).

Dusy (do͞o′ sē) **Branch** and **Meadow** [Kings Canyon NP]. In 1879 L. A. Winchell named this branch of the Middle Fork of Kings River for Frank Dusy (1836–98), a native of Canada and a well-known stockman and mountaineer (Farquhar).

Dutch, Dutchman. With the exception of the Yankees, the Germans seem to be the only settlers whose presence is recorded by their nickname. "Dutch" is an English way of pronouncing *Deutsch* 'German,' and until the 18th century it was common in England to call the Germans "High Dutch" and the Netherlanders "Low Dutch." The American habit of calling Germans "Dutch" and "Dutchmen" became general in the first half of the 19th century, when the "Dutch" began to play an important part in the settlement of the West. In California, as well as in many other states, Dutch has practically replaced the adjective German in geographical nomenclature: there are only two German Creeks, against almost a hundred Dutch Creeks, Flats, Canyons, Valleys, etc. Even these two creeks may not have been named for German settlers but may reflect the Spanish personal name *Germán*. Very often the adjective is combined with a given name: Dutch Bill, Ed, Henry, John, etc. Some of the names were doubtless given for Hollanders, or even Scandinavians: "Europeans . . . save French, English, and 'Eyetalians,' are in California classed under the general denomination of Dutchmen" (Borthwick 1857:329). **Dutch Flat** [Placer Co.] was named for two Germans, the brothers Charles and Joseph Dornbach, who settled there in 1851. The place became known as Dutch Charlie's Flat, and the name was later abbreviated to Dutch Flat. In the 1870s an attempt was made to change the name to German Level. *For* Dutch Corners, *see* Ducor; Newman.

Dutch Creek [Shasta Co.]. Named by Sim Southern in the 1860s for Dr. William Dutch, a San Francisco dentist and frequent visitor in the region (Steger).

Dutschke (duch′ kē) **Hill** [Amador Co.]. The elevation near Ione was named for the pioneer family of Charles Dutschke. Formerly also known as Jones Butte (BGN, Sept.–Dec. 1963).

Du Vrees Creek. *See* Islay: Islais Creek.

Duxbury: Reef, Point [Marin Co.]. The ship *Duxbury* was grounded on the reef on Aug. 21, 1849. The name is shown on the Coast Survey charts of 1851. In Mexican times the cape was known as Punta de Baulenas.

Dyer, Mount [Plumas Co.]. The mountain was named in honor of Assistant U.S. Surveyor General Ephraim Dyer, who came to California in 1850 and surveyed the land along the California-Nevada line from Lake Tahoe to Oregon between 1861 and 1870.

Dyerville [Humboldt Co.]. The town, on Highway 101, is said to have been named for a "tin peddler" called Dyer. It had a post office from 1890 to 1933 (*see* South Fork).

Dyke Ridge. *See* Monarch Divide.

E

Eagle is the favorite bird for place names in the state. There are more than ten Eagle Peaks and as many Eagle Rocks, as well as a large number of Eagle Meadows, Canyons, Lakes, Nests, etc. The name is sometimes used for communities, e.g. Eagleville [Modoc Co.], listed in the Great Register of 1879, and Eagle Tree [San Joaquin Co.]. Not all features are named for the presence of the bird. **Eagle Rock** [Los Angeles Co.] was so named because the shadow of the rock resembles the outline of an eagle. **Eagle Prairie** [Humboldt Co.] was named for a settler nicknamed "Old Eagle Beak." *See* Graeagle. Some names were translations of the Spanish *águila* 'eagle': **Eagle Canyon** [Santa Barbara Co.] is shown as Agalia on copies of the *diseños* of the Dos Pueblos grant (about 1842) and as Arroyo de Agalia on the plat of the same grant (1862).

Eagle Scout Peak and **Creek** [Sequoia NP] were named by Francis P. Farquhar and a group of Boy Scouts from the San Joaquin Valley at the time of the first ascent, July 15, 1926.

Earl [Los Angeles Co.]. Established in 1937 and named after the Earl Estate, formerly owned by E. T. Earl of the Earl Wholesale Fruit Company, Los Angeles.

Earl, Lake [Del Norte Co.]. Although this is in Tolowa territory, it is derived from a Yurok place name that was pronounced *rrł* (Waterman, p. 187; Harrington)—there was no Mr. Earl.

Earlham. *See* El Modena.

Earlimart [Tulare Co.]. When the Southern Pacific reached the place in 1873, the station was called Alila, probably just a "railroad name," which could be spelled forward or backward. In 1909, real estate promoters changed the name to Earlimart to indicate that crops mature early here.

Earp (ûrp) [San Bernardino Co.]. At the request of residents, and of the Santa Fe, the Post Office Dept. gave the name in 1929, for Wyatt Earp—Arizona pioneer, peace officer, and miner, who came to California in the 1860s. The railroad station had been Drennan since 1910.

Eastberne [San Bernardino Co.]. The name, coined from East San Bernardino, was applied to the subdivision about 1880 (Santa Fe).

Easter Bowl [Death Valley NP] was so named because the natural amphitheater is used for annual Easter services, inaugurated in 1927 by H. W. Eichbaum.

East Highlands. *See* High: Highland.

East Lake [Kings Canyon NP] was named by a State Hydrographic Survey party in 1881–82, for Thomas B. East, a hunter, trapper, and cattleman of Eshom Valley (Farquhar).

East Monterey. *See* Seaside.

Easton [Fresno Co.]. The town was first named Covell for A. T. Covell, resident manager of the Washington Irrigated Colony, founded in the 1870s. When the post office was established in 1882, the place was renamed for O. W. Easton, land agent of the colony.

East Palo Alto [San Mateo Co.] was the name given to a community of small subdivisions in 1925. It included the site of the older community of Ravenswood, named in 1854: "The old name has often been proposed as an alternative to East Palo Alto in the event that area incorporates" (Alan K. Brown). Because of the large black population, there was a move in the 1960s to change the name to Nairobi; cf. "Coast area to vote on an African name," *New York Times*, Apr. 4, 1968, p. 43, col. 4; Nov. 26, p. 26, col. 5; Dec. 26, p. 17, col. 2.

East Pasadena [Los Angeles Co.]. Originally named Lamanda Park for Amanda, wife of Leonard J. Rose, on whose property the town was founded. The post office was renamed East Pasadena in 1930; the railroad station retained the old name. *See* Lamanda Park.

East Riverside. *See* High: Highgrove.

Eastwood Grove [Humboldt Co.], in the Prairie Creek Redwoods State Park, was established in 1953 in memory of Alice Eastwood (1859–1953), noted California botanist, who was curator of botany at the California Academy of Sciences for fifty-seven years.

Eaton Canyon [Los Angeles Co.]. Named for Judge Benjamin S. Eaton, who settled at Fair Oaks in 1865 and utilized the water of the canyon for his vineyards. Eaton, a native of Connecticut, had come to Los Angeles in 1852, and was district attorney in 1855–56 (Reid, p. 378).

Ebabias Creek [Sonoma Co.]. The name Ebabais is shown on a *diseño* of the Cañada de Pogolimi grant (1844). The possible Indian origin has not been determined.

Ebbetts Pass [Alpine Co.]. "The pass we called Ebbets' pass, in memory of Major Ebbets, who went over it in the spring of 1851, with a large train of mules, and who found no snow there in April" (Surveyor General, *Report* 1854:90). John Ebbetts came to California in 1849 as captain of the Knickerbocker Exploring Company and died Apr. 15, 1854, in the wreck of the *Secretary* in San Pablo Strait. The Whitney Survey left the pass nameless: "In 1863 this was a simple trail, and it crossed by the route designated on Britton and Rey's map as 'Ebbett's Pass'; this name is, however, one no longer known in that region, as we could find no one who had ever heard of it" (Whitney, *Geology*, p. 446). The old name was restored when the USGS surveyed the Markleeville quadrangle in 1893.

Echo. A favorite term for places where reverberation is heard. The best-known features bearing the name in California are Echo Lake [El Dorado Co.], Echo Rock [Los Angeles Co.], and Echo Peak [Yosemite NP]; the last was named by the Wheeler Survey in 1879.

Eckhard [Yolo Co.]. Probably named for Conrad Eckhardt, a native of Germany and a farmer, who was registered as a voter at West Cottonwood in 1878.

Eckley [Contra Costa Co.]. Commodore John L. Eckley, who came to California as a land agent for D. O. Mills, Thomas O. Larkin, and W. D. M. Howard; he bought the cove for a yacht harbor in the 1870s (Co. Hist.: Contra Costa 1940:732). The name was applied to the station after 1878, when the Central Pacific built the line from Berkeley to Tracy.

Eddy, Mount [Siskiyou Co.]. Named for Nelson Harvey Eddy, a native of New York State, who arrived in 1854 with one yoke of oxen and one cow, lived on the slope of the mountain until 1867, and then became a successful rancher in Shasta Valley (Schrader).

Eden. Of all names alluding to an earthly paradise, Eden is the most popular. The map of the United States is dotted with hundreds of Edens, Edendales, Eden Valleys, etc. California has more than fifteen places or topographic features so named; some may be derived from the well-known English surname. **Mount Eden** town and **Creek** [Alameda Co.]: Eden Landing was established by the Mount Eden Company, an association of farmers, in 1850, and Eden township was formed in 1853. On Hoffmann's map of the Bay region, both Eden Landing and Mount Eden are shown. Since there is no "Mount" to justify the name, either it is a transfer name or members of the large Eden family may have settled in the district as early as 1850 and provided the name. **Eden** [Riverside Co.]: The Eden Hot Springs were named by the owner, John Ryan, in 1890; the post office Eden was established on Mar. 7, 1924. Hoopa Valley [Humboldt Co.] was once called Eden Valley.

Edgar Peak [Yolo Co.]. Probably named for James Edgar, a native of Canada, who had a 1,500-acre farm in the region in 1870.

Edgar Station [Riverside Co.]. *See* Beaumont.

Edgewood [Siskiyou Co.]. The place was first known as Butteville. In 1875 Joseph Cavanaugh changed the name to Edgewood, which he considered more appropriate for its situation at the edge of the forest bordering the Shasta Valley.

Edison [Kern Co.]. Named by the Southern California Edison Company when it built a substation there in 1905. **Lake Thomas A. Edison** [Fresno Co.], near the Inyo Co. line, was named in the 1950s directly for the great inventor (1874–1931).

Edith Lake [Yosemite NP] was named in 1910 by Maj. William W. Forsyth, acting superintendent of the park, for the daughter of Col. John T. Nance (Farquhar). The present spelling, instead of Edyth, was made official by the BGN in 1990.

Ediza (ē dī′ zə) **Lake** [Madera Co.]. Origin unknown. Formerly in Yosemite NP (Browning 1988).

Edna Lake [Yosemite NP]. Named (about 1900) by R. B. Marshall for Edna Bowman, later Mrs. Charles J. Kuhn of San Francisco (Farquhar).

Edom. *See* Palm: Thousand Palms Canyon.

Edson. *See* Essex.

Edwards: Air Force Base, post office [Kern Co.]. The name of old Muroc Air Field was changed to Edwards Air Force Base on Jan. 27, 1950, in memory of Captain Glenn W. Edwards, who was fatally injured in the crash of an experimental YB-49 "Flying Wing." The name of the post office was changed to Edwards on Nov. 1, 1951.

Edwards Canyon [Riverside Co.]. The canyon in the Cleveland National Forest was named in 1960 in memory of Boyd M. Edwards, a USFS employee who lost his life in the Decker fire, in Aug. 1959.

Edyth Lake. *See* Edith.

Eel: River, Canyon, Rock [Mendocino, Humboldt Cos.]. The river was named in Jan. 1850 by the Gregg party because L. K. Wood obtained a large number of eels from the Indians in return for small pieces of a broken frying pan. The name was placed on the map by C. D. Gibbes in 1852. Eel Rock was the name applied to the station of the Northwestern Pacific when the section was built in 1910. A few other features, including Eel Point [San Clemente Island], were probably so named because eel fishing was good there.

Egg Lake [Modoc Co.]. The lake, in the Modoc National Forest, was so named because it is a favorite nesting place for various species of waterfowl (USFS).

Ehrnbeck (ûrn´ bek) **Peak** [Yosemite NP]. Named for Lt. Arthur R. Ehrnbeck, a native of Wisconsin, who made a report in 1909 on a comprehensive road-and-trail project for the national park (Farquhar).

Eiler Lake [Shasta Co.]. Named for Lu Eiler, who discovered Thousand Lake Valley (Steger).

Eiler Mountain [Alameda Co.]. Named for the Eiler brothers who settled here in 1853 (Still).

Eisen, Mount [Sequoia NP], was named in 1941 in memory of Dr. Gustavus A. Eisen (1847–1940), a native of Sweden and a well-known scientist. He accompanied the German geographer Friedrich Ratzel on his trip to California in 1874 and later was instrumental in the establishment of the Sequoia and General Grant National Parks.

Eisenecke (ī´ zə nek): **Valley, Creek** [San Diego Co.]. The name was applied by Mr. and Mrs. Billie Bloch to their ranch and to the creek that flows through it. The name, German for 'iron corner', appeared on the Cuyamaca atlas sheet (1903) and was probably applied because the huge granite cliffs form a natural gateway.

Eisenhower Mountain [Riverside Co.]. For Pres. Dwight D. Eisenhower (1890–1969), supplanting the name Ikes Peak (BGN, 1973, list 7302).

El. Spanish names preceded by the masculine article *el* that are included in cluster, group, or folk names will be found in the alphabetical order of the name proper. Example: El Cerrito will be found under Cerrito. Many names of this type, e.g. **El Bulto** ('the bundle') [Kings Co.], appear on maps of the western San Joaquin Valley; they were apparently bestowed by English-speaking surveyors, with the help of a Spanish dictionary and a good deal of imagination. In this sparsely populated area, the names have been little used, and they have no customary English pronunciations (J. Crosby). For the most part they are omitted from this dictionary.

Elanus Canyon (i lā´ nəs) [San Diego Co.]. For a bird, the white-tailed kite (*Elanus leucurus*); approved by the BGN, list 7702.

El Cajon (kə hōn´): **Valley, Mountains**, town [San Diego Co.]. The name is mentioned on Sept. 10, 1821, as another name for the *sitio rancho Santa Mónica* (Arch. MSB 4:210). *Cajón*, Spanish for 'box', was applied because the place is boxed in by the hills (*see* Cajon). The name appears in the *expediente* of a provisional grant of Aug. 22, 1844, and is mentioned by B. D. Wilson as the name of a Diegueño village in 1852. The valley is called Cajon Valley on the von Leicht–Craven map of 1873, and the ranch was known in 1875 as Cajon Ranch. When the modern town developed, the original name El Cajon was restored. The name of the post office was written Elcajon until 1905, when the Post Office Dept. separated the two words at the insistence of Z. S. Eldredge.

El Camino Real. *See* Camino.

El Capinero (kap i nâr´ ō) [Tulare Co.]. This name has been the subject of some controversy (*WF* 8:370, 9:155–56). Perhaps it is a misspelling of Spanish *caponera*, which in Spanish California referred to a herd of horses (Blanco).

El Capitan (kap´ i tan, kap i tan´, kä pi tän´) [Yosemite NP]. Spanish for 'the captain, the chief'. The name was applied by the Mariposa Battalion in 1851; Bunnell (p. 211) understood that the Paiute name for this cliff contained the word for 'chief', and similar information was obtained later by Brewer (Notes, June 16, 1863). The name El Capitan is used also for a towering rock in Siskiyou Co., a state park in Santa Barbara Co., and a lake in San Diego Co. (*see* Capitan Grande).

El Casco (käs′ kō) [Riverside Co.]. The name was applied to the Southern Pacific station before 1891. The Spanish term means 'skull', but the application is unclear (cf. Gunther).

El Centro [Imperial Co.]. When the town was platted in 1905, the real estate company chose the Spanish term to indicate that the town was in the center of the Imperial Valley. The railroad station had previously been called Cabarker for C. A. Barker, a friend of W. F. Holt, owner of the land on which the town developed.

El Cerrito. *See* Cerrito.

Elder. Although several species of the common shrub or tree (*Sambucus glauca*) are native to various sections of the state, the name is applied to only ten creeks and canyons, plus two inhabited places, Elderwood [Tulare Co.] and Elder Creek [Sacramento Co.]. The Elder Creek that gave the name to the latter settlement appears on Bidwell's map of 1844 as Arroyo de los Saucos 'creek of the elders'.

El Dorado. The name, meaning 'the gilded one', appears in the tableland of Bogotá, Colombia, at the beginning of the 16th century, referring to a legendary Indian chief who was said to have been covered with gold during the performance of religious rites. This chief was eagerly sought by the Spanish conquerors of northern South America. Subsequently, the name designated one of the golden utopias that played such an important role in the conquest of America. With the discovery of gold in California, the name assumed a new significance. Charles Preuss, who was finishing his map of 1848 when news of the discovery of gold reached Washington, placed the legend "El Dorado or Gold Region" along the Plumas River and the South Fork of the American River. **El Dorado** (də rä′ dō) **County** was one of the original twenty-seven counties, created and named by the legislature on Feb. 18, 1850. The town of **El Dorado** [El Dorado Co.], at its inception in 1849, was known as Mud Springs; but at its incorporation on Apr. 16, 1855, the name was changed to El Dorado, after the county. **Eldorado National Forest** was created and named by presidential proclamation in 1910. The name El Dorado was applied to a number of communities in the 1850s, and its popularity soon spread to other states. There is another El Dorado in Calaveras Co., and there are Eldorado Creeks in Mariposa, Placer, and Santa Barbara Cos. **El Dorado Hills** [Sacramento Co.] is a modern real estate development, established in 1962.

Eldridge [Sonoma Co.]. The post office, established about 1895, was named for James Eldridge, to whom part of Rancho Cabeza de Santa Rosa had been patented on Jan. 5, 1880.

Eleanor, Lake; Eleanor Creek; Lake Eleanor Reservoir [Yosemite NP]. The Whitney Survey named the lake in the 1860s, for Josiah D. Whitney's daughter Eleanor. The lake was converted into a reservoir for the water system of San Francisco in 1913 (Farquhar).

Electra Peak [Yosemite NP]. The name appears on the Mount Lyell atlas sheet, published in 1901, and was perhaps given because the anticlerical drama *Electra*, by the Spanish author Benito Pérez Galdós, attracted worldwide attention in the same year. **Electra** [Amador Co.]: The post office and the Pacific Gas and Electric Company station, built in 1902, may have been named for the same reason.

Elena (el′ ə nə) [Shasta Co.]. Named in memory of Elena Haggen, one of the first women to settle in Big Bend (Steger).

Elephant. Several mountains in the state bear the name because of their resemblance to the animal: Elephant Back [Alpine, Tulare Cos.], Elephants Head [Santa Clara Co.], Elephant Rock [Yosemite NP]. **Elephant Hill** [Fresno Co.] was named because "fossil remains of an elephantine, Pliomastodon, were found on the top of this hill" (BGN, 1932). **Elephants Playground** [Plumas Co.] was suggested by the elephantine scale of the boulders in the meadow (Stewart). "To see the elephant" was a common expression for having mined gold during the Gold Rush, and a number of mines were named Elephant.

El Estero (ə stâr′ ō) [Monterey Co.]. A lake in Monterey, originally an estuary; from Spanish *estero* 'estuary' (Clark 1991).

Elftman. *See* Dominguez.

El Granada (grə nä′ də) [San Mateo Co.]. The post office was named in 1910. If it was named for the city in Spain, it does not need the article; if the name chosen was the Spanish word for 'pomegranate', then the feminine article *la* should have been used.

Elizabeth is the most popular of all given names in the state's toponymy. There are about ten Elizabeth Lakes, several Creeks, and at least one Mountain. The oldest is apparently **Elizabeth Lake** [Los Angeles Co.], mentioned in 1853 in the Pac. R.R. *Reports* as Lake Elizabeth. The lake in Tuolumne Co. was named in 1909 by R. B. Marshall of the USGS, for the daughter of Dr. Samuel E. Simmons of Sacramento. **Elizabeth Pass** [Kings Canyon NP] was named by Stewart Edward White for his wife, when they crossed the divide in 1905. **Elizabethtown** [Plumas Co.] was named in 1852 for Elizabeth Stark Blakesley and was popularly known as Betsyburg.

Elizabeth, Lake [Alameda Co.]. This lake in the city of Fremont was named in 1968 for Fremont's sister city—Elizabeth, South Australia (Mosier and Mosier).

El Jarro. *See* Jarro.

Elk; Elkhorn. Although not as popular as the deer, the elk (*Cervus elaphas*) is represented in more than fifty geographic features in the state. Once near extinction in California, the elk has now been reintroduced in a number of locations. In 1837 Belcher referred to the mountains of Solano and Napa Cos. as the Elk Range and even spoke of an Elk station (1:119 ff.). The names Elkhorn Peak and Elkhorn Slough [Solano Co.] are apparently not derived from the name given by Belcher but offer additional evidence that the animal once abounded in this region. **Elk River** [Humboldt Co.] was named by the Gregg party on Christmas Day, 1849, after they had celebrated the day with a dinner of elk meat. **Elk Grove** [Sacramento Co.] also dates from pioneer days. In 1850 James Hall opened his Elk Grove House, with an elk's head painted over the door. The name is found in a number of other post offices and communities: Elk [Mendocino and Fresno Cos.], Elk Creek [Glenn Co.], Elkhorn [Monterey Co.], Elk Horn [Yolo Co.], Elks Retreat [Butte Co.], Elk Valley [Del Norte Co.]. *For* Elk Cove, *see* Alcove Canyon; *for* Elkhorn Station, *see* Burrel.

Ellery Lake [Mono Co.]. Named for state engineer Nathaniel Ellery, who built the road between Tioga Pass and Mono Lake in 1909 (Farquhar).

Elliotta (el ē ō′ tə) **Springs** [Riverside Co.]. Named La Elliotta Plunge Baths by the developer, W. E. Elliott (Gunther).

Ellis Landing [Contra Costa Co.]. Named for Capt. George Ellis, who operated a schooner on San Francisco Bay in the 1850s. It was best known for its shell mound.

Ellis Meadow [Tulare Co.]. Named for Sam L. N. Ellis, for many years head ranger of the USFS in the region and one-time supervisor of Tulare Co. (Farquhar).

Ellwood [Santa Barbara Co.]. The name commemorates Ellwood Cooper, one of California's notable pioneers in horticulture, who was instrumental in introducing the eucalyptus tree and in developing olive culture in southern California. He served as president of the State Board of Horticulture from 1883 to 1903.

El Macero (mə sâr′ ō) [Yolo Co.]. Perhaps a variant of Spanish *la masera* 'the kneading-trough'.

Elmdale. *See* Denair.

Elmira (el mī′ rə) [Solano Co.]. When the California Pacific Railroad from Vallejo to Sacramento was built in 1868, the name Vaca, after Vacaville 3 miles west, was applied to the station. When the extension to Rumsey was built about 1875, Vacaville itself became a railroad station and the station Vaca was renamed Elmira, after the city in New York.

El Mirage: Dry Lake, Valley, settlement [San Bernardino Co.]. The lake west of Victorville was named for the optical illusion frequently occurring in the desert, with the Spanish definite article added for good measure. The post office existed from 1917 to 1934.

El Modena (mō dē′ nə, mə dē′ nə) [Orange Co.]. The settlement was named Modena after the Italian city (which, however, is pronounced with the accent on the first syllable), or more likely after one of several American towns by that name. When the post office was proposed about 1887, the Post Office Dept. objected that Modena was too similar to Madera, and the settlers substituted Earlham, the name of a Quaker college in Indiana; an Earlham post office operated from 1887 (Brigandi). But in 1888 the Spanish article El was introduced, and the post office was renamed El Modena. In 1910 the Post Office Dept. realized that in Spanish a masculine article cannot be followed by a feminine noun and changed the name to El Modeno. In 1965 the BGN confirmed the form El

Modeno; but in 1970 it went back to El Modena (list 7001).

El Monte (mon′ tē) [Los Angeles Co.]. The town was settled in 1852 by squatters who believed that there was a flaw in the title to the land. The place was called El Monte 'the thicket' because of a dense stand of willows. It is mentioned as Monte in the *Statutes* of 1854 (p. 223) and in the Pac. R.R. *Reports* (vol. 3, Prel., p. 27). The original post office name was also Monte; this was changed to El Monte before 1880. About 1895 the Post Office Dept. contracted the two words into one, but in 1905 the name was changed back to El Monte at Z. S. Eldredge's insistence.

Elms Canyon [Los Angeles Co.]. Named in 1882 or 1883 for Dr. Henry Elms, owner of the land at the mouth of the canyon.

Elmwood Colony. *See* Denair.

El Paso de Robles. *See* Paso: Paso Robles.

El Pinal (el pi näl′) [San Joaquin Co.] is Spanish, meaning 'the pine grove', derived from *pino* 'pine tree'.

El Piojo Creek. *See* Piojo.

El Polin (pō lēn′) **Spring** [San Francisco Co.]. This place in the Presidio has been associated with a Spanish legend which said that all maidens who drank from it during the full moon were assured of many children and eternal bliss; another version suggests an Indian origin (Loewenstein). However, Spanish *polín* means 'wooden roller, for moving great guns or any other heavy object'—a plausible item of furniture for the military presidio.

El Pomar (pō mär′) [San Luis Obispo Co.] means 'orchard', especially an apple orchard (Dart). However, the usual word for 'apple orchard' in Californian Spanish is *manzanar*, from *manzana* 'apple'.

El Portal (pôr tal′) [Mariposa Co.]. In 1907 the Yosemite Valley Railroad gave the Spanish name to its terminus at the 'gateway' to the park.

El Rio (rē′ ō) [Ventura Co.]. The town was founded in 1875 by Simon Cohen and called New Jerusalem. About 1895 the Post Office Dept. changed the name to Elrio, and in 1905, on Eldredge's insistence, to El Rio. The Spanish name, *El Río* 'the river', refers to the nearby Santa Clara River. El Rio is now part of the city of Oxnard. *See* Rio.

El Segundo (sə gŏŏn′ dō, sə gun′ dō) [Los Angeles Co.]. The name, meaning 'the second', was applied in 1911 by Col. Rheem, of the Standard Oil Company, to the company's second refinery in California.

Elsinore: Lake, Mountain, Valley [Riverside Co.]. In 1883 Donald M. Graham, William Collier, and F. H. Heald purchased and subdivided Rancho La Laguna, which doubtless had been so named because of the lake previously called La Laguna Grande 'the big lake' or (tautologically) Lake Laguna. The township of one section was called Laguna to preserve the old name. When a post office was established in 1884, the Post Office Dept. insisted on a new name because there was already a Laguna post office in the state; hence Donald Graham chose the name of the Danish castle (in Danish, Helsingør) made famous by Shakespeare's *Hamlet*. The town is now officially named Lake Elsinore (BGN, list 7601).

El Sobrante. *See* Sobrante.

El Sombroso (sōm brō′ sō) [Santa Clara Co.]. The name of this prominence is Spanish for 'the shady one'.

El Sueno (el swān′ yō) [Santa Barbara Co.]. From Spanish *el sueño* 'the dream'; the reason for the name is not known.

Elterpom (el′ tər pom) **Creek** [Trinity Co.]. The name of the tributary of the South Fork of the Trinity, also spelled Eltapom, is from Wintu *eltipom* 'place on the other side', referring to the Weaverville area, from *elti* 'inside, on the other side' plus *pom* 'place' (Shepherd).

El Toro, Eltoro. *See* Toro.

El Venado. *See* Venado.

El Verano (və rä′ nō) [Sonoma Co.]. About 1890, George H. Maxwell, a lawyer and later a well-known irrigationist, whose parental home adjoined El Verano, chose the Spanish name, meaning 'the summer', because "the climate here was considered about perfect" (Ruth Denny).

Elverta [Sacramento Co.]. The town was named in 1908 for Elverta Dike, whose husband had donated a building lot for a community church (Lizzie Silver). The post office is listed in 1915.

Elysian (ə lē′ zhən) **Valley** [Lassen Co.]. The beauty of the valley inspired Daney H. Keatley and L. N. Breed to bestow upon it the name Elysian when they settled there in the summer of 1856—after Elysium, the classical abode of blessed souls. There is also a public park in Los Angeles called Elysian Park.

Embarcadero (em bär kə dâr′ ō). The Spanish term for a boat landing (from *embarcar* 'to embark') was commonly used in Spanish times and has survived in a number of places. A land grant in Santa Clara Co., dated June 18, 1845, was named Embarcadero de Santa Clara.

Emerald. The name of the bright green precious stone is repeatedly used for place names, mostly for bodies of water [Los Angeles Co., El Dorado Co.], but also for mountains [Kern Co., Fresno Co.] and settlements [San Mateo Co., Los Angeles Co.]. **Emerald Peak** [Kings Canyon NP] was named by T. S. Solomons in 1895 on account of its color (Farquhar). **Emerald Lake** [Trinity Co.] was formerly called Lower Stuarts Forks Lake; it was renamed by Anton Weber in the 1920s, perhaps because of its color (Stang). *See* Sapphire Lake.

Emeric: Creek, Lake [Mariposa Co.]. Named by Lt. N. F. McClure in 1895, for Henry F. Emeric, president of the State Board of Fish Commissioners and a charter member of the Sierra Club (Farquhar).

Emerson, Mount [Inyo Co.]. John Muir wrote in 1873: "I have named a grand wide-winged mountain on the head of the Joaquin Mount Emerson. Its head is high above its fellows and wings are white with ice and snow." Muir had become acquainted with Ralph Waldo Emerson (1803–82) when the latter visited Yosemite in May 1871. Muir's Mount Emerson was probably the present Mount Humphreys, about 2.5 miles northwest (Farquhar).

Emeryville [Alameda Co.]. Named in 1897 for Joseph S. Emery, a native of New Hampshire; he came to California in 1850, and in 1859 he bought the land on which Emeryville now stands. Emery School District was organized and named in 1884 (Knave, Oct. 6, 1946).

Emigrant. There are about twenty Emigrant Passes, Lakes, Meadows, etc., along the routes by which trains of immigrants (usually called emigrants) reached the state in the early days. Among the oldest and best known of these are Emigrant Gap [Placer Co.], Emigrant Pass [Tuolumne Co.], and Emigrant Canyon [Death Valley NP].

Emma Lake; Mount Emma [Mono Co.]. Two small lakes were presumably named for Anna and Emma Mack, sisters of Sen. Maurice Mack, whose parental home was nearby in the 1870s (Maule).

Emory, Fort [San Diego Co.]. Named on Dec. 14, 1942, by the War Dept., General Order no. 67, in honor of William H. Emory, topographical engineer of Kearny's detachment in 1846–47, a member of the commission to establish the boundary between the United States and Mexico, and a major general in the Civil War. The name was suggested by Col. P. H. Ottosen of Fort Rosecrans.

Empire [Stanislaus Co.]. Founded in 1850 as Empire City (probably named after New York, the "Empire City") but shown on Gibbes's map (1852) as Empire. Although twice almost destroyed and deserted (1852 and 1855), the place has survived with the name. **Empire Creek** [Sierra Co.]: Before 1950 the stream was officially known as the Little North Fork of the Middle Fork of the North Fork of the Yuba River. Then the BGN abolished this monstrosity by naming the stream after the Empire Ranch, through which it flows.

Encanto [San Diego Co.]. The Spanish word for 'enchantment' was selected for the town by Alice Klauber because of the charming climate and the fascinating view (Sanchez). The post office is listed in 1910.

Encapa Islands. *See* Anacapa Islands.

Enchanted Gorge [Kings Canyon NP] was named by T. S. Solomons in 1895 for the gorge between Scylla and Charybdis (Farquhar).

Encina (en sē′ nə); **Encinitas** (en si nē′ təs). English speakers tend to use the word 'oak' indiscriminately for any type of that tree, but Spanish speakers distinguish the *encina* or *encino* 'coast live oak' (*Quercus agrifolia*) from the *roble* 'deciduous oak'. The words *encina, encinitas* 'little oaks' and *encinal* 'oak grove' were included in the titles of seven land grants and have survived in the names of a number of communities and geographic features. The site of part of the present city of Oakland was once called Encinal de Temescal. **Encinitas Creek** and town [San Diego Co.]: On July 16, 1769, the Portolá expedition gave the name Cañada de los Encinos to the valley through which the creek runs. The name Los Encinitos was applied to the land grant on July 13, 1842. The modern community dates from 1881, when Nathan Eaton, "hermit, keeper of bees, and brother of General Eaton of Civil War fame" (Santa Fe), built the first house. The Santa Fe station was named Encenitos in 1881; the Postal Directory of 1892 and the Official Railway Map of 1900 have the modern spelling. **Encino** (en sē′ nō) post office, **Park, Reservoir** [Los Angeles Co.]: The Portolá expedition found many live oaks in the vicinity and called San Fernando Valley Santa Catalina de Bononia de los Encinos on Aug. 5, 1769. A provisional grant, Encino, was made about 1840 and was regranted to three Indians on July 18, 1845. The name was applied to the station when the Southern Pacific extension from Burbank was built in the 1890s. The post office is listed in 1939.

Encinal del Temescal. *See* Oak: Oakland; Temescal.

Enderts Beach [Del Norte Co.]. Named for Fred W. Endert of Crescent City.

Enecapa, Eneeapah. *See* Anacapa Islands.

Engels, Engelmine [Plumas Co.]. The place was named for the Engel brothers, who developed a big copper mining company around 1900. When a post office was established on July 22, 1916, it was called Engelmine.

Englebright: Dam, Reservoir [Nevada Co.]. The dam was named by the California Hydraulic Miners Association in 1945 in memory of Harry L. Englebright of Nevada City, congressman, 1926–43.

English. About ten geographic features in California bear the name English, sometimes perhaps from the family name. Not only were English miners and settlers less numerous than either the Scots or the "Dutch," but their presence was less conspicuous.

Enterprise [Butte Co.]. Probably named for the Union Enterprise Company, which built flumes there in 1852 (*Grizzly Bear*, June 1921).

Epidote (ep′ i dōt): **Peak** [Mono Co.]. The name was apparently applied by the USGS when the Bridgeport quadrangle was surveyed in the years 1904–9, because of the presence of the mineral—a composition of calcium, aluminum, and iron.

Ericsson, Mount [Sequoia NP]. When Prof. Bolton C. Brown and his wife made the first ascent in 1896, they named the peak Crag Ericsson in honor of John Ericsson (1803–89), designer of the *Monitor*, the ironclad that fought the *Merrimac* in the Civil War (Farquhar). The generic term was later changed to Mount, probably when the USGS mapped the Mount Whitney quadrangle in 1906.

Esalen (es′ ə lən) **Institute** [Monterey Co.]. Formerly Slates Hot Springs, then Big Sur Hot Springs; reopened in 1962 as an educational foundation, named for the local Indian tribe—otherwise spelled "Esselen" (Clark 1991).

Escalon (es′ kə lon) [San Joaquin Co.]. James W. Jones, so the story goes, came upon the Spanish word meaning 'step of a stair' in a book in the Stockton Public Library. Pleased with the sound of the word, he reserved it for the town that he laid out on his land with the coming of the railroad in 1895–96. **Escalona** (es kə lo′ nə) **Gulch** [Santa Cruz Co.] may have the same Spanish source (Clark 1986).

Escarpado (es kär pä′ dō) **Canyon** [Fresno Co.]. The Spanish term means 'craggy', from *escarpe* 'crag, bluff'.

Escarpines. The word appears in the name of the Bolsa de los Escarpines grant [Monterey Co.], dated Oct. 7, 1837. *Escarpín* is Spanish for 'sock; light shoe', but the word may have something to do with *escarpe* 'bluff, declivity, sloped bank', or *escorpina* 'a small saltwater fish'. On some maps the name appears as Bolsa de las Escorpinas. *See* Espinosa Lake.

Escondido; Escondida. The Spanish word for 'hidden' was used in a geographical sense mainly with watering places. The Anza expedition camped "at the place called Agua Escondida" on Feb. 22, 1776 (Anza; Font). There are Escondido Creeks in San Diego, Los Angeles, and Santa Barbara Cos. **Escondido** (es kən dē′ dō) [San Diego Co.]: In 1885 a syndicate of Los Angeles and San Diego businessmen bought the Rancho El Rincón del Diablo, laid out the town at the crossroads called Apex, and named it Escondido after the creek. Lake Wohlford was formerly known as Lake Escondido.

Escorpion [Los Angeles Co.]. The Spanish word for 'scorpion' is mentioned by Fages as a place name as early as Sept. 1783 (Prov. Recs. 3:130). In July 1834 it is recorded as the name of a sheep and horse ranch (DSP Ben. Mil. 79:89). On Aug. 7, 1845, it was applied to a grant of half a league, conveyed to three Indians, Urbano, Odón, and Manuel. Whitney (*Geology*, p. 120) says: "The Sierra Santa Susanna . . . is also familiarly known as the 'Scorpion Hills.'"

Eshom (ish′ əm): **Creek, Valley, Point** [Tulare Co.]. Named for John Perry Eshom, a native of Illinois, who settled in the Tule River precinct before 1866.

Esken. *See* Esquon.

Esmeralda (ez mə ral′ də) [Calaveras Co.]. The name, probably derived from the nearby Esmeralda Mine, is shown on the Land Office map of 1891; the word is Spanish for 'emerald'. The name was very popular in the 19th century, partly because of Esmeralda, the heroine of Victor Hugo's novel *Notre-Dame de Paris*. The spelling Esmerelda is also found.

Espada (es pä′ də) **Creek** [Santa Barbara Co.]. "At this place a soldier lost his sword; he allowed it to be stolen from his belt, but he afterwards recovered it as the Indians who had seen the act ran after the thief. . . . For this reason the name of Ranchería de la Espada stuck to the village" (Costansó, Aug. 27, 1769).

Esparto (es pär′ tō) [Yolo Co.]. The Vaca Valley Railroad (now the Southern Pacific) reached the place in 1875 and called the station Esperanza. When the post office was established in the 1880s, the name had to be changed because there was another Esperanza, in Tulare Co. *Esparto* is the Spanish name for a type of grass native to southern Europe.

Esperanza. The Spanish word for 'hope' is found as an optimistic name for settlements in various parts of the world; in Spanish it can also be a woman's name. California has two railroad stops so named [Orange, Siskiyou Cos.]. *For* Esperanza [Yolo Co.], *see* Esparto.

Esperanza Creek [Calaveras Co.] is mentioned in Hutchings' *Illustrated California Magazine* (3:490) and was probably named after the mine between Mountain Ranch and Railroad Flat. A land grant in Monterey Co. dated Nov. 29, 1834, was named Encinal y Buena Esperanza ('oak grove and good hope').

Espinosa (es pi nō′ sə) **Lake** [Monterey Co.]. Perpetuates the names of José and Salvador Espinosa, who owned adjoining ranchos, the boundary between them running through the lake. *El rancho de los Espinosas llamado* [called] *San Miguel* is mentioned as early as 1828 (Registro, 14).

Espíritu Santo. The Spanish phrase meaning 'Holy Ghost' is found in the names of two land grants. Lomerías Muertas 'dead or bare ridge' or Lomerías del Espíritu Santo [San Benito Co.] was granted on Aug. 16, 1842. Loma del Espíritu Santo [San Benito Co.], a grant dated Apr. 15, 1839, but not confirmed by the U.S. government, may have been named for the grantee, María del Espíritu Santo Carrillo, and not directly for the Holy Ghost.

Esquon (es′ kwon) [Butte Co.]. From Konkow (northwestern Maidu) *eskeni* (Shipley, p.c.). The name of the land grant, dated Dec. 22, 1844, was apparently derived from that of an Indian rancheria, shown as Ra. Esque on a *diseño* (about 1844) of the Aguas Frias grant. Kroeber (AAE 29:267) gives it as Esken. The grant was also known as Neal's Rancho for the co-grantee, Samuel Neal, a Pennsylvania German who came with Frémont in 1844. The station Esquon of the old Sacramento Northern Railroad was so named because it was on land included in the grant.

Essex [San Bernardino Co.]. The station had been called Edson by the Atlantic and Pacific Railroad in 1883. In 1906 the Santa Fe changed the name for operating convenience. According to local tradition, Essex was the name of a pioneer miner (Santa Fe). There is another Essex in Humboldt Co.

Estanislao. *See* Stanislaus.

Estero (es târ′ ō). The Spanish word corresponds to the English "estuary" but was also used for an inlet or lagoon near the sea. It was frequently applied as a geographical term in Spanish times and has survived in the names of several coastal features. **Estero Bay** and **Point** [San Luis Obispo Co.]: The *estero* was mentioned by Costansó on Sept. 9, 1769, and Punta del Estero is shown on his map. The bay appears on later Spanish maps as Los Esteros. Vancouver (2:446) has Ponto del Esteros [*sic*], and Wilkes's map has Esteros Point. Point Esteros and Esteros Bay are shown on the Coast Survey chart of 1852, and Estero Bay is mentioned in the *Report* of 1858 (p. 322). In 1854 Parke referred to the Carrizo Plains [San Luis Obispo Co.] as the Estero, "another remarkable feature . . . a broad, smooth plain" (Pac. R.R. *Reports* 8:1.8), and the Parke-Custer map shows Llano Estero; but here the name refers not to a coastal feature—a salt marsh can also be called *estero* in Spanish. **Estero Americano** [Sonoma Co.]: The name arose in Spanish times, probably because American ships engaged in the fur trade sometimes anchored at the mouth of the *estero*. In 1810 Gabriel Moraga came across three Americans from such a ship, who were hunting deer a few miles from the inlet (PSP 19:279). The name was applied to a land grant dated Sept. 4, 1839, and in the *Statutes* of 1855 the stream is mentioned as forming the boundary between Marin and Sonoma counties. Duflot de Mofras (1844) and Ringgold (General chart, 1850) have Estero Americano for Tomales Bay. *See* Creek. **Estero de Limantour** [Marin Co.]: The name was first applied to what is now Drakes Estero for José Yves Limantour, supercargo of the *Ayacucho*, which was wrecked at the entrance to the *estero* in 1841. Limantour was a French trader, naturalized in Mexico, who gained wide notoriety in the 1850s by presenting to the authorities no fewer than seven claims for grants, totaling more than 124 leagues—all rejected as fraudulent. On the chart of Drake's Bay (1860) the Coast Survey renamed the estuary Drake's Estero. Two years later the survey officials tried to restore the original name in spite of their belief that the wreck of Limantour's vessel had been a fake too: "[Limantour] asserted that in trading upon this coast in 1841 he lost the Mexican vessel Ayachuco [*sic*] at the entrance to this estero" (*Report* 1862:326). But Davidson in his 1869 edition of the *Coast Pilot* kept Drake's name for the large estuary, and Limantour's name was transferred to the smaller inlet farther east.

Estrada (es trä′ də) **Creek** [Santa Cruz Co.]. Probably for José Antonio Estrada, who farmed in the area in 1867 (Clark 1986).

Estrella (es trā′ yə): **Creek**, town, **Army Air Field** [San Luis Obispo Co.]. The name Estrella 'star' is shown north of the creek, a tributary to the Salinas, on the *diseño* of San Miguel, 1840. On July 16, 1844, the name was given to part of the land grant conveyed to the Christian Indians of San Miguel; the grant was not confirmed by the U.S. government. The town is shown on G. H. Derby's map of 1850 (misspelled Estrelta, although it is spelled correctly in his report).

Estuary. *See* Creek.

Ethanac (ē′ thə nak) [Riverside Co.]. The post office was established June 25, 1900; the name was coined from that of Ethan A. Chase, a landowner and political leader. When the Santa Fe reached the place in 1905, the name was applied to the station. In 1926 the post office was renamed Romoland (Brigandi).

Eticuera (et i kwâr′ ə) **Creek** [Napa Co.]. Etécuero is shown on a *diseño* of Las Putas grant (175 ND) about 1843. Perhaps transferred from Michoacán, Mexico, where *etúcuero* is Tarascan for 'salt place'.

Etiwanda (et i wän′ də) [San Bernardino Co.]. George and William Chaffey started the settlement in 1882 and named it for the chief of a supposed tribe near Lake Michigan. The Santa Fe station is Etiwanda, and the nearby Southern Pacific station is Etiwa.

Etna [Siskiyou Co.]. The original name, Rough and Ready, was changed by statute in 1874 to the more conventional name of the nearby flour mill (spelled Aetna). *See also* Aetna Springs.

Eto (ē′ tō) **Lake** [San Luis Obispo Co.]. For Japanese-American farmer Tamaji Eto, who owned land here (Hall-Patton).

Ettawa (et′ ə wə) [Lake Co.]. Perhaps from Lake Miwok *éetaw* 'hot' (Callaghan). But another story is that the name was coined some time before 1900 by the owner of the springs, from the name of his mother, Etta Waughtel.

Ettersburg [Humboldt Co.]. In 1894 Albert F. Etter took up a homestead here, and two years later he started the Ettersburg Experimental Place, where he created a large number of fruit varieties by his unique methods. The post office is listed in 1916. The name commemorates Ettersburg, the castle on the Rhine that the Etter family left two hundred years before.

Eureka (yo͞o rē′ kə). The Greek expression for 'I have found it', associated with the great mathematician Archimedes (287?–212 B.C.E.), has common currency in Western civilization. As a geographical term it apparently originated in California: the gazetteers of the 1840s mention no such place in the United States, but in 1880 there were forty of them. The motto became popular after the California constitutional convention approved the great seal of the state on Oct. 2, 1849, with the inscription "Eureka." The expression became popular also as a trade and mining term, and is still retained in the names of about fifteen features in the state. **Eureka** [Humboldt Co.]: The settlement was established by the Union and Mendocino companies, headed by C. S. Ricks and J. T. Ryan of San Francisco, and named on May 13, 1850 (Doyle). *For* Eureka [Nevada Co.], *see* Granite: Graniteville. **Eureka Valley** [Death Valley NP]: The name has been used since 1888 and was either applied in derision or because something, perhaps water, was actually found. Lt. D. A. Lyle of the Wheeler Survey called it Termination Valley, since he could find no source of water in it.

Evelyn, next to Elizabeth and Helen, is the most popular girl's name used in California geography. **Evelyn Lake** [Sequoia NP] was named for Evelyn Clough, sister of William O. Clough, who discovered Cloughs Cave (Farquhar). **Evelyn** [Inyo Co.]: When F. M. ("Borax") Smith built the Tonopah and Tidewater Railroad in 1907, he named this station for his wife. **Evelyn Lake** [Yosemite NP] was named for a daughter of Maj. William W. Forsyth, acting superintendent of the park from 1909 to 1912 (Farquhar).

Evergreen [Santa Clara Co.]. The popular American place name was applied about 1875 to the post office, now discontinued.

Everitt Hill [Siskiyou Co.]. The elevation on the slope of Mount Shasta was named for John Samuel Everitt, supervisor of the Shasta National Forest, who lost his life in a forest fire on this spot, Aug. 25, 1934 (Schrader).

Evolution: Group, Creek, Valley, Lake, Meadow, Basin [Kings Canyon NP]. Six peaks in this group were named in July 1895 by T. S. Solomons, in honor of Charles Darwin, Thomas Huxley, Herbert Spencer, and Alfred Wallace—British exponents of the theory of evolution—and for Ernst Haeckel, the German scientist, and John Fiske, the American historian and philosopher (Farquhar). The creek was named by J. N. LeConte in Jan. 1905: "The locality . . . is drained by what is known as the Middle Branch of the South Fork of the San Joaquin River, a name clumsy and objectionable to the last degree. I therefore propose to reject it, the sheepmen to the contrary notwithstanding, and refer to it as Evolution Creek" (*SCB* 5:229). The other features were named when the USGS mapped the district, 1907–9. Two other peaks of the group have since been named in honor of distinguished scientists: one in 1912 for Jean Lamarck, the French naturalist, and the other in 1942 (by David R. Brower) for Gregor Mendel, the German-Austrian author of the Mendelian law of heredity.

Exeter [Tulare Co.]. The town was founded in 1880 by D. W. Parkhurst for the Pacific Improvement Company and was named after his home city in England (I. G. Simmons). The post office was established on June 17, 1889.

Explorers Pass [Imperial Co.]. Named by Lt. Joseph C. Ives on Jan. 12, 1858, while he was exploring the Colorado River: "below the pass, which we call after our little steamboat, the Explorer's Pass" (Ives, p. 47).

Eyese (ī′ ēz) **Creek** [Siskiyou Co.]. The name of this affluent of the Klamath is derived from that of the Karok village *áyiith* at the mouth of the creek.

F

Fagan [Butte Co.]. Named by the Southern Pacific for Edward Fagan, on whose land a siding was constructed about 1903.

Fairbank Point [San Luis Obispo Co.]. The point south of the town of Morro Bay was named in memory of Dr. and Mrs. Charles Oliver Fairbank, who had lived nearby (BGN, 1962).

Fairfax [Marin Co.]. Named for Charles Snowden Fairfax, of Fairfax Co., Virginia, popularly known as Lord Fairfax, who settled in 1856 on what was once a part of Rancho Cañada de Herrera (or Providencia).

Fairfield [Solano Co.]. In 1859 Robert H. Waterman (1808–84), a famous clipper ship captain, lived here in a house that he modeled after the prow of a ship. He gave the land for a new city, which he named after Fairfield, Connecticut, his former home.

Fairfield Peak [Lassen NP] was named for Asa M. Fairfield, a pioneer educator and historian of Lassen Co. The name appears on the atlas sheet of the USGS made in 1925–26 and in the 6th Report of the BGN.

Fair Oaks [Sacramento Co.]. The name was applied to the post office about 1895. Other combinations with the adjective "fair," a favorite in American place names, include Fairmead [Madera Co.], Fairmont [Los Angeles Co.], Fairport [Modoc Co.], Fairview [Fresno Co.], and Fairville [Sonoma Co.]

Fair Play [El Dorado Co.]. N. Sisson and Charles Staples settled here in 1853; the post office is listed in 1862. According to local tradition, the name arose from an incident in which an appeal for fair play forestalled a fight between two miners. However, Fair Play was a popular place name in the days before the Civil War; Lippincott's *Gazetteer* of 1854 lists eight towns of that name in various states.

Faith Valley. *See* Hope Valley.

Falda (fäl′ də) [San Diego Co.]. The name was applied to the station of the San Diego Central after 1890. It was probably chosen because the place is near the steep slope of Loma Alta, since in Spanish geographical terminology *falda* 'skirt' is used for the lower part of a slope.

Fales (fālz) **Hot Springs** [Mono Co.]. Samuel Fales, a native of Michigan, settled in Antelope Valley before 1868 and developed the springs in 1877 (Maule).

Falk [Humboldt Co.]. Named for the brothers Noah H. and Elijah H. Falk, of Pennsylvania German stock, who came to California in 1852 and 1878, respectively. Both were prominent lumbermen, and Elijah Falk was mayor of Eureka in 1915.

Fallbrook [San Diego Co.]. Named after Fallbrook, Pennsylvania, the former home of Charles V. Reche, who settled in the district in 1859 and became its first postmaster in 1878. After the town was washed out in the early 1880s, a new settlement developed around the railroad station, which had been named North Fallbrook by the California Southern Railroad in 1882.

Fallen Leaf Lake [El Dorado Co.]. Apparently named by the Whitney Survey because, from the surrounding heights, the outline of the lake resembles a fallen leaf. The name appears on the von Leicht–Hoffmann map of Lake Tahoe (1874).

Fallen Moon, Lake of the [El Dorado Co.]. Named by Frank Ernest Hill, who made the lake the subject of a poem, "The Lake of the Fallen Moon," published in 1923 in the *Sierra Club Bulletin* (Farquhar).

Fallon [Marin Co.]. The station on the North Pacific Coast Railroad was named in the 1890s, probably for a member of the pioneer Fallon family. Luke Fallon settled in the Tomales township in 1854; James L. Fallon settled there five years later (Co. Hist.: Marin 1880:405–6).

Fall River; Fall River Mills [Shasta Co.]. The river was so named by Frémont in 1846 because of its cascades. The name appears on the Frémont-Preuss map of 1848 and is repeatedly mentioned in the reports of the government surveys in the 1850s. Eddy's map also shows a Fall Lake. The town took the name Fall River Mills from the mills built in 1872 by W. H. Winters, although on the county map of 1884 it is shown as Fall City (Steger). **Fall River** [Butte Co.]: The stream was so named because of the falls that are near its junction with the Middle Fork of the Feather River. **Fallsvale** [San Bernardino Co.]: When the post office was established on Nov. 25, 1929, it was given an abbreviated version of the name Valley of the Falls; *see* Forest Falls. The term "falls" is sometimes locally used in northern California for what would elsewhere be called rapids (*see* Ikes Falls; Katimin).

Falsa Vela. *See* Anacapa Islands.

False. The adjective is often used, especially by navigators, to describe an unnamed geographic feature that resembles and may be mistaken for another feature already named. **False Cape** [Humboldt Co.]: It was probably this cape on which Hezeta in 1775 bestowed the name Punta Gorda 'fat point'. In 1851 it is mentioned as False Cape Mendocino, a term used by American sailors. Davidson applied the old Spanish name to the present Punta Gorda south of Cape Mendocino in 1854 and renamed the False Cape as Cape Fortunas—a name originally bestowed on Point Arena by Ferrer in 1543: "We have ventured to call this headland Cape Fortunas, to avoid the repetition of Mendocino, and to commemorate Ferrelo's [Ferrer's] discoveries" (*Coast Pilot* 1858:68). The present version, long used by navigators, was officially established by decision of the BGN in 1940. *See* Punta: Punta Gorda. **False Klamath Cove** and **Rock** [Del Norte Co.]: The cove was so named because, from a ship, it resembles the cove at the mouth of Klamath River. The rock was named by the Coast Survey after the cove, although there is no "true" Klamath Rock. *For* False Bay [San Diego Co.], *see* Mission: Mission Bay.

Famoso (fə mō′ sō) [Kern Co.]. Spanish for 'famous'. When the Southern Pacific reached the place in the 1870s, it called the station Poso after Poso Creek. When the post office was established about 1890, the Post Office Dept. applied the name Spottiswood because the old name was too much like Pozo in San Luis Obispo Co. The residents did not like the Anglo-Saxon name and requested that the Post Office Dept. change the name to Famoso.

Fandango: Valley, Creek, Mountain, Pass [Modoc Co.]. The valley was named by members of the "Wolverine Rangers," an immigrant party of 1849: "After camping one night the weather grew so terribly cold that the men had to dance to keep warm, and named their wild camping place 'Fandango Valley'" (Oliver Goldsmith, a forty-niner, quoted by Bruff 1:580–81). The often-repeated story that a band of Indians surprised and massacred a party of immigrants in the early 1850s, while they were dancing and making merry in celebration of their successful crossing of Lassen's Pass, is more romantic but less authentic. The name is recorded in early reports and on county maps, and was given currency by the Wheeler atlas, in which sheet 38-B shows Valley, Creek, and Peak.

Fanita. *See* Santee.

Fannette (fan et′) **Island** [Lake Tahoe, El Dorado Co.]. Earlier called Coquette Island, named for a visitor because of her beauty and stony heart (Lekisch).

Fannie Creek [Shasta Co.]. For Fannie Brown, a Wintu shaman (Shepherd).

Farallon (fâr′ ə lon, fâr′ ə lōn): **Islands; Gulf of the Farallones** [San Francisco Co.]. The Spanish word for 'small rocky island in the sea' is repeatedly shown on maps of the 16th and 17th centuries in different latitudes. The rocky islands off the Golden Gate were called Islands of Saint James by Drake (1579) and appear with that name on English maps of the 18th century. The Vizcaíno expedition (1602–3) referred to them as *farallones* and named them Los Frayles 'the friars'. Gonzáles referred to them as los Farallones in 1734, and Bodega in 1775 called them Farallones de los Frayles. Most of the Spanish and Mexican maps used Farallones de San Francisco, or simply Farallones, but Davidson and the Coast Survey used Bodega's Farallones de los Frayles until the 1870s. On modern maps the islands are usually designated individually as North, Middle, and South Farallon (or Farallone and Farallones).

Faria (fə rē′ ə) **Beach** [Ventura Co.]. The beach park was founded in 1915 with a gift of land from the family of Manuel Faria, a native of the Azores, who arrived here in 1908 (Ricard).

Farley [Mendocino Co.]. When the Northwestern Pacific was extended north from Ukiah in 1911, the station was named for Jackson Farley, a native of Virginia who had settled in the district about 1857.

Farmersville [Tulare Co.]. The convenient descriptive name came into being when the post office was established in the 1870s.

Farmington [San Joaquin Co.]. A common American place name applied about 1859 by W. B. Stamper. The settlement is in the center of a rich farming area.

Farnsworth Bank [Santa Catalina Island]. Named for Samuel S. Farnsworth, who built a road on the island, 1893–1902 (BGN, 1936).

Farquhar (fär′ kwär), **Mount** [Kings Canyon NP]. For Francis P. Farquhar (1887–1974), a conservationist, writer, and Sierra Club director (BGN, list 8901); *see* Glossary.

Farwell [Butte, Glenn Cos.]. Rancho Arroyo Chico, granted to Edward A. Farwell of Boston, Massachusetts, on Mar. 29, 1844, was also known as Rancho de Farwell.

Faucherie (fôsh ə rē′): **Lake, Creek** [Nevada Co.]. The lake was named for B. Faucherie, a pioneer hydraulic engineer of the 1850s and 1860s who also created French Lake and the Magenta flume.

Fauntleroy Rock [Humboldt Co.]. The rock south of Cape Mendocino was named for the Coast Survey brig *R. H. Fauntleroy*, Davidson's survey vessel in 1854–58. The brig, in turn, was named in memory of Robert H. Fauntleroy of the Coast Survey, son-in-law of the Scottish reformer Robert Owen and father-in-law of George Davidson. There is another Fauntleroy Rock in Crescent City harbor.

Fawnskin Meadows [San Bernardino Co.]. Skins were stretched on trees by a party of hunters in 1891 and allowed to remain there for many years; hence the name (V. W. Bruner). The settlement Fawnskin that developed on Highway 18, north of Big Bear Lake, received a post office on May 18, 1918.

Feather Falls [Butte Co.]. The post office was established in 1921. The name was apparently suggested by the Feather River and the well-known waterfalls of Fall River, several miles north.

Feather River [Butte, Plumas Cos.]. The river was known by the present name to the trappers of the Hudson's Bay Company in the 1830s. The term Río de las Plumas may have been used in Mexican times. However, the Spanish and the English versions probably became current because Sutter used both: "I named Feather River, 'Rio de las Plumas,' from the fact that the Indians went profusely decorated with feathers, and there were piles of feathers lying around everywhere, out of which the natives made blankets" (p. 62). Wilkes's map of 1841 shows Feather River; a *diseño* of the Arroyo Chico grant (1843) and the Frémont-Preuss map of 1845 have Río de las Plumas. Gibbes's map of 1852 still has the Spanish version, but Eddy's official map and most of the other American maps have Feather River. Jedediah Smith's name for the stream in 1828 had been Ya-loo (*Travels*, pp. 70, 176). *See also* Plumas. *For* Feather Lake [Fresno Co.], *see* Vermilion Valley.

Felis. *See* Los Feliz.

Feliz [San Mateo Co.]. This land grant, also called Cañada de las Auras, dated Apr. 30, 1844, was known by the name of the grantee, Domingo Féliz.

Feliz (fā′ liz): **Creek** [Mendocino Co.]. Named for Fernando Féliz, grantee of the Sanel rancho on Nov. 9, 1844.

Fellows [Kern Co.]. Named in 1908 for Charley A. Fellows, a building contractor, by the Sunset Western Railroad Company, which built the line now jointly owned and operated by the Santa Fe and the Southern Pacific.

Felton [Santa Cruz Co.]. The town was founded in 1878 and named for John Brooks Felton (1827–77), philanthropist, mayor of Oakland, regent of the University of California, and twice a candidate for the U.S. Senate (Clark 1986).

Fender Mountain [Shasta Co.]. Named for W. H. Fender, who had a furniture factory near Kenyon Sulphur Springs in the 1880s (Steger).

Fenner. *See* Amboy.

Fermin (fûr′ min), **Point** [Los Angeles Co.]. Vancouver, in 1793, gave this name to the western point at the entrance of San Pedro Bay, to honor his friend Padre Fermín Francisco de Lasuén. The name is spelled Firmin on some maps.

Fernandez (fər nan′ dəz) [Butte Co.]. The name of a rancho granted to the brothers Dionisio and Máximo Fernández on June 12, 1846.

Fernandez Pass [Yosemite NP] was named for Joseph Fernandez, a first sergeant of the 4th Cavalry—who, with Lt. Harry C. Benson, explored the headwaters of the Merced in 1895–97 (Farquhar).

Ferndale [Humboldt Co.]. The town was founded about 1870 and appropriately named because of the luxuriant growth of ferns in the valley. Other communities bear similar names: Fern [Shasta Co.], Fernbridge [Humboldt Co.], Fern Cove [Sonoma Co.], and Fern Canyon [Marin Co.]. Some fifteen Fern Creeks, Canyons, Hills, and Mountains are shown on topographical atlas sheets, and many more such names are used locally.

Ferris Canyon [Mono Co.]. Named for Andrew Ferris, who obtained a patent to land near Sweetwater (Maule).

Ferrum (fâr′ əm) [Riverside Co.]. The Latin word for 'iron' was applied to the railroad station in 1948, when Henry J. Kaiser built a branch line to haul iron ore from Eagle Mountains to the steel mill at Fontana (John Hilton, *Desert Magazine*, Mar. 1949, p. 7).

Fetters Hot Springs [Sonoma Co.]. George Fetters, a native of Pittsburgh, Pennsylvania, struck flowing hot water when prospecting for mineral water here, on land which he had bought in 1907 (Co. Hist.: Sonoma 1911: 618–19).

Fiddler Creek [Shasta Co.]. Named for Capt. Hercules Fiddler, who settled here in the 1860s (BGN, list 8201); earlier spelled Fidler.

Fiddletown [Amador Co.]. "In the days of the Gold Rush the mining settlement got its name Fiddletown because a large number of miners were from Missouri and had brought their fiddles with them. It was said to be a common sight to see one miner working on the claim while his partner played the fiddle for their enjoyment" (Doyle). According to another version, an old lady claimed that her family were the first settlers at the place and called it Violin City because four members of the family played the violin (*CHSQ* 24:167). On May 24, 1878, the name of the post office was changed to Oleta, but on July 1, 1932, the old name was restored. There is a Fiddle Creek [Sierra Co.] and a Fiddler Gulch [Kern Co.]. A Violin Canyon runs parallel to Highway 99 in Los Angeles Co.

Fieldbrook [Humboldt Co.]. The place was originally known as Bokman's Prairie. When the town was laid out in 1900, the name was changed to Fieldbrook because there were brooks east and west of it (Regina Schneitter).

Fields Landing [Humboldt Co.]. The post office was named in 1889, having previously been called Adele (Turner). The name is mentioned in the business directory of the county in 1890 and appears on the Mining Bureau map of 1891 as a station of the Eel River and Eureka Railroad. The station was named for Waterman Field, who had been a resident of the county since the 1860s.

Fifield (fī′ fēld): **Ridge, Valley** [San Mateo Co.]. Albert B. and Winfield J. Fifield settled here in 1859 (Alan K. Brown).

Figarden [Fresno Co.]. Named for the large fig orchards there. The name appeared originally on the Santa Fe Railroad map of 1921 as Fig Garden. On succeeding maps it has its present form.

Fig Tree John Springs [Riverside Co.]. The springs were so named because Juan Razón, a Cahuilla Indian popularly known as "Fig Tree John," lived near them. The place was submerged by the Salton Sea in 1907. The springs reappeared when the water subsided, but John had moved to a safer place, Agua Dulce.

Figueroa (fig ə rō′ ə) **Mountain** [Santa Barbara Co.]. The mountain was named for a member of the well-known southern California family, the most eminent of whom was José Figueroa, governor of California in 1833–35.

Fillmore [Ventura Co.]. Named by the Southern Pacific in 1887 for J. A. Fillmore, general superintendent of the company's Pacific system.

Fin Dome [Kings Canyon NP] was named in 1899 by Bolton C. Brown. He compared the ridge above Rae Lake to a sea serpent and on his sketch map marked "The Head," "The Fin," "The Tail" (*SCB* 3:136).

Finger. About ten mountains and rocks, including the high peaks in Kings Canyon NP and Yosemite NP, are so named because of their shape. The name also occurs as a generic term: Columbia Finger [Yosemite NP]. However, Finger Creek [San Mateo Co.] was named for Theodor Finger, who settled here in 1855 (Alan K. Brown).

Finley [Lake Co.]. The town was started by John Syler on Apr. 20, 1889. From a list of names he submitted, the Post Office Dept. selected the name Finley for the post office established on Oct. 25, 1907. It was the middle name of his father, Samuel Finley Syler (Mauldin).

Finney Creek [San Mateo, Santa Cruz Cos.]. Named for Seldon J. Finney, state assemblyman 1869–70 and state senator 1871–74, whose home was on the creek.

Fir. Although true firs (genus *Abies*) grow in the state and the Douglas fir (*Pseudotsuga taxifolia*) is especially common, this word is seldom found in place names. **Fir Top Mountain** [Sierra Co.] has a bare appearance except at its rather flat top, which is covered with trees, presumably firs (Stewart).

Firebaugh (fī′ ər bô) [Fresno Co.]. In 1854 A. D. Fierbaugh established a trading post and a ferry across the San Joaquin River. The name was misspelled as Firebaugh's Ferry as early as 1856, and this version was used when the place became a station on the Gilroy-Sageland stage line in the 1860s.

Firmin, Point. *See* Fermin, Point.

First Garrote. *See* Groveland; Second Garrote.

Fiscalini (fis kə lē′ nē) **Creek** [San Luis Obispo Co.]. For the family of Joseph Fiscalini, who came here from Switzerland in 1876 (Hall-Patton).

Fish. There are some twenty-five Fish Creeks, Valleys, and Canyons in California, a small number for a state in which fish are abundant. The word appears occasionally in the names of mountains: Fish Valley Peak [Mono Co.], Fish Spring Hill [Inyo Co.], Fish Creek Mountain [Madera Co.], and in several settlements, Fish Rock [Mendocino Co.], Fish Camp [Mariposa Co.], Fish Spring [Imperial Co.]. Fish Creek and Fish Creek Mountains in Anza-Borrego Desert State Park [San Diego Co.] were probably named for the small desert pupfish that once lived in the lower reaches of the creek (Brigandi). *For* Fish Creek Mountain [San Bernardino Co.], *see* Grinnell: Grinnell Peak. *For* Fishpond [San Bernardino Co.], *see* Barstow.

Fishermen. About twenty geographic features, mainly Bays and Coves along the coast, or Canyons and Creeks in the mountains, were so named because they were favorite fishing places. Mount Whitney was once known as Fisherman's Peak.

Fiske, Mount [Kings Canyon NP]. For John Fiske (1842–1901), American historian and writer on evolution (Browning 1986); *see* Evolution.

Fitch Mountain [Sonoma Co.]. The mountain was known to Indians and earlier settlers as Sotoyome and later was named for Henry Delano Fitch (1799–1849) of San Diego, who in 1841 was granted the Rancho Sotoyome, on which the mountain is situated. Fitch's romantic elopement with Josefa Carrillo in 1829 is famous in California literature.

Fitzhugh Creek [Modoc Co.]. Named for the Fitzhugh family, pioneer settlers of the 1870s (Irma Laird).

Five-and-Ten Divide [Siskiyou Co.]. The ridge west of Happy Camp forms the divide between Fivemile Creek and Tenmile Creek (BGN, July–Sept. 1965).

Five-Finger Falls [Santa Cruz Co.]. For the five-finger fern, also called western maidenhair fern, *Adiantum pedatum* (Clark 1986).

Five Points [Fresno Co.]. The post office was established on July 28, 1944, and so named because five roads converge at this point.

Flatiron. About ten Hills, Buttes, and Ridges are so named because of their shape. The best-known features are the Ridges in Mono Co. and in Lassen NP.

Flea Valley; Flea Valley Creek [Butte Co.]. Apparently the only place in the state with the American name of this obnoxious insect; at any rate, it is the only survivor from earlier days. Flea Valley House is shown on the Mining Bureau map of 1891. The nearby Western Pacific station and post office, established about 1907, bear the Spanish word *pulga* 'flea' (*see* Pulgas).

Fleener Chimneys [Lava Beds NM]. Named for Sam Fleener, a homesteader in the vicinity (BGN, Dec. 1948). A post office named Fleener is listed between 1889 and 1893.

Fleischmann Lake [Plumas Co.]. Named for Maj. Max C. Fleischmann in appreciation of his support of the Boy Scout movement in America (BGN, May 1954).

Fleming Point [Alameda Co.]. The site of the racetrack Golden Gate Fields was named for J. J. Fleming, who settled in the vicinity in 1853 (Ferrier, p. 116). His name is shown on Hoffmann's map of the Bay region.

Fletcher Lake and **Creek** [Yosemite NP] were named in 1895 by Lt. N. F. McClure, for Arthur G. Fletcher, a state fish commissioner, who was instrumental in stocking the streams of Yosemite.

Flinn Springs [San Diego Co.]. Possibly named for W. E. Flinn, a native of North Carolina, who was a farmer at Ranchita in 1866.

Flint. The name has been given to several geographic features where obsidian deposits (rather than true flint) provided the Indians with arrowheads. However, the origin of the name of the most prominent feature, **Flint Rock Head** [Del Norte Co.], is shrouded in mystery. The name was applied by the Coast

Survey and is apparently an interpretation of the Yurok *okne'get* 'where they get arrow points'. When Waterman pointed out to the Yurok that they could not possibly get arrowheads there, they told him that some spirit had "tried" to change the granite into obsidian for their benefit and that he "almost did it" (Waterman, pp. 232–33, 272). *See* Pedernales.

Flintridge [Los Angeles Co.]. When part of Rancho la Cañada was subdivided in 1920, the tract was named by Frank P. Flint, U.S. senator from California, 1905–11. In 1979 a merger with La Cañada produced the combined name La Cañada Flintridge.

Florence Lake [Fresno Co.]. Named in 1896 for Florence Evelyn Starr by her brother, Walter A. Starr—who, with Allen L. Chickering, camped here on a trip through the High Sierra from Yosemite to Kings River Canyon. The lake, enlarged by a dam, is now a reservoir of the Southern California Edison Company's system (W. A. Starr). **Florence, Mount; Florence Creek** [Yosemite NP]: The mountain was named at the suggestion of B. F. Taylor in 1886, in memory of Florence, a daughter of James M. Hutchings. She was born in Yosemite Valley on Aug. 23, 1864, and died there on Sept. 26, 1881 (Farquhar). There is a Florence Peak of almost the same height in Sequoia NP.

Flores (flôr′ əs). The word for 'flowers' was used repeatedly for place names in Spanish times and appears in the titles of a number of land grants and claims. **Las Flores** [San Diego Co.]: Font (*Compl. diary*, p. 189) mentions, on Jan. 9, 1776, a lake called Las Flores. This was probably in the valley that the Portolá expedition had called Santa Praxedis de los Rosales ('of the rose gardens') because of "innumerable Castilian rosebushes and other flowers" (Crespí, p. 133). Las Flores, apparently the name of a rancheria, is recorded in 1778 (PSP 2:1). Later, an *asistencia* of Mission San Luis Rey and an experimental Indian pueblo were established at Las Flores. On Oct. 8, 1844, the land was purchased by the Pico family, and their Rancho Santa Margarita y San Onofre became the Rancho Santa Margarita y las Flores (Brigandi). In 1846–47, Emory (*Notes of a military reconnoissance*, p. 117) mentions Las Flores as "a deserted mission." About 1888 the name was given to the station of the Santa Fe. **Las Flores** [Tehama Co.] was laid out in 1916 and named after the Las Flores land grant (of Dec. 24, 1844), on which it is situated. The name is also preserved in Los Angeles and Santa Barbara Cos. In Orange Co. there is a Miraflores, intended to mean 'floral view'.

Flores Peak [Orange Co.]. The name of the peak in Santiago Canyon commemorates the Mexican outlaw Juan Flores, part of whose band was captured here in 1857. *See* Barton Mound; Sheriffs Springs.

Florin [Sacramento Co.]. The name was given to the station by E. B. Crocker of the Central Pacific Railroad when the line was constructed in 1863–69. The florin is a silver coin long used in England (a two-shilling piece); but the coin may not be the source of the name. According to the County History (1880:235), the name was applied because of the large number of flowers in the region.

Floriston [Nevada Co.]. The railroad station established in the 1870s bore the name Bronco, after the creek a mile to the south. When the post office was established in 1891, the new name was chosen. It is probably coined from the Latin or Spanish stem *flor* 'flower'. There are other fanciful names on the Truckee-Verdi section: Polaris, Boca, Iceland, Mystic, Calvada.

Flournoy (flûr′ noi) [Tehama Co.]. The post office, listed in 1910, was named for the Flournoy family, farmers in the Henleyville district since the 1870s.

Flugge, Rancho de. *See* Boga.

Flying Dutchman Creek [Kern Co.]. The name appears on the Breckenridge Mountain atlas sheet of the Corps of Engineers for a tributary to Havilah Canyon Creek. It is not known whether it was named for the fabled Dutch mariner, or for Wagner's opera, or for some fleet-footed German.

Foerster (fôr′ stər): **Peak, Creek** [Yosemite NP] was named in 1895 by Lt. N. F. McClure, for Lewis Foerster (1868–1936), a soldier and mountaineer, and a native of Germany: "His service was outstanding, and it was in recognition of his achievements and because of his close association with the particular region that I gave his name to a prominent peak" (*SCB* 22:102).

Folger Peak [Alpine Co.]. Named either for Robert M. Folger, proprietor of the *Alpine Chronicle*, founded in 1864, the first newspaper on the eastern slope of the Sierra, or for his brother, Alexander M. Folger, first postmaster at Markleeville (Maule).

Folsom (fōl′ səm) [Sacramento Co.]. The town was laid out in 1855 by Theodore D. Judah as the temporary terminus of the Sacramento Valley Railroad, the first in California. It was named for Captain Joseph L. Folsom, who came to California as assistant quartermaster of Stevenson's New York Volunteers and bought Rancho Rio de los Americanos, part of the vast Leidesdorff estate, on which the railroad station was built. The settlement of 1849 had been called Negro Bar because the first placer miners there were African Americans, according to the County History (1880:222). This, however, was another camp near Folsom.

Fontana (fon tan′ ə) [San Bernardino Co.]. A town was laid out here by the Semi-Tropic Land and Water Company in the "boom year" 1887, but it failed to develop. Before 1905 the Fontana Development Company bought the interests, and a new town was started by A. B. Miller, under the name of Rosena. It is uncertain whether the company took its name from a family or from the Spanish poetic word for 'fountain'. In 1913 the name Rosena was changed to Fontana in a solemn ceremony, Judge B. F. Bledsoe presiding and Mrs. E. B. Miller, mother of the "father of Fontana," officiating (Santa Fe).

Fontanillis Lake [El Dorado Co.]. Earlier spelled Fontinalis; apparently a fanciful derivation from 'fountain' (Lekisch).

Fonts Point [San Diego Co.]. This had been called High Point; but in 1940 local residents decided to name it for Padre Pedro Font, diarist of the 2d Anza Expedition (1775–76), who visited this point in Borrego Badlands (Brigandi).

Foolish Lake [Fresno Co.]. Named in 1948 by a DFG survey party. Scott M. Soule, the biologist who named Crazy Lake (*q.v.*), continued in the same vein with this lake, stating that "it would be foolish for anyone ever to revisit it" (Browning 1986).

Forbestown [Butte Co.]. Named for Ben F. Forbes, a native of Wisconsin, who established a general merchandise store at this place in 1850. Mentioned in the *Statutes* of 1853 (p. 313). *For* Forbestown [Lake Co.], *see* Lake: Lakeport.

Fordyce Lake [Nevada Co.]. The reservoir was built in 1873–75 by the South Yuba Canal Company and was named for the hydraulic engineer who began building flumes and canals in the region in 1853. Jerome Fordyce of Fordyce Valley is listed in a directory of 1867.

Forest. This word—usually combined with City, Grove, Hill, Home, Knoll, etc.—is an extremely popular place name in the United States. California has several places so named, the oldest of which is probably **Forest Hill** [Placer Co.], settled in the early 1850s and so named because of the dense pine forest surrounding the place. **Forest** [Sierra Co.]: In 1862, William Hughes of Forest City, answering Bancroft's questionnaire, stated that the town was settled in 1852 and that it was given the name Forest City by a woman (he crossed out 'lady') named Mrs. Moody (Bancroft Scrapbooks 13:73). Another source states that the name came into use when a Mrs. Forest Mooney "dated" her newspaper articles "Forest City" (Theodore H. Hittell 3:100). The name of the post office was abbreviated to Forest on Jan. 17, 1895. The names of several other post offices include the word: Forest Ranch [Butte Co.], June 7, 1878; Forest Home [San Bernardino Co.], May 1, 1906; Forest Knolls [Marin Co.], Mar. 22, 1916; and Forest Glen [Trinity Co.], Mar. 9, 1920. **Forest Falls** [San Bernardino Co.] was a post office established July 1, 1960; the name is a combination of two discontinued post offices, Fallsvale and Forest Home. **Foresta** settlement and **Falls** [Yosemite NP] were established as a summer resort in 1913, with the name presumably coined from "forest" (Browning 1988). *For* Forest Hill Divide [Placer Co.], *see* Divide, The. *For* Forestville [Humboldt Co.], *see* Scotia.

Forester Pass [Sequoia NP]. The pass was discovered on Aug. 18, 1929, by a party that included S. B. Show, Jesse W. Nelson, and Frank Cunningham, all of the USFS. At the suggestion of Cunningham, then supervisor of Sequoia National Forest, the new route to the headwaters of Kern River was named Forester Pass for the men who discovered it. The name was approved by the BGN on Jan. 24, 1938 (W. I. Hutchinson).

Forestville [Sonoma Co.]. This place at the edge of the timber country was originally named Forrestville for its founder, A. J. Forrester. By the end of the 1880s the present spelling was generally used. Andrew Jackson Forrester, a native of Illinois and saloonkeeper at Russian River, was a registered voter in 1866.

Fork. The practice of designating as "forks" the branches or important tributaries of a river system seems to have prevailed in California more frequently than in other parts of the United States and is especially noticeable on the west slope of the Sierra Nevada. Stewart, pp. 266–67, has explained this predilection for "forking" rivers to the rapidity of settlement after the discovery of gold and to the circumstance that the gold seekers (and later the surveyors) proceeded upstream. Most of the specific names that modify the term "fork" indicate the relative location: north, east, middle; but some are descriptive or are the names of other features: the Triple Peak Fork of the Merced, the Wheatfield Fork of the Gualala, the Fish Fork of the San Gabriel. Others are names of men: the Stewart Fork of the Trinity, the Dana Fork of the Tuolumne, the Lewis Fork of the Fresno. In many places the practice is carried to the second degree, e.g. the North Fork of the North Fork of the Navarro River, the Silver Fork of the South Fork of the American River, the East Fork of the West Fork of the Mojave River. The Yuba River actually has an East, a Middle, a North, and a South Fork of the North Fork. In a few places the "forking" was carried to a third degree and has led to amusing monstrosities such as the East Branch of the East Fork of the North Fork of the Trinity River, or the West Fork of the South Fork of the North Fork of the San Joaquin River. Although such combinations are never used locally, the streams have been so designated by the USGS. A more sensible naming is found in a few places where the term "fork" is avoided: East and West Walker River, Natchez and Rocky Honcut Creek.

Forks, The. *See* Downieville.

Forks of Salmon [Siskiyou Co.]. So named when the post office was established in 1859, because of its situation at the confluence of the North and South Forks of the Salmon River. A place called Salmon is shown on Eddy's map (1854), not here but at the junction of the East Fork and the main stream.

Forks-of-the-Roads. *See* Alta: Altaville.

Forlorn Hope. *See* Hopeton.

Forrestville. *See* Forestville.

Forsythe Creek [Mendocino Co.]. Named for Benjamin Franklin Forsythe, a native of Pennsylvania, who settled in Calpella township in 1857.

Forsyth Peak [Yosemite NP]. Named for Col. William W. Forsyth (1856–1933), U.S. Army, who was acting superintendent of the national park, 1909–12.

Fort. Military reservations are listed under their specific names.

Fort. The word is found not only as a generic term in the names of military reservations and as part of the names of communities that developed around old forts, but also as a descriptive term in orographic features. **Fort Mountain** [Calaveras Co.]: Whitney referred to the mountain near Railroad Flat as Fort Hill and said that it was so called because of its "fancied resemblance to a fortress" (*Geology*, p. 266). It was Phil Schumacher of the Whitney Survey who named the mountain. In a letter datelined Railroad Flat, June 29, 1872, he states that the top is "so even and horizontal that it will give room to measure a base-line" (Davidson papers). **Fort Mountain** [Shasta Co.] was so named because it seems to stand guard over the site of old Fort Crook (Steger).

Fort Benson [San Bernardino Co.]. Jerome Benson, a former Mormon, joined by other settlers, fortified his house when the court decided that he had settled (1856) within the boundaries of the San Bernardino Rancho, which belonged to the Mormons.

Fort Bidwell. *See* Bidwell.

Fort Bragg [Mendocino Co.]. When the town was founded in 1885, it was named after the military post that had been established in 1857 by Lt. H. G. Gibson and named in honor of Lt. Col. Braxton Bragg, a hero of the Mexican War. In the Civil War battles of Murfreesboro and Chickamauga, Bragg (1817–76), then a general, was the Confederate opponent of Rosecrans, whose name is commemorated in Fort Rosecrans [San Diego Co.].

Fort Crook [Shasta Co.]. The old fort was named in 1857 for Lt. George Crook, who later served with distinction as a general in the Civil War.

Fort Dick [Del Norte Co.]. Fort Dick Landing is first mentioned in the Civil War records. The "fort" was a log house, built by the citizens for defense against the Indians, and was probably named for a settler. In 1888 the four brothers Bertsch moved their shake-and-shingle mill to the place and called it Newburg. With the establishment of the post office in 1896, the old name was revived.

Fort Jones [Siskiyou Co.]. The fort was established on Oct. 16, 1852, by Companies A and E, 1st Dragoons, and named for Col. Roger Jones, adjutant general of the army (F. B. Rogers). The garrison evacuated the post on June 23, 1858, but in 1860 the settlers of Wheelock renamed their settlement Fort Jones, in gratitude for the military protection they had received. On Nov. 19, 1860, the name was also adopted for the post office, until then called Ottitiewa for the Scott Valley branch of the Shasta Indians. On Goddard's map, 1857, Fort Jones is shown, but not Ottitiewa.

Fort Point [San Francisco Co.]. The name was applied to the point by the Coast Survey in 1811 because the ruins of an old Spanish fort—the Castillo de San Joaquin—were still to be seen on it. Ayala had called it Punta de San José in Aug. 1775. On Cañizares's Plano of 1776 it is shown as Punta de San Josef o Cantil Blanco ('white cliff'). It is the site of modern Fort Winfield Scott.

Fort Reading. *See* Reading Peak.

Fort Ross [Sonoma Co.]. The fort was the nucleus of Russian activities in California, 1812–41. The name Ross is an obsolete, poetic name for 'Russian'; it was selected from lots placed at the base of an image of Christ bestowed on the settlement when it was dedicated on Sept. 11, 1812 (*CHSQ* 12:192). The fort was sometimes called by the Spanish Presidio Ruso or Presidio de Bodega (*see* Presidio). A post office was established and named Fort Ross on May 23, 1877.

Fort Seward [Humboldt Co.]. The town takes its name from the military post that was established on Sept. 26, 1861, and named for William H. Seward (1801–72), Lincoln's secretary of state. The post office is listed in 1913.

Fort Sutter. *See* Sutter.

Fort Tejon. *See* Tejon.

Fortuna (fôr to͞o′ nə) [Humboldt Co.]. Latin for 'fortune'. The place was opened for settlement in the late 1870s by a minister named Gardner, who owned the land. He named it Fortune because he believed it was an ideal place in which to live. Later, for the sake of euphony, he changed its appellation to Fortuna, at that time a unique name in the United States. Names that did not survive were Springville, for the many springs nearby, and Slide, for the landslide northwest of the town (Marie Linser).

Fortunas, Cabo de; Fortunas, Cape. *See* Arena, Point; False: False Cape.

Fort Yuma. *See* Yuma.

Foss Lake [San Diego Co.]. Probably named for David R. Foss of New Hampshire, who was a merchant in San Luis Rey in 1871.

Foster City [San Mateo Co.]. Established in 1965 and named for T. Jack Foster, who deeded a large parcel of real estate to the county.

Foster Lake [Trinity Co.]. Named for William Foster (1869–1950), because he had assisted in maintaining sport fishing in the area (BGN, May 1954).

Four Brothers. *See* Ship.

Four Creeks. *See* Kaweah River.

Four Gables [Fresno, Inyo Cos.]. Probably named by the USGS when the Mount Goddard quadrangle was surveyed, 1907–9, because of the four gable-like formations, two extending northward and one each eastward and westward. The Seven Gables are 4 miles west.

Fourth Crossing [Calaveras Co.] was the place where travelers forded the fourth fork of the Calaveras River on the road between Stockton and Murphys Diggings. A post office of this name was established on June 2, 1855, and again on Feb. 15, 1892. The present settlement is not on the Calaveras River but east of it, on San Antonio Creek. There was also a Second and a Third Crossing along the same route.

Fouts Springs [Colusa Co.]. Named for John F. Fouts, who discovered the springs in 1873.

Fowler [Fresno Co.]. Named for Thomas Fowler, state senator from Fresno, 1869–72 and 1877–78. The name Fowler's Switch was applied to the railroad switch that was built on the Fowler ranch in 1872 and around which the present town developed.

Foxen Canyon [Santa Barbara Co.]. The canyon that cuts through the Tinaquaic rancho preserves the name of the grantee, Benjamin ("Don Julian") Foxen, an English sailor who came to California in 1828.

Fox Meadow [Kings Canyon NP]. For John Fox, for many years a hunter, packer, and guide in the Kings River region (Farquhar).

Frances. *See* Irvine.

Frances, Mount. *See* Liberty Cap.

Frances, River. *See* Kaweah River.

Francisca. *See* Benicia.

Francisco Draco, Puerto Sir. *See* Drakes Bay.

Franklin [Sacramento Co.]. Named after the Franklin House, built by Andrew George in 1856. Until 1887 the place was also known as

Georgetown, although the post office name has been Franklin since 1862.

Franklin Canyon [Contra Costa Co.]. Named for Edward Franklin, who bought Vicente Martinez's share of the Ignacio Martinez estate on Oct. 6, 1853, and lived in the canyon until 1875.

Franklin City. *See* Schilling.

Franklin K. Lane Grove [Humboldt Co.]. Dedicated in 1924 and named in honor of the secretary of the interior in Woodrow Wilson's cabinet; he was the first president of the Save-the-Redwoods League.

Franklin Lakes and **Pass** [Sequoia NP]. The name for the two lakes was "derived from the 'Lady Franklin' mine located in this vicinity in the seventies" (BGN, 6th Report).

Franklin Point [San Mateo Co.]. The cape appears as Middle Point on the early charts of the Coast Survey. It was changed after the ship *Sir John Franklin* was wrecked there on Jan. 17, 1865. *For* a probable early name, *see* Ano Nuevo, Point.

Frayles, Los. *See* Farallon Islands.

Frazier: Mountain [Ventura Co.]; **Game Refuge, Park** [Kern Co.]. The mountain was named for William T. Frazer, who mined in this area from 1852 (Ricard). The town was named after the mountain by Harry McBain in 1926. This family name has several alternative spellings.

Frazier's Station. *See* Carlsbad.

Freaner (frē′ nər) **Peak** [Shasta Co.]. The name of Stoney Peak was officially changed to Freaner Peak by the BGN in 1947, at the instigation of the Shasta Historical Society. Col. James L. Freaner undertook to build a road between the Upper Sacramento and Oregon in 1852. In July of the same year, he was killed by Indians while on his way from Yreka to the Democratic state convention at Benicia. In 1854 Beckwith mentions a "lake [now Cornez Lake] called Freaner, the name of an unfortunate gentleman who is supposed to have been killed by the Indians in its vicinity" (Pac. R.R. *Reports* 2:2.57).

Fredericksburg [Alpine Co.]. The town was started in 1864 and may have been named for Frederick Frevert, who operated a sawmill nearby (Maule). In view of the strong Confederate sentiment in this region, the Union defeat at Fredericksburg, Virginia, in Dec. 1862 may have had something to do with the naming.

Fredonyer (frə don′ yər): **Pass, Peak** [Lassen Co.]. Named for Dr. Atlas Fredonyer, who discovered the pass in 1852 and built the first house in Mountain Meadows.

Freedom [Santa Cruz Co.]. Originally known as Whiskey Hill, doubtless because of the quenchless thirst of the residents, served by eleven saloons in 1852. In 1892 or 1893 an enterprising purveyor of alcoholic beverages had a huge sign put across the front of his place of business, sporting two American flags and the legend "The Flag of Freedom." The place became known as Freedom, a name that was finally applied to the entire town (J. E. Gardner, "The flag of freedom," *California Folklore Quarterly* 5:199).

Freel Peak [Alpine, El Dorado Cos.]. When William Eimbeck of the Coast Survey used the highest point of Jobs Peaks for triangulation in 1874, he called it Freel's Peak, for James Freel, a settler at the foot of the mountain. *See* Jobs Peak.

Freeman Island [San Pedro Bay, Los Angeles Co.]. For astronaut Theodore C. Freeman, 1930–64 (BGN, list 7902).

Freeport [Sacramento Co.]. The place was established as a shipping center in 1862 by the Freeport Railroad Company. It was really a "free port" compared with Sacramento, which at that time levied a tax on all transit across the levee.

Freestone [Sonoma Co.]. The town developed around Ferdinand Harbordt's store and was so named because of a quarry of easily worked, free sandstone. The post office is listed in 1870.

Freezeout: Creek, Gulch, Flat [Sonoma Co.]. These places near Duncans Mills are the only survivors of a number of early-day Freezeouts. They were probably all named after the card game so called, in which a player drops out as soon as his capital is exhausted. *See* Ione.

Fremont (frē′ mont). One of the most spectacular and most controversial characters in American history, John C. Frémont is honored in more than a hundred place names throughout the West. In California the name Fremont was especially popular soon after the Mexican War, but its popularity did not last: Fremont Co. became Yolo Co.; Fremont Canyon is now Sierra Canyon; the peak in San Benito Co. where Frémont raised the Stars and Stripes on Mar. 6, 1846, is officially Gabilan Peak; the town Fremont in Yolo Co., founded by Jonas Spect on Mar. 21, 1849, vanished soon after the post office was discontinued in 1864; Fremont Pass, north of Los Angeles, was replaced by San Fernando Pass. At the end of the 19th century the name was remembered in California geography only by a moderately high peak and a railroad siding in San Bernardino Co., but current maps show the following features: **Fremont Peak State Park** [San Benito Co.] was created in 1934, and the former Gabilan Peak was named Fremont Peak by the BGN in 1960. Frémont did not raise the flag on the peak in 1846, but on Hill 2146 at the head of Steinbach Canyon, 2 miles distant, and not within the park area (Rogers). **Fremont** [Alameda Co.]: The towns of Centerville, Niles, Irvington, Mission San Jose, and Warm Springs—all listed separately in this book—united to form this city, incorporated on Jan. 24, 1956. The name was selected by the Incorporation Committee. There is a Fremont Ford in Merced Co.

Fremont Canyon [Orange Co.]. Named for Fremont Smith, a stock herder of the 1870s.

French; Frenchman. About seventy-five places in the state have names that include the word French. Most of these were so named because of the presence of French prospectors or settlers; some, because French-Canadian trappers of the Hudson's Bay Company were active in those parts; others, doubtless for other foreigners who were thought to be French; and still others, probably for persons surnamed French. **French Camp** [San Joaquin Co.] was the southernmost regular camp site of the trappers after La Framboise established his headquarters here in 1832. By the Spanish Californians it was called Campo de los Franceses, a name preserved in the land grant of Jan. 13, 1844. When Charles M. Weber had the townsite plotted in 1850, he called it Castoria (from Latin *castor* 'beaver'); but popular usage retained the old name, French Camp, and it was applied to the post office in 1859. **French Corral** [Nevada Co.]: A stock pen built by a French settler provided the name for the town that developed after a rich gold placer was discovered in the vicinity in 1849. **French Gulch** [Shasta Co.]: The post office was established in 1856 and named after the gulch where French miners had prospected in 1849 or 1850. **Frenchmans Lake** [Santa Clara Co.] was named for Peter Coutts (Paulin Caperon), a colorful Frenchman who settled near it in the 1870s (Hoover, pp. 539 ff.). *For* French Bar [Stanislaus Co.], *see* La Grange.

Freshwater Lagoon [Humboldt Co.]. The Yurok name for the lagoon south of Orick was *pe'gwi*, which means, according to Waterman (p. 264), 'freshwater lagoon'. Men came from the Indian village at the site of Orick to bathe here after a sweat bath. **Freshwater: Slough, Creek, Corners**, town [Humboldt Co.]: The name is derived from the "fresh water" in the slough as compared with the salt water in Eureka Slough. The name occurs elsewhere in the state. Suisun Bay was sometimes called Freshwater Bay on early maps, after the Spanish Boca del Puerto Dulce (Font, *Compl. diary*, p. 369).

Fresno (frez´ nō): **River, Flats, Dome, Grove of Big Trees** [Madera Co.]; **Fresno: County,** city. Fresno is Spanish for 'ash tree' and was doubtless applied because the Oregon ash, *Fraxinus oregona*, was native here. Goddard's map shows Ash Slough north of the river. Río Fresno is mentioned in the *San Francisco Alta California* of Apr. 2, 1851, and the name appears as Fresno River on Tassin's map of the same year. The stream is shown as Fresno Creek on Gibbes's map of 1852 and is mentioned as Frezno River in the Indian Report. This phonetic spelling was also used when the county was created and named on Apr. 19, 1856 (*Statutes*, p. 183). Before 1860 an attempt was made to establish a Fresno City at the site of the present station of Tranquility (Goddard's and Hoffmann's maps). When the Central Pacific reached the site of the present city on May 28, 1872, the name was applied to the station. In 1874 Millerton, county seat and all, moved to the new station. In Mexican times the name was used in other regions: Bear Creek [Shasta Co.] is shown as Arroyo de los Fresnos on a *diseño* of the Breisgau grant (1844). It is not impossible that *el monte redondo* mentioned by Sebastián Rodríguez on Apr. 23, 1828, when he was in the vicinity of the present city of Fresno, was a round grove of *fresnos* and may have been the origin of the name. *See* Roeding Park.

Friant (frī´ ənt): town, **Dam** [Fresno Co.]. The town is situated at the landing of the old Converse (later Jones) Ferry, established in 1852 by Charles Converse. When the place became the terminus of the Southern Pacific branch from Fresno in 1891, the station was named

Pollasky for Marcus Pollasky, an agent of the railroad and its promoters. In the early 1920s the town was renamed for Thomas Friant of the White-Friant Lumber Company.

Frijoles, Arroyo de los. *See* Bean Hollow; Lucerne: Lake Lucerne.

Front. *See* Mount: Mount Owen.

Frontera (frun târ′ ə) [San Bernardino Co.]. The site of the state prison for women is named with the Spanish word for 'border, boundary'.

Fruitland [Humboldt Co.]. The name of a Dutch colony sponsored by David P. Cutten. The post office is listed in 1891.

Fruto [Glenn Co.]. The Spanish word for 'fruit' or 'produce' was given to the post office about 1890.

Frys Point [Sequoia NP] was named by R. B. Marshall in 1909 for Walter Fry, a native of Illinois. Fry worked for a lumber company in the 1890s, revolted at cutting down more of the big trees, entered the government service, and was superintendent of Sequoia and General Grant National Parks, 1914–20 (Farquhar).

Fuegos, Bahía de los. *See* San Pedro: Bay.

Fuller: Canyon, settlement [Lake Co.]. Named when the family connected with the well-known W. P. Fuller paint company settled in the vicinity about 1905 (Mauldin).

Fuller Lake [Nevada Co.]. The reservoir was created before 1871 by the Meager Mining Company and, like other lakes of the company, was probably named for one of its officials.

Fullerton [Orange Co.] was founded in 1887 by Edward and George Amerigue, the Wilshire brothers, and the Pacific Land and Improvement Company. George H. Fullerton, president of the land company and "right-of-way man" of the California Central (Santa Fe), routed the new railroad through the property of the Amerigues and received in return an interest in the townsite and the honor of having the new city named for him.

Fulton [Sonoma Co.]. Founded and named in 1871 by Thomas and James Fulton, natives of Indiana, who were residents of Santa Rosa in 1867.

Fulton Sulphur Springs [Los Angeles Co.]. *See* Santa Fe: Santa Fe Springs.

Fumos, Bahía de los. *See* San Pedro: Bay.

Funeral Mountains and **Peak** [Death Valley NP]. The name has been applied both to the Black Mountains and to the Amargosa Range. Chalfant (*Death Valley* 1936:49) quotes a passage from R. H. Stretch's diary, published in the *Virginia Enterprise* (1866): "The name of the mountains is probably due to their peculiar appearance. They are principally light-colored rocks, but are frequently capped with heavy masses of black limestone or basalt, the débris of which, running down the slopes, gives them the appearance of being fringed with crape, a species of natural mourning." There is no evidence for the story that the mountains were named by J. B. Colton of the "Jayhawkers" because four members of the party died there.

Funks Creek [Colusa Co.]. The slough was named for John Funk, who owned a large parcel of land there in the 1850s.

Funston, Fort [San Francisco Co.]. Named by the War Dept. in 1917, in honor of Maj. Gen. Frederick Funston (1865–1917), a hero of the Spanish-American War, who had been in command of the Dept. of California at the time of the San Francisco earthquake of 1906. It had previously been known as Laguna Merced Military Reservation.

Funston Meadow, Creek, and **Camp** [Sequoia NP]. Named for James Funston, who grazed sheep in the vicinity about 1870 (Farquhar).

Furnace Creek Wash, Ranch, and **Pass** [Death Valley NP]. Probably so named because of the extreme heat. According to Hanks, Dr. Darwin French and his party named the creek in 1860. The tradition that Mormons had built a furnace there in the 1850s to extract lead from galena is apparently without foundation. Furnace Creek Ranch, established in 1870, was called Greenland in the 1880s and sometimes called Coleman, for the owner, William T. Coleman, "Lion of the Vigilantes." It was given its present name by the Pacific Coast Borax Company after 1889. The summit southwest of Pyramid Peak is locally known as Furnace Creek Pass. According to a statement by William L. Manly made in 1893, Asahel Bennett and his party built a furnace there in 1860. (Manly's journalism, 1877–1904, was published as *The Jayhawkers' oath and other sketches* [Los Angeles: W. F. Lewis, 1940].) **Furnace Canyon** [San Bernardino Co.], although within the territory of the Mormon settlements, was so named because of the extreme heat (Fisk).

G

Gabb, Mount [Fresno Co.] was the name given to a peak by the Whitney Survey, after William M. Gabb (1839–78), paleontologist of the party. It is uncertain whether this peak is the Mount Gabb shown on LeConte's map of 1907 and on the later Mount Goddard atlas sheet.

Gabbro (gab′ rō) **Peak** [Mono Co.] was named because of the rock of granitic texture that occurs here. The name was probably applied when the Bridgeport quadrangle was surveyed, 1905–9.

Gabilan, i.e. Spanish *gavilán* 'sparrow hawk' (*Falco sparverius*), was repeatedly used for names of mountains in Spanish times; it is also a family name. **Gabilan** (gab′ i lan): **Peak, Creek, Range** [Monterey, San Benito Cos.] are named for *un gran cerro llamado del Gavilán* 'a high hill called that of the hawk', mentioned in 1828 (Registro, p. 14; *see* Fremont). Note that the letters *b* and *v* are often interchanged in Spanish. The creek is shown as Arroyo del Gavilán on a *diseño* of San Miguelito (1841). A land grant, Ciénega del Gabilán, was granted on Oct. 26, 1843. The range was called Sierra de Gavilán in the 1840s (Castro 2:44); it is mentioned as the San Juan or Gavilan or Salinas Range by Blake in 1857 (Pac. R.R. *Reports* 5:2.139) and is shown as Sierra Gabilan on Goddard's map. The three geographic features now called Gabilan, and a town of that name, appear on the maps of the Whitney Survey. Riverside Co. has both a Gavilan Peak and Mountain. *See also* Gavilan.

Galena (gə lē′ nə) **Canyon** [Death Valley NP] and several other geographic features in the state were named because of the occurrence of galena (lead sulphide), a common ore of lead.

Gale Peak [Yosemite NP] was named in 1894 by Lt. N. F. McClure for Capt. (later Col.) George H. Gale (1858–1920), a native of Maine and acting superintendent of the national park in 1894 (Farquhar).

Galero, Cabo de. *See* Conception, Point.

Gallagher Beach and **Canyon** [Santa Catalina Island] were named for Tom Gallagher, who lived on the beach as a squatter for many years. When the squatters were asked to leave the island, Tom moved to San Clemente Island, where another beach was named for him (Windle, p. 118).

Gallina, Spanish for 'hen', appears in **Cañada de la Gallina** (gä yē′ nə) 'valley of the hen' [Santa Barbara Co.] . The plural is found in **Gallinas** (gə lē′ nəs) **Valley** and **Creek** [Marin Co.], named for the *sitio de las Gallinas* 'place of the hens' mentioned by Padre Payeras in 1819 (Docs. Hist. Cal. 4:341). On Feb. 12, 1844, the name was incorporated into the San Pedro, Santa Margarita y las Gallinas land grant. The creek is recorded on Hoffmann's map of the Bay region.

Galt [Sacramento Co.] was named in 1869 by John McFarland, an early settler, after his former home in Ontario, Canada, which had been named for John Galt (1779–1839), the Scotch novelist.

Garapito (gâr ə pē′ tō) **Creek** [Los Angeles Co.] is probably a variant of Spanish *garrapata* 'tick' (*q.v.*).

Garberville [Humboldt Co.] was named for Jacob C. Garber, a native of Virginia; he was a resident of Rohnerville in 1871 and soon afterward settled at the place that bears his name. Garberville is shown on the Land Office map of 1879 and is listed as a post office in 1880.

Garcia (gär′ sē ə, gär′ shə, gär sē′ ə) **River** [Mendocino Co.] was probably named for Rafael García, who was granted nine leagues on the coast north of Presidio Ruso (Fort Ross) on Nov. 15, 1844. The same García was grantee of Tamales y Baulenes [Marin Co.].

Garden occurs as a generic term in place names all over the United States. In California there are a dozen such communities; **Gardena** (gär dē′ nə) [Los Angeles Co.] and **Garden Grove**

[Orange Co.] are the most important. **Garden Valley** [El Dorado Co.], according to local tradition, received its name because people realized that it was more profitable to raise vegetables than to mine for gold. However, the post office was established as Garden Valley on Dec. 16, 1852, while it was still a rich gold camp (as late as 1857 a nugget worth $525 was found there). Note also Garden Canyon Creek [Stanislaus Co.], Garden Valley [Yuba Co.], and Garden Acres [San Joaquin Co.].

Gardiner, Mount, and **Gardiner Creek** [Kings Canyon NP] were named after James T. Gardiner (1842–1912) in 1865 by the Whitney Survey, of which he was a member from 1864 to 1867. The Whitney Survey spelled the name Gardner in accordance with Gardiner's own spelling of the name at that time. At the instigation of the Sierra Club, and with the approval of the BGN, the spelling was changed to conform to the original form of the family name as later adopted by Gardiner himself.

Garey [Santa Barbara Co.] was established in 1889 and named for Thomas A. Garey, who came to California in 1852 and settled first in Pomona Valley. Garey was a well-known nurseryman and horticulturist, specializing in the culture of citrus fruit.

Garfield: Grove, Creek [Sequoia NP]. Named by the USGS in 1902 in memory of James A. Garfield (1831–81), who was U.S. president until his assassination.

Garlock [Kern Co.], an old railroad station west of Randsburg, was named for Eugene Garlock, who set up the first stamp mill of the Randsburg mining district in 1895 at the place known as Cow Wells.

Garnet, a common red mineral, occurs in various parts of the state; it has given its name to orographic features in Calaveras, Madera, Mariposa, and Riverside Cos. **Garnet** [Riverside Co.] is the name applied to a Southern Pacific station in 1923, after the hill west of it.

Garrapata (gâr ə pä′ tə) **Creek** [Monterey Co.] is shown in 1835 on a *diseño* (Docs. Hist. Cal. 1:484) as Arroyo de las Garrapatas; it was probably named because of the presence of wood ticks, which thrive especially where wild lilac (*Ceanothus*) grows; it is misspelled Garrapatos on some maps. *See also* Garapito.

Garrote. *See* Groveland; Second Garrote.

Garus (gâr′ əs) [Santa Barbara Co.] was coined by transposing the letters of the word "sugar." The place is in a sugar-beet region (Knave, Mar. 17, 1935); *see* Barsug.

Garvanza (gär van′ zə) [Los Angeles Co.] represents Spanish *garbanzo* 'chickpea'.

Garvey [Los Angeles Co.]. Previously Garvalia, after Richard Garvey, Sr., an Irishman who homesteaded five thousand acres here in 1911 (Patricia A. Warren).

Garza, the Spanish for 'heron', occurs in **Garza** (gär′ zə) **Peak** [Kings Co.]. The plural is found in **Garzas** (gär′ zəs) **Creek** [Stanislaus, Merced Cos.], which takes its name from a *parage llamado . . . las Garzas* 'a place called the herons', mentioned on July 4, 1840 (DSP Mont. 3:86). The creek is shown as Arroyo de las Garzas on a *diseño* of the Orestimba grant (1843).

Gaskill Peak [San Diego Co.]. Christopher B. Gaskill, born in England, homesteaded here in 1887 (Stein).

Gas Point [Shasta Co.] was a name applied because the old prospectors gathered here to "gas and spin yarns" (Steger). From 1875 to 1933 it was the name of the post office.

Gasquet (gas′ kē) [Del Norte Co.] was named for a member of the Gasquet family. Horace Gasquet, a native of France, came to Del Norte Co. before 1860 and had a ranch in Mountain township. Howard Gasquet is listed in McKenney's Directory of 1883 as the postmaster and hotelkeeper at Gasquet.

Gate of the Antipodes (an tip′ ō dēz) [Sierra Co.]. The term "antipodes" refers to a place diametrically opposite to one's position on the earth. The name "China," shown west of this narrow pass, may have suggested this fanciful place name to a surveyor, since China was often considered to be just opposite America on the globe (Stewart).

Gato, the Spanish word for 'cat' (including the bobcat, *Felis rufus*), is usually pluralized in geographical names, including two land grants: Los Gatos or Santa Rita [Monterey Co.], dated 1824 and regranted on Sept. 30, 1837, and Rinconada ('corner') de los Gatos [Santa Clara Co.], dated May 21, 1840. Names of creeks in Santa Clara, San Benito, and Fresno Cos., and Cañada del Gato in Santa Barbara Co., still include the word. **Los Gatos** (gat′ əs) **Creek** and town [Santa Clara Co.] preserve the name of the land grant Rinconada de los Gatos. The Santa Cruz Mountains were called Cuesta de los Gatos by Frémont (*Geog. memoir*, p. 33) and are so designated on the German edition of Eddy's map (1856), whereas Eddy's original map has the modern name Santa Cruz Mountains. The Creek is shown on the maps of the Whitney Survey. The town was laid out in 1850 by J. A. Forbes (Co. Hist.: Santa Clara 1876:16). The post office is listed in 1867, and the station was named when the railroad from San Jose reached the place on June 1, 1878. **Los Gatos Creek** [Fresno Co.], which flows past Coalinga, once bore the names Arroyo Pasajero and Arroyo Poso de Chane; it was officially named Los Gatos by decision of the BGN (May–Aug. 1964).

Gavilan (gav′ i lan) **Creek** [Santa Clara Co.] reflects the name of nearby Gavilan College. Spanish *gavilán* means 'sparrow hawk' but is also a family name; *see* Gabilan.

Gaviota (gav ē ō′ tə): **Canyon, Pass, Creek, Peak**, town [Santa Barbara Co.]. The Portolá expedition camped in the valley on Aug. 24, 1769. Crespí says that he "called this place San Luis, King of France, and the soldiers know it as La Gaviota, because they killed a seagull (*gaviota*) there" (Crespí, p. 172). This name appears frequently in documents of the succeeding years. The Cajón de la Gaviota is mentioned in 1795 (PSP 14:76), and Gabiota (for the canyon) is shown on the *diseño* of the Rancho Nuestra Señora del Refugio, 1838. Parke's report and map misspell it as Gaviote Pass. The correct spelling appears again on Goddard's map (1860).

Gayley, Mount [Inyo Co.]. For Prof. Charles Mills Gayley (1858–1932), classicist at the University of California, Berkeley (Browning 1986).

Gaylor: Lakes, Peak [Yosemite NP]. For Andrew J. Gaylor, national park ranger, 1907–21 (Browning 1988).

Gazelle (gə zel′) [Siskiyou Co.], the name of a small African antelope, was given to the post office in 1870; many places had earlier been named for the native antelope. In the 1850s a steamer *Gazelle* plied between San Francisco and the upper Sacramento River.

Gazos (gä′ zōs, gaz′ ōs) **Creek** [San Mateo, Santa Cruz Cos.] may be a mangling of *garzas* 'herons'. The Portolá expedition found an Indian village here (Crespí, p. 219).

Gefo (gef′ ō) [El Dorado Co.]. Supposed to be derived somehow from the name of George Goss, who planted fish here (Lekisch).

Gem: Lake, Pass [Mono Co.]. The lake was named Gem-o'-the-Mountains by Theodore C. Agnew before 1896 (Farquhar). When the USGS mapped the region in 1898–99, it shortened the name and also gave it to the pass north of the lake.

Gemini (jem′ i nī) [Kings Canyon NP], as a name for the twin-peaked mountain north of the Pinnacles, was proposed by Chester Versteeg of the Sierra Club and approved by the Board on Geographical Names in Mar. 1957. Gemini is the name of the constellation that includes the twin stars Castor and Pollux.

General Creek [El Dorado Co.] was named for Gen. William Phipps, for whom Phipps Lake and Peak were also named. The general, a native of Kentucky and veteran of Indian wars, came to Georgetown in 1854 (W. T. Russell, Knave, July 23, 1944).

General Grant Grove [Kings Canyon NP]. The largest *Sequoia gigantea* in the park was christened General Grant by Mrs. L. P. Baker of Visalia in Aug. 1867, in honor of Ulysses S. Grant (1822–85), U.S. president 1869–77. The name of the tree was given to General Grant National Park when it was established in 1890, at the suggestion of David K. Zumwalt, the attorney of the Southern Pacific Railroad. When General Grant NP was incorporated into Kings Canyon NP in 1940, its identity and name were preserved in the name of the grove.

Genesee (jen′ ə sē): **Valley**, settlement [Plumas Co.]. The Postal Guide of 1867 spells the name Geneseo, but that of 1880 has Genesee; Genesee Valley is mentioned in Jan. 1863 (Bancroft Scrapbooks 8:114). The older gazetteers of New York State likewise have Geneseo for the town, but Genesee for the valley. The name may have been brought from New York by the Ingalls family, who settled in (and perhaps named) Genesee Valley in the 1860s. The name is supposedly Iroquoian for 'beautiful valley' (Stewart). There is also a Geneseo in San Luis Obispo Co.

Geneva. *See* Planada.

Genevra (jə nēv′ rə) [Colusa Co.] was the name, of unknown origin, given by the Southern Pacific Railroad to replace the term Berlin, originally given by the Northern Railroad. The post office, established on Aug. 18, 1876, retained the name Berlin until it was discontinued on June 30, 1934. **Mount Genevra** [Sequoia NP] was named for Genevra Magee in 1899 by J. N. LeConte and his party, who were with Mrs. Magee on the summit of Mount Brewer (Farquhar).

Gente Barbuda, Isla de. *See* Santa Cruz: Santa Cruz Island.

Gentiles, Laguna de los. *See* Caslamayomi.

George J. Hatfield State Park [Merced Co.] was created in 1953 and named for the former district attorney and lieutenant governor.

Georges Creek [Inyo Co.] was named for George, leader of the band of Paiute Indians that fought the white intruders in the mid-southern part of Owens Valley in 1863. His rancheria was on the creek that now bears his name.

Georges Gap [Plumas Co.] was named for George Geisendorfer, who lived just south of it. Geisendorfer's settlement is shown on Bowman's map (1873), but the pass is labeled Georgia Gap. There is another Georges Gap [Los Angeles Co.] on the Angeles Highway, on the saddle between Arroyo Seco and Tujunga watersheds (Wheelock).

George Stewart, Mount. *See* Stewart in the Glossary.

Georgetown [El Dorado Co.] was founded in 1849 by George Ehrenhaft (Bancroft 6:482); but it is uncertain whether it was named for him or for George Phipps, a sailor who led a party of prospectors to the place in 1849 (Rensch-Hoover, 104–5). It was repeatedly mentioned in 1850 as a major mining center. An earlier name, or nickname, was Growlersburg. *For* Georgetown [Sacramento Co.], *see* Franklin.

Georgia Gap. *See* Georges Gap.

Gerber (gûr′ bər) [Tehama Co.] was named in 1916 by the Southern Pacific Railroad for H. E. Gerber, of Sacramento, who sold the land to the railroad. The post office is listed in 1917.

Gerle (gûr′ lē) **Creek** [El Dorado Co.] was named for Christopher C. Gerle, a native of Sweden, who settled in El Dorado Co. before 1860. The name has also been spelled Gurley.

German. There are two German Creeks in the state [Los Angeles, Mendocino Cos.], and in the 1850s there was a mining town, German Bar [Nevada Co.]. Artois [Glenn Co.] was formerly called Germantown. Elsewhere the nickname Dutch is used in place of German in California toponymy. Even the two creeks may not have been named for German settlers; in Spanish, Germán is a common given name.

German [Sonoma Co.] was the name of a land grant dated Apr. 8, 1846. This is probably a Spanish phonetic rendering of the German name Hermann, which Ernest Rufus, the grantee, apparently intended to apply to his rancho. The letter *h* is silent in Spanish, and the aspirate sound is (before *e* and *i*) written with *g*. The name Rancho de Hermann appears on the *diseño*, but the *expediente* has "German." The name of Hermann, victor over the Romans in the Teutoburg Forest in the year 9 C.E., was favored in place naming by German immigrants to the United States in the middle of the 19th century. *See* Gualala.

Germanos, Los. *See* Juristac.

Geyser is found as a name in Sonoma Co. and elsewhere where hot springs appear in the form of small geysers. **Geyserville** [Sonoma Co.] was founded in 1851 by Elisha Ely and so named to advertise the nearby "geysers." Originally the place was known as Clairville [*sic*] for John Clar, for many years engaged in translating and transcribing Spanish archives (Bowman). **The Geysers** was approved in 1980 by the BGN (list 8004) as a name for an area of geothermal activity [Lake, Sonoma Cos.].

Ghirardelli (gē rär del′ ē) **Square** [San Francisco Co.]. In 1897 Domingo Ghirardelli moved his chocolate and spice factory to this location on North Point Street. The factory complex was transformed into the present array of shops and restaurants by William Matson Roth in the late 1960s (Loewenstein).

Gianone (jə nō′ nē) **Hill** [Santa Cruz Co.]. For Ambrogio Gianone, a dairyman here in the late 19th century (Clark 1986).

Giant [Contra Costa Co.]. Named when the Giant Powder Company of Wilmington, Delaware, built its west coast plant in 1880. "Giant powder" was an early name of dynamite.

Giant Forest [Sequoia NP] was discovered in 1858 by Hale Tharp and was named in 1875 by John Muir: "This part of the Sequoia belt seemed to me the finest, and I then named it 'the Giant Forest'" (Muir, *Our national parks* [Boston: Houghton Mifflin, 1901], p. 300).

Gibbs, Mount [Yosemite NP] was named by F. L. Olmsted on the first ascent, Aug. 31, 1864, for Oliver Wolcott Gibbs (1822–1908), professor of science at Harvard and a lifelong friend of J. D. Whitney.

Gibson [Shasta Co.] was the name applied to

this Southern Pacific station when the extension from Redding was built in 1886. The name honors Reuben Gibson, who led whites and Wintus against Modocs at Battle Rock in June 1855 (Steger).

Gibsonville [Sierra Co.], an old mining town on Little Shasta Creek, was first called Gibsons New Diggings, probably named for the same man who discovered one of the Secret Canyons in Placer Co.

Gigling (gig′ ling) [Monterey Co.], a Southern Pacific station, was named about 1920 for the Gigling (or Geigling) family. Valentine Geigling, a native of Germany, settled in the county as a farmer before 1857.

Gilbert, Mount [Kings Canyon NP] was named for Grove K. Gilbert (1843–1918), geologist of the USGS and author of numerous monographs. The name was confirmed by the BGN on July 19, 1911.

Gillette [Tulare Co.] was named for King C. Gillette (1855–1932), of safety razor fame, who owned a citrus orchard of five hundred acres here (Santa Fe).

Gillett Mountain [Tuolumne Co.] was named in 1909 by R. B. Marshall for J. N. Gillett, congressman, 1903–6, and governor of California, 1907–11 (Farquhar).

Gilman (gil′ mən) **Hot Springs** [Riverside Co.], already known to the Indians, were developed by the Branch family in the 1880s under the name San Jacinto Hot Springs. The present name originated when William E. and Josephine Gilman bought the place in 1913. The post office was established on June 1, 1938.

Gilman Lake [Mono Co.] was named in 1905 for Robert Gilman Brown, vice president and general manager of the Standard Consolidated Mining Company, by one of the company's engineers, who was engaged in mapping the district (Farquhar).

Gilroy (gil′ roi): town, **Hot Springs** [Santa Clara Co.]. John Gilroy, a Scottish sailor and the first permanent non-Spanish settler in California, was left ashore in Monterey in 1814 by the Hudson's Bay vessel *Isaac Todd* because he was sick with scurvy. His real name was Cameron, but he changed it to his mother's family name because he had left home as a minor and was in danger of being sent back. He settled in the Santa Clara Valley, where he married María Clara Ortega, grantee (June 3, 1833) of part of the San Isidro land grant. The settlement that developed on the rancho became known as San Isidro, and later as Gilroy. Bowen's *Post-office Guide* of 1851 lists Gilroy's. After the coming of the railroad in 1869, the name Gilroy appears on the map for the station, and Old Gilroy (San Isidro post office) for the older settlement (von Leicht–Craven). The springs were discovered in 1865 by a Mexican sheepherder while he was hunting for some of his flock and were named after the town years later. The Spanish name for Gilroy Valley was Llano de las Llagas (*see* Llagas).

Girard (ji rärd′): **Creek, Ridge** [Shasta Co.]. For Louis Girard, a French Canadian, who discovered gold on the creek (Steger).

Giraud (ji rō′) **Peak** [Kings Canyon NP] was probably named for Pierre Giraud, sheepman of Inyo Co., known as "Little Pete." The name is spelled Giroud on some maps. *See* Little: Little Pete Meadow.

Glacier. Of the numerous geographic features so called, **Glacier Divide, Lake**, and **Creek** [Kings Canyon NP] were named because of the actual presence of glaciers. On the crest of the divide are eight small glaciers; the lake and creek are just south of Palisade Glacier. **Glacier Ridge** and **Monument** in the same park, as well as **Glacier Point** in Yosemite, were so named because of the evidence of glacial action in former periods. Glacier Point was named by the Mariposa Battalion in 1851. **Glacier Canyon Creek** [Mono Co.] has its origin in a glacier on the north slope of Mount Dana. *For* Glacier Pass, *see* Sawtooth.

Glade, an old Germanic term for an open space in the forest, has become a common generic name in North America. Most of the Glades in California are in northern counties. According to Mauldin, some sixty open spaces or clearings are so called in Lake Co. alone.

Glamis (glam′ is) [Imperial Co.] is listed as a post office in 1887, and the station is shown on the Official Railway Map of 1900. Glamis (or Glammis) is a castle in Scotland, made famous by Shakespeare's *Macbeth*.

Glass Mountain [Modoc, Siskiyou Cos.] is chiefly of obsidian ("volcanic glass"), varying in shades from crystal clear through milky white to jet black. **Glass Mountain** and **Creek** in Mono Co. were probably named also for the presence of obsidian.

Glaucophane (glô′ kə fān) **Ridge** [San Benito Co.] was named because of the presence of glaucophane, probably by the USGS when the

Panoche quadrangle was mapped, 1908–11. In the United States this mineral—a silicate of aluminum, sodium, iron, and magnesium—appears chiefly in the Coast Ranges of California.

Gleason Mountain [Los Angeles Co.] was probably named for George Gleason, a resident of the county in 1872.

Glen, a Celtic generic term for 'narrow valley', became the most popular of the archaic geographical names revived by the Romantic movement of the 19th century. It is seldom used as a true generic term replacing "canyon," but it has provided the first element of numerous composite place names. California has not only its share of Glenbrooks, Glenburns, Glendales, and Glenwoods, but also a number of unique combinations, e.g. Glen Alder [Placer Co.], Glen Arbor [Santa Cruz Co.], Glen Frazer [Contra Costa Co.], and Glen Una [Santa Clara Co.]. **Glen Alpine Creek** [El Dorado Co.] was named from a reference in Sir Walter Scott's *Lady of the lake* (Lekisch). **Glen Ellen** [Sonoma Co.] was named in 1869 by Charles V. Stuart, a native of Pennsylvania and a pioneer of 1849, for his wife, Ellen Mary. The post office was established and named on July 19, 1871. **Glencoe** [Calaveras Co.] is shown on Hoffmann's map (1873) as Mosquito. When the post office was established about 1878, a new name was chosen because there was a Mosquito post office in El Dorado Co. At that time there were eighteen other Glencoes in the United States, directly or indirectly named after the valley in Scotland. **Glendale** [Los Angeles Co.] developed soon after the railroad was built from Los Angeles to San Fernando in 1873–74 and was named Riverdale (Co. Map, 1881). The name was refused by the Post Office Dept. in 1886 because of the Riverdale already existing in Fresno Co., so the post office was called Mason. Soon afterward it was changed to Glendale; this name is shown on the Land Office map of 1891. There were twenty-five other Glendales in the United States at that time. **Glendora** [Los Angeles Co.] is a name created in 1887 by George Whitcomb from his wife's name, Ledora. **Glen Blair** [Mendocino Co.] was named after the Glen Blair Mill Company, founded by Captain Blair "soon after Fort Bragg was in operation" (Co. Hist.: Mendocino 1914:67). **Glenavon** [Riverside Co.] contains another Celtic term, *avon* 'river'; it was chosen by L. V. W. Brown in 1909 (A. C. Fulmor). **Glen Aulin** [Yosemite NP] contains *glen* as a true generic: "It was probably in the winter of 1913–14 . . . I . . . suggested Glen Aulin 'beautiful valley or glen' [to R. B. Marshall], and wrote it for him in this way, that it might be correctly pronounced—the 'au' as in author. The correct Gaelic (Irish) orthography is *Gleann Alainn*" (James McCormick to F. P. Farquhar). **Glenwood** [Santa Cruz Co.] was founded by Charles C. Martin, who came round the Horn in 1850. It was known as Martinville until the post office was established on Aug. 23, 1880. The state also has a Glen Valley Slough [Colusa Co.], a Glen Anne Creek [Santa Barbara Co.], a Glenshire [Nevada Co.], and a Glenhaven [Lake Co.].

Glenn: County, station. The county was formed by act of the legislature on Mar. 11, 1891, from the northern part of Colusa Co.; it was named for Dr. Hugh J. Glenn (1824–82), whose estate gave financial backing to a proposal for creating and naming the new county. Dr. Glenn came to California from Missouri in 1849, bought Rancho Jacinto in 1867, and became the most important wheat grower of the state. The station was named after the county when the Southern Pacific branch from Colusa was built in 1917.

Glenn Ranch [San Bernardino Co.] was established as a resort on the ranch where Jerry Glenn, an immigrant from Texas, settled in the 1860s.

Glennville [Kern Co.] was established as a post office in the 1870s; it was named for James Madison Glenn, a native of Tennessee, who had settled in Linn's Valley in 1857.

Glen Pass [Kings Canyon NP] was named for Glen H. Crow, a ranger in the USFS and an assistant in the USGS, when the Mount Whitney quadrangle was surveyed in 1907 (Farquhar).

Glorietta is the name of communities in Contra Costa and Fresno Cos. and of a bay in San Diego harbor. A site in Orange Co. is spelled Gloryetta. The name was probably chosen for its sound. In Spain and Mexico *glorieta* refers to a small plaza from which several streets radiate.

Goat. Since no goats are native to California, the various Goat Mountains, Buttes, and Rocks were probably so named because antelope or mountain sheep were mistaken for goats.

Yerba Buena Island in San Francisco Bay was called Goat Island for many years because, at the time of the Gold Rush, it was populated by several hundred goats, descendants of half a dozen that had been placed on the island in 1842–43 by Nathan Spear and John Fuller, merchants in Yerba Buena (Davis, p. 266).

Gobernadora (gō bər nə dôr′ ə) **Canyon** [Orange Co.] was called Cañada de la Gobernadera on a *diseño* of the Trabuco grant. The canyon probably received its name because of the presence of the gobernadora, or creosote bush (*Larrea divaricata*). The name also appears as Cañada Gubernadora. **Gobernador** (gō bər nə dôr′) **Creek** [Santa Barbara Co.], spelled Governador on a map of El Rincon grant, was perhaps also named for the shrub; there seems no reason to connect it with the Spanish word for 'governor'.

Goddard, Mount [Kings Canyon NP], was named in 1865 by the Whitney Survey for George H. Goddard (*see* Glossary): "Thirty-two miles north-northwest is a very high mountain, called Mount Goddard, in honor of a Civil Engineer who has done much to advance our knowledge of the geography of California, and who is the author of Britton and Rey's Map" (Whitney, *Geology*, p. 382). Nearby Goddard Divide, Canyon, and Creek appear on the 1912 atlas sheet of the USGS.

Goethe (gûr′ tə), **Mount** [Fresno Co.]. The highest peak of Glacier Divide was named by the USFS and the Sierra Club in 1949, the bicentennial of the birth of Johann Wolfgang Goethe (1749–1832), German poet, philosopher, and scientist. The name was suggested by David R. Brower and Erwin G. Gudde, and was confirmed by the BGN in Aug. 1949. Goethe Cirque, Glacier, and Lake are named after the peak. The pronunciation (gā′ tē) also occurs, as for the street of this name in San Francisco.

Goethe (gā′ tē) **Grove** [Humboldt Co.]. The Mary Glide Goethe Grove in the Prairie Creek Redwoods State Park, established through the efforts of the Save-the-Redwoods League in 1948, memorializes Mary Glide Goethe, wife of Charles M. Goethe, of Sacramento. The area below the Grove is called Goethe Addition. Mr. and Mrs. Goethe were well-known philanthropists and outstanding leaders in nature conservation.

Goffs. *See* Amboy.

Golconda (gōl kon′ də) **Ravine** [Nevada Co.]. The term refers to a rich mine, after the ancient diamond mines near Hyderabad, India; the term is Telugu for 'round hill'.

Gold, the precious metal so intimately connected with the history of the "Golden State," is mentioned in the names of more than a hundred geographic features (*see* Oro; Placer). Kern Co. has a Goldpan Canyon; Madera Co. has a Coarse Gold Creek, a Fine Gold Creek, and a Little Fine Gold Creek. In Death Valley NP are Goldbelt Springs and Goldbar. The stream **Gold Run** [Nevada Co.] is mentioned on Aug. 26, 1849: "On the banks of Deer Creek and Gold Run—as they have ever since been called—they struck some of the richest and most famous diggings ever known in California" (Wistar 1:126). **Gold Lake** [Plumas Co.] commemorates what seems to have been the first large-scale hoax perpetrated after the discovery of gold. Early prospectors, noting that the gold became coarser as they proceeded upstream, believed there must be a lake or other source of gold high in the mountains. The first rumor of a lake rich with golden pebbles seems to have started in 1849. In the summer of 1850 some traders utilized these rumors by deliberately spreading the news of the actual discovery of such a gold lake—which, they said, could unfortunately not be exploited by small parties because of hostile Indians. All doubts vanished when, in July, Peter Lassen, who knew the region between Honey Lake and Sacramento Valley better than any other man, fell for the hoax. From Lassen's Ranch alone four expeditions set out about Aug. 1, and by the end of the year the floating population of the mines swarmed through the northern Sierra Nevada in search of the lake. The traders reaped rich profits, but the gold hunters "didn't diskiver nare a Gold Lake" (Bruff 2:825). The name apparently became attached to the lake in Plumas Co. because it fits the general description of the mythical lake: deep-seated, near several buttes, two days' travel from Honey Lake. The range northeast of it was known as Gold Lake Mountains in the 1850s. The name of Gold Lake in Sierra Co. may also have some connection with the "great Gold Lake hunt," and some writers think that this is the lake. In fact, neither this nor the Gold Lakes in Sierra and Butte Cos. have anything to do with the "Gold Lake hunt" or the "Goose Lake hunt," as Schaeffer sarcastically calls it in his account. The Gold Lake of 1849 remains a phantom in spite of the numerous modern accounts, including the fascinating story in Joseph C. Tucker, *To the golden goal* (San Francisco: Doxey, 1895), and the tales about Thomas R. Stoddart, which were started by the ghost writer of William Downie's *Hunting for gold* (San Francisco: California Publishing Co., 1893). The best direct contemporary account is given by Schaeffer (pp. 81 ff.). He and his party left Washington City, a mining camp on the South fork of the Yuba River, on June 7, 1850, but gave up in disgust after four days. This did not prevent other parties from joining the "Goose Lake hunt." Bruff's *Gold Rush* gives many interesting but somewhat confusing data (*see* Last Chance). **Gold Bluffs** [Humboldt Co.] commemorates a rush almost contemporary with the Gold Lake hunt, but this time it was no hoax. The Gold Bluffs were discovered in Apr. 1850 by Hermann Ehrenberg, in later years a famous pioneer of Arizona. They were so called because gold had been washed from them, but the yield from washing the sand proved to be unprofitable. **Gold Run** [Placer Co.] refers to a run of auriferous gravel worked there by early miners. The present town developed around O. W. Hollenbeck's trading center in 1862; the post office was established and named on July 15, 1863. **Gold Discovery Site** [El Dorado Co.] is the location of Sutter's sawmill, in the tailrace of which James W. Marshall discovered gold on Jan. 24, 1848; it became a state historical monument in 1942. Excavation in 1947 showed that the actual site of the discovery is not on state property (*CHSQ* 26:129–30).

Golden names several geographic features, including **Golden Valley** [San Bernardino Co.], because of the golden glow at sunset, or the presence of fields of golden poppies; *see* Cuddeback Lake. **Golden Canyon** [Death Valley NP] was named because of the yellowish clay there. *For* Golden Rock [San Francisco Bay], *see* Red: Red Rock.

Golden Gate was the name given to the entrance of San Francisco Bay by Frémont in the spring of 1846. He chose this name because he foresaw the day when the riches of the greater Orient would flow through the Golden Gate, just as the riches of the lesser Orient had once flowed into the Golden Horn. In a footnote to his *Geographical Memoir* (p. 32), Frémont justifies his choice: "Called *Chrysopylae* (golden gate) on the map, on the same principle that the harbor of *Byzantium* (Constantinople afterwards) was called *Chrysoceras* (golden horn). . . . The form of the entrance into the bay of San Francisco, and its advantages for commerce (Asiatic inclusive,) suggests the name that is given to this entrance." Frémont was apparently determined to fix the Greek name on the entrance, but Preuss cautiously put "Chrysopylae or Golden Gate" on his map of 1848, and the latter was naturally the name accepted. **Golden Gate Park** [San Francisco Co.] was created and named by act of the legislature on Apr. 4, 1870: "The land . . . is hereby designated and shall be known as the 'Golden Gate Park.'" **Golden Gate Bridge** was a name spontaneously used when the project was first discussed in 1917 by M. H. O'Shaughnessy, city engineer of San Francisco, and Joseph B. Strauss, who later constructed the bridge. It became official with the incorporation of the Golden Gate Bridge and Highway District by act of the legislature on May 25, 1923. **Golden Gate National Cemetery** [San Mateo Co.] was named by General Orders no. 4, War Dept., 1939. The name Golden Gate is found for a Creek in Amador Co., a Hill in Calaveras Co., and a site in Colusa Co.

Golden Trout Creek [Tulare Co.], a tributary of the Kern River, was once called Whitney Creek, then Volcano Creek. The present name was applied by the USGS (probably when the Olancha quadrangle was surveyed in 1905) because in 1903 Dr. B. W. Evermann identified a new species of trout here, *Salmo roosevelti* (Farquhar).

Goldstein (gōld′ stēn) **Peak** [Tulare Co.] was named for Ike Goldstein of Visalia, who ran hogs on the mountain at one time (A. L. Dickey).

Goldtree [San Luis Obispo Co.] is an Americanization of the surname of Morris Goldbaum, who settled here in the 1890s. (*Baum* is German for 'tree'.)

Goler Wash, Canyon, and **Heights** [Death Valley NP] are reminiscent of one of the strangest incidents in the history of Death Valley. Among the Argonauts who crossed Death Valley in 1849 were two Germans, John Galler (or Goler or Goller) and possibly Wolfgang Tauber. Tauber apparently discovered gold in either the Canyon or the Wash (which are 50 miles apart). In 1912 Carl Menge discovered a rich pocket of gold at his Oro Fino claim in Goler Wash (Death Valley Guide, p. 62). If at all, it was probably here that Tauber found the gold. Both W. Tauber and John Galler or Goller are listed as members of sections of the Jayhawker party (*HSSC:Q* 22:102 ff.; John Ellenbecker, *The Jayhawkers of Death Valley*). Tauber died at sea in 1850, and in later years Galler or Goller tried to find the mysterious gold deposits; hence the find became known as Goler's mine or nuggets (Weight, pp. 31 ff.). The well-known pioneer carriage-maker of Los Angeles, John Goller, is apparently identical with the gold seeker—although Harris Newmark, who knew Goller very well, does not mention the Death Valley episode in his *Sixty years in southern California*. The story of Goller's nugget was still alive enough in 1893 to provide the name of Goler Gulch, miles away from Death Valley, in Kern Co., where the discovery of gold led to the development of the Johannesburg-Randsburg district.

Goleta (gō lē′ tə) [Santa Barbara Co.] developed on the land grant La Goleta, dated June 10, 1846. In 1875 the town was laid out and named after the grant. According to Michael C. White (MS, 1877, Bancroft Library), the place was originally so named because a schooner (*goleta*) was built there in 1829. According to other and probably more authentic sources, the name originated because an American schooner, stranded in the estuary, lay there for many years (AGS Santa Barbara, p. 168). A *diseño* of La Goleta grant shows a wreck at the mouth of the estuary.

Golgones, Sierra de. *See* Diablo, Mount.

Gomez (gō′ mez) [Sutter Co.] was named in 1906 by the Northern Electric Railroad for Nathaniel Gomez, a native of Portugal, because the right-of-way cut through his property (H. H. Harter).

Gonzales (gon zal′ əs) [Monterey Co.] is named for Teodoro Gonzales, *regidor* and acting *alcalde* at Monterey, who was granted Rincón de la Punta del Monte de la Soledad at this site on Sept. 20, 1836. Duflot de Mofras's map of 1844 shows a Gonsales south of the present location—perhaps indicating San Miguelito, the rancho of José Rafael Gonzales. The present Gonzales may have been named for Teodoro's sons, Alfredo and Mariano, who were prominently associated with the Monterey and Salinas Railroad (Bancroft 3:761).

Goodale: Creek, Mountain [Inyo Co.] were named for Ezra and Thomas Goodale, who settled here in the 1870s. **Goodale Pass** [Fresno Co.] may be named for Gus G. Goodale (son of Thomas), at one time a ranger in the USFS.

Goode, Mount [Kings Canyon NP]. In mapping the Mount Goddard quadrangle, 1907–9, the USGS inadvertently placed a second name on a peak already known as Black Giant. This name, which honors Richard U. Goode (1858–1903), geographer in charge of the USGS in the western United States, was transferred to the present peak by decision of the BGN in 1926.

Goodwin Dam [Stanislaus Co.]. The two diversion dams on the Stanislaus River were named for A. D. Goodwin, Manteca rancher and manufacturer of farm equipment, who was one of the principal boosters of the Tri-Dam Project, completed in 1957.

Goodyears: Bar, Creek, and **Hill** [Sierra Co.] take their name from the river bar opposite the present settlement, where Miles and Andrew Goodyear discovered gold in 1849. In 1851 the name was applied to the settlement, which previously had been known as Slaughters Bar.

Goose is a term found in more than fifty geographic features in the state, named because of the presence of the wild Canada goose (*Branta canadensis*), so important as a food for travelers in the early days. **Goose Lake** [Modoc Co.] was probably the body of water called Pit Lake (after Pit River) by John Work in 1832 (*CHSQ* 22:205). However, Pit Lake on Hood's map (1838) has no connection with what was later called Goose Lake. Pitts Lake on Wilkes's map of 1841 was apparently intended for Goose Lake but is placed more than one degree of longitude too far east. The name Goose Lake probably arose spontaneously in the 1840s, because the honkers were as numerous in the region then as they are now. The name appears on Williamson's map (1849) and in his report (Tyson, pt. 2, p. 22). Miguel Costansó of the Portolá expedition mentions a Llano de los Ánsares 'plain of the geese' in what is now San Mateo Co. on Oct. 28, 1769 (p. 102). **Goosenest Mountain** [Siskiyou Co.] is the crater of an extinct volcano and supposedly resembles a goose nest; it was named around 1874 (Luecke 1982).

Goose Creek [Modoc Co.], which empties into Upper Alkali Lake near Lake City, was named for a German homesteader of the late 1860s whose name was Goos (William S. Brown).

Gordo, Gorda, Spanish for 'fat, plump', is used in a geographical sense for massive promontories along the coast, usually with Punta 'point'. **Punta Gorda** (gôr′ də) was the name given by Hezeta in 1775 to a bold headland north of Cape Mendocino. It was shown on the maps in the latitude of present False Cape as late as 1846 (Tanner). In the first edition of his *Coast Pilot*, Davidson applied the name to the point south of Mendocino: "Punta Gorda is 17 miles . . . from Shelter Cove, and, as its name implies, is a large, bold rounding point" (1858:67). **Gorda Rock** and **Seavalley** take their names from the point. **Gorda** [Monterey Co.], as the name of a settlement, mine, and school, preserves the old name of the nearby cape, Punta Gorda, which the Coast Survey changed to Cape San Martin in 1862. The names of **Cerro Gordo Mine, Spring**, and **Mountain** [Inyo Co.] take their names from Cerro Gordo Peak (tautological: 'peak fat peak'). The Cerro Gordo Mountains (for the Inyo Mountains, northeast of Owens Lake), are shown in 1871 on Wheeler atlas sheet 65 (*see* Cerro).

Gordon Creek and **Hill** [Mendocino Co.] were named by the Coast Survey after Alexander Gordon, a native of Canada; he came to California in 1863, settled first at Caspar, and in 1875 purchased a ranch north of Westport.

Gordons Ferry [Kern Co.] is the place where roads and trails met at the Kern River in the 1850s and 1860s, and where there was a station of the Butterfield stages in 1858–60. Major Gordon operated the ferry.

Gordon Valley [Yolo, Napa Cos.] was named for William ("Julian") Gordon, the first white settler in what is now Yolo Co., grantee of the Quesesosi or Jesús María grant on Jan. 27, 1843 (*see* Quesesosi).

Gorge of Despair [Kings Canyon NP] was named in 1879 by Lilbourne A. Winchell, who spent five months of that year exploring the High Sierra (Farquhar).

Gorman [Los Angeles Co.] was established as a post office, Gorman's Station, on Dec. 18, 1877, and reestablished as Gorman on Sept. 29, 1915. It was obviously named for H. Gorman, the first postmaster.

Goshen (gō′ shən) [Tulare Co.] was named by an employee of the Southern Pacific when the railroad reached this point in 1872. The name of the "best of the land" in Egypt, given by Pharaoh to Jacob, has been a favorite for places throughout the United States.

Gossage Creek [Sonoma Co.] was probably named for Jerome Bonaparte Gossage, from Ohio, who is listed in the Great Register of 1867.

Gottville [Siskiyou Co.] was a mining town named for William N. Gott, whose family owned one of the large mines in the 1880s. The post office name was Klamath River after 1934.

Gouge Eye. *See* Pleasant: Pleasant Grove.

Gould, Mount [Kings Canyon NP] was named University Peak, in honor of the University of California, by J. N. LeConte and a party at the first ascent in 1890. On July 12, 1896, LeConte transferred that name to the present University Peak—when, with Helen Gompertz and Belle and Estelle Miller, he made the first ascent. The following day he climbed the old University Peak and named it for his companion, Wilson S. Gould of Oakland.

Goumaz (g o͞o′ məz) [Lassen Co.] was probably named for Philip J. Goumaz, a native of Switzerland, who came to California in 1863 and settled in Lassen Co. in 1866.

Governador Creek. *See* Gobernadora Canyon.

Government is a term designating a number of places because they were at some time occupied by a federal agency. **Government Well** [Shasta Co.] was named when, in the late 1850s, a detachment of soldiers from Fort Crook established a camp here to protect the freighters from the Indians and dug a well to provide water for the camp. Nearby **Government Lake** was named after the well (Steger).

Grabners (grab′ nərz) [Fresno Co.] was established as a post office in 1915 and named for F. Grabner, on whose land the office was originally situated (Marie Goodrich).

Grace Lake [Shasta Co.] is one of two artificial lakes near H. H. Noble's "castle"; they were named for his daughters Grace and Nora (Steger).

Grace Meadow [Yosemite NP], in the upper Jack Main Canyon, was named for Grace Sovulewski, daughter of Gabriel Sovulewski, for many years an official of the park (Farquhar).

Graciosa (gras ē ō′ sə): **Ridge, Valley** [Santa Barbara Co.]. Named when the Portolá expedition camped in a valley southwest of the canyon on Aug. 31, 1769. Since it was the first time the soldiers saw native women dance, they called the lagoon Baile de las Indias 'dance of the Indian women'. Other members of the party called it La Graciosa because a soldier, through a slip of the tongue, said that he had seen *una laguna graciosa* 'a graceful lagoon'. As this adjective is ordinarily used only for persons, the soldier may have been thinking of one of the Indian dancers. However, on Feb. 29, 1776, Font mentions the Laguna Graciosa and says it was so called because "it is small and of very fine water" (*Compl. diary*, p. 266). The ridge, la Cuesta de la Graciosa, is mentioned on Mar. 19, 1824 (DSP 1:166), and Cañada de la Graciosa is shown on a *diseño* of 1841. A station was named La Graciosa when the Pacific Coast Railroad was built in the early 1880s.

Graeagle (grā ē′ gəl) [Plumas Co.] was the site of a post office established in 1919 and named Davies Mill. In 1921 the name was changed to the present one, a contraction from nearby Gray Eagle Creek. This name may have some connection with Edward D. Baker, the "Gray Eagle of Republicanism," who was in the mining region in 1856 while stumping the state for Frémont.

Grafton. *See* Knights Landing.

Gragg Canyon [San Luis Obispo Co.]. For George Gragg (d. 1914), an early settler (BGN, 1992).

Granada (grə nä′ də, grə nä′ də) [San Mateo Co.]. The name of the onetime Moorish city in Spain has been given to places in San Mateo Co. and in Los Angeles Co.; the latter was changed to Granada Hills in 1942 (Atkinson). *See also* El Granada; Grenada.

Grand is a word frequently used in geographical nomenclature to mean 'great, high' or 'majestic, imposing'. Usually the adjective describes an orographic feature: Grand Mountain [Yosemite NP], Grand Sentinel [Kings Canyon NP], Grand Bluff [Fresno Co.], Grand Dike [Kings Canyon NP]. There is a city named Grand Terrace [San Bernardino Co.]; and Black Point [Marin Co.] was for some time called Grandview Terrace.

Grande, the Spanish word for 'great, large' is found in a number of geographical names as a descriptive adjective; the best known are Arroyo Grande [San Luis Obispo Co.] and Mesa Grande [San Diego Co.]. *See* Arroyo.

Grandview Terrace. *See* Black Mountain: Black Point.

Grangeville [Kings Co.], a popular American place name derived from French *grange* 'barn', was applied to the post office (now discontinued) in the 1870s when the town was the center of the Lucerne district.

Granite, referring to a rock formation, is a term used to name more than fifty geographic features in mountainous areas of the state. The cluster name between the Middle and South Forks of the Kings River, where large masses of granite are found, originated with the Whitney Survey. Placer Co. has a Granite Chief, Inyo Co. a Granite Park, and Madera Co. a Granite Stairway. Sometimes the name has been applied to settlements: Granite Gate [Los Angeles Co.], Granite Creek [Madera Co.], and Granite Hill [El Dorado Co.]. In some cases the places may have been named for the "Granite State," New Hampshire. **Graniteville** [Nevada Co.] was known as Eureka as early as 1850. When the post office was established on Aug. 27, 1867, the new name was chosen to avoid confusion with other post offices named Eureka.

Grape, Grapevine, in the earlier cases referring to the native genus *Vitis*, is a popular fruit mentioned in place names (*see* Uva: Uvas). The most important geographic feature named for it is the long range of the **Grapevine Mountains**, extending from Inyo Co. into the state of Nevada; "Grape Vine Canon" in this region is mentioned by the Nevada Boundary Survey in the *Sacramento Daily Union* of Aug. 10, 1861. The names of **Grapevine Canyon** and **Creek** [Kern Co.] can be traced back to Spanish times: Fages noticed the wealth of wild grapes in 1772, and an expedition gave the name Cajón de las Uvas 'canyon of the grapes' on July 29, 1806 (Arch. MSB 4:49 ff.); **Grapevine Grade** refers to the road that descends this canyon into the San Joaquin Valley. The railroad stations Grape [Imperial Co.], Grapegrowers [Fresno Co.], and Grapeland [San Bernardino Co.] were named because of grape culture in those areas.

Grapit (grap′ it) [Glenn Co.], a Southern Pacific station, was coined because of a nearby gravel pit (Knave, Mar. 17, 1935).

Grass Valley [Nevada Co.] has been described as "a little valley among the hills, whose verdure, surrounded by the red, dry mountain earth of that region, suggested a name to the first discoverer, that of Grass Valley" (Delano, p. 66). The settlement was started in the fall of 1849, when a sawmill was erected. The first log cabin was built by a man named Scott, and the first quartz gold was discovered accidentally on Gold Hill in 1850 by a German (Vischer, pp. 244–45). This started the town on its phenomenal development as a gold mining center. In the early days the names Grass Valley and Centerville were used alternately. On July 10, 1851, the post office was established with the name Centerville, but on Aug. 20, 1852, it was changed to Grass Valley. The name Grass Valley became very popular in California; it is found in Amador, Plumas, and Trinity Cos., as well as elsewhere. Kings Co. has a Pepper Grass Valley, named for a native herb (*Lepidium* sp.).

Graton (grā´ tən) [Sonoma Co.] was founded in 1904 by James H. Gray and J. H. Brush, of Santa Rosa, and was named for the former by abbreviating Graytown to Graton (F. L. Perkins).

Grave Creek [Trinity Co.], a tributary of the Trinity River, was named because of the nearby gravestone of a young Englishman, with this inscription: "In memory of Joseph Martin. Drowned in Trinity River, June 15, 1862. Age 33 years. Native of Darbyshire, England" (R. H. Cross, in Knave, Dec. 15, 1946).

Gravelly Valley [Lake Co.] is so called because the valley is strewn with gravel and debris brought by winter floods (Doyle). The names of more than twenty other geographic features contain the adjective "gravelly" or the noun "gravel."

Gravenstein (grā´ vən stēn) [Sonoma Co.]. For a variety of apple grown in the area.

Graves Grove [Del Norte Co.], a stand of redwood, was presented to the state by George F. Schwarz, of New York; it was named in honor of Henry S. Graves, of the Yale University School of Forestry and former chief forester, USFS (Doyle).

Graveyard Meadows [Fresno Co.] was named by sheepmen because two of their number were murdered and buried here (Farquhar); **Graveyard Peak** is nearby. **Graveyard Canyon** [Los Angeles Co.] and several other places were so named because of the presence of Indian burial grounds.

Gray is a term found in the names of more than twenty-five geographic features, mainly Rocks, Peaks, and Buttes, but also several Creeks and Valleys. Most of them were doubtless named because of the color; some were named for persons. Placer Co. has a Gray Horse Valley and Canyon; Plumas Co. has a Gray Eagle Valley and Creek.

Grayback was once a familiar term for the body louse; it is a common folk name for orographic features that look gray. The popular name of San Gorgonio Peak [San Bernardino Co.] is still Grayback.

Gray Eagle Creek. *See* Graeagle.

Grayson [Stanislaus Co.] was named for Andrew J. Grayson, a native of Louisiana, who came to California in 1846; he was active in the Mexican War and in later years became a well-known authority on Pacific Coast birds. The town that he founded in 1850 is shown on Gibbes's map of 1852 and is often mentioned in the Pac. R.R. *Reports*.

Grays Well [Imperial Co.] was a camp for men constructing the plank road to Yuma; it was named for Newt Gray, a Holtville pioneer and road builder.

Great is a term used with the names of a few physical features to describe their magnitude and importance. **Great Basin**, as a name for the arid plains covering most of Nevada, parts of Utah, and parts of eastern California, was applied by Frémont in 1843: "As most of the streams known to exist in it are known to lose themselves, it has been called, very properly, the 'Great Basin'" (Williamson, in Pac. R.R. *Reports* 5:1.9). **Great Central Valley** is the commonly accepted designation of the inland basin formed by the Sacramento and San Joaquin Valleys; it is often called simply "The Valley." The first to use the word "great" as a common name for both valleys was probably Lansford W. Hastings, who referred to the Great California Valley in his *Emigrants' Guide* (1845:75). Other designations were Valley of California and Great Interior Valley. The **Great Western Divide** was named by J. N. LeConte: "The Great Western Divide lies parallel with the Main Crest west of Mt. Whitney. The southern end of it is between the Kern River water shed and the headwaters of the Kaweah River. Further north it is between the Kern and Roaring River, a tributary of the Kings, and still further north between two tributaries of the Kings, ending in Mt. Brewer. I named it [1896] because it was nearly as high as the Main Crest, parallel with, and about 15 to 20 miles west of it." **Great Cliffs** [Kings Canyon NP] is the name of the mighty monolith on the Middle Fork of the Kings River, probably given by the

USGS when the Mount Goddard quadrangle was mapped in 1907–9. *For* Great Dome [Madera Co.], *see* Balloon Dome.

Greco (grek′ ō) **Island** [San Mateo Co.]. V. C. Greco's salt works were here from around 1910 (Alan K. Brown).

Green is a term applied to more than fifty geographic features. Most of them are valleys so named because of the luxuriant vegetation; some are springs and sloughs that look greenish; others were named for persons named Green. The name is occasionally a translation from Spanish *verde:* **Green Valley** [Contra Costa Co.] is shown in 1833 as Cañada Verde on the *diseño* of the San Ramon grant (322 ND). **Green Valley** [Santa Cruz Co.] is Cañada Verde on the *diseño* (1844) of the Los Corralitos grant. In combination with -field, -ville, -wood, -spot, -view, -dale, etc., "green" is the most popular adjective of color applied to communities in the United States, and California has more than twenty such combinations. **Greenwater Springs** and **Valley** [Death Valley NP] were so named because the waters have a greenish tinge. The place has been known by this name since 1884, but it did not become famous until 1905, when the discovery of copper deposits caused a short-lived boom. **Greenspot** [San Bernardino Co.] was named in the early 1900s by the Cram family, early settlers, because it was the only green spot at the upper end of the San Bernardino Valley. **Greenfield** [Monterey Co.] was laid out on the Arroyo Seco Rancho by the California Home Extension Association between 1902 and 1905; it was called Clarke City for John S. Clarke, one of its principal officers. In 1905 the Post Office Dept. declined to use the name and selected the present name from a large number submitted by residents (R. W. Dunham). There are places called Greenacres [Kern Co.], Greenbrae [Marin Co.], Green Valley [Los Angeles Co.], and Green Valley Lake [San Bernardino Co.]. *For* Green Flat [Tuolumne Co.], *see* Relief Valley; *for* Greenland [Death Valley NP], *see* Furnace Creek Ranch.

Green Creek [Shasta Co.] was named for Myron Green, who was in charge of the U.S. trout hatchery at Baird Station, 1879–83 (Steger).

Greenhorn, a common designation for a newcomer, was a favorite for place names in California in mining days; it has survived for some fifteen geographic features and for a railroad station in Plumas Co. **Greenhorn River** [Nevada Co.] is probably the oldest feature so named: "There we found a small camp of overlanders [new arrivals from the East] washing successfully for gold. They called the creek 'Greenhorn,' and showed us quite a lot of bright, shining, yellow scales" (Wistar 1:127; Aug. 25, 1849). The following story about **Greenhorn Creek** in Siskiyou Co. is still current: "A greenhorn came to Yreka in the mining days, and being a greenhorn asked around and about as to where was the best place to mine. The old-timers of course sent him off to a place where no one had found gold and where there seemed to be no likelihood. He struck it rich, and had the laugh on them—and so it was known as Greenhorn" (Stewart). Similar stories have been told about the prominent Greenhorn Mountains in Kern and Tulare Cos., and about other places so named that are still on maps or on the atlas sheets of the USGS and the USFS. Most of these names can be traced back to the Gold Rush period.

Greenlead Creek [San Bernardino Co.] was probably so named because of the presence of pyromorphite, or lead chlorophosphate.

Green Valley Creek [Sonoma Co.] may be the stream mentioned by Padre Mariano Payeras as Arroyo Verde on Oct. 21, 1822, when he accompanied a *comisionado* to Fort Ross. The official became sick and vomited "green" at this place (Arch. MSB 12:416).

Greenville [Alameda Co.] was named for John Green, a native of Ireland, who came to California in 1857; for many years he had a store and a farm at Dublin.

Greenville [Plumas Co.] was named for a man named Green who built a house here in the mid-1850s. Because his wife served meals to the miners, the house became known as Green's Hotel, and the settlement as Greenville.

Greenwater Spring [Death Valley NP] is the only name on the maps reminiscent of the once bustling town of Greenwater. The town was established in 1906 by Arthur Kunze and for some years was the center of extensive copper mining in the area.

Greenwich. *See* Tehachapi.

Greenwood [El Dorado Co.] is named for John Greenwood, son of the famous trapper and guide Caleb Greenwood. John guided a party overland to California in 1845, served in the

Mexican War, and established a trading post here before 1850. Greenwood Valley is mentioned in 1850. The place became an important mining center, and on Oct. 9, 1852, the post office was transferred from Louisville.

Greenwood: Creek, Ridge [Mendocino Co.]. Named after the settlement Greenwood, founded about 1862 at the mouth of the creek by Britton Greenwood, a member of the second Donner relief party and brother of John (of El Dorado Co.).

Grenada (grə nä′ də) [Siskiyou Co.] refers to a region originally known as Starveout because of its poor land (Ella Soulé). After the formation of the irrigation district it was renamed, possibly after Grenada Co. in Mississippi, a region known for its fertile soil. *See also* Granada.

Grider (grī′ dər): **Creek, Ridge** [Siskiyou Co.]. For William T. Grider, who settled in Seiad Valley in 1870. The creek is recorded on the county map of 1883.

Gridley [Butte Co.] was named in 1870 by the Southern Pacific for George W. Gridley, owner of the land on which the town was built.

Griffith Park [Los Angeles Co.] was named by the Los Angeles city council on Dec. 16, 1896, for its donor, Griffith J. Griffith.

Grigsby Creek [Plumas, Lassen Cos.] was named for Challan Grigsby of the USFS (Stewart).

Grigsby Soda Springs [Napa Co.] became known by this name in 1946, when T. B. Grigsby purchased the Samuel Soda Springs.

Grimes [Colusa Co.] was named about 1865 for Cleaton Grimes, who settled there in 1851.

Grindstone Creek [Glenn Co.] was named for the first commercial industry of the county: the making of grindstones as early as 1845 by Peter Lassen, Ezekiel Merritt, and other pioneers (*see* Stone). The creek is shown on Bidwell's 1844 map and on *diseños* as Rio de Capay.

Grinnell (gri nel′): **Lake** [Fresno Co.], south of Red and White Mountain, was named in Dec. 1946 by Leon A. Talbot of the Division of Fish and Game, in memory of Joseph Grinnell, professor of zoology and director of the Museum of Vertebrate Zoölogy at the University of California, Berkeley, 1908–39 (Browning 1986). **Grinnell Peak** [San Bernardino Co.], a mountain northeast of San Gorgonio Peak, was labeled Fish Creek Mountain by the USGS. Upon petition of the Sierra Club it was renamed for the zoologist who had made studies here at the beginning of the 20th century (Wheelock).

Grissom Island [San Pedro Bay, Los Angeles Co.]. For astronaut Virgil I. Grissom, 1926–67 (BGN 7902).

Griswold: Hills, Canyon, and **Creek** [San Benito Co.]. Named for "Griswold, a hard-looking customer, [who] expatiated on the qualities of his 'ranch'—squatter claim of course" (Brewer, p. 137). There is also a Griswold Creek in Tuolumne Co.

Grizzly refers to the grizzly bear, *Ursus arctos*, which has now disappeared from California but played an important role in the lore of the state; about two hundred place names preserve its memory. The majority of these names are found in the Sierra Nevada and the foothills. This does not necessarily prove that the grizzly was less common in the coast ranges: prospectors often used the term for any bear, whereas farmers distinguished the different types. Three settlements bear the name: Grizzly [Plumas Co.], Grizzly Bluff [Humboldt Co.], and Grizzly Flats [El Dorado Co.]. A number of geographic features were named Grizzly because of an incident; e.g. **Grizzly Mountain** [Trinity Co.]: "Jim Willburn, my wife's grandfather, wounded a bear on this mountain. The bear charged him before the gun could be reloaded and caught Willburn in a hug. Willburn stabbed the bear with a homemade hunting knife and finally killed it after being badly mauled. Local people then came to use the name Grizzly Mountain" (L. P. Duncan). About **Grizzly Camp** [Humboldt Co.], D. Dartt writes: "Back in 1880, Bob Pratt camped at this site with his mules and cargo destined for the New River miners. Feed being plentiful the mules were turned out to graze on the nearby slope. Sometime during the night the Packer was awakened by distress bawling of a mule. Rushing to the rescue of the animal with rifle in hand, expecting to find a Grizzly Bear tearing the mule to bits, the Packer found instead the mule down and under a large Sugar Pine log where it had slipped from the rolling ground. The animal was extricated and being somewhat disgusted the Old Packer decided to name the place 'Grizzly Camp.'" **Grizzly Creek Redwoods State Park** [Humboldt Co.] was established in 1943 and named after a tributary of the Van Dusen River.

Grossmont [San Diego Co.] was promoted as an artists' colony about 1900 by Col. Ed Fletcher and William B. Gross, realtors, and was named for the latter. The post office is listed in 1912.

Grouse, referring to the blue grouse (*Dendragapus obscurus*), is a term used for about thirty geographic features, including two high mountains in Inyo and Mono Cos., a meadow in Kings Canyon NP, and a ridge in Nevada Co.; the bird's drumming is a familiar sound in the woods.

Groveland [Tuolumne Co.]—a peaceful name, perhaps reminiscent of a town "back home"—was chosen by residents to replace the earlier name, Garrote, applied in 1850 when miners hanged a thief here. Garrote post office is listed from 1854 to about 1875. When a nearby place was called Second Garrote for a similar reason, the original Garrote became First Garrote. Hoffmann's map of 1873 still has 1st and 2nd Garrote; but on the Land Office map of 1879, the former appears as Groveland.

Grover Beach [San Luis Obispo Co.] was first named Grover, for Henry Grover, in 1892. In 1937 H. V. Bagwell renamed it Grover City.

Grover Hot Springs [Alpine Co.] was named for Alvin M. Grover, on whose homestead the springs were situated, and who was county assessor for many years (Knave, Aug. 19, 1956). Grover Hot Springs is now a state park.

Groves Prairie [Humboldt Co.] was named for Dave Groves, who squatted on the area in the late 1860s with the intention of raising horses and cattle (USFS).

Growlersburg. *See* Georgetown.

Grub Gulch [Madera Co.] and **Grub Flat** [Plumas Co.] are apparently the only survivors from Gold Rush days of several place names that contained the miners' term "grub," meaning food earned the hard way.

Grullas, Laguna de las. *See* Crane.

Grunigen (grin′ i gən) **Creek** [Tulare Co.] was so named because it runs by the home of the Grunigen family, who formerly spelled their name von Grueningen (Farquhar).

Guachama (gwä chä′ mə) **Rancheria** [San Bernardino Co.] preserves the name of the Guachama Indians, a division of the Serrano tribe.

Guadalasca (gwä də läs′ kə) [Ventura Co.]. Perhaps from Ventureño Chumash *shuwalaxsho*, a village in Big Sycamore Canyon; the meaning is unknown, but cf. *xsho* 'sycamore' (Applegate 1975:41). It was a land grant dated May 6, 1836, and Apr. 6, 1837. The Spanish name seems to contain *guad-*, from Arabic *wâdî* 'river,' found in numerous place names of Spain and Mexico, such as Guadalupe. A transcript of the *expediente* (Legis. Recs. 3:13) has the version Guadalaesa, probably a misspelling.

Guadalupe, as a name associated with the patron of Catholic Mexico, the Virgin of Guadalupe, was an extremely popular place name in early California. The modern pronunciation varies: (gwä də lo͞op′, gwä də lo͞o′ pē, gwä də lo͞o′ pā). The **Guadalupe River** [Santa Clara Co.] was named Río de Nuestra Señora de Guadalupe by the Anza expedition on Mar. 30, 1776, in honor of the Virgin. The name is frequently mentioned in documents. It appears on Font's map of the Bay region (1777), again on Eld's sketch of 1841, and on Duflot de Mofras's map of 1844. **Guadalupe Lake** [Santa Barbara Co.] was called Laguna Grande de San Daniel and Laguna Larga by the Portolá expedition on Sept. 1, 1769. The lake appears with its present name on the maps of the Land Office in the 1850s, and on the Parke-Custer map of 1855 it is shown as Guadalupe Largo. As the name of a settlement, Guadalupe appears on Narváez's Plano of 1830, and a rancho of that name is mentioned in 1834 (PSP 79:106). On Apr. 8, 1837, the name was applied to a land grant. Wilkes's map of 1841 shows the Santa Maria River as Río Guadalupe. The name was adopted for the post office in 1872. **Guadalupe Valley** [San Mateo Co.] was apparently called Cañada de Guadalupe y Visitación by Manuel Sánchez on Feb. 8, 1835, in a petition for a grant (Docs. Hist. Cal. 1:482). The name was included in the name of the land grant to Jacob P. Leese on July 31, 1841. **Guadalupe Valley** and **Mountains** [Mariposa Co.] are named for the vanished mining town, Guadalupe, shown on Gibbes's map of 1852 and again on Hoffmann's map of 1873. The name has no connection with the river named, in 1806, Nuestra Señora de Guadalupe; this was probably the Stanislaus. Guadalupe y Llanitos de los Correos, a land grant in Monterey Co.—dated Sept. 23, 1831, May 18, 1833, and Feb. 10, 1835—has left no trace in current toponymy. A now-vanished town in Nevada Co. was called Walloupa by the miners in 1852, for an Indian chief, Guadalupe (Co. Hist.: Nevada 1880:71). For Río de Nuestra Señora de Guadalupe, *see also* Stanislaus River.

Guajome (wä hō´ mē) [San Diego Co.], as the name of a land grant dated July 19, 1845, is derived from Luiseño *wakhaumai* according to Kroeber. Engelhardt (*San Juan Capistrano*, p. 244) lists Guajaumere and Guajaimeie among the rancherias mentioned in the baptismal register of Mission San Juan Capistrano. The origin is apparently Luiseño *waxáawu-may* 'little frog', from *waxáawut* 'frog'.

Gualala (wä lä´ lə): **River**, post office, **Point** [Sonoma, Mendocino Cos.]. The origin of the name has been discussed in detail by Robert Oswalt (1960, 1964:8). The earliest recording of the name is from 1813, when a Russian explorer used the term "Wallalakh" to refer to a group of Indians north of Fort Ross. The origin is Southwestern (Kashaya) Pomo *walaali*, an abbreviated form of *qhawalaali*, the name of the Indian village that formerly stood at the site of the town of Gualala; it is derived from *ahqha walaali* 'water go-down-place'. The name was apparently applied to the river by Ernest Rufus, captain of Sutter's Indian Company in the Micheltorena campaign and grantee of Rancho German; the *diseño* of the grant shows the name Arroyo Valale (1846). English speakers subsequently folk-etymologized the name to Walhalla, and the Coast Survey used the spelling Walalla until recent years. The Spanish spelling Gualala was used for the name of the post office on Sept. 9, 1862: the name-giver wrongly assumed that the name was Spanish and hence used this spelling because Spanish *gua* is pronounced something like English *wa*. The spelling Gualala was accepted by the Land Office maps in the 1870s and is now generally used.

Guapa (wäp´ ə) [Riverside Co.]. Now a part of Prado Basin County Park, this place was mentioned in Hugo Reid's 1852 letters as a Gabrielino rancheria; earliest spellings were Guapia (1821) and Haupa (1845). There have been many subsequent spellings and speculations about etymology (Gunther).

Guaslay. *See* Cuaslui Creek.

Guasna. *See* Huasna River.

Guasti (gwä´ stē) [San Bernardino Co.] was a railroad station called South Cucamonga when the Southern Pacific line was built in the 1890s. When the Italian Vineyard Company began its operations here, soon after 1900, the new name was given for Secondo Guasti, the founder of the company and for many years a leading viticulturist in the state.

Guatay (gwä´ tī): **Creek, Valley**, town, and **Indian Reservation** [San Diego Co.]. Probably from Diegueño *wataay* 'big house', i.e. 'ceremonial house', from *'ewaa* 'house' plus *'etaay* 'big' (Langdon).

Gubernadora, Cañada. *See* Gobernadora Canyon.

Gubserville [Santa Clara Co.] was established as a post office on the stage route from San Jose to Saratoga on July 5, 1882, and named for Frank Gubser, the first postmaster.

Gudde (go͞od´ ē) **Ridge** [Contra Costa Co.]. Five miles of wooded ridge northeast of Oakland were dedicated in 1970 to the memory of Erwin G. Gudde (1889–1969), professor of German at Berkeley and author of *California Place Names;* cf. BGN, list 7003, and *Names* 19:33 (1971).

Guejito (wā hē´ tō) [San Diego Co.]. Guejito y Cañada de Palomea was the name of a land grant dated Sept. 20, 1845; Sierra de Guejito, shown on the *diseño*, seems to refer to Roderick Mountain and the surrounding hills. *Guejito* is perhaps a diminutive of *guijo* 'pebble'; or it may be for *güejita*, a diminutive of *güeja*, which in northwest Mexico refers to a gourd vessel (from Yaqui *bueha;* Santamaría). Palomea is probably a misspelling, since Cañada Paloma 'pigeon valley' is mentioned in 1845 (DSP Ben. P and J 2:73).

Guenoc (gwen´ ok) [Lake Co.]. The name was given to a land grant dated Dec. 8, 1844, and Aug. 8, 1845. It appears on the *diseños* as the name for a small stream and for a *laguna* in the territory of the Lake Miwok, who called it *wénok pólpol* 'Guenoc lake'. This name may contain Lake Miwok *wéne* 'medicine'; or it may be a borrowing from Wappo *wénnokh* 'Southern Pomo' (Callaghan). *See* Novato; San Anselmo; San Quentin.

Guerneville (gûrn´ vil) [Sonoma Co.] was settled in 1860. When the post office was established in the 1870s, it was named for George E. Guerne, a native of Ohio, who had built a saw- and planing-mill there in 1864.

Guernsey [Kings Co.] was named for James Guernsey, landowner. The name first appeared on the Santa Fe railroad map of 1902.

Guesesosi. *See* Quesesosi.

Guijarral (gē hə ral′) [Fresno Co.] is Spanish for a 'place where cobblestones are found', from *guijarro*.

Guijarros, Punta de los. *See* Ballast Point.

Guilicos (wil′ i kōs) [Sonoma Co.]. From Lake Miwok *wíilok* 'dusty' (Callaghan) plus the Spanish masculine plural suffix *-os*. This is the name of a land grant dated Nov. 13, 1837; it was derived from the name of an Indian tribe, mentioned on Apr. 10, 1823, by Padre Amorós as *la nacion Guiluc* (Arch. Arz. SF 4:2.84). The name appears repeatedly in mission records, with various spellings. The name of the grant was spelled Guilucos until 1859, when the spelling Los Guilicos is shown on a surveyor's plan of the rancho. There was a Lake Miwok village, *wíilok-yomi* 'dusty place', south of Middletown. The name Los Guilicos for the station of the Northwestern Pacific is shown on the maps after 1890.

Guillemot (gil′ ə mot) **Island** [Monterey Co.]. This island offshore of Point Lobos is named for the pigeon guillemot (*Cepphus columba*), a bird that nests there (Clark 1991).

Guinado. *See* Quinado Canyon.

Guinda (gwin′ də) [Yolo Co.] is the Spanish word for the morello, a variety of sour cherry; it was given to the station by the Southern Pacific in the early 1890s because an old cherry tree was then standing at the southeast corner of the townsite. The Spanish word is in fact pronounced (gēn′ dä).

Gulch evolved as a term for a ravine in America early in the 19th century. It was rarely used as a true generic term until the discovery of gold in California, when nearly every ravine, particularly those that yielded gold, became a "gulch." Now it is in general use throughout the state for a small, short canyon. The variant form "gulsh" also occurs.

Gull, referring to the state's most common aquatic bird (genus *Larus*), is the name of an island south of Santa Cruz Island and a number of rocks and coves along the coast, as well as Gull Lake in Mono Co. *See* Gaviota.

Gulling [Plumas Co.] was named for Charles Gulling, an early lumberman and organizer of the Grizzly Creek Ice Company (Myrick).

Gulnac (gul′ nak) **Peak** [Santa Clara Co.] was named either for William Gulnac or for one of his sons. Gulnac came to California in 1833 and later was the grantee of Campo de los Franceses.

Gunther Island [Humboldt Co.]. *See* Indian: Indian Island.

Guntleys. *See* Christine.

Gurley Creek. *See* Gerle Creek.

Gushee (gush′ ē) **Gulch** [San Mateo Co.]. Horace Gushee's ranch was here in the 1860s (Alan K. Brown).

Gustine (gus′ tēn) [Merced Co.] was laid out by land baron Henry Miller and named in memory of his daughter Augustine, who had been thrown from a horse and killed. The station is shown on the Official Railway Map of 1900.

Guthrie Canyon [Riverside Co.], in Cleveland National Forest, was named in memory of John D. Guthrie, a California Division of Forestry employee who lost his life in the Decker fire of Aug. 1959.

Guthrie Creek and **Gulch** [Humboldt Co.] were probably named for Alexander P. Guthrie, who settled in the Eel River district in the 1860s.

Guyapipe. *See* Cuyapipe.

Guyot (gē′ ō), **Mount** [Sequoia NP], had its name suggested by F. H. Wales; the term was applied by Capt. J. W. A. Wright and his party on Sept. 3, 1881, in honor of Arnold H. Guyot (1807–84), a native of Switzerland, explorer of the Appalachian Mountains, and for many years professor of geography and geology at the College of New Jersey (later Princeton). **Guyot Pass, Creek**, and **Flat** were named after the mountain by the USGS in 1907.

Gwins Peak. *See* Liberty Cap.

Gypsum, a common mineral, calcium sulphate, gives its name to several places in the state—including two railroad stops, Gypsum [Orange Co.] and Gypsite [Kern Co.], as well as Gyp Hill in Death Valley, named in 1937 by Don Curry, park naturalist.

H

Ha Amar (hä ə mär′) **Creek** [Humboldt Co.]. From Yurok *ho'omer* (Waterman, map 11), with no known etymology.

Habra. *See* La Habra.

Hacienda (hä sē en′ də), a Spanish word for 'estate, farm', is found as a place name in Alameda, Kings, Los Angeles, and Sonoma Cos.

Hackamore [Modoc Co.] is a western American word for a halter used mainly in breaking horses. It is from Spanish *jáquima* 'headstall of a bridle', and in this form (also spelled Jaquina) it was applied to the station of the Nevada-California-Oregon Railroad in 1910. It was possibly a folk-etymological rendering of an Indian word. In 1928 the Southern Pacific changed the spelling to the American form, which had always prevailed in speech.

Haeckel (hek′ əl), **Mount** [Kings Canyon NP], was named in 1895 by T. S. Solomons in honor of Ernst Haeckel (1834–1919), the best known of the German Darwinists and author of the law of recapitulation. *See* Evolution.

Hagata (hag′ ə tô) **Canyon** [Lassen Co.]. For John Hagata, rancher and sheepman (Morrill).

Hagerman (hā′ gər mən) **Peak** [Santa Clara Co.], north of Pacheco Pass, was probably named for one of the descendants of George "H. F." Hageman, a native of Germany and a hotel-keeper in Gilroy township in 1875.

Hahn, Mount [Death Valley NP], was named for C. F. R. Hahn, who guided the Lyle detachment of the Wheeler Survey into Death Valley and disappeared in Aug. 1871 near Last Chance Spring.

Haiwee (hā′ wā): **Creek**, post office, **Reservoir** [Inyo Co.]. From Panamint *heewi* 'dove' (Dayley), or from a related form in some other Numic language. Haiwee Meadows are mentioned by the Nevada Boundary Survey (*Sacramento Union*, Aug. 10, 1861), and Haiwee Meadows as a settlement is recorded on the von Leicht–Craven map (1874), several miles east of the road. The spelling Haway is also found, but the USGS, when it surveyed the Olancha quadrangle in 1905, accepted the spelling used by the Whitney Survey map.

Halagow (hal′ ə gou) **Creek** [Humboldt Co.]. From Yurok *helega'a*, referring to the boat dance that formed part of the annual world-renewal ceremony centered at Pecwan (Waterman, p. 240).

Halcyon (hal′ sē ən) [San Luis Obispo Co.] was established in 1903 by a group of theosophists (Hine, p. 54). Halcyon means 'calm, peaceful' and is found repeatedly as a place name in the southeastern United States.

Half Dome [Yosemite NP] was the name given by the Mariposa Battalion in 1851 to the split mountain that is also known as South Dome.

Half Moon Bay [San Mateo Co.] was the name given by the Coast Survey to the bay because of its shape. It appears first on the sketch of 1854 as the name for a triangulation station on the hill east of the town. The settlement was first known as Spanish Town and appears thus on the maps of the Whitney Survey. The post office is listed as Halfmoon Bay in 1867, and this name for the town appears on the Land Office map of 1879. Since about 1905 the Post Office Dept. has spelled the name Half Moon Bay. By a decision of the BGN (May–Aug. 1960), this spelling in three words is now official for both the town and the bay.

Half Moon Lake [El Dorado Co.] was so named because of its shape, probably by the USGS when the Pyramid Peak quadrangle was mapped in 1889.

Half Potato Hill. *See* Potato: Potato Hill.

Halleck [San Bernardino Co]. *See* Oro: Oro Grande.

Halleck Creek [Marin Co.] commemorates Henry W. Halleck, co-claimant in 1852 of the part of the Nicasio grant through which the creek flows. Halleck was California's secretary of state and custodian of the Spanish archives after the Mexican War, and was general-in-chief of the Union armies, 1862–64.

Halloran Springs [San Bernardino Co.] is a name of unknown origin, but it was in existence in 1875, when Eric Bergland led a detachment of the Wheeler Survey through the region (Belden).

Halls Flat [Lassen Co.]. Named by the USFS for W. G. Hall and sons, who had a summer cattle camp here about 1880.

Hambre. *See* Alhambra Valley.

Hambright Creek [Glenn Co.]. For Robert Hambright, a Mexican War veteran, who had settled on the bank of the stream. It is shown as Hambright Slough on the Mining Bureau map of 1891.

Hamburg [Siskiyou Co.] was established as a mining camp in the fall of 1851 and named by Sigmund Simon, probably after the seaport in Germany. It was also known as Hamburg Bar (Co. Hist.: Siskiyou 1881:218).

Hamilton, Mount [Santa Clara Co.]. Named by William Brewer and Charles Hoffmann of the Whitney Survey, for the Rev. Laurentine Hamilton, who accompanied the two surveyors to the top of the mountain on Aug. 26, 1861. Brewer, in a letter dated Jan. 31, 1888, states: "Hoffmann and I were encumbered with our instruments, and, as we neared the summit, Mr. Hamilton pushed on ahead of us, and reaching it, swung his hat in the air and shouted back to us: 'First on top—for this is the highest point.' . . . When we worked up our notes in the office, and failing to find any old name in use, Hoffmann and I, after discussing two names, agreed to call it Mt. Hamilton." The date of naming is confirmed in Hoffmann's field notes: on Aug. 26 he calls the mountain the highest peak of the Mount Diablo range; on Aug. 27 he calls it Mount Hamilton. Hamilton was at that time preaching in San Jose. In 1864 he moved to Oakland, where he died in the pulpit of his Independent Church on Easter Sunday, 1882 (F. J. Neubauer). By a strange irony of fate it was later discovered that the peak climbed by Brewer and Hoffmann not only had an old name, Isabel (*q.v.*), but also that the peak 2 miles southeast of the site of the Lick Observatory was 14 feet higher. The USGS promptly (1895) placed the old name on this peak; but the entire mountain remains known as Mount Hamilton.

Hamilton City [Glenn Co.] was started by a sugar company in 1906 and named for J. G. Hamilton, promoter of the town. The railroad name is Hamilton. The Hamilton mentioned in the *Statutes* of 1851 as the county seat of Butte Co. was 30 miles southeast on the Feather River.

Hamilton Field [Marin Co.] was named by the War Dept. in 1932 in honor of Lt. Lloyd A. Hamilton, of the 17th Aero Squadron, U.S. Army, who won the Distinguished Service Cross for heroism in Belgium and was killed in action in France in 1918. Before the site was given to the government by the county, it was known as Marin Meadows.

Hamilton Lake. *See* Precipice Lake.

Hammarskjold (ham′ ər shōld) **Grove** [Humboldt Co.], a stand of redwoods on Pepperwood Flat, was selected by the Save-the-Redwoods League in 1962 and named for Dag Hammarskjöld (1905–61), the Swedish secretary-general of the United Nations.

Hammer Field [Fresno Co.], an army air base, was activated on Aug. 1, 1941, and named in memory of Lt. Earl M. Hammer; he was the first aviator from California to achieve an air victory in World War I and the first to be killed in action while on a mission over Germany.

Hammerhorn Mountain [Mendocino Co.] was a name given in the late 1860s, by a man named Fleming, to the mountain on which he had shot a deer that had round knobs at the tips of the horns (USFS).

Hammonton [Yuba Co.] was named in 1905 for W. P. Hammon, an official of a gold-dredging company.

Hampshire Rocks [Placer Co.] were named after New Hampshire, the "Granite State," because they were of "real hard granite" (*Sacramento Union*, July 11, 1864).

Hanaupah (hə nōō′ pä, han′ ə pä): **Canyon, Spring** [Death Valley NP], is from the local Panamint Indian language, apparently a combination of *hunuppin* 'canyon' and *paa* 'water' (McLaughlin).

Hanford [Kings Co.]. Named in 1877 by the Central Pacific (now the Southern Pacific) for its treasurer, James Hanford.

Hangtown Creek [El Dorado Co.] preserves the nickname of Placerville, which was called Hangtown in the 1850s because its citizens had hanged several robbers in 1849. *For* Hangtown Crossing [Sacramento Co.], *see* Mill: Mills.

Hanksite. *See* Westend.

Hanna Mountain [Mono Co.]. For Thomas R. Hanna, son-in-law of John Muir. Hanna made the preliminary survey of the Hetch Hetchy project in 1905 and was owner of the May Lundy Mine from 1920 to 1940.

Hansen Dam [Los Angeles Co.]. Named by the U.S. District Engineer's Office after Hansen Hill, which was at the site of the spillway. The name had been used locally because Dr. Homer A. Hansen had his home on the hill for many years (N. B. Hodgkinson).

Hansonville. *See* Rackerby.

Hapaha (hä′ pə hä) **Flat** [San Diego Co.]. Presumably an Indian name, but the source has not been established.

Happy, a typical folk name, was as popular with the prospectors in the Gold Rush days as with modern real estate developers. **Happy Camp** [Siskiyou Co.], an old mining town, was named in 1851. H. C. Chester, who interviewed Jack Titus in 1882 or 1883, states that Titus named the camp because his partner James Camp, upon arriving there, exclaimed: "This is the happiest day of my life." Redick McKee mentions the camp on Nov. 8, 1851, as "Mr. Roache's 'Happy Camp,' at the place known as Murderer's Bar" (Indian Report 1853:178). **Happy Isles** [Yosemite NP] was named in 1885 by W. E. Dennison, guardian of Yosemite Valley: "I have named them the Happy Isles, for no one can visit them without for the while forgetting the grinding strife of his world and being happy" (Farquhar). **Happy Gap** [Kings Canyon NP] was named before 1896. According to J. N. LeConte, anyone who succeeds in getting a pack train to this point at once perceives the appropriateness of the name (Farquhar).

Haraszthy (hä′ rəs tē): **Creek, Falls** [Sonoma Co.]. For Agoston Haraszthy (ca. 1812–69), a Hungarian immigrant who pioneered the cultivation of wine grapes in Sonoma Co.

Harbin: Hot Springs, Creek, and **Mountain** [Lake Co.]. For James M. Harbin, who came overland in 1846 and was claimant of part of Salvador Vallejo's Napa grant in 1853. About 1857 he settled at the springs that became known as Harbin Hot Springs.

Harbison Canyon [San Diego Co.]. Named for John S. Harbison, apiarist, who came to San Diego Co. in 1869 and settled in this canyon (Co. Library).

Harbor City [Los Angeles Co.], in the "shoestring strip," was named Harbor Industrial City when it was laid out by W. I. Hollingsworth in 1912. When the expected connection with Los Angeles harbor (then being developed by dredging the slough) did not materialize, the less pretentious form of the name was adopted.

Harden Lake [Yosemite NP] was apparently named by the Whitney Survey for the owner of Harden's Ranch on the South Fork of the Tuolumne. Both ranch and lake are spelled Hardin on the Hoffmann-Gardner map of 1867.

Hardin Butte [Lava Beds NM]. Named in memory of Maj. Charles B. Hardin, who died on June 7, 1939; the name was approved by the BGN in June 1940.

Harding Canyon and **Creek** [Orange Co.]. For Ike Harding, who had a goat farm in the canyon (Stephenson).

Hardin Mountain [Del Norte Co.]. Named for Silas T. Hardin, one of the pioneer settlers of Summit Valley (Co. Hist.: Del Norte 1953).

Hardins: Flat and **Hill** [Tuolumne Co.]. For an eccentric Englishman, "Little Johnny 'Ardin," who had a sawmill on the flat in the 1850s (Paden and Schlichtmann).

Hardscrabble, an old American place name, came to be used on the frontier wherever the scrabble for daily grub was hard. It was repeatedly used for mining camps in the Gold Rush days. Today only two Hardscrabble Creeks are left, in Mendocino and Del Norte Cos. The once famous mining town in Siskiyou Co. was named in the 1850s by two former residents of Hardscrabble, Wisconsin (Luddy). *For* Hardscrabble in Amador Co., *see* Ione.

Hardwick [Kings Co.]. Named for a traffic official of the Southern Pacific when the branch road was built in the 1890s. The post office is listed in 1910.

Harina, Arroyo de la. *See* San Lorenzo: Creek.

Harkless Flat [Inyo Co.]. Named for a pioneer rancher near Big Pine, who cut and sold wood in this area (Brierly).

Harkness, Mount [Lassen NP], originally bore the name Mount Juniper because of a single fine specimen of the Sierra juniper on its slope; the lake north of it is still called Juniper Lake. The present name, shown on the Mining Bureau map of 1891, was given in honor of Dr. Harvey W. Harkness, president of the California Academy of Sciences in the 1890s.

Harlan Canyon [Riverside Co.], in Cleveland National Forest, was named in 1960 in memory of Nelson D. Harlan, a USFS employee who lost his life in the Decker fire of August 1959.

Harlem Springs [San Bernardino Co.] was so named because, in 1887, the promoters of the resort accepted the proposal of Mrs. Crawford, a property owner nearby, to name "their" town and hotel in remembrance of her alma mater, the Harlem School for Young Ladies, New York City.

Harmony, always a favorite name for pioneer settlements, was especially popular in Civil War days. Lippincott's *Gazetteer* of 1868 lists twenty-four towns with this name. **Harmony** [San Luis Obispo Co.] was the name applied to the valley by settlers in the 1860s. When the post office was established in 1915, the name was adopted at the suggestion of Marius G. Salmina, of the Harmony Valley Creamery Association. **Harmony Ridge** [Nevada Co.] was named after the two Harmony Mines on the south slope.

Harper. *See* Costa Mesa.

Harper Lake [San Bernardino Co.]. For J. D. Harper, who lived at the edge of the lake, on what was later known as Black's Ranch.

Harrill's Mill. *See* Mill: Millville.

Harris [Humboldt Co.]. Probably for William C. Harris, a native of Wisconsin, who is listed as the postmaster and hotelkeeper at Harris in McKenney's Directory of 1883–84.

Harrisburg Flat [Death Valley NP], a place name long in use locally, commemorates "Shorty" (Frank) Harris, who was one of the most colorful and best-known "single-blanket jackass prospectors" of the Death Valley region from the 1890s until his death in 1934. This particular place became associated with Shorty's name because he found a rich deposit of gold here in 1906.

Harrisburgh. *See* Warm Springs.

Harrison, Mount [San Bernardino Co.], was named when, in the 1880s, Myron H. Crafts named three peaks in the region in honor of three presidents: James Garfield, Abraham Lincoln, and Benjamin Harrison (1833–1901). Only the last is still shown on maps (*see* Crafton).

Harrison Gulch [Shasta Co.]. For W. H. Harrison, the first judge of Shasta Co., who settled on the North Fork of Cottonwood Creek in 1852 (Steger).

Harrison Pass [Sequoia NP], probably used by sheepmen as early as 1875, came in the 1880s to be known by its present name, which is for Ben Harrison, part Cherokee Indian, who herded sheep in the upper Kern River region (Farquhar).

Harter [Sutter Co.]. Named in 1908 by Clyde B. Harter, owner of the land, for his father, George Harter, a pioneer. The name replaced the name Las Uvas ('the grapes'), which the Northern Electric had given to the station in 1906 because the right-of-way crossed a vineyard.

Hartoum [San Bernardino Co.] was formerly called Khartoum, after the city in the Sudan; it fitted into the scheme of fanciful railroad names in the desert—Bengal, Siam, Java, Nome, Klondike, etc. The Santa Fe Railroad has used the present form since 1903.

Harvard, Mount [Los Angeles Co.], formerly called South Gable Promontory, was named in honor of Harvard University when, on Apr. 7, 1892, the university's president, Charles W. Eliot, visited the nearby point where the Harvard photographing telescope had stood in 1889–90 (Reid, 364 ff.).

Harwood, Mount [San Bernardino Co.], in the San Gabriel Mountains, is named in memory of Aurelia Squire Harwood, educator and conservationist, who died in 1928 (BGN, July–Sept. 1965).

Haselbusch. *See* Hazelbusch.

Haskell: Peak, Creek [Sierra Co.]. Named for Edward W. Haskell, father of Burnette G. Haskell, founder of the Kaweah Cooperative Colony in Tulare Co. (*see* Colony). Edward Haskell had a ranch for many years at the foot of the peak (Oscar Berland).

Hastings: Creek, Cut, Slough, and **Tract** [Contra Costa Co.]. Probably named for Lyman H. Hastings, a native of Ohio and a farmer at Marsh Creek, who discovered quicksilver on the east side of Mount Diablo in 1860.

Hatchatchie. *See* Hetch Hetchy.

Hatchet: Creek, Mountain, and **Mountain Pass** [Shasta Co.]. Named, according to Steger, because Indians "slyly appropriated" a hatchet from a party of immigrants.

Hat Creek [Lassen NP], but not Hat Mountain, is mentioned in early reports (Beckwith, Brewer). According to Steger, the stream was named when a member of the party blazing the trail for the Noble route in 1852 lost his hat in the stream; but the Achumawi name for the creek was *hatiwīwi* (David L. Olmsted, *Achumawi dictionary, UCPL* 45.24 [1966]), and this may point to an Indian origin of the name. Hat Creek post office was established in 1884, but the name was changed to Cassel in 1888; the later Hat Creek post office was established in 1909. **Hat Mountain** may have been named independently because of its shape. For mountains the name is used elsewhere: there are Hat Mountains in Lassen and Shasta Cos.

Haupt (hôpt) **Creek** [Sonoma Co.] commemorates Charles Haupt, a German immigrant, who settled here in the 1860s, married a Pomo girl, and was held in high esteem by the Indians of the district (Leland). It is shown as Charley Haupt Creek on the county map of 1879.

Havasu (hav′ ə so͞o) **Lake** [San Bernardino Co.], an artificial lake on the Colorado River, was named by John C. Page; the name was approved by the BGN in 1939. The origin is Mojave *havasúu* 'blue', *'ahá havasúu* 'Colorado River south of Parker Dam' (lit. 'blue water'). The name was given because of the contrast with the otherwise reddish color of the Colorado River (Munro).

Havens Anchorage [Mendocino Co.] was a name given to the anchorage by the Coast Survey before 1855. According to the County History (1880:380–81), it honors a "commander of a Government coast surveying vessel by the name of Haven." In fact, in 1855 Lt. Edwin J. De Haven was an assistant in the Coast Survey and in charge of the *Arago*, then off the coast of Texas; but since his name cannot be connected with the California place, it is more likely that the anchorage was named for H. W. Havens, a member of a detachment of the *Laura Virginia* party (*see* Humboldt Bay), which explored along the coast north of the Golden Gate in 1850 (cf. Co. Hist.: Mendocino 1880:96 ff.).

Havilah (hav′ i lə) [Kern Co.] was named in 1864 by the founder, Asbury Harpending, a native of Kentucky, after the Biblical gold land, Havilah, mentioned in Genesis 2:11.

Haway Creek. *See* Haiwee Creek.

Hawkins, Mount [Los Angeles Co.], was named by the USFS for Nellie Hawkins, a popular waitress at nearby Squirrel Inn about 1890.

Hawkins Peak [Alpine Co.] was named for John Hawkins, who squatted on a ranch east of the peak in 1858 (Maule).

Hawthorne [Los Angeles Co.] was named about 1906 for Nathaniel Hawthorne (1804–64), the American novelist, by Mrs. Laurine H. Woolwine, daughter of H. D. Harding, who was one of the founders of the town.

Hay, an important term to immigrants and settlers, occurs in numerous place names. **Hayfork: River, Mountain, Bally**, town [Trinity Co.], is a cluster name dating from the 1850s. When the Ruch family settled in the region in 1852, it was known as Hayfields because it was the largest farming section of the county. Later the name Hayfork was applied to a branch of the Trinity River, and this in turn gave the name to the town. The mountain called Hayfork Bally appears as Hayfork Baldy on some maps (*see* Bally). **Hayfield Reservoir** [Riverside Co.], a site established by the Metropolitan water district, had been humorously named the Hayfields because a thin growth of grass there was used by cattlemen for pasturage. **Haypress Creek** and **Valley** [Sierra Co.] recall a handmade hay press of "early days," which can still be seen about half a mile from the Forest Service station (Stewart). The name occurs elsewhere in the state: Haypress Flat [Alpine Co.], Haypress Meadow [Siskiyou Co.].

Hayden Hill [Lassen Co.] was named for J. W. Hayden, who located several mines here in 1869. A post office named Hayden was established on Apr. 13, 1871; after 1878 it was called Haydenhill.

Haystack. About twenty orographic features resembling the outline of a haystack are so named, usually in combination with Hill, Peak, or Mountain; two elevations in Siskiyou and Del Norte Cos., however, are named simply Haystack.

Hayward [Alameda Co.] was the site where William Hayward opened a hotel in 1852, after settling inadvertently on another part of Guillermo Castro's Rancho San Lorenzo. When Castro laid out the present town in 1854, he retained the name. The post office was established on Jan. 6, 1860, with the name Haywood, although William Hayward is listed as postmaster. On Mar. 22, 1880, the name was changed to Haywards, and on Jan. 15, 1911, to the present spelling.

Hazard [Los Angeles Co.] was established as a post office on Oct. 2, 1950, and named for Henry G. Hazard, who had come to California in 1853 and was mayor of Los Angeles in 1889.

Hazel, i.e. the native hazel bush or hazelnut (*Corylus cornuta*), occurs in the name of only a few geographical features, the oldest of which is probably Hazel Green Creek just west of Yosemite NP, named by the Mariposa Battalion in 1851 (Bunnell, p. 316). **Hazel Creek** post office [Shasta Co.] was named in 1877, after having been called Portuguee since 1870. A site in Santa Cruz Co. was called Los Avellanos ('the hazel bushes') de Nuestra Señora del Pilar by the Portolá expedition on Nov. 23, 1769 (Crespí, p. 239).

Hazelbusch [Butte Co.] was the name applied to the station by the Northern Electric Company about 1903, for one of the Hazelbusch families from Germany who had settled there in the 1870s. On some maps it is spelled, probably correctly, Haselbusch.

Healdsburg [Sonoma Co.] was the name given by the Post Office Dept. in 1857, for Harmon G. Heald, who had a trading post there beginning in 1846 and built the first store in the town in 1852. From 1854 to 1857 the post office name was Russian River.

Hearst [Butte Co.] was named as a Southern Pacific railway station in the 1890s, for George Hearst (1820–91), U.S. senator from 1887 to 1891. **Hearst** [Mendocino Co.] was established as a post office in 1892; it was probably also named in memory of George Hearst, who had died the preceding year. **Hearst San Simeon State Historical Monument** [San Luis Obispo Co.] is the site of the castle built by William Randolph Hearst (1863–1951), newspaper magnate and son of George Hearst.

Heaton Station. *See* Cisco.

Heber (hē′ bər) [Imperial Co.] was founded in 1903 by the California Development Company and named for its president, A. H. Heber, a leader in the development of the Imperial Valley.

Hechicera (ā chi sâr′ ə), **Cerro de la** [San Diego Co.]. Spanish for 'hill of the sorceress'.

Hecker Pass [Santa Clara Co.] was named for Henry Hecker, a county supervisor, when the highway was opened in 1928. The monument erected on the pass bears this legend: "This testimonial dedicated to Henry Hecker, whose foresight made possible the completion of the Yosemite-to-the-Sea Highway, May 27, 1928." Henry Hecker was the nephew of Friedrich Hecker, leader of the Baden insurrection of 1848, "Latin farmer" at Belleville, Missouri (*see* Hilgard), and colonel of a Union regiment in the Civil War.

Hector [San Bernardino Co.] was named as a Santa Fe station in 1902, after the Trojan warrior featured in the *Iliad*. Other names on this sector were also chosen from classical literature.

Hedges. *See* Tumco.

Hedionda (hed ē on′ də) **Creek** [Santa Clara Co.] contains a Spanish adjective meaning 'fetid, stinking', which was repeatedly applied to creeks or ponds with unpleasant odors. In the Las Animas land grant case, in 1861, Manuel Larios testified that the creek which empties into Pescadero Creek was so named "from the fact of being stinking water." Two days later, John Gilroy made the more refined statement that "the Hedionda Creek derives its name from a sulphur spring." But two days later still, in the Juristac case, he too labeled it "stinking spring" (*WF* 6:372); *see also* Agua.

Heenan Lake [Alpine Co.]. Probably named for a Mr. Heenan who worked in the Leviathan Mine nearby and was killed in the 1860s by a blast (Maule, p. 5).

Heffernan Honor Grove [Humboldt Co.], a stand of redwoods in Prairie Creek State Park, was established in 1966 in honor of Helen Heffernan, former chief of the California Bureau of Elementary Education.

Helen is probably the second most popular feminine name for lakes, after Elizabeth. R. B. Marshall of the USGS was so infatuated with the name that he bestowed it on three lakes in the Sierra: one east of Kuna Crest, for the daughter of George Otis Smith, director of the USGS; another at the East Fork of Cherry Creek, for Helen Keyes, the daughter of Col. William W. Forsyth; and a third northeast of Muir Pass, for Muir's daughter, Helen Funk. This is all the more astonishing because there was already a Helen Lake (now Starr King Lake) in Yosemite NP. **Helen Lake** [Lassen NP] was named in 1864 for Helen Tanner Brodt, the first white woman to ascend Lassen Peak. **Helendale** [San Bernardino Co.] was originally a Santa Fe station called Point of Rocks; it was named Helen on Dec. 15, 1897, for the daughter of A. G. Wells, vice president of the company. On Sept. 22, 1918, the name was changed to Helendale (Santa Fe). **Mount Helen** [Contra Costa Co.] was named for Helen Lillian Muir Funk, the younger daughter of John Muir (BGN, list 1995). *For* Mount Helena, *see* Saint Helena, Mount.

Helena (hə lē′ nə) [Trinity Co.]. Called North Fork until 1891, when, to avoid confusion with another North Fork post office in the state, the town was renamed after Helena Hall Meckel, the wife of pioneer businessman Christian Meckel (Alice Goen Jones, p. 205).

Heliograph Hill. *See* Telegraph: Telegraph Peak.

Helisma. *See* Burson.

Helix (hē′ liks), **Mount** [San Diego Co.]. Named for the trail that winds around to the summit of the cone-shaped mountain; the term was applied by Capt. Rufus K. Porter in the early 1870s. The boom town conceived and named Helix in 1887 did not flourish.

Hell, referring to the realm of fallen angels, was never as popular in place names as the name of His Satanic Majesty. A few mining camps that included the word in their names dropped it when the country became respectable; the only survivor is **Helltown** [Butte Co.], shown on the Chico atlas sheet 5 miles northeast of Paradise. **Hellhole**, a canyon between Montezuma and Borrego Valleys [San Diego Co.] was named by W. Helm, an early cattleman. The name **Hell-for-Sure Pass** [Fresno Co.] was applied to the Baird trail by J. N. LeConte in 1904; he may have thought of the old Hell-for-Sartain Pass in the southern Appalachians (Drury). **Hell's Hollow**, near the town of Bear Valley [Mariposa Co.], was so called because of the many accidents to men and animals traversing the rough trail in early days (Doyle). **Hell Gate** [Death Valley NP] was so named because travelers are struck by the marked change in temperature when crossing the pass on a hot day (Death Valley Survey). **Hell Hole** [Placer Co.], a deep gorge on the Rubicon River, is a favorite of deer hunters (Morley). Devils Half Acre in Shasta Co. is matched by Hells Half Acre in Lake Co. *For* Hell-Out-For-Noon, *see* Alpha. (*See also* Bumpass Hell.)

Hellen, Mount. *See* Saint Helena, Mount.

Hellgrammite (hel′ grə mīt) **Lake** [Plumas Co.]. For the larva of an insect, *Corydalus cornutus*, used as fish bait (BGN list 8104).

Helm: Meadow and **Creek** [Fresno Co.] were named for William Helm, a native of Canada, who came to California in 1859, settled at Big Dry Creek in 1865, and for many years raised more sheep than anyone else in central California.

Hemet (hem′ ət): **Valley, Dam, Lake, Butte,** town [Riverside Co.]. The name appears as early as 1879 and was often spelled Hemmet. This may well be from Scandinavian *hemmet* 'the home'; there is a town so named in Denmark. The dam was built between 1891 and 1895, the post office opened in 1895, and the station is shown on the Official Railway Map of 1900.

Hemlock: The mountain hemlock (*Tsuga mertensiana*) and the Douglas fir (*Pseudotsuga taxifolia*), sometimes called hemlock, give this name to a number of places in the Sierra Nevada and the Coast Ranges. **Hemlock Creek** [San Bernardino Co.] and some other places were probably named for the growth of water hemlock (genus *Cicuta*), a deadly poisonous plant of the carrot family, highly dangerous to cattle that feed on the tubers.

Hendy Redwood Grove [Mendocino Co.]. Named for Joshua Hendy, its former owner.

Heneet. *See* Yuba.

Henley [Siskiyou Co.] was originally a mining settlement of the 1850s named Cottonwood, after the creek on which it was situated. The post office is listed as Henley as early as 1858. It was named for a prominent citizen named Henly [*sic*] (Co. Hist.: Siskiyou 1881:210).

Henleyville [Tehama Co.]. Probably named for William N. Henley, a native of Indiana, who registered as a voter at Henleys on Aug. 3, 1866.

Hennerville Peak. *See* Hunewill.

Henness Pass [Nevada Co.] was discovered by Patrick Henness (or Hanness) and his partner Jackson. The former returned to the Atlantic seaboard, but Jackson remained on his ranch and built a wagon road in 1852 across the pass that continued to bear his partner's name. The name is mentioned in a report of the state surveyor general (1856:191): "the route known as the Henness, or Downieville, Route."

Henness Trail [Yosemite NP] was named for James A. Hennessy, a native of Ireland, whose residence at the confluence of Indian Creek with Merced River is shown on the Hoffmann-Gardner map of 1867. The name was misspelled Henness by the Wheeler Survey, and this spelling is now generally accepted. The 1932 Yosemite map of the USGS showed a place called Henness west of El Portal.

Hennigers (hen′ i gərz) **Flat** [Los Angeles Co.], on the west slope of Mount Harvard, was so named because William K. Henniger had a squatter's claim here in the 1880s (Reid, p. 365).

Henry, Mount [Kings Canyon NP], was named by J. N. LeConte in honor of Joseph Henry (1797–1878), an eminent physicist, who was professor of natural history at the College of New Jersey (now Princeton University), 1832–78, and president of the National Academy of Sciences, 1868–78 (Farquhar). *For* Mount Henry [Sequoia NP], *see* Kaweah Peaks; *for* Mount Henry [Napa Co.], *see* Saint John Mountain.

Henshaw, Lake [San Diego Co.], was created in 1924 and named for William G. Henshaw, who owned the part of the old Warner's Ranch that is now flooded.

Hercules [Contra Costa Co.]. Named for the ancient Greek hero in the 1890s when the Hercules Powder Company was established there. The post office is listed in 1915.

Herlong [Lassen Co.]. Named by the War Dept. in memory of Capt. Henry W. Herlong (1911–41), the first American ordnance officer to lose his life in World War II, when the Sierra Ordnance Dept. was established there.

Hermann, Rancho de. *See* German [Sonoma Co.].

Hermit is a term repeatedly used in the state to name isolated peaks or rocks. The best known are The Hermit [Kings Canyon NP], named by T. S. Solomons in 1895, and Hermit Butte [Modoc Co.], named by the USFS.

Hermosa (hər mō′ sə) **Beach** [Los Angeles Co.] was created as an advertising name, from the Spanish for 'beautiful'; it was used by the Hermosa Beach Land and Water Company when the original subdivision was laid out in 1901. The name has been used elsewhere in southern California; in San Bernardino Co. it was once combined with the name Iowa to form the unique name Ioamosa for adjoining tracts of land, formerly called the Hermosa and Iowa Tracts, and now called Alta Loma.

Hernandez (hər nan′ dəz) [San Benito Co.] recalls two farmers, Rafael and Jesús Hernández, listed in the Great Register of 1879 as residents of the county; but they cannot now be identified with this place. *For* Hernandez' Spring [Inyo Co.], *see* Resting Spring.

Herndon [Fresno Co.] was originally a station of the Central Pacific (Southern Pacific) Railroad, called Sycamore in 1872; but the railroad abandoned its plan to lay out a townsite here and chose instead the present site of Fresno. Sycamore station was renamed Herndon about 1895, for a relative of the promoter of a local irrigation project (B. R. Walker).

Herpoco (hûr′ pə kō) [Contra Costa Co.] was coined from Hercules Powder Company and applied to the station in 1989 by the Santa Fe to distinguish it from the Southern Pacific station Hercules.

Herrera, Cañada de [Marin Co.], was the name of a land grant dated Aug. 10, 1839. One Francisco Herrera, a *soldado de cuero* 'leather-jacket soldier', was the sponsor of four Indians baptized on Dec. 14, 1817, at San Rafael; but it is not certain whether the valley was named for him.

Herscheys Hollow [Plumas Co.] recalls a murderer named Herschey, who is said to have buried $20,000; he was killed in the early 1870s and buried in the hollow, 2 miles south of Vinton (R. F. Ramelli).

Hesperia (hes pēr′ ē ə) [San Bernardino Co.] was named in 1885 as a California Southern (Santa Fe) railroad station, possibly after Hesperia, Michigan. Greek and Roman poets used the name in the sense of "The Western Land."

Hester Lake [Kings Canyon NP] was named in 1960 in memory of Robert M. Hester and his father, Clinton Hester. Robert Hester was the co-pilot of a B-24 bomber that crashed here in 1943. His body was not discovered until 1960 (BGN, 1960).

Hetch Hetchy (hech hech′ ē): **Valley, Falls, Dome, Reservoir, Junction** [Yosemite NP]. Powers (p. 357) maintains that the original form was Hatchatchie, which is perhaps from Southern Sierra Miwok *aččačča* 'magpie' (Callaghan). The reservoir was constructed by the city of San Francisco from 1914 to 1923.

Hetten [Trinity Co.]. This former mining town in the southwest corner of the county had a post office between 1890 and 1900. The name is from Wintu *xetin* 'camass', an edible root (*Camassia* sp.); the English word is from Chinook Jargon *qamash*. **Hettenshaw: Valley, Peak** [Trinity Co.] are from Wintu *xetin č'aaw* 'camass sing'; the plants were said to sing in that location (Shepherd). *See* Kettenpom.

Hewes Park [Orange Co.] was named about 1890 for David Hewes of San Francisco, on whose ranch the park is situated (Stephenson). Hewes was Leland Stanford's brother-in-law and supplied the golden spike for the ceremonies of the meeting of the Union Pacific and the Central Pacific in May 1869.

Hexie Mountain [Riverside Co.] was named for the mine at its base, originally called Hexahedron and sometimes shortened to Hexie (Wheelock).

Hiampum. *See* Hyampom.

Hickey Grove State Park [Mendocino Co.] was created in 1921 and named in memory of Edward R. Hickey.

Hickman [Stanislaus Co.] was named in 1891 by the Southern Pacific for Louis Hickman, an early settler and onetime mayor of Stockton, who owned the ranch adjacent to the station.

Hicks [San Bernardino Co.]. *See* Hodge. *For* Hicksville [Amador Co.], *see* Ione.

Hi-corum (hī′ kôr əm) [San Bernardino Co.], a mining district at the west end of the Providence Mountains, was named for Hi-corum, a Chemehuevi Indian, discoverer of the ore deposits of the Orange Blossom Mine near Bagdad and probably discoverer of Mitchells Caverns (O. J. Fisk). The name is apparently *haiku-rïmpa* 'white-man mouth', from *haiku* 'white man' plus *tïmpa* 'mouth' (Munro).

Hidden is a term sometimes applied to new real estate developments to suggest a secluded location: Hidden Hills [Los Angeles Co.], Hidden Meadows [San Diego Co.], Hidden Valley [Lake Co., Los Angeles Co.].

Higgins Mountain [Del Norte Co.] was named for Lew Higgins, who discovered the first copper deposits in the district.

High is often used with a generic term to form a place name. In addition to the seven communities listed in the gazetteers, a number of physical features include the word in their names: High Dome and High Plateau Mountain [Del Norte Co.], Highland Peak [Alpine Co.], High Valley [Colusa Co.]. The phonetic spelling that is sometimes used produces a comical effect: Hydril Hill [Kings Co.] (from "high drill"), Hipass [San Diego Co.], Hytree [Del Norte Co.], Hi Vista [Los Angeles Co.]. **Highgrove** [Riverside Co.]: The townsite was mapped and recorded in 1887 by A. J. Twogood and S. H. Herrick and named because there were orange groves on the hillside there. The name of the California Southern railroad station here had been East Riverside since 1882 (Santa Fe). **Highland Lake, Creek, Peak**, and **Reservoir** [Alpine Co.]: The cluster name commemorates the short-lived Highland City, which was on the high land of the divide between the Carson and the Stanislaus. The city, Lake, and Creek are shown on the maps of the Whitney Survey. **Highland; East Highlands** [San Bernardino Co.]: Both names were applied to stations when the Santa Fe was built through the valley in 1894–95. The two places are actually in the lowlands; they were named after the narrow fertile tableland several hundred feet above the valley, which had been called Highland when the school district was organized in 1883. The name of the post office was Messina until 1898. East Highland was formerly Cramville, named for Louis Cram, the first settler. **Hipass** [San Diego Co.], where a post office was established on Nov. 12, 1917, was named because of its situation; in 1956 the name was changed to Tierra del Sol. **Hi Vista** [Los Angeles Co.] was named in 1930 by Mrs. M. R. Card, wife of the man who developed the place, because it commands a beautiful view of the Sierra Madre and the San Bernardino Mountains. **High Sierra** is a commonly accepted name for the part of the Sierra Nevada that includes the high peaks; the term was used by the Whitney Survey to designate the highest regions, as distinguished from the mining and stock-running regions.

Highway Highlands [Los Angeles Co.] was so named in 1923 by one of the promoters, Mark S. Collins, because of its situation on the state highway (Foothill Boulevard). The gazetteers also list a Highway in Monterey Co.; and a station of the old Sacramento Northern, called Highway Crossing, in Solano Co.

Hilarita (hil ə rē′ tə) [Marin Co.]. Spanish diminutive of the woman's name *Hilaria*.

Hildreth Mountain [Madera Co.] was named for either Jonathan or Emphrey Hildreth, farmers from Missouri, who settled in the region about 1870. There is also a Hildreth Peak in Santa Barbara Co. *For* Hildreth's Diggings [Tuolumne Co.], *see* Columbia.

Hilgard, Mount [Fresno Co.], was named before 1896 by T. S. Solomons, at the suggestion of Ernest C. Bonner, an admiring former student of Prof. Hilgard (Farquhar). Eugene W. Hilgard (1833–1916), the father of scientific agriculture in California, was a scion of a family of "Latin farmers" (i.e. professors, doctors, lawyers, etc., who tilled the soil) who had left Germany for political reasons and settled in Belleville, Missouri.

Hill, The. *See* Mokelumne Hill.

Hillsborough [San Mateo Co.] was incorporated in 1910 and named after Hillsboro (or

Hillsborough), New Hampshire, the ancestral home of W. D. M. Howard, the former owner of the site. California has several other place names with the word Hill as the first element.

Hills Ferry. *See* Newman.

Hillyer, Mount [Los Angeles Co.], was named by the USFS for Margaret Hillyer, an employee of Angeles National Forest.

Hilmar (hil′ mär) [Merced Co.], established in 1917, was named after the Hilmar Colony, founded as a colony of Swedes by Nels O. Hultberg and probably named for his son Hilmar—or perhaps for Hilmar A. Carlson, a pioneer citizen.

Hilton [Sonoma Co.] was a name chosen by the Post Office Dept. in 1894 from a list submitted by the residents. Hilton Ridenhour was the son of the oldest pioneer, around whose ranch the settlement grew up.

Hilton: Creek, Lakes [Mono Co.]. Named for Richard Hilton, a blacksmith from Michigan, who settled in Round Valley about 1870 and later operated a "milk ranch" in Long Valley.

Hilton, Mount [Trinity Co.], in the Trinity Alps, is named for James Hilton, author of *Lost horizon* (BGN, 1966).

Hilts [Siskiyou Co.]. For John Hilt, who came to the area in 1855 and built a sawmill on Cottonwood Creek. In 1903 a group of lumbermen bought the estate, naming their company and the town for its original owner. The name of the station is Hilt; the post office, transferred from Coles on July 6, 1903, is Hilts.

Hinkley [San Bernardino Co.] was named upon the arrival of the railroad in 1882, by D. C. Henderson, of Barstow, for his son Hinckley. Formerly spelled Hinckley.

Hiouchi (hī o͞o′ chē) **Redwoods** and **Bridge** [Del Norte Co.] are named with an Indian word supposedly meaning 'blue waters', according to Drury (p. 356).

Hipass. *See* High; Tierra del Sol.

Hirz (hûrts) **Creek** [Shasta Co.], a tributary of the McCloud River, was named for Christian Hirz, who mined here in the late 1860s (Steger).

Hitchcock, Mount [Sequoia NP], was named by the Rev. F. H. Wales, of Tulare, on Sept. 7, 1881, when he ascended nearby Mount Young. Charles H. Hitchcock (1830–1919) was a professor of geology at Dartmouth and the first U.S. scientist to conduct a high mountain observatory.

Hitchens' Spring. *See* Jayhawker Spring.

Hites Cove [Mariposa Co.] was named for John R. Hite, a lucky miner whose yields from his claims in the Sweetwater district are said to have run into millions. The mine at the cove was opened in 1864. The name appears on the Hoffmann-Gardner map (1867) and is listed as a post office until about 1890. Some maps use the form Hite Cove.

Hi Vista. *See* High.

Hoaglin [Trinity Co.] was named for a member of the Hoaglin (or Hoaglen) family, which settled in the district before 1900.

Hobart Mills [Nevada Co.] was established as a post office about 1900; it was named after the Hobart Mills, which had operated there since 1897. Walter Scott Hobart was a leading lumberman of the Lake Tahoe district from the 1860s until his death in 1892.

Hobergs [Lake Co.] was named for Gustave Hoberg, who settled here in 1885; his summer resort became known by his name. In 1929 the Post Office Dept. accepted the name for the post office.

Hobo Hot Springs [Kern Co.] was first known as Clear Creek Hot Springs. In the early 1900s, when the crew of a compressor camped there and helped themselves to cattle and sheep on the fields, the farmers complained to Sheriff Fred McCracken: "After hearing these men with their tales and woes, / Said sheriff dubbed the campers a lot of Hoboes. / Thereafter when conversation hit on such things / There was always some reference to 'Hobo Hot Springs'" (Earl E. Lambert). This name was applied to the post office when it was established on Feb. 4, 1934, and was changed to Miracle Hot Springs on Nov. 1, 1947.

Hockett Meadows, Lakes, and **Trail** [Sequoia NP]. For J. B. Hockett, a pioneer of the region as early as 1849, who built the trail in 1862–64 (Farquhar).

Hock Farm [Sutter Co.]. Established by Sutter in 1841 and supposedly named after an Indian village; but such a source has not been confirmed. It was Sutter's residence from 1849 to 1865.

Hodge [San Bernardino Co.] was a Santa Fe station earlier called Hicks. At the suggestion of Arthur Brisbane, a well-known journalist in the early part of the 20th century and owner of a ranch in the Mojave Desert, the Santa Fe changed the name to Hodge—for Gilbert and

Robert Hodge, of Buffalo, New York, owners of a ranch in the desert since 1912. The name Hicks appears with Hodge on some modern maps.

Hodges, Lake [San Diego Co.], was created in 1922, when the Santa Fe dammed the San Dieguito River at the Carroll reservoir site to supply water for its park development. The lake was named for W. E. Hodges, vice president of the company.

Hodges Peak [Los Angeles Co.] was named for Dr. J. S. Hodge, who purchased a tract in 1888 and built a wagon road to the summit.

Hoffmann, Mount [Lassen NP], was named for George J. Hoffmann, who camped and hunted near it for many years. He was the son of Charles F. Hoffmann, Whitney's topographer. The name of the mountain is misspelled Hoffman on some maps.

Hoffmann, Mount [Modoc Co.]. For John D. Hoffmann, a brother of Whitney's topographer. This Hoffmann was one of the group of German civil engineers who played an important part in the scientific delineation and recording of the geography of California. He was co-author of the first topographical map of the Tahoe region.

Hoffmann, Mount [Yosemite NP], was named for Whitney's topographer, Charles F. Hoffmann: "June 24 [1863] we climbed a peak over eleven thousand feet high, about five miles from camp, which we named Mount Hoffmann, after our topographer. It commanded a sublime view" (Brewer, p. 407). *See* Glossary.

Hoffmeister Creek [Shasta Co.] was named for Charles Hoffmeister, who had a cattle range here (Steger).

Hog. The maps show some twenty Hog Creeks, Canyons, Mountains, and Islands in California, and doubtless many more such names are used locally. The state had no native hogs, but descendants of domestic animals, originally escaped from missions and ranchos, have existed from early days down to the present. *See* Coches. **Hog Flat Reservoir** [Lassen Co.]: The meadow now covered by the reservoir was called Hog Flat when the farmers of the irrigation district fattened huge herds of hogs on the luxuriant vegetation that grew there after the first dam built to create the artificial lake broke. The name is perpetuated in the name of the reservoir (A. G. Brenneis). *For* Hog Canyon [San Bernardino Co.], *see* Wildwood; *for* Hog Root [Tulare Co.], *see* Alpaugh.

Hogback was a term used repeatedly as a descriptive term for sharp ridges, invoking a similarity with the wild "razorback" hog. A Hogback Ridge is found in Death Valley NP, a Hogback Peak in Kings Canyon NP, and a plain Hogback in Sonoma Co. There is a Hogback Creek west of Haiwee Reservoir [Inyo Co.], but the peak after which it was apparently named is now Round Mountain.

Holbrook Grove [Humboldt Co.] became a part of the state park system in 1933 and was dedicated by Mrs. Silas H. Palmer to the memory of her father, Charles Holbrook (1830–1925), a pioneer dealer in stoves and iron and a leading businessman and philanthropist of Sacramento and San Francisco.

Holcomb Valley [San Bernardino Co.] was named for William F. Holcomb, a native of Indiana, who came to California in 1850. In May 1860, while employed as a bear hunter for prospectors in Bear Valley, he discovered gold in the valley that bears his name. *For* Holcomb Village, *see* Deadman: Deadman's Hole.

Hole. Although this word is not a generally accepted generic term, it is sometimes used for depressions in the ground, for water holes, and instead of the term "hollow" or "cove." In the 1870s a landmark along the road from Yreka to the Lava Beds [Siskiyou Co.] was called Hole in the Ground, and the same name appears on maps just west of Lassen Volcanic NP. A depression in the land on the boundary between Fresno and Kings Cos. is called The Dark Hole, and Death Valley NP has a Hole in the Rock Spring. *See* Hollow; Jolla.

Hollenbeck: Flat and **Butte** [Modoc Co.]. For Asa Hollenbeck, a cowman who used the nearby range from the 1880s to the 1920s (William S. Brown).

Hollenbeck Park [Los Angeles Co.] was a tract given to the city of Los Angeles by William H. Workman and Mrs. J. F. Hollenbeck, when Workman was mayor (1887–88); it was named at his request in memory of Mr. Hollenbeck, his "old and cherished friend" (Workman, pp. 221 ff.).

Hollister [San Benito Co.]. Named in 1868 by the San Justo Homestead Association of farmers, for Col. W. W. Hollister, who drove the first flock of sheep across the continent and acquired the San Justo grant on which the new community was established. When someone suggested at the first town meeting that the name of the land grant be retained, one citizen protested violently against the adding of another to the long list of saints' names in the state, and San Justo was rejected.

Hollister Peak [San Luis Obispo Co.]. Named by the Coast Survey in 1884 for the Hollister family, which had established a ranch at the base of the mountain in 1866. It appears as Cerro Alto 'high hill' on older maps and was also known as Morro Twin.

Hollister Valley [San Diego Co.]. *See* Thing Valley.

Hollow. The descriptive generic term is repeatedly used in geographical nomenclature, especially for depressions for which the terms "valley," "canyon," and "glen" are not suited. In Butte Co., north of Chico, four canyons through which tributaries to Mud Creek flow are called Grizzly, Sheep, Cabin, and Sycamore Hollows. *See* Corral: Corral Hollow; Herscheys Hollow.

Hollywood [Los Angeles Co.], the area that was destined to become the film center of the world, was laid out and named by Horace H. Wilcox in 1886. The name is a transfer name. There is a Hollywood in County Wicklow, Ireland, and the word occurs as a place name in the states of Arkansas, Georgia, Maryland, and North Carolina, where the American holly (*Ilex opaca*) is native; and in Minnesota, where it is apparently a transfer name, like our Hollywood. According to Tom Patterson, Mrs. Wilcox heard the name from a fellow passenger on a trip east. Mr. Wilcox imported two holly trees to justify the name, but they did not survive in the temperature of southern California. The name and derivations of it became very popular after Hollywood attained its fame.

Holmby Hills [Los Angeles Co.]. Named after Holmby, England, birthplace of Arthur Letts, Sr., founder of the Broadway Department Store, Los Angeles. In 1919 he bought Rancho San Jose de Buenos Ayres, on which the place is situated.

Holmes [Humboldt Co.] was named Holmes Camp in 1908, for the head man of the logging company operating there. The generic term was dropped when the post office was established in 1912 (Ruth Nalander).

Holt [San Joaquin Co.]. When the Santa Fe took over the San Francisco and San Joaquin Railroad in 1900, the station and town were established and named for Charles Parker Holt, who had extensive farming interests on Roberts Island and was the original builder of caterpillar tractors (Santa Fe).

Holtville [Imperial Co.]. Established and named Holton in 1903 by W. F. Holt, president of the Holton Power Company and one of the organizers of an irrigation project for the Imperial Valley in 1899. At the request of the Post Office Dept., the name was soon changed to the present form.

Holy City [Santa Clara Co.]. Founded by W. E. Riker in 1920 as a religious community for people of the "white race" and named so "because of the principles revealed for an indisputable solution for the Economic, Racial and Spiritual problems of this world" (E. Allington). The post office is listed in 1927.

Holy Jim Canyon [Orange Co.]. James ("Cussin' Jim") Smith had an apiary here, and the name was applied in irony (Stephenson).

Home is a term favored by realtors, instead of "house," because of its connotations of warmth and security; it has been used in place names such as Home Garden [Kings Co.], Home Gardens [Riverside Co.], Homeland [Riverside Co.], and Homewood [Placer Co.].

Homer. *See* Amboy.

Homers Nose [Sequoia NP] was named in 1872 for Joseph Homer, a veteran of the Mexican War: "As my father and the two government surveyors were looking at the mountain Mr. Powell laughingly remarked, 'Homer, that south projection looks like your nose.' 'All right,' said Mr. Orth, 'I am marking it on my map as Homers Nose,' and so it was named" (E. B. Homer to F. P. Farquhar).

Homestead Valley [Marin Co.]. Named in 1903 by the Tamalpais Land and Water Company, after the country home of S. M. Throckmorton.

Honcut (hon´ kut): **Creek**, town [Butte, Yuba Cos.]. The name comes from *hoan'kut*, the Maidu village on the Yuba just below the mouth of the creek (Powers, p. 283). Theodor Cordua, grantee of the land grant dated Dec. 22, 1844, mentions the name in his memoirs (p. 6): "This name I had given my ten-league grant, because the Honcut River formed the northern boundary of my . . . holdings. . . . The names of the rancherias or Indian villages I found rather pretty, for instance . . . Honcut." The creek was apparently the one that Jedediah Smith had called Red Bank Creek (*Travels*, p. 71). It appears on a *diseño* of Rancho Honcut as Arroyo Honcut, and in the present form in the *Statutes* of 1850 (p. 62) and on Gibbes's map of 1852. The post office is listed in 1867.

Hondo, Honda, Spanish for 'deep', is combined with Creek or Arroyo in about ten California place names and is also found as the name for a number of communities. **La Honda** (hon′ də) **Creek** and town [San Mateo Co.]: The creek is shown as *Arroyo ondo* on several *diseños* of land grants, and as Arroyo Hondo on a map of Rancho Cañada de Raymundo (1856). On Hoffmann's map of the Bay region (1874), the valley of San Gregorio Creek west of La Honda Creek is designated as Honda. The post office La Honda is listed in 1880. After 1895 the name was spelled Lahonda, but the original form was restored in 1905, owing to Z. S. Eldredge's efforts. **Hondo** (hon′ dō) [Los Angeles Co.]: The post office was established in 1919 and named after Rio Hondo, the creek on which it is situated. On *diseños* of several land grants the stream is called Río de San Gabriel, Río Hondo de Asuza, and Zanja Onda ('deep ditch'). The San Gabriel River took a new course in 1867; the old channel became known as Rio Hondo after July 1888 (Bowman). **Honda** [Santa Barbara Co.] was named after the Cañada Honda Creek when the last link of the Southern Pacific coast line between Surf and Ellwood was completed in 1900.

Honey Lake [Lassen Co.]. In 1850 Bruff named the lake for his friend Capt. George H. Derby, of the U.S. Topographical Engineers; but he reported that earlier explorers had called it Honey Lake, "from the sweet substance which they found exuding from the heads of wild oats in the basin." The latter name was placed on the map by Blake in 1853. However, the "honey" does not exude from the grass; the sweet liquid is deposited by the honeydew aphis and is said to have been gathered by Indians as food. It is also possible that the name is a popular etymology from Maidu *hanïlekim*, mentioned in the native creation myth as the name for the lake and its valley; it is from *hanïlek* 'to carry (an object) swiftly away' (Shipley). Most of the other physical features named Honey (including a Honey Run in Butte Co.) probably received their names because wild honey was found there. In Humboldt Co. there is a site called Honeydew.

Honolulu. *See* Klamath.

Hood [Sacramento Co.]. Named in 1910 by Madison P. Barnes, for William Hood, at that time chief engineer of the Southern Pacific.

Hood: Creek, Mountain [Sonoma Co.]. Named for William Hood, a Scottish carpenter, who came to California in 1846; in the 1850s he bought the Guilicos rancho, which contained the mountain and creek.

Hoodoo. A typical miner's term, referring either to bad luck or to an unusual rock formation. The name is preserved in Hoodoo Creek [Lake Co.] and probably in some other topographic features.

Hooker: Canyon, Creek [Sonoma Co.]. Named for Joseph ("Fighting Joe") Hooker (1814–79), who came to California in 1849, after serving as an officer in the Mexican War. Hooker purchased part of Rancho Agua Caliente in 1851 and lived there for several years. In the Civil War he commanded the Army of the Potomac, Jan.–June 1863 (*CHSQ* 16:304–20).

Hooker: Creek, station [Tehama Co.]. The creek was named for J. M. Hooker, who settled near its mouth in 1852; the railroad station took the name of the creek (McNamar). A post office was established on Nov. 20, 1885, reestablished on May 31, 1889, and discontinued on Oct. 31, 1928.

Hookton: Channel, Slough [Humboldt Co.]. These names commemorate Hookton, an important shipping center of the lower Eel River region in the 1860s and 1870s.

Hoopa (ho͞o′ pä): **Valley, Valley Indian Reservation** [Humboldt Co.]. The name is not from the language of the resident Hupa (Athabaskan) tribe, but from the name for the valley in the language of the neighboring Yurok tribe, i.e. *hup'oo* (Robins). The present spelling was used by Heintzelman in 1858. The spelling variant Hupa is now usually used to designate the tribe and is also used in the place name Hupa Mountain. Hoopa Valley was formerly called Eden Valley.

Hooper [Siskiyou Co.]. The place was known as Hoopersville for Frank Hooper, a miner (Co. Hist.: Siskiyou 1881:217). The shorter form was apparently used when the name was applied to a station of the McCloud River Railroad in the 1890s.

Hooper, Mount; Hooper Creek [Fresno Co.]. R. B. Marshall of the USGS named the mountain (probably at the time the Mount Goddard quadrangle was surveyed, 1907–9) in memory of Maj. William B. Hooper, a native of Virginia and at one time owner of the Occidental Hotel in San Francisco, who died in 1903. The BGN (June 7, 1911) called the mountain Hooper Peak; but since there is already a Hooper Peak in Yosemite NP, the USGS and common usage prefer Mount Hooper.

Hoosimbim (hōō´ sim bim) **Mountain** [Trinity Co.]. From Wintu *huusun meem*, lit. 'buzzard's water' (Shepherd).

Hoover Lake [Mono Co.]. Named in 1905 by an engineer of the Standard Consolidated Mining Company, for Theodore J. Hoover, who was at that time manager of the company at Bodie. Hoover, a brother of the former president of the United States, was professor of mining and metallurgy at Stanford University, 1919–41.

Hoover Wilderness Area [Mono Co.]. The name was applied in 1928 by the USFS, in honor of Herbert Hoover (1874–1964), thirty-first president of the United States, in the year of his election (G. B. Doll).

Hope Ranch [Santa Barbara Co.]. On June 10, 1870, a patent was issued to Thomas W. Hope for the land grant La Calera y las Positas or Cañada de Calera. The name was changed to Hope Ranch and transferred to the modern subdivision by the heirs.

Hopeton [Merced Co.]. The place is an old mining camp and was named Forlorn Hope, apparently after the name of a mine, when it was still in Mariposa Co. A post office was first established as Forlorn Hope on Aug. 17, 1854. The name was then changed to Hopetown; and when the post office was reestablished on Oct. 2, 1866, the present name was used.

Hope Valley [Alpine Co.]. The Mormons crossed the summit here on their trek from Sutters Fort to Salt Lake City, and Bigler recorded on July 29, 1848, "Campt at the head of which we called Hope Valley as we began to have hope." In analogy, O. B. Powers and a party of surveyors in Nov. 1855 called the two beautiful grassy valleys to the south Faith Valley and Charity Valley.

Hopkins, Mount. *See* Crocker, Mount.

Hopland [Mendocino Co.]. The town was started in 1859, when Knox, Willard, and Conner opened a saloon and R. (or Thomas) Harrison established a store. The place was called Sanel after the land grant on which it was situated and is so shown on the von Leicht–Craven map. In 1874, when a toll road was built on the east side of Russian River, the town (with the exception of a brick store) was moved across the river. The name was changed to Hopland because Stephen Knowles's experiment of growing hops there had proved successful. This name appears on the Land Office map of 1879. In 1890, when the California Northwestern Railroad was extended to Ukiah, the tracks were laid on the west side of the river. Business flowed back to Sanel, and the old Indian name was given to the railroad station and the post office. In 1891–92 the Post Office Dept. discontinued the old Hopland office but gave the name Hopland to the one at Sanel. *See* Feliz Creek; Sanel. Attempts to grow hops on a large scale were made in other parts of the state, but the only name reminiscent of them is Hopyard Road in the Livermore Valley, where the Pleasanton Hop Company had four hundred acres in cultivation in 1903 (E. T. Planer).

Hoppaw (hop´ ou) **Creek** [Del Norte Co.]. Also spelled Hoppow; named after the Yurok village *ho'pew* (Waterman, map 9; Robins).

Horestimba, Arroyo de. *See* Orestimba Creek.

Horicon (hôr´ i kon) [Sonoma Co.]. The name of the school district was probably chosen by an admirer of James Fenimore Cooper—who, in *The Last of the Mohicans*, had called Lake George in New York "The Horicon" (Leland).

Hornblende Mountains [El Dorado Co.]. When the USGS mapped the Placerville quadrangle in 1887, the name for this mineral, a variety of amphibole, was applied apparently because of its presence here.

Hornbrook [Siskiyou Co.]. Named in 1886 by the Southern Pacific, after the brook that ran through the property of David Horn.

Hornitos (hôr nē´ təs, hôr nē´ tōs) [Mariposa Co.]. Spanish for 'little ovens', diminutive plural of *horno* 'oven'. The town was first settled in 1852 by Mexican miners who had been driven out of Quartzburg (*see* Kern: Kernville). The post office was established as Hornitas on June 18, 1856, and changed to the present spelling on Aug. 20, 1877. In 1858 the place gained wide recognition when the Mount Gaines Quartz Mill built two arrastras (*see* Arrastre) driven by an engine of 30 h.p. In the volcanic districts of Latin America, *hornito* describes a low oven-shaped mound. But the name is doubtless a transfer name, probably from Los Hornitos in the Mexican state of Durango. The stories that the shallow graves of the Mexican miners looked like *hornitos*, or that German miners built little bake-ovens of stones and mud, belong obviously in the field of folk etymology. The *hornitos* of brick on and near the cemetery give the impression of having been built in later years to justify the name.

Horse. In Spanish times the horse was so common a sight and was held in such little esteem that the word *caballo* rarely appears in place names (*see* Cavallo). When the first Americans and Europeans filtered into the Mexican province, they found not only a wealth of domestic horses, but large herds that had gone wild. The name Horse was given to numerous Creeks, Canyons, Flats, and Lakes from the very beginning of American occupation. Today there are probably some five hundred Horse names in the state, including numerous places named for dead and blind horses, gray and black horses, wild and stud horses. Sometimes the name was transferred to post offices and communities: Horse Creek [Siskiyou Co.], Horse Lake [Lassen Co.], White Horse [Modoc Co.]. About twenty Horsethief Creeks, Canyons, Points, etc., commemorate the hideouts of horse thieves or places where they were caught and punished (*for* Horsethief Bend, *see* Bend). **Horsetown** [Shasta Co.] preserves the name of one of the richest diggings in the northern mining district in the 1850s. The original name, One-Horse Town, was probably given to the camp in derision. According to an item in the Bancroft Scrapbooks (13:30), "it was named in honor of a favorite horse, the only one at the time about the camp." The abbreviated name is shown on Lapham and Taylor's map of 1856. **Horsehead Mountain** [Trinity Co.] was given this name because the skull of a horse that had fallen and died there could be seen for many years from the trail (K. Smith). **Horse Grotto** [Lava Beds NM]: When J. D. Howard explored the cave about 1918, he found six horses belonging to a settler, Frank Adams, seeking shelter from the wind.

Horse Linto (hos lin′ tən) **Creek** [Humboldt Co.]. From Hupa *xahslin-ding* 'riffles-place', the Indian village originally at the mouth of this creek. Given by Gibbs (Schoolcraft 3:539) as Has-lintah; by McKee (Indian Report, Oct. 7, 1851) as Kas-lin-ta; by Goddard (AAE 1:12) as Xaslindiñ; and by Kroeber as Haslinding. The mapmakers' spelling Horse Linto has given rise to a current spelling-pronunciation (hôrs′ lin tō).

Horseshoe. A popular descriptive term applied to about fifteen lakes resembling a horseshoe and to as many U-shaped river bends. The name is also found combined with Bar, Bay, Cove, Creek, Point, and Slough. The finding of a horseshoe may sometimes have been the reason for the naming.

Horton: Creek, Lakes [Inyo Co.]. Named for William Horton, who settled in Round Valley in 1864 (Robinson).

Hospital. The names of about ten geographic features in the state include this word. **Hospital Creek** [San Joaquin, Stanislaus Cos.]: The name perhaps refers to the Indian sweathouses near the springs in the canyon (Still). The creek is shown on Hoffmann's map of 1873, but the name may have resulted from a misunderstanding: on a *diseño* of El Pescadero (1843), *Ospital* seems to be a stream flowing from the hills, whereas Goddard's map of 1860 shows the stream as Arroyo del Osnito (not identifiable as a Spanish word, but perhaps an error for *osito* 'little bear'). **Hospital Rock** [Sequoia NP]: The name was given to the huge boulder by Hale D. Tharp in 1873, when Alfred Everton, accidentally shot in a bear trap, found a temporary hospital in the shelter formed by the overhanging rock. Others had found shelter here under similar circumstances, and the Potwisha Indians had used it for their sick (Farquhar). **Hospital Rock** [Modoc Co.]: The shelter formed by the rock at the south end of Tule Lake was used as a hospital by the soldiers in the Modoc War, 1872–73 (Doyle).

Hosselkus (hos′ əl kus): **Creek, Valley** [Plumas Co.]. Named for a large family of Palatinate Germans from the Mohawk Valley, New York, who settled here in the early 1850s. Edwin D. Hosselkus kept a general store with the Blood brothers in Indian Valley in 1855.

Hot Rock [Lassen NP]. The rock erupted from Mount Lassen in 1915 and was thrown a distance of 3 miles. It retained its heat for more than a week, but the name still stuck to the rock after it cooled (Doyle).

Hot Sigler Springs. *See* Seigler Springs.

Hot Spring Valley. *See* Drakesbad.

Hough (huf) **Creek** [Plumas Co.]. The stream was probably named for Peter Lassen's bookkeeper and associate in the Gold Lake hunt (1850), a man by the name of Hough (or Huff).

Howards. *See* Occidental.

Howards Bluff [Humboldt Co.]. Named for Maj. E. H. Howard, who owned the land on which the bluff is situated (Co. Hist.: Humboldt 1881:128).

Howard Springs [Lake Co.]. Named for C. W. Howard, who opened the springs to the public in 1877 (Co. Hist.: Napa 1881:153).

Howells. *See* Cosumnes.

Howlands Landing [Santa Catalina Island]. Named for William Howland, of Los Angeles. His sons Frank and Charlie acquired a sheep concession on San Clemente Island, where another Howlands Landing was named for them (Windle, p. 124).

Hoya. *See* Jolla.

Huasna (wäz′ nə): **River, Creek, Valley** [San Luis Obispo Co.]. From the Purisimeño Chumash village name *wasna,* of unknown origin (Applegate 1975:45, after Harrington); transmitted through the Huasna land grant, dated Dec. 8, 1843. *Lomería colindante con Guasna* 'hills adjacent to Guasna' is shown on a *diseño* of the Arroyo Grande grant (1842), and Arroyo del Huasna on a *diseño* of the Huasna grant. The name appears as Río Wasna on the Parke-Custer map (1855); and on the von Leicht–Craven map, Huasna Creek is shown for what is now the Huasna River.

Huckleberry. Although the native huckleberry (genus *Vaccinium*) is widespread in northern and central California, only a few geographic features are named for it, including Lakes in Tuolumne and Shasta Cos. The name is doubtless used locally for other features.

Huddart Park [San Mateo Co.]. The mountain tract north and west of Woodside was named for J. M. Huddart, who bequeathed the area to the city of San Francisco in 1935 (Wyatt).

Huecos [Santa Clara Co.]. The name of a land grant dated May 6, 1846: Spanish *hueco* means 'hole, gap, hollow'. The name is found in other parts of the state. On Mar. 28, 1822, the Los Nietos grant [Los Angeles Co.] is described as consisting of the *parages llamados Los Coyotes, Las Bolsas, y Huecos y Naciós* (DSP Ben. P and J 6:25).

Hueneme (wī nē′ mə, wī nē′ mē): **Point, Canyon**, town [Ventura Co.]. From Ventureño Chumash *wene'mu* 'sleeping place' (Applegate 1975:45, after Harrington). The name was applied to the point in 1856 by James Alden, in charge of the Coast Survey steamer *Active*. The settlement was founded and named after the point by W. E. Barnard and his partners in 1870. The name of the post office in 1870 was Wynema; in 1874 it was changed to Hueneme, and in 1940 to Port Hueneme. Cf. Madison Beeler, "Hueneme," *Names* 14.36–40 (1966); Elaine K. Garber, "Hueneme: Origin of the name," Ventura County Historical Society *Quarterly* 12.3.11–15 (1967).

Huerhuero (wâr wâr′ ō): **Creek, Soda Springs** [San Luis Obispo Co.]. Possibly from the Obispeño Chumash village name *elewexe* (Harrington). The land grant Huerhuero was recorded in 1843 (Legis. Recs. 4:58 ff.). Rancho Huerhuero is recorded by Parke-Custer (1855), Huer Huero Creek by von Leicht–Craven (1874). Another possible origin is Mexican Spanish *huero* 'putrid, rotten', especially referring to eggs, and here perhaps referring to the odor of sulphur water. There is a tendency among locals to pronounce the name (wâr ō wâr′ ō).

Huerta. The Spanish word for 'garden, orchard', once a popular generic term that appeared in the principal or secondary name of about ten land grants or claims, does not seem to have survived as a place name. Three of the grants so named were confirmed by the U.S. government. *See* Cerro: Cerro Romualdo; Cuate; Noche Buena.

Hueso (wā′ sō), **Arroyo** [San Diego Co.]. Spanish for 'bone creek'.

Hughes: Valley, Mountain [Fresno Co.]. Named for John R. Hughes, the first settler in the valley.

Hughes Lake [Los Angeles Co.]. The lake is recorded as Hughe's Lake in 1898 and was doubtless named for G. O. Hughes, who owned the adjoining land (R. Flickwir).

Hughson [Stanislaus Co.]. Named for Hiram Hughson, owner of the land on which the firm of Flack and Jacobson laid out the town in 1907.

Huichica (wi chē′ kə): **Creek** [Napa Co.]. Arroyo de Huichica is shown on a *diseño* of El Rancho de Huichica, granted to Jacob Leese on July 6, 1844. According to Barrett (*Pomo*, p. 312), the name was apparently derived from that of an Indian village, *hū'tci*, near the plaza of the town of Sonoma. However, on a *diseño* of the Entre Napa grant (1835) a locality called Huichica is shown quite a bit southeast of Sonoma.

Huling (hyo͞o′ ling) **Creek** [Shasta Co.]. The tributary of Eagle Creek was named for William Huling, who settled here in 1851 (Steger).

Hull Mountain [Mendocino Co.]. In 1856 James

Hull, a settler in the Sacramento Valley, built a hunting cabin near the top of the mountain. In a fight with a grizzly he killed the animal but was so badly mauled that he died of his wounds (USFS).

Humboldt: Bay, Point, County, Creek, Hill, State Park. The bay was discovered in the summer of 1806 by Jonathan Winship, an American in the employ of the Russian-American Company. It was described in Russian documents as a "bay of Indians" (Davidson, *Humboldt Bay*, pp. 10 ff.), and its entrance was named for the Russian explorer Nicolai Rezanof, who was in San Francisco in the spring of 1806. When the Gregg party reached the bay on Dec. 20, 1849, they believed that they had found Trinidad Bay and gave it the English version of the name, Trinity. On Apr. 8, 1850, Captain Hans Bühne of the *Laura Virginia* party entered it in a boat and a few days later piloted the company's schooner through the channel. He and Douglas Ottinger, commander of the expedition, are responsible for the present name, which honors Alexander von Humboldt (*see* Glossary), who was then at the height of his fame. A short-lived Humboldt City existed in the early 1850s. Fort Humboldt was established in 1853. The county was created and named by act of the legislature on May 12, 1853.

Humbug. A favorite prospector's term to express disappointment when a claim did not yield as much as expected or when a creek was found to be dry. Two cluster names in Placer and Siskiyou Cos. originated at the time of the Gold Rush; the camp in Siskiyou Co. disproved its name by yielding richly. About ten other places still bear this name, including Big Humbug Creek [Tuolumne Co.] and Little Humbug Creek [Sierra Co.]. The place now called North Bloomfield [Nevada Co.] flourished in the 1850s as Humbug or Humbug City.

Hume: post office, **Lake** [Fresno Co.]. The lake was impounded from the flow of Tenmile Creek by the Hume and Bennett lumber firm, which operated a mill there at the turn of the century (Co. Hist.: Fresno 1933:85). A recreation area has been developed at Hume Lake by the USFS.

Humphreys [Fresno Co.]. The station was named for John W. Humphreys, a pioneer lumberman and stock raiser of the Tollhouse and Pine Ridge area (Co. Hist.: Fresno 1956).

Humphreys, Mount; Humphreys Basin [Fresno Co.]. The name was applied by the Whitney Survey in honor of Andrew A. Humphreys (1810–73), who played an important role in the topographical survey of the western United States and distinguished himself as a general in the Civil War.

Hunewill (hun′ i wil): Peak, Hills [Mono Co.]. Named for N. B. Hunewill, who operated a sawmill in Buckeye Canyon in the 1860s to supply lumber for the mining towns. The name is misspelled Hennerville on the Bridgeport atlas sheet.

Hungry. More than ten Valleys, Creeks, and Hollows include this word in their names, among them Hungrymans Gulch [Santa Cruz Island]. All were probably named because of failure to find food at a certain place. Some may go back to Spanish times, when *hambre* 'hunger' was used in place names for the same reason. **Hungry Creek** [Siskiyou Co.]: "The place referred to delights in the significant cognomen of 'Hungry Creek,' so called doubtless from the fact of its discoverers having been hemmed in by the snows without any provisions" (*Shasta Courier*, Mar. 19, 1853; Boggs, p. 157). *See* Rag: Raggedyass Gulch. **Hungry Bills Ranch** [Death Valley NP] was named for a Panamint Indian, who served as a scout in the Modoc War and received the ranch for his services (Death Valley Survey). Inyo Co. has a Hungry Packer Lake.

Hunter Liggett, Fort [Monterey Co.]. Established in 1941 and named for Maj. Gen. Hunter Liggett, a commander during World War I (Clark 1991).

Hunters Point [San Francisco Co.]. The place was named for Robert E. Hunter, who participated in an 1849 project to found a city called South San Francisco on the point. On Hoffmann's 1874 map of the Bay region, the promontory is called Hunter's Point, while the point itself bears the old name Avisadero. Other maps make the same distinction but spell the name Hunter Point. At present, Hunters Point is the generally accepted designation for the entire promontory. *See* Avisadero.

Hunters Valley [Mariposa Co.]. Hunter's Valley is mentioned by Browne (pp. 30–31) in 1869 as a site of mines. It was probably named for William W. Hunter, an engineer from Pennsylvania; he was a resident of Mariposa in 1867 and of Hornitos in 1879.

Huntington. Henry E. Huntington (1850–1927), a nephew of the railroad magnate Collis P. Huntington and promoter of most of the electric railroads in southern California, has been honored in more place names than any other of the later industrial pioneers. He will best be remembered as the donor of the Huntington Library and Art Gallery in San Marino. **Huntington Park** [Los Angeles Co.] was laid out in 1903 and named by the subdivider, E. V. Baker, for his friend Huntington. **Huntington Beach** [Orange Co.] was originally called Pacific Beach but the name was changed by the townsite company in 1903 (Stephenson). **Huntington Lake** [Fresno Co.]: The reservoir was named in 1912 by the Pacific Light and Power Corporation for its president, Henry E. Huntington.

Huntington, Mount [Fresno, Mono Cos.]. Named for the railroad magnate Collis Huntington (1821–1900). *See* Crocker, Mount.

Huntoon Valley [Mono Co.]. Named for R. S. Huntoon, who settled here in the 1870s, or for some one of his relatives (Maule); *see* Yaney Canyon.

Hupa. *See* Hoopa.

Hurd Peak [Inyo Co.]. The peak bears the name of the engineer H. C. Hurd, who in 1906 made the first known ascent.

Hurdygurdy. This was a typical miner's term, applied to hand organs and to the "hurdygurdy girls." It was also the trade name of a widely used water wheel. The name is preserved in Hurdygurdy Butte and Creek in Del Norte Co., and in some local toponyms.

Hurleton (hûrl′ tən) [Butte Co.]. The name was given in 1880 to the post office; the first postmaster was Smith H. Hurles, a native of Ireland. He had named his place Boston Ranch after Boston, Massachusetts, where he formerly lived.

Huron [Fresno Co.]. The name was applied by the Southern Pacific in 1877 to a station on the proposed route from Goshen to Tres Pinos. This was the name given by the French to an Iroquoian group in Ontario, and it became a popular place name in the United States.

Hurricane Deck [Santa Barbara Co.]. Since early days the name has been used for this remote and rugged section of Los Padres National Forest because of the heavy winds that frequently blow there (William S. Brown).

Hutchings, Mount [Kings Canyon NP]; **Hutchings Creek** [Yosemite NP]. Named for James M. Hutchings (1818–1902), a native of Northamptonshire, England, who came to California in 1849. He is best known for his *Illustrated California Magazine* (1856–61) and for his enthusiastic books on Yosemite.

Hutchinson Meadow [Fresno Co.]. For James S. Hutchinson, a native of San Francisco and an explorer and a mountain climber in the Sierra for many years (Farquhar).

Huxley, Mount [Kings Canyon NP], was named by T. S. Solomons in 1895, in honor of the English biologist Thomas H. Huxley (1825–95). *See* Evolution.

Hyampom (hī′ əm pom) [Trinity Co.]. Perhaps from Wintu *xayiin-pom* 'slippery place', from *xay-* 'to slip' (Pitkin, p. 728). The name is sometimes spelled Hiampum. According to the County History (1858:28), the place was settled on Jan. 12, 1855, by Hank Young. The post office was established on Oct. 22, 1890.

Hyatt Lake [Tuolumne Co.]. Named in 1909 by R. B. Marshall for Edward Hyatt, at that time topographic engineer of the USGS, after 1927 the state engineer.

Hyde Park [Los Angeles Co.]. The post office preserves the name of one of the "ghost towns" of the "boom year" of 1887, laid out by Moses L. Wicks and named for the owner of a lumber yard at the site.

Hydesville [Humboldt Co.]. Named in 1858 for John Hyde, owner of the land on which the town was built (Ruth Stover). The post office is listed in 1867.

Hynes [Los Angeles Co.]. The original name of the station and town was South Clearwater. When the post office was established on July 15, 1898, the name was changed to Hynes, for C. B. Hynes, superintendent of the Salt Lake Railroad. The town is now a part of Paramount.

I

Ian Campbell, Mount [Fresno Co.]. For Dr. Ian Campbell (1899–1978), onetime state geologist (BGN list 8201).

Iaqua (ī´ ə kwä) **Buttes** [Humboldt Co.] recall a native salutation still heard in Humboldt Co. Bruff writes, on Feb. 12, 1851 (2:951): "He [the leader of a band of Indians] came up, and saluted me, as usual with them, saying 'Ay-a-qui-ya?' (How do you do?) which I reciprocated and shook hands." The greeting is preserved in Yurok *oyekwi'*, Karuk *ayukii*, and has been said to mean 'friend' in the now-extinct Wiyot language. The present spelling was used in Nov. 1864, when a detachment from Fort Humboldt went to Camp Iaqua (Pico Docs. 3:171). The town "at old Fort Iaqua" is mentioned in the County Business Directory of 1895–96, and the post office is listed in 1910.

Ibex: Spring, Hills, Pass, and **Wash** [Death Valley NP] are named for the bighorn sheep (*Ovis canadensis*), which desert dwellers often called ibex, the name of the European mountain goat (Death Valley Survey).

Ibis Mountain [San Bernardino Co.]. For the Ibis Mine on its east slope; the ibis was a sacred bird of the ancient Egyptians. In 1896 the Santa Fe named a nearby siding Ibex (the wild goat of the Old World), to fit the alphabetic order of stations in this section; but in 1904 the name was changed to Ibis, after the mountain.

Icaria (ī kâr´ ē ə) [Sonoma Co.] was the name given by Armand J. DeHay to a French socialist community, established here in 1881 (Morley). The Greek place name had been used by Étienne Cabet in 1840 and was adopted by the Fourierists for their utopian settlements.

Iceberg, The [Alpine Co.], a prominent grayish rock, has the appearance of a mountain of ice (J. Ellis). **Iceberg Meadow** and **Peak** were named after the rock when the USGS mapped the Dardanelles quadrangle in the 1890s. **Iceberg Lake** [Madera Co.] was so named because the Clyde Minaret glacier at one time discharged bergs into it. Ice still remains on the lake well into summer, owing to its elevation (D. R. Brower). **Iceberg Cave** [Lava Beds NM] was named by J. D. Howard in Jan. 1917 because it contained a 20-foot ice pinnacle, formed by water dropping through the ceiling.

Iceland [Nevada Co.] was the name given to the Southern Pacific station about 1902. Like Polaris, Floriston, and Mystic on the same sector, it is apparently just a railroad name, although appropriate enough for the place in wintertime. *For* Iceland Lake [Tuolumne Co.], *see* Lewis Lakes.

Ickes, Mount [Kings Canyon NP] was named in honor of Harold L. Ickes (1874–1952), who was secretary of the interior, 1933–46 (BGN, 1964).

Idria (id´ rē ə) [San Benito Co.] was named after the New Idria quicksilver mine, which in turn had been named in the early 1850s after Idrija, a production center of mercury in Slovenia. Brewer (Notes, July 19, 1861) calls the place New Idria, and this was the name of the first post office, established on Mar. 22, 1869. When the new post office was established on Dec. 22, 1894, the abbreviated name was used.

Idyllwild (ī´ dəl wīld) [Riverside Co.], a name suggested by Mrs. Laura Rutledge as descriptive of the timbered resort area, was accepted by the Post Office Dept. in 1899. Places called Idlewild are in Del Norte and Placer Cos.

Igerna (ī gûr´ nə) [Siskiyou Co.]. Apparently a name coined for a freight station on the railroad; origin unknown (Hendryx).

Ignacio (ig nā´ shō) [Marin Co.] was named for Ignacio Pacheco, a soldier of the San Francisco company after 1827, and in 1840 a grantee of the San Jose grant on which the settlement is situated. The place is shown as Pacheco on Hoffmann's map of the Bay region; but with the establishment of the post office about 1895, the name was changed because there was a Pacheco in Contra Costa Co.

Ignasio Valley [Contra Costa Co.]. *See* Ygnacio Valley.

Igo (ī′ gō) [Shasta Co.], according to Steger, was a name suggested in 1868 by Charles Hoffmann because he heard the little son of George McPherson say: "Daddy, I go, I go," whenever the latter left for the mine. However, the von Leicht-Craven map of 1874, revised by Hoffmann himself, has the old name, Piety Hill. The post office was established and named Igo in 1873, and the place is shown on the Land Office map of 1879. In folk etymology the name is associated with nearby Ono: Igo was named when a Chinese man driven from his claim there said, "I go," and Ono was named when the same Chinese said, "Oh, no!" because American prospectors tried to drive him from this site too. It has not been ascertained whether such a story also accounts for the names of two places, Igo and Ono, in San Bernardino Co., within 20 miles of each other.

Ikes: Creek, Falls [Humboldt Co.] were named for a Karuk Indian who lived by the falls, who was called "Little Ike," from his Karuk name, *éehkan*. The term "falls" is used in local English for what might be called rapids elsewhere.

Illilouette (il il o͞o et′): **Canyon, Creek**, and **Fall** [Yosemite NP] are named because of a misunderstanding: "Tululowehäck. The cañon of the South Fork of the Merced, called the Illilouette in the California Geological Report, that being the spelling given by Messrs. King and Gardner—a good illustration of how difficult it is to catch the exact pronunciation of these names. Mr. Hutchings spells it Tooluluwack" (Whitney, *Yosemite book* 1870:17). The Southern Sierra Miwok form is *ṭiṭilwiyak*, lit. 'something shiny'; cf. *ṭululli* 'shiny' (Broadbent; Callaghan).

Immanuel Peak [Monterey Co.] was named for Immanuel Innocente, a neighbor of Michael Pfeiffer, for whom Pfeiffer Redwoods State Park was named (AGS Monterey Peninsula 1941:184).

Imola (ī mō′ lə) [Napa Co.], the name of an ancient city in Italy, was applied in the 1920s to the post office and to the station for the state mental hospital. The name Imola was apparently chosen because of a large hospital for the mentally ill in the Italian city.

Imperial: Valley, city, **County**, and **Dam** got their name when the California Development Company organized a subsidiary to colonize the newly reclaimed southern part of the Colorado Desert. The leading men of the project—George Chaffey, Anthony H. Heher, and L. M. Holt—considered Colorado Desert a name unlikely to attract settlers; so they called the promoting organization the Imperial Land Company, and the region Imperial Valley. The "mother town," Imperial, was platted early in 1901; the county was organized on Aug. 15, 1907. **Imperial Beach** [San Diego Co.] was given this name apparently because of its advertising value; the post office is listed in 1910. *For* Imperial Junction, *see* Niland.

Ina (ī′ nə) **Coolbrith, Mount** [Sierra Co.], was named by the BGN in Feb. 1932, at the request of the State Geographic Board, acting on behalf of the Ina Coolbrith Circle and other organizations. The mountain was formerly called Summit Peak. Ina Donna Coolbrith (1842–1928) came to California in 1849 with a party led by Jim Beckwourth (for whom a nearby pass is named). In 1915 she was named poet laureate of California by the legislature.

Iñaja (in′ yə hä) **Indian Reservation** [San Diego Co.]. From Diegueño *'enyehaa* 'my water' (or 'my tears'), from *'iny-* 'my' plus *'ehaa* 'water' (Langdon). *See* Anahuac.

Inam (i näm′) [Siskiyou Co.]. From Karuk *inaam*, referring to the site of the world-renewal ceremony held each year in the autumn.

Incline [Mariposa Co.] was named in 1923 by the Yosemite Valley Railroad for the Incline Logging Railroad, an engineering feat of lumber transportation over a 68 percent grade (J. Law). The post office is listed in 1924.

Inconstance Creek [Siskiyou Co.]. The name of a stream that flows 6 miles and then disappears (BGN, list 7904). *For* Inconstant River, *see* Mojave River.

Independence has been a popular term in U.S. place names since the days of the Revolution. In California there are about twelve features and places so named. Some may be transfer names from the East: Independence, Missouri, was the starting point of many immigrant trains. **Independence** [Inyo Co.] began in 1861, when Charles Putnam built a stone house at the site; the place was first known as Putnam's, later as Little Pine. The present townsite was laid out by Thomas Edwards in 1866 and named after nearby Camp Independence, which had been established by Lt. Col. George S. Evans, of the 2d Cavalry, on Independence Day, 1862. **Independence Lake** [Nevada Co.] was named on Independence Day of 1852 or 1853 by Lola Montez, the actress, who was living in nearby Grass Valley at the time (*see* Lola, Mount). However, according to his manuscript in the Bancroft Library, Augustus Moore claims to have named the lake on the Fourth of July, 1862.

India Basin [San Francisco Co.]. Probably named for cargo ships arriving from India in the 19th century; *see* China: China Basin.

Indian, as a general racial designation of America's original inhabitants, is understandably used more frequently in place names than any other ethnic term. More than three hundred names in the state contain the term, including some ten settlements and one fanciful Indianola [Humboldt Co.]. Some names go back to early mining and exploration days: Indian Valley [Plumas Co.], Indian Diggins [El Dorado Co.], Indian Creek [Siskiyou Co.], Indiangulch [Mariposa Co.], Indian Springs [San Diego Co.]. (*For* Indian Gulch [Calaveras Co.], *see* West Point.) Several Mountains and Rocks are so called because of their assumed or real resemblance to an Indian: Indian Head [Tulare Co.], Indian Rock [Yosemite NP]. There are also combinations like Dead Indian Creek [Riverside Co.], Lake of the Lone Indian [Fresno Co.], and Indian Joe Spring [Inyo Co.]. Trinity River was once Indian Scalp River, and the site of Concord [Contra Costa Co.] was known as Drunken Indian. **Indian Canyon** [Yosemite NP] was once notable as a place where Indians gathered arrowwood (Bunnell, p. 131). **Indian Island** [Humboldt Co.] is the site where more than 200 Indians were massacred in 1860. Before 1970 it was called Gunther Island, after the landowner Friedrich Gunther (Turner 1973). **The Indians** [Monterey Co.] is not named for anything resembling Indians but is short for earlier "The Indian Reservation" or "The Indians' Ranch," referring to the Encinales family and other Salinan Indians of San Antonio Mission who settled here (Clark 1991). **Indian Ranch** [Inyo Co.] is the site of a present-day Indian reservation.

Indiana Colony. *See* Pasadena.

Indio (in′ dē ō): town, **Hills, Mountain** [Riverside Co.]. When the Sunset Route of the Southern Pacific reached the place in May 1876, the station was named after the nearby (still existing) Indian Wells. Before 1879 the name was changed to the Spanish word for 'Indian'. (*See also* Inyo.)

Indios, Cañada de los. *See* Black: Black Star Canyon.

Industry. *See* City of Industry.

Infant Buttes [Fresno Co.] was named by T. S. Solomons (Farquhar). The name was perhaps applied because the buttes, although more than 10,000 feet high, look like infants when compared with the giants east of them.

Infernal Caverns [Modoc Co.] is the site of a battle, fought on Sept. 26–27, 1867, between U.S. soldiers, commanded by Gen. George Crook, and a band of Indians.

Ingalls: Bluffs, Creek [Sonoma Co.]. Probably named for Timothy A. Ingalls from New York, who had a farm in Knights Valley in the 1870s.

Ingalls, Mount [Plumas Co.], was named for the Ingalls family, from New York State, which settled in the Genesee Valley before 1870.

Inglewood [Los Angeles Co.] was founded in 1887 on the part of Rancho de la Centinela owned by Daniel Freeman. The town was probably named by a visitor from an Inglewood in Canada—the sister-in-law of N. R. Vail, one of the promoters (Bertha H. Fuller). *For* Inglewood Rancho, *see* Lennox.

Ingomar (ing′ gə mär) [Merced Co.] was a railroad name applied to the station in 1889, and to the post office on Jan. 28, 1890. The name of the successful play *Ingomar, the barbarian*, a translation of *Der Sohn der Wildnis* by Friedrich Halm (E. F. J. Münch-Bellinghausen), probably suggested the name (Raymond F. Wood). The play had been revived on the American stage in 1885.

Ingot [Shasta Co.], a mining town, has been called this since about 1900, because of the foundry there in which metal was cast into convenient forms (ingots) for shipping. Ingot post office was established in 1919 and discontinued in 1940 (Steger).

Injerto, Arroyo del. *See* San Ramon Creek.

In-ko-pah (ing′ kō pä) **Gorge** and **Mountain** [Imperial Co.], on Highway 80, were named by the Division of Highways on the basis of Diegueño *'enyaak 'iipay*, lit. 'east people' (Langdon).

Inskip [Butte Co.] was established in 1857 and named for a "Dutchman" (probably a Pennsylvania German) named Enskeep, who had discovered gold nearby. **Inskip Hill** and **Little Inskip Hill** [Tehama Co.] may have been named for the same man. The *Marysville Weekly Express*, July 2, 1859, reported that a Dr. Inskip had been on a trapping expedition in the Mount Lassen country that year.

Interlaken (in tər lä′ kən) [Santa Cruz Co.]. Named for the town of Interlaken in Switzerland, supposed to mean 'among the lakes'.

Inverness [Marin Co.]. Judge James McMillan Shafter subdvided the area in 1889 and chose the name of the ancient Scottish city in honor of his family's roots (Teather 1986).

Inwood [Shasta Co.], a combination suggesting 'hidden in the woods', was the name applied to the post office on Jan. 6, 1887 (Steger).

Inyo (in′ yō): **Mountains** [Inyo Co.]. The mountain range east of Owens Lake is clearly shown on the maps of the 1850s, but no name is attached to it. In Apr. 1860 some twenty men led by Col. H. P. Russ and Dr. S. G. George organized the Russ Mining District in this region. When they inquired of the Indians about names, "Chief George," later a leader in the Indian war fought in that district, informed them that the name of the mountains was "Inyo" and that its meaning was "dwelling place of a great spirit" (Chalfant, *Inyo*, p. 83). However, Madison Beeler, "Inyo," *Names* 20.56–59 (1972), and "Inyo once again," *Names* 26.208 (1978), argued for a derivation from Spanish *indio* 'Indian'. **Inyo County** was created on Mar. 22, 1866. **Inyo National Forest** was established and named by executive order of President Theodore Roosevelt on May 25, 1907.

Inyokern [Kern Co.] was earlier called Magnolia, but the name was changed at the request of the Post Office Dept. because there was another town of the same name. The present name was adopted in 1913 by a meeting of the residents; the place is near the boundary of Inyo and Kern Cos.

Ioleo, Rancho del. *See* Yolo.

Iomosa. *See* Alta: Alta Loma; Hermosa.

Ione (ī ōn′) [Amador Co.]. The actual origin and meaning of the name have never been established. According to an often-repeated story, the name was applied in 1848 by Thomas Brown for one of Bulwer-Lytton's heroines. It may also have been named after Ione, Illinois. On Baker's map of 1856 the valley is called Lone Valley, and one would be inclined to seek here the origin of the name. But Ione City was already mentioned in the *San Francisco Alta California* of Mar. 3, 1852, and Ione Valley in the *Statutes* of 1854. The *Placerville Herald* of Sept. 24, 1853, has this story concerning the town: "First named 'Ione City'; then to please some, it was changed to Rickeyville, but as that name was not funny enough for the gamblers . . . they named it 'Freeze Out,' then 'Hardscrabble,' and lately it was called 'Woosterville.'" Another story is told in a publication of the Book Club of California entitled *Heraldry of New Helvetia* (1945: 75): William Hicks of Tennessee came to California with the Chiles-Walker party of 1843 and settled on the Cosumnes River, where the town Hicksville was named for him. The town of Ione was also situated on his land and was reportedly so named because he often used the phrase "I own," indicating his great wealth. The town was originally in Calaveras Co. The post office was established on Sept. 3, 1852, with the name Ione Valley. On Feb. 16, 1857, it was called Jone City; then Ione City; in 1861 Ione Valley; and in 1880 finally Ione.

Iowa Hill [Placer Co.] is the only survivor of several places named by miners from the state of Iowa. The state name is of Indian origin and also designates a Siouan Indian tribe.

Ireland: Lake, Creek [Yosemite NP] were named by Lt. Harry C. Benson for Dr. M. W. Ireland, of Indiana, who was on duty in the park in 1897; during World War I, Ireland was surgeon general of the army (Farquhar).

Irish. Maps of today show about fifteen geographic features so named—comparatively few in view of the strong Irish element among miners and settlers, and the number of place names that they transferred from Ireland to California. Irishtown [Amador Co.] was an old mining settlement.

Irma. *See* Island Mountain.

Irmulco (ûr mul′ kō) [Mendocino Co.], a station of the California Western Railroad, was named in 1908 from "Irvine and Muir Lumber Company" (Borden).

Iron. The word occurs in the names of about thirty Mountains, probably all applied because of the presence or supposed presence of iron deposits. Some fifteen Iron Creeks and Springs probably owe their names to the iron content of the water. **Iron Mountain** [Shasta

Co.]: The settlement developed around the iron mines located by William Magee, deputy to the U.S. surveyor general for California, and by Charles Camden. The claim for 320 acres of unsurveyed land, including the iron deposits, was published in the *Shasta Courier* on Oct. 21, 1871 (Boggs, p. 564).

Ironwood Mountains [Imperial Co.] were so named because of the growth of *Olneya tesota*, known as desert ironwood, the *árbol de hierro* of the Spanish Californians.

Irrigosa (ēr i gō′ sə) [Madera Co.]. Apparently derived, on a Spanish model, from *irrigate*.

Irvine (ûr′ vīn), **Mount** [Sequoia NP]. Named by Norman Clyde on the first ascent in June 1925, in honor of Andrew Irvine, who was lost on the British Mount Everest expedition in June 1924 (Farquhar).

Irvine (ûr′ vīn): town, **Park, Dam, Lake** [Orange Co.]. About 1870 James Irvine, of San Francisco, purchased Rancho San Joaquin and established his famous orchards. When the Santa Fe built the line to Fallbrook Junction in 1888, it named the station for him. In 1913 it gave the name Myford, after Irvine's son, to a siding on the spur to Venta; and in 1923 it named another siding in memory of Irvine's wife, Frances (*see* Kathryn). The post office, established May 20, 1899, was named Myford, because there was a post office in Calaveras Co. named Irvine at that time; not until Mar. 17, 1914, was the name Myford changed to Irvine. When the University of California at Irvine was established in 1965, the Irvine post office became East Irvine, and the same year the railroad name was changed to Valencia.

Irvington [Alameda Co.] was known in the 1870s as Washington Corner (or Corners), after the township in which it is situated. On Mar. 24, 1884, the Post Office Dept. changed the name to Irving in order to avoid confusion with Washington in Nevada Co. In spite of the protests of citizens, the Central Pacific changed the name of the station to Irvington; on Mar. 15, 1887, this name was applied also to the post office. The name of the writer Washington Irving (1783–1859) doubtless had something to do with the change. Irvington became part of the city of Fremont in 1956.

Irwin [Merced Co.] was laid out and named by W. A. Irwin in 1907 (Co. Hist.: Merced 1925: 372–73).

Irwindale [Los Angeles Co.] was named for a citrus grower in the area (Co. Hist.: Los Angeles 1965:3.256). The post office was established May 20, 1899, and the place was incorporated in 1957.

Isaac Walton Grove. *See* Izaak Walton, Mount.

Isabel, Mount; Isabel: Creek, Valley [Santa Clara Co.]. The peaks of what is now called Mount Hamilton were known in Spanish times as Sierra de Santa Isabel, as shown on the *diseños* of several land grants. Mount Isabel is mentioned by Edward S. Townsend on Apr. 23, 1850: "On your right rises the middle range, Mount Diablo and Mount Isabel the crowning point of the ridge" (*CHSQ* 9:360). When William Brewer and Charles Hoffmann of the Whitney Survey climbed the mountain on Aug. 26, 1861, they named what they believed to be the highest peak Mount Hamilton, unaware that it already had a name. The valley east of the peak, however, is called Isabel Valley on the maps of the Whitney Survey. When the USGS in 1895 discovered that the peak 2 miles southeast of the site of the Lick Observatory is really 14 feet higher, this point was properly labeled Mount Isabel, although the mountain group itself retained the name Hamilton. *See* Hamilton.

Isabella [Kern Co.] was named in 1893, the year of the Columbian Exposition, by Stephen Barton, the founder of the town, "because no town on the continent of North America has been named after the good queen who financed the expedition of Columbus" (A. O. Suhre). There were at that time at least nine Isabellas in the United States, but Mr. Barton was apparently certain that none had been named for the Spanish queen. It is now the site of the Lake Isabella reservoir.

Isabel Valley [Humboldt Co.] was named in 1850 for Capt. R. V. Warner's brig, *Isabel* (Co. Hist.: Humboldt 1881:101).

Ishberg Pass [Yosemite NP] was named by Lt. N. F. McClure for a soldier in his detachment, a native of Norway, who in 1895 discovered the route while prospecting for sheepherders' trails (Farquhar).

Ishi Caves [Tehama Co.]. Named for Ishi, a member of the Yahi tribe (*q.v.*), the last "wild Indian" in California, who came to the attention of whites in 1911.

Ishi Pishi Falls [Humboldt Co.]. From Karuk *íshipish*, the former village on the west bank of the Klamath River. The name means 'extending down', referring to the end of a trail that descends from the mountains here.

Isinglass Lake [Siskiyou Co.]. The term refers to the mineral muscovite, a form of mica (Stang).

Islais Creek. *See* Islay.

Island Mountain [Trinity Co.] was named in the 1850s by the first settlers because it is nearly encircled by a river and two creeks. John Rockwell of the Coast Survey mentions Island Peak in 1878. The post office, established in 1905, was called Island until 1907, and then Irma until Aug. 16, 1915, when it was given its present name.

Isla Plana. *See* Mare Island.

Isla Vista (ī′ lə vis′ tə) [Santa Barbara Co.]. Supposed to mean 'island view', with reference to the offshore Channel Islands; better Spanish would be *vista de las islas*.

Islay. This name for the so-called hollyleaf cherry, *Prunus ilicifolia*, is from Salinan *slay'* (Harrington 1944:38). It is mentioned as *yslay* by Fages in 1775, and this soon became the Californians' name for the shrub and its fruit, a favorite Indian food. **Islais** (is′ lis) **Creek** [San Francisco Co.]: The name Los Islais is recorded on the *diseño* of the petition for the grant Salinas y Visitación in 1834. El Arroyo de los Yslais is mentioned in a petition of Manuel Sánchez on Feb. 8, 1835. In American times it is recorded as Islar in 1851 and as Islais in 1853. The early Coast Survey charts call the creek Du Vrees, for a settler, but in 1859 V. Wackenreuder restored the old name on his map. It is still used for the estuary and the bridge (*CFQ* 4:283–84, 5:298–99). **Islay** (iz′ lā) **Creek, Hill** [San Luis Obispo Co.] preserve the name through the grant Cañada de los Osos y Pecho y Islai ('Valley of the Bears, Breast [a rock], and Islay'), dated Sept. 24, 1845.

Isleton (ī′ əl tən) [Sacramento Co.]. This unique name was given by Josiah Pool and John Brocas to the town that they built on Andrus Island in 1874.

Islip (ī′ slip) **Mount; Islip Canyon** [Los Angeles Co.]. The peak was named for George Islip, who settled about 1880 on land now included in the Angeles National Forest (USFS). He was probably the same George Islip, a Canadian, who had a place north of Stanislaus River in the 1850s.

Isthmus: Cove, Landing [Santa Catalina Island, Los Angeles Co.]. The name was applied by the Coast Survey because of the isthmus formed by the cove and Catalina Harbor.

Italian. This adjective of nationality is found in the names of about twenty physical features, most of them applied in modern times when Italian settlers played an important role in California agriculture, especially in viticulture.

Italy, Lake [Fresno Co.]. Named by the USGS about 1907. With a little imagination one can see the resemblance to the outline of Italy on the topographical map.

Ivanhoe [Tulare Co.]. The name of Walter Scott's novel was first applied to the school district in 1885 at the suggestion of Mrs. Ellen Boas. The original name for the settlement was Klink, for George T. Klink, auditor of the Southern Pacific. The name Venice Hill, proposed by the Venice Hill Land Company, was refused by the residents. The present name was bestowed in 1924 through the efforts of the Ivanhoe Farm Bureau (Edith Williams).

Ivanpah (ī′ vən pä): **Valley, Mountain**, station [San Bernardino Co.]. From Chemehuevi *avimpa* or *aavimpa* 'white clay water', from *avi* 'white clay' plus *pa* 'water' (Fowler). Ivanpah Valley and a settlement, Ivanpah, just east of Clark Mountain, are shown on von Schmidt's boundary map (1872). This settlement, obviously a trading camp for the mines that developed around 1870, has vanished. When the Santa Fe built a spur into the valley, it appropriated the name for the terminal. The spur was discontinued in 1920, but the name was not lost: the Union Pacific used it for its station that was formerly called Leastalk (an anagram for "salt lake").

Ivawatch. *See* Avawatz.

Izaak Walton, Mount [Fresno Co.]. The name for the peak at the head of Fish Creek Canyon was proposed in 1919 by Francis P. Farquhar. It honors Izaak Walton (1593–1683), author of *The compleat angler*. The Isaac Walton League of America Grove in the Prairie Creek Redwoods State Park was dedicated on Oct. 3, 1954. The league is one of the oldest conservation organizations in the United States.

J

Jabu (jä′ bōō) **Lake** [El Dorado Co.]. A composite of letters from the name of Jack Butler, who planted fish here (Lekisch).

Jacalitos (jak ə lē′ təs): **Creek, Hills** [Fresno Co.]. *Jacalito*, a diminutive of Mexican Spanish *jacal*, means 'little hut'. The Spanish term is derived in turn from Aztec *xacalli* 'hut'. The name probably does not go back to Spanish times; but like many others in the region, it may have been applied by American surveyors.

Jacinto (jə sin′ tō) [Glenn Co.]. The place preserves the first name of Jacinto Rodríguez, who was granted land on Sept. 2, 1844, that was sometimes called Rancho de Rodríguez (Bidwell map, 1844) and sometimes Rancho Jacinto. *See also* San Jacinto.

Jackass. More than twenty-five place names in the state testify to the wide use of jackasses as pack animals in the mining days. Most of them originated from an incident in which a jackass played a role; some were probably applied in derision. Jackass Creeks are the most numerous, but there are also Jackass Buttes, Peaks, Rocks, Meadows, etc.; the best known is probably the cluster name west of the junction of the Middle and South Forks of the San Joaquin River [Madera Co.]. **Jackassville** [Siskiyou Co.]: A mining camp called City of Six by its first six inhabitants was known in the neighboring camps as the "camp of half a dozen jackasses," later as Jackassville (C. L. Canfield, *The City of Six*, Chicago, 1910). **Jackass Hill** [Tuolumne Co.] was the home of James W. Gillis, who is immortalized in Mark Twain's writings as "The Sage of Jackass Hill." **Jackass Dyke** [Fresno Co.] is one of the few places in California where "dyke" (i.e. an igneous rock wall that has resisted erosion) is used as a generic term. It was not named after the animal, but after Jackass Meadow. **Jackass Flat** [Shasta Co.]: The sad story of a jackass that kicked against progress is told in the *Redding Republican Free Press* of Mar. 7, 1885 (Boggs, p. 709): "The locomotive started ahead, when the animal threw up his tail, kicked up his heels, and blazed away. But alas for the jack, he landed partially on the track, and the remorseless wheels of the locomotive cut off poor jack's fore and hind leg. The poor fellow rolled down the bank, and his kind master came along and put an end to his misery. The railroad boys have named the terrible scene of disaster, 'Jackass Flat.'"

Jack London Historic State Park [Sonoma Co.]. The park, honoring one of California's famous literary men (1876–1916), was created in 1960 on London's ranch near Glen Ellen.

Jack Main Canyon [Yosemite NP]. Named for Jack Main (or Means), who ran sheep in this canyon for many years, beginning in the early 1870s (Farquhar).

Jacks Backbone. *See* Backbone.

Jacksnipe [Solano Co.]. The term refers to various species of aquatic birds similar to the snipe.

Jackson: town, **Gate, Valley, Creek, Butte** [Amador Co.]. The mining camp was first known as Bottileas because of the many bottles (Sp. *botellas*) that had accumulated near the spring where travelers stopped on their way to the southern mines. When "Colonel" Alden M. Jackson, a lawyer from New England, opened an office in the camp and became generally liked for settling quarrels out of court, the grateful miners named the place for him. The post office was established on July 10, 1851, while the town was still in Calaveras Co. It is said that the original name of the butte was Polo Peak, named for an Indian chief. The "Gate," a fissure in the reef of rocks that crosses the creek, was discovered by a boy from Sacramento in 1849. **Jacksonville** [Tuolumne Co.] was named for the same Jackson; both towns appear on Gibbes's map of 1852. *For* Mount Jackson, *see* Shasta, Mount.

Jacks Valley [Lassen Co.]. Named for John

Wright, known as "Coyote Jack," who settled here in 1864 (Doyle).

Jacoby (jə kō′bē) **Canyon** [San Bernardino Co.]. For Frank Jacoby, a miner; the name was established by BGN, list 7501, replacing the variant Chicopee Canyon (Chicopee is a town in Massachusetts).

Jacoby (jə kō′ bē) **Creek** [Humboldt Co.]. Named for A. Jacoby, an early settler. In the 1850s and 1860s, Jacoby's brick store in Union (now Arcata) was used as a refuge during Indian attacks (Coy, *Humboldt*, p. 199). The name was placed on the map by von Leicht–Craven as Jacoby River.

Jacumba (hə kum′ bə, hə ko͞om′ bə): **Mountains, Valley, Hot Springs**, town [San Diego Co.]. Probably originated as a Diegueño village name, of unknown etymology. *La ranchería llamada en su idioma Jacom* 'the village called Jacom in their language' is mentioned on May 13, 1795 (PSP 13:222). The name is repeatedly found in documents, with various spellings. It is recorded for the mountains as la Sierra de Jacum on June 8, 1841 (DSP Ben. P and J 4:14), and Jacum Pass is mentioned in the Pac. R.R. *Reports* (5:1.41). The name was preserved in the old Jacumba House, around which the town developed when the San Diego and Arizona Eastern was built in 1917.

Jago (jā′ gō) **Bay** [Lake Co.]. For Louis Jago, founder of Jago's Resort (Mauldin).

Jail Canyon [Inyo Co.]. The name appears on Wheeler atlas sheet 65-D (Topography); it may have been given because the canyon might easily be converted into a jail by guarding the entrance. The "Land Classification" of the same atlas designates it as Tail Canyon. *See* Calaboose Creek.

Jalama (hə lam′ ə): **Creek**, station [Santa Barbara Co.]. From the Purisimeño Chumash village name *xalam*, meaning 'bundle' (Applegate 1975:28). This rancheria of Mission La Purisima is mentioned as early as 1791 (Arch. MPC, p. 10), and *una cañada . . . Jalama* is recorded on July 25, 1834 (DSP Ben. Mil. 79:105). The name also appears on several *diseños* with the same spelling.

Jamacha (ham′ ə shô) [San Diego Co.]. From Diegueño *hemechaa* 'a type of gourd used for soap' (Langdon). A rancheria Xamacha is mentioned on Nov. 30, 1775 (PSP Ben. Mil. 1:5); this is called Ranchería de Jamacha the next year (SP Sac. 8:72), and in a later record Jamocha or San Jacome de la Marca. The Indians of the village participated in the attack on San Diego Mission on Nov. 4, 1775. The land grant Jamacha or Jamacho is first recorded on Dec. 1, 1831. The spelling Jamacao of the BGN (6th Report) was changed to Jamacha in list 7003.

Jamesburg [Monterey Co.]. Named for John James, who founded the town in 1867.

Jamestown [Tuolumne Co.]. The name was given to the town founded in 1848 by Col. George F. James, a San Francisco lawyer. It appears on Gibbes's map of 1852. After a series of disputes with Mexican settlers, which ended in James's ruin and departure, the name of the town was changed to American Camp, but later the old name was restored. Locally the place is known as "Jimtown."

Jamul (hä mo͞ol′): **Creek, Butte, Mountains**, town [San Diego Co.]. From a Diegueño village name, *hemull* 'foam, lather' (Langdon). A *capitanejo de Jamol* is mentioned by Comandante Rivera, San Diego, Aug. 10, 1776 (PSP 1:227). The *sitio de Jamul* is repeatedly mentioned in records of the 1820s; and on Apr. 20, 1831, Pío Pico was given permission to occupy provisionally *el parage llamado* ('the place called') *Jamul* (Dep. Recs. 9:98). In 1837 the rancho was attacked by Indians, and the caretakers were slain. It was regranted to Pío Pico on Dec. 23, 1845.

Janes Creek [Humboldt Co.]. Named for H. F. Janes, formerly of Janesville, Wisconsin, who came to California in 1849.

Janesville [Lassen Co.]. The post office was named in 1864 by L. N. Breed, the first postmaster—either for Jane Bankhead, wife of the village blacksmith, or for Jane Hill, who was born here on May 17, 1862; possibly for both.

Japacha (hä pə chä′): **Creek, Peak** [San Diego Co.]. The name reflects Diegueño *hapechaa* 'handstone for grinding', Spanish *mano de metate* (Langdon). "Japacha" and "Jepeche" Indian rancherias were listed in mission records (Stein).

Japatul (hä′ pə to͞ol) **Valley** [San Diego Co.]. The name probably preserves the name of the Indian village Japatai, which was still in existence in the early 1880s. Perhaps from Diegueño *hatepull* 'woodpecker' (Langdon).

Jaquima. *See* Hackamore.

Jara (här′ə) **Canyon** [San Mateo Co.]. Sebastián Jara, a woodcutter, lived here from the 1860s (Alan K. Brown).

Jarbo (jär′ bō) **Gap** [Butte Co.]. Said to be the family name of early settlers.

Jarro. The Spanish word for 'jug, pitcher' was used repeatedly in Spanish place names and is preserved in two names. **El Jarro** (hä′ rō) **Creek** [Santa Barbara Co.]: A *cañada* called el Yarro is mentioned on July 25, 1834 (PSP Ben. Mil. 79:105), and an Arroyo del Jarro is shown on the 1837 *diseño* of the San Julián grant. **El Jarro Point** [Santa Cruz Co.]: The Agua Puerca y las Trancas grant was first made on Oct. 12, 1839, as El Jarro (also written as Tarro). The *diseño* of 1843 shows Arroyo de Jarro (probably Scott Creek). The Coast Survey did not apply the name to the point until after 1910.

Jastro (jas′ trō) [Kern Co.]. Named for a Southern Pacific railroad employee.

Javon (hä vōn′): **Canyon, Creek** [Ventura Co.]. The name of the canyon is derived from Spanish *jabón* 'soap' and is reminiscent of one of the strangest of the numerous California mineral booms. It was started in 1875 when H. L. Bickford began to work a "rock soap" mine in the canyon. Within a short time a number of companies, including the Pacific Soap Company, were organized to take soap right out of the earth (Co. Hist.: Santa Barbara 1883:423 ff.). Extravagant claims were made for the miracle product, which was made into scrubbing soap, toilet soap, salt-water soap, and even tooth powder. In the same year, samples of this mineral were shown at the Paris Exposition and attracted considerable attention (Hanks 1884:345). It turned out, however, that the mineral was infusorial earth suitable only for polishing jewelry and silverware, and California's soap bubble burst.

Jawbone Canyon [Kern Co.]. The name was applied by prospectors hecause the outline of the canyon suggests a jawbone. Other physical features bear this name; most of them were probably so named because the jawbone of some animal was found at the place.

Jayhawker Spring and **Well** [Death Valley NP]. The spring, by which the Jayhawkers (*see* Glossary) camped in 1849, was so named in 1936 by the Park Service; it was formerly known as Hitchens' Spring for James Hitchens, who camped there in 1860. The well, by which they also camped, was long known as McLean's Spring. The term "Jayhawker" originally referred to anti-slavery guerrillas who operated in border states during Civil War days.

Jedediah (jed ə dī′ ə) **Smith: Redwoods State Park; Memorial Grove** [Del Norte Co.]. These prominent features of the Smith River–Mill Creek State Park honor one of the greatest pathfinders of the West (1799–1831). The grove was established at the suggestion, and with the support, of Charles M. Goethe of Sacramento (*see* Smith River). Jedediah Smith Mountain, 15 miles northeast of Gasquet, was for some reason changed to Jedediah Mountain by a decision of the BGN, July–Dec. 1965.

Jeff Davis: Peak, Creek [Alpine Co.]. The peak appears on Wheeler atlas sheet 56-B as Sentinel Rock. The present name was apparently not recorded on maps until the district was mapped by the USGS in 1889; however, it may have long been used locally, as many of the inhabitants of nearby Summit City (now abandoned) were Confederate sympathizers during the Civil War. Jefferson Davis (1808–89) was president of the Confederacy, 1861–65.

Jefferson. *See* Lennox.

Jellico (jel′ i kō) [Lassen Co.]. The name was applied to the section-crew camp of the Western Pacific about 1930. The nearby Jelly Camp may have suggested the name.

Jelly Camp [Lassen Co.]. The name was given to a sheepherder's camp in the early 1890s, probably by Victor Ayle, a homesteader. It is shown on the maps of the USFS, but the reason for its application is not known.

Jenkinson Lake [Mono Co.]. Named for Walter E. Jenkinson, who was chiefly responsible for creating the reservoir by damming Sly Park Creek (BGN, Mar. 1957).

Jenner: Creek, town [Sonoma Co.]. Probably named for Elijah K. Jenner, a native of Vermont, or for his son Charles, a native of Wisconsin; both were living in the county in the 1860s.

Jennie Lake [Kings Canyon NP]. Named by S. L. N. Ellis for his wife, Jennie, in 1897 (Farquhar).

Jenny Lind [Calaveras Co.]. Established as a mining camp in the early 1850s and named in honor of the Swedish singer (1820–87), whose tour of eastern cities in 1850–52, managed by the great P. T. Barnum himself, created a furor of excitement throughout America. Despite the persistent legend, Jenny Lind never visited California. The columns of *Las Calaveras* contain various stories concerning the origin of the name: Jenny Lind had come to California during her tour through the United States; a man by the name of Dr. J. Y. Lind founded the place; the braying of mules prompted the miners to use the name of the great singer in sarcasm. The name was popular and often used, because the "Swedish nightingale" was so widely known. Big Arroyo in Tulare Co. was formerly called Jenny Lind Creek, after a mine near its banks.

Jepson, Mount [Kings Canyon NP]. For Willis L. Jepson (1867–1946), professor of botany at the University of California, Berkeley, and charter member of the Sierra Club (Browning 1986).

Jepson Hole. *See* Willis Hole.

Jerusalem Creek [Shasta Co.]. The branch of Cottonwood Creek was named by Jewish settlers (Steger). **Jerusalem Valley** [Lake Co.] was named by Saphonia, daughter of Charles Copsey, who came from Missouri in 1856 with an emigrant train (Knave, June 30, 1957). El Rio [Ventura Co.] was originally called New Jerusalem by its founder, Simon Cohen; and Brewer (Notes, June 7, 1862) mentions an "Old Jerusalem," apparently in Alameda Co.

Jesmond Dene [San Diego Co.]. Named in the 1920s after the public park in Newcastle-upon-Tyne, northern England.

Jesse Canyon [Santa Barbara Co.]. *See* Jesus.

Jess Valley [Modoc Co.]. Named for Jonathan Jess and his brother, pioneers of the 1860s.

Jesus (hā so͞os′) represents Spanish *Jesús*, a common given name for men. There is a Jesus Creek in Santa Cruz Co. (Clark 1986). **Jesus Canyon** [Santa Barbara Co.] was named officially by the BGN in 1978 (list 7802), replacing the name Jesse Canyon.

Jesús de los Temblores, Rio de. *See* Santa Ana: Santa Ana River; Temblor.

Jesús María. Río Jesús María, as the name for a stream, was used in 1808 by Gabriel Moraga for the part of the Sacramento that is north of the junction with the Feather River. Within the next forty years the name was applied to various rivers, real or imaginary: the upper Sacramento, a branch of the San Joaquin, an eastern branch of the lower Sacramento, a river between the Sacramento and the San Joaquin, a river rising near Cape Mendocino and emptying into the northwestern part of San Francisco Bay, and finally to Cache Creek in Yolo Co. (Bowman). On Oct. 21, 1843, the name Río de Jesús María was applied to a land grant in what is now Yolo Co. (*see* Quesesosi). **Jesús María** [Santa Barbara Co.] was the name of a land grant dated Apr. 8, 1837.

Jesus Maria Creek [Calaveras Co.]. The creek and a settlement were named for a Mexican by that name who raised vegetables there in the mining days. The place is shown on Lapham and Taylor's map of 1856. The local pronunciation is (so͞os′ mə rē′ ə).

Jim Crow: Creek, Ravine [Sierra Co.]. "While we were camped at Slate Range [in the fall of 1849], one of our men went back for a Kanaka and an Indian. The Kanaka he returned with, was Jim Crow, whose name still lives in those regions" (William Downie, *Hunting for gold* [San Francisco: California Publishing Co., 1893], p. 35). "Our next meeting [with Jim Crow] was in the following spring, at Crow City, at the head of Jim Crow Canyon, as these places are now called" (ibid., p. 52).

Jim Dollar Mountain [Lake Co.]. Probably named for James Henry Dollar, a native of Indiana, who settled in the county in 1867 or earlier.

Jimeno (hi mē′ nō) [Yolo Co.]. Named after the Jimeno land grant, dated Nov. 2, 1844, which was so called for its grantee, Manuel Jimeno Casarín, a well-known political figure in the last decade of Mexican rule.

Jimgrey (jim′ grā) [San Bernardino Co.]. A contraction of the name of Jim Grey, a road foreman of engineers, given to a railroad siding established by the Santa Fe in 1900.

Jim Jam Ridge [Trinity Co.]. "Jim jams" was a term used by the miners of the district for the result of overindulgence in alcohol, now commonly called the "jitters." In the 1890s three miners on a "sobering-up" hike had so many "sobering-up" drinks that one of them rolled into the campfire, with his pocket full of 30–30 shells. His companions were so affected by "jim jams" that they named the spot for it (Bill Noble to W. E. Hotelling). The name was put on the map of the Trinity National Forest in 1935.

Jimtown. *See* Jamestown.

Joaquin Rocks. *See* Murieta.

Jobs (jōbz): **Peak, Sister** [Alpine Co.]. In the early 1850s, Moses Job opened a store in Sheridan at the foot of the mountain that soon became known by his name. The state surveyor general's report (1856:141) speaks of Job's Group of Mountains, and Hoffmann's map of 1873 refers to the peaks of the range collectively as Job's Peaks. In 1893 the USGS applied the name Jobs Sister to the middle peak, and Freel Peak to the western peak.

Jobs Peak [San Bernardino Co.]. The peak northwest of Lake Arrowhead was so named because a cow (or a mule) called Job wandered up the slope and got lost (Wheelock).

Jofegan (jō′ fā gən) [San Diego Co.]. For Gen. Joseph Fegan, the first commander of Camp Pendleton (Stein).

Johannesburg [Kern Co.]. Named about 1897 by the founders, Chauncey M. Depew and associates, after the famous mining center in the Transvaal, South Africa. It is popularly known as Joburg.

John Little State Park [Monterey Co.]. In 1953 Elizabeth Livermore Schmidt presented the park to the state and named it for a friend, John Little, a pioneer of the area.

John Muir Trail. *See* Muir.

Johnson, Mount [Kings Canyon NP]. The name was originally bestowed by R. B. Marshall on the present Mount Lewis, for Willard D. Johnson (1861–1917), a member of the USGS. To avoid confusion with the nearby Johnson Peak, the name was transferred to its present location.

Johnson Canyon [Riverside Co.]. The canyon in Cleveland National Forest was named in 1960 in memory of Steven W. Johnson, a USFS employee who lost his life in the Decker fire of Aug. 1959.

Johnsondale [Tulare Co.]. The mill town was named by the Mount Whitney Lumber Company in 1938 for Walter Johnson, one of its officials (Mitchell).

Johnson Peak [Yosemite NP]. Named by R. B. Marshall for a teamster and guide with his survey party in the 1890s (Farquhar).

Johnsons Ranch [Yuba Co.]. The land grant, dated Dec. 22, 1844, was first known as Rancho de Pablo, for Pablo Gutiérrez, the grantee. Gutiérrez, Sutter's old Mexican employee, was killed in the Micheltorena campaign; and the title to part of the grant was subsequently acquired by William Johnson, a native of Boston who came to California in 1840. For several years after 1845, Johnson's Ranch was well known as the first settlement reached by the overland immigrants after crossing the Sierra.

Johnson's River [Los Angeles Co.]. *See* San Gabriel River.

Johnstone Peak [Los Angeles Co.]. Named by the USFS for the late W. A. Johnstone, for many years a leader in forest conservation. Approved by the BGN in 1940.

Johnstonville [Lassen Co.]. The settlement was originally known as Toadtown because, according to tradition, little toads covered the ground after a rainstorm. In Dec. 1864 the Board of Supervisors changed the name to Johnstonville, for Robert Johnston, a pioneer farmer in the valley.

Johnsville [Plumas Co.]. Founded in 1876 and known as Johnstown, for William Johns, manager of the Plumas Eureka Mine. The present name was adopted when the post office was established in 1882.

Jolla (hoi′ ə). *Joya*, also spelled *hoya* or *jolla*, is a common Mexican Spanish geographical term referring to a hollow in the mountains, a hollow worn in a river bed, or a hollow on the coast worn by waves. Peñafiel calls it a *corrupción de la palabra castellana Hoya que significa concavidad o excavación de la tierra* 'corruption of the Castilian word *hoya*, which means hollow or excavation in the ground'. The name occurs in three different sections of San Diego Co., but is also found in other parts of the state: there is a La Jolla Creek in Fresno Co.; an intermittent creek rising in the Lompoc Hills [Santa Barbara Co.] is called La Hoya; and a third spelling variant is found in La Joya Peak [Ventura Co.]. On Sept. 17, 1769, the Portolá expedition camped in a "hollow" in what is now Monterey Co., and called it La Hoya de la Sierra de Santa Lucía (Crespí, p. 192), a name that did not survive. Bunnell mentions the use of the word by the Yosemite Indians in the 1850s: "Ho-yas . . . referred to certain holes in detached rocks west of the Sentinel, which afforded 'milling privileges' for a number of squaws, and hence, the locality was a favorite camp ground. The 'Sentinel' or 'Loya' . . . marked the near locality of the Hoyas or mortars" (Bunnell, p. 212 ff.). **La Jolla Valley** [San Diego Co.]: An Indian rancheria called La Joya is mentioned in 1828 (Registro, p. 37). The name for the valley, 3 miles southwest of Lake Hodges, is shown as La Hoya on a *diseño* of the San Bernardo grant (1841); and two *diseños* of the Encinitas grant (1845) show Joya about where Olivenhain now is. The *diseño* of the grant shows an Arroyo de la Hoya as a tributary of what is now San Luis Rey

River. The name is spelled La Joya on some maps. **La Jolla** town, **Point, Bay, Caves, Mesa**, and **Canyon** [San Diego Co.]: It is not known whether the name at the coast has any connection with La Jolla Valley on the San Bernardo grant, or whether it was applied independently. The present town was laid out in 1869. The Pascoe map of 1870 clearly shows lot numbers 1286 and 1288 in a hollow called La Joya, just back of the Beach and Tennis Club (H. F. Randolph). On the maps of the Coast Survey and the Land Office, the name is shown variously spelled, in one or in two words. The first post office, established on Feb. 29, 1888, was called La Jolla Park; the second, established on Aug. 17, 1894, Lajolla. On June 19, 1905, this was changed to La Jolla, on Z. S. Eldredge's insistence. The popular tradition that the name is derived from the Spanish word *joya* 'jewel' sounds plausible, since *y* and *ll* were interchangeable, but it is supported by no evidence. An alternative etymology is from Diegueño *matku-laahuuy* 'place that has holes or caves', related to *llehup* 'hole' (Langdon).

Jollabollas. *See* Bally.

Jolon (hə lōn′): **Creek**, post office [Monterey Co.]. From Antoniano Salinan *xolon* 'it leaks; a leak; a channel where water cuts through' (information from Dr. Katherine Turner, University of California, Berkeley). Bancroft shows the name of the locality near San Antonio Mission on his map (1885) of the Monterey district as it existed in 1801–10 (2:145). Cañada de Jolón is recorded on a *diseño* of the Los Ojitos grant (1842). The name does not seem to appear on early American maps, but Jolon ranch and post office are mentioned by Brewer, May 8, 1861 (p. 93). A local folk etymology claims that the name originated when bandits would stop stagecoaches by shouting, "Hol' on! Hol' on!"

Jonata (hō nä′ tə, hä′ nə tə) [Santa Barbara Co.]. Spanish Jonjonata, from the Chumash village name *xonxon'ata* 'tall oak' (John Johnson 1988:98). *See* San Carlos de Jonata.

Jonathan Rock. See Brother Jonathan Rock.

Jones Butte. *See* Dutschke Hill.

Jones Valley [Shasta Co.]. The valley near Bear Mountain was named for a Mrs. Jones who was killed there by Modoc Indians in 1864 (Steger).

Jonive (hō nē′ və) **Hill** [Sonoma Co.]. The hill west of Sebastopol preserves the name of the grant Cañada de Jonive, one of James Black's grants, dated Feb. 5, 1846. A Cañada de Jonive is shown on the *diseño*. The term is probably the result of a misspelling or misunderstanding of an Indian name.

J.O. Pass [Sequoia NP] was named in 1889 by S. L. N. Ellis, using the first two letters of the name of John W. Warren, which had been carved in a tree by Warren a few years before.

Jordan, Mount [Sequoia NP]. Named in 1925 by the Sierra Club in honor of the educator and scientist David Starr Jordan (1851–1931), president and chancellor of Stanford University. Jordan himself had proposed the name Crag Reflection for a part of the peak.

Joseph, Mount. *See* Lassen Peak.

Joseph Grinnell, Lake. *See* Grinnell.

Josephine, Mount [Los Angeles Co.]. Named for the daughter of Phil Begue, a ranch owner in the vicinity.

Josephine Lake [Kings Canyon NP]. Named by S. L. N. Ellis for Josephine Perkins (Farquhar).

Josephine Peak [Los Angeles Co.]. The peak was named for Josephine Lippencott, whose husband used the elevation for a triangulation station (W. C. Mendenhall).

Joshua Tree National Monument [Riverside, San Bernardino Cos.]. Created in 1936 by proclamation of President Franklin D. Roosevelt. A post office, Joshua Tree, was established on July 16, 1946. The desert tree (*Yucca brevifolia*) was named Joshua tree by the Mormons, to whom it seemed to be a symbol of Joshua leading them to the promised land; it is sometimes called "yucca palm." The names Twenty-nine Palms Oasis, Fortynine Palms Canyon, Lone Palm Oasis, and Lost Palm Canyon—although they are within the monument—derive their names not from the Joshua tree, but from the fan palm, *Washingtonia filifera* (E. N. Fladmark). *See* Palm.

Jota [Napa Co.]. The land grant La Jota, conveyed to George C. Yount on Oct. 21, 1843, contains the Spanish name for the letter *j;* it may also refer to the Spanish dance *la jota. See* Polka.

Jovista (jō vis′ tə) [Tulare Co.]. When a spur of the Southern Pacific was built in 1923, railroad officials coined the name for the station from the two initial letters of Joseph Di Giorgio's given name plus Spanish *vista* 'view'. *See* Di Giorgio.

Joya, La. *See* Jolla.

Juan Rodriguez Island. *See* San Miguel.

Juaquapin (wä kwə pēn′) **Creek** [San Diego Co.]. Perhaps from Diegueño *hakupin* 'warm water' (Langdon).

Jubaville. *See* Marysville.

Jubilee Pass [Death Valley NP]. This name, placed on the Avawatz atlas sheet by the USGS in 1933, was suggested by that of the Jubilee Mine.

Judah, Mount [Placer Co.]. The name, approved by the BGN in 1941, was given, at a rather late date, in memory of the great railroad construction engineer Theodore D. Judah (1826–63), who built the first California railroad and was the guiding spirit in building the Central Pacific across the Sierra Nevada.

Judson Reservoir [San Diego Co.]. Named in the 1880s, probably for Lemon Judson, who owned a ranch in the district.

Jughandle Gulch [Mendocino Co.]. So named because the old road across it made a turn in the shape of a jughandle (Stewart).

Julian [San Diego Co.]. Laid out in 1870 by Drew Bailey and named for his cousin, Mike S. Julian, a mining recorder on whose government claim gold quartz had been discovered.

Julius Caesar, Mount [Fresno, Inyo Cos.]. Named in 1928 because of the peak's proximity to Lake Italy (Browning 1986).

Jumbo Knob [Los Angeles Co.]. The terminal knob of the Linda Vista crescent of hills was named in 1884 because of a fancied resemblance to Barnum's elephant, named Jumbo (Reid, p. 374).

Jumel (jə mel′) **Gulch** [San Mateo Co.]. Named for the Jumel family, ranchers around 1860 (Alan K. Brown).

Jump-Off Creek [Mendocino Co.]. Named after a former road, which was so steep at this point that wagons descending it were braked with drags of hewn trees (F. Price).

Juncal (jung′ kəl) **Canyon** [Santa Barbara Co.]. The Spanish term means 'a place where reeds grow', from *junco* 'reed'.

Junction. This geographical term is commonly applied to places where railroads, highways, mountain ranges, rivers, and county borders meet. **Junction City** [Trinity Co.]: The post office was transferred from Messerville on Aug. 19, 1861, and was named Junction because of the old mining center that had developed at the junction of Canyon Creek with the Trinity River about 1850. **Junction Meadow** and **Peak** [Sequoia NP]: The meadow at the junction of the Kern and Kern-Kaweah Rivers was named by W. B. Wallace in 1881, and the peak by J. N. LeConte in 1896, because it is at the junction of the Kings-Kern Divide with the main Sierra crest (Farquhar). **Junction Butte** and **Bluffs** [Madera Co.]: Applied because butte and bluffs are situated at the junction of the North and Middle Forks of the San Joaquin River. The name was probably given by the USGS when the Mount Lyell quadrangle was surveyed in 1898–99. **Junction Ridge** [Kings Canyon NP]: The continuation of Monarch Divide west of Wren Peak was first called Junction Dike when the USGS mapped the Tehipite quadrangle in 1903. At the western end of the ridge is the junction of the Middle Fork and South Fork of Kings River.

June: Lake, Mountain [Mono Co.]. Origin not known (Sowaal 1985). The name of the lake may be taken from that of a woman, like that of other lakes in the region.

Juniper. The presence of at least one of the four native species of juniper in almost all parts of the state has given the name to more than twenty-five physical features and settlements. **Juniper Lake** [Lassen NP] was named after the peak south of it, which was once called Juniper Peak because a beautiful specimen of *Juniperus occidentalis* was on its slope. The peak was renamed Mount Harkness in the 1890s.

Junipero Serra (hə nip′ ə rō sâr′ ə): **Peak** [Monterey Co.]. The name honors Padre Junípero Serra (1713–84), founder of nine missions in Alta California during his presidency of the missions from 1769 until his death. The Native Daughters of the Golden West had first (June 1905) bestowed the name on a peak in the High Sierra; but the name was transferred to the former Santa Lucia Peak on the recommendation of the Sierra Club, and the transfer was affirmed by the BGN in 1907. "He was familiar with the Mountain [range] of Santa Lucia . . . and it is believed that he never saw the Sierra Nevada . . . The name of Junípero Serra in the Sierra Nevada will be simply a geographical record; his name upon one of these coast peaks that barred the expedition of 1769 and 1770 will be a living designation to some marked and well-known landfall, appealed to every day by the mariner and traveller" (letter of George Davidson, Oct. 5, 1905).

Junta de los Quatro Evangelistas. *See* Suisun Bay.

Juntas [Contra Costa Co.]. A land grant dated Oct. 20, 1832, and Feb. 12, 1844. Since the grant is just south of Suisun Bay, it is probable that the name was derived from Junta de los quatro Evangelistas 'joining of the four evangelists', a name probably referring to the "meeting" of the waters, shown on the Ayala-Cañizares map of 1775 and on later maps. For many years the name was preserved in Las Juntas station of the old San Ramon branch of the Southern Pacific.

Juristac [Santa Clara Co.]. The name of the land grant, dated Oct. 22, 1835, includes a Costanoan place name with the locative ending *-tak;* it may be derived from Chocheño *huri* 'to take off, let go' (Callaghan). The grant was also known as La Brea and as Los Germanos (Antonio and Faustino Germán were the grantees).

Jurupa (hə ro͞op′ ə): **Mountains**, station [Riverside, San Bernardino Cos.]. The name was applied because the range and the railroad station are situated on the territory of the Jurupa land grant, dated Sept. 28, 1838. Jurupa, even earlier, had been a rancho of Mission San Gabriel. The origin must be Gabrielino *horúvnga* 'sagebrush-place' (Munro), or a similar form in a related language, with a 'place' suffix *-pa*. Jurupa is mentioned as a place by Ord in 1849 (p. 119) and is shown on his sketch of the Los Angeles plains. Riverside was apparently first called Jurupa (*see* River: Riverside); however, the von Leicht–Craven map of 1874 shows the two names in different locations. The name Jurupa Mountain was given by the BGN (list 6901) to the highest peak in the range.

K

Kabyai (kä′ bē ī) **Creek** [Shasta Co.]. From a Wintu place name *xebeyay'*, meaning 'unknown' (Harrington).

Kadota (kə dō′ tə) [Merced Co.]. A variety of fig, perhaps originally Japanese.

Kagel (kā′ gəl) **Canyon** [Los Angeles Co.]. Named for Henry Kagel, who took up a claim on government lands at the mouth of the canyon (T. R. Wilson).

Kaiser [San Bernardino Co.]. The station was named by the Southern Pacific for Henry Kaiser (1882–1967), industrialist of World War II, one of whose war plants was in nearby Fontana.

Kaiser: Creek, Creek Diggings, Pass, Peak, Peak Meadows, Ridge [Fresno Co.]. According to L. A. Winchell, Kaiser or Keyser Gulch was mentioned as early as 1862 (Farquhar). It appears as Kaiser Gulch on Hoffmann's map of 1873, and as Kaiser Creek on Bancroft's map of 1882. The other features were named by the USGS and are shown on the Kaiser atlas sheet of 1904. It is possible that the gulch was named for Elijah Keyser, a native of Pennsylvania and argonaut of 1849 (*CFQ* 4:92–93). He struck it rich somewhere in the mines, but neither he nor any other person named Keyser has been identified with this particular location. It is also not impossible that the gulch was originally named for Richard Keyes, a successful miner of 1853, and that the present version is the result of a misunderstanding. At the time of World War I, many places in the United States named Kaiser were changed; but the three other Kaiser Creeks in California, all probably named for local residents, likewise weathered the storm.

Kalmia (kal mē′ ə) **Lake** [Trinity Co.]. For a shrub, the American laurel or alpine laurel, *Kalmia polifolia*, which grows here (BGN, list 8404). The name of the plant, commemorating the 18th-century Finnish botanist Peter Kalm, is usually pronounced by botanists as (kal′ mē ə). The name also occurs in El Dorado Co.

Kalorama (kal ə ram′ ə) **Barranca** [Ventura Co.]. For the steamship *Kalorama*, which was wrecked near here in 1876 (Ricard). Kalorama was the name of the home of Joel Barlow, American poet and diplomat (1754–1812), in the District of Columbia; the name was later applied to a neighborhood and a street in the nation's capital.

Kanaka (kə nak′ ə). The common designation for a native of the Hawaiian Islands (the Hawaiian word for 'human being') is contained in the names of about twelve geographic features, apparently all in northern counties, including Kanaka Glade [Mendocino Co.]. (Glade is a generic term rarely used in California.) Kanakas were widely employed as sailors, and some accompanied early trapping parties even before Sutter brought islanders to work in California in 1839. All Kanaka place names probably originated in the mining days, but the term was loosely used and some names may have been applied for reasons other than the presence of Pacific Islanders. *See* Jim Crow.

Kanawyers (kə nô′ yərz) [Kings Canyon NP]. Napoleon ("Poly") Kanawyer maintained a camp at Copper Creek for many years near its confluence with the South Fork of the Kings River (Farquhar).

Kane: Springs, Lake [Kern Co.]. The springs northeast of Cantil were apparently named for Grover Kane, repeatedly mentioned in Marcia Wynn's *Desert bonanza* (1963). Nearby Koehn Lake was also known as Kane Lake.

Kane Spring [Imperial Co.]. According to an article in the *Overland Monthly* (Nov. 1904), the spring was named for the first owner. Since the name is also spelled Cane, the origin may perhaps be found in the growth of canes at the place.

Kangaroo. Several place names in northern counties, including Kangaroo Mountain

[Siskiyou Co.], were probably given because of the presence of the kangaroo rat (genus *Dipodomys*). "Nice, harmless animals . . . on my ranch where they danced around in large numbers on beautiful summer evenings" (Cordua, p. 26).

Kanick (kā´ nik) [Humboldt Co.]. Also spelled Kenick, from the Yurok village Kenek (Waterman, p. 137; Harrington).

Karnak [Sutter Co.]. Named for the village in Egypt where part of the ruins of ancient Thebes are located.

Karuk (kä´ ro͞ok) **Indian Reservation** [Siskiyou, Humboldt Cos.]. The name of the tribe on the upper Klamath River is from the word *káruk* 'upriver', contrasting with *yúruk* 'downriver'—which term was applied by whites to the Yurok tribe on the lower Klamath.

Kaseberg (kās´ bûrg) **Creek** [Placer Co.]. Named for James W. Kaseberg, a native of Germany, who had a large ranch adjoining Roseville.

Kashia (kə shī´ ə) **Indian Reserve** [Sonoma Co.]. The name refers to the Southwestern Pomo Indians; it is from their self-designation *k'ahšáaya* 'agile people', from *k'ahša* 'agile, nimble' (Oswalt).

Kathryn [Orange Co.]. Named in 1923 by the Santa Fe, for Kathryn Irvine, daughter of James Irvine, owner of the San Joaquin rancho (Gertrude Hellis). *See* Irvine.

Katimin (kä´ ti mēn) [Siskiyou Co.]. From Karuk *ka'tim'íin*, lit. 'upper edge falls', referring to a rapids in the Klamath River and to the adjacent Indian village, a major site for the annual world-renewal ceremony (*see* Pick-aw-ish Campground). Alternative spellings found include Cottamin, Cottamain. Downriver a few miles in Humboldt Co. are Ike's Falls, called *yu'tim'íin* 'lower edge falls' in Karuk, also a ceremonial site.

Kavanaugh Ridge [Mono Co.]. The ridge was named for Stephen ("Steve") Kavanaugh, who was employed about 1900 by M. P. Hayes to drive a tunnel along a gold vein high on the ridge from the East Fork of Green Creek. Kavanaugh never became rich, although the Chemung Mine, which he located and named after his home town in Illinois, has produced more than a million dollars' worth of ore (Maule).

Kaweah (kə wē´ ə): **River, Peaks**, post office, **Basin, Gap** [Sequoia NP]. The name is derived from the name of the Yokuts tribe Kawia or Ga'wia (Kroeber), which lived on the edge of the plains on the north side of the river. The tribal name appears as Cah-was in the Johnston Report (Aug. 31, 1851), as Cahwia in the Indian Report (p. 254), and with many spelling variants in other documents of the 1850s and 1860s. The name seems to have no relation to that of the Cahuilla Indians in Riverside Co., although it is similarly pronounced. The river itself had been discovered and named San Gabriel by a Spanish expedition in 1806 (Arch. MSB 4:40). It is mentioned by Padre Juan Cabot in 1814 and shown on the Estudillo (1819) map as Río de San Gabriel. Derby's map (1850) records "River Frances or San Gabriel." In the Indian Report, the first of the four creeks forming the delta is called Cowier on Apr. 28, 1851; and in the Johnston Report the Cahuia River is mentioned on Sept. 26, 1851. It is shown as Cahwia on Tassin's map (1851). In the Pac. R.R. *Reports* the stream is mentioned as the "Pi-pi-yu-na, or Kah-wée-ya, . . . very commonly known as the Four Creeks" (5:1.13). Most maps of the 1850s label the river system Four Creeks. The modern version, Kaweah, appears on Goddard's map (1860). Mount Kaweah was named in Sept. 1881 when J. W. A. Wright, F. H. Wales, and W. B. Wallace made the first ascent. They named the other three peaks of the group Mount Abert, Mount Henry, and Mount LeConte, but these are now known as Black, Red, and Gray Kaweah (Farquhar). The post office was named Kaweah in 1889 because it was at the headquarters of the Kaweah Co-operative Colony, founded by Burnette G. Haskell (*see* Colony). Kaweah was once commonly called Ragtown (*see* Rag).

Keane Wonder; Keane Spring Canyon [Death Valley NP]. The names recall the Keane Wonder Mine, near Chloride Cliff, discovered by Jack Keane and associates in 1903.

Kearney [Fresno Co.]. Named for M. Theodore Kearney (1838?–1906), a wealthy landholder, who came to the county about 1873 (Co. Hist.: Fresno 1919:218 ff.).

Kearny, Camp [San Diego Co.]. The reservation was named by the War Dept. at the time of World War I, in memory of Gen. Stephen W. Kearny, whose name has been associated with this area since he led the "army of the West" from New Mexico to San Diego in 1846. A town on Bear River [Yuba Co.], doubtless named in his honor, is mentioned in the *Statutes* of 1850 (p. 101).

Kearsarge [kēr′ särj]: **Peak**, station [Inyo Co.]; **Pass, Lakes, Pinnacles** [Kings Canyon NP]. After the Confederate raider *Alabama* sank the Union warship *Hatteras* off the coast of Texas on Jan. 11, 1863, the range north of Owens Lake was named Alabama Hills by Southern sympathizers. When the Union man-of-war *Kearsarge* in turn destroyed the *Alabama* off the coast of France on June 19, 1864, Thomas May and his partners called their claims the Kearsarge mining district (organized Sept. 19, 1864). Kearsarge is mentioned as a stage station on June 23, 1866 (*San Francisco Alta California*), and Kearsarge Mountains is shown on Hoffmann's map. Since the Union warship was named after Mount Kearsarge, New Hampshire, we have (as with Dunderberg) a California peak named after an eastern peak via a warship and a mine. The names were applied to the other features by the USGS when the Mount Whitney quadrangle was mapped in 1905.

Keddie: Ridge, Peak [Plumas, Lassen Cos.]. Named for Arthur W. Keddie, a native of Scotland who came to California in 1863; he was surveyor of Plumas Co., 1870–71 and 1874–77, and made the original survey for the Western Pacific route. A post office in Plumas Co. is listed in 1910.

Keeler [Inyo Co.]. The station was named in 1882 for J. M. Keeler, manager of the Inyo County Marble Quarry. Keeler was a forty-niner who went back east to fight in the Civil War, where he earned a captaincy.

Keeler's Needle [Sequoia NP]. Named in 1881 by S. P. Langley, for his assistant in the Mount Whitney expedition, James E. Keeler (1857–1900). Keeler was later a well-known astronomer and director of Lick Observatory (Farquhar).

Keen Camp [Riverside Co.]. Named in 1880 for the Keen family, the first white settlers of the district (H. D. King). The post office is listed in 1910.

Keene [Kern Co.]. The post office was first established on Feb. 13, 1879, and was doubtless named for a member of the Keene family, prominent in the district.

Keith, Mount [Kings Canyon NP]. Named by Helen M. Gompertz in July 1896 in honor of the California landscape painter William Keith (1838–1911), a native of Scotland and frequent companion of John Muir in his High Sierra travels (Farquhar).

Kekawaka (kik′ ə wä kə) **Creek** [Trinity Co.]. From Wintu *kiki waqat* 'frozen creek', containing *kik-* 'to freeze' and *waqat* 'creek' (Shepherd). The creek is shown on the Mining Bureau map of 1892; the station was named when the last link of the Northwestern Pacific was built in 1910.

Keller: Peak, Creek [San Bernardino Co.]. Apparently named for Francis D. Keller, a native of Illinois and pioneer farmer in the district after 1854.

Kellogg, Mount. *See* Rodgers Peak.

Kelly Mountain [Plumas Co.]. Named for the head of the Kelly family, the first settlers in the vicinity.

Kelsey [El Dorado Co.]. Named for Benjamin and Samuel Kelsey (brothers of Andrew of Kelseyville), who opened the diggings here in 1848. The place is shown on Trask's mining district map of 1853.

Kelseyville, Kelsey Creek [Lake Co.]. The town was established in the 1860s and called Kelsey Town in memory of Andrew Kelsey, the first white settler in the county and a troublesome character, who was killed by Indians in 1849 in revenge for his mistreatment of them. The post office name was Uncle Sam at first, after the mountain of the same name. About 1885 the names of the town and the post office were changed to Kelseyville. *See* Konocti.

Kelso [San Bernardino Co.]. Named for a railroad official in 1906 when a railroad siding was established there by the San Pedro, Los Angeles, and Salt Lake Railroad (now the Union Pacific).

Kelso Creek [Kern Co.]. Named for John W. Kelso, who brought goods by ox team from Los Angeles during the Kern River gold rush in the 1860s.

Keluche (kə lo͞o′ chē) **Creek** [Shasta Co.]. After Charlie Klutchie or Keluche, a Wintu shaman; his Wintu name was *tl'učuheres* (Harrington). It is derived from Wintu *tl'uči* 'sticking, stabbing', from *tl'u-* 'to stick, stab' (Shepherd).

Kendall's City. *See* Boonville.

Kendrick Peak and **Creek** [Yosemite NP]. The peak was named in 1912 by Col. W. W. Forsyth in honor of Henry L. Kendrick (1811–91), veteran of the Mexican War and professor of chemistry at the U.S. Military Academy, 1857–80.

Kenebec Creek [Butte Co.]. Probably for Kennebec, the name of a river in Maine, said to mean 'long reach' in an Algonquian language.

Kenick [Humboldt Co.]. *See* Kanick.

Kennedy: Lake, Peak [Tuolumne Co.]. The name was applied because the lake was included in the strip of land patented to Andrew L. Kennedy on Aug. 27, 1886, for grazing purposes (USFS).

Kennedy Peak [Inyo Co.]. The mountain near the Nine Mile Canyon Road was named for a local rancher (Wheelock).

Kennedy Table [Madera Co.]. The plateau northwest of Kerckhoff Powerhouse was probably named for Alexander Kennedy, a native of Maryland and a stock raiser at Millerton in the 1860s.

Kennett [Shasta Co.]. The town, the site of which is now covered by the waters behind Shasta Dam, was named in 1884 by the Central Pacific Railroad for Squire Kennett, a stockholder. During World War I, it was a thriving copper-smelting and shipping center. The name was later preserved in the Kennett Division of the U.S. Bureau of Reclamation.

Kenshaw Spring [Shasta Co.]. From Wintu *ken-č'aaw*, from *ken-* 'down, inside' + *č'aaw* 'sing' (Shepherd).

Kensington [Contra Costa Co.]. Named after Kensington, England, by Robert Bousefield, when the tract was opened in 1911 (L. Ver Mehr).

Kent Canyon. *See* Quinado Canyon.

Kentfield [Marin Co.]. Albert Emmett Kent established his home here in 1872 and called it Tamalpais. The adjacent station of the North Pacific Coast Railroad was also called Tamalpais until the 1890s, when it was changed to Kent to avoid confusion with the name of the newly built Tamalpais Scenic Railway. When the post office was established about 1905, the name was changed to Kentfield (Elizabeth Kent).

Kent Grove [Humboldt Co.]. Named in memory of William Kent (1864–1928), who had donated the grove for public use; he was a U.S. congressman, 1911–17. Kent had previously given Muir Woods to the federal government but had declined Pres. Theodore Roosevelt's suggestion to name it Kent Woods. He was the son of Albert Emmett Kent, of Kentfield.

Kenwood [Sonoma Co.]. Named before 1887 by a Mrs. Yost, after the town in Illinois. *For* Kenwood [San Bernardino Co.], *see* Devore.

Kerckhoff: Dam, Lake, Dome [Fresno, Madera Cos.]. The lake and dam derive their names from the Kerckhoff plant, which was put into operation by the San Joaquin Power Company on Aug. 5, 1920. William G. Kerchoff (1856–1929)—note the spelling!—of Los Angeles, was a well-known promoter and philanthropist, and one of the organizers of the company. Kerkhoff Canyon [Los Angeles Co.] was probably named for the same man.

Kerman [Fresno Co.]. The Southern Pacific station was named Collis (for Collis P. Huntington) when the connecting line from Fresno was built in 1895. In 1906, when W. G. Kerckhoff and Jacob Mansar of Los Angeles established a colony of Germans and Scandinavians from the Middle West, the settlement and the station were called Kerman, a name coined from the first three letters of each promoter's last name.

Kern: River, County, City; Kernville [Kern Co.]. **Kern: Canyon, Flat, Hot Springs, Lake, Peak, Point, Ridge; Little Kern River, Little Kern Lake, Kern-Kaweah River** [Tulare Co.]. The name of the river was given by Frémont in 1845 for his topographer and artist, Edward M. Kern, of Philadelphia, who narrowly escaped drowning while attempting to cross the stream (*see* Rio: Rio Bravo). Francisco Garcés had named the river Río de San Felipe on May 1, 1776 (Coues, *Trail* 1:280–83). In Aug. 1806 Padre Zalvidea renamed it La Porciúncula for the day of the Porciuncula Indulgence, Aug. 2 (*San Francisco Bulletin*, June 6, 1865). The name continued to be used locally and elicited this statement from Williamson in 1853: "This stream was, and is now, known to the native Californians as the *Po-sun-co-la*, a name doubtless derived from the Indians" (Pac. R.R. *Reports* 5:1.17). As late as 1860 "Kern or Porsiuncula R." is found on Goddard's map. A Kerns Pass is given on Gibbes's map, apparently for the now-nameless pass at Kern Flat. The same map shows a Kern Lake (now disappeared) in Kern Co. The lake in Tulare Co. was formed by a landslide in 1867–68 and was named after the river. Kernville was known as Whiskey Flat in the early 1860s. The county was created from parts of Tulare and Los Angeles Cos., and was named by act of the legislature on Apr. 2, 1866. The name Kern-Kaweah River was bestowed by a Stanford University party in July 1897 upon the west branch of the river, which had been known since 1881 as Cone Creek.

Kerrick Canyon [Yosemite NP]. Named for

James D. Kerrick, who took sheep into the mountains around 1880 (Farquhar).

Kessler: Spring, Peak, Range [San Bernardino Co.]. The names perpetuate the memory of a settler named Kessler, who was supposedly killed by Indians about 1890. One of the suspected Indians was tried and was hanged from an oak tree at Kessler Spring (Gill).

Keswick (kes′ wik) [Shasta Co.]. Named by the Mountain Copper Company for its president, Lord Keswick, of London, when operations were begun in 1896 (M. E. Dittmar).

Kettenpom (ket′ ən pom) **Valley** [Trinity Co.] is from Wintu *xetin-pom* 'camass place', containing *xetin* 'camass', an edible bulb of genus *Camassia* (Shepherd). The Indian name was apparently preserved by the Whitney Survey: von Leicht–Craven record Ketten Pom and Ketten Chow Valleys (*see* Hetten).

Kettinelbe. *See* Phillipsville.

Kettle. A favorite descriptive term for kettle-like depressions as well as for kettle- or dome-shaped mountains. **The Kettle** [Kings Canyon NP] was the name applied by the Whitney Survey to the natural amphitheater at the upper course of Sugarloaf Creek: "From this camp, and the next . . . two miles farther up the divide, an examination was made of an interesting and characteristic feature in the topography of this granitic region, and to which the name of 'The Kettle' was given" (Whitney, *Geology*, p. 374). The mountain southwest of this amphitheater is Kettle Peak. There are also a Kettle Dome and a Kettle Ridge in the park, north of the Middle Fork of the Kings River. Another Kettle Peak is found in Mono Co., a Kettle Mountain in Shasta Co., and a Kettle Rock in Plumas Co.

Kettleman: Plain, Hills, City [Kings Co.]. Named for David Kettleman, who came to California in 1849, went back to the Missouri River, and returned with a herd of cattle, which he pastured in the hills west of Tulare Lake (J. W. Beebe). Kettleman Plains and Kettleman School House are shown in the county atlas of 1892. The BGN approved the name Kettleman, "not Kittleman," for the plain and hills on May 6, 1908. Kettleman City was laid out in 1919 by A. Mansford Brown as the first "oil town" in the county.

Keyes (kēz) [Stanislaus Co.]. The name was given to the Southern Pacific station about 1897, probably in memory of Thomas J. Keyes, who died in 1895. Keyes had represented the county in the state senate from 1871 to 1874.

Keyes Peak [Yosemite NP]. Col. W. W. Forsyth named the peak in 1912 for his son-in-law, Lt. Edward A. Keyes (Farquhar).

Keyesville [Kern Co.]. The ghost town was named for Richard Keyes, whose discovery of gold in Key's or (Keyes) Gulch about 1853 caused the Kern River gold rush. *See* Kaiser Creek.

Keys, Point. *See* Reyes, Point.

Keys Creek [Marin Co.]. Named for John Keyes, a native of Ireland, who came to California in 1849, used this once-important waterway in 1850, and took up land along the creek (Co. Hist.: Marin 1880:402 ff.).

Keyser Gulch. *See* Kaiser [Fresno Co.].

Keystone: Mountain, Ravine; Ravine Creek [Sierra Co.]. These names go back to the rich Keystone Mine, which was developed before 1857 and had a twelve-stamp mill in 1866.

Kezar (kē′ zär) **Stadium** [San Francisco Co.]. Mary A. Kezar gave $100,000 to the city in 1923 to erect the stadium (Loewenstein).

Khartoum. *See* Hartoum.

Kiavah (kī′ ə vä) **Mountain** [Kern Co.]. The name of the mountain west of Walker Pass was apparently not placed on the map until the USGS mapped the Kernville quadrangle in 1906. According to Crites (pp. 225, 267), the peak was known as Scodie Mountain but was renamed for the chief of a tribe that came from Panamint Canyon and took up residence in Sage Canyon. It could also contain Panamint *kiya-* 'to play' plus *pa* 'water, spring' (Fowler).

Kibbie: Creek, Lake, Ridge [Yosemite NP]. Named for Horace M. Kibbe, who owned land in the vicinity (Browning 1988).

Kibesillah (kib ə sil′ ə) [Mendocino Co.]. The name of the region and of the bare rock north of Fort Bragg is from the Northern Pomo for 'flat rock': *khabé* 'rock' plus *silá* 'flat' (Oswalt). Kibesillah Mountain is mentioned by John Rockwell of the Coast Survey in 1878, and this name appears on the maps of the survey.

Kilarc Reservoir [Shasta Co.]. The term "kilarc" designated a high-voltage switch oil used in the power plant built here in 1903–4 (Stang).

Kilbeck. *See* Chubbuck.

Kimball Island [Sacramento Co.]. The island opposite Antioch was named for George W. Kimball, who came from Maine to California in 1850 as captain of a sailing vessel. Settling at the site of Antioch, he became the first postmaster, first notary public, and first justice of the peace in New York township. When his land was declared part of Los Medanos grant, he bought the island that now bears his name.

Kimball Plains [Shasta Co.]. The plains were named for "Kimball Bill," who was in charge of the sheep range of Cone and Kimball of Red Bluff (Steger).

Kimshew (kim′ shōō): **Creek, Point** [Butte Co.]. The name is from Konkow (Northwestern Maidu) *kiwim sewi* 'the stream on the other side', referring to the North Fork of the Feather River; it is from *kiw* 'behind, on the other side' and *sew* 'creek, river' (Shipley). Kimshew is mentioned by Browne (p. 160) as a place where a tunnel had been run to strike the old channel under Table Mountain (*see* Nimshew).

King, Mount. *See* Clarence King, Mount; Starr King, Mount.

King City [Monterey Co.]. Named for Charles Henry King, who laid out the town on his Rancho San Lorenzo when the Southern Pacific was extended to that place in 1886 (Clark 1991).

Kingdon [San Joaquin Co.]. The place was called West Lodi when the Western Pacific reached it in 1909. In 1915 the railroad renamed the place for Kingdon Gould, a grandson of Jay Gould (Amy Boynton).

Kings: River, River Canyon, County; Kings Canyon NP; Kings-Kern Divide. The river was named Río de los Santos Reyes 'river of the Holy Kings' by a party of Spanish explorers in 1805, according to Padre Muñoz's diary of the Moraga expedition of 1806 (Arch. MSB 4: 1–47). The first party probably reached the river on Jan. 6, the festival of the three magi; hence the name. Padre Cabot used the name Río de Reyes in 1814 (Arch. MSB 6:67–72); this form appears on Estudillo's map of 1819 and on later maps. Jedediah Smith in 1827, ignorant of the Spanish name, called it the Wimmel-che, for the Indians living on the river (Merriam, *SCB* 11:376). Frémont reached the river on Apr. 8, 1844, at a time of the year when most of the flood flow passes southward to Tulare Lake; hence he called it "River of the Lake" (*Expl. exp.*, p. 363). After he had learned of the Spanish name, he spoke of "Rio Reyes of Tulare Lake" (*Geog. memoir*, p. 30). Preuss labeled the stream on both of his maps "Lake Fork," a name also used by Frémont (ibid., p. 18). Derby's map (1849) and Williamson's report (Pac. R.R. *Reports* 5:1.13) have the translation King's [*sic*] River; but the *Statutes* of 1852 (p. 240)—as well as Gibbes's, Blake's, and Eddy's maps—have the modern version. Kings River Canyon was explored and named by the Whitney Survey in 1864. Kings Co. was created from a part of Tulare Co. and named on Mar. 22, 1893; the national park, in 1940. **Kingston** [Kings Co.] is the name of the old ferry station opposite Laton, first known in 1856 as Whitmore's Ferry, then as Kings River Station, finally as Kingston. **Kingsburg** [Fresno Co.]: The name was adopted in 1875 for the settlement that was originally known as Kings River Switch, then as Drapersville, and later as Wheatville. *For* Kings River (town) [Fresno Co.], *see* Centerville; *for* Kings River Canal, *see* Main Canal.

Kings Beach [Placer Co.]. Named by the residents for Joe King, in recognition of his gifts to the community. The post office was established on Mar. 25, 1937.

Kingsburg. *See* Kings: River.

Kings Meadows, Creek, and **Falls** [Lassen NP]. The meadows were named for James W. King, who ran horses and mules in the mountains near Lassen Peak and owned a race track at Pine Grove in the 1860s (Steger).

Kings Mountain [San Mateo Co.]. The highest elevation of the Sierra Moreno is commonly called Kings Mountain, from the Kings Mountain House, which was operated for many years by Mrs. Honora King (Wyatt). Honora was probably the widow of Frank King, who had come from Indiana to California in 1852, and in the 1860s acquired the Mountain Brow House and 80 acres of land (Co. Hist.: San Mateo 1878:39).

Kings Peak. *See* Disaster: Peak.

King Spur. *See* Clarence King.

Kings Ridge [Sonoma Co.]. Doubtless named for William King, a native of Canada, who settled at West Austin Creek after 1876 and was for many years a supervisor and assessor of the county (Co. Hist.: Sonoma 1911: 941 ff.).

Kingston: Springs, Peak, Range, Wash, settlement [San Bernardino Co.]. Kingston Spring is mentioned by the Nevada Boundary Survey on Feb. 20, 1861 (*Sacramento Union*, June 29, 1861). It is said that the name was applied by a member of Frémont's second expedition, after Kingston, New York (Fisk). Frémont's party traversed the region in spring 1844, but Preuss fails to record the name on his maps. However, on June 2, 1854, Peg-leg Smith told S. N. Carvalho (*Travels and adventures* [New York, 1859], p. 234) that the name Kingstone Springs had long been in use. The peak is shown on the von Schmidt boundary map of 1872, and the name was affixed to the mountains, including the Nopah Range, by the Whitney and Wheeler surveys. *See* Nopah.

Kinney Lakes [Alpine Co.]. The lakes near Ebbetts Pass, known in the 1860s as Silver Lakes, were named for David Kinney, a native of Iowa, who was a farmer at Silver Mountain in 1873.

Kinto Creek. *See* Quinto Creek.

Kinyon [Siskiyou Co.]. The old post office White Horse, across the line in Modoc Co., was given the name of the boss of the logging camp on Aug. 1, 1952.

Kirker Creek, Pass [Contra Costa Co.]. The name of the pass originated about 1852 when James "Don Santiago" Kirker—a native of Belfast, Ireland, and a well-known Indian fighter in New Mexico and Chihuahua—lived in a cabin at Oak Springs, near the pass. The stream through the old coal-mining district was known as Quercus Creek; but when the Mount Diablo quadrangle was surveyed in 1898, the name Kirker Creek was applied.

Kirkville, Kirk Lake [Sutter Co.]. Named for T. D. Kirk, who laid out a town in 1874 at the place known as Colegrove Point, which he called Kirksville (Co. Hist.: Sutter 1879:98). *See* Colegrove.

Kirkwood [Tehama Co.]. When the section of the Central Pacific from Willows to Tehama was built in 1882, the station was named for Samuel J. Kirkwood, secretary of the interior in 1881 (Ruth L. Zimmerman).

Kirkwood: post office, **Lake** [Alpine, El Dorado Cos.]. Kirkwood Lake was named for Zachariah S. Kirkwood, owner of the adjoining land in the early 1860s. The post office was established as Lake Kirkwood in 1941; it is now called simply Kirkwood.

Kirman Lake [Mono Co.]. The lake was named for Richard Kirman, Sr., an early cattleman (Maule). The name is misspelled Carmen on some maps.

Kit Carson: Pass [Alpine Co.]; post office [Amador Co.]. The pass was named by Frémont for the frontiersman Christopher "Kit" Carson (1809–68), his guide in crossing the Sierra Nevada in Feb. 1844 (*see* Carson [Alpine Co.]). The post office was established on June 16, 1951.

Kittinelbe (kit ə nel′ bē) [Humboldt Co.]. The place derives its name from an Indian village. The site is in Athabaskan territory, but the name apparently derives from the neighboring Wintu language, in which *xetin-elba* means 'camas eating' (Shepherd). It is now known by the name of the post office, Phillipsville (*q.v.*).

Kittleman. *See* Kettleman.

Klamath (klam′ əth): **River, National Forest** [Siskiyou, Humboldt, Del Norte Cos.]. The name is derived from *tlamatl*, the Chinook name for the Klamath tribe (living around Klamath Lake in Oregon). The first person to use the derivative in a geographical sense was apparently Peter Ogden, who refers to "Claminitt Country" in a letter of July 1, 1826 (McArthur), and to the "waters of the Clammitte" in his diary of Nov. 27, 1826 (*OHQ* 10:210 [1910]). An English-speaking person easily substitutes initial *kl* for *tl* because the latter does not occur in English at the beginning of a word. On the maps the name of the river appears variously with *tl, kl,* or *cl* as initial letters, and with many spelling variants. Duflot de Mofras's map of 1844 shows the position of the river fairly accurately. The lower Klamath was apparently the stream on which Jedediah Smith's name had been bestowed in 1828 (Wilkes's map, 1841); *see* Smith River. The present spelling of the name became current when the short-lived Klamath Co. was established on Apr. 25, 1851, with its seat at Orleans. Klamath National Forest was established and named by presidential proclamation in 1905. The bay 4 miles north of the mouth of the Klamath River is named False Klamath Cove, probably because, when seen from the ocean, its outline resembles that of the bay at the river mouth; False Klamath Rock was obviously named (or misnamed) after the cove, for there is no "true" Klamath Rock. The name has been repeatedly applied to settlements and post offices. Two communities in Del Norte Co. are called Klamath and Klamath Glen. The settlement called Klamath River [Siskiyou Co.] was first called Honolulu, then (until 1834) Gottville, for William N. Gott. **Klamathon** [Siskiyou Co.]: In 1890

James McLaughlin built a mill here and named it Pokagama, an Indian name that he brought from Wisconsin. When John R. Cook ran the mill in 1892, he renamed it Klamathton, the name of the river plus *-ton* 'town'. The name Pokagama was then applied to the logging camp near Shovel Creek Springs (Schrader). Klamathton appears as Klamathon on modern maps.

Klinefelter. *See* Amboy.

Klink. *See* Ivanhoe.

Klondike [San Bernardino Co.]. The choice of this name by the Santa Fe in 1897, for one of the hottest spots in the Mojave Desert, was doubtless influenced by the discovery of rich gold deposits on the river in Canada. Klondike Lake [Inyo Co.] and Klondike Canyon [Monterey Co.] probably owe their origin likewise to the Klondike gold rush in 1897.

Kneeland [Humboldt Co.]. The post office was established before 1880 and named after Kneeland's Prairie, a place that had been named for a settler of the early 1850s, John A. Kneeland (Co. Hist.: Humboldt 1881:148).

Knightsen [Contra Costa Co.]. Founded in 1898, when the San Francisco–San Joaquin Valley Railroad (now the Santa Fe) was built, and named for G. M. Knight, a prosperous farmer who had donated the right-of-way to the railroad. The name was suggested by Mr. Knight and includes the suffix of his wife's maiden name, Christensen.

Knights Landing [Yolo Co.]. The name commemorates William Knight, of Indiana, who settled here in 1843. The town was laid out in 1850 and is shown on Gibbes's map. The post office name was Grafton for many years, for the man who established the mail route from Benicia to Yolo in 1855 (Co. Hist.: Yolo 1940:235). **Knights Ferry** [Stanislaus Co.]: After the discovery of gold, William Knight moved to this place, where he operated a ferry. When he died in 1849, the business was taken over by John C. and Lewis Dent (General Grant's brothers-in-law) and James Vantine. The *Statutes* of 1850, the Indian Report (1851), and Gibbes's map (1852) all have the present name.

Knights Valley [Sonoma Co.]. Named for Thomas Knight, the first permanent white settler in the valley in the 1850s. Knight's Creek and Knight's Valley are mentioned by Brewer in 1861.

Knob. The American generic term for a round hill or small mountain is found in California in the names of about fifty orographic features. Most of these are in the southern half of the state; in the north, the term "butte" is more often used for this type of elevation. Since the word ordinarily refers to an isolated hill, or at least one jutting out above others, it is often combined with the word Pilot to designate a landmark [Fresno, Imperial, Kern, San Bernardino Cos.]. Kern and Mariposa Cos. each have a tautological Knob Hill. The site of Knob [Shasta Co.] is named after a nearby knob. The Southern Pacific station called Knob [Imperial Co.], formerly Pilot Knob, is named after the old landmark near the corner where California, Arizona, and Mexico join.

Knocti, Mount [Lake Co.]. *See* Konocti.

Knopf (nôf) **Canyon** [San Mateo Co.]. The Knopf ranch was here from the 1870s (Alan K. Brown).

Knopki (kə nop′ kē) **Creek** [Del Norte Co.]. Perhaps of Indian origin, but this has not been confirmed.

Knotts Berry Farm [Orange Co.]. The famous berry farm was developed by Walter Knott in the 1920s. In 1940 he added a "ghost town" to his farm, consisting of relics from real ghost towns in southern California and Arizona. Since then Knotts Berry Farm has grown into one of southern California's most popular theme parks (Brigandi). *See* Calico.

Knowland State Park [Alameda Co.]. The park and arboretum, opened on Sept. 9, 1951, were named for the late Joseph R. Knowland (1873–1966), for many years publisher of the *Oakland Tribune*.

Knowles [Madera Co.]. Named by the Southern Pacific in 1890 for F. E. Knowles, the founder of the town.

Knownothing Creek [Butte and Siskiyou Cos.]. The names of the two creeks are reminiscent of the American Party, a political movement also called the Knownothing Party—because its members, when questioned, replied that they "knew nothing" about it. Its short but spectacular career shook the country in the 1850s. In the 7th session of the state assembly (1856), the American Party had the majority, though only a few members were rabid "Know-Nothings"; most of them were Whigs, whose party was disintegrating. Knownothing was a favorite name for mines, and the creek in Siskiyou Co. (a tributary to the Salmon River) was named after the Knownothing Mine at Gilta.

Knox Mountain [Modoc Co.]. The mountain in the Modoc National Forest was named for Robert Knox, who built a sawmill here in 1874 (USFS).

Knoxville [Napa Co.]. Named for one of the proprietors of the Manhattan Quicksilver Mine. A post office was established on Nov. 30, 1863. Tbe place is mentioned in a letter from Hoffmann to Whitney on Nov. 29, 1870, and in the reports of the Mining Bureau until recent times. It was in Lake Co. until 1872.

Knuedler (nēd′ lər) **Lake** [San Mateo Co.]. John Knuedler operated a ranch here around 1900 (Alan K. Brown).

Knulthkarn (kə nulth′ kärn) **Creek** [Humboldt Co.]. From Yurok *knułkenok* (Waterman, map 11, item 82).

Koehn (kān) **Lake** [Kern Co.]. The now dry lake south of the road from Mojave to Randsburg was named for Charley Koehn, a German homesteader at Kane Springs. His place was a stopping point before the discovery of the gold deposits in 1893, and it had a post office from 1893 to 1898 (Wynn, p. 60). The lake is also known as Kane Lake or Desert Lake.

Kohler, Camp [Sacramento Co.]. Named in 1942 by the War Dept. for Lt. Frederick Kohler, the first Signal Corps officer killed in World War II.

Koip Peak and **Crest** [Yosemite NP]. The name for the mountain was fixed by Willard D. Johnson of the USGS about 1883. It is probably from Northern Paiute *kóipa* (Fowler) or from Mono *koippï* 'mountain sheep', lit. 'that which is killed', from *koi-* 'kill plural' (McLaughlin).

Kokoweep (kō kō wēp′) **Peak** [San Bernardino Co.]. Perhaps from Southern Paiute *kogo* 'gopher snake' plus *uippi* 'canyon' (Fowler). The name is misspelled Kokoweef on some maps.

Kolana (kō lä′ nə) **Rock** [Tuolumne Co.]. This supposedly Indian name was mentioned by John Muir in 1873 (Browning 1988). Perhaps from Southern Sierra Miwok *kuluunaw*, the name of the location now called Coarsegold (Callaghan).

Kolmer: Creek, Gulch [Sonoma Co.]. Named for Michael Kolmer, who settled here in 1848. *See* Coleman Valley.

Kongsberg; Konigsberg. *See* Silver: Silver Creek.

Konocti (kə nok′ tī), **Mount** [Lake Co.]. George Gibbs had named the isolated peak Mount M'Kee for Colonel Redick McKee, who explored the Clear Lake region in 1851 (Schoolcraft 3:109–10). In 1854 Martin Hammack and Woods Crawford named it Uncle Sam Mountain, and it appears with that name on the von Leicht–Craven map. The BGN (6th Report) decided for the local Indian name of the peak, which, according to Barrett (*Pomo*, p. 183), was Kno'ktai, derived from *kno* 'mountain' and *xatai* 'woman'. A local interpretation of Konocti is 'thrown horse'; it is said that the name arose because of the peculiar shape of an open space on the mountainside—later covered but not obscured by a walnut orchard—which looks like a fallen horse with its head held down (Ruth Lewis). An alternative spelling is Knocti.

Korbel (kôr bel′) [Humboldt Co.]. Korbel Rock is mentioned in Davidson's correspondence as early as 1872, but the place was known as North Fork when the Korbel brothers built their mill there in 1882. On June 24, 1891, the post office was established under the name Korbel (Borden). Nearby Korblex is obviously derived from Korbel.

Kosk (kosk) **Creek** [Shasta Co.]. The tributary of Pit River was named for John Kosk or Kosh, a native of Russia who was superintendent of the Silver City Mine in the 1860s and later operated a ferry across the Pit River (Steger).

Kramer: station, **Hills** [San Bernardino Co.]. The Southern Pacific gave the name in 1882 to the junction from which a branch led to the Johannesburg and Randsburg mines. The Santa Fe retained the name for the present station. One Moritz Kramer, a native of Germany, is listed in the Great Register of 1879.

Kreyenhagen (krā′ ən hā gən) **Hills** [Fresno, Kings Cos.]. The range was named for the Kreyenhagen brothers, large-scale cattle breeders, whose headquarters were on Zapato Creek at the foot of the hills. Their father, Gustave Kreyenhagen, a native of Germany, had come to Merced Co. in 1867 and settled at the place that became known as Kreyenhagen's (*see* Los Banos). Kreyenhagen Shale of the oil-producing geological formation around Coalinga was named after the original Kreyenhagen Ranch.

Kruse (kro͞oz) **Rhododendron Reserve State Park** [Sonoma Co.]. This part of the old Rancho Germán was given to the state by Edward, Emil, and Mollie Kruse, and was made a state park on Jan. 8, 1934.

Kuna (ko͞o′ nə): **Peak, Crest, Ridge, Creek** [Yosemite NP]. The name of the peak was fixed, about 1883, by Willard D. Johnson of the USGS; the word *kúna* means 'fire' in several Numic languages of eastern California, including Southern Paiute, Chemehuevi, and Kawaiisu (Munro).

Kunkle Reservoir [Butte Co.]. The reservoir was named after the mining place Kunkles, shown on the county map of 1862 and named for a settler of the region.

Kyburz (kī´ bərz) [El Dorado Co.]. The post office was established in 1911 and named for the first postmaster, Albert Kyburz—the son of Samuel Kyburz, a native of Switzerland and an important figure at Sutters Fort before and during the Gold Rush.

L

La. Spanish names preceded by the feminine article *la* 'the' that are included in cluster, group, or folk names will be found in the alphabetical order of the name proper; e.g., La Jolla will be found under Jolla. Many names of this type, such as **La Arena** ('the sand') [Kings Co.], can be found on the western side of the San Joaquin Valley; they were apparently bestowed by English-speaking surveyors, with the help of a Spanish dictionary and a good deal of imagination. In this sparsely populated area, the names have been little used, and they have no customary English pronunciations (Crosby). For the most part, they are omitted from this book.

Laborda (lə bôr′ də) **Canyon, Creek** [Riverside Co.]. From Jacques LaBorde, of France, who farmed in the area from 1883 (Gunther).

Labranza (lä brän′ zə) [Merced Co.]. The Spanish term means 'cultivation'.

Labrea Creek. *See* Brea.

Lac [Sonoma Co.]. A place called Lac by Indians (tribe unknown) is mentioned in the petition of May 21, 1844, for the land grant of this name, dated July 25, 1844 (Bowman).

Lachusa Canyon [Los Angeles Co.]. Probably for Spanish *lechuza* 'owl'.

Lacjac (lə jak′) [Fresno Co.]. Coined by Daniel J. Ellis in 1899 from the names of the firm Lachman and Jacobi, for whom he built a winery and distillery here, at that time the largest in the world.

Lack Creek [Shasta Co.]. Named for De Marcus Franklin Lack, who settled at the creek, a tributary of Bear Creek, about 1860 (Steger).

Lacosca (lä kōs′ kə) **Creek** [Santa Barbara Co.]. California Spanish for 'scrub oak' (John Johnson).

La Costa. *See* Ponto.

La Costa (kos′ tə) **Creek** [Alameda Co.]. First called San Antonio Creek in 1857. In 1873 changed to La Costa, modified from the name of the landowner Juan F. LaCoste, a native of France (Mosier and Mosier).

La Crescenta (krə sen′ tə) [Los Angeles Co.]. This name is *not* Spanish for 'the crescent', which would be *la creciente*. The place was settled in the early 1880s and named by Dr. Benjamin B. Briggs. From his home he could see three crescent-shaped formations, which suggested to him the artificial name. The name was accepted for the post office in 1888 (H. A. Scheuner).

La Cumbre, Spanish for 'the peak', is a phrase found in the names of several locations; *see* Cumbre.

Ladder Butte [Lassen Co.]. The USFS gave the name about 1929, when it built a lookout station there.

Laddville. *See* Livermore.

Ladera (lə dâr′ ə) [San Mateo Co.]. Spanish for 'slope, hillside'. There is also a **Ladera Heights** [Los Angeles Co.].

Ladoga. *See* Lodoga.

Ladrones (lä drō′ nās), **Cañada de los** [Santa Barbara Co.]. Spanish for 'valley of the thieves'.

Ladybug. A number of canyons, as well as a peak in Sierra Co., bear this name because they are hibernation refuges of the brightly colored beetle (properly a ladybird, the *Vedalia*).

Lady Franklin Rock [Yosemite NP]. Named for Lady Jane Franklin, widow of John Franklin, the British naval officer whose attempt to discover the Northwest Passage in 1847 led to one of the worst Arctic disasters. Lady Franklin visited Yosemite in 1863 and admired the view of Vernal Fall from this rock (Farquhar).

Lafayette: town, **Creek, Ridge, Reservoir** [Contra Costa Co.]. The town stands among the fifty or more other places in the United States named in honor of the great Frenchman who fought in our War of Independence, the Marquis de Lafayette (1757–1834). The first settler was Elam Brown, who bought the Acalanes rancho and settled near the site of Lafayette on Feb. 7, 1848 (Knave, Feb. 8, 1948). The settlement was named in 1853 by Benjamin Shreve, the owner of the store; the post office was established on Mar. 2, 1857, with the name spelled La Fayette originally.

La Fresa (frā′ sə) [Los Angeles Co.]. Spanish for 'the strawberry'.

Lagoon. Both English 'lagoon' and Spanish *laguna* are derived from Latin *lacuna* 'pool, pond', but in English-speaking countries the term is now almost entirely restricted to a sheet of water near or communicating with the ocean. The most prominent feature for which the generic term is used in this sense is Big Lagoon [Humboldt Co.]. But since *laguna* was used in Spanish and Mexican times for any kind of lake, inland lakes in California are sometimes called lagoons. The swampy lake north of Irvington [Alameda Co.] is called The Lagoon; it was originally called La Laguna, and its overflow passes through Arroyo de la Laguna.

La Grange [Stanislaus Co.]. French for 'the farm'; also a family name. The place was known in the early 1850s as French Bar, but it appears on Baker's map of 1856 as Owens Ferry. In the same year, the settlement was made the county seat, and the present name was applied. La Grange was the name of Lafayette's country home and was already very popular as a place name in the United States.

Laguna (lə gōō′ nə). The Spanish geographical term generally designates a small lake, but in Spanish California it referred to a lake of any size; Clear Lake, a large body of water, was formerly known as the Laguna (Bryant, p. 278), and the shallow, marshy Tulare Lake was once called Laguna de Tache. The word *laguna* appears in the names of more than thirty land grants and claims. It has survived as a semi-generic term and has also become a specific term in the names of mountains, canyons, creeks, and subdivisions, also occurring tautologically as Laguna Lake [Imperial, San Benito, San Luis Obispo Cos.]. **Punta de la Laguna** [Santa Barbara, San Luis Obispo Cos.] was a land grant, dated Dec. 26, 1844, referring to the lake that the Portolá expedition called Laguna Grande de San Daniel on Sept. 1, 1769 (Crespí, p. 181). Costansó records on the same day that the valley was named el Valle de la Laguna Larga (p. 54). **Laguna Seca** [Santa Clara Co.]: La Laguna Zeca 'the dry lake' is mentioned on Oct. 31, 1797 (PSP 15:158), at a time of the year when many California lakes are dry. In 1823 the name was used for the Laguna Seca (or Refugio de la Laguna Seca) provisional grant, which was definitely granted on July 22, 1834. **Laguna Seca** [Monterey Co.]: *La laguna seca* is mentioned on July 24, 1830 (Legis. Recs. 1:147), and a name Laguna is shown on a *diseño* of the Laguna Seca (or Cañadita) grant, dated May 18, 1833, and Jan. 9, 1834. **The Lagoon; Arroyo de la Laguna** [Alameda Co.]: The lake north of Irvington is shown as *Laguna Permanente que es nacimiento de este Arroyo* 'the permanent lake which is the source of this creek' on a *diseño* of the San Ramon grant (1834). **Laguna Peak** [Ventura Co.] was obviously named after the nearby Laguna (now called Mugu Laguna), shown on a *diseño* of the Guadalasca grant (1836). **Laguna Creek** [Santa Cruz Co.] is shown as Arroyo de la Laguna on a *diseño* (1836) of the land grant bearing the same name. **Laguna Salada** [San Mateo Co.], meaning 'salt lake', here actually a lagoon, is shown on a *diseño* (1838) of the San Pedro grant. **Arroyo de la Laguna** [Alameda Co.], a tributary of Alameda Creek, was named after the Laguna shown on a *diseño* of the Valle de San Jose grant (1839). **Laguna Honda** [San Francisco Co.]: At the site of the reservoir at the edge of Sutro Forest there was formerly a 'deep lake', shown on several *diseños*. **Laguna Canyon** and **Beach** [Orange Co.]: The canyon is shown as Cañada de Las Lagunas on a *diseño* (1841) of the Niguel grant; the town of Laguna Beach was named after the canyon. In recent years, other names have capitalized on the fame of Laguna Beach, such as Laguna Niguel (*see* Niguel) and Laguna Hills. **Laguna Mountains** [San Diego] have been known since the 1870s by this name because two lakes, called Laguna Lakes, are on the summit of the mountain range. *For* Lake Laguna [Riverside Co.], *see* Elsinore; *for* Point Laguna [Ventura Co.], *see* Mugu, Point; *for* Laguna Grande [San Mateo Co.], *see* Raimundo; *for* Laguna Larga [Santa Barbara Co.], *see* Guadalupe: Lake; *for* Laguna Merced Military Reservation [San Francisco Co.], *see* Funston, Fort.

Laguna Canyon [Monterey Co.] is not named for a lake, but for Antonio C. Laguna, who acquired land here in 1892 (Clark 1991).

Lagunita. This diminutive, meaning 'little lake', was used in Spanish times much less often than the term *laguna*. **Lagunitas** (lä gōō nē′ təs) **Creek, Lake,** and town [Marin Co.]: The origin of the name is found in the little lakes north of Mount Tamalpais, now converted into reservoirs. The one now called Lagunitas Lake ('little lakes lake') is shown as Laguna on a *diseño* of the Cañada de Herrera grant (1839). The canyon through which Little Carson Creek flows is labeled Puerto Zuelo Lagunitas (*puertezuelo* 'little mountain pass') on Hoffmann's map of the Bay region (1873). On the county map of the same year, the stream on the north slope of Mount Tamalpais, emptying into Lagunitas Lake, is called Lagunitas Creek. On modern maps the name Lagunitas Creek is applied to the entire stream as far as Tomales Bay. This watercourse appears on a *diseño* of Rancho Nicasio (1844) as Arroyo de San Gerónimo and is designated on American maps as San Geronimo Creek. It is also known as Paper Mill Creek because Samuel P. Taylor built a paper mill at the site of Taylorville in 1856. When the USGS mapped the Tamalpais quadrangle in 1895, the name San Geronimo was restricted to the tributary of Lagunitas Creek. The post office is listed in 1908. **Lagunita Lake** [Monterey Co.]: The 'little lake lake' on the Salinas–San Juan Bautista highway is shown as Laguna on a *diseño* of the Vergeles grant (1834). *For* Lagunita [Los Angeles Co.], *see* Bixby Slough.

La Habra (hä′ brə): **Valley**, town [Orange Co.]. *Habra* (which in standard Spanish is spelled without the initial *h*) means 'a gorge, a pass through the mountains'; although it is feminine, nouns beginning with accented *a* take the article *el*, so 'the pass' would be *el (h)abra*. The term refers here to the pass through the Puente Hills, traversed by the Portolá expedition on July 30, 1769. The term was used in the name of a land grant, La Cañada del Habra (Legis. Recs. 3:62), dated Oct. 22, 1839. In American land grant papers the name was given as La Habra, and this form was applied to the valley and in 1912 to the post office.

La Jolla. *See* Jolla.

Lake. The shores of a lake have been most conducive to settlement ever since primitive times, and the word "lake" is probably found in the names of more American communities than any other geographical generic term with the exception of "mount" and "mountain." In California it is used in combination with "shore," "side," "view," "wood," etc. for about fifteen communities, and at least seven additional lake names were copied in naming a town or a post office. Some of the name-giving lakes have long since disappeared. Several mountains and creeks are likewise named after nearby lakes. **Lake City** [Nevada Co.] was founded in 1857 and named after the numerous small lakes in the vicinity (Co. Hist.: Nevada 1880:59, 172). **Lake County** was created on May 20, 1861, from part of Napa Co., and so named because Clear Lake is the principal feature of the area. **Lakeport** [Lake Co.] was settled in 1859 and first called Forbestown, for William Forbes, the owner of the land. When the community became the seat of the newly created Lake Co., the name was changed to Lakeport because it is on the edge of Clear Lake. **Lakeville** [Sonoma Co.] is shown on Hoffmann's map of the Bay region (1873); its situation on the shore of the lake-like expansion of Petaluma Creek doubtless suggested the name. **Lake City** [Modoc Co.] was established as a post office in the 1870s and so named because it is situated between Upper and Middle Alkali Lakes (locally known as Surprise Valley Lakes). **Lakeside** [San Diego Co.] was laid out on the shore of Lake Lindo in 1886 by El Cajon Valley Company and named after the lake (Hunzicker). **Lakeview**: post office, **Mountains** [Riverside Co.], was established as a post office about 1895 and given this name because of the view of Lake Moreno. The lake, long since drained, was formerly called Brown's Lake (*moreno* 'brown'); *see* Moreno. **Lake Hughes** [Los Angeles Co.]: The post office was named in 1925 after the lake; *see* Hughes Lake. **Lakewood Village** [Los Angeles Co.]: The subdivision was laid out in 1934 on the land of Rancho Los Cerritos by Clark J. Bonner of the Montana Land Company and Charles B. Hopper, promoter of the development, and was named Lakewood Village because it is near Bouton Lake (C. B. Hopper). **Lake Mountain** [Trinity Co.]: The post office, established on Apr. 18, 1878, and later discontinued, was reestablished on Mar. 25, 1936 (Leona Miller). **Lakehead** [Shasta Co.] was the name applied to the station at the northern "head" of Shasta Lake by the Southern Pacific after completion of Shasta Dam. The post office was established on Nov. 1, 1950, and named after the station (Edith Ramey). The term was established by the BGN in list 6902, to replace the former name, Loftus. Note also Lake of the Pines [Nevada Co.], Lakeland Village [Riverside Co.], Lake Los Angeles [Los Angeles Co.], and Lakeshore [Fresno Co.]. *For* Lake Elsinore, *see* Elsinore; *for* Lake Fork, *see* Kings River.

Lake Arrowhead [San Bernardino Co.]. When the post office was established on Apr. 29, 1922, at Arrowhead Lake, the generic name was placed first to avoid confusion with other Arrowhead names in the vicinity. *See* Arrowhead.

Lamanda Park [Los Angeles Co.]. In 1886 Leonard J. Rose laid out the subdivision on his famous Sunnyslope Farm and named it by prefixing the initial letter of his given name to his wife's name, Amanda. In 1905, the zealous Z. S. Eldredge, scenting a Spanish name, persuaded the Post Office Dept. to "restore it" to La Manda. Fortunately, the Santa Fe station kept the name Lamanda, and in 1920 the post office restored the old form.

Lamarck, Mount; Creek, Lakes [Kings Canyon NP]. For French biologist Jean Baptiste Pierre Antoine de Monet de Lamarck, 1744–1829 (Browning 1986). *See* Evolution Peaks.

La Mesa. *See* Mesa.

La Mirada (mi rä′ də) [Los Angeles Co.]. The name was applied to the station of the Santa Fe when the line to San Diego was built in 1888. It is a Spanish word meaning 'glance, gaze'. In 1953 an olive grove in the area was subdivided and the settlement was named after the station (Co. Hist.: Los Angeles 1965: 2.132–33).

Lamoine (lə moin′) [Shasta Co.]. Named in 1898 by the Coggins brothers, owners of the Lamoine Lumber and Trading Company, after their native home in Maine (Steger). The name is spelled La Moine by the Post Office Dept.

Lamont [Kern Co.]. Named by the MacFadden family, landholders, after their former home in Scotland (Santa Fe).

Lanare (lə nâr′) [Fresno Co.]. Coined from the name of L. A. Nares, chief promoter of a colonization project here, and applied in 1911 when the Laton and Western Railway was built.

Lancaster [Los Angeles Co.]. Named in 1877 by settlers, after their former home in Pennsylvania.

Lancaster Mountain [San Diego Co.]. Named for A. W. Lancaster, who purchased a farm near the mountain in 1872 (Hunzicker).

Lancha Plana (län′ chə plä′ nə) [Calaveras Co.] was a mining town, settled probably in 1848. In 1850 two prospectors, Kaiser and Winter, built a raft of casks lashed together for a ferry across the Tuolumne River; hence the name 'flat boat'.

Lane Memorial Grove. *See* Franklin K. Lane Grove.

Lanfair [San Bernardino Co.]. The station was named for E. L. Lanfair, an early settler, when the Southern Pacific branch from Needles to Mohave was built in 1884. A post office was established in 1912.

Lang [Los Angeles Co.]. Named for John Lang, a native of New York who was a farmer at Soledad in the 1870s. In the State Gazetteer of 1888, he is listed as postmaster, hotelkeeper, and real estate agent at Lang.

Langille (lan′ jil) **Peak** [Kings Canyon NP]. Named by the USGS, at the suggestion of Charles H. Shinn, then head ranger, for Harold D. Langille, a forest inspector for the U.S. General Land Office, after his tour of inspection in 1904 (Farquhar).

Langley, Mount [Sequoia NP]. Named in 1905 in honor of Samuel P. Langley (1834–1906), astronomer and physicist, who organized an expedition to Mount Whitney in 1881 for research in solar heat and who was well known for his experiments in the problem of mechanical flight. In 1864 Clarence King had named the peak Sheep Mountain. In 1871 Albert Bierstadt made a painting of the peak and named it for William W. Corcoran, donor of the Corcoran Art Gallery, Washington, D.C., where the painting later hung. The same year Clarence King mistook his Sheep Mountain for Mount Whitney, which he and his companions had named seven years before. The mistake was corrected by W. A. Goodyear on July 27, 1873, and Whitney's name was restored to the peak for which it was intended. The BGN approved the name Mount Corcoran until 1943, when the Sierra Club argued that "Langley" had become established through usage; the BGN then reversed its decision and made the present name official.

Langley Hill [San Mateo Co.]. Named for Frank and Laban Langley, who owned five sections of land here in 1877.

Langville. *See* Capay.

La Olla (ō′ yə) [Santa Barbara Co.]. The name of this valley is Spanish for 'the pot'.

La Paleta (pə lā′ tə): **Valley, Creek** [San Diego Co.]. Spanish *paleta* has several meanings: 'shovel; shoulder blade; artist's palette'; in Mexico it means 'frozen dessert on a stick'. It may have been applied to the valley to indicate its shape or coloring—or, like many other place names in southern California, for the pleasant sound of the name.

La Palma (päl′ mə) [Orange Co.]. Spanish for 'the palm tree'.

La Panza (pän′ zə): settlement, **Range** [San Luis Obispo Co.]. According to Sanchez, the name arose because hunters used the paunch (*la panza*) of a beef as bait to catch bear. It appears to be the same place recorded as *paraje la Panza* by Sebastián Rodríguez on Apr. 27, 1828. It is shown as La Pansa on the von Leicht–Craven map, and as La Panza on the Land Office map of 1879, at the site of present La Panza Ranch. The discovery of gold in the canyon in 1878 precipitated a small-scale gold rush and was followed by the establishment of a post office. For some years the residents suffered the name to be spelled in one word, until Z. S. Eldredge in 1905 prevailed on the Post Office Dept. to restore its rightful form. The present spelling was established by BGN in list 6802.

La Patera (pə târ′ ə) [Santa Barbara Co.]. *Patera* means 'place where ducks congregate' and is shown on a *diseño* of the Dos Pueblos grant (1842) as the name for a stream that flows into the *estero*. The name was applied to the Southern Pacific station when the line westward from Santa Barbara was built in the 1880s.

La Porte (lə pôrt′) [Plumas Co.]. Named in 1857 after La Porte, Indiana, the birthplace of Frank Everts, a local banker. Previously called Rabbit Creek (*q.v.*).

La Posta (pōs′ tə): town, **Creek, Indian Reservation** [San Diego Co.]. Spanish for 'relay stage or post', where mail riders used to stop (Stein).

La Presa (prā′ sə) [San Diego Co.]. Spanish for 'the dam', named in 1887 for nearby Sweetwater Dam (Stein).

La Puente (po͞o en′ tē, pwen′ tē) [Los Angeles Co.]. The post office was established as Puente (*q.v.*), meaning 'bridge', on Sept. 15, 1884. The place was so known until the BGN decided for La Puente in 1955. In modern standard Spanish, the word is masculine, hence *el puente*.

La Quinta (kēn′ tə) [Riverside Co.]. The post office was established in 1931 and named after La Quinta Hotel, which had been opened by a Mr. Kiener (Mary Brooks). *Quinta* means 'country estate'.

Laquisimes, Río de los. *See* Stanislaus River.

Larabee: Creek, station, **Buttes** [Humboldt Co.]. Named for a stock raiser, Larabee, whose ranch was burned by Indians in the spring of 1861. The stream is shown on the von Leicht–Craven map as Lariby Creek, and Laribee Buttes are mentioned by Rockwell in 1878. A post office, Laribee, was established on Apr. 6, 1888. The present spelling was used for the station in 1908 and for a new post office on Sept. 30, 1921.

Largo [Mendocino Co.]. The name was applied to the station of the Northwestern Pacific in 1908. It is a Spanish word meaning 'long' and may have been chosen because the station was on the land of Lemuel F. Long, resident in the county since 1858, assemblyman 1877–78.

Laribee; Lariby. *See* Larabee.

La Riviera [Sacramento Co.]. The name refers to the resort area on the Mediterranean coast of France and Italy; it is Italian for 'the coast'.

Larkin's Children [Colusa, Glenn Cos.]. The grant, dated Dec. 14, 1844, was made to the three minor children of Thomas O. Larkin, a native of Massachusetts, who came to California in 1832 and was U.S. consul at Monterey after 1843. The land could not be granted to Larkin himself, because he never became a Mexican citizen.

Larkspur [Marin Co.]. Subdivided by Charles W. Wright in 1887. Wright's British-born wife, Georgiana, is credited with naming the town for a flower blooming on the hills; she thought the lupine was larkspur (Teather 1986).

Las. Names preceded by the Spanish feminine plural article *las* that are not listed here will be found under the first letter of the name proper; e.g., Las Posas is listed under Posas. The pronunciation (läs) often merges, as (lôs), with that of the masculine article *los*.

Las Aguilas (ä′ gi ləs): **Creek, Canyon, Mountains, Valley** [San Benito Co.]. In 1854 Manuel Larios testified in the Real de las Águilas land case that "the name . . . was given to said place in 1826 in consequence of two persons by the name of Aguila having encamped there while lassoing cattle" (*WF* 6:371). The two Aguilas, however, were not the grantees of the land grant that was called Real de las Águilas and dated Jan. 16, 1844.

Las Chiches (chē′ chās) [San Luis Obispo Co.]. The name of this flat is Mexican Spanish for 'the nipples'.

Las Chollas; Las Choyas. *See* Chollas.

Las Cruces (krōō′ səs) [Santa Barbara Co.]. The name, 'the crosses', is derived from a land grant, dated July 12, 1836, and May 11, 1837. A Rancho de las Cruces, probably belonging to Mission Santa Ynez, is mentioned on July 21, 1822 (Guerra y Noriega 5:266), and Lomas de las Cruces is shown on a *diseño* of Rancho Santa Rosa (about 1835). The canyon labeled Cañada de las Cruces on the Lompoc atlas sheet is shown as Arroyo de las Cruces on a *diseño* of the Las Cruces grant. A post office Las Cruces is listed in 1880.

La Selva Beach [Santa Cruz Co.]. This was originally a resort; the name is Spanish for 'the forest' (Clark 1986). However, California Spanish would normally use *monte* to mean 'forest'.

Las Flores (flôr′ əs) [San Diego Co.]. Spanish for 'the flowers'. The Portolá expedition in 1769 named the place Santa Praxedis de los Rosales 'St. Praxed of the Rose Gardens'. Crespí wrote of abundant flowers here, with a large Indian settlement under San Luis Rey Mission from 1835 onward (Stein).

Las Garzas (gär′ zəs) **Creek** [Monterey Co.]. Spanish for 'the herons'; also called Garzas Creek (Clark 1991).

Las Llajas (yä′ həs) **Canyon** [Ventura Co.]. Probably a misspelling of *llagas* 'wounds', *q.v.* (Ricard).

Las Peñas (pān′ yəs) [San Mateo Co.]. Spanish for 'the crags' (Alan K. Brown).

Las Pilitas (pi lē′ təs) [San Luis Obispo Co.]. Spanish diminutive of *pila* 'baptismal font', referring to natural bowls or basins in a stream (Dart).

Las Plumas (plōō′ məs) [Butte Co.]. This community, named from the Feather River on which it is situated, was inhabited mainly by the employees of the large Pacific Gas and Electric Company Big Bend powerhouse. The first post office was established on Nov. 24, 1908. Since 1966 it has been under the waters of the Oroville Dam project.

Las Posadas (pō sä′ dəs) **Forest** [Napa Co.]. In 1938 Mrs. Anson S. Blake of Berkeley donated the Moores Creek Ranch to the State Board of Forestry to preserve and study the unique flora of the forest. The ranch had been a portion of the La Jota land grant, and the name owes its origin to the fact that *las posadas* 'the guest houses' were apparently located at this beautiful spot (*CHSQ* 35:1 ff.). *See* Moores Creek.

Lassen: Peak, County, National Forest, Volcanic NP; Lassen: Meadow, View [Plumas Co.]; **Lassen Creek** [Modoc Co.]. In a geographical sense, the name of the noted Danish pioneer Peter Lassen (*see* Glossary) was probably applied first to the mountain. In 1827 Jedediah Smith, who often called a range a "mountain," applied the name Mount Joseph to the entire chain between latitude 39° north and 41° north (Burr's map). Wilkes's map of 1841 limited the term to what is now Lassen Peak; but influenced by the holy names farther south, he placed "St." before it. This name, Mount St. Joseph, is shown on most maps until the middle 1850s. Bruff calls it Snow-Butte or Mount St. Jose in 1850 (2:810). The people, however, attached the name of the generally respected pioneer to the peak, as is shown by the reference to "Mount Saint Joseph (sometimes called Lassen's Peak)" in the Pac. R.R. *Reports* (2: 2.51). H. L. Abbot, of the same survey, speaks of "Lassen's butte" in 1851 (ibid., 6:1.128); and Blake, the geologist of the survey, used the form Lassens Butte on his map, an early example of the omission of the apostrophe—a simplification that is now required by the BGN. The now generally accepted name is Mount Lassen, although officially it is Lassen Peak (BGN, June 2, 1915). The county was created by the legislature from parts of Plumas and Shasta Cos. and named on Apr. 1, 1864; the national forest was created by presidential proclamation in 1908, and the national park by act of Congress in 1916. Fandango Pass [Modoc Co.] was formerly Lassen Pass or Lassens Horn or Lassen Cut-off (*see also* Cape Horn); and the meadows in Nevada where the Humboldt River changes its course from west to south were Lassen Meadows. Former misspellings: Lawson, Lasson, and Lassin.

Lassic, Mount; Black Lassic: Peak, Creek; Red Lassic: Peak, Creek [Trinity Co.]. According to P. E. Goddard, the mountain preserves the name of an Athabaskan tribe, named after its last chief, Lassik (Hodge). The spelling is that of the USFS. The name is probably not Athabaskan, however (Golla); it may be from the neighboring Wintu language, in which *lasik* means 'bag'. Black Lassic Creek is a typical absurdity caused by thoughtless naming after other features. Neither Lassic nor the creek could be called black, but only the peak after which the creek was named.

Last Chance. A name favored by prospectors and explorers, especially for places where they found water when about to give up from thirst. The name may sometimes have been applied because the place was the last at which one could obtain supplies of water or liquor before entering a desert or a mountain fastness. E. J. McKenney tells the story of three men working a mine near Coloma [El Dorado Co.] who were ready to give up when a Sacramento firm staked them with a few months' supplies. They took up work, struck it rich, and called their claim the Last Chance (Knave, Dec. 29, 1946). There are more than twenty "last chance" names in the state, including a Last Chance Ditch [Kings Co.]. On the Caliente Road in Sequoia National Forest, two water-supply places are designated as Big and Little Last Chance Can. **Last Chance Spring** and **Range** [Inyo Co.]: In Aug. 1871 the Lyle detachment of the Wheeler Survey, completely exhausted after its guide had disappeared, fortunately reached the spring. The name appears on the von Schmidt boundary map (1872) and on Wheeler atlas sheet 65. **Last Chance Creek** [Plumas Co.]: A prospector led the first party to discover Gold Lake from Nevada City about June 5, 1850. After a long fruitless search, the leader was given a "last chance" to find the lake within forty-eight hours or else. Schaeffer (p. 87) relates that the threat was not carried out because some believed that his "reason was dethroned." **Last Chance** [Placer Co.]: According to local tradition, the miners applied the name to their claim when they shot a deer with their last bullet after all their provisions were gone. **Last Chance Creek** [Santa Cruz Co.] is in an area with abundant water; Clark 1986 suggests that the name may be derived from that of nearby Las Trancas Creek (*q.v.*).

Las Trampas (träm′ pəs): **Ridge, Peak, Creek** [Contra Costa Co.]. On the *diseños* of several ranchos of the district the ridge is labeled *Sierras pequeñas del . . . Arroyo San Ramón alias Las Trampas* ('little ridge of the . . . Arroyo San Ramón alias The Traps'). The abbreviated versions Las Trampas and Arroyo de las Trampas appear on the *diseño* of Laguna de los Palos Colorados. According to the testimony of José Martínez, in land grant case 276 N.D. (1862), traps were set in the chaparral of these hills to catch elk (Cutter).

Las Trancas (trang′ kəs) **Creek** [San Mateo, Santa Cruz Cos.]. The Spanish word for 'barriers' often refers to cattle guards; it is also found as Los Trancos (Alan K. Brown; Clark 1986); *see also* Tranca.

Lasuen (lä sōō ān′), **Point** [Los Angeles Co.]. In Nov. 1792 Vancouver named the southeast point of San Pedro Bay for his friend, Padre Fermín Francisco de Lasuén. *See* Fermin.

Las Virgenes (vûr′ jə nəs) **Creek** [Ventura, Los Angeles Cos.]. *El paraje de las Vírgenes* 'the place of the virgins' (referring to the virgin martyrs of the church calendar) was the name of a land grant in 1802 (Docs. Hist. Cal. 4:121). This grant was later abandoned, but the name was preserved through a new grant, dated Apr. 6, 1837.

Las Yeguas (yā′ gwəs) **Canyon** [Santa Barbara Co.]. Shown on the *diseño* of the Cañada del Corral grant as Cañada de las Yeguas 'valley of the mares'. On the *diseño* of the Dos Pueblos grant the same valley is labeled Arroyo de las Llagas 'arroyo of the wounds [stigmata]'. The apparent confusion results in part because *ll* is pronounced like *y* in Mexican Spanish and often alternates with it in non-standard spelling (*see* Las Llajas; Llagas; Yegua).

Latache. *See* Lemoore.

Lathrop [San Joaquin Co.]. Laid out in 1887 and named by Leland Stanford for his brother-in-law, Charles Lathrop.

Laton (lā′ tən) [Fresno Co.]. Named for Charles A. Laton of San Francisco, who acquired a part of the Laguna de Tache ranch with L. A. Nares in the 1890s.

Latona. *See* Redding.

Latour Butte [Shasta Co.]. The elevation was named for James La Tour, whose homestead is now included in La Tour State Park (Steger).

Latrobe [El Dorado Co.]. When the Placerville and Sacramento Valley Railroad reached the place in Aug. 1864, F. A. Bishop, the chief engineer, named the station for Benjamin H. Latrobe, who had constructed the famous

Thomas Viaduct over the Patapsco River for the Baltimore and Ohio Railroad in the 1830s.

Laurel. The presence of the California laurel, *Umbellularia californica* (Spanish *laurel*) in almost all sections of the state, either as a tree or a shrub, has given rise to more than twenty-five Laurel Creeks, Lakes, Mountains, and Hills. Most of the places named for the laurel tree are in the central and southern parts of the state. In the northern counties the laurel is usually called pepperwood, bay tree, or spice tree.

Laureles. The Spanish word for 'laurel trees' appears in the names of two land grants in Monterey Co., dated Sept. 19, 1839, and Mar. 4, 1844. It has also survived in **Los Laureles** (lôr el′ əz) **Canyon** [Santa Barbara Co.]. However, Laureles Arroyo [San Mateo Co.], shown on a *diseño* of Las Pulgas (1835), was translated into Laurel Creek.

Laurel Lake [Yosemite NP]. For the "mountain laurel" around the shore (Browning 1988). This is a popular name for the flowering *Ceanothus* bush, unrelated to the *Umbellularia*.

Lauterwasser (lô′ tər wô sər) **Creek** [Contra Costa Co.]. The tributary to San Pablo Creek was formerly a large clear stream for which the name, meaning 'pure water', was highly descriptive; but in fact it was named for F. P. Lauterwasser, a German butcher from San Francisco, one of the earliest settlers in the Orinda district.

Lava. The presence of lava flow is indicated in the names of about ten orographic features, including Lava Top on the boundary line of Plumas and Butte Cos. **Lava Beds National Monument** [Modoc, Siskiyou Cos.] was created and named by presidential proclamation in 1925. The name Lava Beds for the volcanic formations had long been in use, and it gained prominence after the Modoc War of 1872–73, in which the Indians used them as strongholds.

La Verne [Los Angeles Co.]. Named in 1916 after La Verne Heights, a subdivision west of the town that had been called by the given name of the promoter. The former name was Lordsburg, for I. W. Lord, who laid out the town in 1886 (Santa Fe).

Lavezzola (lav ə zō′ lə) **Creek** [Sierra Co.]. This name of an early settler was applied by the BGN in 1950 to the stream formerly known as the North Fork of the North Fork of the Yuba River.

Lavic (lä′ vik): **Lake, Mountain**, settlement [San Bernardino Co.]. These places west of Ludlow were apparently named for a prospector. The lake is dry; the mountain is known for its manganese deposits. The settlement had a post office from 1902 to 1909.

Lavigia (lä vig′ ē ə) **Hill** [Santa Barbara Co.] is mentioned as *el cerro de la Vigia* 'lookout hill' in a description of Rancho Nuestra Señora del Refugio in 1828 (Registro, p. 31). When W. E. Greenwell of the Coast Survey charted the Santa Barbara Channel between 1856 and 1861, he used the hill for a triangulation point and kept the Spanish name.

Lawlor, Mount [Los Angeles Co.]. The mountain was named by the USFS about 1890 for Oscar Lawlor, a prominent Los Angeles attorney.

Lawndale. This popular American place name, which preserves the old English generic term "dale" (valley), is represented in California by Lawndale [Los Angeles Co.], established and named by Charles Hopper in 1905; Lawndale [San Mateo Co.], named in 1924 by the Associated Cemeteries at the suggestion of M. Jensen, later mayor of the town; and Lawndale [Sonoma Co.], a station of the Northwestern Pacific.

Lawrence Creek [Humboldt Co.]. The stream was probably named for either Asa or William H. Lawrence, farmers at Rohnerville, whose names appear in the Great Register of 1879.

Laws [Inyo Co.]. The Carson and Colorado Railroad built the narrow-gauge road in 1883 and named the station, which was near the site of old Owensville, for R. J. Laws, assistant superintendent.

Lawson Peak [Sequoia NP] was named in 1976 for Andrew C. Lawson (1861–1952), professor of geology at the University of California, Berkeley (Browning 1986). *See also* Lassen Peak.

Laytonville [Mendocino Co.]. Named for Frank B. Layton, who came to California from Nova Scotia, Canada, in 1867 and settled here in 1875 (Co. Hist.: Mendocino 1880:583 ff.).

League. Some Mexican land grants were described as a number of *sitios*, each *sitio* being a parcel of land 5,000 *varas* square. In American land grant papers the *sitio* was translated as "league." There were three grants called Four Leagues [Napa, Sacramento, San Francisco Cos.]; two Five Leagues [Yolo Co.]; and six Eleven Leagues [Butte, Fresno, Glenn, Madera, Stanislaus Cos.]. Since the *vara* (yard) was not uniform in practice, the size of a *sitio,* or league, differs. The provincial league was 4,438.19 acres; the league around 1850 was 4,340.28 acres; and the present league, used since June 1857 in the patent surveys for land grants, is 4,438.68 (Bowman).

Leastalk. *See* Ivanpah.

Leavitt: Meadow, Peak, Creek [Mono Co.]. In 1863 Hiram L. Leavitt built a hotel at the east end of Sonora Pass to accommodate the traffic between Aurora and Sonora (Maule). Leavitt's is recorded on Hoffmann's map, and Levitt Peak on the Mining Bureau map of 1891.

Lebec (lə bek′) [Kern Co.]. The place was named for a man who was killed here by a grizzly bear, probably a member of a trapping party. His comrades buried him and cut this inscription on an oak tree: "Peter Lebeck, killed by a X bear, Oct. 17, 1837." The inscription, which is now in Kern County Museum, was first seen and recorded by Bigler on July 31, 1847. On Sept. 29, 1853, William Blake of the Pacific Railroad Survey recorded it again but misspelled the name as Le Beck. This gave rise to the stories that Lebeck was a veteran of Napoleon's armies, a settler in the region, and that his name was Pierre Lebeque—all unproven fantasies. When the post office was established on Sept. 6, 1895, the Post Office Dept. further corrupted the name to Lebec.

Lecheria (lā chə rē′ ə), **Canyon de la** [Kern Co.]. Mentioned by BGN, list 6702. Spanish *lechería*, as observed by Malcolm Lowry in *Under the volcano*, does not mean 'brothel' as one might guess, but rather 'dairy' (from *leche* 'milk').

Lechler (lek′ lər) **Canyon** [Los Angeles Co.]. The canyon with the intermittent stream was named by the USGS after the Lechler Ranch in Oak Canyon when the Camulos quadrangle was surveyed in 1900–1901. The name was also spelled Leckler. George W. Lechler died in Piru City on Dec. 10, 1906 (*HSSC:P* 7:85).

Lechusa (li cho͞o′ sə) **Canyon** [Los Angeles Co.]. *Lechuza* is Spanish for the 'barn owl' (*Tyto alba*). The canyon is on the Malibu grant; the name probably goes back to Spanish times.

Leckler Canyon. *See* Lechler Canyon.

LeConte (lə kont′) was the name of two eminent Berkeley professors, father (Joseph) and son (J. N.); *see* Glossary. **LeConte Canyon** [Kings Canyon NP], **Point** [Yosemite NP]: The point was named for J. N. LeConte (*see* Glossary) by R. B. Marshall in 1909. The name for the canyon was applied before 1908 and approved with the spelling Leconte by the BGN on June 7, 1911. **LeConte Falls** [Yosemite NP] was the name originally given to what is now called Waterwheel Falls. Through confusion by mapmakers, the name LeConte Falls is now applied to what was earlier called California Falls (Browning 1988). **Mount LeConte** [Sequoia NP]: "Forms one of the most striking points of the whole range. . . . Some time ago those residents of the Lone Pine district who are interested in the mountains decided upon naming this peak LeConte, in honor of Professor Joseph LeConte. . . . It was then determined to make the ascent of the mountain and erect a monument. . . . In a small can we put a photograph of the Professor, with the following memorandum: 'To-day, the 14th of August, 1895, we, undersigned, hereby named this mountain LeConte, in honor of the eminent geologist Professor Joseph LeConte. . . . A. W. de la Cour Carroll, Stafford W. Austin'" (*SCB* 1:325–26); *see also* Kaweah Peaks. **LeConte Divide** [Kings Canyon NP], between the San Joaquin and Kings Rivers, was approved (with the spelling Leconte) by decision of the BGN on June 7, 1911. There is also a Mount LeConte in Death Valley.

Ledro. *See* Lerdo.

Leech: Lake, Mountain [Mendocino Co.]. It is not known whether this is a family name or refers to the aquatic animal, *Hirudo* spp. (Stang).

Leesville [Colusa Co.]. Named about 1876 for "Lee" Harl, the owner of the land, who acquired his nickname (according to local tradition) because he was an ardent admirer of the great Southern commander, Robert E. Lee.

Lee Vining (lē vī′ ning, lə vī′ ning): **Creek, Peak**, settlement [Mono Co.]. From the name of Lee (Leroy) Vining, of Laporte, Indiana, who came to California in 1852 in search of gold. Vining settled on the creek named for him and operated a sawmill (Maule). The creek is shown on maps as early as 1863; the post office was established as Leevining on Mar. 1, 1928. In 1957 the BGN decided that the name should be spelled in two words: Lee Vining.

Leggett [Mendocino Co.]. The place was long known as Leggett Valley, for an early pioneer.

When the post office was established on Oct. 16, 1949, the name was abbreviated by the Post Office Dept.

Le Grand [Merced Co.]. Established in 1896 by the San Francisco and San Joaquin Valley Railroad (now the Santa Fe) and named for William Legrand Dickinson, a Stockton resident who owned large tracts of land there. The name Dickinson itself was given to the railroad siding 3 miles south.

Lehamite (lə hä´ mi tē) **Creek** and **Falls** [Yosemite NP]. "The cliff along the east side of Indian Cañon was known as Le-ham-ite, and designated the place of the arrow wood" (Bunnell, *Biennial report of the Commission to Manage the Yosemite Valley* [Sacramento: State Printing Office, 1889–90], p. 11). The word could have been confused with the name of one of the Indian villages, about 0.4 mile east of Indian Canyon, which was given as Le-sam´-ai-ti (Powers, p. 365); cf. Browning 1988. The Southern Sierra Miwok origin is *leeha'-mite* 'there are several syringa bushes', from *leeha* 'syringa, mock orange' (Callaghan).

Leidig (lī´ dig) **Meadow** [Yosemite NP] commemorates the Leidig family, Pennsylvania Germans who came to Yosemite Valley in 1866. George F. Leidig owned the Leidig and Davaney Hotel at the foot of Sentinel Rock in the 1860s and 1870s, and had a winter house in the meadow that now bears his name. His son, Charles, the first white boy born in the valley (1869), was later a ranger in the park service.

Leidy (lī´ dē) **Creek** [Mono Co.]. Named for George Leidy, of Fort Madison, Iowa, who settled in Fish Lake Valley in 1882 (Robinson).

Lembert Dome [Yosemite NP]. In 1885 John B. Lembert took up a homestead at the foot of the dome to which his name was applied in the 1890s. The name was placed on the map by the USGS in 1901. The Wheeler Survey had named the prominent landmark Soda Spring Dome (atlas sheet 56-D).

Lemon. The name of this fruit has been used for a number of places in southern counties: Lemon Grove [San Diego Co.]; Lemoncove [Tulare Co.]; Lemon Heights [Orange Co.]; a fancy Lemona [Riverside Co.]; and two plain Lemons, one in Ventura Co., the other in Los Angeles Co.—a post office established Dec. 11, 1895, and renamed Walnut on Dec. 22, 1908.

Lemoore [Kings Co.] was settled by John Kurtz in 1859. In 1871 Dr. Lovern Lee Moore arrived, called the settlement Latache, and founded the Lower Kings River Ditch Company. When the post office was established on Sept. 21, 1875, its name was coined from Moore's name.

Lempom [Tulare Co.]. The name, given to the railroad station in 1917, was coined from lemon and pomegranate (Santa Fe).

Lennox [Los Angeles Co.]. Named after Lenox, Massachusetts, the former home of a resident of the settlement, who met with a group of other citizens to choose a name for their community. It had been known in the early days as Inglewood Rancho and later for a short time as Jefferson. In 1921 the Los Angeles County Board of Supervisors officially recognized the name by changing the name of Olivian Avenue to Lennox Avenue, now Lennox Boulevard.

Lenwood [San Bernardino Co.]. Coined from the name of Ellen Woods, whose husband, Frank Woods, subdivided land and planned a town here in the early 1920s (Santa Fe).

Leodocia. *See* Red: Red Bluff.

Leonard Lake [Mendocino Co.]. Named for John Leonard, who, according to local tradition, obtained title to the lake from Jim Patrick, a bear hunter, in exchange for a pinto horse.

Leonis (lē ō´ nis) **Valley** [Los Angeles Co.]. Named for Miguel Leonis, a Basque sheepherder, who settled in the San Fernando Valley before 1868 and later became the owner of Rancho El Escorpión.

Le Perron (lə pə rōn´) **Peak** [Humboldt Co.]. Named for Yves Le Perron, an early prospector and hunter; also spelled Le Parron (Turner 1993).

Lerdo (lâr´ dō) [Kern Co.]. The name was applied to the station by the Southern Pacific when the railroad reached the place in the fall of 1872. According to Sanchez, it is a Spanish surname. Sometimes spelled Ledro (Land Office map, 1895). *See also* Cawelo.

Leucadia (lo͞o kā´ dē ə) [San Diego Co.]. Established in 1885 by a group of English colonists with a predilection for Greek. Leucadia (or Santa Maura) is one of the Ionian Islands; from one of its cliffs Sappho is said to have leaped into the sea. All street names in the California town are Greek: Hygeia Street, Athena Street, etc. (C. E. Thrailkill). The post office was established on Mar. 27, 1888.

Leviathan: Creek, Peak [Alpine Co.]. A Biblical term for a huge marine animal, originally applied to the mine that was worked here in the 1860s; like many mines, it was given a name suggestive of tremendous riches.

Lewis. *See* Mococo.

Lewis, Mount [Yosemite NP]. Named in memory of Washington B. Lewis, a native of Michigan, who for many years was connected with the USGS and was superintendent of Yosemite NP. The former name was Mount Johnson. **Lewis Creek** [Yosemite NP], also known as Maclure Fork, was named for the same man.

Lewis Creek and **Canyon** [Kings Canyon NP]. Named for the brothers Frank M. and Jeff Lewis, pioneer stockmen, prospectors, and hunters (Farquhar).

Lewis Lakes [Tuolumne Co.]. A group of three lakes was named in memory of Bert Lewis, of the USFS, who was killed in World War I on May 27, 1918. Another nearby lake, also called Lewis Lake, was named Iceland Lake by decision of the BGN, 1965.

Lewiston: town, **Lake** [Trinity Co.]. The town was settled by B. F. Lewis in 1853 and had a post office from May 24, 1854. The lake was created by impounding the Trinity River in the late 1950s.

Lexington. The scene of the first bloody conflict between colonists and British troops, on Apr. 19, 1775, is one of the most popular place names in the United States. For Southern sympathizers the name gained new significance when the siege of Lexington, Missouri, in Aug. 1861, ended in one of the early Confederate victories. The name is currently represented by Lexington Hills [Santa Clara Co.].

Liberty Cap [Yosemite NP]. "Owing to the exalted and striking individuality of this boldly singular mountain . . . , it had many godfathers in early days; who christened it Mt. Frances, Gwin's Peak, Bellows' Butte, Mt. Broderick, and others; but, when Governor Stanford (now U.S. Senator) was in front of it with his party in 1865, and inquired its name, the above list of appellatives was enumerated, and the Governor invited to take his choice of candidates. A puzzled smile lighted up his face and played about his eyes, as he responded, 'Mr. H., I cannot say that I like either of those names very much for that magnificent mountain; don't you think a more appropriate one could be given?' Producing an old-fashioned halfdollar with the ideal Cap of Liberty well defined upon it, the writer suggested the close resemblance in form of the mountain before us with the embossed cap on the coin when the Governor exclaimed, 'Why! Mr. H., that would make a most excellent and appropriate name for that mountain. Let us so call it.' Thence forward it was so called; and as everyone preferentially respects this name, all others have been quietly renunciated" (Hutchings, *Sierras*, p. 445).

Liberty Settlement. *See* River: Riverdale.

Lick. Several Lick Creeks and a Licking Fork (of the South Fork of the Mokelumne) were probably so named after salt licks for cattle or game. *For* Lick Springs, *see* Tuscan Springs.

Lick Observatory [Santa Clara Co.]. The name commemorates the donor of the observatory to the University of California, James Lick (1796–1876), a Pennsylvania German real estate speculator and public benefactor. The name was suggested by George Davidson in 1873 after Lick had accepted his plan to establish the observatory on what is still called Observatory Point at Lake Tahoe (von Leicht–Craven map, 1874). In 1874 the choice of site was changed to Mount Hamilton. The road to the peak was built in 1876 by Santa Clara Co. and named Lick Avenue. Lick came to California shortly before the discovery of gold with a small fortune acquired in South America, which he augmented by clever real estate speculation. In addition to giving the observatory to his adopted state, he endowed the California Academy of Sciences and made numerous other bequests. Lick Mills in San Jose is the site of his mahogany-finished flour mill, built in 1855.

Liddell (li del′) **Creek** [Santa Cruz Co.]. For George Liddell, who had a sawmill here in 1851 (Clark 1986).

Liebel (lē′ bel) **Peak** [Kern Co.]. The mountain was named in the 1930s for Michael Otto Liebel, a prospector of German descent who came into the Paiute Mountain country in 1876. He married an Indian woman, settled at the foot of the mountain, and brought up a large family.

Liebre (lē ā′ brē): **Mountain, Gulch** [Los Angeles Co.]; **Liebre Twins** [Kern Co.]. La Liebre land grant, dated Apr. 21, 1846, was evidently named after *el paraje que llaman la Cueba de la liebre* 'the place that is called the burrow of the hare' (i.e. the jackrabbit), mentioned in Oct. 1825 (Docs. Hist. Cal. 4:767). Sierra de la Liebre is shown on Wheeler atlas sheet 73-C.

The modern names (all of which are outside the boundaries of the land grant) were used when the USGS mapped the Tejon quadrangle. By mistake the name Liebre Twins was transferred on the atlas sheet to a single-top higher peak to the west (Wheelock).

Lieutenants, The [Siskiyou Co.]. The two peaks in the Siskiyou Mountains were named as lieutenants to the towering El Capitan, close to the southwest of them (BGN, 1965).

Lights Creek [Plumas Co.]. Named for Ephraim Light, a pioneer rancher north of Indian Valley. He later was one of the first men to commercialize hot springs in Santa Rosa (R. Batha).

Lignite. *See* Carbon: Carbondale.

Likely [Modoc Co.]. In 1878 the settlers had to find a new name for their place because the old name, South Fork, had been rejected by the Post Office Dept. They suggested three other names, and each was rejected because the name existed elsewhere. The settlers met again, and one remarked dejectedly that it was not likely that they would ever get a name. Another spoke up and said, "What's the matter with 'Likely'? It is not likely that there will be another post office in the state called 'Likely.'" The suggestion was adopted, and the Post Office Dept. approved the name. The mountain southwest of the town is now called Likely Mountain instead of South Fork Mountain (BGN, 1965).

Lilac [San Diego Co.]. For the "California lilac" or *Ceanothus*, unrelated to the domestic lilac (Stein).

Lillis [Fresno Co.]. Named for Simon C. Lillis, a superintendent of the Laguna de Tache cattle ranch before 1917.

Lily: Lake, Pond. Bodies of water have been given these names in several areas; some of them may refer to the yellow water lily or cow-lily, *Nuphar polysepalum* (Stang).

Limantour. *See* Estero: Estero de Limantour.

Lime; Limestone. Some fifteen geographic features are named from the presence of limestone. None seems to have been named for the citrus fruit. Lime Saddle is in Butte Co., and Lime Point in San Francisco Co. There are also several Limekiln Creeks and Canyons, and three settlements: Lime [Tuolumne Co.], Limestairs [Trinity Co.], and Limestone [El Dorado Co.] (*see* Calera). *For* Lime Point Military Reservation, *see* Baker, Fort.

Limoneria (līm ə nâr′ ə, lim ə nâr′ ə) [Ventura Co.]. For the Limoneria Company, which established a citrus packing house here (Ricard); from Spanish *limonería* 'place of lemons'.

Limpia Concepción, Punta de la. *See* Concepcion, Point.

Lincoln [Placer Co.]. "The TOWN of LINCOLN, at Auburn Ravine, Placer county, was laid out about two years ago, under the auspices of C. L. Wilson, projector of the California Central Railroad, and soon afterward a sale of town lots at auction took place at the St. George Hotel in this city. The town does not take its name from the President of the United States, but from its founder, whose middle name is Lincoln" (*Sacramento Union*, Nov. 4, 1861). Lincoln Acres [San Diego Co.] and Lincoln Park [Los Angeles Co.], like scores of other communities in the United States, may have been named for Abraham Lincoln (1809–65; president, 1861–65). Note also Lincoln Village [San Joaquin Co.].

Linda [Yuba Co.] was laid out by John Rose in Jan. 1850, as a rival of Marysville. Named after the miniature steamer *Linda*, launched on the Sacramento River in Dec. 1849. The steamer had been named after the seagoing *Linda*, which had brought the engine, scow, and wheel for its smaller counterpart round Cape Horn (Ramey).

Lindemann Lake [Modoc Co.] was named for J. B. Lindemann, a native of Germany and a resident of Eagleville in the 1870s.

Linden [San Joaquin Co.]. In 1860 the place was known as Linden Mills. The name was shortened to the present form when the post office was established in 1867. Since the linden tree (*Tilia* genus) is not native in the state, this is probably a transfer from the eastern United States, where the name is popular. *For* Linden [Tulare Co.], *see* Denlin.

Lindero (lin dâr′ ō) **Canyon** [Los Angeles Co.]. Cañada del Lindero 'boundary valley' is shown on a *diseño* of El Conejo grant, on the boundary of Las Virgenes grant.

Lindo; Linda. This Spanish adjective, meaning 'handsome, pretty', has been repeatedly used in place naming. There is a Lindo Lake in San Diego Co., and a Linda Creek in both Placer and Sacramento Cos. **Linda Vista** [San Diego Co.], meaning 'pretty view', was mentioned as a new settlement in the *San Diego Union*,

Apr. 22, 1886. During World War II the name was applied to a large housing project on Kearny Mesa, 5 miles east of the old place (San Diego Public Library). There is another Linda Vista in Santa Clara Co.

Lindsay [Tulare Co.]. Capt. A. J. Hutchinson chose the maiden name of his wife for the town that he founded in 1888.

Lingo Canyon [San Luis Obispo Co.]. For George Lingo, an early settler.

Linn, Mount [Tehama Co.]. The name was given by Frémont, probably to the highest peak of the Yolla Bolly group, in honor of Lewis F. Linn (a senator from Missouri, 1833–43), who took an active part in the struggle for the acquisition of the Oregon Territory. Mount Linn and "Mt. Tsashtl" (Shasta) are the only California mountains named on Preuss's map of 1848. Von Leicht–Craven (1874) show "South Yallo Balley or Mt. Linn." The Mining Bureau map of 1891 transferred the name to a peak farther east, and Douglas's gazetteer places it at latitude 40°03' north, longitude 122°46' west (*see also* Bally).

Linns Valley [Kern Co.]. Named for William Lynn, who settled here in 1854 (Co. Hist.: Kern 1914:188).

Linora (li nôr′ ə) [Merced Co.]. Applied to the station by the Southern Pacific when the line was built about 1890. A typical railroad name, probably coined like other fanciful names on this section.

Lion. The presence of the feline variously called cougar, puma, panther, or mountain lion (*Felis concolor*) is indicated in the names of more than fifty geographic features, mainly Canyons. **Lion Rock** and **Lake** [Sequoia NP]: The rock was so named because Mansell Brooks, a sheepman, killed a mountain lion near there in 1883 (Farquhar). **Lion Rock** [San Luis Obispo Co.] was named because of the presence of sea lions; it was already known in Spanish times as El Lobo 'the [sea] wolf', as shown on a plat of part of the land grant Cañada de los Osos y Pecho y Islai.

Lippincott Mountain [Sequoia NP]. Named by the USGS in 1903, for Joseph B. Lippincott, at that time hydrographer for the survey.

Lisque (lē′ skā) **Creek** [Santa Barbara Co.]. Probably for a Spanish spelling *lisgüey*, from the Ineseño Chumash village name *aliswey* 'in the tarweed' (Applegate 1975:25, after Harrington).

Litchfield [Lassen Co.]. Named for the Litch family, pioneer settlers, by Mrs. B. F. Gibson when the townsite was laid out in 1912 (Clara Litch). The post office is listed in 1915.

Little. The adjective is used frequently for geographic features that resemble another more prominent feature nearby, e.g. Little Yosemite Valley. This unimaginative method of naming often produces a comic effect: Little Shuteye Peak [Madera Co.] is not named for a little shut-eye but is a peak smaller than the real Shuteye Peak; likewise, Little Bear Creek [Los Angeles Co.] is not named for a little bear, nor Little Rattlesnake Mountain [Del Norte Co.] for a little rattlesnake. A tributary of the San Gabriel River is called Little Santa Anita Canyon Creek. Sometimes, however, the specific name is actually modified: Little Lake Valley [Mendocino Co.] is named after a diminutive lake near its southern border; Little Lakes Valley [Inyo Co.] because there are some forty little lakes in it; Little Pete Meadow [Kings Canyon NP] for Pierre ("Little Pete") Giraud (Farquhar); and Little Claire Lake [Sequoia NP] for the seven-year-old daughter of Ralph Hopping (Farquhar). Often, similar features in the same locality receive the adjectives "big" and "little" at the same time: Big and Little Maria Mountains [Riverside Co.], Big and Little Five Lakes [Sequoia NP]. Sometimes a feature is modified by "little" because it is thought to resemble a well-known feature elsewhere, e.g. Little Gibraltar [Santa Catalina Island]. A number of mining towns and settlements were named after states or cities: Little Texas [San Diego Co.], Little York [Nevada Co.]. "Little" geographic features often have namesakes in nearby post offices and communities: Little Lake [Inyo Co.], Littleriver [Mendocino Co.], Little River [Humboldt Co.], Littlerock [Los Angeles Co.], Little Shasta [Siskiyou Co.], and Little Valley [Lassen Co.]. **Little Norway** [El Dorado Co.] was so named in 1961, replacing the former name Vade. **Little Sur River** [Monterey Co.] is adapted from Spanish El Río Chiquito del Sur 'the little river of the south', as contrasted with the Big Sur River, El Río Grande del Sur (Clark 1991); *see* Sur. *For* Little Pine [Inyo Co.], *see* Independence; *for* Little Truckee Lake, *see* Webber Lake.

Live Oak. The common designation of several species of California's evergreen native oak (especially *Quercus agrifolia*) is contained in some twenty geographical names, including a number of communities and post offices: the city so called in Sutter Co. was named in 1874 by H. L. Gregory. **Live Oak Acres** [Ventura Co.] was named about 1925 by W. C. Hickey. Other examples are Live Oaks [Sonoma Co.] and Live Oak Springs [San Diego Co.] (*see* Encina; Oak). *For* Live Oaks [Sacramento Co.], *see* Michigan Bar.

Livermore: city, **Valley, Pass** [Alameda Co.]. Robert Livermore (1799–1858), an English sailor, came to California in the 1820s and became co-grantee of Rancho Las Positas on Apr. 10, 1839. After the American occupation his residence became known as Livermore's and is so designated on maps of the 1850s. The town did not come into existence until 1864. It was first named Laddville, for Alphonso S. Ladd, who erected the first building. In 1869 the post office was established and named Nottingham, after Livermore's home town. However, in the same year William H. Mendenhall, a member of the Bear Flag party, platted the townsite and named it Livermore in memory of the pioneer. The name was accepted for the post office and the railroad station.

Livingston [Merced Co.]. When the valley route of the Southern Pacific reached the place in 1871, the station was named Cressey for the owner of the land. In 1872 the postmaster, O. J. Little, also a landholder, objected to the name (according to Co. Hist.: Merced 1925: 366); and the name of the explorer of central Africa, David Livingstone (minus the final *e*), was chosen instead. The old name was not lost but was applied to a Santa Fe station, 3 miles east, about 1900.

Llagas (yā′ gəs, yä′ gəs) **Creek** [Santa Clara Co.]. On Nov. 25, 1774, Padre Palou named a place near the creek Las Llagas de Nuestro Padre San Francisco 'the wounds [stigmata] of Our Father Saint Francis'. Anza refers to it as Las Llagas in 1776, and Josef Moraga calls the creek Arroyo de las Llagas de Nuestro Padre San Francisco (Anza 3:411). Llano de las Llagas was the Spanish name for Gilroy Valley (Beechey 2:48). A land grant, San Francisco de las Llagas, is dated Feb. 3, 1834. Trask's *Report* of 1854 (p. 8) mentions "Llagos River." The present version is recorded in Hoffmann's notes on Aug. 20, 1861. A local English pronunciation (i lä′ gəs) is documented (Shafer p. 241). *See also* Las Llajas; Las Yeguas Canyon.

Llajome. *See* Yajome.

Llanada (yə nä′ də) [San Benito Co.]. A post office, now discontinued, was established here about 1900 and given the Spanish name, meaning 'plain; level ground', because it is situated on the wide expanse of Panoche Valley. A local pronunciation (i lan′ də) is documented (Shafer, p. 241).

Llano. The Spanish word means 'level field, even ground' and was commonly used as a generic geographical term. Six land grants include the word in their titles; another has the diminutive, Llanitos. **Llano** (yä′ nō) [Los Angeles Co.]: A post office with this name was established in 1890. It was discontinued in 1900; but after the well-known socialist leader Job Harriman of Los Angeles founded the colony Llano del Rio, the post office was reestablished on Jan. 23, 1915. The socialist colony was moved to Louisiana in 1917, but the post office remained. A local pronunciation (i lan′ ō) is documented (Shafer, p. 241).

Lleguas (yā′ gwəs) **Pass** [Alameda Co.]. Probably a misspelling of Spanish *yeguas* 'mares'.

Llokaya. *See* Ukiah.

Llorones, Ensenada de los. *See* Mission: Creek, Rock.

Lobdell Lake [Mono Co.]. Named for J. B. Lobdell, who is said to have hidden out here in the 1860s to escape military service (Maule).

Lobitos (lō bē′ təs): **Creek**, settlement [San Mateo Co.]. The name is the diminutive of *lobos* 'wolves', referring in California to sea lions (Alan K. Brown). The stream is shown as Arroyo de los Lovitos on a *diseño* of the Cañada Verde grant (about 1838), and as Arroyo de los Lobitos on a *diseño* of the San Gregorio grant (about 1839).

Lobo. The geographical names containing the Spanish word for 'wolf' do not refer to the member of the dog family, which was never common (and is now extinct) in California; they occur only along the coast and refer to the *lobo marino* or 'sea wolf', i.e. the seal or California sea lion (*Zalophus californianus*). **Point Lobos** (lō′ bəs, lō′ bōs); **Lobos Creek** [San Francisco Co.]: The name of the westernmost point of San Francisco is as appropriate now as it was when it originated. Punta de los Lobos is mentioned by Chamisso in 1816 (Rurik, p. 73), and on June 25, 1846, the name Punta de Lobos was given to a land grant (later rejected by the U.S. courts) that included this point as well as the presidio.

The English version, Point Lobos, is used by Beechey in 1826 (2:424) and by the Coast Survey in 1851. **Point Lobos; Lobos Rocks** [Monterey Co.]: "Innumerable sea wolves" were observed in the vicinity by Crespí and Fages in May 1770 (Palou 2:284). Punta de Lobos is shown on a *diseño* of the San José y Sur Chiquito grant (1839). On American maps the entire cape was generally designated as Point Carmel; the map of the Coast Survey of 1855 has Point Lobos, but the survey's special map of Monterey Bay of 1878 has Pyramid Point because the projecting rock, now called Pinnacle Point, is pyramid-shaped. The name Point Lobos for the entire cape has been used since 1890; it was made permanent in 1933 by the creation of the Point Lobos Reserve, a state park. *For* El Lobo, *see* Lion: Lion Rock.

Locatelli (lō kə tel′ ē) **Creek** [Santa Cruz Co.]. Giuseppe Locatelli acquired land here in 1898 (Clark 1986).

Loch Leven (lok lev′ ən) [Kings Canyon NP]. A lake named for Loch Leven in Scotland, from which the Loch Leven trout was introduced to California (Browning 1986).

Loch Lomond (lok lō′ mənd) [Santa Cruz Co.]. A reservoir named for the lake in south central Scotland (Clark 1986). There is also a Loch Lomond post office in Lake Co. *See* Ben Lomond.

Lockeford [San Joaquin Co.]. The name was first applied to the ford that crossed the Mokelumne River on land of Dr. D. J. Locke, a graduate of Harvard, who settled there in 1851. In 1860 John A. Clapp used the name for his hotel, and shortly thereafter the town was laid out and named. The post office is listed in 1867.

Lockwood [Monterey Co.]. Named in 1888 for Belva Ann Lockwood, a feminist who ran for president of the United States in 1884 (Clark 1991).

Locoallomi [Napa Co.]. The name of the land grant, dated Sept. 30, 1841, appears to be Lake Miwok, perhaps *lakáh-yomi* 'cottonwood place' or *lakáa-yomi* 'goose place' (Callaghan). The forms Aloquiomi (apparently *alók-yomi* 'ear place'), Loaquiomi, Liaquiomi, and Loaquiomis also occur in the records of Mission San Francisco Solano (Arch. Mis., vol. 1). There has been further confusion with Collayomi and Loconoma (*qq.v.*). *See also* Lokoya.

Loconoma (lō kō nō′ mə) **Valley** [Lake Co.]. The name apparently reflects Wappo *lóknoma* 'goose place' (*see* Collayomi, Locoallomi).

Loco Siding [Inyo Co.]. When the California-Nevada Railroad was built in 1910, the name Loco, probably a local Indian name, was applied to the station. Other stations of this sector bear Indian names—Coso, Haiwee, Olancha. Loko Indians are mentioned by Hodge (1:932) as a division of the Mono-Paviotso group.

Lodi (lō′ dī) [San Joaquin Co.]. The station was called Mokelumne when the Central Pacific reached the place in 1869. To avoid confusion with similar names, the present name was chosen in 1874. Lodi in Italy was the scene of Napoleon's first spectacular victory on May 10, 1796, and in the 1870s more than twenty U.S. communities were already so named. A famous race horse called Lodi may have had some connection with selection of the name in California.

Lodoga (lə dō′ gə) [Colusa Co.]. The place is shown on a map of 1902 as Ladoga. It may have been named after the large lake between Finland and Russia, or after Ladoga in Indiana or Wisconsin. The post office was established on Feb. 1, 1898, with the spelling Lodoga.

Loftus [Shasta Co.]. A post office was established in 1922 at Steinaker's service station and named after Pollock Bridge, which had been built across the Sacramento in 1916 by George Pollock. In 1939, when the bridge and settlement became victims of the Shasta Dam project, the post office, together with the old name, was moved 5 miles north, and in 1940 another 2 miles north (Stella Woolman). As the post office was often confused with Pollock Pines in El Dorado Co., it was renamed in 1944 for Charles T. Loftus, a resident of the canyon and son of Tom Loftus, creator of the famous Banner strawberry (Steger). The BGN changed the name to Lakehead (*q.v.*, under Lake) in list 6902.

Logan Lake [Shasta Co.]. Named for P. B. Logan, who had his summer range here (Steger).

Log Cabin [Yuba Co.]. The settlement was named in 1927 when a post office (now discontinued) was established in a log cabin (Maudelene Cleveland).

Loghry (lou′ rē) **State Forest** [Santa Cruz Co.]. This land was deeded by the J. P. Loghry family to the state in 1944 (Clark 1986).

Log Meadow [Sequoia NP]. A huge, hollow, fallen Sequoia at the edge of a meadow was first seen by Hale D. Tharp in 1858 and later used by him as a summer cabin.

Lokoya (lō koi′ ə) [Napa Co.]. Probably abbreviated from Locoallomi (*q.v.*). *La tribu de Locaya* is mentioned on Aug. 6, 1845 (DSP Ben. 5:384), and the name Lokoya was applied in 1925 to the post office formerly called Solid Comfort.

Lola, Mount [Nevada Co.]. The mountain preserves the memory of the great lover and lesser actress Lola Montez, who precipitated a revolution in Munich in 1848. In the 1850s she found a fruitful field for her endeavors in Grass Valley and other mining towns of California (*see* Independence: Lake).

Loleta (lō lē′ tə) [Humboldt Co.]. When the local railroad was built in 1883, the station was named Swauger for Samuel A. Swauger, the owner of the property. In 1893 the residents, wishing a change, accepted Mrs. Rufus F. Herrick's choice of the present name, supposed to be from the local Wiyot Indian language. The Indian name was in fact *katawółoʼt*, but an elderly Indian played a joke on Mrs. Herrick by telling her that the name was *hóš wiwítak* 'let's have intercourse!'—the latter part of which she interpreted, baby-talk-wise, as Loleta (Teeter 1958).

Loma (lō′ mə). The Spanish geographical term designates a low, long elevation or hill, but apparently even in Spanish times it was occasionally applied to higher hills or mountains. It has survived as a semi-generic term in the counties that were in the Spanish domain: with Alta 'high' [Marin, Monterey, Santa Barbara, San Diego Cos.], Pelona 'bald' [San Luis Obispo, Santa Barbara Cos.], Atravesada 'oblique' [Fresno Co.], Chiquita 'little' [Santa Clara Co.], and Verde 'green' [Los Angeles Co.]. **Las Lomas** [Kings Co.], unlike other fancy map names in the Kettleman Hills, is really descriptive of the feature. Las Lomas Muertas [San Diego Co.] is a name that was repeatedly used in Spanish times for bare hills. Loma has also become a favorite specific term, often used for a "Spanish" effect without regard to meaning or idiomatic usage: Del Loma [Trinity Co.], Loma Rica [Yuba Co.], Loma Mar [San Mateo Co.], Oro Loma [Fresno Co.], Casa Loma [Orange Co.], Loma Portal and Loma Alta Mountain [San Diego Co.]. The name **Point Loma** [San Diego Co.], earlier Punta de la Loma 'point of the hill', was applied to the tip of the peninsula because of its promontory. It appears first on Juan Pantoja's *plano* of 1782 (Wagner, p. 395). Wilkes 1841 has Loma Point; the Coast Survey chart of 1851 has Point Loma. The post office was established about 1895. **Loma Prieta** (prē et′ ə, prē ā′ tə) [Santa Clara Co.]: In 1854 the peak was used as a primary triangulation station by the Coast Survey and named in honor of their chief, Alexander D. Bache, superintendent of the survey from 1843 to 1867. The Whitney Survey and the State Mining Bureau accepted the Coast Survey name; but the Land Office maps continued to use the old Spanish name, and this was accepted by the USGS when the New Almaden quadrangle was surveyed in 1916. "The name 'Loma Prieta' is very commonly given by the Spanish-Mexican population to any high chaparral-covered point which looks black in the distance. Mount Bache is one of these 'Black Mountains'" (Whitney, *Geology*, p. 65). A short-lived post office, Loma Prieta, is listed in 1892. In 1989 the peak was the epicenter of an earthquake that had a devastating effect in Santa Cruz, San Francisco, and Oakland. **Loma Linda** [San Bernardino Co.]: When the Colton-Indio sector of the Sunset Route was built in 1875–76, the name Mound Station was applied to the stop, doubtless because of the slight elevation south of it; it is so named on the Land Office map of 1879, but later Southern Pacific maps have Mound City. When the Seventh Day Adventists built a sanatorium in the 1890s and a post office was established on Jan. 14, 1901, the name Loma Linda 'pretty hill', first spelled Lomalinda, was chosen. *For* Laguna de Loma Alta, *see* Mountain.

Lomerias Muertas (lō mə rē′ əs mo͞o âr′ təs) [San Benito Co.]. The elevations north of San Juan Bautista were given this name, meaning 'dead hills', because they are almost entirely without trees. On Aug. 16, 1842, the name was applied to a land grant. Lomerias Muertas are often mentioned in Hoffmann's notes in the summer of 1861.

Lomita (lō mē′ tə). The diminutive of Spanish *loma* 'low, long hill' was rarely used before 1846. One of the names that can be traced to Spanish times is Lomita de las Linares, a hill one mile slightly south of Rucker in Santa Clara Co. From about 1780 to the middle of the 19th century it was a well-known landmark. In 1875 the origin of the name was revealed in the testimony of Justo Larios in Circuit Court case 1397, San Francisco: "The

Lomita de las Linares was named after my grandmother. They were on the road from Monterey to San Jose and passing this spot they saw wild animals, elks, and they went to chase the deer and left my grandmother upon this spot and in that way it came to be named the Lomita de las Linares. . . . They left her upon the lomita so she could not be frightened while they went to chase the deer. Her name was Linares" (*WF* 6:374). Lomita Mountain, 8 miles west of this hill, means literally "little hill mountain." **Lomita** [Los Angeles Co.] also had its origin in Spanish times. The surveyor's plat of the San Pedro Rancho shows a Lomita del Toro, a few miles east of the modern city. Lomita Park [San Mateo Co.] and the many other Lomitas are modern realtors' applications.

Lompico (lôm pē´ kō): settlement, **Creek** [Santa Cruz Co.]. Supposedly coined from Sp. *loma* 'hill' plus *pico* 'peak' (Clark 1986).

Lompoc (lôm´ pōk, lōm´ pok): town, **Hills** [Santa Barbara Co.]. From the Purisimeño Chumash place name *lompo', 'olompo'*, perhaps meaning 'stagnant water' (Applegate 1975:34, after Harrington). A *ranchería Lompoc o Lompocop* under the jurisdiction of Mission La Purisima is recorded in 1791 (Arch. MPC 10). A rancho, Lompoc, is mentioned in 1835 (DSP 4:48–51); the land was granted to Joaquín and José Carrillo on Apr. 15, 1837. In 1874 the California Immigrant Union purchased a portion of the rancho, laid out the town, and sold lots with the condition that the sale and consumption of alcohol be prohibited.

Lone. There are about twelve Lone Rocks, Mountains, Lakes, etc., on the maps of the state, and as many Lone Tree, Pine, Rock (etc.) combinations. There is also a Lonely Gulch near Lake Tahoe. **Lake of the Lone Indian** [Fresno Co.]: In 1902 a party that included Lincoln Hutchinson gave the name to the lake because the mountain above it distinctly shows the profile of the face of a reclining Indian (*SCB* 4:197). **Lone Pine** [Inyo Co.]: The Hill party camped here in 1860 while prospecting the Iowa silver mine, and named the place for a tall pine, which was a landmark for many years until it was undermined by the creek. The name and a picture of the pine are shown on Farley's map of 1861. Other inhabited places having names that include the adjective are Lone Star [Fresno Co.], possibly for Texas, the "Lone Star State"; and Lone Hill [Los Angeles Co.]. The names of the post office Lonoak in San Benito Co. and the Southern Pacific siding Lonoke in Santa Clara Co. seem to have been coined from "lone oak." **Lone Point** [Santa Catalina Island]: The Coast Survey named the cape Long Point, but most maps label it Lone Point. The atlas sheet of the Corps of Engineers (1943) has Long Point, but Lone Point Light. *For* Lone Valley, *see* Ione.

Long. The descriptive adjective "long" is included in the names of more than a hundred features, mainly Valleys, but also Bars, Beaches, Gulches, Canyons, and Lakes. There are a great number of Long Valley Creeks, even a Long Grade Canyon Creek [Ventura Co.] and a Long Pine Canyon Creek [San Bernardino Co.]. There is a Long Dave Canyon and Valley in Ventura Co. There are settlements called Long Barn [Tuolumne Co.] and Longvale (from Long Valley) [Mendocino Co.]. However, Long Lake [Plumas Co.] is probably based on a family name, because it is not at all long. **Long Beach** [Los Angeles Co.] was founded in 1882 by William E. Willmore and named Willmore City. In the "boom year" of 1887, the Long Beach Land and Water Company acquired the interests and applied the new, truly descriptive name.

Longley Pass [Kings Canyon NP] was discovered and named in Aug. 1894: "As the writer [Howard Longley] had been first to reach its summit, the party concluded to call it Longley's Pass, as a means of identification in the future" (*SCB* 1:190).

Lookout. The term is often given to elevations that command a wide view. The state has more than twenty Lookout Mountains, about ten Peaks and ten Points, and a few Ridges, Hills, and Rocks. **Lookout** [Modoc Co.] was named about 1860 after the hill above the town on which, as the story goes, the Pit River Indians had a lookout when the Modocs were on wife-stealing expeditions (Doyle); *see* Lavigia.

Loomis [Placer Co.]. Named in 1884 by the Southern Pacific for Jim Loomis, the local railroad agent, express agent, postmaster, and saloonkeeper. Twenty years before, the station had been named Pino to distinguish it from the nearby settlement of Pine Grove; but the railroad and the Post Office Dept. found that the name was confused with Reno and therefore changed it.

Loomis Peak [Shasta Co.]. Named for Benjamin Franklin Loomis, an early settler, according to the 6th Report of the BGN. He was the author of *Pictorial Lassen* (Steger).

Loon Lake [El Dorado Co.]. Recorded on Hoffmann's map of 1873 and doubtless so named because the loon (*Gavia immer*), an aquatic bird, was found there.

Loraine [Kern Co.]. A settlement was established here by French miners in 1902 and was first named Paris; the name was changed in 1912 to commemorate the Lorraine area of France (Darling 1988).

Lord-Ellis Summit [Humboldt Co.]. For William Lord and Edward Ellis, who were active in having the road built over this summit in the 1880s (Turner 1993). The spelling with the hyphen was prescribed by the BGN in list 7502.

Los (lôs, lōs). Names preceded by the Spanish masculine plural article *los* that are not listed here will be found under the first letter of the name proper; thus Los Coyotes is listed under Coyote. The pronunciation tends to merge with (läs), representing the feminine article *las*.

Los Altos (al′ tōs) [Santa Clara Co.]. The post office was established in 1908 and took the Spanish name, meaning 'the heights', which the subdivider had chosen for the site in 1907.

Los Amoles (ä mō′ lās) **Creek** [Santa Barbara Co.]. Mexican Spanish for 'the soaproot plants' (probably *Chlorogalum pomeridianum*), from Aztec *amulli*.

Los Angeles (an′ jə ləs, ang′ gə ləs): **River**, city, **County**. Hugo Reid (Dakin, p. 220) says that the name of the rancheria on the site of the city was Yang-na; this is apparently Gabrielino *iyáanga'* 'poison oak place'. The Portolá expedition camped on the bank of the river on Aug. 2, 1769, and named it in honor of Nuestra Señora de los Ángeles de Porciúncula, whose feast day they had celebrated the preceding day. The Porciuncula chapel, the cradle of the Franciscan order, is in the basilica of Our Lady of the Angels near Assisi, Italy. Crespí and Costansó simply called the stream Porciúncula. This abbreviated form was commonly used; but Palou, in Dec. 1773, gives the full name: Nuestra Señora de los Ángeles de Porciúncula (Hist. Mem. 3:220). The name was not preserved through the river, but through the name of the pueblo, the projected establishment of which is mentioned on Dec. 27, 1779 (PSP 1:59): *la erección de un pueblo con el título de Reina de los Ángeles sobre el río de la Porciúncula* 'the founding of a town with the name Queen of the Angels on the river of the Porciuncula'. On Aug. 26, 1781, Gov. Neve issued the final instruction for the founding of the town, which took place on Sept. 4, 1781. Although the place was not named for the Angels, but for the Virgin, the most common designation of the future metropolis seems to have been Pueblo de Los Angeles. After the American occupation there was some confusion about the proper use of the name. Stockton datelined his general order of Jan. 11, 1847, from "Ciudad de los Angelos"; and Emory gives the English equivalent of this version, City of the Angels (Mil. Rec. 121). Ord shows the City of Los Angeles on his sketch of Aug. 1849. The present abbreviated version was definitely established when the county was organized on Feb. 18, 1850, and the city incorporated, Apr. 4, 1850. The controversy over the correct pronunciation of the name of the city was carried on in years past with a fierceness worthy of a better cause but seems to have been settled by the younger generation. There are still some enthusiasts who insist upon a "Spanish" pronunciation, not realizing, however, that the Spanish *g*, correctly pronounced, is an entirely foreign sound to Americans. The city will doubtless continue to flourish with its name pronounced euphoniously and rhythmically: (lôs an′ jə ləs). In fact, on Sept. 12, 1952, a jury appointed by the mayor decided that this is the official pronunciation of the metropolis (D. A. Stein, "Los Angeles: A noble fight nobly lost," *Names* 1:35–38 [1953]); however, an alternative pronunciation (an′ jə lēz) is still occasionally heard. The history of the name is discussed by Weber and by Treutlein. Treutlein argues that the name given at the founding in 1781 was "La Reyna de los Angeles." However, it was founded on the Río de la Porciúncula (now the Los Angeles River), which had been discovered by the Portolá expedition in 1769; the full name of the river was the "Río de Nuestra Señora de los Ángeles de Porciúncula," whose feast day they had celebrated the previous day. Some recent writers, says Treutlein, have confused the name of the town with that of the river, erroneously stating that the town was called "El Pueblo de Nuestra Señora la Reina de los

Angeles (de Porciúncula)." He also disputes the arguments of Weber that "la reina" was not originally part of the name. *For* Isla de Los Ángeles or Los Angeles Island, *see* Angel Island.

Los Banos (ban′ əs): **Creek**, town, **Game Refuge** [Merced Co.]. The creek took its name from the pools near its source, called Los Baños ('the baths') del Padre Arroyo, for Padre Felipe Arroyo de la Cuesta, who was at San Juan Bautista Mission from 1808 to 1833. According to tradition, he used to refresh himself in the pools when on missionary trips to the San Joaquin Valley. The pools are called Baños del Padre Arroyo on *diseños* of the San Luis Gonzaga (1841) and Panocha de San Juan (1844) grants. It is possible that these were the "baths" mentioned on Oct. 20, 1798, as los Baños de S. Juan (Prov. Recs. 5:282). About 1870, Gustave Kreyenhagen, a German immigrant, built a store at the old wagon road about 3 miles west of the creek on land leased from Henry Miller (of Miller and Lux). The place became known as Kreyenhagen's and is shown on the maps of the Whitney Survey. A post office was established at the store in 1874 and named Los Banos after the creek. After the railroad was built in 1889, Miller and Lux laid out the present town, which became the headquarters of that famous cattle firm.

Los Bueyes (bo͞o ā′ yəs) **Creek** [Monterey Co.]. Spanish for 'the oxen' (Clark 1991). The same term is reflected as **Los Buellis Hills** [Santa Clara Co.], a respelling of Los Buelles, shown on the *diseño* of the Pueblo Lands of San Jose, 1838.

Los Coyotes (kä yō′ tēz) **Indian Reservation** [San Diego Co.]. The name *coyotes* was applied by the Spanish to Cahuilla Indians in this mountainous area west of the Borrego Desert. *See* Coyote: Coyote Canyon.

Los Feliz (fē′ lis, fə lēs′) [Los Angeles Co.]. The name of the land grant, which included the site of modern Griffith Park, preserves the name of José Féliz (or Félix), mentioned as having cultivated a garden here in 1813 (Bancroft 2:349). After his death the widow married Juan Diego Berdugo; but when the land was granted to her on Mar. 22, 1843, the grant was named Los Felis, referring to señora Féliz and her children of the first marriage. The spelling Los Felis is still found.

Los Gatos. *See* Gato.

Los Machos: Creek, Hills [San Luis Obispo Co.]. Spanish for either 'the masculine ones' or 'the mules'.

Los Molinos. *See* Molino.

Los Nietos (nē et′ ōs): town, **Valley** [Los Angeles Co.]. Manuel Nieto was the grantee of one of the first California land grants, dated Nov. 20, 1784. It is called *rancho de Nieto* (PSP 16:250) and *rancho de los Nietos* (Prov. Recs. 5:262) in 1797. On May 22, 1834, the large rancho (33 leagues) was regranted to his five heirs (*los Nietos*) in five parcels: Los Alamitos, Los Cerritos, Santa Gertrudis, Los Coyotes, Las Bolsas. The name is shown as Nieto on Narváez's Plano (1830) and on Duflot de Mofras's map (1844), and as Los Nietos on the von Leicht–Craven map. The first post office bearing the name was established on June 14, 1867; it was reestablished on May 29, 1891, after the Santa Fe had given the name to the station.

Los Olivos (ō lē′ vōs) [Santa Barbara Co.]. When the Pacific Coast Railroad built the extension from Los Alamos about 1890, the terminal was called Los Olivos because of the extensive plantings of olive cuttings made there by the Hayne brothers. The immense olive orchard disappeared, but the name remained.

Los Padres National Forest [Monterey, San Luis Obispo, Santa Barbara, Ventura Cos.]. Named in 1936 by order of President Franklin D. Roosevelt, to commemorate the Franciscan padres who founded the California missions, eight of which are in or near the forest area. The present name replaces the name Santa Barbara National Forest, which had been given by order of President Theodore Roosevelt in 1903 to a combination of forest reserves; it was extended in 1910 to include San Luis Obispo National Forest, and in 1919, Monterey National Forest.

Lospe (lōs′ pē) **Mountain** [Santa Barbara Co.]. From Purisimeño Chumash *lospe* 'flower' (Applegate 1975:34, after Harrington).

Los Penasquitos (pen əs kē′ tōs): **Canyon, Creek** [San Diego Co.]. The name Peñasquitos 'small crags' appears in the *expediente* of the land grant of Santa María de los Peñasquitos, June 15, 1823, and is mentioned as Panasquitas by Emory in 1849. Von Leicht–Craven (1873) have Penasquito Creek, but the Coast Survey and USGS used the plural form.

Los Serranos (sə rä′ nōs) [San Bernardino Co.]. Spanish for 'the mountain people' (from *sierra* 'mountain range'); the term is applied to an Indian tribe of the area.

Lost. The adjective "lost" in a place name generally indicates an incident or story behind the name. There are almost a hundred such names in the state, mostly Creeks and Rivers that suddenly go underground, or Meadows, Hills, Mines, Canyons, etc., that were discovered, "lost" for a time, and later found again. Sometimes a feature because of its isolation or seclusion may seem to be lost. There are also a number of names referring to something that was actually lost, e.g. Lost Horse (Cow, Man, etc.) Creeks. In Humboldt Co. there is a Little Lost Man Creek as well as a Lost Man Creek. El Dorado Co. has a Lost Corner Mountain; Riverside Co., where the groups of Washington palms have provided many place names, a Lost Palm Canyon. **Lost Cannon Creek, Canyon**, and **Peak** [Mono Co.] were so named on the erroneous assumption that Frémont abandoned his howitzer here in 1844 (Maule). **Lost Wagons** [Death Valley NP]: In the summer of 1889, C. B. Zabriskie was obliged to abandon one or two wagons of the Pacific Coast Borax Company here. The place and name have been associated with the burning of the wagons of the Jayhawker party of 1849, which actually occurred much farther south. **The Lost Arrow** [Yosemite NP] has been said to be a translation of the Miwok Indian *ummo, hammo,* or *ummoso*. As Catherine Callaghan points out (p.c.), this is in fact probably a man's proper name in Central Sierra Miwok, transcribed by Gifford (AAE 12:150) as *Omusa* 'missing things when shooting with arrows'. **Lost Hills** [Kern Co.]: In 1910 the town took its name from the so-called Lost Hills, slight elevations that seem to belong to the Kettleman Hills but look as if they were "lost." **Lost River** [Modoc Co.] is literally lost between Clear Lake and Tule Lake; reclamation engineers have tried in vain to find an underground flow. **Lost Islands** [San Diego Co.] is the common name for the two submarine mountains west of the Coronados, called Cortez Bank and Tanner Bank by the Coast Survey (*see* Cortes Bank). **The Lost Coast** [Humboldt Co.] was the name given to the area around Shelter Cove when the area became depopulated around 1937 (Turner 1993). Note also Lost Lake [Riverside Co.]. *For* Lost Valley [Fresno Co.], *see* Blaney Meadows.

Los Trancos. *See* Las Trancas; Tranca.

Lotus [El Dorado Co.]. The place was first named Marshall for James W. Marshall, the discoverer of gold at Sutter's mill. After the admission of the state into the Union in 1850, the name was changed to Uniontown. When the post office was established on Jan. 6, 1881, the new name was chosen, probably to avoid confusion with Union House [Sacramento Co.]. There was no other Uniontown in the state at that time; Uniontown in Humboldt Co. had been Arcata since 1860. According to W. T. Russell (Knave, July 23, 1944), the first postmaster, George E. Gallaner, suggested the new name because, in his opinion, the inhabitants were as easygoing as the lotus-eaters in the *Odyssey*.

Louisville. *See* Greenwood.

Love. A number of geographic features in various parts of the state bear names like Love, Lovejoy, Loveland; but usually they were named for persons (as Lovelady and Lovelock were) and have nothing to do with tender emotions. However, El Dorado, Santa Clara, and Placer Cos. each have a Lovers Leap. This is a favorite American name for a cliff, not because a lover ever jumped from one so named, but because it would be "a good place to take off, if anyone wanted to" (Stewart, p. 129). By decision of the BGN, 1961, the name of Point Aulon (*see* Abalone) was changed to Lovers Point.

Love Creek [Santa Cruz Co.]. Named for Henry (Harry) Love, captain of the California Rangers who killed one of the bandits called Joaquin Murieta in 1853.

Lovelady Ridge [Mendocino Co.]. Named for Joshua Lovelady, who settled at the foot of the ridge in the 1860s.

Lovelock [Butte Co.]. The place was originally named Lovelocks, for George Lovelock, who founded the mining town in 1855 by opening a hotel and store (Co. Hist.: Butte 1882:260).

Lovitos, Arroyo de. *See* Lobitos.

Low Divide. *See* Alta: Altaville.

Lowe, Mount [Los Angeles Co.]. The peak was named in honor of Thaddeus S. C. Lowe by his companions on the first horseback ascent, Sept. 24, 1892. Lowe (1832–1913), a versatile scientist and inventor, was chief of the Aeronautic Corps of the U.S. Army in the Civil War. "It was discovered that until this time this giant peak, the monarch of the Sierra Madres, was unnamed. One of the party suggested, that whereas Prof. T. S. C. Lowe, the great scientist, had first ridden to the top, had made the first trail to its lofty summit, was the first man to have planted the stars and stripes on its highest point, and was the first man to conceive the project of reaching its dizzy height with a railroad and with courage and means to put such a project into execution, as was now being done, no more fit and appropriate name could be given this mountain than the name of 'Mount Lowe'" (Auglaize, Ohio, *Republican*, Dec. 1, 1892; quoted by Reid, p. 445).

Lower Lake [Lake Co.]. The town was established in 1860 and named from its proximity to the southeastern tip of Clear Lake, at that time generally known as Lower Lake. The post office is listed in 1859. The adjective "lower" is used elsewhere in the state: Lower Coon Mountain [Del Norte Co.], Lower Alkali Lake [Modoc Co.], Lower Salmon Lake [Sierra Co.], Lower Sardine Lake [Sierra Co.], Lower Soda Springs [Shasta Co.].

Lower Otay Dam. *See* Savage Dam; Otay.

Loyalton [Sierra Co.]. Named by the Post Office Dept. in 1863, at the suggestion of the Rev. Adam G. Doom, the first postmaster, who considered the name expressive of the strong Union sentiment of the place. The settlement was previously known as Smiths Neck, perhaps because the Smith Mining Company was operating there at that time.

Loyd Meadows [Tulare Co.]. Named for John W. Loyd, who ran sheep there in the 1870s (Farquhar).

Lubeck Pass [San Bernardino Co.]. Named for Bill Lubeck, who settled on the Colorado River in pre-railroad days (*Desert Magazine*, Apr. 1941).

Lucas Valley [Marin Co.]. Probably named for John Lucas, who inherited the rancho of which the valley is a part from his uncle, Timothy Murphy.

Lucerne. The name of the medieval town and the beautiful lake in Switzerland, the locale of Wilhelm Tell's exploits, is a popular place name in the United States. There are fourteen post offices (often spelled Luzerne) and a great number of lakes and other features so called. California has its share with at least four Lucernes—one of which, however, was named for the European name of our alfalfa, rather than for the Swiss city. **Lake Lucerne** [San Mateo Co.]: The name of the lake in Switzerland replaces the former name Bean Hollow Lagoon, an American version of the Spanish name, Laguna del Arroyo de los Frijoles (Wyatt). Bean Hollow Lake is now the name of the reservoir farther upstream. **Lucerne Valley** [Kings Co.]: The reclaimed district was known by the name of the stream that traverses it, Mussel Slough. Frank L. Dodge, editor of the *Hanford Weekly Sentinel*, and other boosters disliked the name "slough" because to Easterners the name still suggests a muddy backwater. In an article of Apr. 21, 1887, the district was rechristened Lucerne Valley because the distant mountains, the glittering Tulare Lake, and the richness of the soil made the reclaimed tule land "eminently worthy to be a namesake of that old, rich and venerable Lucerne of Europe" (*see* Mussel: Mussel Slough). **Lucerne Valley** [San Bernardino Co.]: James E. Goulding settled in the valley in 1897. When questioned about a name for it, Goulding said, "Well, if we get water, I know from past experience that we can raise plenty of alfalfa in this country, or lucerne, as the Mormons called it in southwest Colorado. Why not call it Lucerne Valley?" (quoted in *CHSQ* 27:120). In 1912 Dr. F. J. Gohar and his family homesteaded at the site of the present town and found that lucerne, or alfalfa, actually could be grown successfully in the valley (Belden, July 26, 1963). A post office was established on Sept. 9, 1912. **Lucerne** [Lake Co.]: The shore of Clear Lake, which presumably resembled the Swiss lake, suggested the name for the hotel and post office established on July 2, 1926.

Lucia (lo͞o sē′ ə) [Monterey Co.]. The post office, now discontinued, was established about 1900 and was obviously named because the settlement is at the foot of the Santa Lucia Mountains.

Ludlow [San Bernardino Co.]. Named in the 1870s by the Central Pacific (now Southern Pacific), for William B. Ludlow, master carrepairer of the Western Division. The name of the post office was Stagg from 1902 to 1926.

Luffenholtz: Creek, settlement [Humboldt Co.]. Named for the mill owner, who settled here in the spring of 1851. His name was originally spelled Luffelholz. On the county map of 1886 the stream is labeled Lutfenholts Creek.

Lugo [Los Angeles Co.]. The station of the Pacific Electric commemorates Antonio María Lugo (1778–1860), on whose Rancho San Antonio the place is situated.

Lugo [San Bernardino Co.]. It is not certain for which of the numerous Lugos the Santa Fe station was named. Three members of the Lugo family were grantees of San Bernardino Rancho, some 18 miles south of the station. The district formerly called Lugonia was on part of the San Bernardino grant; the name was coined from the family name in 1877 (*see also* Red: Redlands).

Lukens (lo͞o′ kənz) **Lake** [Yosemite NP]. Named in 1894 by R. B. Marshall, for Theodore P. Lukens, mayor of Pasadena at that time; he was a noted advocate of reforestation. **Mount Lukens** [Los Angeles Co.], referring to the same man, is the common designation for Sister Elsie Peak (*q.v.*).

Lunada (lo͞o nä′ də) **Bay** [Los Angeles Co.]. The Spanish word, meaning 'formed like a half-moon', was applied in modern times to the crescent-shaped bight.

Lundy: Lake, Canyon, settlement [Mono Co.]. William O. Lundy obtained a timber patent here in 1880. The settlement near the May Lundy Mine was first known as Mill Creek.

Lutfenholts Creek. *See* Luffenholtz Creek.

Luther Pass [El Dorado, Alpine Cos.]. Named for Ira M. Luther, who crossed the pass in a wagon in 1854 and was active later in an attempt to route the Central Pacific over it.

Luzon (lo͞o zon′) [Contra Costa Co.]. The name of the major island of the Philippines was given to the station of the San Francisco and San Joaquin Valley Railroad at the time of the Spanish-American War, when Philippine place names were popular (Santa Fe).

Lyell (lī′ əl): **Mount; Lyell Canyon, Fork** (of the Merced), **Fork** (of the Tuolumne), and **Glacier** [Yosemite NP]. The peak was named in 1863 for the English geologist Charles Lyell (1797–1875) by Brewer and Hoffmann of the Whitney Survey: "As we had named the other mountain Mount Dana, after the most eminent of American geologists, we named this Mount Lyell, after the most eminent of English geologists" (Brewer, p. 411).

Lynwood [Los Angeles Co.]. The name was first applied to a dairy, for Lynn Wood Sessions, the wife of the owner. The station and subdivision were named after the dairy (W. E. Wellinger). The place was incorporated on July 16, 1921.

Lyons: Valley, Peak [San Diego Co.]. The names of these features preserve the name of Gen. Nathaniel Lyon, who was instrumental in keeping Missouri in the Union at the beginning of the Civil War and was killed in the battle of Wilson's Creek on Aug. 7, 1861. In the early 1850s he was stationed in San Diego and acquired a section of land in the valley. *See* Battle: Battle Island.

Lyons Peak [Siskiyou Co.]. Named by the USFS for George Washington Lyons, supervisor of the Modoc National Forest, who died of injuries incurred while he was on duty (William S. Brown).

Lyons Springs [Ventura Co.]. Named for Gertrude A. Lyons, owner of the property (WSP 338:63).

Lytle (lī′ təl) **Creek** [San Bernardino Co.]. Named for Capt. Andrew Lytle, a former officer of the Mormon Battalion, who led a division of the Latter-Day Saints into San Bernardino Valley in 1851.

Lytton [Sonoma Co.]. Named for Captain Litton, who developed the place as a resort in 1875. The name was misspelled when the Northwestern Pacific applied it to a station on the railroad built in the 1880s.

M

Maacama. See **Mayacmas**.

McAdams Creek [Siskiyou Co.]. Named for a Scot who mined at the creek in 1854 (Co. Hist.: Siskiyou 1881:216).

McAdie (mə kā′ dē), **Mount** [Sequoia NP]. Named in 1905 in honor of Alexander G. McAdie, who was forecast official (1895–1902) and in charge (1903–13) of the U.S. Weather Bureau, San Francisco. He was professor of meteorology, Harvard University, and director of the Blue Hill Observatory from 1913 to 1931. "Our party had the honor of naming the peak directly south of Lone Pine Pass Mt. McAdie, to commemorate your services in advancing the science of climatology" (J. E. Church, *SCB* 5:317).

McAfee (mak′ ə fē): **Creek, Meadow** [Mono Co.]. Named for A. G. McAfee, a rancher who settled in Fish Lake Valley in 1864 (Robinson).

McAlpine. *See* Calpine.

McArthur [Shasta Co.]. Named about 1896 by the John McArthur Company for John McArthur, who had settled in the Pit River Valley in 1869 and owned most of the land in the district.

MacArthur, Fort [Los Angeles Co.]. On Jan. 10, 1914, the War Dept. named the fort in honor of Lt. Gen. Arthur MacArthur (1845–1912), a veteran of the Civil and Spanish-American Wars and father of Gen. Douglas MacArthur.

McArthur-Burney Falls State Park [Shasta Co.]. The original 160 acres were given to the state in 1920 by Frank McArthur, in memory of his mother, Catherine Clark McArthur (Steger). *See* Burney.

McCabe: Lakes, Creek [Yosemite NP]. Named in 1900 for Lt. Edward R. W. McCabe, 17th Infantry.

McCall Creek [Shasta Co.]. The tributary of Dog Creek was named for John McCall, who owned 60 acres of mining claims on this creek (Steger).

McCann [Humboldt Co.]. The name was applied in 1881 to a station on the stage route, for an old settler who had a mill here (L. M. Armstrong). The settler was probably Willard O. McCann, who came to the county in 1869.

McCarthysville. *See* Saratoga.

McClellan Air Force Base [Sacramento Co.]. Named in 1939 by the War Dept. in memory of Maj. Hez McClellan of the Army Air Corps.

McCloud: River, town [Siskiyou, Shasta Cos.]. The river took its name from Alexander R. McLeod, leader of a Hudson's Bay Company brigade that trapped in California in 1828–29. The name is spelled McLoud on Reading's map and in Trask's Report (1854:45), and McCloud by Beckwith in 1854 in the Pac. R.R. *Reports* (2:2.54), doubtless because the name McLeod is ordinarily so pronounced. The maps of Butler 1851 and Eddy 1854 have the correct spelling, McLeod's Fork. After Ross McCloud settled here in 1855 and became prominent in the development of the region, the river was usually associated with his name. Various spellings were used until the Whitney Survey decided for McCloud (von Leicht–Craven 1874). Nearby McCloud Reservoir was renamed McCloud Lake by the BGN in list 6903.

McClure Lake [Madera Co.]. Named for Lt. Nathaniel F. McClure, 5th Cavalry, stationed in Yosemite National Park 1894–95 (Farquhar). Probably named by R. B. Marshall in 1899.

McClure Meadow [Kings Canyon NP]. Named for Wilbur F. McClure, state engineer, 1912–26, in recognition of his assistance in building the John Muir Trail (*SCB* 10:86). **McClure, Lake** [Mariposa Co.]: Named in 1927 by the Merced irrigation district, in memory of the same McClure, who had died in the preceding year.

McClure, Mount [Yosemite NP]. *See* Maclure, Mount.

McColl [Shasta Co.]. The station was named in memory of John McColl, state senator from Shasta and Trinity Cos., 1932–38 (Steger).

McComber Lake [Shasta Co.]. In the late 1850s George W. McComber settled on the meadow along Battle Creek that became known as McComber Flat. A dam was built in 1906; the lake formed by the inundated meadow kept the name McComber.

McConnell [Sacramento Co.]. The station was named by the Central Pacific in the 1870s for Thomas McConnell, a native of Vermont who had settled in the county in 1855 and engaged in raising sheep.

McConnell State Park [Merced Co.]. In 1949 the land for the park was given to the state by Mr. and Mrs. Warren F. McConnell.

McCoy: Mountains, Spring [Riverside Co.]. Probably named for James McCoy, a native of Ireland who came to San Diego in 1850 as a soldier, was sheriff of San Diego Co. (1861–71), and state senator (1871–74).

MacCullough (mə kul′ ə): **Lake, Valley** [Lake Co.]. Named for H. V. S. MacCullough, who settled in the county and had a homesite in the valley before 1900 (Mauldin).

McDairmids Prairie. *See* Metropolitan.

Macdoel (mak dō′ əl) [Siskiyou Co.]. When the Southern Pacific extension from Weed to Klamath Falls was built in 1906, the station was named for the owner of the land, William MacDoel.

McDonald Creek [Sonoma Co.]. Named for William McDonald, who came to Napa Valley in 1847 and built a house on the banks of this creek in 1850 (*WF* 6:372).

McDonald Lake [Trinity Co.]. Named for Warren P. McDonald because of his aid in the conservation of wildlife (BGN, May 1954).

McDowell, Fort [Angel Island] was named by the War Dept. in 1900 in honor of Irvin McDowell, commander of the Union army at the first battle of Bull Run, and later of the Dept. of the Pacific.

McDuffie, Mount [Kings Canyon NP] was named by the Sierra Club in memory of Duncan McDuffie, mountaineer and conservationist, who died on Apr. 21, 1951. The Duncan and Jean McDuffie Grove in Prairie Creek Redwoods State Park was established in 1951 through the efforts of the Save-the-Redwoods League; the Sierra Grove bearing their names was established in the Calaveras Big Trees State Park in 1959.

McFarland [Kern Co.]. Named in 1908 for J. B. McFarland, one of the founders of the town.

McGee: Creek, Canyon, Mountain, Meadow [Mono, Inyo Cos.]. The names commemorate the four McGee brothers, early cattlemen in Mono and Inyo Cos., of whom Alney ("Allie"), John, and Bart played important roles in pioneer days. The tributary of Owens River in Long Valley [Mono Co.] became known as McGee Creek because the brothers had their headquarters there. McGee Creek west of Bishop received its name because John McGee, at one time sheriff of Inyo Co., owned a ranch where the creek debouches upon the valley (Brierly).

McGee, Mount; McGee Lakes [Kings Canyon NP]. Named in memory of William J. McGee (1853–1912), American geologist and anthropologist.

McGill's. *See* Miguel.

McGinty: Point, Reservoir [Modoc Co.]. The point was named about 1917 for J. B. McGinty, a nearby homesteader (William S. Brown).

McGonigle Canyon [San Diego Co.]. Named in the early 1870s for Felix McGonigle, an early settler and landholder in this district. It was formerly known as Cordero Valley and Canyon, after the Cordero brothers, Spanish soldiers in the Portolá expedition.

Machado (mə chä′ dō) **Creek** [Santa Clara Co.] was named for a pioneer family.

Machesna (mə ches′ nə) **Mountain** [San Luis Obispo Co.]. Apparently a Spanish version of the name of the McChesney family, early settlers (Hall-Patton).

Macho. The Spanish word for 'masculine' also means 'mule'; it is rare in California place names, but *see* Mocho.

McIntyre Creek [Tulare Co.]. Named for Thomas McIntyre, a pioneer of the county, who ran sheep here in the 1880s (Farquhar).

McKee, Mount. *See* Konocti.

MacKerricher Beach State Park [Mendocino Co.]. The land for the park was a gift from the MacKerricher family, whose forebears Duncan and Jessie came from Canada and settled on it in 1864.

McKinley Grove [Fresno Co.]. When the USGS mapped the Kaiser quadrangle in 1901–2, R. B. Marshall named the grove of big trees in memory of President William McKinley (1843–1901), who had died the victim of an assassin. **McKinleyville** [Humboldt Co.] was named in honor of President McKinley after his assassination.

McKinney: Creek, Bay [Placer, El Dorado Cos.]. Named for John McKinney, miner, hunter, and trapper, who came to El Dorado Co. in the early 1850s and later opened one of the first Lake Tahoe resorts, on the southern shore of McKinney Bay (W. T. Russell in Knave, July 23, 1944).

McKinstry: Peak, Lake, Meadow [Placer Co.]. The mountain is labeled McKinstry's Mountain on Bowman's map (1873), and M'Kinstry Peak on sheet 47-D of Wheeler's atlas. It has been suggested that the name may commemorate George McKinstry Jr., who was one of the best known of Sutter's men and the first sheriff of the northern district (1846–47); it seems more likely, however, that the places were named for Elliott McKinstry, a miner at Oregon Bar in 1866, or for Lee McKinstry, a miner in Georgetown township [El Dorado Co.] in 1867.

McKittrick: town, **Valley, Field, Summit** [Kern Co.]. When the Southern Pacific built the extension from Bakersfield to the asphaltum beds in the early 1890s, the terminal was named Asphalto. When the oil fields were developed, the railroad extended the tracks in 1900 and named the new station for Capt. William H. McKittrick, son-in-law of Gen. William Shafter, the owner of the land.

McLaren Park [San Francisco Co.]. Named on Apr. 14, 1927, by action of the Board of Park Commissioners, in honor of John McLaren (1846–1943), superintendent of San Francisco parks from 1890 to 1943. **McLaren Meadows** [Contra Costa Co.] was named in 1943 for John McLaren by resolution of the East Bay Regional Park Board, in recognition of his assistance in creating the local regional parks.

McLean's River. *See* Clearwater.

McLeod's Fork; McLoud River. *See* McCloud River.

Maclure, Mount; Maclure: Fork, Creek, Glacier, Lake [Yosemite NP]. The Whitney Survey named the mountain in 1868, in honor of the Scottish-American geologist William Maclure (1763–1840): "To the pioneer of American geology . . . one of the dominating peaks of the Sierra Nevada is very properly dedicated" (Whitney, *Yosemite book*, 1869: 101). In 1901 the USGS applied the name to this branch of the Lyell Fork of Tuolumne River. The other features were named in 1932 by the BGN. On some maps the name is erroneously spelled McClure. *For* Maclure Fork, *see* Lewis, Mount: Lewis Creek.

McNears: Landing, Point [Marin Co.]. This point was known as Point San Pedro from 1811. When an important fishing industry was developed here by Chinese in the 1870s (Co. Hist.: Marin 1880:346–47), the landing and later the point became known as McNears, for the firm of McNear and Brothers, who owned the land.

Macomb (mə kōm´) **Ridge** [Yosemite NP]. For Lt. Montgomery M. Macomb of the Wheeler Survey, who mapped the region in 1879.

McPherson [Orange Co.]. Named for its founders, the brothers Robert and Stephen McPherson. The post office operated from 1886 to 1900 (Brigandi).

McQuaide, Camp [Santa Cruz Co.]. Established by the War Dept. on Sept. 17, 1940, as a special training center and named in honor of Joseph P. McQuaide, a chaplain in the Spanish-American War and in World War I (Shirley Potter).

Mad. Among miners and settlers this adjective was a favorite descriptive term in place naming, used in the sense of 'crazy, enraged, angry'; it was usually applied because of an incident. Several of these names have survived. **Mad River** [Humboldt, Trinity Cos.] was named in Dec. 1849 by members of an exploring party led by Dr. Josiah Gregg. The leader became very angry with his companions for not waiting for him when he wished to determine the latitude of the mouth of the river: "As the canoes were about pushing off, the Doctor . . . hastily caught up his instruments and ran for the canoe, to reach which, however, he was compelled to wade several steps in the water. His cup of wrath was now filled to the brim; but he remained silent until the opposite shore was gained, when he opened upon us a perfect battery of the most withering and violent abuse. Several times during the ebullition of the old man's passion, he indulged in such insulting language and comparisons, that some of the party . . . came very near inflicting upon him summary punishment by consigning him, instruments and all, to this beautiful river. Fortunately for the old gentleman, pacific councils prevailed. . . . This stream, in commemoration of the difficulty I have just related, we called Mad River" (Wood 1932:12). Mad River is mentioned in the *Statutes* of 1851 (p. 179) and was apparently put on the map by Gibbes in 1852. **Mad Canyon Creek** [Placer Co.]: The tributary of the Middle Fork of American

River was named after the Mad Cañon Diggings, a mining camp mentioned in the *Sacramento Daily Transcript* of Mar. 15, 1851. **Mad Mule Canyon, Mad Ox Canyon** [Shasta Co.]: The incidents that led to the naming of the two canyons on Whiskey Creek are not known. The former was known as a goldmining site in 1851 and was a voting precinct in 1853 (Steger); the latter is mentioned in the *Shasta Courier*, May 29, 1852.

Maddox. *See* Robbins.

Maddox, Mount [Kings Canyon NP]. Named in memory of Ben M. Maddox of Visalia, who died in 1933 (Farquhar).

Madeline: Plains, post office [Lassen Co.]. The name commemorates a little girl by that name, killed when a party of immigrants were attacked by Indians in the early 1850s (H. E. Risdon). It was first applied to the pass west of Mud Lake [Nevada Co.] and appears in the Pac. R.R. *Reports* as Madelin Pass. The plains are shown by von Leicht–Craven with the present spelling. The post office was named in the 1870s.

Madera (mə dâr′ ə): city, **County, Peak**. When the California Lumber Company built a flume from the forest area to the railroad in 1876, the pleasant-sounding Spanish word for 'wood' was chosen as an appropriate name for the new lumber town. On Mar. 11, 1893, the part of Fresno Co. north and west of the San Joaquin River was organized as a new county and named after the town. The name of Black Peak was changed to Madera Peak by the BGN (6th Report) at the request of various organizations of the county. **Madera Creek** [Santa Clara Co.] is shown as Arroyo del Matadero 'slaughtering place' on *diseños* of the 1830s and 1840s, and on American maps until the 1870s. The name may have been changed accidentally because of the similarity in sound. *See also* Corte Madera.

Madison [Yolo Co.]. In 1877, when the Vaca Valley and Clear Lake Railroad extended its line from Winters to a point one mile beyond the old trading center of Cottonwood, a new townsite was laid out at the terminus and named Madison (Co. Hist.: Yolo 1879:71). The name was not given in honor of our fourth president; it was applied in 1870 by Daniel Bradley Hulbert, a native of Madison, Wisconsin, according to William E. Asiston.

Madonna, Mount [Santa Clara Co.]. The Italian designation for the Virgin Mary was applied to the peak west of Gilroy by Hiram Wentworth, a pioneer of the region (Doyle).

Madora (mə dôr′ ə) **Lake** [Plumas Co.]. The origin has not been discovered.

Madrone (mə drōn′). The common designation for one of our most beautiful native trees (*Arbutus menziesii*) is derived from its Spanish name, *madroño*, and is found in numerous place names, chiefly in the mountains, foothills, and gravelly valleys of the Coast Ranges, the principal habitat of the tree. "Madrona" is a common variant of the name.

Madulce (mə do͞ol′ sē) **Peak** [Santa Barbara Co.]. Named with a California Spanish term for 'wild strawberry'; formerly also called Strawberry Peak (John Johnson).

Magalia (mə gāl′ yə): post office, **Reservoir** [Butte Co.]. The settlement was started in 1850 by E. B. Vinson and Charles Chamberlain, and became known as Dogtown. In 1862, A. C. Buffum, a citizen, supposedly proposed the Latin word for 'cottages' as a more fitting name for a town that, he declared, was after all a town of cottages and not dog houses (Doyle). The new name was probably applied when the post office was transferred from Butte Mills on Nov. 14, 1861.

Magee Peak [Shasta Co.]. Named for William Magee, a deputy to the U.S. surveyor general, who surveyed this district in the 1860s.

Maggie, Mount [Tulare Co.]. Named in the 1870s by Frank Knowles, for Maggie Kincaid, a schoolteacher in Tulare Co. (Farquhar).

Magnesia Spring Canyon [Riverside Co.]. The canyon was named after the magnesia spring in the canyon that opens into Coachella Valley.

Magnolia [San Bernardino Co.] refers to a flowering tree native to the southeastern United States but widely planted as an ornamental in sourthern California. The name Magnolia Villa was formerly applied to Upland (*q.v.*). *For* Magnolia in Kern Co., *see* Inyokern.

Magra (mag′ rə) [Placer Co.]. The origin of this place name has not been discovered.

Mahala (mə hä′ lə) **Creek** [Humboldt Co.]. The term *mahala* was once used in California English to mean 'Indian woman'; it may be from an Indian pronunciation of Spanish *mujer*, or from a Yokuts term such as Chowchila *mokheelo* 'woman', Yawelmani *mokhelo* 'old woman' (Gamble).

Mahews [Sacramento Co.]. *See* Mayhew.

Mahnke (mang′ kē) **Peak** [Lake Co.]. The mountain on which Bear Creek rises was named for the pioneer family Mahnke, who killed three bears there, one as recently as 1946 (Mauldin).

Mahogany Flat [Death Valley NP]. Named by Col. J. R. White for the desert mahogany (perhaps *Cercocarpus intricatus*), a shrub or scraggy tree native in our desert regions.

Maidens Grave [Amador Co.] is believed to be the burial site of Rachel Melton, a girl from Iowa, who came to California in a covered wagon in 1850.

Mailliard [mä′ yärd] **Redwoods State Park** [Mendocino Co.] was named in 1954 for John Ward Mailliard Jr., a distinguished philanthropist and conservationist, who died on July 11, 1954.

Main Canal [Stanislaus, Merced, Fresno Cos.]. By decision of the BGN, 1963, this is the official name for the important irrigation canal hitherto known as Old, San Joaquin, or Kings River Canal.

Malacomas. *See* Mayacmas; Saint Helena, Mount.

Malaga (mal′ ə gə) [Fresno Co.]. The post office was established about 1885 and named for the malaga grape, which is grown commercially in the district; this is named in turn for the city of Málaga in Spain. *For* Malaga [Los Angeles Co.], *see* Malibu.

Malakoff (mal′ ə kôf) **Diggins State Park** [Nevada Co.]. The state park, created in 1966, commemorates one of the richest gold diggings in the state. The original mine was opened in 1855, during the Crimean War, and was named after the Malakoff Tower near Sebastopol, Russia.

Malapai (mal′ ə pī) **Hill** [Joshua Tree NM]. The name is an Americanism derived from *malpais*, American Spanish for 'badlands' in lava-bed country. Miners used the word also for basaltic formations, and this hill was apparently named for its basaltic rocks. **Malapi Spring** is in Inyo Co.; **Malpais Creek** is in Tulare Co.

Malibu (mal′ i bo͞o): **Creek, Point, Lake, Beach** [Los Angeles Co.]. The origin of the name is in the name of a rancheria (probably Chumash) Umalibo, which was under the jurisdiction of Mission San Buenaventura (Engelhardt, p. 166). The present spelling appears in the name of the Topanga Malibu Sequit grant, dated July 12, 1805. A spelling variant is recorded on Dec. 31, 1827, *la cierra de Maligo* (DSP 2:49), and the rancho is called Malago in the *Statutes* of 1851 (p. 172). Similar variants persist until about 1880 (Goddard 1860: Malico; Wheeler atlas sheet 73-C, 1881: Malaga). On the county map of 1881 the spelling Malibu is restored. The Chumash original may have been *(hu-)mal-iwu* 'it makes a loud noise all the time over there', referring to the surf (Beeler 1957:237–38).

Malki (mäl′ kē) [Riverside Co.]. The Wanikik or Pass Cahuilla Indian name for the Banning district, applied to the Morongo Indian Reservation since 1908 (Gunther), where Malki Museum now attracts many visitors.

Mallacomes. *See* Mayacmas; Saint Helena, Mount.

Mallory, Mount [Sequoia NP]. Named in memory of George H. L. Mallory, a member of the British Mount Everest expeditions of 1921, 1922, and 1924, who was lost in June 1924 after reaching a height above 28,000 feet. The name was proposed by Norman Clyde, who made the first ascent of Mount Mallory in July 1925.

Malpaso (mal pas′ ō): **Creek, Canyon** [Monterey Co.]. Applied by the Coast Survey from Arroyo de Mal Paso 'creek difficult to cross', shown on a *diseño* of the Sur Chiquito grant, dated Apr. 2, 1835. **Mal Pass, Mallo Pass Creek** [Mendocino Co.]: When the Coast Survey charted the sector in 1869–70, the name Mal Passo was applied to the steep gulch, probably because the surveyors knew the creek in Monterey Co. and because this one was really "mal paso."

Mama Pottinger Canyon. *See* Pottinger.

Mammoth. The descriptive name of about fifteen geographic features and places in the state, including Mammoth Peak [Tuolumne Co.]. A number are survivals of mining-boom days, when the name was a favorite. **Mammoth Peak** [Yosemite NP]: "A very high and massive peak was seen to the east of Mount Lyell, which it nearly equalled in altitude; it was called Mammoth Mountain" (Whitney, *Geology*, p. 401). This is probably the peak later named Mount Ritter. It is possible that the USGS, when the Mount Lyell quadrangle was mapped, transferred the name to the present Mammoth Peak, which is due north of Mount Lyell. **Mammoth: Mountain, Lakes, Creek, Pass, Crest** [Mono, Madera, Fresno Cos.]: The cluster recalls the big boom town Mammoth City, which flourished briefly after the organization of the Mammoth Mining Company on June 3, 1878.

Manchester [Mendocino Co.]. The post office was established in the 1870s and was probably named after one of the thirty-five Manchesters that already existed in the United States.

Mandeville Island [San Joaquin Co.]. Named probably for James W. Mandeville, former assemblyman, state senator, U.S. surveyor general, and state controller.

Man Eaten Lake [Siskiyou Co.]. This is probably the lake called in Karuk *ára u'ipamvâanatihirak* 'the place where a person ate himself long ago'. The name refers to a story about a man who, in a fit of hunger, ate his entire family and finally devoured his own flesh; at the end he became a roving skeleton, still seeking food.

Mangalar (mang gə lär′) **Spring** [San Diego Co.]. For Mexican Spanish *manglar* 'place of the sumac', from *mangle* 'sumac' (Stein).

Manhattan Beach [Los Angeles Co.]. Named in 1902 after New York's Manhattan Island at the suggestion of Stewart Merrill, founder of the town. The place was formerly known as Shore Acres, a name given to the station established by the Santa Fe (Edna Alterton). The New York name is Algonquian, referring to a local tribe.

Manikin: Creek, Flat [Tulare Co.]. The places were named for an early settler by that name (Jesse Pattee). The Great Register of 1872 lists James Henry Mankins, a native of Arkansas, farmer at Venice.

Maniobra (mä nē ō′ brə) **Valley** [Imperial Co.]. The name is Spanish for 'handiwork; manoeuvre'.

Maniz, Rancho de. *See* Muniz.

Manly: Peak, Pass, Fall [Inyo Co.]; **Peak, Beacon** [Death Valley NP]. The features in Inyo Co. commemorate William Lewis Manly, who played a prominent role in one of the Death Valley expeditions of 1849. In 1936 the National Park Service, dissatisfied with having Manly's name on only a minor peak, changed the name of the double peak marked Baldy on the Ballarat atlas sheet. It designated the north peak as Manly Peak, for Manly, and the south peak as Rogers Peak, for John Rogers, as a fitting tribute to the two men who led their starving party out of Death Valley.

Mann: Ridge, Gulch [Alameda Co.]. Named for George Mann, who settled in the district in 1855. Also misspelled as Man.

Manresa (man rē′ sə) [Santa Cruz Co.]. Once the site of Villa Manresa, a religious retreat operated by the College of Santa Clara. Named after Manresa, Spain, where St. Ignatius Loyola developed ideas that led to his forming the Society of Jesus (Clark 1986).

Manteca (man tē′ kə) [San Joaquin Co.]. In 1904 or 1905 the Southern Pacific supposedly named the station after a local creamery, which had taken its name from the Spanish for 'butter' or 'lard'. The former name of the station was Cowell, for Joshua Cowell, who had given the railroad the right-of-way in 1870.

Manton [Tehama Co.]. Named in 1892 by J. M. Meeder, probably after the town in Rhode Island (G. L. Childs).

Mantor Meadow [Kern Co.]. This High Sierra meadow was named for a pioneer sheepherder (Crites, p. 269).

Manuel Peak [Monterey Co.]. For Manuel Inocente, a Chumash neophyte from Mission San Buenaventura who settled in the Big Sur area in the 18th century (John Johnson).

Manvel [San Bernardino Co.]. When the railroad spur was built north from Goffs, Isaac Blake named the station for A. A. Manvell, president of the Santa Fe (Myrick). A post office was established on Mar. 30, 1893, and changed to Barnwell on Feb. 21, 1907.

Manzana (män zä′ nə) **Creek** [Santa Barbara Co.]. The Spanish word for 'apple' was bestowed on the creek in the 1870s because a large apple orchard was adjacent to it at that time (William S. Brown).

Manzanar (man zə när′) [Inyo Co.]. Before its acquisition by Los Angeles for the Owens Valley water project, the site was the center of an important fruit-growing industry, whence it received its Spanish name, meaning 'apple orchard' (from *manzana* 'apple'). A post office is listed in 1912. The place was later notorious as the site of a relocation camp for Japanese Americans during World War II.

Manzanita (man zə nē′ tə). The common English name for the genus *Arctostaphylos* is the Spanish word *manzanita*, diminutive of *manzana* 'apple'. This beautiful shrub, with some thirty-nine species, grows abundantly in all parts of the state (Rowntree). More than a hundred geographic features are named for it, including Manzanita Chute [Shasta Co.], a smooth slope covered with manzanitas.

Maple. Many creeks are so named because of the presence of this tree (genus *Acer*). Most of them are in the northern Coast Range, where the tree thrives best; but a number appear in other sections less favorable to its growth and where its presence is more notable when it is found.

Maranda Indian Reservation [Riverside Co.]. Although this has been called the smallest Indian reservation in the United States, it is actually only a burial site, 24 feet square, near the west portal of the Metropolitan water district's San Jacinto Tunnel. When the Colorado River Aqueduct was planned in 1930, it was stipulated that this grave site would be preserved, and the route of the channel was eventually changed to respect this guarantee. Named for the family of José Maranda of Mexico, who moved to this area in the mid-19th century and married an Indian woman (Gunther).

Maray (mə rā′), **The** [San Mateo Co.]. A flat in Portola valley, presumably reflecting French *marais* 'marsh' (Alan K. Brown).

Marble. The frequent appearance of marble or similar limestone formations has given rise to more than twenty-five geographical names, including the Marble Mountains [Siskiyou Co.], topped by a summit of limestone that looks like snow from a distance.

March Air Force Base [Riverside Co.]. The air base was named March Field in Mar. 1918 in memory of Lt. Peyton C. March, Jr., who had lost his life in an airplane accident in San Antonio, Texas, on Feb. 13, 1918.

Marconi (mär kō′ nē) [Marin Co.]. In 1914 the Radio Corporation of America installed a station here and named it in honor of Guglielmo Marconi (1874–1937), the famous Italian pioneer in wireless communication (D. J. Steele).

Marcuse (mär kyo͞oz′) [Sutter Co.]. The name for the station is recorded on the Official Railway Map of 1900; it was probably given for Abraham and Jonas Marcuse, who came from Germany before 1870 and acquired large land holdings in the county.

Mareep (mə rēp′) **Creek** [Humboldt Co.]. For the Yurok village *meriip* (Waterman, p. 250; Harrington).

Mare Island [Solano Co.]. The island (or rather part of a peninsula) was called Isla Plana 'flat island' by Ayala in 1775. According to a widely circulated story, the island was named Isla de la Yegua 'isle of the mare' by General Mariano G. Vallejo (*see* Glossary) when his favorite white mare saved herself from a capsized boat by swimming to the island, from which she was later retrieved (Co. Hist.: Solano 1879:247). A different story is told by Joseph W. Revere: "This island is famous for being the resort of a large herd of these animals [elk], which are invariably accompanied by a wild mare, who has found her way thither. But although we saw this beautiful band, feeding in company with their equine friend, we could not get near enough for a shot" (*Tour of duty* 1849:67). This report is confirmed by another reliable witness, Bayard Taylor (p. 215). Vallejo's mare and the mare that joined the elks were perhaps one and the same, not a favorite horse but an ordinary animal not worth the trouble of retrieving. Isla de la Yegua was the name of the grant of the southern tip of the peninsula, dated Oct. 31, 1840, and May 2, 1841; litigation over this grant continued until May 25, 1942, when the U.S. Supreme Court rejected the Navy's claim for the entire peninsula. In the *Statutes* of 1850 (p. 60), and on Gibbes's map of 1852, the hybrid Yegua Island occurs, but the Coast Survey established the English version. Mare Island Strait is shown on the Coast Survey sketch of San Francisco Bay in 1850. Mare Island Navy Yard was established and named by act of Congress on Aug. 31, 1852.

Margarita. *See* Ranch House.

Maria Ygnacia (ig nä′ sē ə) **Creek** [Santa Barbara Co.] is named for an Indian woman who received a land allotment there and died in 1864 (John Johnson).

Maricopa (mâr i kō′ pə) [Kern Co.]. The name was applied to the new terminal when the spur of the Southern Pacific was extended from Sunset (now Hazelton) in 1903–4. Sunset Valley had apparently once been called Maricopa Valley (BGN, 6th Report). It is doubtless a transfer name from Arizona, where the name of the Maricopa Indians on the Gila River is included in the names of a county and several other features. As a name for the tribe, Maricopa is an abbreviation of the term Cocomaricopa, used earlier; the origin of the name is unknown (*HNAI* 10:83).

Marie Lake [Fresno Co.]. The lake was named by R. B. Marshall when the Mount Goddard quadrangle was mapped, 1907–9, in honor of Mary Hooper, later Mrs. Frederick L. Perry, eldest daughter of Maj. William B. Hooper (Farquhar).

Marin (mə rin´): **Islands, County, Peninsula**. The bay between San Pedro and San Quentin points, in which the two islands lie, was named Bahía de Nuestra Señora del Rosario la Marinera ('bay of Our Lady of the Rosary, she of seafarers') by Ayala in 1775. This name is shown on the maps of Ayala and Cañizares (1775, 1776, and 1781) and Dalrymple (1790). Although another map shows the abbreviated version Bahía del Rosario (Wagner, p. 492), it is possible that a different abbreviation—Bahía de la Marinera—survived locally, and that the name Marinera was also applied to the islands. Ya. [Ysla] de Marin is shown near the two islands on a *diseño* of the Corte Madera del Presidio grant (about 1834), and in 1850 the islands are designated as Marin Islands on Ringgold's charts and in the *Statutes* (p. 60). The larger island may be the one called del Oro by Payeras on May 28, 1819 (Docs. Hist. Cal. 4:341). Vallejo's Report (1850) told the story of a great Indian chief and military leader named Marin, for whom he said the islands had been named. That the Indian was a chief is questionable, and the stories of his prowess are doubtless fictional; but apparently there actually was an Indian, a boatman, by the name of Marin at Mission San Rafael (*CFQ* 4:166–67), who quite possibly lived for a time on one of the islands. Whether the islands were named for this Indian or whether Ayala's "Bahía de . . . la Marinera" was the origin of the modern name is still an unanswered question. *For* Marin Meadows, *see* Hamilton Field.

Marina (mə rē´ nə). The name means 'shore' or 'seacoast' in Spanish and is now widely used to designate a harbor for small boats. The term is used for a site in Monterey Co. and for a district in San Francisco; it does not seem to have been used in California as a place name in Spanish times.

Marinera, Isla de la. *See* Marin.

Marion. *See* Reseda.

Marion: Lake, Peak [Kings Canyon NP]. The lake was named in 1902 by J. N. LeConte, for his wife, Helen Marion, who was with him on a pioneering trip up Cartridge Creek (Farquhar).

Mariposa (mâr i pō´ zə, mâr i pō´ sə): **Creek**, town, **County, Grove** [Mariposa Co.]; **Mariposa Peak** [Merced Co.]. Padre Muñoz, who accompanied Gabriel Moraga on his expedition through the San Joaquin Valley in 1806, records in his diary (Arch. MSB 4:1–47) on Sept. 27, 1806: "This place is called [the place] of the *mariposas* [butterflies] because of their great multitude, especially at night and morning. . . . One of the corporals of the expedition got one in his ear, causing him considerable annoyance and no little discomfort in its extraction." It is, of course, not certain that the arroyo at which the expedition camped is identical with present Mariposa Creek. The name Las Mariposas was given to two land grants, dated Sept. 19, 1843, and Feb. 22, 1844. Frémont speaks in his *Memoirs* (p. 444) of a Mariposas River in connection with his third expedition (1845–46). In 1847 Frémont acquired the claim for the rancho that had been granted to Juan B. Alvarado in 1844. He expresses his own ideas (p. 447) on the origin of the name: "On some of the higher ridges were fields of a poppy which, fluttering and tremulous on its long thin stalk, suggests the idea of a butterfly settling on a flower, and gives to this flower its name of Mariposas—butterflies—and the flower extends its name to the stream." Frémont's version of the origin of the name of the stream is hardly acceptable even without Muñoz' direct evidence. However, the Mariposa lily (*Calochortus* spp.) takes its name from the California toponym. When the county (one of the original twenty-seven, and at first including most of southern California) was created on Feb. 18, 1850, the singular form, Mariposa, was chosen. The town sprang up when gold was discovered on Mariposa Creek in 1849; it was moved to the present site early in 1850. **Mariposa Grove** [Yosemite NP] was discovered by Galen Clark and Milton Mann in May 1857 and was so named because it was in Mariposa Co. It is mentioned as Mammoth Grove of Mariposa in Hutchings' *Illustrated California Magazine* of Dec. 1858. **Mariposa Peak** [Merced, San Benito, Santa Clara Cos.] was apparently named by the USGS when the Quien Sabe quadrangle was mapped in 1917–18.

Maristul. *See* Moristul.

Marjorie, Lake [Kings Canyon NP]. Named for Marjorie Mott, later Mrs. David C. Berger, daughter of Ernest J. Mott of San Francisco (Farquhar).

Markham [Sonoma Co.]. The place near the mouth of the Russian River was named for

Andrew Markham, who built a sawmill there in the late 1880s (Borden).

Markham, Mount [Los Angeles Co.]. Named by the USFS for Henry H. Markham, governor of California from 1891 to 1895.

Markleeville: town, **Creek, Peak** [Alpine Co.]. The post office was established on Oct. 21, 1863, and named for Jacob J. Marklee, a settler of 1861 who was later killed in a quarrel over the land on which the town was built. The town was incorporated by act of the legislature in 1864 and kept the name of the post office.

Markleville. *See* Clover: Cloverdale.

Mark West: Springs, Creek, station [Sonoma Co.]. The places preserve the name of Mark (William Marcus) West, an Englishman who came to California from Mexico in 1832, was naturalized in 1834, and received the San Miguel grant on Nov. 2, 1840. The post office at the station and the old mill existed from 1865 until 1917. In 1894–95 the short-lived utopian colony Altruria existed here.

Markwood Meadow [Fresno Co.]. The meadow southeast of Shaver Lake was named for William Markwood, a sheepman of the 1870s (Farquhar).

Marliave (mär′ līv), **Mount** [Nevada Co.]. Named for Elmer C. Marliave and Burton H. Marliave, engineering geologists (BGN, 1994).

Maronga. *See* Morongo.

Marshall [Marin Co.]. Originally called Marshalls, for Alexander S. Marshall and his brothers, James, Hugh, and Samuel, who settled in the county in the 1850s and built a hotel here in 1870 (Co. Hist.: Marin 1880:413–14, 503).

Marshall Historical Monument [El Dorado Co.]. The monument at Coloma was erected by the state in 1890 in honor of James W. Marshall, the discoverer of gold at Sutter's mill on Jan. 24, 1848.

Marsh Creek [Contra Costa Co.]. The only feature that preserves the name of John Marsh, one of the foremost pioneers of central California (*see* Glossary.) The creek on which Marsh built his home was formerly called Arroyo de los Poblanos. His shipping point, 4 miles west of Antioch, was for many years known as Marsh's Landing.

Mars Hill [Death Valley NP] was named "to recognize the unusual similarity of the feature to the surface of Mars, and to commemorate the May 1992 testing of the Russian Marsokhod Rover on this site in preparation for 1997 Mars landing" (BGN, list 1995).

Martell [Amador Co.]. The post office was established about 1906; it was probably named for a descendant of Louis Martell, who came from Canada before 1866 and settled in nearby Jackson.

Martha Lake [Kings Canyon NP] was named in 1907 by George R. Davis, topographer of the USGS, for his mother (Farquhar).

Martin Creek [Sonoma Co.]. Probably named for Martin E. Cook, co-patentee of part of the Mallacomes grant, through which it flows.

Martinez (mär tē′ nəs) [Contra Costa Co.]. Named in 1849 for Ignacio Martínez, born in Mexico City in 1774, *comandante* at the Presidio of San Francisco, 1822–27. The town was laid out in 1849 by Col. William M. Smith on Rancho El Pinole, which had been granted to Martínez in 1823 in recognition of military service (*see* Pinole).

Martinez [Riverside Co.]. A Cahuilla Indian settlement, now part of Torres-Martinez Indian Reservation. First mentioned in 1862 as named for an Indian surnamed Martínez (Gunther).

Martins Beach [San Mateo Co.]. Named for Nicholas Martin, a prosperous rancher, who came to California in 1850 and at one time owned this beach (Wyatt).

Martins Ferry [Humboldt Co.]. John Frederick Martin operated the ferry here from 1858 (Turner 1993).

Martin Station. *See* Neponset.

Martinville. *See* Glen: Glenwood.

Mártires, Arroyo de los. *See* Mojave River.

Marvin Pass [Kings Canyon NP]. The pass between Mount Maddox and Mitchell Peak was named by Sam L. N. Ellis of the USFS for his son Marvin (Farquhar).

Mar Vista [Los Angeles Co.]. The original name, Ocean Park Heights, was changed by the community in 1904 to avoid confusion with the nearby Ocean Park (E. A. Johnson). Mar Vista is pseudo-Spanish, suggesting 'view of the sea' (for a general discussion of such pseudo-Spanish place names in California, *see* Diament).

Mary Austin, Mount [Inyo Co.]. The mountain west of Independence was named for Austin (1868–1934), the author of *Land of little rain* and many other books and articles on the

Southwest; she was a longtime resident of the area (BGN, 1966).

Mary Blaine Mountain [Trinity Co.]. The mountain was named before 1870 after the mine named for Mary Blaine, who operated a roadhouse at the junction of the old Trinity and Klamath trails (J. D. Beebe).

Marysville [Yuba Co.]. Theodor Cordua, a native of Mecklenburg, Germany, established a ranch at the site of modern Marysville in the fall of 1842, on land that he had leased from John Sutter. He was the first settler in the Sacramento Valley north of New Helvetia: "I called my whole settlement New Mecklenburg, hoping that I would be able to share it with many of my own countrymen" (p. 7). Cordua gave the name Honcut (*q.v.*) to the grant that he received on Dec. 22, 1844, but on most maps it is labeled Mecklenburg, or New Mecklenburg, or Cordua's Rancho. The present city was laid out in the winter of 1849–50 by Auguste Le Plongeon, a French surveyor, for Covillaud and Company, who had acquired Cordua's land grant. Sutter, in conveying his equity in the townsite to Covillaud and his partners, used the name Jubaville (Yubaville), and many would have favored this name had it not been for Yuba City across the river. The names Sicardoro (for Theodore Sicard), Circumdoro, and Norwich were proposed. Finally, at a public meeting in Jan. 1850, the town was named Marysville, in honor of Mary Murphy Covillaud, a survivor of the Donner party and the wife of Charles Covillaud, the principal owner. The name Marysville had already been used in advertisements prior to the meeting. The prominent peaks west of the city are called Marysville Buttes or Sutter Buttes.

Mason. *See* Glen: Glendale.

Mason, Fort [San Francisco Co.]. The name was given to the fort at Point San Jose by the War Dept. in 1882 in memory of Gen. Richard B. Mason, who had been military governor of California, 1847–49.

Masonic: Gulch, Mountain, settlement [Mono Co.]. Named after the mines, which were first worked in 1862 by members of the Masonic Lodge (Maule).

Massacre Canyon [Riverside Co.]. The name was applied to the canyon a few miles north of San Jacinto because it had once been the scene of a battle between two Indian groups (Drury, p. 127). **Massacre Cave** [Monterey Co.] is an Indian burial site discovered in 1962; the exaggerated term "massacre" was apparently applied by the press (Clark 1991).

Matadero, Arroyo de. *See* Madera: Madera Creek.

Matagual (mä′ tə wäl) **Valley** [San Diego Co.]. Matagua is mentioned as the name of a rancheria in the Valle de San José in 1795 (SP Mis. 2:55 ff.). On an 1844 map of the valley, the name Matajuai is shown near the southeastern end of what is now Matagual Valley (Joseph J. Hill, *Warner's Ranch* [Los Angeles: Young and McCallister, 1927], pp. 29, 207). The Diegueño name may have been *mak aahway* 'place-kill', i.e. 'battleground', or *mat-iihway* 'ground-red', related to *'ehwatt* 'red' (Langdon).

Matanzas (mə tan′ zəs) **Creek** [Sonoma Co.]. Spanish for 'a slaughter'. The stream is labeled Matanza Creek on a map of part of Cabeza de Santa Rosa Rancho (1859), and Matanzas Creek on a map of Los Guilicos Rancho. "At the killing season, cattle were driven from the rodeo ground to a particular spot on the rancho, near a brook or forest. It was usual to slaughter from fifty to one hundred at a time. . . . The occasion was called the matanza" (William Heath Davis, pp. 45–46).

Mather (math′ ər): **Pass** [Kings Canyon NP], station [Yosemite NP], **Grove** [Humboldt Co.]. These names commemorate Stephen T. Mather (1867–1930), a native of San Francisco and first director of the National Park Service (1917 to 1929). The pass was named on Aug. 25, 1921, by what was probably the first group to cross with a pack train. The name was applied to the station on the Hetch Hetchy Road by M. M. O'Shaughnessy, city engineer of San Francisco.

Mather (mā′ thər) **Air Force Base** [Sacramento Co.]. Named Mather Field by the War Dept. in 1936, in memory of Lt. Carl Mather, Aviation Section, Signal Officers' Reserve Corps.

Matheson (math′ ə sən) [Shasta Co.]. Named in 1920 by the Mountain Copper Company, a subsidiary of Matheson and Company of London, in memory of the founder of the famous firm, James Matheson (W. F. Kett).

Mathews, Lake [Riverside Co.]. The major reservoir of the Colorado River Aqueduct was named about 1940 for W. B. Mathews (1865–1931), the first general counsel of the Metropolitan Water District of southern California, Los Angeles, and a leader of the building project (Riverside Public Library).

Mathles (math′ ləs) **Creek** [Shasta Co.]. Probably from Wintu *maałas* 'baked salmon' (Pitkin, p. 342).

Matilija (mə til′ i hä): **Canyon, Hot Springs**, station [Ventura Co.]. Matilija was one of the Chumash rancherias under the jurisdiction of Mission San Buenaventura and is mentioned in its archives (1:27). Arroyo de Matilija is recorded in 1827 (Dep. Recs. 5:74), and Ranchería de Matilija is shown on a *diseño* of El Rincon. The name is mentioned by Taylor with the phonetic spelling Matiliha on Oct. 18, 1861, and with the Spanish spelling Matilija on July 24, 1863. The name was applied to the Southern Pacific station when the extension from Ventura to Nordhoff (Ojai) was built in 1898. The meaning of the Chumash name is unknown. The matilija poppy (*Romneya trichocalyx*) takes its name from the canyon.

Matilton (mə til′ tən) [Humboldt Co.]. The name of this former village site of the Hupa (Athabaskan) Indians, also called Captain John's Ranch, comes from the Hupa name *me'dil-ding* 'canoe place' (Golla).

Matterdome, The [Riverside Co.]. This mountain was named after Matt Smithson, a pioneer of the area, on the pattern of "Matterhorn."

Matterhorn: Peak, Canyon [Yosemite NP]. The name of one of the grandest Alpine peaks was applied in 1877 by John Muir, perhaps to Banner Peak near Mount Ritter (Farquhar), and by the Wheeler Survey in 1878 to the peak and the canyon that still bear the name: "That the name is a poor one there can be no doubt, for . . . there is only the barest suggestion of resemblance to the wonderful Swiss mountain after which it is called" (Lincoln Hutchinson, "The ascent of 'Matterhorn Peak,'" *SCB* 3:159–63). The name Matterhorn Peak on LeConte's map and the USGS map is tautological: *Horn* is a German generic term for 'peak'.

Matthes (math′ əs) **Crest** [Yosemite NP]. The name was proposed in 1946 by Reid Moran, then a Yosemite ranger-naturalist, in honor of Dr. François Emile Matthes, senior geologist, USGS. The crest is the cotype of a geological form first generically described by Matthes as a cockscomb (David Brower); *see* Cockscomb Crest.

Mattole (mə tōl′): **River, Canyon** [Humboldt Co.]. The name refers to an Athabaskan Indian tribe, called *bedool* by themselves (Golla) and *me'tuul* by the Wiyot (Teeter 1958); its original source is unknown. The canyon is submarine, one mile offshore at latitude 40°17' north, and was named by the BGN in 1937.

Maturango (mat ə rang′ gō) **Peak** [Inyo Co.]. Perhaps from Spanish *maturrango* 'bad horseman; clumsy, rough person' (Sowaal 1985); or from a Panamint form such as *mattootoongwïnï* 'stand braced, as against the wind' (McLaughlin); or from Panamint *muatangga*, the name for Koso Hot Springs (Fowler). The Whitney Survey left the highest peak of the Argus Range nameless, but the name appears in 1877 on sheet 65-D of the Wheeler atlas. Later maps show various spellings. The USGS restored Wheeler's version when the Ballarat quadrangle was mapped in 1905–6 and 1910–11.

Mawah (mä′ wä) **Creek** [Humboldt Co.]. From Yurok *mawa* (Waterman, map 17, item 106).

Maxon: Resort, Dome, Meadow [Fresno Co.]. Named for Charles N. Maxon, an early settler who operated a hotel at the site of the modern resort (Co. Hist.: Fresno 1956).

Maxwell [Colusa Co.]. Established in 1878 and named for an early resident, George Maxwell.

Maxwell Creek [Mariposa Co.]. The tributary to Merced River was named for a forty-niner called George Maxwell. The post office at Coulterville (*q.v.*) was called Maxwell Creek from 1852 to 1872.

Mayacmas (mā yak′ məs, mə yak′ ə məs) **Mountains; Maacama Creek** [Sonoma, Lake Cos.]. The mountain chain, forming the divide of the headwaters of Russian River and Clear Lake, was named for the Indians on the west slope, probably a division of the Yuki. According to Barrett (*Pomo*, p. 269), there was a Yukian Wappo village, *maiya'kma*, one mile south of Calistoga. Serro de los Mallacomes (Mount Saint Helena) is shown on a *diseño* of the Caymus grant (1836). Later the name appears in the title and on the *diseños* of a land grant Mallacomes y Plano de Agua Caliente or Moristul, dated Sept. 3, 1841, and Oct. 11 and 14, 1843. The present spelling is used in the *Statutes* of 1850 (pp. 60–61). Although this version was also used by the Whitney Survey, confusion persists to the present day. The BGN (5th Report) decided for Miyakma, but in 1941 it reversed this decision in favor of Mayacmas ("not Miyakma, Cobb Mountain

Range, Malacomas, Mayacamas, nor St. Helena Range"). The name of the stream is often spelled Maacama.

Mayfield [Santa Clara Co.]. In 1853 Elisha O. Crosby, who had been a member of the California Constitutional Convention of 1849, bought a tract of land that he called Mayfield Farm. The name Mayfield was given to the post office in 1855, to the railroad station in 1863, and to the town that was laid out by William Paul in 1867. In 1925 the town was annexed to Palo Alto.

Mayfield Canyon Battleground [Inyo Co.]. The place was registered on June 20, 1935, in commemoration of the battle fought by California cavalry and settlers against the Indians in Apr. 1862. One leader of the settlers, named Mayfield, was killed.

Mayhew [Sacramento Co.]. The name is shown on the Land Office map of 1879. According to the County History (1890), Mayhew Station was one of the first stations on the Sacramento Valley Railroad built by Theodore D. Judah in 1856 and was named for the station agent. In McKenney's Directory of 1880 it is listed as Mahews, with L. Mahew as agent.

May Lake [Yosemite NP] was named by Charles F. Hoffmann, for Lucy Mayotta ("May") Browne, who became his wife in 1870 (Farquhar).

Maywood [Los Angeles Co.]. This popular American place name was chosen for the town by a vote of the citizens, probably at the time of the incorporation in 1924 (Myrtle Reed).

Mazourka (mə zo͞or′ kə) **Canyon** [Inyo Co.]. The name, also spelled Mazurka, refers to a dance of Polish origin.

Mc. This prefix is alphabetized here as if spelled "Mac" (*q.v.*).

Meachim Hill [Sonoma Co.]. Probably named for Alonzo Meacham, who established a general store and trading post in Santa Rosa in 1853.

Meadow. The common generic term has also been repeatedly used as a specific term, especially in connection with Creek and Lake. A number of Meadow names have been transferred to towns and stations. **Meadow Valley** [Plumas Co.] was once a rich gold-mining center; its post office was established on Oct. 3, 1855. **Meadow Lake** [Nevada Co.] is a reminder of the even richer mining town of the 1860s, Meadow Lake City.

Meadows Canyon [Monterey Co.]. For James Meadows, who arrived in the area in 1842 and married Loreta Onésimo de Peralta, a Rumsen (Costanoan) Indian woman. Their daughter Isabella Meadows was the last surviving speaker of the Rumsen language (Clark 1991).

Meads [Shasta Co.] was named for Dr. Elwood Mead (1858–1936), late commissioner of the U.S. Bureau of Reclamation (Steger).

Meamber (mē am′ bər) **Creek** [Siskiyou Co.]. Named for a family of settlers (Hendryx).

Mears Creek [Shasta Co.]. The tributary of the Sacramento River was named for Henry Mears, a trapper, who built a rock fort on this creek in the winter of 1862, as a protection against the Indians (Steger).

Mecca: town, **Hills** [Riverside Co.]. When this part of the desert was reclaimed through irrigation for date culture, the name of the Arabian city was considered more appropriate than Walters, the name by which the settlement had been known since 1896. The name was changed to Mecca on Sept. 26, 1903, at the suggestion of R. H. Myers, founder of the Mecca Land Company (*Westways*, Feb. 1951).

Mecklenburg. *See* Marysville.

Medanos (mə dä′ nōs) means 'sand dunes', with a shift of accent from Spanish *médanos*. **Los Medanos** [Contra Costa Co.]: A *paraje que llaman los Méganos* 'place called the sand dunes' (with a variant spelling) is mentioned in Durán's diary on May 24, 1817. The name Meganos was used for a land grant, dated Oct. 13, 1835, which was finally patented to John Marsh's daughter. Another land grant in the district, with the spelling Los Medanos, was dated Nov. 26, 1839. The name was applied to the Southern Pacific station in 1878 because it was situated on the latter land grant. **Medanos Point** [San Diego Co.]: On a *diseño* of the Pueblo Lands of San Diego, some *méganos* are shown near Punta Falza (False Point). Medanos Point was named by the Coast Survey in modern times.

Meder (mē′ dər) **Creek** [Santa Cruz Co.]. For Moses A. Meder, a Mormon who had come to San Francisco in 1846 with Samuel Brannan and later acquired the property through which the creek runs.

Medicine is a colloquial term referring to the religious rites of American Indians. **Medicine Lake** [Siskiyou Co.] was supposedly the site of puberty ceremonies of the Shasta tribe. The name does not seem to appear on older maps. The Land Office map of 1890 has the present name, but the Mining Bureau map of 1891 has Crystal Lake. There are about twenty-five other Medicine Lakes, Creeks, and Gulches in the state. Some names may go back to Indian times, others doubtless were applied because the water or nearby herbs proved of healing value to exhausted travelers.

Meeks: Bay, Creek [El Dorado Co.]. The name appears on the maps of the 1870s: Micks Bay and Meadow (von Leicht–Hoffmann, Wheeler), Meeks Bay (Bowman). Several Meeks were registered in the county, but none can be definitely connected with the place.

Meganos. *See* Medanos.

Mehlschau (mel′ skow) **Creek** [San Luis Obispo Co.]. For a family of early settlers (Hall-Patton).

Meiners (mī′ nərz) **Oaks** [Ventura Co.]. Named in 1925 for Carl Meiners, a landowner in the district (H. M. Rider).

Mellen. *See* Topock.

Meloland (mel′ ō land) [Imperial Co.]. The place was named about 1910 by the author Harold Bell Wright, because of the mellow nature of the loamy soil (L. G. Goar).

Melones (mə lō′ nēz): post office, **Reservoir** [Calaveras Co.]. The present name of the place on the Stanislaus River, between Angels Camp and Sonora, did not appear on the maps until the name of the post office was changed from Robinsons to Melones on Feb. 15, 1902. The name itself, however, goes back to 1850. The *San Francisco Alta California*, June 16, 1851, refers to a Meloneys Diggings, and other newspapers to a Melones claim. A rich mine on the slope of Carson Hill, called Melones, is frequently mentioned. This mine once had a mill with 120 stamps and is claimed to have produced $4.5 million by 1934. In 1896 it was combined with five other quartz mines to form Melones Consolidated Mines. This group of mines was called the largest mining town of the state by Browne (p. 59). Edward Vischer describes the origin but not the meaning of the name: "Only a few miles from the southern end of the town [Angels Camp] . . . is the famous Carson Hill, called 'el Cerro de Melones' [hill of melons] by the Mexicans. . . . That was in the summer of 1851 when the wealth of Carson Hill seemed inexhaustible. . . . There were several productive dry diggings . . . called 'Meloncitos' by the Mexicans to differentiate them from the mountain" (p. 324). That Carson Hill was called Cerro de Melones has not been proved, although this statement has often been repeated. Likewise unproved is the generally accepted story that the Mexicans found gold flakes in the shape of melon seeds there. Flakes that looked like melon or cucumber seeds were in fact actually found in various places (Browne, p. 51; John S. Hittell, *Mining in the Pacific states of North America* [San Francisco: Bancroft, 1861], p. 46). This does not prove, however, that Mexicans called a camp, not to mention a hill, *melones*—the Spanish word for melons, not melon seeds. Melones as a place name is found in other Spanish-speaking countries and may be a transfer name. It is also possible that Melones is a misreading of the name McLeans or McLanes, applied to a camp or mine; such a name was repeatedly mentioned in old diaries and was recorded on Gibbes's, Goddard's, and other early maps close to Robinsons Ferry. The rare compound of tellurium and nickel is called Mellonite, after the Melones mine where it was found. The idea that Melones was once called Slumgullion existed only in the vivid imagination of Bret Harte.

Melville, Mount [San Mateo Co.]. Named in memory of Melville B. Anderson, professor of English literature at Stanford University.

Mendel, Mount [Kings Canyon NP]. For Johann Gregor Mendel, pioneering geneticist (1822–84). *See* Evolution Peaks.

Mendenhall Peak [Los Angeles Co.]. Named by the USFS for Frank Mendenhall, a hunter.

Mendenhall Valley [San Diego Co.]. Named for Enos Mendenhall, who settled here in 1870.

Mendico (men′ di kō): **Creek, Hill, Lake** [San Mateo Co.]. For Juan Mendico, a Basque, who settled here in 1859 (Alan K. Brown); also spelled Mindego.

Mendocino (men də sē′ nō), **Cape** [Humboldt Co.]; **Mendocino: County**, town, **National Forest, Canyon**. The origin of the name cannot be satisfactorily explained. A Cabo Mendocino in the general region appears on the maps of Ortelius in 1587 (Wagner, pp. 396–97). Padre Antonio de la Ascensión wrote the following account a few years after his return from the Vizcaíno expedition (1602–3): "There may be some curious person who may wish to know why this cape or point of land came to be named 'Mendocino.' The reason was that when Don Antonio de Mendoza was viceroy of New Spain in 1542, he sent two ships to the Philippines. . . . The first land seen [by them] returning by that latitude was this Cabo Mendocino, to which they gave the name in honor and remembrance of the viceroy" (Wagner's translation, *CHSQ* 7:366). This story remained current and was repeated, somewhat garbled, more than two centuries later by Duflot de Mofras (p. 97). It has never been substantiated, but neither can it be refuted. Since the name apparently does not appear on maps until 1587, it is of course possible (and more plausible) that the cape was named for Lorenzo Suárez de Mendoza, viceroy of New Spain from 1580 to 1583. If one of the two viceroys was thus honored, the place name was created by the relatively rare method of using the adjective form of the personal name, comparable to Smithsonian or Wagnerian (in Argentina, a Mendocino is a man from the city of Mendoza). It is also not impossible that some European cartographer simply placed the name on the map. It is the oldest name of a cape that has survived the various phases of real and imaginary California geography, with the same spelling and in the same general location, although it was not definitely identified with the cape at latitude 40°27' north until Malaspina (1791) placed it at 40°29' north (Wagner, ibid.). The county, one of the original twenty-seven, was created on Feb. 18, 1850, and named Mendocino after the cape—although the latter was and is outside the county. The town was settled and probably named by William Kasten in 1852; the post office is listed in 1853. The national forest was created as Stony Creek Forest Reserve in 1907, was named California National Forest in 1908, and received its present name in 1932. The submarine canyon, 2 miles off the coast, was named by the Coast Survey, and in 1938 the name was approved by the BGN.

Mendota (men dō´ tə) [Fresno Co.]. The name was given to the station in 1895 when the Southern Pacific built the extension from Fresno. Like many railroad stations, it was probably named after a town "back home": there were a number of Mendotas in the Midwest at that time. The term is from a Siouan language, perhaps referring to the confluence of two streams.

Menifee: post office, **Valley** [Riverside Co.]. For Menifee Wilson, a miner around 1880. The post office was established in 1887 (Gunther). A new master-planned community in the valley is called Lake Menifee (Brigandi).

Menlo Park [San Mateo Co.]. In Aug. 1854, D. J. Oliver and D. C. McGlynn, brothers-in-law from Menlough, County Galway, Ireland, erected an arched gate at the joint entrance to their ranches with the inscription "Menlo Park" and the date. When the San Francisco and San Jose Railroad reached the place in 1863, it adopted the name for the station. The gate stood until July 7, 1922, when an automobile struck and destroyed the landmark (Stanger, p. 152). The town of this name in New Jersey is named after the one in California.

Mentone (men´ tōn) [San Bernardino Co.]. The land was purchased by the Mentone Company in 1886, and the town was laid out in 1887 (Santa Fe). The name was probably chosen for advertising purposes; the asserted resemblance to the Riviera resort is imaginary.

Merced (mûr sed´), **Lake** [San Francisco Co.]. When the Anza expedition camped here on Sept. 22, 1775, they named the lagoon for Nuestra Señora de la Merced 'Our Lady of Mercy' (Palou 4:40). It is shown as Laguna de la Merced on Cañizares's 1776 map. In 1798 some of the cattle of the San Francisco presidio were kept in *el parage de la Laguna de Merced* (SP Mis. and C. 1:74). On Sept. 25, 1835, the name was applied to a land grant. The hybrid name, Lake Merced, came into use in the early 1850s, although Eddy's and other maps retained the all-Spanish name. The name La Merced was also applied to a land grant in Los Angeles Co., dated Oct. 4, 1844.

Merced: River, County, Falls (post office), city [Merced Co.]; **Grove, Peak, Pass, Lake** [Yosemite NP]. The name Nuestra Señora de la Merced was given to the river by an expedition headed by Gabriel Moraga on Sept. 29, 1806 (Muñoz), five days after the feast day of Our Lady of Mercy. Frémont speaks of Río de la Merced in his *Expl. exp.* (p. 360), and Tuolumne River is labeled Río de la Merced on Preuss's map of 1845. The modern name seems to have come into general use with the Gold Rush. A short-lived settlement, Merced City, on the San Joaquin north of the mouth of Merced River, is mentioned in the *San Francisco Alta California* on Feb. 7, 1850. The mining town Merced Falls was founded soon afterward; the post office is listed in 1858. Merced Co., carved out of Mariposa Co., was named on Apr. 19, 1855. The modern city came into existence after the Southern Pacific reached the place on Jan. 15, 1872. The grove of Big Trees was discovered (or rediscovered) in 1871–72 by surveyors for the Coulterville Road and was named by John T. McLean, president of the Turnpike Company (Farquhar). Merced Peak is shown on Wheeler atlas sheet 56-D. Merced Pass was discovered by Corporal Ottoway in 1895 and named by Lt. H. C. Benson (Farquhar). The lake above Little Yosemite Valley was discovered by John Muir in 1872 and named Shadow Lake, but it has been designated as Merced Lake by the BGN (list no. 30). *For* Rio de los Merced [*sic*], *see* Tuolumne River.

Mercur (mûr′ kər) **Peak** [Yosemite NP]. Named in 1912 by Col. W. W. Forsyth, for James Mercur (1842–96), professor of civil and military engineering at West Point from 1884 until his death.

Mercy Hot Springs [Fresno Co.]. Probably named for John N. Mercy, a native of France, a stock raiser in the county in the 1860s and 1870s.

Meridian [Sutter Co.]. The name was given to the post office in 1860 because the place is only a quarter mile west of the Mount Diablo meridian. *For* Meridian District, *see* Bostonia.

Merriam Mountains [San Diego Co.]. The range near Escondido was named for Maj. G. F. Merriam, one of the earliest settlers in the area.

Merriam Peak [Fresno Co.]. Named by the California State Geographic Board in 1929, in honor of C. Hart Merriam (*see* Glossary).

Merrill (mâr′ il): **Creek, Mountain; Merrillville** [Lassen Co.]. A post office (now discontinued) was established before 1880 and probably named for C. A. Merrill, a native of Maine, who came to Lassen Co. in 1874 (Co. Hist.: Plumas 1882:504). "Merrill and Marker, tunneling Eagle Lake," are mentioned in McKenney's Directory, 1883–84.

Merrill (mâr′ il) **Ice Cave** [Lava Beds NM]. This cave with permanent ice was named for Charles H. Merrill, who homesteaded the land on which it is situated (BGN, Dec. 1948).

Merrimac [Butte Co.]. The now-discontinued post office was established about 1885 and given the popular American place name, which originated in New England, perhaps an Algonquian term meaning 'deep place'.

Merritt [Yolo Co.]. The Southern Pacific station was named for Hiram P. Merritt, who came to California from Vermont in 1852 and settled in the county before 1866.

Merritt, Lake [Alameda Co.]. The slough of San Antonio Creek became a "lake" through the efforts of Dr. Samuel J. Merritt, mayor of Oakland in 1869. It was named Lake Peralta but was renamed Lake Merritt when it and the surrounding land became a city park in 1891.

Merritt Island [Yolo Co.]. Probably named for Ezekiel Merritt, who is said to have had a hunting or trapping camp here. Merritt was a tough character connected with Sutters Fort and participated in every fight in the last troublesome years of Mexican rule.

Mesa (mā′ sə). The Spanish word for 'table', referring to a plateau-like hill, is generally used in the American Southwest as a generic term, but it has not wholly replaced the corresponding English term "table." Five of the twenty-odd Mesa names in the state are land-grant names; others may also have originated in Spanish times, but most of them are probably modern applications. Monterey Co. has a Mesa Grande, Las Mesas Potrero, and a Mesa Coyote; Riverside Co. has Mesas de Burro, de Colorado, and de la Punta. Some Mesa names are hybrids: The Mesa [San Diego Co.], Burton Mesa [Santa Barbara Co.], and Mount Mesa [Los Angeles Co.]. Mesa Peak [Los Angeles Co.] is actually a peak and not a mesa; it was perhaps so named because of the real but unnamed mesa on Malibu Creek to the southeast. Some settlements derive their names from nearby mesas: Mesa Grande Indian Reservation [San Diego Co.], Mesaville [Riverside Co.]. **La Mesa Battlefield** [Los Angeles Co.] is the site of one of the last engagements in the Mexican War, on Jan. 9, 1847. **La Mesa** [San Diego Co.] is found first as La Mesa Heights in 1886. In 1894 a new town was started at Allison Springs and called La Mesa Springs. When the town was incorporated in 1912, the abbreviated form was used. The post office name had always been La Mesa; it was spelled as one word for a number of years until Z. S. Eldredge came to the rescue in 1905.

Mescal (mes kal′). The Mexican Spanish term is from Aztec *mexcalli* and designates the fleshy edible parts of several desert plants, including species of agave and yucca. Indians used it chiefly for food, but also to make rope, baskets, etc. As a place name, Mescal occurs in Los Angeles, Monterey, Santa Barbara, and San Bernardino Cos. **Mescal Range** and **Springs** [San Bernardino Co.] refer to *Yucca mojavensis*, which grows plentifully in the Ivanpah Mountains. The Indians prepared the plant by roasting it in pit ovens, the common native method of preparing the food (Gill).

Mescalitan (mes kal′ i tən) **Island** [Santa Barbara Co.]. The Portolá expedition camped in the vicinity on Aug. 20, 1769, near several fairly rich Indian villages: "The soldiers named these towns Mescaltitlan, but others call them the towns of La Isla; I christened them with the name of Santa Margarita de Cortona" (Crespí, p. 168). The Aztec suffix *-titlan* means 'among'—hence the meaning 'among the mescal'; the term is probably a transfer name from Mexico. The abbreviated form Mescal Island is also used.

Mesitas, Las. *See* Anacapa.

Meskua. *See* Moscow.

Mesquite (mes kēt′). Mesquite is the common name for several species of the genus *Prosopis*. The word is of Aztec origin (*mizquitl*), and designates the leguminous tree also called *algarroba*. The two species native to California are the only trees that grow, without traceable water supply, in the arid regions of the Great Basin. They are valued by Indians and desert dwellers for their edible beans and their wood. A number of places testify to their presence or former presence, e.g. Mesquite Valley (northwest arm of Death Valley); Mesquite Flat and Well [Death Valley NP]; and Mesquite Dry Lake [San Bernardino Co.].

Messelbeck Reservoir [Shasta Co.]. The reservoir was built in 1875 and named for Frank Messelbeck, the former owner of the property (Steger).

Messerville. *See* Junction: Junction City.

Messick [Sutter Co.]. Apparently named for Charles C. Messick, a native of Woodland, who came to the Meridian district in the 1870s (Co. Library).

Messina. *See* High: Highland.

Metate (mā tä′ tā) **Hill** [San Diego Co.]. From the Mexican Spanish word for a grinding slab, derived in turn from Aztec *metlatl*.

Metralla (mi trä′ yə) **Canyon** [Kern Co.]. The Spanish word means 'grapeshot'; the name was made official by BGN, list 8104. The reason for the naming is not known.

Metropolitan [Humboldt Co.]. Originally known as McDairmids Prairie, the place was renamed in 1904 when the Metropolitan Redwood Lumber Company built its mill there (Borden).

Mettah (mā′ tə) **Creek** [Humboldt Co.]. From the Yurok village name *metaa* (Waterman, p. 245; Harrington).

Metz [Monterey Co.]. When the Southern Pacific reached the place in 1886, the station was named Chalone, after Chalone Peaks (now The Pinnacles). In 1891 a post office was established and named for the first postmaster, William H. H. Metz, a native of Ohio, who had settled here as a stock raiser in 1871.

Meyers [El Dorado Co.]. The post office was established on Oct. 6, 1904, and named for a homesteader who had settled on the land before 1860. Brewer mentions the settlement in his Notes, Nov. 7, 1863.

Meyers Canyon [Imperial, San Diego Cos.]. Named for a Doctor Meyers, cattleman and rancher near Descanso (*Desert Magazine*, June 1939).

Mezes (mā′ zəs, mē′ zēz) **Hill** [San Mateo Co.]. The estate of Simón Monserrate Mezes was here from the late 1850s (Alan K. Brown). *For* Mezesville, *see* Redwood: Redwood City.

Miami Mountain [Mariposa Co.]. The name is perhaps the local Indian (Yokuts) name *mē-ah-nee* (W. M. Sell); association with the well-known eastern place name, Miami (itself of Indian origin), may account for the present form. **Miami Creek** [Madera, Mariposa Cos.]: By decision of the BGN, 1964, this became the new name for the North Fork of the Fresno River.

Mica Butte [Riverside Co.]. Probably named because a deposit of the mineral mica was found there. There is a Mica Gulch near Igo in Shasta Co.

Michie Peak [Yosemite NP]. Named in 1912 by Col. W. W. Forsyth, in memory of Peter S. Michie (1839–1901): a native of Scotland, a general in the Civil War, and professor of natural and experimental philosophy at West Point from 1871 until his death.

Michigan: Bar [Sacramento Co.], **Bluff** [Placer Co.]. Miners from the state of Michigan gave both names in the early 1850s; the term is originally Algonquian, 'big lake'. Michigan

Bar was originally applied to the river bar and was soon transferred to the settlement known as Live Oaks. Michigan Bluff started as Michigan City one-half mile away and assumed its new name and new location when mining operations threatened the foundations of the houses (Doyle). There is also a Michigan Flat in Lassen Co.

Micks Bay. *See* Meeks Bay.

Mid, Middle. The map of the United States is dotted with hundreds of names containing these adjectives. The name is applied for a place either halfway between two other places, or in the center of a valley, or amid trees, etc. For physical features it is often purely descriptive: Middle Fork, Middle Palisade, Middle Alkali Lake. **Midway: Valley, Peak; Midoil; Midland** [Kern Co.]: The valley probably received its name because it is situated between the San Joaquin Valley and the Carrizo Plains. The name became widely known when C. A. Canfield and his associates developed the Midway oilfields in the 1890s. **Mid Hills** [San Bernardino Co.] refers to the range connecting the New York Mountains with the Providence Mountains. **Middle Waters** [Sierra Co.], with a unique generic, is a creek on Henness Pass Road. **Middletown** [Shasta Co.], which vanished in the 1860s, was named because it was about midway between Shasta and Horsetown; the new **Middletown** [Lake Co.] was named in the 1860s when it was the stage stop halfway between Lower Lake and Calistoga. **Midway City** [Orange Co.] derived its name from being exactly midway on Bolsa Avenue between Santa Ana and the beach; the post office is listed in 1930. **Midpines** [Mariposa Co.] was named by N. D. Chamberlain in 1926 because it is "amidst the pines and midway between Merced and Yosemite." Many other settlements and towns were named for similar reasons: Middle Creek [Shasta Co.]; Middle River [San Joaquin Co.]; Midlake [Lake Co.]; Midvale [Madera Co.]; Midway Well [Imperial Co.]; Midway [Alameda Co.]. However, **Midland** [Riverside Co.] is a transfer name from Michigan, given in 1928 by O. M. Knode of the U.S. Gypsum Company. *For* Middle Point [San Mateo Co.], *see* Franklin Point.

Miguel (mi gel′): **Meadow, Creek** [Yosemite NP]. "Next day we proceeded . . . to McGill's, where I again camped. . . . The ranch belongs to Mr. Miguel D. Errera, but his American friends have corrupted Miguel into McGill and by that name is his house known" (N. F. McClure, *SCB* 1:184–85). The original spelling has been restored by the BGN.

Milagra (mi lä′grə) **Valley** [San Mateo Co.]. For *milagro*, Spanish for 'miracle', but the name was applied apparently in modern times; it is shown neither on the *diseños* of Rancho San Pedro nor on early maps of the county.

Mile. In early California geography the word "mile," modified by a number, was a convenient and widely used specific term, and it has survived in many place names. It was combined mainly with House or Creek, sometimes with Canyon, Hill, Point, River, Ridge, Rock, or Slough. With creeks, the number of miles seems usually to have designated the distance along a trail from the crossing of one stream to that of another; in travel by foot, the distance covered was naturally small, and the longest distance indicated in this manner seems to be Twelvemile Creek [Modoc Co.]. In only a few names is the length of the stream itself indicated. Fortymile Creek [El Dorado Co.], a small stream, was probably named in jest or through a misunderstanding. Running off from Klamath River in Township 6 N, Range 6, 7 W, is a trail having Four Mile, Five Mile, and Ten Mile Creeks. The first two distances are approximately correct, but the distance between the last two is something over 2 miles, not 5 miles. Such discrepancies are found elsewhere and may indicate that the miles sometimes get longer after one has walked a while in the mountains (Stewart). Larger numbers are found when places are designated by distance of miles along a road: Fifteen Mile Point, a projection north of Bear Lake Road [San Bernardino Co.], is exactly 15 miles east of Victorville. On Hoffmann's map of the Bay region (1873), Seven, Eight, Thirteen, Fourteen, Fifteen, Eighteen, and Twenty-One Mile Houses are indicated along the road from San Jose to Gilroy. On the old country road to Gasquet [Del Norte Co.], Ten Mile, Eleven Mile, Twelve Mile, and Eighteen Mile Creeks are shown, counting from the Oregon line or a point just above it. **Mile Rocks** [San Francisco Co.] were called One-Mile Rocks by Beechey in Nov. 1826 because they were one mile south of the channel (half a mile south of Point Bonita) by which ships entered the Golden Gate. The Coast Survey applied the present name (*Coast Pilot*, 1869: 59). **Ten Mile River** [Mendocino Co.] was so named in the 1850s because it is 10 miles north of Noyo. People now usually assume that the name refers to the distance from Fort Bragg, which is only about 8 miles (F. F. Spalding). **Five Mile Gulch** [Shasta Co.]: The Irish Placer Mining Company dug a ditch 5 miles long from this gulch to bring water to French Gulch for placer mining. The ditch itself is called Five Mile Ditch (Steger).

Milestone: Mountain, Plateau, Bowl, Creek [Tulare Co.]. Prospectors used this name because the abrupt mountain summit resembles a milestone. It is recorded on Hoffmann's map of 1873. The bowl and plateau were named after the mountain in 1902 by Prof. W. R. Dudley of Stanford.

Miley, Fort [San Francisco Co.]. Named in 1900 by the War Dept. in memory of Lt. Col. John D. Miley, who died in Manila, Sept. 19, 1899.

Milford. *See* Mill.

Mill. There are more than a hundred Mill Creeks in the state—all named, as far as can be ascertained, because a mill once existed there. In addition to several Mill Gulches, the maps show Mill Spur [Yolo Co.], Mill Potrero [Kern Co.], Chino Mill Creek [Riverside Co.], and Burnt Mill Creek [San Bernardino Co.]. The name-giving mill was usually a gristmill or a sawmill, but some places may have been named after a stamp mill for crushing ore, and at least two were named after a windmill: Windmill Creek [San Luis Obispo Co.] and Windmill Canyon [Monterey Co.]. Some of the Mill Creeks were translations of the Spanish *molino*. **Mill Valley** [Marin Co.] was locally so known because John Reed (Reid, or Read), grantee of Rancho Corte de Madera del Presidio, built a sawmill there in 1834 and operated it for many years. The place itself is shown as Read on Hoffmann's map of the Bay region (1873). In 1889 the Tamalpais Land and Water Company acquired the land, built a branch of the North Pacific Coast Railroad to it, and laid out the town of Mill Valley. *For* Mill Valley Junction, *see* Almonte. **Mill Creek** [Tehama Co.] is shown on *diseños* as Arroyo and Río de los Molinos; the latter name was applied to the land grant dated Dec. 20, 1844, and granted to Albert G. Toomes, the pioneer of Tehama. It was, however, called Mill Creek by Bidwell and others as early as 1843, and is so designated on Reading's map of 1849. The resort was first called Mill Creek Homesite, after the creek. When the post office was established in 1937, the present name was adopted at the suggestion of E. J. and W. H. Foster. (*For* Mill Creek [Mono Co.], *see* Lundy.) **Millville** [Shasta Co.] had one of the first gristmills of Shasta Co., built here by D. D. Harrill. The place was known as Harrill's Mill in 1855, as Buscombe (after Harrill's birthplace in North Carolina) in 1856, and under the present name in 1857 (Steger). **Milford** [Lassen Co.]: When J. C. Wemple built a gristmill here in 1861, he thought the name appropriate for the settlement (Co. Hist.: Lassen 1916:239). **Mills** [Sacramento Co.], earlier known as Hangtown Crossing, was named after a gristmill (Co. Hist.: Sacramento 1913:321). The station is shown on the Official Railway Map of 1900. **Millseat Creek** [Shasta Co.], formerly also called Millsite, was so named because several early mills were built on this stream (Steger). **Mill Creek Redwood State Park** [Del Norte Co.] had as its nucleus the Franklin D. Stout Memorial Grove, given to the state in 1929.

Millard (mi lärd′) **Canyon** [Los Angeles Co.]. Named for Henry W. Millard, a native of Missouri, who settled at the mouth of the canyon in 1862, raised bees, and hauled wood to Los Angeles.

Millbrae (mil′ brā) [San Mateo Co.]. In the 1860s Darius O. Mills (1825–1910), a San Francisco banker and promoter, acquired part of the Buri Buri Rancho and built his residence south of the townsite. The name Millbrae (*brae* is Scottish for 'hill slope') was applied first to the railroad station, and in 1867 to the post office. The San Francisco Municipal Airport was originally called Mills Field, because it was acquired from the estate. **Mount Mills, Mills Creek** [Fresno Co.]: At the suggestion of the Sierra Club, of which Mills was a charter member, the name was applied by the BGN after his death in 1910.

Miller [Marin Co.]. The name preserves the memory of James Miller, a native of Ireland, who came to California with the Stevens-Murphy-Townsend party in 1844 and settled in the county in 1845.

Miller [Santa Cruz Co.]. The Southern Pacific station was named before 1900 for the land baron Henry Miller (1827–1916), whose Bloomfield Farm was here (Hoover, p. 494).

Miller, Fort. *See* Millerton Lake.

Miller Lake [Yosemite NP] was named by Lt.

N. F. McClure in 1894, for a soldier in his detachment (Farquhar).

Miller Peak [Riverside Co.]. Named by the California State Park Commission on Nov. 9, 1935, in memory of Frank A. Miller (1858–1935), proprietor of the Mission Inn, Riverside, and a civic leader.

Millerton Lake [Fresno, Madera Cos.]. In 1851 Lt. Tredwell Moore established a fort at the river and named it in honor of Maj. Albert S. Miller, a Mexican War veteran and at that time commanding officer at Benicia. In 1854 the mining town of Rootville, about one mile below, was renamed after the fort; it appears as Millerstown in the *Statutes* of 1854 (p. 222) but as Millerton on the maps. It was the seat of Fresno Co. from 1856 to 1874. The reservoir, formed by Friant Dam, has now obliterated the site of the historic town, but it preserves the name.

Millikin Corners [Santa Clara Co.]. Named for John Millikin, a native of Pennsylvania, who settled in Santa Clara Co. in 1852 and lived on his farm until his death in 1877. The name is also spelled Millican.

Mills. *See* Mill.

Mills, Mount. *See* Millbrae.

Millsaps [Glenn Co.]. Named for George W. Millsaps, of Kentucky, who came to California in 1854.

Mills College [Alameda Co.]. In 1871 Dr. Cyrus Mills moved the Young Ladies Seminary from Benicia to Oakland. It became generally known and was incorporated under the present name in 1886. The post office was named Mills Seminary before 1880 and was changed to Mills College in 1888. Dr. Mills had intended to give the name Alderwood Seminary and had erased his name from all blueprints, except the one that was in the hands of the architect and shown by the latter to a reporter (Wolfe).

Mills Creek. *See* Millbrae.

Millseat Creek. *See* Mill.

Millsholm [Glenn Co.]. The station was named for Edgar Mills when the Southern Pacific spur to Fruto was built through his land in the 1880s. "Holm" is a generic term meaning 'island' in English and Scandinavian, but 'hill' in German. The latter meaning is indicated by the surrounding hills.

Millux [Kern Co.]. The name was applied to the station when the Sunset spur of the Southern Pacific was built in 1901. It is coined from that of the stock-raising firm Miller and Lux, which had large holdings nearby (Santa Fe).

Mill Valley. *See* Mill.

Millville. *See* Mill.

Milo (mī′ lō) [Tulare Co.]. The post office in Mountain View Valley was established in 1882 and named Cramer, for Eleanor Cramer, one of the first white settlers. When the Post Office Dept. requested a change of name in 1888, Henry Murphy sent in a list of names, from which the department selected Milo (Mitchell).

Milpitas (mil pē′ təs) [Santa Clara Co.]. The word is a diminutive of Mexican Spanish *milpas* 'cornfields'. *Milpa* is of Aztec origin, from *milli* 'field' plus locational *-pan*. The name was preserved through the Milpitas grant, dated Sept. 28 and Oct. 2, 1835. Máximo Martínez testified in the U.S. District Court on Oct. 13, 1861, that the place was so called because his father "sowed, cultivated, and lived there, and after raising the crop, left for the pueblo (San Jose). Some Indians were living with us" (Bowman). The town was founded in the 1850s. Milpitas Village is shown on a plat of the Rincon de los Esteros grant in 1858, and the post office was established on May 31, 1856. **Milpitas Ditch** and **Las Milpitas** [Monterey Co.], near San Antonio Mission, are reminiscent of the Milpitas grant, dated May 5, 1838.

Milton [Calaveras Co.]. The name was applied to the terminus of the Stockton and Copperopolis Railroad in 1871. It is not known whether the station name was given for Milton Latham, governor of California for five days in 1860, U.S. senator, successful banker, and unsuccessful railroad builder; or whether it had been intended for W. J. L. Molton, a director of the railroad, and was misspelled on maps.

Mimulus (mim′ yə ləs) **Spring** [Contra Costa Co.]. The genus name for the monkey-flower (diminutive of Latin *mimus* 'comic actor').

Minarets, The; Minaret Creek [Madera Co.]. "To the south of [Mount Ritter] are some grand pinnacles of granite . . . to which we gave the name of 'the minarets'" (Whitney, *Yosemite book* 1868:98).

Mindego. *See* Mendico.

Mineral [Tehama Co.]. When the post office was established on June 4, 1894, it was so named because it was near the Morgan Mineral Springs. When the post office was moved to the new location, the old name was retained.

Mineral King [Tulare Co.]. The settlement, which grew up around the mine that opened here in 1872, was first called Beulah but was changed to Mineral King when a mining district was organized and proclaimed "the king of mineral districts" (Farquhar).

Mingo Creek [Humboldt Co.]. For a dog lost here around 1910, according to the BGN (list 7704).

Minkler [Fresno Co.]. When the Santa Fe spur from Reedley to Delpiedra was built in 1910, the station was probably named for Charles O. Minkler, a farmer at Sanger, or for a member of his family.

Minneola (min ē ō′ lə) [San Bernardino Co.]. In 1902 the Santa Fe named the siding after the now-vanished boom town that had been built at the site as the terminus of an irrigation canal about 1895. The name was given for Minnie Dieterle, the wife of an official of the Southern California Improvement Company, which had built the canal (Santa Fe). But there was probably influence of Minneola in Minnesota and other states, from a Siouan name meaning 'much water' (Stewart 1970).

Minor Creek [Humboldt Co.]. The name may have been given for Isaac Minor, who settled in Humboldt Co. in 1853.

Mira. One of the most popular Spanish words used in coining pleasant-sounding place names. The word does not mean 'view', as is generally assumed, but is a 3d person singular present tense form; a place name like Miramar, as found in Spanish-speaking countries, means literally '(it) looks at (the) sea'. The pattern has been extended in California. **Miramar** (mēr′ ə mär) [San Diego Co.] was the name given by E. W. Scripps to his Linda Vista Ranch. It appears in the *San Diego Union* on Aug. 5, 1891, and was applied to the post office in Apr. 1892. **Mira Loma** (mēr′ ə lō′ mə) [Riverside Co.] is intended to mean 'hill view'. The original name, Stalder, for an old settler, was changed to Wineville when the Charles Stern Winery was built there. The present term replaced the alcoholic name when the post office was established on Nov. 1, 1930 (C. E. Faulhaber). **Miraleste** (mēr ə les′ tē) [Los Angeles Co.] is from Spanish *mira al este* 'it looks to the east'. The name was applied in 1924 by the Palos Verdes Project, a land development company, after consultation with several Spanish scholars. **Miramonte** (mēr ə mon′ tē) [Fresno Co.] and **Mira Monte** [Ventura Co.] were intended to mean 'mountain view'. Spanish *monte* historically meant 'mountain', but the usual modern sense is 'bushes, chaparral'; *see* Monte.

Mirabel [Lake Co.] was once the site of a cinnabar mine owned by Messrs. Mill, Randall, and Bell; there was a post office here in 1892–93 (Feltman, p. 60).

Miracle Hot Springs. *See* Hobo Hot Springs.

Mirador (mēr′ ə dôr) [Tulare Co.]. The name, meaning 'balcony' or 'gallery,' was applied to the Santa Fe station in 1923.

Miramontes (mēr ə mōn′ təs): **Point**, ridge [San Mateo Co.]. The name can be traced to the family of Candelario Miramontes, who settled south of Pilarcitos Creek in 1840 and became grantees of Arroyo de los Pilarcitos on Jan. 2, 1841. The name was first applied to the northern headland of Halfmoon Bay by the Coast Survey in 1854. When Halfmoon Bay was charted in 1862, the name Pillar Point was chosen for this cape and the name Miramontes was transferred to the nameless point 5 miles south, which was actually on the property of the rancho. The name was later given (without a generic term following) to the chain of hills north of Pillar Point, the cartographer probably assuming that *montes* meant 'mountains'.

Miranda [Humboldt Co.]. The name was applied to the post office about 1906; it is not known whether the name-giver had in mind a girl, or the well-known Spanish place name and family name.

Mirror Lake [Yosemite NP] was named for its reflections by C. H. Spencer, of the Mariposa Battalion.

Mission. The importance of the Franciscan missions in the early history of the state is reflected in many place names, not all of them close to an old mission. The names of the twenty-one California missions will be found under the specific names. **Mission: Bay, Bay State Park, Beach, Valley** [San Diego Co.]: Crespí (p. 122) referred to the bay as the second but closed harbor of San Diego, and Font (p. 237) called it Puerto Anegado 'overflowed port'. On Pantoja's map of 1782 it is shown as Puerto Falso (Wagner, p. 453), and this name it retained—later translated to False Bay by the Coast Survey—until the BGN changed the name to Mission Bay by decision of June 2, 1915. The state park was named in 1929. **Mission: Creek, Rock** [San Francisco Co.]:

The tidal channel originally extended to Mission Dolores, making possible communication by water with downtown San Francisco. It is shown on a *diseño* of Limantour's land claim (1842) as Estero de la Misión. The bay between Steamboat and Potrero points, long since filled in, had been named Ensenada de los Llorones 'bay of the weepers' by Ayala in 1775 but became known to Americans as Mission Bay. Twin Peaks were sometimes known as Mission Peaks. **Mission San Jose** (san hō zā′, san ō zā′) [Alameda Co.] was founded on June 11, 1797, as the Misión del Gloriosísimo Patriarca San Joseph. In Spanish times it was commonly known as the Misión de San José. This became Mission San Jose when the post office was established on Apr. 9, 1850. The community became part of the city of Fremont in 1956. **Mission Hills** is a post office in Los Angeles Co.; there are communities of the same name in San Diego and Santa Barbara Cos. **Mission Viejo** (vē ā′ hō) [Orange Co.] takes its name from a Rancho Mission Viejo, granted in 1845 from former lands of Mission San Juan Capistrano. In the 1960s the name was applied to a new master-planned community, which today is the incorporated City of Mission Viejo (Brigandi). If the name is supposed to mean 'old mission', then there is a gender error; the correct Spanish would be *misión vieja*, or better *misión antigua*. But in Rancho Mission Viejo, Viejo 'old' can be interpreted as referring to the masculine noun Rancho.

Mitchell: Canyon, Creek [Contra Costa Co.]. In 1853 Captain Mitchell located a claim in the canyon south of the present town of Clayton (Co. Hist.: Contra Costa 1882).

Mitchell: Meadow, Peak [Tulare Co.]. The meadow was named for Hyman Mitchell, of White River, and the peak for Susman Mitchell, his son (Farquhar).

Mitchells Caverns [San Bernardino Co.]. The caves were discovered before 1895—probably by Hi-corum, a Chemehuevi Indian—and were named for J. E. Mitchell, who later developed them.

Miter, The [Sequoia NP]. The peak was named by Chester H. Versteeg, about 1935, because its shape resembles the ornamental headdress worn by bishops.

Miwok (mē′ wok, mē′ wuk) **Lake** [Yosemite NP]. So named because there are Miwok Indian sites here (Browning 1988); from Central Sierra Miwok *miwwïk* 'people; Indians'. The same term occurs in **Mi-Wuk Village** [Tuolumne Co.]: When Harry Hoefler developed the community in 1955, he selected the name after consultation with Chief Fuller of a nearby Miwok village (Mary L. Storch).

Moaning Cave [Calaveras Co.]. The large domelike vault was so named because of a curious sound heard at the entrance. The "moaning" disappeared after the circular stairway was built in the main chamber (Doyle). The original name was Solomons Hole. It was believed among gold seekers that it had been worked for gold in Mexican, or even Spanish, times, until Trask disproved this in 1851.

Moavi (mō ä′ vē) **Park** [San Bernardino Co.]. Possibly from Mojave *muu'wáav* 'kinsperson' (Munro).

Moccasin (mok′ ə sən). This word was originally introduced to English (from an east-coast Algonquian source) as *mekezin*, by Marc Lescarbot in 1609 (Cutler, p. 21), to designate the footgear of American Indians. It is found in the names of several California communities and physical features. The term also refers to a kind of snake, though not one native to California. **Moccasin: Creek, Peak**, post office [Tuolumne Co.]: The creek is shown as Mocosin on Gibbes's map of 1852. The stream was so named because miners mistook the numerous water snakes for moccasin snakes, found in swampy regions in the South.

Mocho (mō′ chō), **Arroyo; Mocho Mountain** [Alameda, Santa Clara Cos.]. Arroyo Mocho is shown on several *diseños* of the 1830s. According to Still, "It was on account of this creek having no outlet, but sinking into the ground (except in wettest weather) after spreading out into many smaller streams between Livermore and Pleasanton, that it was given the name Arroyo Mocho, meaning 'cut-off creek'." The hybridization Mocho Creek is recorded as early as 1852 (*Statutes*, p. 178). In 1875 William Eimbeck of the Coast Survey established a triangulation station on the nearby mountain, but he was not sure of the meaning of the name and changed the spelling to Macho 'mule'. In 1887 Davidson restored the original spelling of the name of the creek and the mountain (E. K. Gudde, "Mocho Mountain," *Names* 5:246–48 [1957]). Los Machos Hills [Monterey Co.] may refer to 'mules' or may be the result of a similar misunderstanding. **Mocho Creek** [Monterey Co.]: Clark 1991 suggests that this may have been named for someone referred to as *mocho* 'maimed, lacking a body part'.

Mococo (mō kō′ kō) [Contra Costa Co.]. The name was coined in 1912 from Mountain Copper Company, the name of the company that owned the copper smelter at Bulls Head Point. From 1910 to 1912 the Southern Pacific station was called Lewis, for the general manager of the company.

Mocosin Creek. *See* Moccasin: Moccasin Creek.

Modena. *See* El Modena.

Modesto (mō des′ tō, mə des′ tō) [Stanislaus Co.]. The Central Pacific Railway reached the place on Nov. 8, 1870, and the station was supposed to be named for William C. Ralston, one of the railroad's directors and the most colorful of San Francisco's financiers. It is said that Ralston, on hearing of the honor, modestly declined, whereupon the name was changed to the Spanish word for 'modest'.

Modin (mō′ din) **Creek** [Shasta Co.]. The branch of Squaw Creek was named for Jim Modin, whose home was on the stream (Steger).

Modjeska (mō jes′kə): **Canyon, Island, Peak** [Orange Co.]. The Polish actress Helena Modjeska (1840–1909) and her husband, Count Bozenta, financed a farming project at Anaheim during the 1870s; later they purchased most of the canyon (Brigandi). The island (also called Bay Island) was named Modjeska in 1907, when the actress bought a house there. After her death in 1909, J. B. Stephenson, forest ranger, named the mountain in her memory.

Modoc (mō′ dok): **County, National Forest**. The county was created from a part of Siskiyou Co. by act of the legislature on Feb. 17, 1874, and was named for the Indian tribe, which had been subdued after severe fighting in the Modoc War of the preceding year. The name is derived from *moowataakknii*, which means 'southerners' in the language of the neighboring Klamath tribe of Oregon (M. A. R. Barker, *UCPL* 31:242).The modern spelling was used in the Indian Report of 1854 (p. 471). In a geographical sense the name was apparently not applied until the creation of the county, although immigrants, traveling over the bloody trail from Lassens Pass, used the term "Modoc country." Modoc National Forest was created and named in 1908.

Moffett Creek [Siskiyou Co.]. This tributary of the Scott River preserves the name of a prospector who settled at the creek briefly in 1850 (Schrader).

Moffett Field [Santa Clara Co.]. The air field was transferred from the Navy to the Army in 1935. The following year, the War Dept. named it in honor of Rear Admiral William A. Moffett.

Mogneles River. *See* Mokelumne.

Mohave. *See* Mojave.

Mohawk: Valley, Creek, settlement [Plumas Co.]. The region was settled by migrants from New York State (*see* Hosselkus). The settlement had a post office from 1881 to 1926. Mohawk, one of the most popular place names of Indian origin in the United States, is the name given by whites to an Iroquoian tribe of New York State; however, the term is Algonquian in origin.

Mohr Station. *See* Bethany.

Moigne Mountain. *See* Pinto: Pinto Peak.

Mojave (mō hä′ vē, mə hä′ vē): **River, Valley, Sink** [San Bernardino Co.]; **Desert** [San Bernardino, Kern, Los Angeles Cos.]; town [Kern Co.]; **Mohave: Mountains, Canyon, Rock, Wash, Indian Reservation** [San Bernardino Co.]. The name is derived from the populous and warlike Yuman tribe on the Colorado where California, Arizona, and Nevada now meet. These Indians are mentioned in 1775–76 as Jamajabs in the diaries of Font and Garcés. The Spanish and English names are derived from the tribe's name for itself, *hamakháav*, which may be related to *haa-* 'water' (*HNAI* 10:69). It does not mean 'three mountains', as has been often reported; this refutation is argued by Lorraine M. Sherer, "The name Mojave, Mohave," *Southern California Quarterly* 1967:49.1–36, 455–58, and confirmed by Pamela Munro. In literature the name appears with more spelling variants than any other Indian name in California. In Whipple's report the Yuman version is given as Mac-há-vès and the Paiute version as A-mac-há-vès (Pac. R.R. *Reports* 3:3.16). A spelling approximating the modern version, Mohawa, was used as early as 1833 (Pattie, p. 93). In a geographical sense the name was first applied by Frémont to the river on Apr. 23, 1844: "The two different portions in which water is found had received from the priests two different names [Arroyo de los Mártires by Garcés on Mar. 9, 1776];

and subsequently I heard it called by the Spaniards the Río de las Ánimas, but on the map we have called it the Mohahve river" (*Expl. exp.*, p. 377). Frémont's reason for applying this name to the river in the Great Basin, separated by several mountain ranges from the territory of the Mojave Indians, has been a puzzle to scholars. Kroeber believes that it arose from the "erroneous impression that this [river] drained into the Colorado in the habitat of the Mohave." The explorer, however, applied the name although he knew the river was not in Mojave territory. On the day he named the stream, he had met a party of six roving Mojaves, one of whom told him that "a short distance below, this river finally disappeared" (ibid.) and also gave him interesting information about his tribe. This meeting induced Frémont to name the stream Mohahve. Mojave River had been called Inconstant River by Jedediah Smith in 1826 (*Travels*, p. 15), and it is thus designated on the maps of Burr 1839 and Wilkes 1841. Mojave Valley is repeatedly mentioned in the Pac. R.R. *Reports*. The term Sink of Mohave is recorded on Goddard's map (1860); it is now also known as Soda Sink. The term Mohave Desert seems to have been applied by the Wheeler Survey in 1875 (atlas sheet 73-C), although it was doubtless used before that date. The town in Kern Co. came into existence when the Southern Pacific reached the place on Aug. 8, 1876, and called the station Mojave because it was at the western end of the Mojave Desert. The names of the places straddling the Colorado in San Bernardino Co. and in Mohave Co. (Arizona) arose because they were all in or near the habitat of the Mojave Indians. The mountains east of the Colorado are called "Hamook Häbî" on the Whipple-Ives map (1854). The BGN (6th Report) decided for the spelling Mojave for the California names in the Great Basin but left the spelling Mohave for the names on the Colorado, thus accentuating the difference in the origin and application of the two name clusters, despite their common source. The spelling with the *j* is official in the use of the Mojave Tribe, but *h* is official for the Arizona county.

Mokelumne (mō kel′ ə mi, mō kol′ ə mē): **River** [San Joaquin Co.]; **Hill** [Calaveras Co.]; **Peak** [Amador Co.]. According to Barrett (*Miwok* 1908:340), the name is derived from a Plains Miwok village near Lockeford. The ending *-umne* means 'people' (see Cosumnes, Tuolumne); the stem may be related to Plains Miwok *moke* 'fish net', or perhaps Central Sierra Miwok *moke* 'red paint' or *mokólkine* 'manzanita berry' (Callaghan). The Indians are called Muquelemnes by Durán on May 23, 1817, and their name appears with similar spellings elsewhere (Arch. Arz. SF 3:2.83–84, 104, etc.). On the *Plano topográfico de la Misión de San José* (about 1824) a village of "heathen Indians," Muguelemnes, is indicated near the site of the present city of Lodi. The name was applied to the river by the Wilkes party, probably at the suggestion of Sutter: Rio Mokellemos (Eld), Mogneles River (Wilkes map, 1841). The name Sanjón ('ditch') de los Moquelemes was used for a land grant on Jan. 24, 1844, and the name of the river appears in the titles of several other grants. The present spelling, Mokelumne River, was used by Frémont (*Geog. memoir*, p. 16). Mokelumne Hill, first called Big Bar, then known as Mok Hill or The Hill, developed as a mining camp in 1848 and became one of the important centers of the southern mines. The post office was established on July 10, 1851. **Mokelumne City** [San Joaquin Co.] became deserted after the Central Pacific built a station about 10 miles southeast, which it called Mokelumne until 1874, when the name was changed to Lodi.

Molaine Corrals [Trinity Co.]. The place has been known by this name since 1887, when Jim Molaine had a holding pasture here for his cattle (W. E. Hotelling).

Molasky (mə las′ kē) **Creek** [Santa Cruz Co.]. Probably for Henry Molaskey, who lived in the area in the 1870s (Clark 1986).

Molate (mō lä′ tē): **Point, Reef** [Contra Costa Co.]. The name Moleta was applied in Mexican times to the island now known as Red Rock, probably because its shape resembles the muller (called *moleta* in Spanish) used for grinding pigments. Beechey misspelled the word as Molate in 1826, and this version was adopted in 1851 by the Coast Survey when it used the island as a secondary triangulation station. Molate Point was named by the Survey in 1854, and Molate Reef in 1864. *See* Red: Red Rock.

Moleta. *See* Molate; Red.

Molino (mō lē′ nō, mə lē′ nō). The Spanish word for 'mill' was repeatedly used as a geographical term; it not only has survived in a number of places but has been used for place naming by Americans. In some cases, however, it may reflect a Spanish family name. **Molino** [Sonoma Co.] was so named because it is on the Molino land grant, dated Feb. 24, 1836. **Los Molinos** [Tehama Co.] was named after the land grant El Río de los Molinos, dated Dec. 20, 1844. The post office is listed in 1908 (*see* Mill). **Molino Creek** [Santa Cruz Co.] is shown as Arroyo del Molino on a *diseño* of the San Vicente grant (1846). On another *diseño* of the same grant it is misspelled Aroyo del Molyllo. **Molino** [San Bernardino Co.] was the name given to the Santa Fe station in 1909 because the Brookings Lumber Company had built a mill here in 1896 (Santa Fe). *For* Arroyo de Molinos, *see* Mill: Mill Creek.

Molybdenite (mə lib′ də nīt): **Creek, Canyon** [Mono Co.]. Named for an outcropping of ore of molybdenum, a metallic element of the chromium group.

Momyer [mom′ yər] **Creek** [San Bernardino Co.]. Named by the BGN (list 7503) for conservationist Joe R. Momyer (1910–73).

Monache (mō nach′ ē, mō′ nə chē). A division of the Numic branch of Uto-Aztecan Indians, also called Mono (*q.v.*). Kroeber (*Handbook*, map) indicates their main habitat along the eastern slope of the Sierra Nevada, from about 40 miles north of Mono Lake to slightly north of Owens Lake. Their western neighbors, the Yokuts, supposedly called them *monachi* 'fly people' because their chief food staple and trading article were the pupae of a fly, *Ephyda hyans*, found in great quantities on the shores of the Great Basin lakes (*CFQ* 4:90–91). It is certainly not accidental that the name is preserved mainly in two clusters, one at Mono Lake and the other at Owens Lake. **Monache: Meadow, Creek, Mountain** [Tulare Co.], **Monachee** [Inyo Co.]: The name was recorded in a geographical sense when the people of Owens Valley petitioned the legislature in Feb. 1864 to create a new county south of Mono Co. and name it Monache (Chalfant, *Inyo*, p. 175). The von Leicht–Craven map (1874) shows the phonetic spelling Monatchay Meadows; the Mining Bureau map of 1891 has the now-accepted spelling. The name Monache was applied to the mountain and the creek when the USGS mapped the region in 1905. With an additional *e* to indicate the pronunciation, the name was applied to the station when the California-Nevada Railway was built in 1908.

Monarch Divide [Kings Canyon NP]. The divide between the Middle and South Forks of the Kings River was named Dyke Ridge by the Whitney Survey (*see* Dike). The new name was apparently applied by the USGS when the Tehipite quadrangle was mapped in 1903, perhaps by analogy to Kings River.

Monica, Sierra de la. *See* Santa Monica.

Monitor: Creek, Pass [Alpine Co.]. The settlement of Monitor was named in 1863 for the ironclad *Monitor*, which fought the *Merrimac* at the outbreak of the Civil War in 1862 (Browning 1986). It is now a ghost town.

Monmouth (mon′ məth) [Fresno Co.]. The station was named in the 1890s by the Santa Fe after Monmouth, Illinois, former home of a settler; that in turn was probably named for the town and county in England.

Mono (mō′ nō), or its variant Monache (*q.v.*), refers to an Indian tribe speaking a language of the Numic family. **Mono: Lake, Pass, County, Valley, Dome, Craters** [Mono Co.]; **Divide, Rock** [Fresno Co.]: As a geographical term, the shorter and more common form of the name is the older. It was applied to the lake by Lt. Tredwell Moore in the summer of 1852 and appears on Trask's maps of 1853. The county was created and named by act of the legislature on Apr. 24, 1861. There was a short-lived mining town, Monoville, after Cord Norst found gold in the hills around Owens Valley in July 1859. The pronunciation "(mon′ ō) Lake" is now popular among airline pilots who fly over the area. **Mono Jim Peak** [Mono Co.] was named for a Mono Indian guide who, along with Robert Morrison, was killed near Convict Lake in 1871 during a gunfight with escaped convicts (BGN, list 8704).

Monocline (mon′ ə klīn) **Ridge** [Fresno Co.]. This ridge near the San Benito Co. line was so named, probably by the USGS when the Panoche quadrangle was mapped in 1908–11, because it appears to have only one inclination or slope, extending into San Joaquin Valley.

Mono Creek [Santa Barbara Co.]. This name may contain Spanish *mono* 'monkey; cute'; there is no reason to connect it with the name of the Indian tribe in east central California.

Monolith [Kern Co.]. The place was named in

1908 by William Mulholland, builder of the Los Angeles Aqueduct, after the Monolith Portland Cement Company.

Monroe Meadow [Yosemite NP]. Named for George F. Monroe, who drove stages on the Wawona Road to Yosemite Valley from 1868 to 1888.

Monroeville [Glenn Co.]. The place was named for U. P. Monroe, who settled here before 1851. When Colusa Co. was organized in 1851, Monroe's Rancho became the first county seat. The post office is listed as Monroeville the same year.

Monrovia: city, **Hill** [Los Angeles Co.]. Named for William N. Monroe, a railroad construction engineer, who with his associates laid out the town in 1886 on sixty acres of Ranchos Santa Anita and Azusa de Duarte. The Latinized form of the place name is like that of Monrovia in Liberia, named for U.S. President James Monroe.

Monserrate (mōn′ sə rät) **Mountain** [San Diego Co.]. The Monserrate land grant dates from May 4, 1846. The peak was supposedly named after Monserrat, a famous mountain and monastery in Catalonia, Spain (Co. Hist.: San Bernardino 1883:180).

Monson [Tulare Co.]. A railroad name applied to the station in 1887. There were at that time places called Monson in Massachusetts and in Maine.

Montague [Siskiyou Co.]. When the Southern Pacific extension was built from Dunsmuir to the state line in 1886–87, the station was named for S. S. Montague, chief engineer of the Central Pacific (Southern Pacific).

Montalvo (mon tal′ vō) [Ventura Co.]. The name was applied to the station when the Southern Pacific reached the point in the summer of 1887. Garcí Ordóñez de Montalvo is thought to be the author of *Las sergas de Esplandián* (about 1510), in which the name California probably appears for the first time. Montalvo is also the name of James Phelan's former estate in Santa Clara Co. (Drury).

Montaña de Oro (mon tan′ ə di ôr′ ō, mōn tä′ nyə . . .) **State Park** [San Luis Obispo Co.]. The name does not stand for a 'mountain of gold' but evokes the blaze of spring flowers on its slope. The name was given to the northern part of the Spooner Ranch when purchased by Irene Starkey McAllister in 1954; the site was purchased by the state park system in 1965 (Hall-Patton).

Montara (mon tär′ ə), **Point; Montara: Mountain**, post office [San Mateo Co.]. The name, spelled Montoro, was used for the mountain and the point by the Whitney Survey in 1867, and in 1869 the present form was used by the Coast Survey (*Coast Pilot* 1869:54). Both are probably misspellings of one of several similar Spanish words referring to forest and mountain: *montuoso, montaraz, montaña*. A *Cañada Montosa* 'valley full of brush' was shown about 1838 on a *diseño* of nearby Rancho San Pedro. Cf. Beeler, "Montara."

Montclair. The name is often used in the United States. Among the many places so named, the most important is the citrus-producing Montclair in San Bernardino Co., named in 1960.

Monte. This Spanish word for 'bushes, brush, woods' was often used as a generic geographical term in Spanish times. Although Spanish dictionaries also list *monte* with the meaning 'mountain', no evidence has been found that it was so used in Spanish-speaking countries except where a hill or mountain was densely covered with trees. A *diseño* of the San Antonio or El Pescadero grant (1833) bears out the use of the word in Spanish California: Butano Ridge is called "Lomas" and "Sierra con Monte" 'hills/ridge with bushes'. However, Anglo-Americans have long believed that *monte* means 'mountain', as it does in Italian; the usual pronunciation is (mon′ tē). Mount Diablo appears as Monte Diablo as late as 1874 (Hoffmann's maps); *for* Monte Diablo Creek, *see* Convict Creek. The word continues to be applied to elevations: Monte de Oro [Butte Co.], Monte Arido [Santa Barbara Co.]. The term has also been used for communities: Monte Rio [Sonoma Co.], intended to mean 'river mountain', contains *río* 'river'. Monte Vista [Santa Clara Co.] contains *vista* 'view'; the same name appears in San Bernardino and Placer Cos. Monte Nido (nē′ dō) [Los Angeles Co.] contains *nido* 'nest'. Where *monte* is used with an Italian specific term, the combination is grammatically correct, as in Monte Bello ('beautiful mountain') Ridge [Santa Clara Co.]. Monte Sereno [Santa Clara Co.] is supposed to mean 'serene mount'. (*See* Del Monte; El Monte; Diablo.)

Montebello (mon tə bel′ ō) [Los Angeles Co.]. In 1887 Harris Newmark, a Los Angeles businessman, purchased part of the Repetto ranch. The entire settlement was called Montebello, an international name of Italian origin ('beautiful mountain'), but the town itself was named Newmark. This name was dropped on Oct. 16, 1920, in favor of Montebello: "another of the many instances in recent years of the lack, among Californians, of proper historic respect for pioneer names," as Newmark's son remarks in the 1926 edition of his father's book, *Sixty years in southern California* (pp. 555, 668).

Montecito (mon tə sē´ tō): town, **Creek, Peak** [Santa Barbara Co.]. A place called Montecito 'little woods' is mentioned by Fages as early as Dec. 5, 1783, as a site suitable for a mission (Prov. Recs. 3:55), and Mission Santa Barbara was founded there in 1844. The name appears repeatedly in mission and land grant papers. The present settlement of Montecito is farther east, on Santa Barbara pueblo lands; a Terreno (portion of land) de Montecito was given to José Rosas, probably a former soldier, on May 15, 1834. It is mentioned as Monticito in the *Statutes* of 1850 (p. 172).

Monte Cristo Channel [Sierra, Nevada Cos.]. The name of the old river bed, stretching from north of Downieville to North Bloomfield, goes back to the early 1850s, when rich gold deposits were discovered along its course. Alexandre Dumas's novel *The Count of Monte Cristo* was then at the height of its popularity.

Monterey (mon tə rā´): **Bay, Harbor, Presidio**, city, **County, Canyon**. The bay was discovered by Cabrillo on Nov. 16, 1542, and named Bahía de los Pinos. Cermeño crossed it on Dec. 10, 1595, and named it San Pedro, in honor of Saint Peter the Martyr, whose feast day is Dec. 9 (Wagner, 398). Seven years later, on Dec. 16, 1602, Vizcaíno anchored in what is now Monterey harbor and named it "Puerto de Monterey", in honor of Gaspar de Zúñiga y Acevedo, 5th Count de Monterrey, then viceroy of New Spain; his ancestral castle was at Monterrey in Galicia, Spain (Clark 1991). The name stands for *monte del rey* 'mountain/forest of the king'; the city of Monterrey, Mexico, shows the modern Spanish spelling of the same name. When Portolá was sent north in 1769, he was commissioned to find this harbor, which had been described as an excellent port by Vizcaíno. On June 3, 1770, a presidio and a mission were established, both named San Carlos Borromeo. The presidio, however, was known by the name Monterey even in Spanish times; and in 1904 the U.S. War Dept. renewed this name in perpetuation of the first Spanish military post in California. The county, one of the original twenty-seven, was named on Feb. 18, 1850. The submarine canyon, probably the same in which Cabrillo had anchored, was named by the Coast Survey, and the name was confirmed by the BGN in 1938. The Salinas River was called Río de Monterey by Font on Mar. 4, 1776, and was called Monterey River as late as 1850 (*Statutes*, p. 59).

Monterey Park [Los Angeles Co.]. A subdivision on the Repetto ranch was developed in 1906 and named Ramona Acres. At the time of its incorporation, the city was renamed after Monterey Pass to the west (Gertrude Shearer). The pass is now often called Coyote Pass.

Montezuma (mon tə zōō´ mə). The name of the Aztec ruler at the time that Cortez invaded Mexico was suggested as a new name for Alta California by José María Echeandía, governor of the province, 1825–31. The Diputación (provincial legislature) adopted the suggestion in 1827, but the Mexican government refused to concur (Chapman, p. 460). The name Montezuma was frequently used as a U.S. place name, and after the Mexican War it became very popular in California, where it still survives in a few places. A variant form of the name, current in Mexico, is Moctezuma; the Aztec original is Motecuzoma, meaning something like 'angry lord'. **Montezuma Landing, Hills, Creek, Island, Slough**, and station [Solano Co.] recall one of the attempts of the Mormons to settle in California. Lansford W. Hastings, their agent, laid out Montezuma City at the head of Suisun Bay in 1847. Although often mentioned in contemporary newspapers and books, the "city" never seems to have developed beyond Hastings's own adobe. Montezuma Hills appear on Ringgold's general chart (1850); and they are mentioned in Hutchings' *Illustrated California Magazine*, Mar. 1858. **Montezuma** [Tuolumne Co.]: A prosperous trading post, Montezuma House, was established here by Sol Miller and P. K. Aurond in 1850 and named after Montezuma Flats.

Montgomery: Creek, Peak [Mono Co.]. The creek is shown on Wheeler atlas sheet 57, but the peak is labeled White Mountain Peak.

The man for whom the creek was named was perhaps the owner of the mill that Hoffmann's map (1873) shows halfway down the course of the stream, near the edge of the talus. In 1917 the USGS renamed the peak and transferred the old name to a peak 16 miles farther south.

Montgomery Creek [Shasta Co.]. The post office was named Montgomery Ferry in 1877, but the name was changed to the present form in 1878. The creek from which the town derives the name was probably so called because Zack Montgomery once made a big haul when fishing there in the early 1850s (Steger).

Montgomery Memorial State Park [San Diego Co.]. Created in 1952 in memory of John J. Montgomery, the first American to experiment with gliders; he died in a fall in 1911. **Montgomery Hill** [Santa Clara Co.] was also named for the pioneer aviator (BGN, 1964).

Monticello (mon ti sel′ ō) [Napa Co.]. An extremely popular place name in the United States since the time of Jefferson, who gave his Virginia estate the name of a town in northern Italy (lit. 'little mountain'). The post office was established on July 8, 1867. The town is now inundated by Berryessa Lake, but Monticello Dam preserves the old name.

Montoro. *See* Montara.

Montosa, Cañada. *See* Montara.

Montpelier (mont pēl′ yər) [Stanislaus Co.]. The station was named in 1891 upon completion of the Southern Pacific from Oakdale to Merced, probably after the capital of Vermont (named in turn for the city in France, spelled Montpellier). The name of the former post office was spelled Montpellier, but the present spelling is shown on the Official Railway Map of 1900.

Montrose [Los Angeles Co.]. The name, which has long been a popular place name in the United States, was chosen, as the result of a contest, for the subdivision established in 1913 on part of the La Crescenta development.

Monument. About fifteen orographic features bear this name, and in addition a number of Monument Creeks are named after them. Most of the peaks so named are at the boundaries of the state where surveying parties had erected a "monument." The name Monument Peak at the intersection of El Dorado and Alpine Cos. with Douglas Co. (Nevada) was apparently placed on the map (atlas sheet 56-B) by the Wheeler Survey because of the proximity of the peak to one of the granite monuments erected by the boundary survey of 1872. Some Monument Mountains were named because of their "monumental" appearance.

Moody Creek [Shasta Co.]. The branch of Stillwater Creek was named for M. G. and Elizabeth Moody, who filed a land claim on this creek in 1852 (Steger).

Moon. The word is often found in the United States in the names of lakes; **Moon Lake** [Lassen Co.] is a body of water formerly called Tule Lake Reservoir (BGN, list 6804). **Moonshine Creek** [Yuba Co.] was probably named because "moonshiners" were active here; **Moonlight Peak** [Plumas, Lassen Cos.] must have been named by a romantic settler or surveyor.

Moon Creek [Shasta Co.]. The tributary of Rainbow Lake was named for Arch Moon, who filed a homestead claim here in 1883 (Steger).

Mooney, Mount [Los Angeles Co.]. Named in memory of John L. Mooney of the USFS, who died in France in World War I.

Mooney Flat [Yuba Co.]. Named for Thomas Mooney, who in 1851, in partnership with Michael Riley, established a hotel and a trading post on the nearby Empire Ranch, which was on the route of the California Stage Company.

Moonstone Beach. Three beaches [Humboldt and San Luis Obispo Cos., Santa Catalina Island] are so named because of moonstones washed in by breakers.

Moorehouse Creek [Tulare Co.]. Named for Gus Moorehouse, an early prospector (Farquhar).

Moorek (mo͞o′ rek) [Humboldt Co.]. Yurok *muurekw*, the name of a village (Waterman, p. 247; Harrington). Also spelled Morek.

Moores Creek [Napa Co.]. The creek and the ranch were named for "Old Lady Moore," a settler in the 1850s who died under suspicious circumstances (*CHSQ* 25:9). *See* Las Posadas.

Moores Flat [Nevada Co.]. Named for H. M. Moore, who built the first house and store here in 1851, after having driven a herd of cattle across the continent (Brown and Dallison's Nevada, Grass Valley Directory, 1856).

Moore's Station. *See* Riverton.

Moorman Meadow [Mono Co.]. Named after the pioneer Moorman Ranch but spelled Mormon on the Bridgeport atlas sheet (Maule).

Moorpark [Ventura Co.]. The town was founded about 1900 and named for a well-known English variety of apricot, long a favorite variety in southern California.

Moosa: settlement, **Canyon** [San Diego Co.]. According to the file of the San Diego Public Library, the name is an abbreviation of Pamoosa, which in turn is derived from Pamusi, a rancheria mentioned on July 23, 1805; a Diegueño origin seems possible but has not been established. A post office, Moosa, is listed in 1887.

Moose, Moosehead. These names, found mainly in northern counties, were applied because large specimens of elk were mistaken for moose, which have never lived in California. **Moose Lake** [Sequoia NP] was so named because its outline on the map resembles the head of a moose.

Moovalya (mo͞o väl′ yə), **Lake** [San Bernardino Co.]. From Mojave *muuvály*, the name for this area; no further etymology is known (Munro).

Mopeco (mō pē′ kō) [Kern Co.]. The name of the railroad siding was coined from Mohawk Petroleum Corporation (Santa Fe).

Moquelumne. *See* Mokelumne.

Morada (mō rä′ də) [San Joaquin Co.]. The Spanish word means 'dwelling place'.

Moraga (mə rag′ ə, mə rä′gə) **Valley**, settlement [Contra Costa Co.]. Preserves the name of Joaquín Moraga, a soldier in the San Francisco Company in 1819 and co-grantee of the Laguna de los Palos Colorados grant in 1835. He was the son of the explorer Gabriel Moraga (1765–1823). The post office was established on May 5, 1886, and reestablished on Dec. 16, 1915, when the town was developed.

Moraine. Although the deposits of ancient glaciers are frequently found in California, the term is included in the names of only about ten features. They include Moraine Lake [Sequoia NP], named by a party of the Sierra Club in 1897, as well as Mount Moraine, Moraine Ridge, and Moraine Creek [Tulare Co.]. There is another Moraine Ridge in Tuolumne Co., a Moraine Dome in Mariposa Co., and a Moraine Mountain in Madera Co.

Moran [Lassen Co.]. The siding of the Southern Pacific was established on Feb. 28, 1953, and on the suggestion of David F. Myrick was named for Charles Moran, builder of the Nevada-California-Oregon Railway.

Morek. *See* Moorek.

Morena (mô rē′ nə): station, **Butte, Valley, Dam** [San Diego Co.]. The name is shown for the Santa Fe station on the Official Railway Map of 1900. It is not known whether its origin is a family name or the Spanish adjective meaning 'brown'.

Moreno (mô rē′ nō, mə rē′ nō) [Riverside Co.]. When Frank E. Brown declined to have his name used for the town that he and E. C. Judson laid out in 1881–82, the Spanish word for 'brown' was substituted (Co. Hist.: Riverside 1912:170). In 1984 the communities of Moreno, Sunnymead, and Edgemont incorporated as the City of Moreno Valley (Gunther 1984).

Morgan, Mount [Inyo Co.]. Named on Aug. 28, 1878, by the Wheeler Survey, for one of its members, J. H. Morgan of Alabama.

Morgan Hill [Santa Clara Co.]. The settlement that developed on the Morgan Hill Ranch was named about 1892, for Morgan Hill, who had acquired the ranch when he married Diana Murphy, daughter of Daniel Murphy, a wealthy landholder and stock raiser.

Morgans Cove. *See* Ayala Cove.

Morgan Territory [Contra Costa Co.]. The district was named for Jeremiah ("Jerry") Morgan, a Cherokee Indian, who came to California in 1849 and claimed 10,000 acres of unsurveyed land east of Mount Diablo in 1856.

Morgan Valley [Lake Co.]. Named for Charles Morgan, who settled in the valley in 1854 (Co. Hist.: Napa 1881:141).

Moristul [Sonoma Co.]. The alternate name of the three Mallacomes grants is derived from that of the former Wappo village *mutistul* (*muti* 'north', *tul* 'large valley'), four and a half miles west of Calistoga (Barrett, *Pomo*, pp. 270–71). However, the name is already shown with the spellings Maristul and Muristul on *diseños* of the grants.

Mormon. The participation of the Latter-Day Saints in the Mexican War, their connection with the discovery of gold, and their various attempts to settle in California have left distinct traces in our toponymy. More than twenty-five places with the name are still recognized, including three inhabited places. Mormon Creek [Tuolumne Co.], Mormon Island [Sacramento Co.], Mormon Bar [Mariposa Co.], and Mormon Point [Inyo Co.] are names with historical implications. Mormon Meadow in Mono Co., however, is a misspelling on the Bridgeport atlas sheet; it should be Moorman, for a pioneer settler (Maule). *For* Mormon Gulch [Tuolumne Co.], *see* Tuttletown.

Moro Cojo (môr′ ō kō′ hō) **Slough** [Monterey Co.]. From Rancho Moro Cojo, established in 1836. *Moro* means 'a Moor', i.e. a Muslim, but also a dark-colored horse ('blue roan'); *cojo* is 'lame' (Clark 1991). *See* Moro Rock.

Moron. *See* Taft.

Morongo (mə rong′ gō): **Valley, Creek, Pass, Indian Reservation** [San Bernardino Co.]. The names are derived from *maronga,* a Serrano Shoshonean village in the valley (Kroeber, AAE 8:35). Marengo [*sic*] Pass is shown on the Land Office map of 1859 and on the von Leicht–Craven map. The historical name was preserved locally by settlers in the valley, and the post office was established on July 1, 1947, as Morongo. **Morongo Indian Reservation** [Riverside Co.], near Banning, was so named when people from Morongo Valley moved here in the mid-19th century. The reservation site was earlier called Potrero Ajenio, from *potrero* 'horse pasture' and Genio, name of a Cahuilla chief; it was often mispronounced and misspelled as The Portrero or The Protero (Gunther 1984).

Moro Rock [Sequoia NP]. "Mr. Swanson of Three Rivers in the [1860s] had a blue roan mustang—the color that the Mexicans call moro. . . . This moro pony of Swanson's often ranged up under the rock and they called it 'Moro's Rock'" (John R. White to Francis P. Farquhar).

Morrell (môr el′) **Gulch** [Santa Cruz Co.]. Hiram C. Morrell bought property near here in 1867 (Clark 1986).

Morrison, Mount [Mono Co.]. Named for Robert Morrison, of Benton, who with a posse pursued a number of convicts escaped from the Nevada State Penitentiary and was killed by one of them on Sept. 24, 1871. *See* Bloody: Bloody Mountain; Convict.

Morro. The Spanish geographical term for a crown-shaped rock or hill, the best known of which is El Morro in the harbor of Havana, Cuba, was repeatedly used in Spanish times and has survived in San Luis Obispo, Ventura, and San Diego Cos. **Morro Rock, Bay, Creek**, and **Beach** [San Luis Obispo Co.]: The Portolá expedition camped in the valley of what is now Morro Creek on Sept. 8, 1769, and Crespí mentions the rock at the entrance of the bay: "a great rock in the form of a round *morro*" (p. 186). In the *expediente* of the land grant, provisionally granted Dec. 28, 1837, the word is spelled with one *r* (Moro y Cojo or Moro y Cayucos), probably because a relation to *moro* ('Moor' or 'blue roan horse') was assumed; *see* Moro. This remained the common spelling until the Coast Survey changed it back to Morro in the 1890s. A town, Moro, is shown on the von Leicht–Craven map of 1874. Morro Bay State Park was named in 1932 and Morro Strand in 1934. **Morro Hill** [San Diego Co.], also spelled Moro and Mora, is shown as Morro on a *diseño* of the Santa Margarita y Las Flores grant. Point Stir [Monterey Co.] had been named Morro de La Trompa by Costansó in 1769 because it looked like a rock in the shape of a trumpet (*APCH:P* 2:125). *For* Morro Twin, *see* Hollister Peak.

Morse's Landing. *See* Moss Landing.

Mortero (môr târ′ ō) **Palms** [San Diego Co.]. From the Spanish for 'mortar' (a container for grinding); named for the Indian bedrock mortars found here.

Mortmar [Riverside Co.]. The Southern Pacific station was named Mortmere in the 1890s because of the proximity of Salton Sea (French *mort* 'dead', *mer* 'sea'). It is shown on the Official Railway Map of 1900, but on modern maps the Anglo-Saxon *mere* is replaced by Spanish *mar*.

Mosaic Canyon [Death Valley NP] is so called from the appearance of the breccia rock, which is exposed on many smooth surfaces and resembles vari-colored mosaic work (Stewart).

Moscow [Sonoma Co.]. The place on the Russian River is mentioned in the County Historical Atlas of 1877 (p. 24). It is uncertain whether the name was chosen because the settlement is on the Russian River, or whether the name is an American rendering of an Indian name, *meskua* (*Overland Monthly,* Oct. 1904), which in turn might have preserved an original "Moscow" applied by the Russians.

Moses, Mount [Tulare Co.]. The nickname of an elderly member of a fishing party was applied to the mountain by Frank Knowles in the 1870s (Farquhar).

Moskowite Reservoir [Napa Co.] is named for

George Moskowite, of a family long resident in the area (Stang).

Mosquito. Some fifty Creeks, Gulches, and Lakes bear the name of this annoying insect. Although the word came into English from the Spanish ('little fly') and although the Spaniards doubtless were bothered as much by mosquitoes as were Americans, apparently none of the names date from Spanish times. The name was very popular in Gold Rush days. There was once a post office Mosquito in El Dorado Co., and Glencoe [Calaveras Co.] was formerly Mosquito Gulch. The creek from which the first water was conveyed for the dry diggings in Nevada City was called Musketo Creek (Ritchie, p. 93).

Moss Beach [San Mateo Co.] was named because of the presence of the marine plant life. There is another Moss Beach in Santa Cruz Co., and a Moss Creek in Mariposa Co.

Moss Landing [Monterey Co.]. According to an often-repeated story, Charles Moss built a wharf here about 1865; the place became an important whaling station but was abandoned in 1888. The anchorage is not recorded on the detailed Coast Survey charts until about 1900. Hoffmann's map of 1873 shows Morse's Landing.

Mother Grundy Peak [San Diego Co.]. The name was supposedly given to the peak because it shows an upward-facing profile of a large nose and a protruding chin (Hazel Sheckler). The reference is to "Mrs. Grundy", a prototype of a narrow-minded person; but there may be a mixture with Spanish *madre grande* 'big mother' (Stein).

Mother Lode. In early mining days it was believed that a huge vein of gold extended from the Middle Fork of the American River to a point near Mariposa and that the known veins were offsprings of this "mother lode." Although this idea has long since been discarded by geologists, the name continues to be used and cherished.

Mott [Siskiyou Co.]. "The new town of Mott, south of Sisson, . . . was named after M. H. Mott, the energetic and popular road master of the railroad company" (*Yreka Journal*, July 16, 1887).

Mott Lake [Fresno Co.]. "Named in honor of Ernest Julian Mott, mountain explorer" (BGN, 6th Report).

Mound City. *See* Loma: Loma Linda.

Mount. California has ten post offices and communities named after a nearby mountain; several others are transfer names. For names with "Mount" as a generic term, see entries under the specific names. **Mount Bullion** [Mariposa Co.] was named in 1850 after the eminence on Frémont's Mariposas estate, which Frémont had named for his father-in-law, Sen. Thomas Hart Benton (1782–1858). The senator from Missouri had been nicknamed "Old Bullion" because he advocated the adoption of metallic currency. Mount Bullion was formerly known as Princeton, named after the Princeton Mine—which in turn had been named, tradition says, for Mr. Prince Steptoe, one of the discoverers of the ore deposit. **Mount Hebron** peak and town [Siskiyou Co.] bear the name of an ancient town in Palestine. It appears as Mount Hebron for a settlement on the Land Office map of 1891. The peak, 6 miles south, may have been named first, which would explain the generic in the name of the town. **Mount Hermon** [Santa Cruz Co.]: In 1905 a group of Christian people purchased "The Tuxedo," a pleasure resort. Five women of the group were entrusted with the selection of a new name, and they chose that of the peak in the Holy Land (Inga Hamlin). **Mount Owen** [Kern Co.]: When a railroad reached the community in 1909, the station was named Front. The name was later changed to Brown, for George Brown, owner of the hotel; and in 1949 to Mount Owen, after the peak which overlooks the valley (*Los Angeles Times*, June 24, 1951). **Mount Baldy** [San Bernardino Co.]: The community was originally known as Camp Baldy, after nearby Old Baldy peak (*see* Old: Old Baldy); but on July 1, 1951, the name of the post office was changed to Mount Baldy on petition of the residents. The town of **Mount Laguna** [San Diego Co.] takes its name from the Laguna Mountains (*q.v.*).

Mountain. Of the generic geographical terms used as the specific part of a place name, "mountain" is the most popular in the United States. In California there are dozens of Mountain Lakes, Passes, Springs, etc., as well as a Mountain Top [Calaveras Co.] and a Mountain View Peak [Madera Co.]. At one time Lake Tahoe was called Mountain Lake, so named by Frémont; Mountain Lake [San Francisco Co.] was Laguna de Loma Alta 'lake of the high hill' in Spanish times, referring to the 400-foot elevation of the Presidio. The term is also found in names of communities: Mountain View [Kern, Nevada, Santa Clara Cos.]; Mountain Ranch [Calaveras Co.]; Mountain King [Mariposa Co.]; Mountain House [Alameda Co.]; Mountain Spring [San Diego Co.]; and Mountain Center [Riverside Co.]. **Mountain View** [Santa Clara Co.]: The settlement that developed in the early 1850s around the stage station was named Mountain View because the Santa Cruz Mountains, Mount Diablo, and Mount Hamilton could be seen from the place (D. M. Burke). In 1864 the name was also given to the railroad station about one mile north, and to the new town that grew up there and eventually merged with the old town. *For* Mountain View in Butte Co., *see* Dogtown; in Orange Co., *see* Villa Park.

Mountain Charlie Gulch [Santa Cruz Co.]. Named for Charles McKiernan, better known as Mountain Charlie, an early pioneer and "noted character." He built the first road in these mountains, still called Mountain Charlie Road, where he collected toll for many years. Near Glenwood on Bear Creek is Mountain Charlie Tree, a redwood 260 feet high (Doyle).

Mountclef Village [Ventura Co.]. The community northwest of Thousand Oaks and Mountclef Ridge is listed in the BGN Decisions (1964). The name is apparently a combination of Mount with French *clef* 'key.'

Moyacino, Mount. *See* Saint Helena, Mount.

Muah (mo͞o′ ə) **Mountain** [Inyo Co.]. Probably from Panamint *mïa* 'moon' (McLaughlin). The name is not shown on older general maps; it was probably applied by the USGS from local information when the Olancha quadrangle was mapped in 1905.

Muckawee (muk′ ə wē) **Gulch** [Trinity Co.]. The origin of this name has not been discovered.

Mud. The term is contained in the names of more than thirty features on the map, mostly Springs, but also Creeks, Sloughs, and Lakes. There is a tautological Mud Run Creek in Fresno Co. Mud Hills [Riverside Co.] seem to be wrought of mud. Mud Town was the name of Watts before it was incorporated in Los Angeles. *For* Mud Springs in El Dorado Co., *see* El Dorado; in Los Angeles Co., *see* San Dimas.

Muerto (mo͞o âr′ tō, mwâr′ tō). The Spanish word for 'dead; dead man' was frequently used in place naming in Spanish times; it has survived in the names of a number of places in the original form, and sometimes in translation. When the adjective is used with a generic orographic term, it usually designates 'barren': Las Lomas Muertas [San Diego Co.], Lomerías Muertas [San Benito Co.]. The noun *muerto* in a name refers to a place where a corpse was found, or to a burial ground; *see* Dead; Deadman; Punta: Punta de los Muertos.

Mugu (mə go͞o′): **Point, Lagoon, Canyon** [Ventura Co.]. According to Kroeber, the name is from Chumash *muwu* 'beach'. The Indian village was mentioned by Cabrillo in 1542 and may thus have the distinction of being the oldest California name in continuous written use (*CFQ* 5:197–98). It is mentioned again in early mission records (Arch. Mis. S.Buen. 1:27). Vizcaíno in 1603 called the cape Punta de la Conversión, and it appears thus on maps until the 19th century. For some unknown reason the Coast Survey did not accept the name but applied it only to a triangulation point and called the cape Point Laguna. The name Point Mugu for the cape appears in the report of Subassistant W. M. Johnson, who surveyed the sector in 1856 (Coast Survey *Report*, p. 100). He doubtless chose the name because the Indian village still existed at that time. It is shown on the Land Office map of 1859, and it replaced the former name on the charts of the Coast Survey in 1863. The canyon is submarine; its name was approved by the BGN in 1938.

Muguelemnes. *See* Mokelumne.

Muir (myo͞o′ ər). The great naturalist and mountaineer, John Muir (1838–1914), has been commemorated in the nomenclature of the state more than any other person. **Muir Gorge** [Yosemite NP]: "We named this gorge Muir Gorge [in 1894], after John Muir, the first man to go through the cañon" (R. M. Price, *SCB* 1:206). **Muir Grove** [Sequoia NP] and **Pass** [Kings Canyon NP] were named by R. B. Marshall in 1909. **Muir, Mount** and **Lake** [Tulare Co.]: The name was given to the peak by Alexander G. McAdie, of the U.S. Weather Bureau. **Muir** [Contra Costa Co.]: The station was named by the Santa Fe in 1904 because of Muir's nearby residence. **Muir Woods National Monument** [Marin Co.] was given to the United States by William and Elizabeth Kent (of Kentfield) to preserve the virgin stand of coast redwood; and the name was bestowed at their request when the national monument was created in 1908. Nearby **Muir Beach** was named after the monument. **John Muir Trail**: At the instigation of the Sierra Club the legislature appropriated the first installment for the construction of the High Sierra trail in 1915. It was named in memory of Muir, who had died the preceding year (Farquhar). **Muirs Peak** [Los Angeles Co.]: The summit of the ridge forming the east wall of Rubio Canyon was so named because John Muir made the first ascent in Aug. 1875 (Reid, p. 369). **Muir Crest** [Sequoia NP]: The culminating crest of the Sierra Nevada between Shepherd and Cottonwood passes was named for Muir in 1937 by François E. Matthes of the USGS: "The small peak that bears Muir's name at present seems hardly commensurate in importance among the features of the Sierra Nevada with the greatness of the man whose love for the 'Range of Light' inspired the movement for the conservation of its scenic treasures" (*SCB* 22:6).

Mulberry. The station of the Sacramento Northern in Butte Co., the town in San Benito Co., and several physical features named Mulberry, named for plantings of the tree (*Mora* spp.), are reminiscent of a widely heralded California industry that failed to develop. Louis Prévost, a native of France and expert in silk culture, planted the first mulberry trees in San Jose and imported silkworm eggs from France in 1856. In 1866 the promoters of the industry prevailed on the legislature to offer handsome premiums for plantings of mulberry trees and production of cocoons. With true California enthusiasm the people covered the land with mulberry groves. The trees flourished, but the silkworms failed to thrive. In Aug. 1869 Prévost died; a few months later, the legislature repealed the law; and in 1871 a heat wave killed most of the surviving silkworms. Today a few geographical names are the only remnants of an industry that once looked so promising.

Mule. Since the mule played a less important part than the jackass as a pack animal in early mining and mountaineering days, only a limited number of features were named for the animal, including a Mule Mountain and a Mad Mule Gulch [Shasta Co.]; a Crazy Mule Gulch Creek [Yosemite NP]; Mule Springs, originally Dead Mule Springs [Nevada Co.]; and several mining camps. Shasta Co. once had a One Mule Town as well as a One Horse Town (Steger).

Mulholland Aqueduct [Los Angeles Co.]. For the engineer William Mulholland (1835–1935), who brought water to the Los Angeles area from the Owens Valley.

Mulholland Hill [Contra Costa Co.]. The name was apparently applied to the elevation south of Orinda when the Concord quadrangle was mapped in 1892–94. It may preserve the name of one of the three Mulhollands, farmers from Ireland, who settled in the San Pablo district in 1867.

Mulligan Hill [Monterey Co.]. The elevation, shown as Cabeza de Milligan 'Milligan's Head' on a *diseño* of the Bolsa del Potrero grant, was long known as Mulligan Head. It was named for John Milligan, an Irish sailor and one of the earliest foreign residents of California. He arrived before 1819, taught the art of weaving to Indians at several missions, and became part owner of Rancho Bolsa del Potrero; he died in 1834 (Bancroft 4:747–48).

Munchville. *See* Capay.

Muniz [Sonoma Co.]. The name of a land grant dated Dec. 4, 1845. It is called Rancho de Maniz on a *diseño*. The Spanish family name is properly Muñiz.

Mupu (mo͞o′po͞o) [Ventura Co.]. The Ventureño Chumash name for the Santa Paula area, appearing in Mission records of 1828 (Ricard), survives in the name of a school.

Murderer. The many acts of violence committed in the Gold Rush days and during Indian fights have left their mark on California's toponymy, though numerous names containing the word have disappeared. **Murderers Bar** [El Dorado Co.] received its name because five Oregonians were killed here (probably in 1849) in retaliation for the murder of three Indians who had been slain in an attempt to protect their women (Hittell 3:77).

Murdock Lake [Tuolumne Co.]. Named by N. F. McClure in 1895, for William C. Murdock, of the State Board of Fish Commissioners (Farquhar).

Murieta (mûr ē et′ ə). Joaquín Murieta was the "John Doe" of five or more Mexican bandits of the early 1850s (not to be confused with Murrieta, *q.v.*). The name has left its mark in several geographical names, two of which are still in use: **Joaquin Murieta Caves** [Alameda Co.], east of Brushy Peak, and **Joaquin: Rocks, Ridge** [Fresno Co.]; both localities were favorite hideouts, according to tradition. **Joaquin Peak** [Calaveras Co.], near San Andreas, was so named because Murieta is said to have fought the Chaparral skirmish here in Jan. 1853.

Muristul. *See* Moristul.

Muro Blanco (mo͞o′ rō bläng′ kō) [Kings Canyon NP] is Spanish for 'white wall'. The west slope of Arrow Ridge, which looks like a solid whitish wall, apparently suggested the name, which was applied by the USGS in 1904.

Muroc (myo͞o′ rok): station, **Dry Lake, Army Air Field** [Kern Co.]. The Santa Fe named the station on Dec. 25, 1910, by spelling backwards the surname of Clifford and Ralph Corum, homesteaders. The station was built in 1882 and had borne several other names. After Jan. 27, 1950, the airfield became known as Edwards Air Force Base in honor of Capt. Glenn W. Edwards, who was killed in a test flight in 1948. Muroc post office, established on Dec. 17, 1910, was renamed Edwards on Nov. 1, 1951.

Murphy Creek [Yosemite NP]. Named for John L. Murphy, an early settler on the shore of Tenaya Lake (Farquhar).

Murphys [Calaveras Co.]. Established as a mining camp in 1848–49 and named for John M. Murphy, a member of Weber's Stockton Mining Company. With his father, Martin Murphy, and other members of the family, he came from Missouri to California in the Stevens party in 1844. In 1849 the family settled in Santa Clara Co., where John later held several county offices and was mayor of San Jose (Bancroft 4:749). *For* Murphys Old Diggings, *see* Vallecito. **Murphys Peak** [Santa Clara Co.] is also named for the family.

Murphy Spring [Mono Co.]. The spring at the site of the old stage station on the Big Meadows and Bodie Toll Road was owned by J. C. Murphy in the 1880s.

Murray [Kings Co.]. The post office was established in 1919 and named for David Murray, who was a leader in introducing olive culture into the region (J. E. Meadows).

Murray Canyon [Riverside Co.]. For Dr. Welwood Murray, who built a health resort and hotel at Palm Springs about 1885.

Murrieta (mûr ē et′ ə): town, **Valley, Creek, Hot Springs** [Riverside Co.]. The post office was established about 1885 and named for John Murrieta, ranch owner and for many years bookkeeper in the sheriff's office. The name should not be confused with Murieta (*q.v.*).

Muscle Creek. *See* Mussel: Mussel Slough.

Muscupiabe (məs kə pi yä′ bē) [San Bernardino Co.]. The name of the land grant, dated Apr. 29, 1843, is derived from *el cajón de Muscupiavit*, mentioned about 1785 (SP Mis. and C. 1:241) and repeatedly in later years. According to Kroeber, the name consists of Serrano *muskupia*, of unknown meaning, and the locative suffix *-vit*.

Musick (myo͞o′ zik) **Peak** [Fresno Co.]. Named for either Charles or Henry Musick, both of whom were connected with the mill company at Shaver (Farquhar). The name is misspelled Music on the Kaiser atlas sheet.

Musketo Creek. *See* Mosquito.

Muslatt (mus′ lat): **Mountain, Lake** [Del Norte Co.]. The mountain was so called by the Big Flat Indians, but its meaning is unknown (J. Endert).

Mussel. About twenty features, chiefly Rocks, are named because of the presence of mussels. Most of these names occur along the coast, where large colonies of marine mussels are a common sight. On Oct. 30, 1769, the men of the Portolá expedition gave the name Punta de las Almejas 'point of the clams or mussels' to what is now Pedro Point [San Mateo Co.], "on account of the large number of mussels which they found on the beach, very good and large" (Crespí, p. 226). The point 5 miles north of it is now called Mussel Rock. **Mussel Slough** [Kings Co.]: The branch of Kings River was named Muscle Creek because of the freshwater mollusks found there. This spelling variant persisted on maps until the 1890s, though the stream as well as the district was popularly known as Mussel Slough. The name gained historical significance through the bitter struggle between settlers and the Southern Pacific from 1876 until the bloody encounter in May 1880—an episode immortalized in Frank Norris's novel *The octopus*. (*See* Lucerne: Lucerne Valley.)

Mustang. The word, from Spanish *mesteño*, formerly designated the small half-wild horse of the Western plains but is now practically synonymous with "horse" throughout the American Southwest. Some of the California place names were probably given because horses, descended from those that had escaped from the missions or ranchos, were seen or captured in the area.

Mustard. The flowering plant *Brassica campestris* was introduced from Europe and now grows wild throughout California. No place seems to have been named for it; but **Mustard Canyon** and **Hills** [Death Valley NP] were so named because of the mustard-yellow rocks in which the canyon is cut.

Musulacon [Sonoma Co.]. Rincón de Musulacon was the name of a land grant dated May 2, 1846. From Southern Pomo *mussaalahkon* 'Long Snake', a supernatural creature; also said to have been the name of a chief (Oswalt).

Mutau (mo͞o′ tô) **Flat** [Ventura Co.]. Will Mutau was an early homesteader here; he is said to have been murdered in 1874 (Ricard).

Myers Flat [Humboldt Co.]. The settlement was known as Myers, after the Grant Myers ranch. When the post office was established on Jan. 1, 1949, the present name was chosen to avoid confusion with Myers in El Dorado Co. (Mary Mosby).

Myford [Orange Co.]. Named in 1923 by the Santa Fe for Myford Irvine, son of James Irvine, owner of the San Joaquin Rancho (Gertrude Hellis). *See* Irvine.

Mynot (mī′ not) **Creek** [Del Norte Co.]. Said to have been named for a settler surnamed Mynot or Minot (Harrington).

Myoma (mī ō′ mə) [Riverside Co.]. This railroad siding was said to have been named for a nearby Indian settlement in 1877 (Gunther).

Mystic Lake [Riverside Co.]. An ephemeral body of water in the bed of the San Jacinto River; it appears during wet winters and later disappears. The present name was applied around 1898 (Gunther).

N

Nacimiento (nä sim ē en′ tō): **River**, station; **Lake Nacimiento** [San Luis Obispo, Monterey Cos.]. The name for the river apparently arose through a misunderstanding. The Portolá expedition camped near here on Sept. 21, 1769, and Crespí (p. 194) wrote of "a very large arroyo, whose source [*nacimiento*], so they said, was not far off." When the Anza expedition came to the same river, Anza apparently assumed that the previous expedition had named the river Nacimiento, perhaps associating the word with 'the Nativity', another meaning of the word, and not with 'source of the river'. Anza mentions the name of the stream on Apr. 16 and 24, 1774. On Aug. 27, 1795, a padre speaks of *el nacimiento* between San Antonio and San Luis [Obispo]; the small *n* may indicate that here again the meaning 'source' was intended (SP Mis. 2:56–57). However, later documents mention the Río del Nacimiento. Parcels of land called Gallinas, Nacimiento, and Estrella were granted to the Christian Indians of San Miguel on July 16, 1844, but the claims were later rejected by the United States. In the *Statutes* of 1850, the Nacimiento River is mentioned as forming part of the northern boundary of San Luis Obispo Co. as originally defined (pp. 59–60). The Southern Pacific station was named after the river in 1905.

Nacional. The livestock ranchos of the presidios, unless they had a specific name, were generally called *rancho del rey* in Spanish times. After Mexico became independent, this designation was changed to *rancho nacional* (Bowman). Two of the ranchos that later became land grants kept the name: that of Monterey Presidio, dated Apr. 4, 1839, and that of San Diego Presidio, dated July 26, 1843; the latter was also known as Rancho de la Nación. *See* National City.

Nacitone [San Luis Obispo Co.]. The soil conservation district takes its name from the Nacimiento and San Antonio Rivers, whose watersheds are covered by the district (Hall-Patton).

Nacko (nä′kō) **Creek** [Humboldt Co.]. This spelling was established by BGN, 1984 (list 8402), instead of Natchko or Natchka. The earlier spellings seem closer to the Yurok original *nohčka* (Waterman, map 11, item 132).

Nadeau (nə dō′) [Los Angeles Co.]. The Santa Fe station was named after the Gernert and Nadeau Beet Sugarie (!) established at this place in 1881. Rémi Nadeau, a French Canadian, had come to Los Angeles in the 1860s and organized the first freight transportation service by mule teams from the silver mines of Cerro Gordo, Calico, and Lookout across the Mojave Desert.

Najalayegua [Santa Barbara Co.]. An Indian rancheria so named is mentioned as early as June 28, 1785 (PSP 5:157), and repeatedly in later years. On Sept. 23, 1845, the name appears in the title of the Prietos y Najalayegua land grant. The word is apparently Chumash, but its meaning is unknown.

Nance Peak [Yosemite NP]. Named for Col. John Torrence Nance, professor of military science at the University of California (1904–27).

Napa (nap′ ə): **Valley**, city, **County, River, Creek, Slough, Junction, Soda Springs**. The name is mentioned in the baptismal records of Mission Dolores after 1795, and again in the diaries of Padre José Altimira and José Sánchez in June 1823, when they were looking for a suitable site for Mission San Francisco Solano. They came upon a large arroyo in the beautiful plain of Napa, *así llamado de los indios que antes lo habitaban* 'so called for the Indians who formerly inhabited it' (SP Sac. 11:30 ff.). The name appears in the titles of two land grants: Entre Napa, May 9, 1836, and Napa or Trancas y Jalapa, Sept. 21, 1838. The Caymus land grant was sometimes called Paraje en Napa. The city was laid out in 1848 by Nathan Coombs, a native of Massachusetts, who came to California in 1843 and bought a portion of Salvador Vallejo's Napa grant. The county, one of the original twenty-seven, was named on Feb. 18, 1850. River and creek are mentioned in the *Statutes* of 1850 (pp. 60–61). In 1858, the Vaca Mountains were called the Napa Range, and Clear Lake was called Laguna Grande de Napa (Hutchings' *California Magazine* 2:397, 3:146). The meaning of the name Napa has never been satisfactorily explained, although there are probably more theories about it than about any other Indian name in California. First, according to Kroeber, *napa* is the Southern Patwin designation for 'grizzly bear'; but it does not necessarily follow that the word is the source of the place name. It is quite certain that the geographical term originated in Northern Patwin territory. Second, Eastern and Central Pomo have a word *naphó* 'people, family' (Oswalt), but there is no clear connection with the modern place name. A third plausible interpretation was advanced by Mariano Vallejo's son, Platón, who had learned the language of the Suisun Patwin Indians. According to him, the meaning was 'near mother', 'near home', or 'motherland' (*San Francisco Bulletin*, Feb. 7, 1914). *For* Napa Wye, *see* Wyeth.

Naphus (nā′ fəs) **Peak** [Mendocino Co.]. Jim Naphus, a Missourian who came with the Asbills in 1854, was an Indian fighter and later a bandit. Once when he returned from a hunt, he had so much red soil of the mountain on him that he claimed the peak as his own (Asbill).

Naples [Santa Barbara Co.]. The Southern Pacific station was named in 1887, after the city in Italy. **Naples** [Los Angeles Co.] was founded in 1905 by A. C. and A. M. Parson, and was named likewise after the Italian city.

Naranjo (nə ran′ hō) [Tulare Co.]. When citrus growing developed in this region in the early 1900s, the Spanish word for 'orange tree' was given to the community and to the station of the Visalia Railroad (the fruit is *naranja*). The post office is listed in 1904.

Narod (nâr′ əd) [San Bernardino Co.]. When the old railroad to Salt Lake City was built in 1904, the station was given the name Narod, the section foreman's name spelled in reverse (Belden).

Nashmead [Mendocino Co.]. The station and post office were named Nash, for the first postmaster, when the Northwestern Pacific reached the place in 1907; the name was changed to the present form in 1916 (D. O. Strock). "Mead" is an old English term for 'meadow', still used in place naming and poetry.

Nashua [Monterey Co.]. Perhaps for Nashua, New Hampshire; named for a local Indian tribe, the Nashaways. The name is said to mean 'between-water'.

Nashville [El Dorado Co.]. The name was applied at the time of the Gold Rush by miners from Tennessee (Co. Hist.: El Dorado 1883:198). The name is not found on older maps but is mentioned in the *Statutes* of 1854 (p. 222).

Nataqua, Territory of. *See* Roops Fort.

Natchka Creek. *See* Nacko.

National City [San Diego Co.]. The town, which was laid out on the part of Rancho de la Nación owned by Frank, Warren, and Levi Kimball, was named National Ranch in 1868 and National City in 1871. The Spanish name lingered for many years: the Santa Fe station was called Nacion as late as 1886. *See* Nacional.

National Tribute Grove [Del Norte Co.]. The grove in Jedediah Smith Redwoods State Park was created in 1945 to represent every state in the union.

Natividad: Valley, Creek [Monterey Co.]. The place called La Natividad 'the Nativity' is mentioned by Font on Mar. 23, 1776, two days before the day of the Annunciation, i.e., the day on which the Archangel Gabriel announced to the Virgin Mary the coming birth of Christ. Anza's diary on the same day calls the camping place "La Assumpcion." The text leaves no doubt that both names refer to the same place, "at the beginning of a canyon." La Natividad as a place is mentioned again on Sept. 15, 1795 (Arch. MSB 4:592 ff.), and repeatedly in later years. The name was given to a land grant dated May 30, 1823, and Nov. 16, 1837. The place gained historical significance when a party of Americans under Capt. Charles Burroughs fought the Californios here in the "Battle of Natividad" on Nov. 16, 1846. The post office is listed in 1858.

Natividad de Nuestra Señora. *See* Oso: Los Osos.

Natoma (nə tō′ mə) [Sacramento Co.]. The name of an Indian village Natomo on the

American River was repeatedly mentioned in the *New Helvetia diary* of 1847–48 in connection with the building of Sutter's gristmill. The name is from Nisenan (Southern Maidu) *notoma* 'east place, upstream place', from *noto* 'east, upstream' (Shipley, p.c.). This is different from the site of Natoma or Notoma founded by Samuel Brannan on Mormon Island in 1848, shown on maps of 1849 (Jackson) and 1851 (Butler), and mentioned by Felix P. Wierzbicki (*California as it is, and as it may be* [San Francisco: Bartlett, 1849], p. 41). A third site—the present town, southwest of Folsom—is shown on Gibbes's map of 1852; the post office was established on Dec. 2, 1884.

Naufus (nā′ fəs) **Creek** [Trinity Co.]. From Wintu *norboos* 'those living to the south', containing *nor* 'south' and *boh-* 'to live' (Shepherd).

Navajo Creek [San Luis Obispo Co.]. Named for a Navajo woman who was traveling with sheepherders to French Camp (Hall-Patton). The name of the Navajo tribe, of Arizona and New Mexico, is derived from a place name in the Tewa (Pueblo) language, *navahuu* 'field valley' (*HNAI* 10:496).

Navalencia (nā və len′ shə) [Fresno Co.]. The name was coined by combining the names of two varieties of oranges: Navel and Valencia. It was applied to the station by the Santa Fe in 1913 because it is situated in one of the best citrus regions of the valley. The post office is listed in 1919.

Navarro (nə vä′ rō): **River, Head, Point**, post office [Mendocino Co.]. The name Novarra (or Novarro) is shown on a *diseño* of the Albion grant (1844). It may have been an Indian name, transcribed to resemble the name of the Spanish province Navarra, or the family name Navarro. The name for the river is on the Land Office map of 1862; the Coast Survey did not use the name until about 1870. The name of the lumber town and the post office was Wendling until about 1914, when it was changed to Navarro.

Navy Point. *See* Army Point.

Nawtawaket (nô′ tə wä kət): **Creek, Mountains** [Shasta Co.]. From Wintu *noti waqat*, lit. 'south creek', containing *nor* 'south' and *waqat* 'creek' (Shepherd).

Neafus (nē′ fəs) **Peak** [Trinity Co.]. For Jim Neafus, who explored parts of the area around 1854 (Jones, ed., 1851:346).

Neall Lake [Yosemite NP]. Named for John M. Neall, 4th Cavalry, who was stationed in the park, 1892–97 (BGN, list no. 30).

Neal's Rancho. *See* Esquon.

Nearys (nē′ rēz) **Lagoon** [Santa Cruz Co.]. For Patrick Neary, a settler from Ireland (Clark 1986).

Nebo (nē′ bō) is the mountain in the Holy Land from which Moses first saw the land of Canaan; it gives its name to **Nebo Center** [San Bernardino Co.]. *For* Mount Nebo, *see* Bismarck Knob.

Needham Mountain [Sequoia NP]. Named by W. F. Dean, for James C. Needham of Modesto, a congressman from 1899 to 1913 (Farquhar).

Needle. The word is sometimes used to describe orographic features. It occurs as a generic—Agassiz Needle, the former name of Mount Agassiz [Fresno Co.], The Needles [Tulare Co.]—and also as a specific term: Needle Peak [Placer Co.], Needle Rock, Point [Santa Cruz Co.].

Needles [San Bernardino Co.]. The post office and the station of the Atlantic and Pacific Railroad (now the Santa Fe) were established on the Arizona side of the Colorado River on Feb. 18, 1883, and named after the nearby pinnacles. On Oct. 11 of the same year, the railroad transferred the name to a new town on the California side, a location that it considered better suited for a projected division point (Barnes). The name was originally applied to the peaks by the Pacific Railroad Survey, in whose reports they are frequently mentioned. The Whipple-Ives map of 1854 labels the peaks on both sides of the river as The Needles.

Neeley Hill [Sonoma Co.]. Probably named for Robert Neely, a native of Pennsylvania, who came to the county in 1866 or earlier.

Neenach (nē′ nəch) [Los Angeles Co.]. The place is shown on the Land Office maps after 1890; it is probably of Numic origin but its meaning is not known.

Negit (neg′ it) **Island** [Mono Co.]. From Eastern Mono *nïkïtta* 'goose' (McLaughlin): "The island second in size we call Negit Island [1882], the name being the Pa-vi-o-osi [Mono Indian] word for blue-winged goose" (I. C. Russell, USGS *Report* 8:279). It has been claimed that the word is Piute for 'darkness'—the opposite of adjacent Paoha Island—but this is not confirmed by studies of the local Indian languages (Browning 1986).

Negro; Nigger. The names have occurred frequently in times past, particularly in mining days, not because there were large numbers of African Americans, but because the presence of a single one was sufficiently conspicuous to suggest calling a place Negro Bar or Nigger Slide. The name is also found in Spanish times: Lytle Creek was known as Arroyo de los Negros, and the pass northwest of San Bernardino as El Puerto de los Negros. The grant El Cajón de los Negros is dated June 15, 1846. Among the features originally named with the form Nigger, now often changed to Negro, are Nigger Bill Bend in the San Joaquin Valley, south of Stockton; Nigger Head [Mendocino Co.] and Niggerhead Mountain [Los Angeles Co.] (which probably reflect the now obsolete term "niggerhead" in the sense of 'a dark-colored boulder'); Nigger Rube Creek [Tulare Co.]; Nigger Jack Slough [Yuba Co.]; and Niggerville Gulch Creek [Siskiyou Co.]. The name of one of the old mining towns is preserved in Nigger Bar [Sacramento Co.]. Among the names that include the word are Negro Butte and Negro Butte Dry Lake [San Bernardino Co.], and a creek with a rare generic, Negro Run [Plumas Co.]. **Negro Canyon** [Los Angeles Co.] was named for Robert ("Uncle Bob") Owen, who came from Texas in 1853, secured a contract to supply firewood to the soldiers, invested his money wisely, and became the richest black man in the county (Reid, p. 386). **Nigger Jack Peak** [Tuolumne Co.]: Named for "Nigger Jack" Wade, a former slave, who owned 460 acres at the base of Table Mountain (Paden and Schlichtmann). *For* Negro Bar, *see* Folsom. Note that the term "Niggerhead" in place names may refer not to the head of a Negro, but rather to a flanged drum on a winch, used for winding in lines—also called a "gypsyhead."

Neighbours [Riverside Co.]. Named for J. E. Neighbors, the first postmaster (Co. Hist.: Riverside 1935:213). The post office is listed in 1908 as Neighbours.

Nelson [Butte Co.]. Established in the early 1870s as a station on the California and Oregon Railroad (now the Southern Pacific) and named for A. D. Nelson and his sons, farmers in the district.

Nelson [Tulare Co.]. Named for John H. Nelson, a pioneer of the Tule River region (Farquhar).

Nelson: Creek, Point [Plumas Co.]. The creek appears on Gibbes's map of California (1852). It was named for the "tall, sandy haired, thin faced, good natured *hombre*" (Delano, p. 19) who had a store at the confluence of the creek and the Middle Fork of Feather River. The bar and the diggings on the creek proved to be the richest gold-producing placers of the region. They were discovered in 1850 and were shown on the maps as early as 1851. The first post office was established on Mar. 30, 1855.

Nemshas. *See* Nimshew.

Neponset (nə pon′ sət) [Monterey Co.]. An Indian transfer name from Massachusetts, perhaps meaning 'good waterfall' (Stewart). The former name, Martin Station, was changed by the Southern Pacific about 1900.

Nestor [San Diego Co.]. The post office was named about 1890 for Nestor A. Young, state assemblyman in 1887–93.

Neuil. *See* Niguel.

Nevada. The name of the mighty range Sierra Nevada has given to California nomenclature two sonorous names, often applied without regard to their meaning: Sierra ('range') and Nevada ('snow-covered'). **Nevada City** [Nevada Co.]: The place at the site of the city was called Deer Creek Dry Diggings in the fall of 1849 by a prospector named Hunt, who struck rich gold deposits here (*see* Deer: Deer Creek). It became known also as Caldwell's Upper Store, for Dr. A. B. Caldwell, who set up his trading post here in Oct. 1849, a short time after he had opened a "lower store" 7 miles down Deer Creek. In May 1850 the settlers adopted the name Nevada at a public meeting, probably suggested by O. P. Blackman, and in Mar. 1851 the town was incorporated as City of Nevada (H. P. Davis). The name Nevada City is shown on Bancroft's map of 1858, but it was used earlier to distinguish the city from the county. The post office is listed as Nevada City as early as 1851. **Nevada County** was established and named after the city by act of the legislature on Apr. 25, 1851. City and county were NOT named after the state of Nevada as has been sometimes assumed; indeed, according to H. P. Davis, the territory (later the state) of Nevada was in all probability named after the California places. The gold rush to the Washoe country in Nevada, which led to the creation of a new political unit carved from Utah Territory, was started in Nevada City. The "blue-black stuff" was brought across the Sierra and assayed by citizens of Nevada City; and after the *Nevada Journal* on July 1, 1859, had published the facts about the amazing gold and silver content of the ore, the people of the city were in the vanguard of the argonauts to the Washoe mines. However, Stewart (pp. 303–4) states that the name which Congress gave to the territory was a shortening of the name of the mountain range. **Nevada Fall** [Yosemite NP] was named by L. H. Bunnell when the Mariposa Battalion entered the valley in 1851: "The white, foaming water, as it dashed down Yo-wy-we [the Indian name for the fall] from the snowy mountains, represented to my mind a vast avalanche of snow" (Bunnell, p. 205).

Nevahbe (nə vä′ bē) **Ridge** [Mono Co.]. From Mono *nïpapi* 'snow' (McLaughlin), pronounced approximately (nə vä′ vē), according to McLaughlin.

Nevares (nə vär′ ēz) **Peak** [Inyo Co.]. A Spanish family name, also spelled Nevárez.

New. The principal use of the adjective "new" is in connection with transfer names that immigrants brought from the home country. California has had comparatively few such names: New Chicago, New Helvetia, New Mecklenburg, New Philadelphia, New Texas, New Vernon—all have vanished. Occasionally the adjective was used in emulation of another place: New Almaden, New Jerusalem, New Albion. Sometimes the word was used to distinguish a newly founded settlement from an older one: New Rio Dell, Nubieber (*q.v.*), New Cuyama [Santa Barbara Co.]. In several names "new" was combined with an abstract noun: New Hope was to have been the name of a Mormon colony that Sam Brannan tried to found on the Stanislaus in 1846; the gazetteers list Newhope Landing [San Joaquin Co.], New Hope Rock [San Diego Co.], Newlove [Contra Costa Co.]. Most of the names of California communities with the prefix "New" are direct transfer names, in which the "New" is no longer descriptive: New England Mills, Newport, etc. The Pacific Electric station Greenville [Orange Co.] was once actually known as Old Newport. Because of the peculiar hydrographic conditions in several sections of the state, there are also a number of New Creeks, Lakes, and Rivers, the best known of which is **New River** [Imperial Co.], named as early as 1849, when the dry river bed was suddenly filled by overflow from the Colorado. *See* Nuevo. The more important names with the adjective "New" are listed separately.

New Albion. From June 17 to July 23, 1579 (June 27 to Aug. 8, New Style), Francis Drake anchored his *Golden Hinde* in a bay that was probably the one still called Drakes Bay. During his stay he took possession of the country for the English crown and named the country Nova [New] Albion: "Our Generall called this countrey Noua Albion, and that for two causes: the one in respect of the white bankes and cliffes, which lie toward the sea; and the other, because it might haue more affinitie with our Countrey in name, which sometime was so called" (quoted in H. R. Wagner, *Sir Francis Drake's voyage around the world* [Glendale: Clark, 1926], pp. 276–77). It was the first challenge to Spain's claims, and "New England" was to be the nucleus of a British empire in North America. The British claim to the coast of northern California was not given up until the Oregon treaty of 1846. On most British and many other non-Spanish maps, the English name appears for the rather undefined territory otherwise often labeled with the mythical name Quivira. The name New Albion gained new significance after the Spaniards started the colonization of Upper California. The renewed British claim for California is reflected on the maps: Arrowsmith (1790) and Cary (1806) place the name New Albion neatly within the present boundaries of the state. It remains unexplained why the Russians, who had planted themselves squarely in the British claim, should retain the English name. A Spanish padre in July 1818 complains that the Russians use the name as dictated to them by Great Britain: *es uno de los dictados de la Gran Bretaña* (Guerra Docs. 5:66), and Kotzebue writes in 1824: "The whole of the northern part of the [San Francisco] bay, which does not properly belong to California, but is assigned by geographers to New Albion" (*New voyage* 2:112). *See* Albion.

New Almaden (al mə den′) [Santa Clara Co.]. The post office was given this name in 1953, replacing early Almaden (*q.v.*).

Newark [Alameda Co.]. The South Pacific Coast Railroad Company named its station in 1876 after the New Jersey home of A. E. Davis and his brother.

Newberry Springs [San Bernardino Co.]. The Southern Pacific named the station Newberry in 1883. In 1919 the Santa Fe changed the name to Water because for a long period the springs there supplied the Santa Fe with all the water it needed. In 1922 the name Newberry was restored (Santa Fe).

Newburg. *See* Fort Dick.

Newbury Park [Ventura Co.]. The post office was established in 1875, with Egbert S. Newbury as postmaster (Ricard).

New Camaldoli (kə mäl′ dō lē) [Monterey Co.]. The site of the Immaculate Heart Hermitage, founded by Camaldolese monks. The name is from Camaldoli, a hermitage near Florence, Italy, where St. Romuald (927–1027) founded this order of barefooted Benedictine hermits (Clark 1991).

Newcastle [Placer Co.]. The station of the Central Pacific was named in 1864 after the old mining town nearby, which in turn had doubtless been named after one of the "old" Newcastles in the eastern United States. The post office is listed in 1867.

Newcomb Pass [Los Angeles Co.]. Named by the Forest Service for Louis T. Newcomb, a ranger in the timberland reserve that is now the Angeles National Forest (USFS).

Newell [Modoc Co.]. The place, at the time of World War II a Japanese relocation center, was named for Frederick H. Newell, the first chief engineer of the U.S. Reclamation Service (State Library).

New England Mills [Placer Co.]. The name was applied to a station of the Central Pacific about 1877 after the nearby mill, which was operated by Captain Starbuck and his partners from New England, and which supplied timber for the railroad (Virginia Major). The post office name is Weimar.

Newhall: town, **Creek, Pass** [Los Angeles Co.]. A station of the Southern Pacific Railroad was established on Oct. 28, 1876, at the present site of Saugus and was named for Henry M. Newhall (1825–82), the owner of the land—a prominent California pioneer, railroad promoter, and resident of San Francisco. On Feb. 15, 1878, the depot and the name were transferred to the new site.

New Helvetia [Sacramento Co.]. The settlement was founded by Sutter on Aug. 13, 1839, and given the Latin name of Switzerland, the home of Sutter's ancestors. On June 18, 1841, and Feb. 25, 1845, the Spanish version, Nueva Helvetia, was applied to Sutter's vast land grant. *See* Sacramento; Sutter: Sutters Fort.

New Hope [Stanislaus Co.]. A Mormon colony was established here by Sam Brannan in 1847. *See* Stanislaus.

Newhope Landing [San Joaquin Co.]. The name is reminiscent of the New Hope Ranch established by Arthur Thornton about 1855, at the site of present Thornton. From 1880 to 1907 New Hope was listed as a post office. *See* Thornton.

New Idria. *See* Idria.

New Jerusalem. *See* El Rio.

Newman [Stanislaus Co.]. The town was established in 1887 when the Southern Pacific reached the place and was named for Simon Newman, who donated the land for the right-of-way of the railroad. The residents of nearby Hills Ferry and of Dutch Corners, a German settlement, moved to the new town.

Newmark. *See* Montebello.

New Mecklenburg. *See* Marysville.

New Orleans Bar. *See* Orleans.

New Owenyo. *See* Owenyo.

New Pass. *See* Soledad: Soledad Canyon.

Newport: Bay, Beach, Heights [Orange Co.]. Named in 1870, shortly after the first commercial ship sailed into the bay, because it was a "new port" between Los Angeles and San Diego (Brigandi). In 1873 the McFadden brothers, of Delaware, bought the dock and warehouse, which had been built the previous year, and established a lumber business. In 1876 they named their steamer *Newport*, and in 1892 they had the townsite Newport platted as a beach resort, in which lots were not sold but leased. Later the property was sold to W. S. Collins and associates, who on Feb. 16, 1904, filed the map for the subdivision Newport Beach (Sherman, pp. 9 ff.). *For* Newport [Solano Co.], *see* Collinsville.

New Redwood Canyon. *See* Canyon.

New River City. *See* Denny.

New San Pedro. *See* Wilmington.

New Town. *See* Nubieber.

Newtown. *See* Wilmington.

New Years Point. *See* Año Nuevo.

New York. Of all names transferred from the East in pioneer days, New York was the most common in California. It has been retained in about fifteen place names, including the mountain range separating Ivanpah and Lanfair Valleys in San Bernardino County. New York Point and Slough and York Island [Contra Costa Co.]

are reminiscent of a proud New York of the Pacific, founded in 1849. *See* Pittsburg.

Nicasio (ni kash′ ō): **Creek, Hill**, town [Marin Co.]. The name was applied to two land grants, dated Mar. 13, 1835, and Feb. 2, 1844. Arroyo de Nicasio, Casa de los Indios de Nicasio, and Roblar ('oak grove') de Nicasio are shown on a *diseño* of the 1844 grant. Nicasio was probably an Indian who was baptized with the name of one of several saints named Nicasius. The place was settled by Noah Corey in 1852; the post office was established on Apr. 13, 1870 (Co. Hist.: Marin 1880:284 ff.).

Nice (nēs) [Lake Co.]. The old name, Clear Lake Villas, was changed to the present name by the citizens in 1927–28 because the general topography of the district was thought to resemble the Riviera landscape near Nice in France (Helen Bayne).

Nicholaus. *See* Nicolaus.

Nichols [Contra Costa Co.]. Established as a siding by the Santa Fe in 1909 and named for the William H. Nichols Syndicate (now the General Chemical Company of California), the principal landholder in the district (Santa Fe). *For* Nichols Peak [Kern Co.], *see* Nicoll Peak.

Nicolaus (nik′ ə ləs) [Sutter Co.]. Nicholaus Allgeier, a native of Germany, came to California in 1840 as a trapper for the Hudson's Bay Company; and in 1842 he received from Sutter a tract of land on the Feather River, where he built a hut and operated a ferry. "The public spirited proprietor of the tract of land heretofore known as Nicolaus' Ranche has responded to the repeated requests of the people, and has caused one mile square of it to be laid off into a town, to which he has given the name of the ranche. The name is not so euphonious as some 'we wot of,' but our friend Shakspeare has told us, and with some truth, that 'a rose by any other name would smell as sweet.' We argue from this that 'Nicolaus' will lose none of the great advantages it possesses in consequence of its name" (*Sacramento Placer Times*, Feb. 16, 1850; Boggs, p. 42). The maps of 1849 record the place as Nicholas Ferry and Nicholas Alleger. The Postal Guide of 1851 and Schaeffer's sketches have the spelling Nicholaus.

Nicoll (nik′ əl): **Peak** [Kern Co.]. The mountain south of Weldon was named for John Nicoll, who came to Southfork Valley in 1852 and took up a homestead. The name is misspelled Nichols on many maps.

Nido (nē′ dō). The Spanish word for 'nest', often used in the applied sense of 'abode' or 'home', is found in three places: El Nido [Los Angeles and Merced Cos.] and Rionido [Sonoma Co.]. The latter is intended to be Spanish for 'river nest'.

Nietos. *See* Los Nietos.

Nigger. *See* Negro; *for* Nigger Slough, *see* Dominguez.

Night Cap [Sacramento Co.]. The name was applied to the mountain near Sonora Pass by the USGS in the 1890s, probably because the outline resembles an old-fashioned nightcap. *See* Liberty Cap.

Niguel (ni gel′) **Hill** [Orange Co.]. A place called Nigüili by the Indians is mentioned in Boscana's *Chinigchinich* before 1831 (pp. 83, 215–16); it is from Juaneño *nawíl*. The name appears in later records with various spellings and is used for the Niguel land grant, dated June 21, 1842. On a *diseño* the hill is shown, spelled "Lomaria de Neuil." The Coast Survey used the hill as a triangulation station in the 1870s but spelled the name Nihail, a spelling that continued on the maps of the survey for many years. The city of **Laguna Niguel** is now located in the area.

Niland (nī′ lənd) [Imperial Co.]. A contraction of "Nileland," selected in 1916 by the Imperial Farm Lands Association, comparing the fertility of the irrigated region to that of the Nile valley in Egypt. When the Southern Pacific built its branch line into the Imperial Valley, the name of the station was Old Beach, and from 1905 to 1916 it was Imperial Junction.

Nilegarden [San Joaquin Co.]. The Western Pacific applied the name about 1915 to the shipping station near the Nile Garden Irrigation Farms, which the promoter C. B. Hubbard had named in analogy to the Nile delta in Egypt.

Niles [Alameda Co.]. The name was given in 1869 by the Central Pacific, for Judge Addison C. Niles, who was elected to the state Supreme Court in 1871. The place had been known as Vallejo Mills in the 1850s for José de Jesús Vallejo, who had built a flour mill on Alameda Creek. It became part of the city of Fremont in 1956.

Nimshew (nim′ shōō) [Butte Co.]. A division of the Maidu Indians, Nemshaw, near the headwaters of Butte Creek, was noted by Horatio Hale (Wilkes 6:631); it appears in various spellings in the *New Helvetia diary* and in the records of the 1850s. The name is from Konkow (Northwestern Maidu) *nem sewi* 'big stream', containing *sew* 'creek, river' (Shipley, p.c.). The name of the Nemshas grant, dated July 26, 1844, probably comes from the same source, although the grant was much farther south. Nimshew mining tunnel, under Table Mountain, is mentioned by Browne in 1868 (p. 160). *See* Kimshew.

Nina, Lake. *See* Tilden Lake.

Nipinnawasee (nip in ə wä′ sē) [Madera Co.]. Edgar B. Landon brought the name from Michigan, where it is said to mean 'plenty of deer' in an Algonquian language. He applied it to his settlement in 1908, and in 1915 it was accepted by the Post Office Dept.

Nipomo (ni pō′ mō): **Valley, Creek, Hill**, town [San Luis Obispo Co.]. From Obispeño Chumash *nipumu'* 'house-place, village', from *(q)nipu* 'house' + *-mu'* 'nominalizer' (Klar 1975). Nipomo is mentioned as the name of a Chumash rancheria in the records of La Purisima Mission between 1799 and 1822 (Merriam). On Apr. 6, 1837, the name was applied to the land grant. The place is mentioned by Brewer on Apr. 9, 1861; the modern town was laid out in 1889 by the heirs of William G. Dana, who was the original grantee of the rancho and a cousin of the writer Richard H. Dana.

Nippen. *See* Nipton.

Nipple, The [Alpine Co.]. The USGS adopted the local name for the mountain when the Markleeville quadrangle was mapped in 1889. A number of hills are so named in various parts of the state, including Nellies Nipple [Kern Co.].

Nipton [San Bernardino Co.]. According to local tradition, the station was named for one of the engineers of the survey party when the San Pedro, Los Angeles, and Salt Lake (now Union Pacific) Railroad was built in 1904. However, the Southern Pacific map of 1905 shows the name Nippen for the station; the present name does not appear until 1910.

Nisene (nī sēn′) **Marks State Park** [Santa Cruz Co.]. The park was established in 1963 and named for Nisene Marks, the mother of the donors of the property. The land was donated through the Nature Conservancy, and the full name is The Forest of Nisene Marks State Park.

Nob Hill [San Francisco Co.]. Perhaps from slang "nob" 'person of wealth and importance', because wealthy people lived there (Loewenstein).

Nobles Pass [Lassen Co.]. The pass was named for William H. Noble, who in the summer of 1852 began to build a wagon road over the cut-off that had been discovered in 1851 (cf. BGN, July–Sept. 1966).

Noche Buena [Monterey Co.]. The land grant, dated Nov. 15, 1835, was also known as Huerta de la Nación. *Nochebuena* is Spanish for 'Christmas Eve' (lit. 'good night').

Noe (nō′ ē) **Valley** [San Francisco Co.]. José Noé was the last *alcalde* (chief magistrate) under Mexican rule, and a city official after the American occupation; he owned a ranch of some 4,000 acres in the center of present-day San Francisco (Loewenstein).

Nogales. The California walnut (*Juglans californica*) was noticed by the very first overland expedition on Aug. 5, 1769 (Crespí, p. 151). Later the word *nogales* 'walnut groves' was used in the names of two land grants in Los Angeles Co.: Nogales, on Oct. 11, 1838, and Mar. 13, 1840, and Cañada de los Nogales, on Aug. 30, 1844. Neither the Pacific Electric station Los Nogales, near Universal City, nor the Union Pacific station Walnut, southwest of Spadra, derives its name from these grants. *For* Arroyo de los Nogales [Contra Costa Co.], *see* Walnut: Walnut Creek.

Nojogui (nō′ hō wē): **Creek, Falls** [Santa Barbara Co.]. Najague or Najajue was one of the rancherias under the jurisdiction of Mission La Purisima (Arch. MPC, 10) and is repeatedly found in land-grant papers with various spellings. The canyon through which the creek flows is shown as Cañada de Na-jao-ui on a *diseño* of Las Cruces grant; Cuchilla ('ridge') de Nojogue and Planecita Nojoque appear on the *diseño* of the Najoqui grant (1842). Kroeber says that the name may come from Chumash *onohwi*, which still finds its echo in the local pronunciation; its meaning, however, is unknown.

Nolina (nō lē′ nə) **Wash** [San Diego Co.]. Nolina is the genus name of a yucca-like plant (Stein). The plant was named for C. P. Nolin, an 18th-century writer.

Nome-Lackee (nōm lä′ kē) [Tehama Co.]. The name of the former Indian reservation is also the name of a tribe, the Nomlaki or Central Wintun; the spelling Nomi-Lackee is also found. The term is probably from *nom* 'west' plus a form of the verb 'speak' (Shepherd).

Nomwaket (nōm′ wä kət) **Creek** [Shasta Co.]. From Wintu *nomwaqat*, lit. 'west creek', from *nom* 'west' + *waqat* 'creek' (Shepherd).

No Name Canyon [Kern Co.]. The canyon north of Inyokern received the unusual name because the surveyors for the Los Angeles Aqueduct could find no local name in use and apparently did not have the imagination to give it a name of their own (Wheelock). *See* Unnamed Wash.

Noonan Gulch [Trinity Co.]. Named for a family of settlers (Morris); sometimes misprinted as Noonah.

Noonday Rock [San Francisco Co.]. This rock off the Farallon Islands was named by the Coast Survey for the clipper *Noonday*, which struck it on Jan. 2, 1863, and sank within an hour.

Nopah (nō′ pä) **Range** [Inyo Co.]. Perhaps 'carried water', from Southern Paiute *noo-* 'to carry' plus *paa-* 'water' (Munro); but local people claim it is a hybrid name, 'no-water' (Fowler). The mountains were considered part of the Kingston Range by the Whitney and the Wheeler Surveys. When the USGS mapped the region (1909–12), the name Kingston was applied only to the southern part; the northern part was designated Nopah, a name doubtless supplied by local sources. *See* Kingston: Springs.

Norco [Riverside Co.]. The name was coined from North Corona Land Company by Rex B. Clark in 1922. Lake Norconian was so named in 1925 because it is part of the Norco development.

Nord [Butte Co.]. Laid out in 1871 by G. W. Colby and named Nord (German, 'north') by his wife (Co. Hist.: Butte 1882:246).

Norden; Lake Van Norden [Nevada Co.]. The dam was built in 1900, and the lake was later named for Dr. Charles Van Norden, of the South Yuba Water Company. The post office was established in 1927 and named after the lake.

Nordheimer Creek [Siskiyou Co.]. The stream was named for a prospector who lived in the vicinity until the 1930s (Ella Soulé). George A. Nordheim, a native of Prussia, is listed in the Great Registers of the early 1870s. Nordheimer Creek and Ditch are mentioned by Browne (1868:202). Another possibility is that the name commemorates B. Nordheimer, who was one of the first to try to wash gold at Gold Bluffs in 1850 (Alex Rosborough, in Knave, Aug. 26, 1956).

Nordhoff Peak [Ventura Co.]. The peak was named by the USGS in 1903 after the town of Nordhoff, which had been named for Charles Nordhoff (1830–1901); he was a native of Germany, a well-known author, and the grandfather of Charles B. Nordhoff, co-author of *Mutiny on the Bounty. See* Ojai.

Norris Butte [Trinity Co.]. The mountain was named for a settler who lived at the foot of it. The name is misspelled Norse on some maps (K. Smith).

Norte, Rio del. *See* Colorado River.

Northam. *See* Buena Park.

North Baldy [Los Angeles Co.]. *See* Baden-Powell, Mount; Mount: Mount Baldy; Throop Peak.

North Beach [San Francisco Co.]. This section of the city was once at its northern end, until landfill expanded the city area.

North Bloomfield [Nevada Co.]. The town that developed after the discovery of gold here in 1851–52 was first called Humbug, or Humbug City. When the post office was about to be established, the popular American town name Bloomfield was chosen at a public meeting. As there was already a Bloomfield post office, in Sonoma Co., "North" was added to distinguish the two places.

North Columbia [Nevada Co.]. The post office was established in the 1860s and named after Columbia Hill, a well-known mining camp. The "North" was added to distinguish it from Columbia post office in Tuolumne Co.

North Dome [Yosemite NP]. Named in 1851 by Major Savage's party of the Mariposa Battalion.

North Edwards [Kern Co.] is named for its proximity to Edwards Air Force Base (*q.v.*).

North Fork [Madera Co.]. The town came into existence after several mining companies opened up the district in 1878 and was named because of its location on the North Fork of San Joaquin River. The post office was established on Dec. 18, 1888, as Northfork, and many maps still spell the name as one word. *For* North Fork in Humboldt Co., *see* Korbel; *for* North Fork in Trinity Co., *see* Helena; *for* North Fork Dry Diggings, *see* Auburn.

North Guard [Kings Canyon NP]. The names North Guard and South Guard, for the two peaks that flank the somewhat higher Mount Brewer, appear on Lt. Milton F. Davis's map of 1896 (Farquhar).

North Highlands [Sacramento Co.]. The name

reflects the location northeast from the city of Sacramento; there is no place named Highlands in the vicinity.

North Los Angeles. *See* Northridge.

Northridge [Los Angeles Co.]. When the Southern Pacific Railroad built the San Fernando branch line about 1908, the place was known by the Biblical name Zelzah because it was actually a 'watering place in the desert'. In 1933 the name was changed to North Los Angeles, and in 1935 to Northridge, a name proposed by Carl S. Dentzel because this district of Los Angeles lies at the base of San Fernando Valley's northern ridge.

North Sacramento [Sacramento Co.]. The North Sacramento Land Company purchased a part of the Rancho Del Paso in 1910, laid out a subdivision, and named it for its location north of the city of Sacramento (H. S. Wanzer).

North San Juan; San Juan: Hill, Ridge [Nevada Co.]. When the rich gold deposits were found in 1853, one of the discoverers, Christian Kientz, a veteran of the Mexican War, named the hill because its shape reminded him of San Juan de Ulloa in Mexico. When the post office was established on May 21, 1857, the adjective North was added to distinguish it from San Juan in San Benito Co.

North Shore [Riverside Co.]. The name refers to the location north of the Salton Sea.

Nortonville [Contra Costa Co.]. Named for Noah Norton, a native of New York, who located the Black Diamond coal mine and in 1861 built the first house in the town. The mine was one of the centers of the Mount Diablo coal boom, the decline of which began with the explosion in this mine on July 24, 1876. *See* Coal; Coalinga; Pittsburg; Somersville.

Norwalk [Los Angeles Co.]. The place was settled by Atwood and Gilbert Sproul in 1877 and called Corvallis, after their former home town in Oregon. In 1879, when the post office was established, it was renamed after Norwalk, Connecticut, the former home of some of the settlers (Co. Hist.: Los Angeles 1965: 4.180).

Nose Rock [Mendocino Co.]. The name was applied to the rock opposite Cuffey Cove by the Coast Survey, probably because of its outline. *See* Homers Nose.

Nosoni (nə sō′ nē) **Creek** [Shasta Co.]. From Wintu *nosono* 'south peak', containing *nor* 'south' and *sono* 'nose, peak' (Shepherd).

Notoma. *See* Natoma.

Nottingham. *See* Livermore.

Nova Albion. *See* New Albion.

Novarro. *See* Navarro.

Novato (nō vä′ tō): **Valley, Creek**, town [Marin Co.]. A Cañada de Novato, where cattle from Mission San Rafael grazed, was mentioned in 1828 (Registro, p. 4). The name is from that of a chief of the Hookooeko Indians, according to Merriam (*Mewan stock*, p. 355); this chief had probably been baptized with the name of Saint Novatus. On Nov. 18, 1836, the name was applied to a land grant, and on Feb. 2, 1856, to a post office.

Noyo: River, Canyon, town, **Anchorage, Seavalley** [Mendocino Co.]. According to Barrett (*Pomo*, p. 134), this was the name of a former Northern Pomo village near the mouth of Pudding Creek; the meaning was 'under the dust or ashes', from *nó* 'dust, ashes' plus *yow* 'under, in' (Oswalt). The name was later transferred to the river south of Fort Bragg, doubtless by the Coast Survey: *Report*, 1855, Noyou River; *Coast Pilot*, 1858, Noyon River. *See* Pudding Creek.

Nubble, The [Sonoma Co.]. This rare topographical term applies to a little knob projecting at the end of a sloping hill. The name was probably applied when the USGS mapped the Tombs Creek quadrangle.

Nubieber (no͞o′ bē bər) [Lassen Co.]. In 1931 the extensions of the Great Northern Railroad and the Western Pacific Railroad met at this place, which was called New Town at that time. When the newly selected name, Big Valley City, was rejected by the Post Office Dept., L. H. Martin of the Chamber of Commerce of Bieber (*q.v.*) christened the settlement after the pioneer town.

Nueces y Bolbones [Contra Costa Co.]. The name Arroyo de las Nueces y Bolbones ('creek of the walnuts and the Bolbones') was the name of the land granted to Juana Sánchez de Pacheco on July 11, 1834. Bolbones was the name of a tribe that spoke a Miwokan language; the name is better known in modern times as Ohlones. *See* Diablo; Walnut: Walnut Creek.

Nuestra Señora. In place naming, the Franciscan missionaries used this title, referring to 'Our Lady', i.e. the Virgin Mary, in preference to Santa María. It was usually combined with one of her special titles: de Altagracia, de los Dolores, de los Ángeles, de la Merced, del Rosario, etc. The Nuestra Señora part of these names was naturally soon dropped for practical purposes, even in Spanish times; but the special title has survived as a place name in numerous instances. *For* Nuestra Señora de Altagracia, *see* Conejo; *for* Nuestra Señora de los Angeles de Porciúncula, *see* Los Angeles; *for* Bahía de Nuestra Señora del Rosario de la Marinera, *see* Marin.

Nuevo, Nueva. The Spanish adjective meaning 'new' was frequently used in Spanish times and has survived in a few places: Nuevo Canyon [Ventura Co.], Nuevo Creek [Santa Barbara Co.]. Land grants included Nueva Flandria, Nueva Helvetia, Nueva Mecklenburg, and Nueva Salem. **Nuevo** (nōō ā′ vō, nwā′ vō) [Riverside Co.] is the name of a post office established in 1915; it took its name from the Rancho San Jacinto Nuevo y Potrero, granted to Miguel de Pedroarena in 1846 (Brigandi). *See also* Ramona. *For* Nueva Helvetia, *see* New Helvetia; *for* Nuevo San Francisco, *see* Solano. *See also* New.

Nurse Slough [Solano Co.]. The tidal channel was named after Nurse's Landing (now Denverton), which in turn had been named for Dr. Stephen K. Nurse. *See* Denverton.

Nutter Lake [Mono Co.]. Named in 1905 by an engineer of the Standard Consolidated Mining Company, of Bodie, for the assistant superintendent, Edward H. Nutter (Farquhar).

Nut Tree [Solano Co.] is a post office named for the popular restaurant on the site.

O

Oak. The seventeen species of native live and deciduous oaks (genus *Quercus*) have left their mark on the nomenclature of the state. Oak is included in almost 150 names on the maps and, next to Willow and Pine, is the most popular name derived from a tree. Although the Spanish distinguished the *encina* (coast live oak, *Q. agrifolia*) from the *roble* (deciduous oak), we have only a few place names modified by "live," "white," or "black." Because especially beautiful specimens grow on level ground, the number of Oak Flats is rather large. The locality of Oak Run [Shasta Co.] gives its name to a tautological Oak Run Creek. There is an Oak Hall Bend in the river south of Sacramento, and a Seven Oaks Creek in Sonoma Co. The GNIS lists dozens of features that bear the name in one form or another, some of which are only conventional applications. **Oakland** [Alameda Co.]: In Spanish times part of the site of the city was called Encinal del Temescal 'oak grove by the sweathouse' because of the luxuriant growth of oaks. The city was laid out for Horace W. Carpentier, Edson Adams, and Andrew J. Moon by a surveyor, Julius Kellersberger, in 1850. When it was incorporated as a town, the present name was spontaneously chosen. **Oakdale** [Stanislaus Co.] was established and named when the Copperopolis and Visalia Railroad reached the place in 1871 (Santa Fe). **Oakville** [Napa Co.]: The name was applied to the station when the Napa-Calistoga sector of the railroad was built in the early 1870s; it is shown on the von Leicht–Craven map of 1874. **Oakley** [Contra Costa Co.] was settled in 1896–97; it was named for the abundance of native oak trees by R. C. Marsh, who became the first postmaster in 1898. The Old English suffix *-ley* 'field, meadow' has been commonly used for places in the United States, although many such names were doubtless transfers from England. **Oak View** [Ventura Co.] was formerly called Oak View Gardens; it was named in 1925 at a public meeting because of the garden-like appearance of the oak groves. **Oak Creek Pass** [Kern Co.] was the route by which Fremont's second expedition crossed the Tehachapis on Apr. 14, 1844. According to the California Department of Parks and Recreation, *California historical landmarks* (1963), this was the pass used by Padre Garcés when he returned from the exploration of San Joaquin Valley in 1776. **Oak Glen** [San Bernardino Co.] was originally called Potato Canyon; when apple growing replaced potato growing in the fertile valley, Isaac Ford substituted the present more euphonious name (Buie, Sept. 22, 1964). **Oakhurst** [Madera Co.] contains the archaic term "hurst" meaning 'wooded hill'. *See* Encina; Roble.

Oakzanita Peak [San Diego Co.]. From "oak" plus "manzanita" (Stein).

Oasis [Mono Co.]. Named after the fertile and productive Oasis Ranch, on which N. T. Piper and his brother Samuel settled in 1864 (Robinson). **Oasis** [Riverside Co.] originated as an orange grove, planted in 1902 (Gunther).

Oat. The term is used for various native wild grasses, in addition to oats (genus *Avena*) introduced from Europe. There are about twenty Oat Hills and Mountains on California maps, together with a number of Oat Creeks and Canyons and an Oat Knob [Tulare Co.]. In Sonoma Co. there are Big and Little Oat Mountains and a Red Oat Ridge. Oat Creek [San Luis Obispo Co.] is an error for Old Creek. *See* Avena.

Oban (ō´ bən) [Los Angeles Co.]. The station is shown on Southern Pacific maps after 1904; it was probably named after the seaport in Scotland or the town in Kansas.

Obed. *See* Bell.

Obelisk [Fresno Co.]. The name was applied to this peak by the USGS in 1903 because of its shape. The Whitney Survey had named the Clark Range "Obelisk Group." The word is occasionally used as a generic.

O'Brien: post office, **Creek, Mountain** [Shasta Co.]. Named for Con O'Brien, who bought the old Conner Hotel on the Sacramento-Yreka road in 1873 (Steger).

Observatory; Observation. Several Peaks, Hills, and Points are so called because they were used for geodetic or astronomic observation. **Observatory Point** [Lake Tahoe] was in 1873 the prospective site of the observatory later established on Mount Hamilton. The von Leicht–Craven map and the von Leicht–Hoffmann Tahoe map of 1874 show Lick's Observatory here. After the site was changed to Mount Hamilton, the abbreviated name clung to the point. **Observation Peak** [Kings Canyon NP] was so named by J. N. LeConte in 1902 because he used the peak for a triangulation point (Farquhar).

O'Byrne Ferry [Calaveras Co.] commemorates Patrick O. Byrne, who established a ferry and later (1852) built a bridge across the Stanislaus River to Mountain Pass. *See* Byrnes Ferry.

Occidental [Sonoma Co.]. The name was first given to a Methodist church erected here in Apr. 1876 on land given by M. C. Meeker, who then laid out the town. The post office was established on Dec. 7, 1876. The railroad station was called Howards until 1891, for William Howard, a settler of 1849.

Ocean. The best known and one of the oldest of the many names along the coast with the stem Ocean is **Oceanside** [San Diego Co.], founded in 1883 and named by J. C. Hayes. **Ocean Beach** [San Diego Co.] was laid out and named by William H. Carlson and Albert E. Higgins in May 1888 (Co. Hist.: San Diego 1922:1:133–34). **Ocean Park** [Los Angeles Co.] was founded and named in 1892 by Abbot Kinney and F. G. Ryan (Newmark 1916:603). **Oceano** (ō sē an′ ō) [San Luis Obispo Co.] was named by real estate promoters around 1893 (Hall-Patton). In Marin Co. there is an Ocean Roar. *For* Ocean Park Heights and Ocean View [Contra Costa Co.], *see* Albany. **Oceanic Creek** [San Luis Obispo Co.] is named for the Oceanic Quicksilver Mine (Hall-Patton).

Ockenden [Fresno Co.]. Named for Tom Ockenden, who had a trading post here (Farquhar).

Ocotillo (ok ə til′ ō, ō kə tē′ yō) [Imperial Co.] is named for the desert shrub ocotillo (*Fouquieria splendens*). **Ocotillo Wells** [San Diego Co.] was earlier called simply Ocotillo (Brigandi); the post office was established on Nov. 16, 1957. The Spanish word is a diminutive of Mexican Spanish *ocote* 'firewood', itself a borrowing from Aztec *ocotl*.

Odessa [Kings Co.]. The name appears on Santa Fe maps after 1905. As the place was in a wheat-growing district, it was doubtless considered appropriate to name it after Odessa in the Ukraine, the greatest wheat-exporting port in Europe.

Ohm [San Joaquin Co.]. The Southern Pacific siding was named about 1900 after the Ohm Ranch, which was established by Thomas Ohm, a native of Germany, who settled there in 1868 and raised grain on a large scale (Co. Hist.: San Joaquin 1923:972, 975).

Oil. The exploitation of California's oil resources has played as important a role in the 20th century as the discovery of gold in the mid-19th century, although less romantic, and has made its mark on our nomenclature. Kern Co. has a whole cluster—Oil City, Center, Junction; and Oildale. A town in Fresno Co. is called Oilfields, a mountain ridge in Ventura Co. is named Oil Ridge, and several minor features bear the name.

Ojai (ō′ hī): **Valley**, town [Ventura Co.] reflect a Chumash place name, *a'hwai*, and means 'moon' (Kroeber). Aujai o Aujay is on the list of rancherias of Mission San Buenaventura (Arch. Mis. S.Buen. 1:28), and a *ranchería de Ojai* is shown on a *diseño* of El Rincón about 1834. On Apr. 6, 1837, the name was applied to a land grant, spelled Ojay and Ojai. In 1874 R. G. Surdam laid out the town and named it for the writer Charles Nordhoff, who had given enthusiastic accounts of the valley in his articles. In 1916 the town was renamed after the valley. *See* Nordhoff.

Ojala (ō häl′ ə) [Ventura Co.]. What was supposedly the smallest post office in the United States, about the size of a phone booth, was once located here. The presumed Chumash origin has not been confirmed.

Ojitos (ō hē′ tōs) [Monterey Co.]. The name Los Ojitos, meaning 'the little springs', was applied to the land grant on Apr. 5, 1842. A *diseño* of the grant shows Ojo de Agua de los Ojitos. There is also a Los Ojitos in San Mateo Co. (Alan K. Brown). *See* Ojo.

Ojo. The Spanish word means 'eye'; *ojo de agua*, lit. 'eye of water', means 'spring (of water)' and is often abbreviated to *ojo*. This form was widely used as a geographical term in Spanish times, and it appears as the principal name of several land grants: Ojo de Agua de Figueroa [San Francisco Co.], Sept. 16, 1833; Ojo de Agua de la Coche ('sow spring') [Santa Clara Co.], Aug. 4, 1835; Tres Ojos de Agua [Santa Cruz Co.], Mar. 18, 1844; Ojo de Agua [Sonoma Co.], Dec. 20, 1844; it is also included in alternate names of other land grants. The peak on the old land grant west of Morgan Hill [Santa Clara Co.], labeled El Toro on the Morgan Hill atlas sheet, was probably one of the "remarkable hills named El ojo del coche" mentioned by Beechey in 1826 (1831:2.48) and recorded as the Peak of "Ojo del Agua de la Coche" in Hoffmann's notes on Aug. 21, 1861. Ojo Grande [Riverside Co.] was the name of a spring at the head of Carrizo Creek mentioned by Emory (1857: 103). None of these names seems to have survived. *For* Ojo Caliente, *see* Caliente.

Ojotska, Rio. *See* American River.

Olancha: Peak, town, **Creek** [Inyo Co.]. The term "Olanches" was applied by Taylor in 1860 to a Panamint village or band south of Owens Lake; however, the name does not appear to come from the Panamint language. The name appears for a settlement, south of Owens Lake, shown as Olanche on Farley's map of 1861, and for the peak shown as Olancha on Wheeler atlas sheet 65. The name of the town is recorded on the Land Office maps of the 1870s.

Old. This descriptive adjective was applied to a number of communities and physical features, usually after another place had been given the original name. There is an Old Inspiration Point in Yosemite, and an Old Mountain View in Santa Clara Co. For some time, Orange Co. had an Old Newport as well as a Newport. In some names the adjective is not used in contrast to a "new" name: Old Barn [Sacramento Co.], Old Man Mountain [Nevada, Santa Barbara Cos.], Oldwomans Creek [San Mateo Co.]. **Old Diggings** [Shasta Co.]: A mining district was so named because abandoned mines of the 1850s were reopened in later years (Steger). **Old Station** [Shasta Co.]: The site of the Hat Creek stage station and military post was so called after it was abandoned in 1858 (Steger). **Old Woman: Mountains, Springs** [San Bernardino Co.]: According to WSP (224:70), the range was so named because the granite pinnacle at its crest resembles an old woman; and *Desert Magazine* (Dec. 1940) indicates that the Indian name for the mountains likewise meant 'old woman'. However, L. Burr Belden states that the place was so named because Col. Henry Washington found two old Indian women at the springs when he surveyed the base line in 1855. **Old Dad Mountains** [San Bernardino Co.]: According to O. J. Fisk, the mountains were actually named because of their appearance. **Old Harmony Borax Works** [Death Valley NP] commemorates the famous works from which the first twenty-mule teams transported borax. The works were built in 1882 by William T. Coleman. **Old Baldy** and **Old Grayback** [San Bernardino Co.] are the local names for San Antonio and San Bernardino (or San Gorgonio) Peaks. *For* Old Beach [Imperial Co.], *see* Niland; *for* Old Canal [San Joaquin Valley], *see* Main Canal; *for* Old Creek [San Luis Obispo Co.], *see* Oat; *for* Old Jerusalem [Alameda Co.], *see* Jerusalem; *for* Old Saddleback [Orange Co.], *see* Santiago Peak.

Olema (ō lē´ mə): post office, **Creek** [Marin Co.]. The approximate site of modern Olema was probably occupied by a Coast Miwok village mentioned repeatedly as Olemos and Olemus in the baptismal records of Mission Dolores after 1802 (Arch. Mis. 1:87 ff., 108), derived from a stem *óle* 'coyote' (Callaghan). Olema post office was established on Feb. 28, 1859. The Olema station of the Wells Fargo Company was named Point Reyes Station on Apr. 1, 1883. *See* Reyes, Point.

Oleta. *See* Fiddletown.

Oleum (ō´ lē əm) [Contra Costa Co.]. The name of the town at the Union Oil Company refinery was created by lopping off the first four letters of petroleum. The post office is listed in 1912.

Olinda [Shasta Co.]. Originally a Portuguese name, preserved in the Brazilian city near Pernambuco, but now international. Samuel T. Alexander, a settler, transferred the name from Maui, Hawaii, in the early 1880s. **Olinda** [Orange Co.] was previously called Petrolia; the name was changed about 1900 by Mr. Bailey, president of the Olinda Oil Company.

Olivas (ō lē´ vəs) **Creek** [Monterey Co.]. Probably for Louis Olivas, who acquired land here in 1908 (Clark 1991).

Olive. The olive tree was introduced into California by the Franciscan padres who planted seed, brought from Mexico, in the mission gardens. Today olive orchards are cultivated commercially in many parts of the state. **Olive** [Orange Co.]: The olive trees on nearby Burruell Point suggested the name Olive Heights when the town was laid out in 1880. The generic was dropped when the post office was established about 1890. **Olivenhain** (ō lē′ vən hīn) [San Diego Co.] was a German cooperative colony, existing in July 1885, according to the journal of J. E. Hauswirth, a member of the colony. He spells the name Olivenheim 'home of the olives', whereas the present name is Olivenhain 'olive grove'. Other places are called Olive View [Los Angeles Co.], Olivehurst [Yuba Co.], Olive Oak Springs [San Diego Co.], and Oliveto [Sonoma Co.]. *See* Los Olivos.

Olive Lake [Tuolumne Co.]. The lake near Huckleberry Lake was named in memory of Olive Hall, a local resident and conservationist, who died in 1964 (BGN, 1965).

Olmsted Grove [Humboldt Co.]. The grove in the Prairie Creek Redwoods State Park was established in 1953 in honor of Frederick Law Olmsted (1822–1903), noted landscape architect, who for many years was consultant to the California state parks system and active in the redwoods conservation program.

Olney: Creek, Gulch [Shasta Co.]. Named for Nathan Olney, of Oregon, who brought Walla Walla Indians to mine here before the Gold Rush of 1849 (Steger).

Olompali (ə lom′ pə lē, ō ləm pä′ lē) [Marin Co.]. The school preserves the name of a Coast Miwok village called *óolum pálli*, apparently containing *ólom* 'south' (Harrington; Callaghan). The Olumpali or Olompalis Indians are mentioned by members of the Kotzebue expedition in 1816, and Padre Payeras records a *cañada de los Olompalis* in 1819 (Docs. Hist. Cal. 4:341 ff.), and *el punto de Sta Lucía Onompali* on Oct. 20, 1822 (Arch. MSB 12:411). *La ranchería de Olimpali o Santísimo Rosario* is mentioned in 1828 in the Registro (p. 4), and the baptismal records of Mission Dolores show the variant Olompalico. The present spelling was used for the name of a land grant on Oct. 22, 1843.

Olympia [Santa Cruz Co.]. The post office was established and named in 1915. Olympus, the mountain throne of the Greek gods, and Olympia, the site of the Olympic games, have repeatedly been used for American place names.

Olympic Valley [Placer Co.]. The name of the post office was changed from Squaw Village to Olympic Valley on Aug. 1, 1960, in anticipation of the Winter Olympic Games at Squaw Valley.

Omega (ō mā′ gə) [Nevada Co.]. Two mining towns, named after the first and the last letter of the Greek alphabet, are listed as post offices in 1858. Omega, founded in 1851 by E. Paxton, John Douglass, and others, is still on the maps, but little is left to indicate the diggings at old Alpha.

Omenoku (om ə nō′ ko͞o) [Humboldt Co.]. The name of the small promontory north of Trinidad Head is from Yurok *o-menokw* 'where it projects' (Waterman, p. 269).

Omjumi (ōm jo͞o′ mē) **Mountain** [Plumas Co.]. The name of this mountain probably contains the Maidu generic term *om* 'rock' (Kroeber).

Omochumnes [Sacramento Co.]. The probable source is Plains Miwok *omuučaïmni*, meaning people of *omuuča* 'winter' (Callaghan). The land grant, dated Jan. 8, 1844, was named after the Ranchería de los Omochumnes, shown on a *diseño* of the grant.

Omogar (om′ ə gär) **Creek** [Humboldt, Del Norte Cos.]. From Yurok *omega'a* (Waterman, map 9, item 60); the meaning of the name is not known. The spelling was made official by the BGN in 1984 (list 8402), supplanting Omagaar.

Omo (ō′ mō) **Ranch** [El Dorado Co.]. The place is shown on the Land Office map of 1891 and is listed as a post office in 1892. According to Merriam (*Mewan stock*, p. 344), it was named after an Indian village, Omo. It may be an abbreviation of Northern Sierra Miwok *oomu' a' koča* 'menstrual hut' (Callaghan).

Ondo, Arroyo. *See* Hondo.

O'Neals [Madera Co.]. The post office, established on Oct. 4, 1887, took its name from the Charles O'Neal's Hotel.

One-Horse Town. *See* Horse: Horsetown; Mule.

Oneida (ō nī′ də) **Lake** [Yosemite NP] is named for Oneida Lake in New York State (Browning 1986). The term is Iroquoian, meaning 'standing rock', from the verb *-ot-* 'to stand' (Mithun); cf. Oneonta.

O'Neill Forebay [Merced Co.]. The reservoir was named for J. E. O'Neill, rancher and business leader in the San Joaquin Valley (BGN, 1966).

O'Neill Lake [San Diego Co.]. Named for Richard O'Neill, who owned a part of the Santa Margarita Ranch (J. Davidson).

One Mule Town. *See* Mule.

Oneonta (ō nē on´ tə, on ē on´ tə) **Park** [Los Angeles Co.]. The Pacific Electric station was named after Oneonta, New York, birthplace of Henry E. Huntington, the promoter of the railroad. The New York name is probably from Onondaga (Iroquoian) *onę̨yǫda'* 'protruding stone', from the verb *-ǫt-* 'to protrude' (Chafe); cf. Oneida.

One Suerte (wun so͞o âr´ tē) [Monterey Co.]. English "one" plus Spanish *suerte* 'chance, luck', which (like English "lot") can mean 'a piece of ground'; used in Mexican California to mean 'a farming lot located near a pueblo' (Clark 1991).

Onion. About twenty-seven native species of the genus *Allium* are found in California, especially in hilly and mountainous sections; their presence has given rise to almost a hundred Onion, Leek, and Garlic place names, including two settlements called Onion Valley, in Plumas and Calaveras Cos. respectively.

Onlauf (on´ lôf) [Ventura Co.]. Named for the Anlauf family of Santa Paula, which owned property in this section (J. H. Morrison).

Ono (ō´ nō) [Shasta Co.]. When the post office was established in 1883 at Eagle Creek, the settlers chose the name of the Biblical town (I Chron. 8:12), at the suggestion of the Rev. William S. Kidder (Steger). There are (or were at that time) several Onos in the Middle West, and there is now another Ono in California [San Bernardino Co.]. *See* Igo.

Onofre. *See* San Onofre.

Ontario: town, **Peak** [San Bernardino Co.]. The town was laid out in 1882 by George B. Chaffey on the old Cucamonga Rancho and was named for his former home province, Ontario, Canada. The term is said to be Iroquoian, translated 'lake-fine' (Stewart 1970).

Onyx: post office, **Peak** [Kern Co.]. The place had been known as Scodie's Store since the 1860s. When the post office was established in 1890, the name Scodie was rejected because of its similarity to Scotia in Humboldt Co. William Scodie then chose the present name because it is short and unique (A. J. Alexander).

Opahwah (ō pä´ wä) **Butte** [Modoc Co.]. The mountain, also known as Centerville or Rattlesnake Butte, was given this name by the BGN (6th Report) in the 1920s. Opahwah was the name in an unidentified Indian language.

Opata (ō´ pə tə) **Creek** [San Diego Co.]. Ópata is the name of an Indian tribe of Sonora in northwestern Mexico.

Ophir (ō´ fər) [Placer Co.]. This is the only survivor of five mining towns named after Ophir, the land of gold mentioned repeatedly in the Bible. The town was known as Spanish Corral in 1849 but took the Biblical name in 1850 after a rich lode was discovered in the vicinity. Ophir City was the original name of Oroville. **Mount Ophir** [Mariposa Co.] was the site of one of the first mints to coin California gold pieces.

Orange. The importance of the orange in California fruit culture is shown in the number of places—including at least seven communities or settlements—that have been named for it. **Orange**: city, **County**: The town, laid out by Glasell and Chapman in the 1870s, was first called Richland; then its name was changed to Orange because there was another Richland in Sacramento Co. (Santa Fe). The Land Office Map of 1879 shows neither name, but Orange post office was established on Sept. 1, 1873. The name was chosen to accent the developing orange culture, but the choice may have been influenced by the fact that the Glassell family had once lived in Orange Co., Virginia (named for the Prince of Orange, the son-in-law of the English king George II). By 1880 there were Orange counties in six other states, as well as numerous towns and post offices so named. The name Orange for the California county was first proposed in 1872, during an early county division attempt. According to county historian Jim Sleeper, "to encourage immigration, the area was 'boomed' by real estate promoters as a semi-tropical paradise . . . The name *orange* has a Mediterranean flavor about it, so for that reason it was selected to suggest our climate." After six division attempts, the new county was finally created from southern Los Angeles Co. in 1889 (Brigandi). **Orange Cove** [Fresno Co.], situated in a "cove" in the Sierra foothills where citrus thrives, was named in 1913 after the Orosi Orange Lands Company, managed by E. M. Sheridan and M. S. Robertson (Santa Fe).

Orchard Spring [Lassen Co.]. The spring in the lower end of Little Tuledad Canyon had various names until the BGN decided for the present name (1964) because it is near an abandoned apple orchard.

Orcutt [Santa Barbara Co.]. The town was laid

out by the Union Oil Company in 1903 and named for the company's geologist, W. W. Orcutt (AGS: Santa Barbara, p. 180).

Ord, Fort [Monterey Co.]. In 1933 the War Dept. named the military reservation Camp Ord, in memory of Gen. Edward Otho Cresap Ord (1818–83), who was stationed in Monterey as a lieutenant of the 3d Artillery in 1847 (Clark 1991). He was a distinguished general in the Civil War and in 1868 became commander of the Dept. of the Pacific. The name was changed to Fort Ord in 1940. **Ordbend** [Glenn Co.], on the extensive Ord Ranch, was owned by Gen. Ord and two of his brothers in the 1850s. **Ord Mountains** [San Bernardino Co.]: According to local information, the mountains were named after the "Ord Group," a number of claims located here by Sandie Lochery about 1876. The mines were named for General Ord, under whom the prospector may have served (O. J. Fisk).

Oregon: Hills, Peak, Gulch [Butte Co.]; **Oregon Bar** [El Dorado Co.]; **Oregon House** [Yuba Co.]; **Oregon Creek** [Sierra Co.]; **Oregon Gulch** [Shasta Co.]. All these names are survivors of the many camps and features named in mining days by gold-seekers who came from or via Oregon. This most disputed of U.S. place names has been explained by Stewart (1982:153–55) as a misreading of "Ouariconsint" on a 1715 map; that form was a variant of "Ouisconsing," recorded in 1673 for the Wisconsin River—the ultimate origin apparently being in an Algonquian language.

Orestimba (ôr əs tēm´ bə) **Creek** [Stanislaus Co.]. An *arroyo de Orestimac* is mentioned in Father Viader's diary of Oct. 1810 (Arch. MSB 4:92). A *diseño* of the Rancho del Puerto (1843) shows an Arroyo de Horestimba, and a *diseño* of the Orestimba land grant (Feb. 21, 1844) shows an Arroyito de Orestiñoc. The name contains Costanoan words, the first being *ores* 'bear' (Kroeber). A tributary is called Oso ('Bear') Creek, and a nearby mountain is Mount Oso. Locally the stream has been referred to as Orris Timbers Creek.

Orick [Humboldt Co.]. From the Yurok Indian village *oo'rekw* (Waterman, p. 262; Harrington).

Oriflamme (ôr´ i fläm): **Canyon, Mountains** [San Diego Co.]. For the Oriflamme quartz mine, active around 1875–82 (Brigandi); sometimes spelled Oroflamme, doubtless because of an association with Spanish *oro* 'gold'. "Oriflamme" is used in English for anything suggestive of the red flag carried into battle by early French kings. According to Wheelock, the first miners during the Julian gold rush boom of 1870 arrived aboard the sidewheeler *Oriflamme*, and this may be the origin of the name.

Orinda (ôr in´ də) [Contra Costa Co.]. About 1880 Theodore Wagner, then U.S. surveyor general for California, gave the name Orinda Park to his estate between Bear and Lauterwasser Creeks. On Mar. 13, 1888, it was favored with a post office, and in 1890 Orinda Park Station became the provisional terminal of the California-Nevada Railroad. From the Wagner estate the name was transferred in 1895 to what is now Orinda Village—and in 1945, as Orinda, to the "crossroads," formerly known as Bryant. Like similar names with a Latin flavor, it was probably coined for its pleasing sound; however, "General" Wagner might have read the name Orinda, by which the poet Katherine Fowler Philips was known. Possible origins of the name in European literature are discussed by Herman Iventosch, "Orinda, California: or, the literary traces in California toponymy," *Names* 12:103–7 (1964).

Orland [Glenn Co.]. The name was given to the station when the Central Pacific built the connecting line between Willows and Tehama. According to the county history (1918), the place was named after Orland, England, the birthplace of an early settler. The name was drawn from a hat in which three proposed names had been placed, including Leland and Comstock.

Orleans: town, **Mountain** [Humboldt Co.]. The place was settled in 1850 and called New Orleans Bar, probably after the city in Louisiana. The name was shortened to Orleans Bar in 1855 when the settlement became the seat of short-lived Klamath Co. The present form was used by the post office in 1859.

Ornbaun (ôrn´bôn): **Valley, Hot Springs** [Mendocino Co.]. Apparently an error for the name of John S. Ornbaum, a native of Indiana, who came to the county in 1854 and operated a large stock ranch in the valley (Co. Hist.: Mendocino 1914:880).

Oro. The Spanish word for 'gold' was naturally a favorite word in naming places in California. Oro was the name intended for what became Tuolumne Co., and it was the name given to the first county seat of Sutter Co., a short-lived town founded in 1850 some 3 miles above the mouth of Bear River (now in Placer Co.). Often the name was first given to a gold mine, and frequently it was applied without regard to idiomatic usage: Oro Loma [Fresno Co.]; Oro Chino [Mariposa Co.]; Oroleeve, Monte de Oro [Butte Co.]. The combination Oro Fino was especially popular; this name is preserved in Monterey, San Luis Obispo, and Siskiyou Cos. **Oroville** (ôr′ ō vil, ôr′ ə vil) [Butte Co.] was settled in 1849 by Col. John Tatam and other miners and became known as Ophir City, after the ancient gold land mentioned in the Old Testament. When the post office was established in 1855, the name had to be changed because there was an Ophir in Mariposa Co. and an Ophirville in Placer County. Judge J. M. Burt preserved the golden glimmer by coining the new name from Spanish *oro*. **Oro Grande** (ôr′ ō gran′ dē) [San Bernardino Co.] goes back to a gold-mining camp of 1878. The post office was established on Jan. 3, 1881, and was named Halleck, for the chemist of the stamp mill. On May 1, 1925, the name of the post office and of the town was changed to Oro Grande, after a nearby mine. **Orocopia: Mountains, Canyon, Spring** [Riverside Co.]. From the Ora Copia Mining Co., coined in 1904 from Spanish *oro* 'gold' and Latin *copia* 'abundance' (Gunther). *For* Isla del Oro, *see* Marin.

Oroflamme. *See* Oriflamme.

Orosi (ôr ō′ sə) [Tulare Co.]. The town was founded in 1888 by Daniel R. Shafer and associates. The name was coined from *oro* 'gold' by Neal McCallan, because the fields around were covered with golden poppies. The post office is listed in 1892.

Oroville. *See* Oro.

Orrs Springs [Mendocino Co.]. The hot sulphur springs are on the Orr ranch, established in 1858 by Samuel Orr, a native of Kentucky who had come to California in 1850 (Co. Hist.: Mendocino 1880:658). The place is shown on the Official Railway Map of 1900.

Ortega (ôr tā′ gə): **Hill**, station [Santa Barbara Co.]. The name for the station appears on maps after 1890 and was given for a member of the Ortega family, well known in the annals of the county. The intermittent creek west of Summerland is labeled Arroyo de las Ortegas on the Santa Barbara atlas sheet. This is doubtless on the property mentioned as *otro parajito nombrado* ('another little place called') *las Ortegas* in July 1834 (DSP Ben. Mil. 79:96). There is another Ortega Hill (4,970 feet!) in Ventura Co., and a station of the Western Pacific called Ortega in San Joaquin Co.

Ortigalita (ôr ti gə lē′ tə): **Creek, Peak** [Merced Co.]. The name is a diminutive of Spanish *ortigal* 'nettle patch', from *ortiga* 'nettle'. A post office Ortigalito is listed in 1880.

Osdick [San Bernardino Co.]. In 1905 the Santa Fe named the station on the recently acquired Kramer-Johannesburg branch for P. J. ("Pete") Osdick, a pioneer in the Rand mining district. After a dispute with the mining interests in 1931, the name was officially changed to Red Mountain, but school and voting district as well as the Osdick Group of Mines preserved the old pioneer's name.

O'Shean's View. *See* Albany.

Osito. *See* Oso.

Osnito, Arroyo del. *See* Hospital Creek.

Oso. The Spanish word for 'bear', the state animal of California, was frequently used for place names in Spanish times and has survived in the names of more than ten geographic features. **Oso Flaco: Valley, Creek, Lake** [San Luis Obispo Co.]: The Portolá expedition camped here on Sept. 2, 1769, and some of the soldiers killed a bear. Crespí gave the place a holy name; some of the soldiers called it Las Víboras (on account of the many rattlesnakes); and others called it El Oso Flaco 'the lean bear', although the bear must have weighed 375 pounds (Crespí, p. 182). After that, Oso Flaco is repeatedly mentioned in Spanish manuscripts as the name for the lake and for a *sitio*. A *diseño* of the Guadalupe grant records *Méganos de Oso Flaco* 'dunes of the lean bear', and one of the Bolsa de Chamisal (1837) shows Terrenos del Oso Flaco. **Los Osos: Valley, Creek** [San Luis Obispo Co.]: "In this valley we saw troops of bears, which kept the ground plowed up and full of holes which they make searching for roots. . . . The soldiers went out to hunt and succeeded in killing one . . . This valley they named Los Osos, and I called it La Natividad de Nuestra Señora" (Crespí, pp. 184–85). The name was preserved by the land grant Cañada de los Osos, dated Dec. 1, 1842, which on Sept. 24, 1845, was combined in a grant with the extraordinary name Cañada de los Osos y Pecho y Islai 'valley of the bears and breast [after a rock] and wild cherry'. **Cañada de los Osos** [Santa Clara Co.]: A Ranchería de los Osos is mentioned on June 7, 1799 (PSP 17:327); it probably did not refer to bears, but to hostile Indians called Osos by the Spaniards (Prov. Rec. 6:120). Cañada de los Osos is shown on a *diseño* of Rancho Ausaymas in 1833; this name was used for a land grant dated Oct. 20, 1844. The diminutive *osito* was occasionally used. Crespí records on Sept. 10, 1769: "They brought a little bear which they had reared and offered it to us, but we did not accept it. From this circumstance the soldiers took occasion to name the spot El Osito, but I called it San Benvenuto." *See* Bear; Poza: Posa.

Ospital. *See* Hospital Creek.

Ossagon (os' ə gən) **Creek** [Humboldt Co.]. From the Yurok village name *osegen* (Waterman, p. 234); also spelled Ossegon.

Ostrander: Rocks, Lake, Trail [Yosemite NP]. Named by the Whitney Survey for Harvey J. Ostrander, descendant of one of the early Dutch settlers of New Amsterdam; he came to California at the time of the Gold Rush, and took up a homestead at the junction of Glacier Point and Old Mono trails in the 1860s. The rocks and the homestead are shown on the Hoffmann-Gardner Yosemite map of 1867. Bret Harte's mother was a member of the Ostrander family.

Oswald [Sutter Co.]. The place was probably named for August Oswald (or Ostwald), a native of Germany, who came to California in 1847 as a member of Company B of Stevenson's Volunteers and was later one of Allgeier's men at Nicolaus.

Otay (ō tī', ō' tī): **River, Valley, Mesa, Mountain**, town, **Reservoirs** [San Diego Co.]. Perhaps from Diegueño *'etaay* 'big' (Langdon). An Indian rancheria Otai is mentioned in 1775 (PSP Ben. Mil. 1:5), and a *gentil de Otay* is recorded in the following year (PSP 1:230). The form Otay was used for two land grants, on Mar. 24, 1829, and May 4, 1846, and the river is shown as Arroyo de Otay on a *diseño* of 1833. The Boundary Survey, 1849, has Rio Otay; the Coast Survey, 1857, has Valley of Ohjia [sic]; von Leicht–Craven, 1874, has Otay Creek. Otay post office is listed in 1870.

Otis. *See* Yermo.

Otterbein [Los Angeles Co.]. In 1910 Bishop William M. Bell established a settlement for retired ministers of the Church of the United Brethren in Christ and named it in honor of the founder of the church, Philip W. Otterbein (C. H. Bell).

Otter Creek [El Dorado Co.]. The name of the tributary of the Middle Fork of the American River is the only reminder of the river otter (*Lutra canadensis*), which once was common in our streams. Wilkes's map of 1841 shows an Otter River, apparently the Merced.

Ottitiewa (ō tē' tə wä) [Siskiyou Co.]. An old name for Fort Jones (*q.v.*), supposedly referring to the Scott River group of the Shasta Indian tribe.

Ottoway: Peak, Creek, Lakes [Yosemite NP]. The peak was named in 1895 by Lt. N. F. McClure, for a corporal in his detachment.

Outside Creek [Tulare Co.]. The creek was so named because it was on the "outside" edge of the swamp area created by the delta of the Saint Johns and Kaweah Rivers.

Ouzel: Basin, Creek [Kings Canyon NP]. Named in 1899 by David Starr Jordan because "here John Muir studied the water ouzel in its home, and wrote of it the best biography yet given of any bird" (Jordan, *The Alps of King-Kern Divide* [San Francisco: Robertson, 1907], pp. 18–19). The bird, *Cinclus mexicanus*, is also called the dipper.

Ovis (ō' vis) **Bridge** [Lava Beds NM]. The Latin word for 'sheep' was suggested to J. D. Howard by a naturalist when they examined the skulls of bighorns lying under the bridge.

Owens: Lake, Valley, Point [Inyo Co.]; **River** [Mono, Inyo Cos.]; **Mountain** [Fresno Co.]; **Peak** [Kern Co.]. John C. Frémont named the lake in 1845 for Richard Owens, of Ohio, a member of his third expedition (1845–46), captain of Company A of the California Battalion, and his "Secretary of State": "To one of the lakes . . . on the east side of the range I gave Owens' name" (Frémont, *Memoirs*, p. 455). The river was also named by Frémont; at least it appears on Preuss's map of 1848. *For* Owensmouth, *see* Canoga Park.

Owens Creek [Merced, Mariposa Cos.]. The stream is labeled Owen's Creek on Hoffmann's map of 1873; it may have been named for Richard H. Owen, a native of Connecticut and a miner in Mariposa Co. in the 1870s.

Owens Ferry [Stanislaus Co.]. *See* La Grange.

Owens Mountain [Fresno Co.]. The mountain was probably named for George W. Owens, a native of Ohio, who came to California in

1862 and had a stock ranch in the foothills in the 1870s.

Owenyo (ō´ wən yō) [Inyo Co.]. A railroad name, coined from the name of the lake and the second syllable of the county and applied to the station of the Carson and Colorado Railroad in 1905. When the extension to Mojave was built in 1910, the station was shifted to the present location and called New Owenyo, but the prefixed adjective was dropped the following year.

Owlshead Mountains [San Bernardino Co.]. The highest peak of the mountains was named Owlshead Peak by the Wheeler Survey (atlas sheet 65), doubtless because of its appearance from a certain angle.

Oxalis [Fresno Co.]. When the line from Los Banos to Fresno was built in the 1890s, the Southern Pacific applied the botanical name for wood sorrel to the station. When another siding was established a few miles to the south, it received the same name spelled backwards: Silaxo.

Oxnard [Ventura Co.]. When the Southern Pacific branch from Ventura to Burbank was built in 1898–1900, the station was named for Henry T. Oxnard, who had established a beet-sugar refinery here in 1897. The name is originally "oxen-herd," i.e. a herder of oxen.

Oystercatcher Point [Monterey Co.]. For a shore bird, the black oystercatcher, *Haematopus bachmani* (Clark 1991).

Ozena (ō zē´ nə) [Ventura Co.]. There is a tradition that this means 'runny nose' (Charles Johnson), apparently from Spanish *ozena*, translated as 'ozaena, an ulceration of the nostrils'.

P

Pablo, Rancho de. *See* Johnsons Ranch.

Pachalka (pə chäl′ kə) **Spring** [San Bernardino Co.]. The spring was named before 1895 for a Paiute Indian named *pichǻka*, who had his camp there (Fowler). He was a leader of a small group of Las Vegas Indians (O. J. Fisk). The name was spelled in various ways until the BGN decided for the present version on Dec. 6, 1911: "Not Pachaca, Pachanca, Pachapa, Pachauca, Pechaca, nor Pechapa."

Pachappa (pə chä′ pə, pə chap′ ə) [Riverside Co.]. The name was applied to the station when the Santa Fe line was built from Riverside to Santa Ana in 1890. It is apparently a Gabrielino name of unknown meaning, possibly confused with Pachalka.

Pacheco (pə chā′ kō): **Pass** [Merced Co.]; **Peak, Creek, Canyon** [Santa Clara Co.]. The names are in the territories of the San Luis Gonzaga and Ausaymas y San Felipe grants, the former granted in 1843 to Juan P. Pacheco, and the latter in 1833 and 1836 to Francisco Pacheco, who had come to California in 1819. The pass is shown as Pacheco's Pass on the Frémont-Preuss map of 1848. Pacheco Peak is mentioned in Trask's Report (1854:39). Hoffmann mentions a Pachecoville (which has since vanished) in his notes of Sept. 1861. Sierra de Pacheco is shown on a *diseño* of the Soberanes land grant, which later became the nucleus of the vast landholdings of Miller and Lux.

Pacheco: Valley, Creek, town [Contra Costa Co.]. Salvio Pacheco, a native of Monterey and a former soldier, settled in 1844 on Rancho Monte del Diablo, which had been granted to him in 1834. The present town was laid out by Dr. J. H. Carothers in 1857 and was named Pacheco in 1858. *For* Pacheco [Marin Co.], *see* Ignacio.

Pacheco Hill [Marin Co.]. Named for Ignacio Pacheco (1808–64), a native of San Jose and grantee in 1840 of the San Jose land grant on which the hill is situated.

Pacific. California's location on the shore of the Pacific Ocean is reflected in a number of place names, including the half-Spanish Pacifico Mountains [Los Angeles Co.]. The oldest of the names still in existence is probably that of the post office in El Dorado Co., which goes back to the early 1870s. **Pacific Grove** [Monterey Co.] was established as a tent city by Methodists in 1875, and the present town was laid out by the Pacific Improvement Company in 1883. **Pacific Beach** [San Diego Co.] was apparently named by the founders of San Diego College of Letters, a short-lived institution established in 1887 (*San Diego Tribune*, Dec. 7, 1934). **Pacific Palisades** [Los Angeles Co.] was founded and named in 1921 by a Methodist church organization. The post office was established on Oct. 6, 1924. **Pacific Valley** [San Luis Obispo Co.]: "The only stretch of level land along Santa Lucia coast" is "dignified with the name of Pacific Valley though there is nothing at all valley like about it" (Chase, *Coast trails*). **Pacific Ridge** [Lake Co.]. bears the name of the now-vanished Pacific City, which consisted mainly of tents during the copper boom of the 1860s (Mauldin). *For* Pacific Congress Springs, *see* Congress Springs.

Pacifica [San Mateo Co.]. On Oct. 29, 1957, the inhabitants of Linda Mar, Sharp Park, Edgemar, Westview, Pacific Manor, Rockaway Beach, Fairway Park, Vallemar, and Pedro Point voted to incorporate as the city of Pacifica, a name indicative of its situation by the shore of the Pacific Ocean.

Pacoima (pə koi′ mə): **Canyon, Creek, Dam, Wash**, city [Los Angeles Co.]. The name is derived from a Gabrielino word and may mean 'running water' (Keffer, p. 63). The town was established about 1887 by Sen. Charles Maclay, Judge Robert M. Widney, and others; the post office is listed in 1915.

Padding River. *See* Pudding Creek.

Padre Barona: Valley, Creek [San Diego Co.]. The name was apparently first applied to one of the mesas in the region: a land grant, dated Jan. 25, 1846, is called Cañada de San Vicente y Mesa del Padre Barona. Father José Barona was a friar at San Diego (1798–1811) and at San Juan Capistrano (1811–31). The area is simply called Barona nowadays. When the Diegueño Indians on the Capitan Grande Reservation were displaced by the El Capitan reservoir in the 1870s, they were moved to the Barona Ranch Indian Reservation (Brigandi).

Pahrump (pä′rump): **Valley** [Inyo Co.]. The valley extending from Nevada into California is shown as Pahrimp in Wheeler's atlas (sheet 66). The name can be referred to Southern Paiute *pa-* 'water' plus *tïmpa* 'mouth', which after the root *pa-* would be modified phonetically to *-rïmpa* (Fowler).

Pahute Peak. *See* Piute.

Paicines (pī sē′ nəs) [San Benito Co.]. Perhaps from Mutsun (Costanoan) *paysen* 'to get pregnant' (Callaghan). The post office was established before 1880 and named after the Rancho Ciénega de los Paicines, granted on Oct. 5, 1842. The specific term refers to a Costanoan village, Paisin, on the San Benito River (Kroeber, *Handbook*, p. 465). This spelling approximates that of *Ciénega de los pasines*, shown on a *diseño* of the rancho. *Ciénega* is the common Spanish term for 'marsh'. *See* Tres Pinos.

Paige [Tulare Co.]. The Santa Fe station was named for Timothy Paige, a San Francisco banker and owner of the Paige and Morton Ranch, when the branch from Tulare to Corcoran was built in the middle 1890s.

Paines Creek. *See* Paynes Creek.

Painted. This participle is used to describe physical features which, because of their surface composition, appear "painted" in a certain light, or which are actually marked with pictographs made by former dwellers. In the former class are Painted Canyon [Riverside Co.], Painted Gorge [Imperial Co.], Painted Rock [Placer Co.], and Painted Hill [Riverside Co.]. The latter class includes Painted Cave [Santa Barbara Co.], in which picture writings were discovered by Dr. W. J. Hoffman in 1884; and Painted Rock [San Luis Obispo Co.], known to the Spaniards as La Piedra Pintada.

Paintersville [Sacramento Co.]. Named for Levi Painter, who came to California in 1853 and laid out the town in 1879 (Doyle).

Pait Valley. *See* Pate Valley.

Paiute. *See* Piute.

Pajaro (pä′ hə rō): **River, Creek, Gap; Mount Pajaro** [Santa Cruz, Monterey Cos.]. The river was named by the soldiers of the Portolá expedition on Oct. 8, 1769. "We saw in this place a bird which the heathen had killed and stuffed with straw; to some of our party it looked like a royal eagle. . . . For this reason the soldiers called the stream Río del Pájaro, and I added the name of La Señora Santa Ana" (Crespí, pp. 210–11). Llano ('plain') del Pájaro is mentioned on Sept. 17, 1795 (Arch. MSB 4:193). The name of the river is repeatedly recorded in mission and state papers, and it appears later in the titles of several land grants. Beechey (followed by Gannett) was apparently unaware of the circumstance of the naming, since he states that the river was "appropriately named Rio de los Paxaros, from the number of wild ducks which occasionally resort thither" (2:48–49). Pajaro Valley is mentioned by Tyson (1850:51), and Pajaro River in the *Statutes* of 1850 (p. 59). The settlement of Pajaro was called a "considerable" town by Cronise (p. 123); it was later renamed Watsonville Junction. The name was used elsewhere in Spanish times: Islas de Pájaro (perhaps The Brothers) are shown on a *diseño* of Rancho de San Pablo [Contra Costa Co.]. The Pajaro River was apparently once called Sanjón ('ditch') de Tequesquite; *see* Tequesquite.

Pajuela (pä hwā′ lə) **Peak** [Kern Co.]. The Spanish term means 'a short straw'.

Pala [Santa Clara Co.]. The name of the land grants, Pala, dated Nov. 5, 1835, and Cañada de Pala, dated Aug. 10, 1839, are derived from an Indian *capitanejo* named Pala, who was mentioned as early as July 29, 1795 (DSP San Jose 1:50).

Pala (pä′ lə): **Mountain, Indian Reservation**, town [San Diego Co.]. From Luiseño *páala* 'water'. A rancheria called Pala appears in mission records as early as Sept. 28, 1781 (Arch. Mont. 7:3). In 1816 the Mission San Luis Rey built a chapel there, and the *asistencia* was called San Antonio de Pala. There were three Mexican land claims called Pala, the first dated Nov. 5, 1835. The name is recorded by Williamson in 1855. The Indian reservation was established and named by executive order of President Grant on Dec. 27, 1875.

Palen (pā′ lən): **Lake, Pass, Mountains** [Riverside Co.]. Perhaps named for Matthew Palen, a miner from Ireland (Gunther).

Palermo (pə lâr′ mō) [Butte Co.]. Named in 1887 after Palermo, the capital of Sicily. The name was chosen because the land and climate are suited for olive growing (H. Huse). *For* Mount Palermo, *see* Tamalpais.

Palisade: peaks, **Creek, Lakes** [Kings Canyon NP]; **Glacier** [Inyo Co.]. The collective name Palisades was applied to the peaks by the Whitney Survey in 1864. The two highest peaks are now called North and Middle Palisade. In 1879 North Palisade, the highest, was named by L. A. Winchell in honor of Frank Dusy, and in 1895 by Bolton C. Brown in honor of David Starr Jordan. Neither name prevailed. Split Mountain was at one time called South Palisade, but it does not really belong to the group, and the name has been discontinued (Farquhar). A steep elevation along Nacimiento River [Monterey Co.] is called The Palisades. The name is also found as a generic term on the ocean shore, where the sea-facing bluffs are usually so designated. *See* Pacific: Pacific Palisades.

Palm. The isolated groups of the California fan palm, *Washingtonia filifera*, in Riverside and San Diego Cos., have suggested a number of place names. Indeed, so conspicuous and so important is the tree that it is quite natural to call a place where it grows Palm Spring, or Palm Canyon, or Palm Valley. "In the talk of desert men the palm figures constantly . . . the names mean to the traveller not only water, but shade, with the chance of grass for his animals, and the relief of verdure for his sorely harassed eyes" (Chase, *Desert trails*, p. 16). The number of trees found in a place often becomes a part of the name: Lone Palm, Dos Palmas, Two Bunch Palms, Seven Palms Valley, Seventeen Palms, Twentynine Palms, Thousand Palms; these names survive even when they lose their appropriateness. In Spanish documents places named for the *palmas* are repeatedly mentioned. None of these seem to have survived, although some were doubtless for the same places now bearing the American form. Dos Palmas [Riverside Co.] was named in American times. The land grant Valle de las Palmas is situated in Lower California. Las Palmas [Fresno Co.] and several other places were not named for the native palm, but either for cultivated palms or for other trees which were called palms. **Palm Springs** [Riverside Co.] was originally Palmetto Spring: "The fine large trees which mark the course of the run have furnished the name by which it is known—'Palmetto Spring'" (Whipple 1849:7–8). The place is shown as Big Palm Spring on the von Leicht–Craven map of 1874. Later it was known as Agua Caliente because of the hot springs. When the post office was established about 1890, the name Palm Springs was chosen because there was already a post office named Agua Caliente, in Sonoma Co. The Southern Pacific station, now called Garnet, was named Palms when the sector from Colton to Indio was built in 1875–76. On the Southern Pacific map of 1889 it appears as Seven Palms, and in 1900 as Palm Springs. In 1923, when Palm Springs station became Garnet, the name was transferred to the old White Water station. **Thousand Palms Canyon** and post office [Riverside Co.]: This large colony of fan palms has been called Thousand (i.e. many) Palms since the "early days," although on the von Leicht–Craven map of 1874 the place is modestly called 100 Palm Spring, and on the Land Office map of 1891 it is likewise labeled 100 Palms. In 1946 there were actually about a thousand palms in the canyon—some seven hundred old trees, and three hundred from 10 to 20 feet tall. Paul Wilhelm's subsequent plantings of 60,000 seedlings gave new significance to the name. The post office was established in 1915 and called Edom after the ancient country in Asia. In 1939 the name was changed to Thousand Palms upon petition of the residents, but the railroad name remained Edom. There is another Thousand Palms Canyon in Anza Borrego Desert State Park. **Two Bunch Palms** [Riverside Co.] is named from two groups of fan palms, one upper and one lower. Both have small springs; they were Indian camps and later resting places for desert prospectors (*Desert Magazine*, May 1939, p. 40). **Twentynine Palms** [San Diego Co.] was named in 1852 by Col. Henry Washington, the surveyor of the San Bernardino base line, who found 29 "cabbage trees" here. This was the common name for the Washingtonia palm before the German botanist Wendland named it in 1879, either for the father of our country or for Henry Washington. There are numerous other Palm names in Riverside and San Diego Cos.: Biskra, Curtiss, Macomber, and Willis Palms;

Pushawalla Canyon Palms, One Palm, Lost Palms, Hidden Palms, and Palm Desert.

Palmas. *See* Palm; *for* Mount Palmas, *see* Tamalpais.

Palm City [San Diego Co.]. When the post office was established on Jan. 13, 1914, it was so named because it was situated on a road lined by palm trees.

Palmdale [Los Angeles Co.]. The tree for which this town and several other places were named was not a true palm but the Joshua tree (*Yucca brevifolia*), sometimes called yucca palm. The place was settled by German Lutherans in 1886 and called Palmenthal. The post office was established on June 7, 1888, and the name was changed to Palmdale on Aug. 13, 1890.

Palmer, Mount [Death Valley NP]. This peak in the Grapevine Mountains was named by the BGN for Dr. T. S. Palmer after his death in 1955. Palmer was a member of the biological expedition to Death Valley in 1891 and the author of *Place names of the Death Valley region*, 1948.

Palmer: Mountain [Kings Canyon NP]; **Cave** [Sequoia NP]. Named for Joe Palmer, a pioneer miner and mountaineer of the region and discoverer of the cave.

Palo is Spanish for 'stick' but was used in Spanish California for 'tree', in combination with purely descriptive adjectives, e.g. *palo seco* 'dry tree' and *palo prieto* 'dark tree', as well as for definite species of trees: *palo colorado* 'redwood' and *palo verde*, a desert tree with green bark (genus *Cercidium*), also called palo verde in English. On Sept. 26, 1769, the Portolá expedition gave the name Ranchería del Palo Caydo 'village of the fallen tree' to an Indian settlement in what is now Monterey Co. because the Indians "lived in the open near a fallen oak tree" (Costansó, p. 234). This, the first place name in California containing the word *palo*, has not remained in use. But *palo* has survived from Spanish times in a number of geographical names and has also been used for place naming in American times. **Palo Alto** (pal′ ō al′ tō) [Santa Clara Co.]: Tradition connects the origin of the name with the tall tree still standing near the railroad station. However, since this redwood had a twin, which fell in 1885–86, it can hardly be the *palo alto* described by the Anza expedition, as is generally assumed by historians. On Nov. 28, 1774, Palou (3:264) records in his diary: "Near the crossing there is a grove of very tall redwood trees, and a hundred steps farther down another very large one of the same redwood, which is visible more than a league before reaching the arroyo, and appears from a distance like a tower." Palou, from a distance, might have taken "twin redwoods" for one large tree, but Anza and Font leave no doubt in their diary entries of Mar. 30, 1776, that the *palo alto* was a single tree. Font's map of the Bay region (*Compl. diary*, opp. p. 302) likewise shows a single tree. Whether this tree was the redwood a mile downstream, carried away by high water in Mar. 1911, or whether it was another tree that has left no trace, will probably remain an unanswered question. In a geographical sense the name was used when the San Francisquito rancho was sold in 1857: "a certain tract of land known as the Rancho of Palo Alto." This name was doubtless used to avoid confusion with the two adjoining ranchos, which included the name San Francisquito in their full names. Various surveyors' plats of these ranchos after 1858 show clearly that the name "Palo Alto" was at that time associated with "Twin Redwoods." Leland Stanford established his country estate in 1876 on the rancho. Gradually he acquired a total of 8,000 acres, which he called Palo Alto Farm. After the founding of Stanford University, Timothy Hopkins laid out the present town in 1888, naming it University Park. At the same time a real estate company developed a new subdivision adjoining Mayfield and named it Palo Alto. Stanford brought an injunction against the company for using "his" name; through an amicable settlement Palo Alto became College Terrace, and University Park was rechristened Palo Alto on Jan. 30, 1892 (cf. Guy Miller, *WF* 6:78–79, 7:284–88). **Laguna de los Palos Colorados** [Alameda, Contra Costa Cos.] was a land grant, registered on Oct. 10, 1835; a lake and some of the *palos colorados* of Redwood Canyon were on the territory. These as well as other stands of redwood in the San Francisco Bay region are shown as Bosques de Palo Colorado on Ayala's Plano (1775). The name is used in a geographical sense as early as Sept. 2, 1797, when the baptismal records of Mission San Jose show that an Indian woman, Gilpae, *de los Palos Colorados*, was christened Josepha. **Palo Colorado Canyon** [Monterey Co.] is shown as Arroyo del Palo Colorado on a *diseño* of Rancho Sur Chiquito, 1835. **Palo Comado** (kə mä′ dō) **Canyon** [Los Angeles Co.] is perhaps from Spanish *palo quemado* 'burnt tree'. **Palo Escrito** (pal′ ō es krē′ tō) **Peak** [Monterey Co.]: A *terreno* called *palo escrito* 'a tree with writing on it' is mentioned on Feb. 21, 1833 (Arch. Mont. 6:56), and Cañada de Palo Escrito is recorded in a petition for a grant in 1840 (Bowman Index). Palo Scrito Hills are mentioned by Whitney (*Geology*, p. 111), and Palo Escrito Mountain is shown on Hoffmann's map. The name probably originated because there was a tree, mentioned as *palo escrito* in 1828 (Registro, p. 11), with carved symbols. *See* Alamo: Alamo Pintado. **Palos Verdes** (pal′ əs vûr′ dēz) [Los Angeles Co.] is named for Cañada de los Palos Verdes 'valley of green trees', now known as Bixby Slough. (Reference is not, however, to the desert tree called the palo verde.) The land had been occupied by José and Juan Sepúlveda since 1827, was mentioned as a rancho named Palos Verdes in 1844 (DSP Ang. 8:2), and was formally granted on June 3, 1846. The subdivision was laid out on part of the rancho in 1922. **Palo Cedro** (pal′ ō sē′ drō) [Shasta Co.]: A townsite was laid out in 1891 and given the Spanish name because of a cedar tree on the place. The original tree is no longer there (Steger). The post office, listed in 1893, was for a number of years spelled Palocedro. **Palo Verde** (pal′ ō vûr′ dē): town, **Lagoon, Mountains** [Imperial Co.] probably refers to the desert tree, which is native to Arizona and Mexico, but introduced to California. The post office was established on Jan. 13, 1903, as Paloverde, and on July 1, 1905, the name was changed to Palo Verde. **Palo Corona** [Monterey Co.] is a summit near San José Creek, lit. 'crown tree', perhaps referring to an unusually tall redwood (Clark 1991).

Paloma; Palomar. The Spanish words for 'pigeon, dove'—and, with the locative ending *-r*, for 'pigeon-roost, dovecote'—were often used for place names in Spanish times. Paloma is mentioned as a place name in San Diego Co. (DSP Ben. P and J 2:73) and in Los Angeles Co. (Bowman Index). Today the name is found in Calaveras, Monterey, Riverside, and San Luis Obispo Cos., in some instances probably applied by Americans because of the pleasant sound: (pə lō′ mə). *For* Cañada Paloma, *see* Guejito.

Palomar [Los Angeles Co.]. The station of the Pacific Electric, on Rancho San Jose, was doubtless named for Ignacio Palomares, one of the patentees of the grant.

Palomar (pal′ ə mär): **Mountain, State Park** [San Diego Co.]. A Cañada de Palomar is shown on a *diseño* of the Camajal y Palomar grant in 1846 (*see* Camajal). From 1859 to 1868 Joseph Smith lived on his Palomar ranch at the side of the mountain; after he met a violent death, the peak became known as Smith Mountain. In Dec. 1901 the BGN, upon a petition of Julia Wagenet and other local residents, restored the old name officially. The post office Palomar Mountain was established on Dec. 19, 1920, and the state park in 1933.

Palomares (pal ə mär′ əs) **Creek** [Alameda Co.]. Probably named for Francisco Palomares (or a member of his family); he was a resident of San Jose after 1833, an Indian fighter, and a judge.

Palomas Canyon [Los Angeles Co.]. The name probably originated by a mistaken association of *paloma* 'pigeon' with *pelonas* 'bald' (fem. sg.)—since *Lomas pelonas* 'bald hills' are shown on a *diseño* of the Temascal grant, through which the canyon runs.

Palowalla (pä′ lō wä lə) [Riverside Co.]. Perhaps of Chemehuevi Indian origin, or just a combination of Spanish *palo* 'tree' with the second part of the name Chuckwalla. The Little Chuckwalla Mountains are a few miles to the southwest.

Pamo (pä′ mō) **Valley** [San Diego Co.]. From Diegueño *paamuu*, of unknown meaning (Couro and Hutcheson). An Indian rancheria called Pamo is mentioned in Spanish records as early as 1778 (PSP Ben. Mil. 1:41). The Valle de Pamo, or Santa María, land grant is dated Nov. 21, 1843.

Pamoosa. *See* Moosa.

Pampa Peak [Kern Co.]. The peak was named after the former Southern Pacific station, shown on the Official Railway Map of 1900. Like Bena and Ilmon on the same sector, this is probably a railroad name, without meaning.

Panama: town, **Slough** [Kern Co.]. The place was settled before 1866 and was probably so named because the settlers considered the land on which they were living an isthmus, formed by two river channels (H. S. Allen). On Nov. 27, 1874, the name was given to the Panama Ranch, established by Miller and

Lux. Panama Slough is mentioned in court records of Aug. 6, 1900 (V. J. McGovern).

Panamint (pan′ ə mint): **Valley, Range, Springs, Dry Lake** [Inyo Co.]. The name refers to a division of Shoshonean Indians, also called Koso, who once occupied this region. It is apparently derived from Southern Paiute *pa-* 'water' plus *nïwïntsi* 'person' (Munro). The term appears as Panamint in the report of the Nevada Boundary Commission in 1861 (*Sacramento Union*, July 13, 1861) and was probably applied by the Darwin French party in the preceding year (*see* Darwin Wash). The orographic identity of the mountains on the maps remained vague. The map of the State Mining Bureau (1891) and again the Ballarat and Avawatz Mountains atlas sheets designate the various chains from Tin Mountain to Wingate Gap collectively as the Panamint Range. On some maps, the Quail Mountains are added for good measure.

Panasquitas. *See* Los Peñasquitos.

Pancho Rico Creek [Monterey Co.]. This tributary of the Salinas River commemorates Francisco ("Pancho") Rico, grantee of the San Lorenzo grant on Nov. 16, 1842. The name was misspelled Poncho Rico until the BGN restored the proper form (1961).

Panocha. A variant of Spanish *panoche* (*q.v.*).

Panoche (pə nō′ chē, pə nōch′): **Creek, Hills, Valley**, town [Fresno, Merced, San Benito Cos.]. *Panoche* or *panocha*, in the Spanish of Mexico, is a kind of raw sugar, sometimes made from wild plant foods (Santamaría, p. 797). Its production is described by Fages (pp. 79 ff.): "Native sugar is made from the olive-like fruit produced by a very leafy, tufted shrub, six feet high with a stem of reddish color and leaves like those of the *mangle* [sumac]. The preparation . . . consists in gathering the ripe fruit, separating the pulp from the seed, and pressing it in baskets to make cakes of sugar." The *Indian Report* of 1853 (p. 57) mentions an Indian dish, *penocha* or *penona*, made of acorns with sugar. The etymology of the word is not entirely clear, but it may be of Aztec origin. *El punto* ('place') *de la Panocha*, 20 leagues east of San Juan Bautista, is mentioned on June 7, 1830 (DSP Ben. Mil. 72:10), and a *parage llamado la Panocha* 'place called the Panocha' is mentioned on July 4, 1840 (DSP Mont. 3:86). It is improbable that the name has any connection with Narciso Panocha, who was given permission to go after neophytes from Monterey in 1827 (Dep. Recs. 4:79). The name appears in the titles of two land grants: Panocha de San Juan y Carrisalitos in Merced Co., dated Feb. 10, 1844, and Panoche Grande in San Benito Co., dated Mar. 14, 1844 (the latter was not confirmed by the U.S. government). "Paneche Pass" is shown on Derby's map (1850), and "Penoche Valley" is mentioned in Hoffmann's notes in July 1861. The name was used elsewhere: Panocha is shown on a *diseño* of the Cuyama grant [Santa Barbara Co.], and the diminutive, shown as La Panochita on several *diseños*, is used for a hill southwest of Mount Hamilton. The place was mentioned by Sebastián Rodríguez on Apr. 22, 1828: *el paraje llamado la Panochita*.

Panther. This name is sometimes used in the northern counties for the mountain lion (*Felis concolor*) and is found in the names of about twenty geographic features, including Panther Den and Panther Beds Ridge [Sonoma Co.]. **Panther Creek, Gap**, and **Peak** [Sequoia NP]: The creek was so named because a panther was killed here by Hale Tharp (Farquhar).

Panum (pan′ əm) **Crater** [Mono Co.]. Of unidentified Indian origin, the name was given by the BGN in 1971 (list 7104), replacing North Crater.

Panwaucket (pan′ wô kət) **Creek** [Trinity Co.]. An alternative name of Readings Creek, this is Wintu *paani waqat; paan-* is a tree similar to the yew, and *waqat* is 'creek' (Shepherd).

Paoha (pä ō′ hə, pā ō′ hə) **Island** [Mono Co.]. From Eastern Mono *pa-ohaa* 'water baby' (McLaughlin), referring to a dangerous supernatural creature supposed to live in bodies of water. The Land Office map of 1879 used the name Anna Herman Island. The present name was given by I. C. Russell in 1882: "We may therefore name the larger island Paoha Island, in remembrance, perhaps, of the children of the mist that held their revels there on moonlit nights in times long past" (USGS, *Report* 8:279). It has been claimed that the word is Piute for 'white' or 'daylight'—the opposite of adjacent Negit Lake—but this is not confirmed by studies of the local Indian languages (Browning 1986).

Papas, Las. *See* Potato; Twin: Twin Peaks [San Francisco Co.].

Paper Mill Creek [Marin Co.]. The first paper mill on the Pacific Coast was built and operated about 3 miles southeast of Olema by

Samuel P. Taylor in 1856. The mill is recorded on Hoffmann's map, on which the creek is still called Arroyo San Geronimo. On maps of the Coast and Geological Surveys, the entire creek is called Lagunitas Creek.

Papoose Lake [Trinity Co.]. The term for an Indian child was first introduced to English in 1633; it is from Narragansett (New England Algonquian) *papoòs*. At present the term is considered offensive (*see also* Squaw).

Paps. *See* Twin: Twin Peaks [San Francisco Co.].

Paradise. In California some fifty features are so named because of their assumed resemblance to the abode of the blessed; to some the name was no doubt applied in irony. In Siskiyou Co. there is a Paradise Craggy and in San Joaquin Co. a Paradise Dam. **Paradise** [Butte Co.] is shown on the Land Office map of 1879, and on the Official Railway Map of 1900 it is recorded as Paradice. It could not be ascertained whether this spelling confirms the old story that the town was named after the "Pair o' Dice" saloon, or whether it was a misspelling that gave rise to the story. Helltown is a short distance away. **Paradise Cay** [Marin Co.] is a point of land on the northeast coast of the Tiburon Peninsula (BGN, list 6704); this is perhaps a unique occurrence in California of the word "cay," used in the West Indies to designate a small island.

Paraiso (pə rī′ zō, pär ə ē′ sō) **Springs** [Monterey Co.]. Spanish *paraíso* 'paradise' was occasionally used for place names in California in Spanish times—far less often than Paradise in American times. The name in Monterey Co. seems to be the only survivor. A Cañada del Paraíso, now the upper San Jacinto Valley [Riverside Co.], is mentioned by Font on Dec. 16, 1775.

Paraje. The common Spanish generic term, sometimes spelled *parage*, for 'place, site'. It is often found in land grant and mission records and seems to have been used interchangeably with *sitio* (*see* Sitio).

Paramount [Los Angeles Co.]. When the cities of Hynes and Clearwater were merged in 1948, Frank Zamboni, president of the Kiwanis Club, proposed the new name because the main street was Paramount Boulevard, named for the motion picture company (Betty Doheney).

Pardee Reservoir [Amador Co.]. Named by the East Bay Municipal Utility District on Oct. 29, 1929, for George C. Pardee, who was mayor of Oakland, 1893–95, governor of California, 1903–7, and president of the board of directors of the utility district, 1924–41.

Parker: Pass [Yosemite NP]; **Creek, Lake, Peak** [Mono Co.]. The tributary to Mono Lake is labeled Cranes Creek on Eddy's map (1854) but is left nameless on other older maps. The present name, for an old settler on the banks of the creek, was apparently not applied until the Mount Lyell quadrangle was mapped in 1898. *For* Parker Valley, *see* Ukiah Valley.

Parker Dam [San Bernardino Co.]. When the new branch of the Santa Fe was built in 1905–7, the station in Arizona was named Parker, both for Earl H. Parker, location engineer of the railroad, and after Parker, an Indian settlement in the vicinity (Barnes). The general area became known as Parker, and the name was transferred to the proposed dam for water diversion into Arizona and then to the dam for Havasu Lake. The post office is listed in 1936.

Parker Mountain [Los Angeles Co.]. The name is shown on the Tujunga atlas sheet of 1900 and probably commemorates James L. Parker, a ranger in the San Gabriel Forest Reserve in 1898 (Co. Surveyor). Older maps designate the peak as Mount Parkinson.

Parkfield [Monterey Co.]. After the Post Office Dept. had rejected the original name, Russelsville, Postmaster Sittenfelt selected the present name in 1883 because of the surrounding natural oak park (Myrtle Flentge).

Parkinson, Mount. *See* Parker Mountain.

Parks Bar [Yuba Co.] was named for David Parks, who settled there in 1848 (Co. Hist.: Yuba 1879:88).

Parlier (pär lēr′) [Fresno Co.]. The post office, established in 1898, was named for its first postmaster, I. N. Parlier, who had settled in the district in 1876.

Parsons: Peak, Memorial Lodge [Yosemite NP]. Named in 1901 by R. B. Marshall for Edward T. Parsons (1861–1914), for many years a director of the Sierra Club. The lodge was erected and named by the Sierra Club in 1915.

Par Value Lake [Mono Co.]. The lake was named after the adjacent mining claim called Par Value (Maule).

Pasadena (pas ə dē′ nə) [Los Angeles Co.]. The community was founded in 1874 and called Indiana Colony because the original promoters came from Indiana. When the post office was established in 1875, another name had to be chosen, and rarely have pioneer settlers gone to more trouble to select a name for their town than the good people of Indiana Colony. Hiram Reid's account of the naming (pp. 338 ff.) sounds more convincing than various other stories: Judge B. S. Eaton, in discussing with another stockholder, Calvin Fletcher, the possibility of finding a suitable Spanish name for the proposed post office, recalled a conversation he had had with Manuel Garfias, the patentee of Rancho San Pascual, on part of which the town was situated. When asked why he had chosen so impractical a place for his house, Garfias replied, "Porque es la llave del Rancho." Fletcher was disappointed, because "yavvey," the only word he caught, would never do for a place name. Judge Eaton then translated Garfias's reply as 'key of the rancho'. This was at least a cue to a suitable name. Dr. T. B. Elliott, the president of the Indiana Colony, then took up the idea. He wrote to a friend who was a missionary among the Chippewa (Ojibway) Indians in the Mississippi Valley for an Indian version of 'Key of the Ranch' or 'Entrance to the Upper Part of the Valley', and received in due course these suggestions: Weoquân Pâ sâ de ná 'Crown of the Valley'; Gish kâ de ná Pâ sâ de ná 'Peak of the Valley'; Tape Dâegun Pâ sâ de ná 'Key of the Valley'; Pe quâ de na Pâ sâ de ná 'Hill of the Valley'. Since Dr. Elliott could not very well propose the name Tapedaegunpasadena or Weoquanpasadena, he quietly dropped the specific part and submitted to the townspeople the pleasing and euphonious name Pasadena. The interpretation that Pasadena alone means 'crown of the valley' has persisted until the present day. As for the original Chippewa word, it can be identified with *passadina* 'there is a valley' (Frederick Baraga, *Dictionary of the Otchipwe language* [Montreal, 1878]).

Pasajero, Arroyo. *See* Gatos: Los Gatos Creek.

Pasatiempo (pas ə tē em′ pō) [Santa Cruz Co.]. A residential golf club development founded in the 1920s; Spanish for 'pastime' (Clark 1986).

Pasines, Cie'naga de los. *See* Paicines.

Paskenta (pas ken′ tə) [Tehama Co.]. From Wintu *phas kenti* 'under the cliff', containing *phas* 'cliff', or from a similar form in the related Nomlaki (Central Wintun) language (Shepherd). The post office was established on Sept. 3, 1872.

Paso. The Spanish word is a common geographical term for 'pass, passage, crossing, ford, narrows, channel'. Often used in Spanish times, it appears as a generic term in the names of six land grants, and as a specific term in one. **Paso Robles** (pas′ ə rō′ bəlz): city, **Creek** [San Luis Obispo Co.]: The deciduous oaks that later provided the name for the place are mentioned by Font on Mar. 4, 1776. Paso de Robles 'passage through the oaks' is recorded as a rancho where the padres of San Miguel sowed wheat in 1828 (Registro, p. 17). On May 12, 1844, the name was used for a land grant. The city was founded on the rancho in 1886 and was incorporated in 1889 with the name El Paso de Robles. In common usage the name is abbreviated to Paso Robles (*see* Bartolo Viejo; Del Paso). **El Paso Creek** [Kern Co.] was named Pass Creek on Apr. 14, 1844, when Frémont and Preuss crossed the Tehachapi Mountains. Later surveyors Hispanicized the name. **El Paso: Peak, Mountains** [Kern Co.]: The pass north of Mojave, which gave the name to the orographic features, was discovered and named by Jacob Kuhrts in 1857. The well-known German pioneer of Los Angeles passed through Bedrock Canyon together with John Searles (of Searles Lake) on his way to Los Angeles (*HSSC:Q* 29:141). *For* Paso de los Americanos, *see* American River.

Pass Creek. *See* Paso: El Paso Creek; Tehachapi.

Passing Island. *See* San Nicolas Island.

Pastoria de las Borregas [Santa Clara Co.]. The name, meaning 'pasture of the ewes', was applied to the land grant on Jan. 15, 1842. **Pastoria** (pä stə rē′ ə) **Creek** [Kern Co.] was so named because the plateau through which it runs was used as the government pasture lands in the days of the Indian reservation at Rancho El Tejon in the 1850s (Rensch-Hoover, p. 132).

Patch [Kern Co.]. The place was known as Weedpatch from the many weeds that grew there because of subirrigation. The Santa Fe Railroad abbreviated the name for the siding, to distinguish it from the settlement Weed Patch farther south. *See* Algoso.

Patchen Pass [Santa Clara, Santa Cruz Cos.]. Patchen post office was established on Mar. 28, 1872, and named for a famous race horse with the improbable name of George M. Patchen (Clark 1986); the spelling Patchin also is found.

Paterson. *See* Cherokee.

Pate Valley [Yosemite NP] is part of the Grand Canyon of Tuolumne River, though the place is not shown on the maps of the Whitney and Wheeler surveys. It may have been named for Francis M. Pate of Alabama, a resident of Indian Gulch in 1867. However, the name appears as Pait in the early 1890s, which suggests a sheepman (Farquhar).

Patricks: Creek, Point, State Park [Humboldt Co.] were called Patrick Point on the county map of 1886. It was doubtless given for Patrick Beegan, whose preemption claim is recorded in the Trinidad Record Book on Jan. 13, 1851, and whose tract 6 miles north of Trinidad was known as Patrick's Ranch (Joseph P. Tracy); cf. BGN, list 6704.

Patterson [Stanislaus Co.]. Thomas W. Patterson, a Fresno banker, laid out the town about 1910 and named it for his uncle, John D. Patterson, who had purchased the land in 1864.

Patterson, Mount [Mono Co.]. The highest peak in the Sweetwater Mountains was perhaps named for James H. Patterson, a native of Ohio, who settled at Big Meadows before 1867.

Patterson: Pass, Run [Alameda Co.]. Andrew Jackson Patterson was one of three brothers who settled in the district in the 1850s. Once when he and his wife were driving through the pass, a heavy windstorm came up; the wagon turned over, and Mrs. Patterson's leg was broken. It is said that the pass received its name from this incident (Lila McKinne).

Patterson Mountain [Fresno Co.]. Probably named for John A. Patterson, a native of Georgia, who brought the first cattle into the upper Kings River region in 1853; he was county supervisor in 1856, and assemblyman in 1875–76.

Patton [San Bernardino Co.]. The Santa Fe station was called Asylum in 1891 because it was the station for a state mental hospital. When the post office was established in 1895, it was named for Henry Patton of Santa Barbara (Santa Fe). In 1909 the station appears on the Santa Fe map as Patton.

Pattymocus (pat′ ē mok əs) [Tehama Co.]. Said to be a Wintu name referring to a tipped-over basket; it apparently contains the element *muk-* 'turtleshell-shaped' (Shepherd).

Pauba (pô′ bə) [Riverside Co.]. The name of a land grant, dated Oct. 18, 1844, and Feb. 4, 1846. It is probably a Luiseño word, but its meaning is not known. A place Pauba is shown on one *diseño*, and Las Paubas on another.

Pauley Creek [Sierra Co.]. The BGN gave this name to the East Fork of the North Fork of the Yuba River in 1950. An early settler named Pauley had built the first sawmill on the stream.

Paulsell [Stanislaus Co.]. Named in 1897 by the Sierra Railroad for A. C. Paulsell, a native of Tennessee, the owner of the property. Paulsell came to Dent Township in 1854; he was president of the Farmers' Co-operative Union, assemblyman in 1873–74, and member of the State Harbor Commission in 1884–90.

Pauma (pô′ mə): **Valley, Creek, Indian Reservation** [San Diego Co.]. The name goes back to a rancheria of Luiseño Indians, mentioned in the records of the 1790s (PSP 15:182). Potreros de Paoma are mentioned on Nov. 27, 1841 (Arch. LA 2:120), and on Nov. 8, 1844, the name was applied to the land grant, Pauma or Potrero de Pauma. The meaning of the word is not known, but since the place is near Pala (from Luiseño *páala* 'water'), a connection with Luiseño *páamay* 'little water' is possible. *See* Pala.

Páxaros, Río de los. *See* Pajaro River.

Paxton [Plumas Co.]. The post office was established on Sept. 3, 1917, and was named for Elmer E. Paxton, general manager of the Indian Valley Railroad and the Engel Mining Company (Helen Strong).

Paynes Creek [Tehama Co.]. A settlement, Payne's Creek, is listed in the Directory of the Marysville Appeal of 1878—apparently named after Paines Creek, shown on the Land Office map of 1879. The post office Paynes Creek is listed in 1892. A Mr. Payne had a sawmill farther south, on the Sacramento River, perhaps as early as 1851; and John Paine (or Payne) was a resident of Red Bluff in the 1860s.

Payson Canyon [Inyo Co.]. The canyon was named for "Old Lew" Payson, who lived for many years at Antelope Springs (Robinson).

Peach. Several physical features bear the name of the fruit: Peaches Creek [Sonoma Co.], Peachtree Canyon [San Bernardino Co.], Peachtree Creek [Stanislaus Co.], and Peachtree Valley [Monterey Co.]. In San Bernardino Co. there is a Peachy Canyon. **Peachton** [Butte Co.] was named in 1907 by the Sacramento Northern because a large peach orchard flourished there (Florence Campbell).

Peanut [Trinity Co.]. In 1898 the settlers around Cuff's store decided to apply for a post office and name it for Cuff's wife. When the matter was submitted to A. L. Paulsen, postmaster at Weaverville, he suggested instead the name Peanut, because it would be a unique name—and because he was very fond of peanuts and was eating them at the time. The petitioners agreed to enter it on the application as a second choice. The Post Office Dept. apparently agreed with Paulsen and chose the name Peanut on Jan. 20, 1900.

Pearblossom [Los Angeles Co.]. Named in 1924 by Guy C. Chase because it was then a center of pear orchards. The name remained after the pears were killed by blight and replaced by peaches (R. M. Yost). **Pearland** [Los Angeles Co.] is an advertising name applied to the subdivision in 1919.

Pearch [pûrch] **Creek** [Humboldt Co.]. Perhaps named for John Adam Pearch, postmaster at Orleans in 1859 (Turner). Established by the BGN (list 7702), instead of Perch Creek.

Pear Lake [Sequoia NP]. The name was given to the lake because, as on the Tehipite atlas sheet, it presents a perfect outline of a pear.

Pearson Springs [Lake Co.]. *See* Saratoga: Saratoga Springs.

Pease [Sutter Co.]. The station of the Sacramento Northern was named for George Pease, through whose property the railroad was built in 1907.

Peavine [Sierra Co.]. Probably named for a flowering wild pea, *Lathyrus* sp. Many creeks and other natural features in the state also bear this name.

Pebble Beach [Monterey Co.]. The name developed locally because there is a beach with pebbles at this point, and it was accepted by the Pacific Improvement Company when it acquired the property in 1880. The post office is listed in 1910. There are several other Pebble Beaches and Creeks in the state.

Pechaco (pə chā′ kō) **Creek** [Sonoma Co.]. Perhaps for the Spanish family name Pacheco (*q.v.*).

Pechanga (pə chäng′ gə) **Indian Reservation** [Riverside Co.]. Of unidentified Luiseño origin.

Pecho: Rock, Creek [San Luis Obispo Co.]. The Spanish word for 'breast' was used in a geographical sense for rocks and hills shaped like a woman's breast. Arroyo del Pecho is shown on a *diseño* of the San Miguelito grant (1839). El Pecho (for the rock) and Cañada and Arroyo del Pecho are shown on several *diseños* of the grants, which were combined in 1845 in the Cañada de los Osos y Pecho y Islai grant. *For* Los Pechos de la Chola, *see* Twin Peaks.

Peckinpah (pek′ in pä) **Creek, Mountain** [Madera Co.]. Charlie Peckinpah and his brothers started a sawmill here in 1884 (Browning 1986; BGN, 1994). The film director Sam Peckinpah is a member of this family.

Pecks Canyon [Tulare Co.]. Named for a man who ran sheep here about 1870 (Farquhar). One James Peck, a native of Kentucky, is registered in Visalia in 1879.

Peconom (pə kō′ nəm) **Creek** [Lassen Co.]. Named for a Maidu Indian woman, Roxie Peconom, who gathered wild foods along the stream and died at an advanced age in 1962 (BGN, 1993). The name is probably from Maidu *pekúnim* 'mountain lion' (Shipley).

Pecwan (pek′ wän): **Creek, Ridge** [Humboldt Co.]. Also spelled *pekwan;* from the Yurok village name *pekwan* (Waterman, p. 243; Harrington). The Pak-wan "band" is mentioned in the Indian Report on Oct. 6, 1851. The spelling Pecwan was established by BGN, list 8204.

Pedernales (ped ər nä′ ləs), **Point** [Santa Barbara Co.]. On Aug. 28, 1769, the Portolá expedition camped somewhere near Point Arguello and found an Indian rancheria: "In this village the soldiers gathered good flints for their weapons; for this reason they named it Los Pedernales" (Crespí pp. 176–77). On Spanish maps, the headland (of which Arguello is the principal point) was called Punta Pedernales; on American and European maps, Arguello was used, in various spellings. In the 1850s the name became attached to Purisima Point: "This is known on the coast as Point Pedernales, signifying Point of Flints, but generally and erroneously printed Pedro Nales" (*Coast Pilot* 1858:24). When the Coast Survey recharted this section in 1873–74, it called a triangulation point Pedernales but designated the present point as Promontory. The name was not attached to the point at latitude 34°36′ north until after 1900 (*Coast Pilot* 1904). *See* Arguello.

Pedley [Riverside Co.]. When the railroad from Ontario to Riverside (now the Union Pacific) was built in 1905, the station was named, probably for Francis X. Pedley, who was at that time engaged in real estate promotion at Arlington. Pedley was a native of England who had come to California in 1882 and to Riverside Co. in 1894 (Co. Hist.: Riverside 1912:598–99).

Pedregoso. *See* Santa Barbara.

Pedro (pē′ drō): **Point, Creek, Valley, Hill** [San Mateo Co.]. San Pedro is mentioned as a rancho of Mission Dolores as early as 1791. On Jan. 26, 1839, the name was used for a land grant. The anchorage San Pedro is shown on Duflot de Mofras's map. The point was originally named Punta del Ángel Custodio 'guardian angel' by Crespí, and Punta de las Almejas 'mussels' by Portolá's soldiers on Oct. 30, 1769; it appears as Point San Pedro on sketch J2 (1850) of the Coast Survey. The "San" has been dropped by common usage and by the Post Office Dept., but is still seen on maps. The Pedro Point post office is listed in 1939. In 1957 the community of Pedro Point was incorporated into the City of Pacifica.

Pedro Nales. *See* Pedernales.

Peeler Lake [Mono Co.]. Named for Barney Peeler, an early settler (Maule).

Pegleg Mountain [Lassen Co.]. J. J. ("Pegleg") Johnson, a trapper and Indian fighter, had taken a homestead near the mountain in the 1850s, and by local usage the mountain was given his nickname (A. G. Brenneis).

Pekwan. *See* Pecwan.

Pelican. Three Points, two Bays, and one Rock on the coast are named for the largest of our aquatic birds (genus *Pelecanus*). The oldest is probably Pelican Bay [Del Norte Co.], shown on Gibbs's map of 1851. *See* Alcatraz.

Pelona. The Spanish word for 'bald' (fem.) was used as a descriptive name for orographic features devoid of trees. It is preserved in Sierra Pelona and Sierra Pelona Valley [Los Angeles Co.] and in Loma Pelona [Monterey, Santa Barbara Cos.]. *See* Bald; Palomas Canyon.

Peñasquitos (pān yə skē′ tōs) [San Diego Co.]. The word is a diminutive plural derived from *peña* 'rock, crag'; it was applied to a land grant dated Aug. 11, 1832, and May 5, 1834. Santa Margarita de los Peñasquitos was one of the ranchos and localities under the jurisdiction of Santiago Argüello as of Dec. 31, 1830 (Engelhardt, *San Diego*, p. 229).

Pences. *See* Pentz.

Pendleton, Camp [San Diego Co.]. Named in 1941 in memory of Marine major general Joseph H. Pendleton, a veteran of the "banana" wars in Central America during the 1920s, who was instrumental in getting Marine Corps units based on the West Coast (Frances Kracha).

Peninsula, The [San Mateo Co.]. The term "San Francisco Peninsula," comprising the area between San Francisco and San Jose, is attested from 1869. In the early 1870s the abbreviation "The Peninsula" was already used (Alan K. Brown).

Peninsular Island. *See* Belvedere.

Penitencia (pen i ten′ sē ə): **Creek, Canyon** [Santa Clara Co.]. This tributary of Coyote Creek was named after the Penitencia adobe house that stood at the highway, which had probably been used as a house of confession and penitence in mission times (Bowman). La Penitencia, apparently still the adobe house, is shown on Eld's sketch of 1841, and Arroyo de la Penitencia is shown on a *diseño* of the Pueblo Lands of San Jose (1840).

Penngrove [Sonoma Co.]. Penn's Grove is mentioned as a station of the San Francisco and North Pacific Railroad in 1880; it was possibly named after Penn's Grove, New Jersey. The contraction was used by the Post Office Dept. in 1908.

Penobscot Creek [El Dorado Co.]. A name transferred from Maine, said to mean 'at the sloping rock' in the Penobscot (Algonquian) Indian language.

Penole; Penoli. *See* Pinole.

Peñon Blanco (pān yōn′ bläng′ kō): **Point, Ridge** [Mariposa Co.]. Named after the Peñon Blanco mine, which "takes in nearly the whole of the prominent Peñon Blanco hill" (Browne, p. 35). *Peñón blanco* is Spanish for 'white crag'. *For* Peñón de la Campana, *see* Picacho.

Penryn (pen′ rin) [Placer Co.]. Established in 1864 by Griffith Griffith, owner of local granite quarries, and named after his home, Penrhyn, Wales. When the Central Pacific Railroad built a station there, the officials struck the *h* from the original spelling.

Pentz [Butte Co.]. The post office was established between 1864 and 1867, and was named for Manoah Pence, the first postmaster; the misspelling was apparently done by the Post Office Dept. (Co. Hist.: Butte 1882:251). The county map of 1861 shows the name Pences north of Oroville.

Peoria: Pass, Basin, Creek, Mountain [Tuolumne Co.]. The pass is shown on Hoffmann's map (1873); it was probably named after the town in Illinois, or another of the ten Peorias then existing in the United States. The term is from Peouarea, a subtribe of the Illinois (Algonquian) Indians, according to Stewart 1970.

Pepperwood. About ten geographic features are so named, apparently all in the northern Coast Ranges, where the California laurel (*Umbellularia californica*) is called "pepperwood". **Pepperwood** [Humboldt Co.]: The post office was established about 1900, so named because of the presence of one of the finest groves of this pepperwood in the state.

Peral (pə ral′) [Tulare Co.]. The Spanish word for 'pear tree' was applied to the Santa Fe station when the line from Visalia to Fresno was built about 1895.

Peralta (pə ral′ tə) [Orange Co.]. The settlement was founded before 1900 and named in memory of Juan Pablo Peralta, co-grantee in 1810 of the Santiago de Santa Ana grant, on which the place is situated. The community is now considered part of Anaheim Hills in the City of Anaheim (Brigandi). The group of hills near the place in the Santa Ana Mountains is now called Peralta Hills (BGN, 1965). *For* Lake Peralta [Alameda Co.], *see* Merritt, Lake.

Perch Creek. *See* Pearch Creek.

Peregoy (pâr′ ə goi) **Meadows** [Yosemite NP]: In 1869 Charles E. Peregoy, a native of Maryland, built a hotel known as the Mountain View House on the old horse trail from Clark's to Yosemite Valley. The remains of the old building were long visible on the meadow. His place is shown as Peregoy's on the Hoffmann-Gardner map, and the name is mentioned repeatedly in Brewer's Notes of the summer of 1864.

Perkins [Sacramento Co.]. The post office was established about 1866 and named Brighton. About 1885 the name was changed to Perkins for Thomas C. Perkins, a native of Massachusetts, who had settled here in 1861 and was the first postmaster. *See* Brighton.

Perkins, Mount [Kings Canyon NP]. Named in 1906 by Robert D. Pike, in honor of George C. Perkins (1839–1923), governor of California in 1880–83, U.S. senator from California in 1893–1915 (Farquhar).

Permanente (pûr mə nen′ tē) **Creek** [Santa Clara Co.]. The creek is shown as Arroyo Permanente on a *diseño* of Rancho San Antonio (1839). Permanente post office, established in 1938, and Henry Kaiser's Permanente Cement Company were named after the stream (J. H. Rogers). *Permanente* is often found on Spanish maps to designate surface water which does not dry up in summer.

Perris (pâr′ is): town, **Valley** [Riverside Co.]. The town was laid out in 1886 and named for Fred T. Perris, chief engineer for the California Southern Railroad and one of the founders of the town. The valley was formerly called San Jacinto Plains, after San Jacinto land grant. Before 1885 the town was 2 miles south of Perris, called Pinacate after a nearby gold mine. Because of some litigation, both the site of the town and the name were changed (Co. Hist.: Riverside 1935:228). *See* Pinacate.

Perrott Grove [Humboldt Co.]. This unit of Humboldt State Park was named for the donor, Sarah T. Perrott.

Perry [Los Angeles Co.]. The station was named in 1905 for the president of the Pacific Electric Railroad, at the suggestion of John Kirsch, owner of the local store.

Perry, Mount [Inyo Co.] was named for John W. S. Perry, 1848–98, who designed the twenty-mule wagons used to haul borax from Death Valley (BGN, 1961).

Perry Aiken Creek [Mono Co.]. Named for Perry Aiken, a settler there in 1870 (Robinson).

Persian: Creek, Canyon [Tulare Co.]. The name of the tributary to Cottonwood Creek is not reminiscent of any Persian settlers on its banks, but of the old John Persian ranch (A. L. Dickey). James Persian, a native of Pennsylvania, and John, Silas, and Henry Persian, natives of Missouri, were residents of the county in 1867.

Persido Bar. *See* Presidio Bar.

Persinger (pûr′ sing ər) **Canyon** [Los Angeles Co.]. Named by the Forest Service for the Persinger family, homesteaders of 1888 (USFS).

Peru, Rio. *See* Piru.

Pesante (pi sän′ tē) **Canyon** [Monterey Co.]. For John Pesante, native of Switzerland, who purchased land here in 1869 (Clark 1991).

Pescadero. In Spanish the word means 'fishing place', from *pescar* 'to fish'. Older maps show various places *de los Pescadores* 'of the fishermen', but the five land grants and the surviving names have only Pescadero. **Pescadero** (pes kə dâr′ ō) **Creek** [Santa Clara Co.]: In 1861 Manuel Larios testified in the Las Animas land grant case that the Castros "had an Indian boy who went to this creek to fish," and so it was called the Pescadero. Two days later John Gilroy testified in the same case that "the Pescadero draws its name from the fact of our catching salmon there." Two days later still, in the Juristac land grant case, Gilroy stated: "the Castros, I and an Indian gave it that name in 1814, being a place where we used to catch salmon" (*WF* 6:371–72). Arroyo del Pescadero is shown on *diseños* of the 1830s. The stream is mentioned as Pescadero River in Trask's Report (1854:8). **Pescadero Creek**, town, and **Point** [San Mateo Co.]: The name is mentioned in the *expediente* of the land grant El Pescadero or San Antonio, Dec. 17, 1833. On the charts of the 1860s the Coast Survey called the creek Pescador River. The town was settled by Spanish-speaking people in the early 1850s and was called Pescadero from the beginning. **Pescadero Point** and **Rocks** [Monterey Co.]: La Punta del Pescadero is mentioned in a letter of Apr. 2, 1835 (Legis. Recs. 3:61). The name El Pescadero was applied to a land grant on Feb. 29, 1836, and Pescadero appears on the early charts of the Coast Survey. **Pescadero Colony** [San Joaquin Co.]: The name of the tract was preserved from the land grant, El Pescadero, dated Nov. 28, 1843. The name may go back to the Río del Pescadero (probably the old channel of the San Joaquin) named by Fernando de Rivera in Dec. 1776 (PSP 14:13–14). **Pescadero Creek** [San Benito Co.] is recorded as Sanjón ('ditch') del Pescadero on a *diseño* of the Paicines grant.

Pescado Creek [Kern Co.]. From the Spanish word for 'fish'; made official by the BGN (list 6702).

Petaluma (pet ə lo͞o′ mə): city, **Creek** [Sonoma Co.]. From Coast Miwok *péta lúuma* 'hillside back', i.e. 'hillside ridge' (Callaghan). *El llano de los Petalumas* 'the plain of the Petaluma Indians' is mentioned by Padre Payeras on May 31, 1819 (Docs. Hist. Cal. 4:341 ff.). On June 21, 1834, the name Petaluma was applied to a land grant, of which Mariano G. Vallejo was grantee. Part of the rancho was preempted in 1850 by G. W. Keller, who laid out the city in 1851 and kept the Indian name. The post office was established and named on Feb. 9, 1852.

Peter, Lake [Tulare Co.]. According to Judge W. B. Wallace, the name was bestowed on the little lake in 1877 by his companion, Joe Palmer, because the dim trail they had followed "petered out" at that place (Farquhar).

Peter Peak [Kings Canyon NP]: Named by the Sierra Club in 1938, in memory of Peter Grubb, who had made the first recorded ascent in July 1936 (Farquhar).

Peters [San Joaquin Co.]. The station of the Stockton and Copperopolis Railroad was named in the 1870s for J. D. Peters, owner of the land. Peters, an associate of Frank Stewart in land speculation, had come to California at the time of the Gold Rush; he later gained a fortune in the grain and shipping business, with headquarters at Stockton.

Petes. *See* Pitas Point.

Petrified Forest. There are several petrified forests in the state, so named because of the presence of petrified trees. The two best known are in Sonoma and Kern Cos.

Petroglyph Point [Modoc Co.]. So named because the rock wall near the Oregon line is covered with petroglyphs carved by aborigines.

Petrolia [Humboldt Co.]. The first California oil deposits to be exploited commercially were found northeast of the town by a U.S. Army officer. Reports of the discovery were published, with reservations, in the *Sonoma Journal* and the *Mining Press* on Feb. 1, 1861, and in the summer of 1865 the San Francisco newspapers reported the first shipment of oil. The name of the post office, established Sept. 13, 1865, was originally spelled Petrolea. *For* Petrolia [Orange Co.], *see* Olinda.

Petticoat Mountain [Lake Co.]. So named because it is shaped like an old-fashioned petticoat (Mauldin). The once well-known mining camp Petticoat Slide was, according to tradition, so named because a lady slid in the mud, fell, and exposed her petticoats.

Pettit Peak [Yosemite NP]: The peak was named by Col. W. W. Forsyth, acting superintendent of the park, 1901–12, for James S. Pettit, colonel of the 4th U.S. Volunteer Infantry in the Spanish-American War.

Pfeiffer (fī′ fər): **Point, Rock, Redwoods State Park** [Monterey Co.]. The name was given to the point for Michael Pfeiffer, a pioneer settler, when the Coast Survey resurveyed this sector, 1885–87. The state park was created and named in 1933. Pfeiffer preempted a claim in Sycamore Canyon, in the Big Sur

Valley, in Nov. 1869 (AGS Monterey, p. 183). The Julia Pfeiffer State Park, made possible by a land gift of 1,700 acres, was created in 1961.

Phelan (fā′ lən, fē′ lən) [San Bernardino Co.]. The post office was named on Nov. 25, 1916, for Sen. James D. Phelan (1861–1930), who had used his influence in establishing it (Ruth Mannigel).

Phillipsville [Humboldt Co.]. George Stump Phillips settled here about 1865, and the place became known as Phillips Flat. When the post office was established on Mar. 12, 1883, the name was changed to Philippsville. It was later changed to Kettinelbe (*q.v.*), after the Indian village at the site; but when the post office was reestablished on Aug. 1, 1948, the old name was again chosen (current spelling confirmed by BGN, list 8402).

Philo (fī′ lō) [Mendocino Co.]. Named after 1868 by Cornelius Prather, landowner and first postmaster, for his favorite girl cousin, whose given name was probably Philomena.

Phipps Peak [El Dorado Co.]. Named for Gen. William Phipps, a native of Kentucky and veteran of Indian wars, who came to Georgetown in 1854 and was famous for his marksmanship (W. T. Russell, in Knave, July 23, 1944).

Picacho (pi kä′ chō): **Peak, Mines, Wash**, settlement; **Little Picacho** [Imperial Co.]. The peak, which has the appearance of an obelisk, is a well-known landmark. Font mentions it as La Campana 'the bell' on Dec. 4, 1775 (*Compl. diary*), and Garcés calls it Peñón de La Campana 'rock of the bell' (Coues, *Trail*, pp. 162, 215). In early American times it became known as Chimney Rock, a name still used locally (*see* Chimney). The present name is tautological: *picacho* itself means 'peak'. The Sutter Buttes were called Picachos in Spanish times, and the hill southeast of Arroyo Grande [San Luis Obispo Co.] is still called Picacho. *For* Picacho Prieto [Marin Co.], *see* Tamalpais.

Picayune (pik′ ə yo͞on) **Valley** [Placer Co.]. The mining district formed in 1864 apparently failed to yield a golden harvest: a picayune was a small coin (five cents) of the mid-19th century, and the term came to mean 'paltry, insignificant'.

Pick-aw-ish (pik ä′ wish) **Campground** [Siskiyou Co.]. From *pikyávish*, a term used to refer to the Karuk world-renewal ceremony held every fall near this site. The term is not originally Karuk, but rather is a whites' simplification of the phrase *ithívthaaneen upikyávish* 'he will remake the world', referring to the ceremonial actions of the Karuk priest.

Pickel Meadow [Mono Co.]. The meadow was named for Frank Pickel, stockman and prospector of the 1860s. It had formerly been known as Duffields Valley (Maule, p. 10). The name is misspelled Pickle Meadow by the USGS.

Pickering Mineral Hot Springs. *See* Urbita Springs.

Picket Guard Peak [Tulare Co.] was named before 1898: "There is a fine pyramidal peak at the eastern end of the third range, which was always in the background of the view as we entered and ascended the narrow cleft of the Kern-Kaweah. This was named the Picket Guard" (William R. Dudley, *SCB* 2:189).

Pickett Peak [Alpine Co.]. The name recalls an early stage station, Pickett Place, which was near the peak and was named for Edward M. Pickett (Maule).

Pickle Meadow. *See* Pickel Meadow.

Pico (pē′ kō) [Los Angeles Co.]. The name was applied to the station of the San Pedro, Los Angeles, and Salt Lake Railroad in 1904 because it was situated on that part of the Paso de Bartolo Viejo grant which was owned by Pío Pico (1801–94), the last governor of Mexican California. **Pico Canyon** and **Oil Field** [Los Angeles Co.] were named for Andrés Pico (1810–76), brother of Pío Pico, and the last commanding Mexican officer to surrender, in 1847. "Andreas Pico knew the locality now called Pico Cañon . . . and had made oil for San Fernando Mission in a small way. He was probably the pioneer coal oil manufacturer of the State" (Hanks, "Report," 1883: 294). **Pico Heights** [Los Angeles Co.] is listed as a post office in 1904.

Pico: Creek, Rock [San Luis Obispo Co.]. The creek was named for José de Jesús Pico, patentee (1876) of the Piedra Blanca grant (1840), through which the creek flows. The Coast Survey applied the name to the bare rock south of San Simeon.

Pico Blanco [Monterey Co.]. "A striking white mountain called Pico Blanco [white peak], the second highest point of the range. It looked strangely white, almost as though it were snow-covered" (Chase, *Coast trails*, p. 208). The geographical term *pico* was used here in the original sense: 'peak of the mountain'. It has survived (or was applied by Americans) in the names of several other orographic features: *see* the specific names.

Pico Rivera [Los Angeles Co.]. In Jan. 1958, the towns of Pico and Rivera incorporated under the new name, which was confusing because the two post offices were not amalgamated.

Picture City. *See* Agoura.

Piedmont [Alameda Co.]. This name, inspired by French *piedmont* or Italian *piemonte* 'foot (of the) mountain', is one of the popular American place names with an aristocratic ring. It is here actually descriptive because the city is situated at the 'foot of the mountain'. About 1876 an organization that called itself the Piedmont Springs Company purchased a tract of land on which there was a sulphur spring and soon afterward erected the Piedmont Springs Hotel (Co. Hist.: Alameda 1928:1.532). The residential district was developed about 1900; the post office is listed in 1904.

Piedra. The Spanish word for 'stone' has survived in many names from Spanish times: Piedras Altas [Monterey Co.], Piedra Azul Canyon [Merced Co.], Piedra Gorda [Los Angeles Co.], Piedra de Lumbre ('flintstone') Canyon [San Diego Co.], Corral de Piedra Creek [San Luis Obispo Co.]. **Piedras Blancas** (pyā´ drəs bläng´ kəs) rocks and **Point** [San Luis Obispo Co.]: The two large, white, pointed rocks were noticed by early navigators, although no name for them is found on the maps. La Piedra Blanca is mentioned on Apr. 27, 1836 (Docs. Hist. Cal. 1:255); and on Jan. 18, 1840, the name, also in the singular, was applied to a land grant. The Coast Survey has used the name Piedras Blancas since the early 1850s. **Piedra** (pē ā´ drə) [Fresno Co.]: In 1911 the Santa Fe constructed a branch line to handle rock from a quarry and gave the siding its Spanish name. The post office was once called Delpiedra. *For* Piedra Pintada, *see* Painted.

Pierce Mountain [Humboldt Co.]. Named by Henry Washington in honor of President Franklin Pierce (1804–69) when the initial point of the Humboldt base line was established: "It has been named Mount Pierce, as a compliment to the President, and the monument . . . was erected . . . on the 6th. of Oct., 1853" (*CHSQ* 34:14).

Piercy (pēr´ sē) [Mendocino Co.]. The name was given to the post office in 1920 for Sam Piercy, the oldest white settler of the district (C. C. Kirk).

Pierpont Bay [Ventura Co.]. The bay was named for Ernest Pierpont, a native of Virginia, who was a resident of Ojai Valley in the 1890s.

Piety Hill [Shasta Co.]. The old mining town was named after Piety Hill, Michigan, the former home of "Grandma" McKenny, a beloved and respected resident (Steger). The place is now called Igo.

Pigeon Point [San Mateo Co.]. The name was applied by the Coast Survey because the clipper ship *Carrier Pigeon* was wrecked here on May 6, 1853. The Spanish name for the point as shown on several *diseños* was Punta de las Ballenas 'point of whales'. Punta Falsa de Año Nuevo on Camancho's map of 1785 was probably modern Pigeon and Franklin Points. *See* Ano Nuevo.

Pika (pī´ kə) **Lake** [Fresno Co.]. Refers to a small mammal of the high mountains, *Ochotona princeps*, also called the cony. The name was borrowed by naturalists from Evenki, a language of Siberia.

Pike. The expression "Pike" or "Pike Countyan" was used in California in the 1860s or earlier, for a person who had supposedly come from Pike Co., Missouri, but actually it was applied to anyone of migratory habits. The name found its way into geographical names, some of which have survived. **Pike County Peak** [Yuba Co.] was named after the Pike County House, a hotel built and named by a Mr. Thompson in 1860 (Co. Hist.: Yuba 1879:92). **Pike** [Sierra Co.]: The post office was established as Pike City in the 1870s and may have been named for a family named Pike. The "City" was dropped when the Post Office Dept. systematically simplified many of its names in the 1890s.

Pilarcitos (pil ər sē´ tōs): **Creek, Lake** [San Mateo Co.]. The Portolá expedition camped at the creek on Oct. 28, 1769, and Crespí named it Arroyo de San Simón y San Judás in honor of the Apostles Simon and Jude, because it was their feast day. The name Arroyo de los Pilarcitos 'creek of the little pillars', i.e. pillar-shaped rocks, was given to a land grant on Jan. 2, 1841, but the name is recorded in land grant papers as early as 1836. The grant was also known as Rancho de Miramontes and as San Benito. In 1862 the Coast Survey placed the name Pilarcitos Creek on its chart of Halfmoon Bay (Alan K. Brown). **Pilarcitos: Canyon, Ridge** [Monterey Co.]: A place, Pilarcitos, about midway between missions San Juan Bautista and Soledad, is shown by Bancroft on his map of the Monterey district, 1801–24 (2:145). A *sitio de los Pilarcitos* is mentioned on July 24, 1830 (Legis. Recs. 1:147) and in later records. On June 23, 1834, the place was made a temporary grant—which, however, was called Chamisal when granted in 1835 (Bowman Index). Clark 1991 believes that two different sites called Pilarcitos may be involved here.

Piletas (pi lē´ təs) **Canyon** [San Luis Obispo Co.]. Also spelled Pilitas; Spanish *pileta* and *pilita* are both possible diminutives of *pila* 'basin'. BGN, list 6802, distinguished Upper and Lower Piletas Canyons.

Pillar: Point, Rock, Reef [San Mateo Co.]. A *parage llamado el 'Pilar'* ('place called the "pillar"') is mentioned on Oct. 18, 1796 (PSP 14:19) and in later records, probably because of a rock formation. It was probably not the place (in Santa Cruz Co.) named in honor of Nuestra Señora del Pilar by Crespí on Oct. 12 and Nov. 23, 1769. Narváez's Plano of 1830 and some of the later maps show a settlement Pilares at the approximate site of the present town of Half Moon Bay. The cape is called Miramontes on the Coast Survey chart of 1854; but when Halfmoon Bay was charted in 1862, this name was replaced by Pillar Point.

Pill Hill is a name given locally to hills on which medical facilities are located, e.g. in Oakland (Mosier and Mosier) and in Monterey (Clark 1991).

Pillsbury Lake [Lake Co.]. The reservoir, filled for the first time in Feb. 1922, was named for E. S. Pillsbury, who had organized the Snow Mountain Water and Power Company with Sen. C. N. Felton in 1906 (*Pacific Service Magazine* 17:344 ff.).

Pilot. The word, usually combined with Peak, Hill, or Knob, was a favored name with overland immigrants, prospectors, and surveyors, for a landmark that would "pilot" them in the right direction. In California there are some forty orographic features called "Pilot." **Pilot Knob** [Imperial Co.]: This black rock was called San Pablo in Anza's diary entry of Feb. 10, 1774. The present name was used as early as 1846 by soldiers, surveyors, and immigrants, and seldom has a name been applied more appropriately. **Pilot Knob** [Siskiyou Co.]: This high peak of the Siskiyou Mountains has been so known since the 1840s, when it "piloted" the caravans using the old California-Oregon trail. **Pilot Hill** [El Dorado Co.], as the name of the elevation between the Middle and South Forks of the American River, goes back to the early mining days. As early as 1849 it was the center of a rich placer-mining region. On Apr. 18, 1854, a post office was established. *For* Pilot River [Alpine Co.], *see* Carson River.

Pinacate (pē nə kä´ tē, pin ə kä´ tē) [Riverside Co.]. Mexican Spanish for a type of black beetle locally called "stink bug," genus *Eleodes* (Gunther), from Aztec *pinacatl*. **Pinecate Peak** [San Benito Co.] represents another spelling. Los Pinacates are mentioned in this area on Apr. 6, 1817 (Arch. Arz. SF 3:2.11). Cañada de los Pinacates is shown on several *diseños* in different locations but was probably the valley through which the road winds. An unconfirmed land grant called Cañada de los Pinacates is dated Apr. 20, 1835.

Pinal (pi näl´) **Creek** [Monterey Co.]. Spanish for 'pine grove', from *pino* 'pine'. The diminutive occurs in Pinalito Canyon and Pinalito Creek, in the same area (Clark 1991).

Pinchot (pin´shō), **Mount; Pass** [Kings Canyon NP]. "There stood a great rounded mass of red slate on the Main Crest, and I allowed myself to change the name Red Mountain given it by Professor Brown, and already applied to scores of the slate peaks of the Sierra, to Mt. Pinchot" (J. N. LeConte, *SCB* 4:262). Gifford Pinchot was at that time chief of the U.S. Division of Forestry, later professor of forestry at Yale and governor of Pennsylvania.

Pincushion Peak [Fresno Co.] was so named because its regular outline resembles a pincushion. The name, which probably had already been in use locally, was applied by the USGS when the Kaiser quadrangle was mapped in 1901–2.

Pine. The existence of almost twenty varieties of true pines (*Pinus* spp.) in the state and the great commercial value of a number of them have made this tree one of the most popular for naming places. About two hundred physical features bear some form of the name Pine on the maps, and many more are so called locally. Some Pine names are translations from the Spanish: Pine Canyon [Monterey Co.] was Arroyo del Pino; Pine Creek [Tehama,

Butte Cos.] was Arroyo de los Pinos; Pine Ridge [Santa Clara Co.] was Pinalitos. *See* Pinos. Occasionally the name of the species is added: Digger Pine Flat, Sugar Pine Hill, Torrey Pines Park. Some names are modified by Lone, Big, or Little, or by a number. "Piney" in Piney Ridge and Creek [Mariposa Co.] is a common term for 'pine-covered'. Dozens of places are named after the tree, including Pinecrest [Tuolumne Co.], Pinedale [Fresno Co.], Pine Grove [Amador Co.], Pineridge [Fresno Co.], Pine Valley [San Diego Co.], Bigpine [Inyo Co.], Lone Pine [Inyo Co.], Pine Knot and Pine Lake [San Bernardino Co.]. *For* Pine Grove [Sonoma Co.], *see* Sebastopol.

Pinecate Peak. *See* Pinacate.

Pinnacles. The name is applied to pillar-like formations created by erosion. The most notable are the **Pinnacle Rocks** [San Benito Co.], mentioned by Vancouver in 1794, and once called Chalone Peaks (*q.v.*); since 1908, they have been included in **Pinnacles National Monument**. With the exception of Pilot Pinnacle [Lassen NP], the word does not seem to have been used as a generic term. Pinnacle Point is the modern name of the rocky promontory of Point Lobos near Monterey; in 1878 the Coast Survey recorded the name Pyramid Point for the projecting rock.

Pino. *See* Loomis.

Pinoche (pi nō´ chē) **Peak** [Mariposa Co.]. Perhaps a variant of Spanish *panoche* (*q.v.*).

Pinole (pi nōl´): **Point, Creek, Ridge**, town, **Shoal** [Contra Costa Co.]. The Spanish word, designating ground and toasted grain or seeds, is derived from Aztec *pinolli*. Fages (p. 79) reports that the California Indians made a chocolate-colored *pinole* from the seeds of the cattail reed, and from the flower in season they made a yellow and sweet *pinole*. It may have been a pinole of this kind that the Indians gave to José de Cañizares in 1775 (Eldredge, *Portolá*, pp. 66–67), and because of this incident the name may have become attached to the region. According to a legend, however, the name was given in Mexican times: A detachment of soldiers, prevented from crossing Carquinez Strait because of high winds, ran out of provisions. "On their march they found a village of Indians, who had corn from which they manufactured meal (Pinole). That camp they named El Pinole" (Co. Hist.: Contra Costa 1878:12). In 1823, possessory rights to Pinole y Cañada de la Hambre were granted to Ignacio Martínez in recognition of his military services. The formal land grant is dated June 1, 1842. Another grant, Boca de la Cañada del Pinole, is dated June 21, 1842. The point was called Penoli on Ringgold's maps, and Penole on the Coast Survey chart in 1850. The Parke-Custer map, in 1855, gives the correct spelling, but the Coast Survey did not correct the mistake until fifty years later. The post office was established and named in the 1870s.

Piñon (pin yōn´) [Riverside Co.]. Spanish *piñón* (also spelled "pinyon" in English) is an augmentative form of *piña* 'pine cone'. It refers to several species of pine, such as *Pinus monophylla*, that have edible seeds. **Piñon Hills** [San Bernardino Co.], about 40 miles northwest of San Bernardino, first bore the name Smithson Springs for the owner of a cattle ranch; it was later called Desert Springs, and in 1962 Piñon Hills (Belden).

Pinos. The wealth of native pines in California made this tree the most popular for place naming, even in Spanish times. The name was repeatedly used by explorers who were struck by the pine-covered mountains and promontories after their passage along the treeless shores of Lower California. Monterey Bay was the Bahía de los Pinos, and Point Reyes was probably the Cabo de Pinos that Cabrillo named in 1542. From Monterey Bay southward there are still about fifteen geographic features called Pinos. **Point Pinos** (pī´ nəs, pē´ nōs) [Monterey Co.]: The name Punta de Pinos was given to the cape by Vizcaíno in Dec. 1602 (Wagner, p. 402). It was applied to a land grant, dated May 24, 1833, and Oct. 4, 1844. The generic part was Americanized by the Coast Survey in 1855. **Potrero Los Pinos; Los Pinos Mountain** [San Diego Co.]: The names originated with the Potrero Los Pinos, a part of the Potreros de San Juan Capistrano land grant, dated Apr. 5, 1845. In Santa Barbara Co., there was a land grant Cañada de Los Pinos, dated Mar. 16, 1844, and the Santa Cruz Mountains are called Sierra Verde de Pinos on Crespí's map of 1772. Spanish *pinos* has often come to be pronounced in English as (pē´ nəs), sounding like English *penis*, and so the term has been modified in various California place names (cf. Shafer pp. 242, 246); in some places it has been hypercorrected to "Piños" (nonexistent in Spanish), and a name like "Mount Pinos" has been changed to "Pine Mountain" [Ventura Co.]. *See* Tres Pinos; Pine.

Pinoso (pi nō′ sō), **Arroyo** [Fresno Co.]. The valley south of Coalinga had several other names before the BGN decided (1964) for the present name, which is Spanish for 'piney'.

Pintado. In Spanish and even in early American times, the Spanish adjective for 'painted' was often used for colored canyons and mountains, as well as for trees or rocks with Indian symbols and figures.

Pinto. The Spanish word means 'mottled' in the Southwest and is now usually applied to a piebald horse. It was used as a name for a mountain chain near the Mojave Sink as early as 1776: "the sierra that I named Pinta for the veins that run in it of various colors" (Garcés, Mar. 9, 1776; Coues, *Trail*, p. 238). **Pinto Peak** [Death Valley NP]: The name was applied to the range north of Towne's Pass and to its highest peak by the Wheeler Survey in 1871, appearing on its atlas sheet 65-D. The name was later extended to the ridge south of the pass (Carson and Colorado Railroad map of the 1880s). On the Ballarat atlas sheet it is transferred, without reference to its meaning, to a drab and colorless peak. The range, left nameless on the atlas sheet, is locally called Moigne Mountain, for an old prospector. Pinto Lake in Santa Cruz Co. was probably named for a person.

Pinyon is an alternative spelling of Piñon (*q.v.*), referring to pine trees that bear edible nuts. But **Pinyon Peak** [Monterey Co.] is probably from *peñón* 'large crag', augmentative of *peña* 'crag', since there are no piñon pines here (Clark 1991).

Piojo is Spanish for 'louse'. **El Piojo** (pē ō′ hō) **Creek** [Monterey Co.] was named after the land grant El Piojo, dated Aug. 20, 1842.

Pioneer [Amador Co.]. Named in honor of early settlers; a post office was established here in 1947. **Pioneer Basin** [Fresno Co.]: When the Mount Goddard quadrangle was surveyed, between 1907 and 1909, R. B. Marshall named four peaks of the Silver Divide for the pioneer financiers Charles Crocker, Collis P. Huntington, Mark Hopkins, and Leland Stanford —the "Big Four" who were responsible for the Central Pacific Railroad. The depression surrounded by the peaks was called Pioneer Basin (Farquhar). **Pioneertown** [San Bernardino Co.] was originally built as a set for Western movies. The late actor Dick Curtis applied the name on Labor Day, 1946 (Hester Downing).

Pipe Creek [Mendocino Co.]. So named by settlers in the early 1860s because they found two corncob pipes nearby (Asbill). *See* Bell Spring Mountain.

Pipe Line Canyon [Lake Co.]. In the 1890s an iron pipeline, 4 miles long, was laid from Chicken Springs to the Highland Springs Resort (Mauldin).

Pi-Pi (pī′ pī) **Creek** [El Dorado Co.] is said to be from an unidentified Indian language.

Piru (pi ro͞o′, pē′ ro͞o, pī ro͞o′): **River, Creek, Canyon**, town, **Lake** [Ventura Co.]. The name is derived from Kitanemuk (Takic) *pi'idhuku*, the name of a plant (cf. W. Bright, "The Alliklik mystery," *Journal of California Anthropology* 2:228–36 [1975]). A place called Piru is mentioned on May 31, 1817 (Arch. Arz. SF 3:1.132), and with various spellings in the following years; the Arroyo de Piruc is shown on a *diseño* of the San Francisco grant, 1838. In American times, the name appears as Piro in the *Statutes* of 1850 (p. 59) and as Rio Peru on the Parke-Custer map of 1854–55. The town was laid out in 1888 and called Piru City after the Piru ranch developed by the Chicago publisher David C. Cook.

Pisgah (piz′ gə): **Mountain**, station [San Bernardino Co.]. The extinct volcanic crater was named after Mount Pisgah in Jordan, from which Moses saw the promised land. The station was named after the mountain in 1905 (Santa Fe).

Pismo (piz′ mō): **Beach, Creek, Lake, State Park** [San Luis Obispo Co.]. From Obispeño Chumash *pismu'* 'tar, asphalt', lit. 'dark place, dark stuff', from *piso'* 'to be black, dark'; cf. Kathryn Klar, *Names* 23:26–30 (1975). The town was laid out in 1891 when the last link of the Southern Pacific coast route was built from San Luis Obispo to Ellwood. It was named Pismo because of its situation on the Pismo land grant (dated Nov. 18, 1840); the generic term Beach was added to the name after 1904. The state park was established and named in 1935. *See* Asphalt; Brea.

Pitas (pē′ təs) **Point** [Ventura Co.]. The point received its name from an Indian village that the Portolá expedition called Los Pitos 'the whistles': "Two leagues beyond is the village of Los Pitos, so called because of the whistle which the men of the first expedition of Commander Portolá heard blown there all night" (Font, *Compl. diary*, pp. 248–49). Crespí had recorded the same incident on Aug. 15, 1769: "During the night they [the Indians] disturbed us and kept us awake playing all night on some doleful pipes or whistles." He adds that he called the village Santa Conefundis but does not say that the name Los Pitos was given by the soldiers. *El citio llamado* ['the place called'] *de los Pitos* is mentioned on July 25, 1822 (Docs. Hist. Cal. 4:583). The Coast Survey left the point nameless until 1868, when it was labeled Las Petes. In 1889 Davidson changed the name to Point Las Pitas, probably assuming that it was named for the American agave or century plant, *pita* in Spanish. The Coast Survey chart still keeps Davidson's name although the point is commonly called Pitas Point.

Pitman Creek [Fresno Co.]. Named for Elias Pitman, who had a hunting cabin on the banks of the creek "in the early days" (Farquhar).

Pitos, Los. *See* Pitas Point.

Pit River. Like the Missouri and the Mississippi, the Pit and the Sacramento present a geographical anomaly. Above the confluence with the Little or Upper Sacramento, the Pit is the main stream; consequently, the entire course from Goose Lake to Suisun Bay should be called either Pit or Sacramento. The Pit River had been known to the trappers of the Hudson's Bay Company since the 1820s and was so called because the Indians dug pits on its banks as traps. A graphic account of these pits is given by Joaquin Miller in his *Life amongst the Modocs* (London: Bentley, 1873), p. 373: "We crossed the McCloud, and our course lay through a saddle in the mountains to Pit river; so called from the blind pits dug out like a jug by the Indians in places where their enemies or game are likely to pass. These pits are dangerous traps; they are ten or fifteen feet deep, small at the mouth, but made to diverge in descent, so that it is impossible for anything to escape that once falls into their capacious maws. To add to their horror, at the bottom, elk and deer antlers that have been ground sharp at the points are set up so as to pierce any unfortunate man or beast they may chance to swallow up. They are dug by the squaws, and the earth taken from them is carried in baskets and thrown into the river." Hood's map of 1838 has the correct spelling, but Wilkes's map of 1841 has Pitts River and Pitts Lake. The observant Eld, to be sure, crossed out one "t" in his manuscript when he learned the reason for its naming, but his journal was not published. In 1850 Williamson tried to rectify the mistaken idea that the river was named for the English prime minister William Pitt (1759–1806): "We passed many pits about six feet deep and lightly covered with twigs and grass. The river derives its name from these pits, which are dug by the Indians to entrap game. On this account, Lieut. Williamson always spelled the name with a single t, although on most maps it is written with two" (Pac. R.R. *Reports* 6:1.64). His efforts were in vain: although the newspapers of the northern counties usually spelled the name correctly, most maps of the 1850s and 1860s have Pitt River. Scholfield applied the name Sacramento to both the Pit and the Sacramento in 1851. The Whitney Survey tried again to call the entire waterway Sacramento River and at the same time restored the proper spelling. The von Leicht–Craven map has Upper Sacramento or Pit River. However, Brewer in his field notes uses consistently the spelling Pitt, and most maps continued the wrong spelling until the USGS issued the Redding atlas sheet in 1901. *For* Pit Lake [Modoc Co.], *see* Goose Lake.

Pittsburg: Landing, city, **Station** [Contra Costa Co.]. Col. J. D. Stevenson, of the New York Volunteers, and Dr. W. C. Parker bought Rancho Los Medanos in 1849 and laid out the "City of New York of the Pacific" at the site of modern Pittsburg. When the coal deposits, discovered in the hills north of Mount Diablo in 1852, were commercially exploited after 1858, two railroads were built, one terminating in this New York, the other at the wharf of the Pittsburg Coal Company at the present site of Pittsburg Landing (plat of Rancho Los Medanos, 1865). The coal company was doubtless named after the industrial city in Pennsylvania. On Hoffmann's map of the Bay region (1873) both terminals of the railroad—the one at the river as well as the one in the hills—are labeled Pittsburgh. The struggle to exploit the deposits of inferior coal was finally given up, and both cities declined. After 1900 the steel industry instilled new life into the sleeping communities: old New York was named Pittsburg (or Pittsburgh) and old Pittsburg became Pittsburg Landing. With the exception of the large city in Pennsylvania, which insists on the retention of the final *h*, the spelling Pittsburg is now generally used in the United States.

Pitts Lake [Modoc Co.]. *See* Goose Lake; Pit River. *For* Pitts River, *see* Pit River.

Pittville [Lassen Co.]. The post office was established in 1878 and named after the river, its name then still incorrectly spelled Pitt. Another town, Pittsburg, obviously named after the river, is mentioned by Beckwith in the Pac. R.R. *Reports* (2:2.55).

Piute (pī′ yo͞ot). In California this spelling is now generally accepted for most of the places named for the tribe. The name Paiute (as anthropologists prefer to spell it) is applied to three groups of the Shoshonean family: the Northern Paiute or Paviotso in northwestern California and adjacent Nevada and Oregon, the Owens Valley Paiute (or Eastern Mono) in California, and the Southern Paiute of southern Nevada and southern Utah. However, the name has often been loosely applied to many Indians of the Great Basin area. The term Paiute has been said to mean 'water Ute' or 'true Ute'. However, the name Ute (a neighboring tribe, after whom the state of Utah is named) enters English from Spanish *yuta*, whereas Paiute comes through Spanish *payuchis*, probably from Southern Paiute *payuutsi* 'Paiute Indian' (Sapir, p. 640). By false analogy, the English term Paiute has been made to resemble the tribal name of the Utes (*HNAI* 11:393). The spellings Paiute and Piute were generally used by the men of the Pacific Railroad Survey in the 1850s. **Piute Creek** [Lassen Co.], a tributary of the Susan River, appears to be the only feature named for what anthropologists now designate as Northern Paiute. **Piute Mountain** and **Creek** [Yosemite NP]; **Piute Mountain** [Mono Co.]; **Piute Pass, Creek,** and **Canyon** [Fresno Co.]; and **Paiute Monument** [Inyo Co.] are all in the territory of the Eastern Mono tribe, locally called Paiute. The Sierra pass on the Fresno-Inyo county line was named by L. A. Winchell because it was used by Owens Valley Indians. The name was applied to the creek by J. N. LeConte in 1904 to avoid the name "North Branch of the South Fork of the San Joaquin River" (Farquhar); *see* Winnedumah. **Piute Range** and station [San Bernardino Co.] are in the territory of the Chemehuevi tribe, closely related to the Southern Paiute. The station was named by the Santa Fe in 1903–4. A nearby Pah-Ute Springs, shown on the Santa Fe map of 1880, doubtless suggested the name. **Piute Butte** [Los Angeles Co.] as well as **Piute Mountains, Peak**, and town, and **Pahute Peak** [Kern Co.] are decidedly outside of Paiute territory, but close enough to justify their naming. When rangers of Sequoia National Forest climbed a mountain near Piute Peak about 1945, they found a tube containing a record left by earlier climbers indicating that this was the original Piute Peak. Since, however, the namegivers had spelled the name Pahute, the rangers left this name on the new peak, and we now have both names (Stewart).

Pixley [Tulare Co.]. The name of Frank Pixley, founder and editor of the *San Francisco Argonaut*, was given to the settlement in the 1880s (Co. Hist.: Tulare 1913:51).

Pizzlewig Creek [Siskiyou Co.], formerly known as Sweet Pizzlewig Creek, now retains the shortened form. The story behind the name was told by "Hardscrabble" in Aug. 1900: "A poor forlorn woman deficient in virtuous ways settled on top of the mountains to be free from the temptations of the world, but her charm seemed to be so catching that the babbling brook that flowed by her door paused for a moment and secreted away that poetical name and shall now for ever be known as Sweet Pizzlewig for its momentary folly" (Luddy). "Pizzle" is old-fashioned slang for 'penis; to copulate'.

Placentia [Orange Co.]. Founded in 1910 and named after the school district. This had been named in 1884 by Mrs. Sarah J. McFadden, who wrote, "Placentia means delightful situation" (Brigandi). She apparently associated it with the Latin verb *placet* 'it is pleasant'; however, there was a city named Placentia in ancient Italy, now known as Piacenza.

Placer (plas′ ər). This western American term of Spanish origin designates alluvial or glacial deposits containing gold particles, which can be obtained by washing. In the "golden days" the word was frequently used for place names. **Placerita** (plas ə rē′ tə) **Creek** [Los Angeles Co.], perhaps for a Spanish diminutive *placerito*, recalls the discovery of gold by Francisco López six years before Marshall's find. For a few years after 1842 a placer on Rancho San Francisco was worked with moderate success. **Placerville** [El Dorado Co.] was first settled in 1848 by William Daylor of Sutters

Fort and became known as Dry Diggings. In 1850 the camp was named Placerville because the streets of the camp were almost impassable on account of the numerous placering holes. The town never bore the name Hangtown, as has often been asserted; it was simply a nickname given to the place because of an incident that occurred on Jan. 22, 1849, when two Frenchmen and a Chileno were hanged, as witnessed by E. Gould Buffum (p. 65). According to the *Sacramento Union* (Apr. 2, 1853), "The name of 'Hangtown' was originally given to the town, in consequence of the carrying into effect in a summary manner, some of Judge Lynch's sentences, and the citizens find it somewhat difficult to get rid of the objectionable soubriquet." **Placer County** was created on Apr. 25, 1851, from parts of Sutter and Yuba Cos. and was so named because of many placers in its territory.

Plainsburg [Merced Co.]. In 1869, after Farley's hotel was opened near the settlement known as Welch's Store, the place was named Plainsburg (Co. Hist.: Merced 1925:363). It is actually a "burg in the plains." *See* Athlone.

Planada (plə nä′ də) [Merced Co.]. From a number of names submitted in a contest held in 1911, this Spanish word for 'plain' was chosen as a suitable name for the town, which is situated in a rich fertile plain (Frances Osterhaut). The former post office name was Geneva; the former railroad name, Whitton.

Plano (plā′ nō) [Tulare Co.]. The name is Spanish for 'level ground, plain, flat' but is not current as a generic term. It was selected when the post office was established in 1871, probably by A. J. Adams, the first postmaster.

Plantation [Sonoma Co.]. The post office was established about 1900 and was named after the old Plantation House, a roadhouse shown on the county map of 1879.

Plaskett (plas ket′): **Creek, Rock** [Monterey Co.]. Probably named for William L. Plaskett, a native of Indiana, or one of his sons, who were farmers in nearby San Antonio in the 1870s.

Plaster City [Imperial Co.]. According to Brigandi, this company town dates from around 1922, when a narrow-gauge railroad was built to the mines of the Portland Cement Company (now the United States Gypsum Company). The post office was established in Aug. 1936.

Platina (plat′ i nə) [Shasta Co.]. The post office was established on Apr. 23, 1921, and was so named because it is in a platinum ore area (Steger). Spanish *platina*, pronounced (plä tē′ nä), is derived from *plata* 'silver' and was the name first given to platinum when it was discovered in South America. An alternative source is Wintu *pathina* 'come out', referring to a place where one comes out of the mountains (Shepherd).

Playa. The generic name means beach but has different meanings in various parts of Spanish America. In the name of the land grant Boca de la Playa [Orange Co.], dated May 7, 1846, and in La Playa [San Diego Co.] the term is obviously used for beach. **Playa del Rey** (plä′ yə del rā′) [Los Angeles Co.] was named in 1902 for the unsuccessful venture of Port Ballona; it was advertised as meaning 'playground of the king' with a certain justification, for in Argentina and Chile *playa* sometimes means 'playground'. In the case of Rancho de La Playa, which belonged to the Mission Santa Barbara before the secularization—and the many other playas found on *diseños* in various places of the state—the meaning is 'dry lake or river bed', as in Mexican regional Spanish.

Pleasant. A favorite American specific term used to describe places of beauty and tranquillity. California has many features so named. More than half the names refer to valleys. **Pleasant Valley** [El Dorado Co.] was discovered and named by Henry W. Bigler on June 8, 1848. On July 3, the Mormons from Coloma and Mormon Island started their trek to Salt Lake from here. It had a post office from Mar. 23, 1864. **Pleasant Grove** [Sutter Co.] had a post office established in the late 1860s; the name was probably applied to offset the earlier term Gouge Eye (Co. Hist.: Sutter 1879:98).

Pleasanton [Alameda Co.]. The origin of the name is not in "pleasant town" as the present spelling would suggest. The town was named in 1867 for Gen. Alfred Pleasonton by John W. Kottinger, a pioneer of 1851 who may have served with Pleasonton in the Mexican War. The name was misspelled, probably through a clerical error, when the post office was established on June 4, 1867. An attempt was made in later years to correct the error (Postal Guide, 1898), but apparently without success, for within a few years the name appeared again as Pleasanton.

Pleasants Peak [Orange Co.]. Adopted in 1933 by the BGN at the suggestion of the Orange County Historical Society, for J. E. Pleasants, a forty-niner whose name had been associated with the peak since 1860. J. E. Pleasants was a son of James M. Pleasants, for whom Pleasants Valley in Solano Co. was named. The original name of the peak was Sugarloaf.

Pleasants Valley [Solano Co.]. This valley, between Vacaville and Winters, was named for James Marshall Pleasants, a native of Kentucky, and his son William James, who arrived here on Dec. 6, 1850, and settled as farmers. William James Pleasants later described his experiences in *Twice across the plains* (San Francisco, 1906).

Pleasant Valley [Inyo Co.]. Named for James ("Cage") Pleasant, a dairyman from Visalia, who was killed by the Indians here in 1862 (Chalfant, *Inyo*, p. 216).

Pleasonton. *See* Pleasanton.

Pleito (plā´ tō): **Hills, Creek** [Monterey Co.]. The place called el Pleito is mentioned as early as Sept. 1796 (Prov. Recs. 6:172). A rancho called San Bartolomé or Pleito is repeatedly mentioned in later years and was made a land grant on July 18, 1845. The variant spelling Pleyto occurs. *Pleito* is Spanish for 'quarrel, dispute' and may have been applied because there was some dispute between the missions of San Miguel and San Antonio concerning the property rights of Rancho San Bartolomé. **Pleito Hills, Creek** and a playful diminutive, **Pleitito Creek**, exist in Kern Co.

Plumas. Feather River was known as Río de las Plumas in Mexican times and appears with this name on American maps as late as 1852 (Gibbes). When **Plumas** (plo͞o´ məs) **County** was formed from a portion of Butte Co. on Mar. 18, 1854, the old name was revived. A flourishing town on the east bank of the Feather River, founded in 1850 and named Plumas City, has vanished. Two Western Pacific stations are called Plumas [Lassen Co.] and Las Plumas [Butte Co.]. A lake in Yuba Co., exceedingly irregular in shape, is called Plumas Lake; the nearby Sacramento Northern station is named after the lake. **Plumas National Forest** was created and named by presidential proclamation in 1905. *See* Feather River; *for* Rancho de las Plumas [Butte, Sutter Cos.], *see* Boga.

Plum Valley [Sierra Co.]. The valley is first mentioned in 1858 and was so named because of the great quantities of wild plums (perhaps *Prunus subcordata*).

Pluto, Mount [Placer Co.]. One of the last volcanoes in the Tahoe region; named for the Greek god of the underworld (Lekisch).

Pluton (plo͞o´ ton) **Creek** [Sonoma Co.]. The name—the Spanish or French equivalent of Pluto, the god of the lower world in classical mythology—may have been applied to the stream by Forrest Shepherd in Feb. 1851, when he saw the hot geysers rising from lower strata. Pluton Geysers, Pluton Valley, and Pluto's Cauldron are mentioned in an article by Shepherd in the *American Journal of Science and Arts*, Nov. 1851 (pp. 153 ff.). Only the north fork now bears the name; the main stream is called Sulphur Creek. There is a Mount Pluto in Placer Co., and Pluto's Cave near Mount Shasta is mentioned by Brewer on Nov. 11, 1863. The presence of plutonic rock, a type of igneous formation, may have some connection with these names.

Plymouth [Amador Co.]. The site of an old goldmining camp called Puckerville or Pokerville. A post office was established there on Sept. 8, 1871, and was named after the Plymouth Mines, which had been in operation since the early 1850s.

Poblanos (pō blä´ nəs) [Contra Costa Co.]. Arroyo de los Poblanos is shown for Marsh Creek on the "Plano topographico de la Misión de San José" (about 1824), and part of it appears on modern maps as Cañada de los Poblanos on the Mount Diablo atlas sheet. In Mexico *poblano* means a person from the city of Puebla; however, in California it may be another version of the name of the Bolbones Indians at the foot of Mount Diablo.

Pochea (pō chē´ ə) [Riverside Co.]. Indian village site near Hemet, said to mean 'where the rabbit went in' (Gunther).

Pocitas, Las. *See* Posita.

Poco, Arroyo [San Diego Co.]. Recognized by the BGN in 1990. This is probably an Anglo attempt to put 'little creek' into Spanish; but *poco* means 'a little bit', not 'small in size'. Better Spanish would have been Arroyo Chiquito.

Poe: Canyon, Reservoir [Plumas Co.]. The canyon was named for the American poet Edgar Allen Poe (1809–49) by Mrs. Virgil Bogue in 1908. The lake and the powerhouse were built by the Pacific Gas and Electric Company in 1958, in anticipation of the construction of the Oroville Dam.

Poggi (pō´ jē) **Canyon** [San Diego Co.]. For Joseph Poggi, an Italian immigrant and cattleman (Stein).

Pogolimi [Sonoma Co.]. The name of the land grant Cañada ('valley') de Pogolimi (or Pogolomi or Pogolome), dated Feb. 12, 1844, is apparently derived from an Indian word of unknown origin and meaning.

Pogonip (pō´gə nip) [Santa Cruz Co.]. The area around a country club of the same name, founded in the early 1930s; however, the name was given to Pogonip Creek much earlier (Clark 1986). Pogonip is a term used in the Great Basin for an icy fog that forms in mountain valleys of the area; the term is *pakï-nappï* 'cloud, fog' in Southern Paiute and other Numic languages (McLaughlin). It is not known how the term arrived in Santa Cruz Co.

Poho (pō´hō) **Ridge** [El Dorado Co.]. Said to be from an unidentified Indian language.

Pohono. *See* Bridalveil Fall.

Point. Names of coastal features preceded by the generic Point will be found listed under the specific name. *For* Point of Rocks, *see* Helen: Helendale; *for* Point Loma, *see* Loma; *for* Point Reyes, *see* Reyes.

Point Arena (ə rē´nə): town, **Creek** [Mendocino Co.]. The town developed around the store built in 1859 and was named after the nearby cape. The post office name was Punta Arenas in 1867, but the von Leicht–Craven map, 1874, has Point Arena. *See* Arena.

Poison. The word is frequently found attached to names of physical features. Poison Rock in Mendocino Co. probably preserves the name of a meadow or creek no longer so known. Poison Meadows and Canyons are often properly so named because poisonous plants grow there that are known to have killed livestock, especially sheep. In relation to surface waters, the term is loosely applied: the arsenic and other "poisons" are only Glauber's and Epsom salts, though of course these may be fatal if drunk in excess by thirsty desert travelers. **Poison Lake** [Lassen Co.] was so named "from the innumerable quantity of animalculae and frogiponiana in its waters, which could only be rendered drinkable by filtration" (Delano, p. 36). **Poison Valley** [Alpine Co.] was shunned by early cattlemen, or herding there was watchfully done, after some of their stock had died from eating water hemlock (*Cicuta*) and larkspur (*Delphinium*) (Maule).

Poker Flat [Sierra Co.]. The site in the rich gold-producing Slate Creek region, still shown on the Downieville atlas sheet of 1907, is the only remnant of several Poker Flats, Bars, and "Villes" named during the Gold Rush. The scene of Bret Harte's "The outcasts of Poker Flat" is purely fictional and has nothing to do with the place. *For* Pokerville, *see* Plymouth.

Pokywaket (pō´ kē wak ət) **Creek** [Shasta Co.]. From Wintu *puki waqat* 'unripe creek', containing *puk-* 'unripe' and *waqat* 'creek' (Shepherd). Unripe acorns were brought here to be "cured" (Steger).

Polaris (pō lâr´ is) [Nevada Co.]. The Latin name of the North Star is one of the fanciful railroad names applied to stations of the Truckee-Verdi section in 1867.

Polemonium (pol i mō´ nē um) **Peak** [Kings Canyon NP]. Named in 1985 for a flower—the sky pilot, *Polemonium eximium* (Browning 1986).

Poleta (pō lā´ tə): **Canyon**, station [Inyo Co.]. The Southern Pacific station, also spelled Polita, was named after the canyon, which had been named after the Poleta Mine. A Mexican or Spaniard named Poleta located the mine in the early 1880s (Robinson).

Polka [Santa Clara Co.]. The land grant was part of the original San Isidro grant (1808); it is dated June 19, 1833. In 1849 Daniel Murphy bought the rancho, and in his claim, filed on Feb. 17, 1852, he calls it La Polka, a name probably suggested by the new dance that was at the peak of its popularity in those years.

Pollard Gulch [Shasta Co.]. Named for John Pollard, a miner who lived here for a short time in the early 1880s (Steger).

Pollasky. *See* Friant.

Pollock Bridge. *See* Loftus.

Pollock Pines [El Dorado Co.]. The post office was established in 1935 and named for the grove of pines belonging to the Pollock family, the first white settlers.

Polly Dome [Yosemite NP]. Named by R. B. Marshall, for Mrs. Polly McCabe, daughter of Col. W. W. Forsyth (Farquhar).

Polo Peak. *See* Jackson.

Polvadero (pōl və dâr´ ə): **Gap, Creek** [Fresno Co.]. Spanish *polvadera* means 'cloud of dust'. The name (spelled with the incorrect *-o* ending) was applied to the gap between Kettleman and Guijarral Hills.

Pomins (pō mēnz´) [El Dorado Co.]. A post office here was named after Pomins Lodge, owned by Frank and Marion Pomin (F. Slade). William Pomin, a pioneer, had built the Tahoe House in 1864.

Pomo [Mendocino Co.]. A post office was named in the 1870s, probably after the Indian village that stood at the site of the Potter Valley flour mill (Barrett, *Pomo*, p. 140); it was called *phóomóo* 'at red-earth hole' in the Northern Pomo language, from *phóo* 'red earth, magnesite'. As the name of the major language family in Mendocino, Lake, and Sonoma Cos., the word Pomo has a different derivation, from Northern Pomo *phó'ma'* 'inhabitant', containing *phó* 'live, inhabit' (Oswalt).

Pomona (pə mō′ nə) [Los Angeles Co.]. The name was applied to a new settlement by the Los Angeles Immigration and Land Cooperative Association on Aug. 20, 1875, as a result of a contest for a name, won by Solomon Gates, a nurseryman. The Roman goddess of orchards and gardens had already given her name to at least six other communities in the United States. Pomona College was originally in Pomona and was named after the city in 1887.

Pomponio (pom pō′ nē ō) **Creek** [San Mateo Co.]; **Cañada Pomponio** [Marin Co.]. Both features were named for José Pomponio Lupugeyum, a Bolinas Indian, captain of a group that called itself Los Insurgentes; he was captured and executed in 1824 (Alan K. Brown). A Cuchilla ('ridge') de Pomponio is shown on a *diseño* of the San Gregorio grant (1839).

Poncho Rico Creek. *See* Pancho Rico Creek.

Ponderosa Way [Mariposa Co.]. A firebreak of some 200 miles, most of which was completed in the 1930s. It was so named because it separates the higher country, where the ponderosa pines grow, from the highly inflammable and not very valuable lower country (Stewart).

Pondosa [Siskiyou Co.]. The post office was established in 1926 and named Pondosa, the trade name of the ponderosa pine, *Pinus ponderosa* (E. Fritz).

Ponto [San Diego Co.]. The name, poetic Spanish for 'sea', was applied to the Santa Fe station in 1919. The place had previously been known as La Costa.

Pony. This small but hardy horse was greatly valued in pioneer days. The name became very popular as a place name when the famous (though short-lived) Pony Express was inaugurated in 1860. About fifteen geographic features bear the name.

Poonkiny (pōōn′ kin ē) **Creek** [Mendocino Co.]. From Yuki *p'unkini*, the aromatic plant called 'wormwood', genus *Artemisia;* the name may contain *kin-* 'to stink' (Shepherd).

Poopenaut (pōō′ pə nôt, pōō′ pə nō) **Valley** [Yosemite NP] is of mysterious origin but is said by one writer to be named after an early settler of German extraction. Also spelled Poopenant, Poopino, Poopeno (Browning 1988).

Poopout Hill [San Bernardino Co.]. The name appears in the Decisions of the BGN, 1966. The name of this hill near San Gorgonio Mountain probably contains the American slang word for 'exhausted'; the elevation is also known as Trail Head Hill.

Poorman Creek [Nevada, Butte Cos.]. According to the journal of John Steele written during his California mining adventure in 1850, the name is that of a pioneer miner surnamed Poorman (Bidwell, *Echoes*, p. 163). San Luis Obispo Co. has a Poorman Canyon, Plumas Co. a Poorman Creek, and Mendocino Co. a Poor Mans Valley. The name was often used in gold-mining days, sometimes because the diggings were really poor, at other times to scare away new prospectors by pretending unprofitable results. *For* Poor Man's Flat, *see* Windsor.

Popcorn Caves [Shasta Co.] are so called because of the volcanic formation there, formed by gas escaping through hot lava (Harris).

Pope: Valley, Creek [Napa Co.]. Valley and creek preserve the name of William (Julian) Pope, a member of Pattie's party in 1828 and grantee of the Locoallomi grant on Sept. 30, 1841. The Indian name for Pope Creek was Nombadjara (D. T. Davis), and this name is shown on a *diseño* of the Las Putas grant. *For* Popes Valley [Fresno Co.], *see* Watts Valley.

Poplar. Besides the town in Tulare Co. and the cluster name on the Middle Fork of Feather River [Plumas Co.], only a few places in the state bear the name of the tree. Three species are native to the state—*Populus fremontii, trichocarpa*, and *tremuloides*—but the first two are locally called cottonwood, and the last, aspen; hence the lack of Poplar place names. In the eastern states the Poplar names outnumber the Cottonwood names. *See* Alamo.

Porciúncula, La. *See* Kern River; *for* Nuestra

Señora de los Ángeles de Porciúncula, *see* Los Angeles.

Porcupine. The name of this prickly animal (*Erethizon dorsatum*) was used repeatedly for geographical names. The best-known feature is Porcupine Flat in Yosemite Valley, mentioned in a letter of Hoffman to Whitney on Sept. 8, 1867.

Porphyry (pôr′ fə rē) [Yosemite NP]: This refers to a dark reddish rock with feldspar crystals in it (Browning 1988).

Portal. A name sometimes used in place of "gate," corresponding to Spanish *portal*. The best-known instances of its use are El Portal [Mariposa Co.] and Portal Ridge [Los Angeles Co.].

Port Animal Depot [Los Angeles Co.]. Established by the War Dept. on Sept. 7, 1944, and so named because the post was a "staging area" for animals (mules, war dogs, carrier pigeons, etc.) to be shipped overseas during World War II. It is commonly called Puente Animal Depot because of its proximity to the town of Puente (Marian Guntrup).

Port Chicago. *See* Chicago.

Port Costa [Contra Costa Co.]. The name was applied to the station of the Southern Pacific in 1878, obviously because it was in Contra Costa Co. and situated at the coast.

Porterville [Tulare Co.]. In 1859 Royal Porter Putnam, who was known by his middle name, operated a stage depot known as Porter's Station; later he had a store called Porter's Trading Post or Store. In 1864 he laid out the town and named it Portersville. The Official Railway Map of 1900 shows the present form.

Port Harford. *See* San Luis Obispo.

Port Kenyon [Humboldt Co.]. The name was applied by the Coast Survey to the shipping point on the land of J. G. Kenyon (Co. Hist.: Humboldt 1881:158).

Portola (pôr tō′ lə). Gaspar de Portolá y de Rovira, leader of the expedition of 1769, is honored in the name of a post office in Plumas Co. (listed in 1910), **Portola Valley** in San Mateo Co., and Portola Redwoods State Park—the last sometimes pronounced the Spanish way, (pôr tō lä′). The name of the state park was suggested by Aubrey Drury at a Park Commission meeting and was adopted on motion of Commissioner Leo Carrillo, a descendant of members of the Portolá expeditions. *See* Glossary.

Portuguese Bend [Los Angeles Co.]. The name was applied for Joseph Clark (Machado), a native of the Azores, who owned a fleet of whalers around 1860 (*CHSQ* 35:238). **Portuguese Flat** [Shasta Co.], locally pronounced (pôr′ tə gē), was named for Portuguese settlers in the mining days. Hazel Creek post office in Shasta Co. was called "Portuguee" from 1870 to 1877. **Portugee Canyon** is in Monterey Co.

Posa. *See* Poza.

Posé Flat. *See* Poza: Poso Flat.

Posey [Tulare Co.]. Earlier spelled as Pose and Posa, probably for Spanish *poza* 'water hole' (*q.v.*). The post office is listed in 1915.

Posita. The name is a Mexican localism for 'pond, water hole,' a diminutive of *pozo* 'well'. **Las Positas** (pō sē′ təs) **Creek** [Alameda Co.]: Padre Viader mentions *una posa de buena agua* 'a pool of good water' in the Valle de San José (Livermore Valley) in Aug. 1810 (Docs. Hist. Cal. 4:74), and Joaquín Piña mentions a *Paraje nombrado de las Positas del Valle* 'called a place of the little pools of the valley' on May 26, 1829. The name appears in the title of the land grant Las Positas or Las Pocitas del Valle de San Jose, granted Apr. 10, 1839, to Livermore and Noriega. It is shown as Los Positos on Eld's sketch of 1841, and as las Pocitas on several *diseños*. The Land Office maps of the 1850s have Posita Creek; Hoffmann's map of 1873 has Las Positas Creek. On the Pleasanton atlas sheet of 1906 it appears as Arroyo las Positas. **Positas** [Santa Barbara Co.] is so named in the *expediente* of the land grant La Calera y las Positas, dated May 8, 1843.

Poso. *See* Poza; *for* Arroyo Poso de Chane, *see* Gatos.

Posolmi y Posita de las Ánimas [Santa Clara Co.]. This land grant, dated Feb. 15, 1844, was also known as Rancho de Yñigo, from the name of the Indian grantee, Lupe (or López) Yñigo. Posolmi may be from Mutsun (Costanoan) *puslubmin* 'person with a big belly' (Callaghan); *posita de las ánimas* means 'little pool of the souls'. *See* Animas.

Post Office: Cave, Bridge [Lava Beds NM]. The presence of many pigeon holes in the walls suggested the name to J. D. Howard. Ballarat [Death Valley NP] was formerly Post Office Springs.

Post Peak [Yosemite NP]. Named for William S. Post, of the USGS, when the Mount Lyell quadrangle was surveyed in 1898–99.

Potato. More than twenty-five geographic features, mainly Hills, bear the name of this staple food. **Potato Canyon** [San Bernardino Co.], now called Oak Glen, was so named because potatoes were once grown there, and **Potato Hill** [Tehama Co.] because it has the shape of a potato. Some hills may resemble the shape of a "potato hill" rather than the potato itself; others may have been named because potatoes were planted on the slope. Santa Cruz Island has a Potato Harbor. Mineral Peak [Tulare Co.] was known as Half Potato Hill in the 1880s. The Twin Peaks in San Francisco were once called Las Papas 'the potatoes' and appeared with this name on many early maps.

Potem (pot´ əm) **Creek** [Shasta Co.]. Perhaps from Wintu *patem* 'mountain lion' (Shepherd).

Potholes. The term is used by geologists to describe regularly shaped holes in rocks, like those found in the canyon of the North Fork of Mokelumne River. It was once believed that they had been made by Indians, but most of them were formed by glacial action or by running water in stream beds, as in The Pothole and Pothole Creek [Tulare Co.], Pothole Valley and Pothole Spring [Modoc Co.]. Sometimes the name is used for a natural or excavated but regular depression in the ground: Pothole Lake [Inyo Co.] seems to be literally in a pothole. **Potholes** [Imperial Co.]: The name was applied to the terminus of the spur built by the Southern Pacific from Yuma in 1907. According to Doyle, it is the site of an old mining town and was probably so named because the gold was found in "pots or pockets." There is no documentary evidence that gold was mined here in the 18th century, as often asserted, but small-scale gold washing was carried on here until the 1930s, mainly by Mexicans.

Potrero (pō trâr´ ō). The Spanish word for 'pasture', derived from *potro* 'colt', was one of the most common generic terms in California; it appears in the names of more than twenty land grants or claims. It has survived in the name of one community in San Diego Co. and a district in San Francisco and in the names of numerous physical features. Although *potrero* has not entered general American English, like "corral" and "canyon," it is still used, especially in southern California, as a true generic term: The Potrero, Big Potrero, Round Potrero [San Diego Co.]; Potrero Chico, Potrero Grande, Potrero de Felipe Lugo [Los Angeles Co.]; Potrero Seco [Ventura Co.]; Mill Potrero [Kern Co.]; La Carpa Potreros, Montgomery Potrero, Pine Corral Potrero, Salisbury Potrero [Santa Barbara Co.]. It is also found as a specific term combined with Creek, Hill, Peak, and Point, and sometimes tautologically with Meadow.

Pots. *See* Devil: Devil's Mush Pot Cave.

Potter Point [Yosemite NP]. The name was given in 1909 by R. B. Marshall for Dr. Charles Potter of Boston (Farquhar).

Potter Valley [Mendocino Co.]. William Potter and his brother were the first white settlers in the Chico region. In 1853 they moved to the valley that bears their name.

Pottinger Canyon [Kern Co.]. The canyon in Santa Maria Valley was probably named for Thomas B. Pottinger, a native of Maryland and a resident of Bakersfield in 1879. Nearby Mama Pottinger Canyon may have been named for his wife or for his mother.

Potwisha (pot wish´ ə) [Tulare Co.]. The locality at the junction of the Marble and Middle Forks of Kaweah River was named by the BGN (6th Report) at the suggestion of George W. Stewart, for a branch of the Monache Indians (cf. *HNAI* 8:426).

Poverty Hills [Inyo Co.]. So named because, although gold was mined in rather large quantities in the nearby Fish Springs Hills in the 1870s, none was found in these hills (Brierly). **Poverty Bar** [Trinity Co.] was probably named for a similar reason, although it is said that Chinese reopened the diggings and made good money. **Poverty Hill** [Sierra Co.] near Scales has a rich gold mine; the name is ironic (Morley).

Poway (pou´ wā, pou´ wī): **Valley**, town [San Diego Co.]. The name is derived from a rancho of Mission San Diego, mentioned in 1828 as Paguay (Registro, p. 37). Cañada y Arroyo de Paguay is shown on a *diseño* of Rancho San Bernardo (1841). The post office is listed in 1880. From Diegueño *pawiiy;* the similarity to *pawiiy* 'arrowhead' may be accidental (Couro and Hutcheson).

Powder Mill Flat [Santa Cruz Co.]. The name is derived from the California Powder Works, which were operated here between 1865 and 1916 (Hoover, p. 575).

Powell, Mount [Kings Canyon NP]. Named in 1912 by R. B. Marshall in memory of John Wesley Powell (1834–1902), who was in charge of the expedition that navigated the Colorado through the Grand Canyon in 1869. Powell was director of the USGS, 1881–94, and director of the Bureau of Ethnology, Smithsonian Institution, 1879–1902 (Farquhar).

Powellton [Butte Co.]. The post office was established in the 1870s and named for R. P. Powell, who had acquired a ranch here before 1853. It exemplifies the use of the traditional English suffix *-ton* 'town' in the formation of a California place name.

Poza; Pozo. In Spanish geographical nomenclature a distinction is made between *poza* 'puddle' and *pozo* 'well'. In Spanish California, however, the words (also spelled *posa, poso*) were apparently used indiscriminately for 'water hole' in the widest sense. The various spellings were found in the names of several land grants and on many *diseños;* some of these names have survived. The word has even been actively used in place naming in American times. **Las Posas** (läs pō′səs) [Ventura Co.]: A place called Las Pozas o Simi, between San Buenaventura and San Fernando, is mentioned on Jan. 28, 1819 (Arch. Arz. SF 3:2.52). The land grant Las Pozas, named after the place, is dated May 15, 1834. Arroyo de las Pozas and Cuchilla ('ridge') de las Pozas are shown on a *diseño* of Calleguas grant (1837). **Posa** [Santa Clara Co.] was the name of a land grant also designated as Posa de San Juan Bautista and Posa de Chaboya, dated Mar. 10, 1839, and of another grant, Posa de los Ositos [Monterey Co.], meaning 'water hole of the little bears' and dated Apr. 16, 1839; but these do not seem to have survived as place names. **Pozo** [San Luis Obispo Co.]: "G. W. Lingo, Esq., a well-known citizen, had the honor of proposing the name of the post-office in the valley. *Pozo,* in Spanish, means a well or hole, whence the likeness of the valley itself to a place of this sort, and the Spanish word *Pozo* was adopted as the name of the post-office" (Co. Hist.: San Luis Obispo 1883:366). In the same county is Poso Ortega, a little lake in the Temblor Range. A tributary of Nacimiento River [Monterey Co.] is called Pozo Hondo 'deep well' Creek. In Spanish times the site of San Miguel mission was known as Las Pozas (SP Mis. 2:56–57). **Poso Creek, Camp**, and **Flat**, as well as **Mount Poso** [Kern, Tulare Cos.] and **Posey** [Tulare Co.] apparently refer to the stream called Rio de Santiago by Garcés on May 1, 1776 (Coues, *Trail*, p. 283). Williamson calls it O-co-ya or Pose Creek in 1853 (Pac. R.R. *Reports* 5:1.14). The maps of the 1850s have either name or both. In the Indian Report of 1854 and on Goddard's map (1860), it is spelled Posa. Brewer mentions Posé Flat and Little Posé Flat on June 7, 1863 (p. 394). The older Land Office maps have Poso Creek for the lower course and Posey Creek for the upper course. The post office is listed in 1915. This name, as well as Poso Slough [Merced Co.] and others, may be derived from *poso* 'sediment'. *See* Famoso.

Prado (prā′ dō): station, **Dam, Reservoir** [Riverside Co.]. The Spanish word for 'meadow' was given to a station of the Santa Fe in 1907 and later was applied to the dam built by the Corps of Engineers. **El Prado** [Fresno Co.] was the name that was given to the station when the San Joaquin and Eastern Railroad was built from here to Huntington Lake.

Prairie. The word, so familiar throughout the American West, is rarely used in California names. In the northwestern section of the state, especially in Humboldt Co., the term is used as a true generic for glade or clearing: Groves, Hancorn, Pitt Place, Bukers, Stevens, Boyes, and Elk Prairies. Sometimes Prairie is found as the specific term of a name: Prairie Creek [Humboldt Co.], Prairie Creek [Yuba Co.], Prairie Fork of the San Gabriel River [Los Angeles Co.]. In Sutter Co., Prairie Buttes was a former name of Sutter Buttes. **Prairie Creek Redwoods State Park** [Humboldt, Del Norte Cos.] was created in 1923.

Prater, Mount [Inyo, Fresno Cos.]. Named in memory of Alfred Prater, who, with his wife, made what was probably the first ascent, in 1928.

Prather (prā′ thər) [Fresno Co.]. The post office, listed in 1915, was so named because it is on the Prather Brothers Lodge Ranch, established about 1912 by Joseph E. and Fred Prather.

Prattville [Plumas Co.]. The settlement, known as Big Meadows, was named for the first postmaster, Dr. Willard Pratt, when the post office was established in the 1870s. The name of the post office is now Almanor.

Precipice Lake [Sequoia NP] was earlier called Hamilton Lake (BGN, list 6801). *For* Precipice Peak, *see* Cotter, Mount.

Precita (prə sē′ tə) **Park** [San Francisco Co.]. For Spanish *presita,* diminutive of *presa* 'dam'.

Prefumo Creek [San Luis Obispo Co.]. For Pietro Benedetto Prefumo, an early resident (Hall-Patton).

Prenda [Riverside Co.]. The optimistic name, a Spanish word meaning 'pledge, security', was given to the terminus when the Santa Fe spur was built into the newly developed citrus district in 1907 (Santa Fe).

Presidio (prə sid′ ē ō, prə sē′ dē ō). The Spanish word means 'garrison, fortified barracks'. In Spanish California there were four presidios, the names of which are still in use: Monterey Presidio was established in 1770; the San Diego mission guard was made a royal presidio on Jan. 1, 1774; that of San Francisco was founded in 1776, and that of Santa Barbara in 1782. The Sonoma garrison (1836) was really a presidio, but it was rarely so called. The Russian establishment that is now Fort Ross was sometimes called by the Spanish *Presidio ruso* or *Presidio de Bodega* (Bowman).

Presidio Bar [Siskiyou Co.] is an Anglicization of *pasirú'uuvree*, a former Karuk Indian village nearby. Also called Persido Bar.

Price Creek [Humboldt Co.]. Preserves the name of Isaac Price, a settler of 1852 (Co. Hist.: Humboldt 1915:828).

Price Creek [Yosemite NP]. Named for Lt. George Ehler Price, a veteran of the Spanish-American War (Farquhar).

Priest Grade [Tuolumne Co.]. Named about 1870 for W. C. Priest, owner of Priest's Hotel near Big Oak Flat. The name is repeatedly mentioned in the Davidson correspondence.

Prieto, Prieta. The descriptive adjective, meaning 'dark-colored, blackish,' was repeatedly used in Spanish times and is preserved in the names of several mountains. The word is used in the name of the Prietos y Najalayegua grant [Santa Barbara Co.] on Sept. 23, 1845. Rancho de los Prietos is mentioned in the Guerra Documents (7:147), and Corral de los Prietos is shown on a *diseño* of San Marcos.

Primer Cañón [Tehama Co.]. The name, meaning 'first canyon', was applied to the land grant on May 22, 1844. This grant is also known as Río (or Arroyo) de los Berrendos.

Primero [Tulare Co.]. The Spanish word for 'first' was applied to the station because it is the first station north of Orosi on the Santa Fe branch to Porterville, built in 1913–14 (Santa Fe).

Prince Island [Del Norte Co.]. The island was named by the Coast Survey about 1900, probably for Francis Prince, a native of New York who settled at Smith River before 1879.

Princeton [Colusa Co.]. The post office, listed in 1858, was named at the suggestion of Dr. Almon Lull, a graduate of Princeton University (E. L. Hemstreet). There is another Princeton in San Mateo Co. *For* Princeton [Mariposa Co.], *see* Mount: Mount Bullion.

Prisoners Harbor [Santa Cruz Island]. The name was put on the map by the Coast Survey, which published a hydrographic sketch of the harbor in 1852. The name may preserve the memory of an interesting historical incident. At one time the Mexican government intended to make California a penal colony; eighty criminals, sent on the *María Ester*, arrived in Santa Barbara in Mar. 1830. When the Californians refused to receive them, many of the prisoners were provided with tools, cattle, fishhooks, and a little grain, and were shipped to Santa Cruz Island. It is most likely that they were landed at this bay because plenty of wood and water could be obtained in its vicinity. When a fire destroyed their possessions, the convicts built a raft and returned to the mainland, landing at Carpinteria (Bancroft 3:48).

Proberta (prō bûr′ tə) [Tehama Co.]. The name was applied to the Southern Pacific station in 1889 after Edward Probert.

Project City [Shasta Co.]. The town started and grew with the Central Valley Project; the name was adopted at a public meeting in the spring of 1939 (W. K. Gaslin).

Promontory. *See* Pedernales, Point.

Prospect Peak [Shasta Co.]. Prospectors for gold have searched hereabouts without success (Steger).

Prospero [Los Angeles Co.]. The land grant is dated May 16, 1843, and was so called because the grantee was Próspero Valenzuela, an Indian.

Prosser Creek [Nevada Co.]. The name is shown on the von Leicht–Hoffmann Tahoe map of 1874. According to Henry T. Williams, *The Pacific tourist* (New York: Williams, 1876), p. 224, a man by that name operated a hotel there "in the early days." One William Jones Prosser, a native of England and resident of Truckee, is listed in the Great Register of 1872, but he could not be identified with the place.

Providence Mountain [San Bernardino Co.]. The name is shown on the maps after 1857 for the entire range of New York Mountains, Mid

Hills, and Providence Mountains. It was probably applied by early travelers and immigrants because they found numerous springs on the range, which is situated between desert valleys. When the USGS mapped the Ivanpah quadrangle in 1909–10, it limited the name to the southern end of the range.

Providencia [Los Angeles Co.]. The land grant is dated Mar. 1, 1843; the name is an example of the rare use of an abstract noun (referring to divine providence) for a land grant. Providencia was also an alternate name of another grant, Cañada de Herrera [Marin Co.].

Prunedale [Monterey Co.]. Presumably named for prune orchards once existing in the area, although these were later replaced by other crops (Clark 1991).

Puckerville. *See* Plymouth.

Pudding Creek [Mendocino Co.]. The Coast Survey charts show Padding River in 1870 and Pudding River in 1871. According to a local story, sailors called Noyo River "Put In Creek" because its mouth provided the only safe anchorage (E. C. Cretser). This name may have been transferred later to the stream north of Fort Bragg and changed to the present form by folk etymology (*see* Noyo River). It is, of course, possible that the name simply arose because of the presence of conglomerate, or pudding stone. There is a Pudding Stone Reservoir in Los Angeles Co.

Pueblo. In Spanish California, *pueblo* was a common generic name corresponding to English "town." The oldest civic community is Pueblo de San José de Guadalupe, founded in 1777, now the city of San Jose. The term *ciudad* 'city' was occasionally used for Los Angeles, and *villa* 'town' was used only once: Villa de Branciforte, modern Santa Cruz (Bowman). The term does not appear among present-day place names. *See* Dos Pueblos.

Puente (po͞o en´ tē, pwen´ tē): town, **Hills** [Los Angeles Co.]. The name, meaning 'bridge,' obviously goes back to the Portolá expedition, which camped at San Jose Creek on July 30, 1769: "In order to cross the arroyo it was necessary to make a bridge of poles, because it was so miry" (Crespí). On the return journey the expedition camped in the same place, on Jan. 17, 1770, and Portolá mentions the plain as Llano de la Puente (in present-day Spanish, 'bridge' is masculine, *el puente*). On Nov. 22, 1819, a *rancho llamado La Puente* 'ranch called The Bridge' is mentioned, and on July 22, 1845, the name was used for the rancho granted to two well-known pioneers of the Los Angeles district, John Rowland and William Workman (*WF* 6:269–70). When the Los Angeles–Colton section of the Southern Pacific was built in 1875, the name was applied to the station. The post office was established Sept. 15, 1884, as Puente. In 1955 the BGN decided for La Puente. The name of the land grant Rincón de La Puente del Monte, in Monterey Co., dated Sept. 20, 1836, is a misspelling of Rincón de la Punta del Monte. *For* Arroyo de las Puentes [Santa Cruz Co.], *see* Cojo.

Puentecita (pwen tə sē´ tə) **Gulch** [Santa Cruz Co.]. Spanish for 'little bridge', diminutive of *puente* 'bridge' (Clark 1986).

Puerta del Suelo (pwâr´ tə del swā´ lō) [Kern Co.]. As the name for a mountain pass, the name was ratified in this form by the BGN in 1981 (list 8103), with the supposed meaning 'passage' or 'entry floor'. This is clearly a hypercorrection of Spanish *puertezuelo* 'little pass', a diminutive of *puerto* 'pass'; *see* Puerto; Puerto Suelo.

Puerto; Puerta. The generic term *puerto* 'port' was frequently used along the coast in Spanish times, but apparently it has not survived in this sense. It also means a 'mountain pass', and Whipple, in 1849, mentions such a Puerto in San Diego Co. **Puerto** (po͞o âr´ tō, pwâr´ tō): **Canyon, Creek** [Stanislaus Co.]: Apparently the feminine form, *puerta* 'door, gate', was used also, since the creek appears as Arroyo de la Puerta on a *diseño* of Rancho del Puerto, granted Jan. 10, 1844, as well as on a *diseño* of the Pescadero grant (1843). Brewer mentions Cañada del Puerto and Puerto Canyon on June 10, 1862: "the canyon comes through by a very narrow 'door,' which gives the name to the valley behind." *For* Puerto Anegado, *see* Mission Bay; *for* Puerto Dulce, *see* Suisun Bay; *for* Boca del Puerto Dulce, *see* Freshwater Lagoon, *also* Carquinez; *for* Puerto Falso, *see* Mission: Mission Bay.

Puerto Suelo (pwâr´ tō swā´ lō) [Monterey Co.]. Name of a pass and a creek, for Spanish *puertezuelo* 'a small pass', from *puerto* 'port, pass'; variously misspelled as Puerto Suello, etc. (Clark 1991); *see* Puerta del Suelo; Puerto. *For* Puerto Zuelo Lagunitas [Marin Co.], *see* Lagunita.

Pujol (po͞o hōl´) [Riverside Co.]. A historic post

office site named for Domingo Pujol, of Spain, a landowner in the 1880s (Gunther). Now also pronounced (po͞o zhōl′).

Pulgas. Although Americans were as much molested by fleas (*pulgas*) as the Spaniards were, they seldom used the word for place names (*see* Flea Valley). In Spanish times, however, quite a number of places were named for the little insect, some of which have survived. **Pulgas** (po͝ol′ gəs): **Ridge, Creek** [San Mateo Co.] were so named because they are on the Pulgas (or San Luis or Cochenitos) land grant, dated in 1795 and again on Nov. 27, 1835. Beechey (2:44) mentions a "farm-house . . . called Las Pulgas" in Nov. 1826: "a name which afforded much mirth to our travellers, in which they were heartily joined by the inmates of the dwelling, who were very well aware that the name had not been bestowed without cause." The soldiers of the Portolá expedition had given the name Ranchería de las Pulgas to a deserted village on Purisima Creek, about 10 miles southwest of the Pulgas rancho, on Oct. 27, 1769 (Crespí, Costansó). **Las Pulgas Canyon** [San Diego Co.] was a *sitio* of Mission San Luis Rey in 1828 (PSP, Presidios 1:98). **Pulga** [Butte Co.] was named when the Western Pacific built the line in 1907. It is a renewal of the name of the old mining camp Pulga Bar, which later was apparently called Big Bar.

Punta. In Spanish times the word *punta* was found frequently in names of less prominent headlands on the coast. The word corresponds exactly to the English geographical term 'point'; hence the Spanish term has survived in only a few places. There is a Punta del Castillo in Santa Barbara Co., a Mesa de la Punta in Riverside Co., a station called Punta in Ventura Co., and a Punta Arena on Santa Cruz Island. The generic term also appears in the names of eight land grants. **Punta Gorda** (po͞on′ tə gôr′ də) [Humboldt Co.]: When Bruno de Hezeta was in Trinidad Bay in June 1775, he gave this Spanish name, meaning 'massive point', to a broad promontory at the mouth. It is not certain whether the name was applied to Table Bluff or to False Cape. On his large map Hezeta places it in the latitude of the former, on his small map in the latitude of the latter (Wagner, p. 458). Later cartographers identify the name with the promontory which is now called False Cape. In 1854 the Coast Survey, trying to save the name Cape Fortunas for False Cape, moved the name Point Gorda southward to the cape that some maps show as Cabo Vizcaino (Wagner, p. 523). In 1870 the Coast Survey restored the Spanish name, Punta Gorda. **Punta de Los Muertos** [San Diego Co.] is a name used since 1782. The term, meaning 'point of the dead', was given because a squadron had anchored at that place and buried a number of sailors carried off by an epidemic. After the American occupation, the morbid name disappeared because "New Town" developed at exactly this place, and its promoters expected it to be very much alive. The old Spanish name was recalled when an explosion on the USS *Bennington*, anchoring off this point, caused the death of sixty-five sailors on July 21, 1905. **Rincón de la Punta del Monte** [Monterey Co.] is the name of a land grant dated Sept. 20, 1836; the word Punta 'point' is given as Puente 'bridge' on the Land Office maps. *For* Punta Falsa de Año Nuevo, *see* Año Nuevo.

Purificacion. The word, referring to the Purification of the Virgin Mary, is found in the name of a land grant, Lomas ('hills') de la Purificación [Santa Barbara Co.], dated Dec. 27, 1844. The word is shown on a *diseño* of the grant, apparently designating a valley.

Purisima (pə ris′ i mə): **Point, Hills, Canyon** [Santa Barbara Co.]. These geographic features were named after Mission La Purísima Concepción (Immaculate Conception of the Virgin Mary), founded by Lasuén on Dec. 8, 1787. *See* Conception; *for* La Purisima River, *see also* Santa Ynez. **Purisima: Creek**, town [San Mateo Co.] bear the name given to an Indian village here in 1786. The name of the land grant La Purísima Concepción [Santa Clara Co.], dated June 30, 1840, does not seem to have survived in a geographical name.

Pushawalla (po͞osh′ wä lə) **Canyon** [Riverside Co.] is described in the *Desert Magazine* of Dec. 1945. The name is perhaps of Indian origin, but its meaning is not known.

Pusher. *See* Dunsmuir.

Putah Creek [Lake, Napa, Solano Cos.]. From Lake Miwok *puṭa wuwwe* 'grassy creek' (Callaghan; cf. Beeler 1974:141). The similarity to Spanish *puta* 'prostitute' is purely accidental. In the records of Mission San Francisco Solano (Sonoma Mission) of 1824, the natives of the place are mentioned with various spellings from Putto to Puttato. In the baptismal records of Mission Dolores an *adulto de Putü* is mentioned in 1817, and the wife of Pedro Putay in 1821 (Arch. Mis. 1:94.81). In 1842 the stream was well known by its name: 'I know that the Rio was called 'Putos.'. . . It is well-known by the name which has been given it" (J. J. Warner, land-grant case 232 ND). The name was probably fixed by William Wolfskill, who named his grant Rio de los Putos on May 24, 1842. In 1843 the name was used in the titles of three other land grants, in one of which the spelling Putas occurs. In the *Statutes* of the early 1850s, in the Indian Reports, and in the Pac. R.R. *Reports*, the spelling of the name is in complete confusion. The present version was applied to a town in 1853, was used in the *Statutes* of 1854, was made popular by the Bancroft maps, and finally was adopted by the USGS.

Putnam Peak [Solano Co.]. The peak was probably named for Ansel W. Putnam, who had settled as a farmer at Vacaville before 1867. *For* Putnam's [Inyo Co.], *see* Independence.

Putos, Río de. *See* Putah Creek.

Pyramid. Some twenty-five features in California—mostly Peaks, Rocks, and Hills—are called Pyramid because of their shape. Among these are two high peaks in Sequoia National Park and in Inyo Co., and Pyramid Head on San Clemente Island. For the application of the name Pyramid Point to Point Lobos, *see* Lobo: Point Lobos [Monterey Co.].

Pywiack (pī′ wē ak) **Cascade** [Yosemite NP]. In Southern Sierra Miwok, *paywayak* is the name for Vernal Falls, from *paywa* 'chaparral' (Callaghan). The name was recorded by John Muir in 1873 and was made official by decision of the BGN (list no. 30).

Q

Quail, of three species, are a familiar sight in California, and the valley quail (*Lophortyx californicus*) is the state bird. It is not surprising that many geographic features are named after them, such as **Quail Valley** [Riverside Co.]. *See* Codorniz.

Quanai (kwə nī′) **Canyon** [San Diego Co.]. From Diegueño *kwa'naay* 'wire grass', a plant used in basket making (Langdon).

Quarry, referring to a source of stone for construction, is a term sometimes found in place names. A quarry on a hill or mountain slope often leads to naming the whole feature after it; but there is also a high mountain, **Quarry Peak**, in Yosemite NP.

Quartz has long been an important mineral in California because gold is often embedded in it. The maps show a number of Quartz Mountains, Hills, and Rocks, as well as several Canyons and Creeks. Two settlements called Quartz are in Butte and Tuolumne Cos., and there is a ghost town, Quartzburg, in Mariposa Co. **Quartzite Peak** [Inyo Co.] is named for a rock composed of quartz.

Quatal (kwə täl′) **Canyon** [Ventura, Santa Barbara Cos.]. For California Spanish *guatal*, a place where *guata* 'juniper' grows (John Johnson); from Luiseño or Gabrielino *wáa'at* (Harrington 1944:38).

Quate; Quati. *See* Cuate.

Quedow (kwə dou′) **Mountain** [Tulare Co.]. Established in 1966 by the BGN (list 6603), replacing the names Credow, Cuidow, and Cuidado Mountain. The origin has not been identified.

Quemado, Spanish for 'burnt', was once used in several place names. The only survivor seems to be **Arroyo Quemado** (kā mä′ dō) [Santa Barbara Co.], mentioned as early as December 1794 and shown on the *diseño* of Rancho Refugio in 1838.

Quentin. *See* San Quentin.

Quercus Creek. *See* Kirker Creek.

Quesesosi [Yolo Co.] was a land grant, sometimes written Guesesosi and also known as Jesús María, dated Jan. 27, 1843. A *diseño* bears the legend TERRENO QUE SE SOLI (on the left side), CITA EN (across the top), and CANTIDAD DE OCHO SITIOS (at the bottom). Put together, this means 'Land that is requested in the amount of eight *sitios*.' It has been suggested that the sequence *que se soli* may have been mistaken for the name of the *terreno*, and that the name Quesesosi arose through this mistake (*WF* 6:82). To be sure, the *diseño* on which this legend was noted was made two years after the name Quesesosi had been used on another *diseño* (*WF* 7:171 ff.); however, earlier maps may have had the same notation.

Quicksilver, i.e. mercury, as a term applied to places in Los Angeles and Sonoma Cos., recalls an era when the discovery of small amounts of quicksilver ore raised hopes of finding large deposits of the valuable metal (*see* Almaden).

Quien Sabe (kyen sä′ bā) **Creek** [San Benito Co.], with the Spanish phrase 'Who knows?', is found in the name of a land grant, Santa Ana y Quien Sabe, dated Apr. 8, 1839. It had been used previously (Apr. 15, 1836) for a grant that was not confirmed. The maps of the Santa Ana y Quien Sabe grant show Cañada de Quien Sabe, and Sierra de Quien Sabe for a part of the Diablo Range. The name is recorded as *arroyo llamado Quien sabe* by Sebastián Rodríguez on Apr. 21, 1828. The name may have been applied in jest.

Quijarral. *See* Guijarral.

Quimby. *See* Denny.

Quinado (kē nä′ dō) **Canyon** [Monterey Co.]. Probably a Spanish adaptation from the name of an Indian village, Mutsun (Costanoan) *kináw* (Clark 1991, after Harrington).

Quincy [Plumas Co.] grew around the hotel that H. J. Bradley built on his American Ranch in the early 1850s. When it became the seat of the newly formed Plumas Co. in 1854, it was called Quincy, after Bradley's native town in Illinois.

Quinliven (kwin′ liv ən) **Gulch** [Mendocino Co.], near Anchor Bay, was named for the Quinliven family, which settled early in the area (BGN, 1962).

Quinn Camp and **Peak** [Sequoia NP] were named for Harry Quinn, a native of Ireland, who came to California in 1868 and became a sheep raiser in Tulare Co. (Farquhar).

Quintin. *See* San Quentin.

Quinto (kin′ tō) **Creek** [Stanislaus Co.], containing the Spanish word for 'fifth', may have referred to a share of land, or it may have been applied here to indicate that this creek was the fifth one from some point on the road. It is spelled Kinto on the Land Office map of 1879.

Quiota (kē ō′ tə) **Creek** [Santa Barbara Co.]. For Mexican Spanish *quiote* 'Yucca whipplei' (John Johnson), from Aztec *quiotl* 'sprout'.

Qui-quai-mungo Range. *See* San Gabriel.

Quito (kē′ tō) [Santa Clara Co.] was the name of a land grant dated Mar. 12, 1841. In 1878 it was given to a siding on the Southern Pacific line between San Jose and Los Gatos. Quito is Spanish for 'quits' (as well as being the capital of Ecuador). However, the name may have been the result of an error: the grant was also called Tito, for an Indian who occupied a part of the former Mission Santa Clara. *For* Arroyo Quito, *see* Campbell Creek [Santa Clara County].

Quivira. *See* California; New Albion.

R

Rabbit. The term is used by many Americans to include not only several species of true rabbits but also hares (genus *Lepus*). Probably more than fifty creeks and canyons, as well as a number of flats and islands, are named for the little animal. The mining camp at the site of modern La Porte [Plumas Co.] had been named Rabbit Creek for the presence of the "snowshoe rabbit" (i.e. the snowshoe hare, *Lepus americanus*); but the settlers became indignant when the Post Office Dept. tried to bestow the name Rabbit Town upon them. Hot Springs Mountain [San Diego Co.] was earlier called Rabbit Peak, adapted from Cupeño *sú'ish péki'* 'rabbit's house' (Brigandi). *See* Conejo.

Raccoon: Strait, Shoal [San Francisco Bay]. The raccoon (*Procyon lotor*) is a common animal in California but seems to have given its name to few places. The channel between Angel Island and Marin Co. was named for the British warship *Raccoon*, which anchored in San Francisco Bay in 1814 to make repairs (PSP 19:368). Raccoon Strait is shown on Tyson's map of 1850 and on the charts of the Coast Survey after 1858. The word "raccoon" was originally introduced to English from Virginia Algonquian by the famous John Smith, as *aroughcun* (Cutler, p. 18).

Racetrack [Death Valley NP]. "The Racetrack in Death Valley is a circular dry lake in the northwest corner of the Monument immediately adjacent to Ubehebe Peak. It is so named because it is almost a true circle in shape and has two rock formations protruding, which by stretch of imagination could appear to be a judges' stand and a grandstand" (T. R. Goodwin).

Rackerby [Yuba Co.]. The place was known as Hansonville for James H. Hanson, who had settled here in 1851. In 1884, William M. Rackerby, a pioneer of 1849, came to Hansonville as a merchant and rancher; when the Hansonville post office was discontinued in 1892, a new post office was established with Rackerby as postmaster, and it was named for him.

Radec (rā´ dek) [Riverside Co.]. The post office was established about 1885 and given the name Cedar spelled backward; it closed in 1901 (Brigandi). *See* Sniktaw.

Rademacher (rä´ də mā kər): station, **Mountains** [Kern Co.]. The Southern Pacific station, on the branch line built in 1909 from Mojave to Owens Lake, was probably named for Alexander Rademacher, a native of Germany and resident at Greenwich in 1879.

Radio Hill [Plumas Co.]. This sharp hill just outside of Quincy used to have no name, except that it was occasionally called Cemetery Hill because of the cemetery there. After the USFS established its key radio station for the region on top of this hill about 1945, it became generally known as Radio Hill (Stewart).

Rae Lake [Kings Canyon NP]. The lake was named by R. B. Marshall in 1906 for Rachel ("Rae") Colby, wife of William E. Colby, the well-known conservationist and mountaineer (Farquhar).

Rafferty: Creek, Peak [Yosemite NP]. Named in 1895 by Lt. Nathaniel F. McClure, for Capt. Ogden Rafferty (1860–1922), an army surgeon (Farquhar).

Rag, Ragged. These have been used as specific terms in place names, either to describe physical features—mainly Peaks, Points, and Canyons—or to indicate the dilapidated condition of a camp or its settlers. Kaweah was once commonly called Ragtown because most of the inhabitants lived in makeshift tenthouses. Ragged(y)ass Gulch [Siskiyou Co.] became so known because the miners in this gulch looked more dilapidated than the ordinary miners of the period. It is shown as Raggedass Creek on the map of Klamath National Forest (Schrader). The nearby Hungry Creek seems to confirm the story of hardship.

Railroad Canyon [Riverside Co.] was so named

because the Santa Fe Railroad passed through it. After the railroad bed was washed out one winter, the line was rerouted via Corona, but the name remained and was still applied to the reservoir (Wheelock). The reservoir is now, however, called Canyon Lake (Brigandi).

Railroad Flat [Calaveras Co.]. The camp was settled in 1849 and later so named because a short track conveyed the lorries with gold ore and waste to and from the diggings. The place had a post office in 1857–58, reestablished on Mar. 17, 1869.

Raimundo, Cañada de [San Mateo Co.]. The name of a land grant dated Aug. 4, 1840, also spelled Raymundo. One Raimundo, a native of Baja California, is mentioned on June 20, 1797, as having been sent out after Indians who had run away from Mission San Jose (PSP 15:16). On a plat of Pulgas rancho (1856), Laguna Grande or Lake Raymundo seems to be what is now called Crystal Springs Lake.

Rainbow Lake [Shasta Co.]. The artificial lake formed from Moon Creek was so named because it is stocked with rainbow trout (Steger).

Rainbow Mountain [San Bernardino Co.]. The mountain west of Ivanpah Range was so named because the stratified volcanic rock resembles the rainbow (Gill).

Raines Valley [Fresno Co.]. The valley east of Centerville was named for James Raines, a settler of 1863—who, according to Doyle, later served a term in a penitentiary and was finally lynched.

Raisin [Fresno Co.]. The post office was established about 1906 and named for the chief product of the district.

Raker Peak [Lassen NP]. Named for John E. Raker, a congressman from 1910 to 1926, who introduced the bill to create Lassen Volcanic National Park.

Raliez Valley. *See* Reliz: Reliez Valley.

Ralston Peak [El Dorado Co.]. The name for the peak above Echo Lake appears on atlas sheet 56-D of the Wheeler Survey, which mapped the region in 1876. It was probably given (not necessarily by the Wheeler Survey) in memory of William C. Ralston, who died on Aug. 27, 1875. *See* Modesto.

Rambaud (ram′ bō) **Peak** [Kings Canyon NP]. Named for Pete Rambaud, a Basque sheepman, who brought the first sheep into the region of the Middle Fork of Kings River from the Inyo side, through Bishop Pass, in 1877 (Farquhar).

Ramola. *See* Romoland.

Ramona [San Diego Co.]. The name was given to this community, as well as to several others, soon after 1884, when Helen Hunt Jackson's sentimental novel *Ramona*, about a young Indian woman, was at the height of its popularity. Jackson used the name of her friend Ramona Wolf, wife of the storekeeper at Temecula, although Mrs. Wolf was not Indian (Gunther). The Spanish name Ramona is the feminine counterpart of *Ramón* 'Raymond'. In 1886 the new townsite of Nuevo was given the name of Ramona for its post office but lost it a few months later to a new tract in Los Angeles Co. When the latter closed in 1895, Nuevo took back the name and has been called Ramona ever since (Brigandi). Two railroad stations are called Ramona [Sacramento Co.] and Ramona Park [Los Angeles Co.]. Riverside Co. has Ramona Hot Springs and Ramona Indian Reservation. *See* Alessandro; Romoland.

Rampart Pass. *See* Siberian Pass.

Ramshaw Meadows [Tulare Co.]. Named for Peter Ramshaw, a stockman in this region from 1861 to 1880 (Farquhar).

Rana. The Spanish word for 'frog' was sometimes used in place naming. Ciénega ('marsh') de las Ranas was the original name of the San Joaquin grant [Orange Co.], dated Apr. 8, 1837. A tributary of Tularcitos Creek [Monterey Co.] is called **Rana** (rä′ nə) **Creek**; and there is a Santa Fe station Rana in San Bernardino Co.

Ranchería (ran chə rē′ ə). The Spanish word, originally designating 'a collection of ranchos or huts', has taken the meaning of 'hamlet' or 'village' in American Spanish. It was invariably used in Spanish California for Indian villages. It is still used to refer to Indian communities; and Prof. Leanne Hinton, a linguist of the University of California, sometimes tells people she is from "Berkeley Ranchería." The term is also preserved in many geographic features, mainly creeks on the banks of which there was once a ranchería; the best known is the cluster Rancheria: Creek, Falls, Mountain, and Trail, in Yosemite NP.

Ranch House [San Diego Co.]. The Santa Fe station was originally called Margarita because it was on the Rancho Santa Margarita y las Flores. Before 1900 the name was changed to Ranch House because it was near the adobe house of the Picos, onetime owners of the rancho (Santa Fe).

Ranchita (ran chē′ tə) [San Diego Co.]. For Spanish *ranchito* 'little ranch', diminutive of *rancho*. When a post office station was requested in 1935, an error resulted in a change of gender (Stein). There is a **Ranchito Creek** in Monterey Co.

Rancho. In American Spanish the word *rancho* was applied originally to a hut or a number of huts in which herdsmen or farm laborers lived. In Spanish times the word was often used in this sense in reference to the farms of the missions, pueblos, and presidios. A presidio rancho was called *rancho del rey*—and later, in Mexican times, *rancho nacional*. In Mexico the word acquired the meaning of 'small farm'; but at the fringe of the Spanish empire, in the present southwestern United States, the term evolved as the designation for a grazing range. When the private land grants were separated from the public domain, the word *rancho* became synonymous with 'landed estate'—called *hacienda* in other Spanish American countries—and was in fact used as a generic geographical term. The western American term "ranch" is adapted from *rancho*. **Rancho Cordova** (kôr dō′ və) [Sacramento Co.]: Earlier named Cordova Village because the Cordova Vineyards (originally named for the Spanish city, Córdoba) were situated in the center of the Rancho de los Americanos grant. The post office was established on May 16, 1955, and given the name Rancho Cordova. **Rancho Murieta** [Sacramento Co.] was perhaps named for Joaquín Murieta; *see* Murieta. **Rancho Palos Verdes** [Los Angeles Co.] was named for the Palos Verdes peninsula (*q.v.*) on which it is located. **Rancho Rinconada** [Santa Clara Co.]: Spanish *rinconada* is an external corner, like that formed by two streets; it is derived from *rincón* 'internal corner, nook'. **Rancho Santa Fe** [San Diego Co.] was a name given in 1906, when the Santa Fe Railroad purchased the San Dieguito Ranch for experimental planting of eucalyptus trees. In 1927 the railroad sold the ranch to promoters, who subdivided it but kept the old name. **Rancho Mirage** [Riverside Co.] received its official name when the post office was established on Feb. 1, 1961. **Rancho California** [Riverside Co.] is a land development name given to the former Vail Ranch in 1964 (Gunther). **Rancho Santa Margarita** [Orange Co.] is the name of a modern community dating from 1986, not of a Spanish rancho. The original name selected was Santa Margarita, in honor of the rancho Santa Margarita y las Flores of Spanish times; but when it was learned that a Santa Margarita existed in San Luis Obispo Co., the present name was chosen (Brigandi).

Randolph. *See* Sablon.

Randsburg; Rand Mountains [Kern Co.]. The name was applied as a good omen to the town and the district in 1895, after Witwatersrand in the Transvaal—commonly called "The Rand"—a rocky ridge that has been known since 1886 as one of the richest gold-mining districts in the world.

Rankin Peak [Los Angeles Co.]. Named for the late Edward P. Rankin, a resident of Monrovia, who was actively interested in the mountains of the area (BGN, 1949).

Ransom Point [Contra Costa Co.]. Named for Leander Ransom, who, as a deputy U.S. surveyor, established the Mount Diablo base and meridian lines in 1851.

Raspberry Lake [Siskiyou Co.] is named for the native berry, *Rubus leucodermis*.

Rattlesnake. Some two hundred features on the maps of the state are named for the venomous serpent (genus *Crotalus*), and many more are so named locally. The name is usually connected with a creek or a canyon, but there are also Rattlesnake Ridges, Mountains, Buttes, Meadows, Gulches, Valleys, and Points; a Rattlesnake Bridge joins El Dorado and Placer Cos. In early mining days the name was also used for communities: Rattlesnake Flat is shown on Gibbes's map of 1852, south of the South Fork of Tuolumne River; and Rattlesnake Bar, on the North Fork of the American River, is mentioned in the *Statutes* of 1854 (p. 222). Rattlesnake Creek [Humboldt Co.] is said to have been named because a winding road near the creek was first called Snake Road, then Rattlesnake Road. *For* Rattlesnake Butte [Modoc Co.], *see* Opahwah Butte; *for* Rattlesnake Island [Los Angeles Co.], *see* Terminal Island.

Raven. The bird *Corvus corax* is common in California. **Ravendale** [Lassen Co.] was established as a post office in 1910; *see also* Ravenswood.

Ravenna [Los Angeles Co.]. This typical railroad name was applied to the station when the last section of the valley route was built in 1876. There were at that time a number of Ravennas in the eastern United States, all probably named after the Italian city.

Ravenswood: Point, Slough [San Mateo Co.]. The names preserve the name of the town of Ravenswood, which had been laid out in the 1850s when the Central Pacific was expected to cross lower San Francisco Bay, and which was probably named after one of the several Ravenswoods in the East. When the railroad bridge was finally built in 1920, the town had disappeared, but it left its name in a railroad siding.

Rawhide [Tuolumne Co.]. The place west of Jamestown seems to be the only survivor of the once-popular name for California gold mines. The rich mining town of the 19th century was earlier called Rawhide Ranch.

Raymond [Madera Co.]. The name was applied to the terminal of the Southern Pacific spur from Berenda about 1885, for Walter Raymond of the Raymond-Whitcomb Yosemite Tours, which started at this point (F. E. Knowles).

Raymond, Mount [Yosemite NP]. Named by the Whitney Survey in 1863 for Israel W. Raymond (1811–87), one of the chief proponents in the campaign to set aside Yosemite Valley for public enjoyment. For many years he was a member of the state commission that managed Yosemite Valley before it became a national park (Farquhar).

Raymond Peak [Alpine Co.]. Named in 1865 by the Whitney Survey for Rossiter W. Raymond, a graduate of the famous mining school of Freiberg, Germany, and commissioner of mining statistics in the U.S. Treasury Dept. Reed's map of the county, 1865, shows a Raymond City north of the peak.

Raymundo, Cañada de. *See* Raimundo.

Raza de Buena Gente, Isla. *See* Terminal Island.

Read. *See* Mill: Mill Valley.

Reading (red′ ing) **Peak** [Lassen NP], earlier known as White Mountain, was named in memory of Maj. Pierson B. Reading by the BGN in 1943 (*see also* Redding; Shasta). Next to John Bidwell and John Marshall, Reading was the best known of Sutter's men and was the real pioneer of Shasta Co. He was grantee of Rancho Buenaventura in 1844, participated in the Bear Flag Revolt, and discovered gold in the Trinity region in July 1848. He is still commemorated by Reading Peak, Reading Adobe, Readings Bar, and Fort Reading; Readings Springs and Diggings have vanished. **Reading Rock** [Humboldt Co.] was also named after Major Reading, by the Coast Survey in 1849 or 1850 (*Report* 1862:341). The present spelling was confirmed by the BGN in 1970 (list 7004).

Real de las Aguilas. *See* Las Aguilas.

Recess Peak [Fresno Co.]. Named because of its proximity to the First Recess of Mono Creek, which was discovered with the other three Recesses by Theodore S. Solomons in 1894 (Farquhar).

Reche (rech′ ē): **Canyon, Mountain, Wells** [San Bernardino Co.]. The places were named for Charles L. Reche, an early homesteader who dug the first well in the area (BGN, 1960).

Reconnaissance Peak [Plumas Co.]. The peak was so named because of the timber reconnaissance work done there in 1914–15 (D. N. Rogers).

Rector Canyon [Napa Co.]. Named for a man who settled here before 1847 (Doyle), possibly John Potter Rector, whose name is recorded in the Great Register of Napa Co., 1867–68.

Red. Next to "black," "red" is the most common adjective of color used in place names. The maps of the state have more than 550 names that include this adjective, and many more are used locally. Usually the word is found to describe hills, peaks, cliffs, and points. Most of the Red Lakes and Creeks (unless they contain the nickname of a person) are doubtless named after red orographic features or because the soil looks red, but there are notable exceptions: the Red River of California, as the Colorado River was sometimes called, was so named because its water appears "reddish." Often the adjective does not describe the feature but modifies another term: Redrock Mountain, etc. [Los Angeles Co. and elsewhere]; Red Bridge Slough [San Joaquin Co.]; Red Fox Canyon and Red Rover Canyon [Los Angeles Co.]; Red Reef Canyon [Ventura Co.]; Red Pass Lake [San Bernardino Co.]; Red Clover Creek [Plumas Co.]; Redbird Creek [El Dorado Co.]; and many others. **Red Rock** [San Francisco Bay] was known in Spanish times as Moleta for its conical shape, referring to a 'muller' (a type of stone pestle) —misspelled "Molate" by Beechey in 1826. It was also known as Golden Rock because pirates were supposed to have buried vast treasures there. It was possibly the island mentioned by Padre Payeras in 1819 (Docs. Hist. Cal. 4:341 ff.): *otra mucho más chica* [isla] *cerca Sn. Rafael llamada del oro* 'another much smaller island near San Rafael called golden'. The name Golden Rock is the official name (*Statutes* 1850:60), but as early as 1848 the present name was in use (Ord's map). The Coast Survey kept the name Molate Rock until 1897. **Red Bank Creek** [Tehama Co.] is shown as Baranca Colorada on Bidwell's map of 1844, a name also applied to the land grant Barranca Colorada 'red ravine', dated Dec. 24, 1844. In the 1850s the creek was sometimes called Red Bluff Creek (*Statutes*, 1851; Gibbes's map, 1852). *For* Red Bank Creek [Butte, Yuba Cos.], *see* Honcut Creek. **Red Cap: Bar, Creek** [Humboldt Co.] were named by Gibbs's party after an Indian elder who wore a red woolen cap (1851; Schoolcraft 3:148–49). **Red Lassic Peak** and **Creek** [Trinity Co.]: The color refers to the peak—not to Lassik, the last chief of an Athabaskan tribe—and is used to distinguish it from Lassik and Black Lassik Peaks (*see* Lassic). **Red Lake Peak** [Alpine Co.] was formerly called Red Mountain, and the small, marshy, half-dry lake at its foot was named Red Lake after the mountain (Surveyor General, *Report*, 1856: 105). The name of the mountain was apparently lost, but it stuck to the lake, which was made into a reservoir. When it became necessary to have a name for the mountain, it was named after the lake: Red Lake Peak! However, according to George Stewart, the lake was originally Reed Lake (*see* Reed). There is another Red Lake and a Red Lake Mountain in Shasta Co. **Red Slate Mountain** [Mono Co.] was named by the Whitney Survey and recorded on Hoffmann's map of 1873. **Red and White Mountain** [Fresno Co.] was named by Theodore S. Solomons in 1894. When Lincoln Hutchinson and his party made the first ascent on July 18, 1902, they kept the name because "it is peculiarly descriptive of the great peak of red slate fantastically streaked with seams of white granite" (*SCB* 4:201). **Red Box Divide** [Los Angeles Co.] marks the boundary between the watersheds of Arroyo Seco and San Gabriel River; it was given this name because a large red box used by forest rangers for storing firefighting equipment was visible from the road (AGS: Los Angeles, p. 298). Many communities have been named after nearby "red" physical features, e.g. **Red Bluff** [Tehama Co.]: Early in May 1850, Bruff (pp. 789, 794) refers to the plans of Sashel Woods and Charles L. Wilson to lay out the town at Red Bluffs or "the Bluffs." In the same year, the town Red Bluffs is mentioned in the *Statutes* (p. 62). According to Bancroft (6:496), the earlier name was Leodocia. In 1853 it was known as Covertsburg (*San Francisco Alta California*, Jan. 15, 1853), but Eddy's map of 1854 shows the town as Red Bluffs. **Redlands** [San Bernardino Co.]: The district formerly known as Lugonia was developed after the California Southern built a line to connect with the Santa Fe in 1885 (*see* Lugo). The present name, descriptive of the soil, was given to the town, which was platted in 1887 by E. G. Judson and Frank E. Brown. **Redbanks** [Tulare Co.] was named by the Santa Fe in 1914 after the Redbanks Orchard Company, so called because the soil in the district was red. **Red Mountain** [San Bernardino Co.]: The post office was named in 1929 after the nearby reddish-colored mountain; *see* Osdick. *For* Red Mountain [Kings Canyon NP], *see* Pinchot, Mount. *For* Red Rock [San Bernardino Co.], *see* Topock.

Redding [Shasta Co.]. A town laid out south of the present city was first called Latona, and later Reading, in honor of Pierson B. Reading (*see* Reading Peak, Reading Rock): "While upon this matter we have to record an objection to the name 'Latona.' It is not a proper name for a town or anything else that we know of. As well as we can remember, Latona was the name of one of the high old goddesses of Grecian mythology, who conducted herself in a very improper manner. We would take the liberty of suggesting the name of 'Reading' as by far the more appropriate" (*Shasta Courier*, Nov. 2, 1861; Boggs, p. 397). After the Central Pacific acquired the right-of-way, the present town was laid out in 1872 by B. B. Redding—land agent of the company, former secretary of state of California, and future state fish commissioner—and named for him. The local people, however, wished to keep the pioneer name, and in Jan. 1874 the legislature enacted solemnly: "That the name of the Town of Redding, Shasta County, shall hereafter be known and spelled Reading, in honor of the late Maj. Pearson [*sic*] B. Reading, the pioneer of Shasta County" (Boggs, p. 595). This did not end the confusion, for the railroad refused to recognize the change. In the end the friends of the living railroad official were more influential than those of the dead pioneer, and in Apr. 1880 the legislature changed the name back to Redding (ibid., p. 666). Mr. Redding showed his appreciation by donating a fine 245-pound bell to the local Presbyterian church (*Sacramento Record Union*, Mar. 9, 1881).

Redding Canyon [Inyo Co.]. Named for John Redding, an old miner who lived here in 1879 (Robinson).

Red Dog [Nevada Co.]. The once famous mining town was named in the early 1850s by Charlie Wilson after his former home, Red Dog Hill, Illinois (H. P. Davis); that town in turn may have been named for the popular card game, or in allusion to "Red Dog" banks and bank notes—a well-known term current about 1850, synonymous with "wildcat." Red Dog is mentioned in Brewer's Notes of Nov. 1861. It had a post office from 1855 to 1869. The former name was Chalk Bluff.

Redinger (red′ ing gər) **Lake** [Fresno Co.]. Named in 1955 for Daniel H. Redinger, for many years resident engineer of the Big Creek hydroelectric project of the Southern California Edison Company.

Redlands. *See* Red.

Redondo (rə don′ dō) **Beach** [Los Angeles Co.]. The city was founded in 1881 and incorporated in 1892 (Santa Fe). Although the city is on the territory of Rancho San Pedro, the name was doubtless derived from the adjacent Rancho Sausal Redondo 'round willow grove'. The post office was established on July 15, 1889. The adjective *redondo* 'round' was once frequent in the names of bays, lakes, and valleys. The part of San Francisco Bay now called San Pablo Bay was originally known as Bahía Redonda (Crespí, 1772; Cañizares, 1776, 1781).

Red Rock. *See* Red.

Reds Meadow. *See* Sotcher Lake.

Redwood. This name is popularly applied to two species: *Sequoia sempervirens*, the coast redwood, and *Sequoiadendron gigantea*, the "big tree" of the Sierra. The term is included in more than fifty place names in the coastal habitat, although in some places the name-giving redwoods have long since disappeared. As early as 1769 the distinctiveness of the tree and its abundance were noticed by Crespí, who already used the term *palo colorado* (pp. 211, 232–33). **Redwood Creek** and **Canyon** [Contra Costa Co.] are among the places that were known in Spanish times as *los palos colorados*. **Redwood City** [San Mateo Co.] was earlier known as Red Woods City (1854), from Red Woods Embarcadero (1850), from The Redwoods, applied in 1830s–1840s to the nearby redwood forest and lumbering area (Alan K. Brown). In the early 1850s it was known as Mezesville, for S. M. Mezes. **Redwood Empire**, a popular name for the area traversed by the Redwood Highway from Marin to Del Norte Cos., was first used by Aubrey Drury in 1915. **Redwood Creek** [Humboldt Co.] became well known in 1963 when Paul A. Zahl of the National Geographic Society discovered here the tallest redwoods ever measured. As for the Sierra redwood, its name occurs frequently on the Tehipite atlas sheet [Tulare, Fresno Cos.]: **Redwood Meadow, Creek, Canyon**, and **Mountain** are shown. *See* Palo; Sequoia.

Redwood Memorial Groves. The efforts of the Save-the-Redwoods League established many groves in state parks and in other redwood reserves. Some of these memorialize organizations, such as the California Federation of Women's Clubs (1931), Garden Clubs of America (1931) and Native Daughters of the Golden West (1930) in the Humboldt Redwoods State Park; California Garden Clubs (1949), California Real Estate Association (1953), National Council of State Garden Clubs (1949), and Rotary Clubs (1952) in the Prairie Creek Redwoods State Park; and Daughters of the American Revolution (1949) in the Jedediah Smith Redwoods State Park. Many other groves memorialize individuals, some of which are listed separately in this book.

Reed. This term for a type of water plant, *Phragmites* sp., is little used in California, and is rarely found in place names; instead, the Mexican Spanish word *tule*, from Aztec *tollin*—which originally refers to the bulrush, *Scirpus* sp.—is used almost exclusively (*see* Tule). The former Reed Lake near Carson Pass in Alpine Co., apparently the only sheet of water named originally because of the broad margin of reeds, was renamed Red Lake (Stewart); *see* Red: Red Lake Peak.

Reed [Marin Co.]. When the Northwestern Pacific Railroad was built in the 1870s, the station was named for John Reed (or Read, or Reid), owner of the Corte Madera Rancho, on which the station is situated.

Reedley [Fresno Co.]. Named for Thomas L. Reed, a veteran of Sherman's march to the sea, who gave half his holdings to the city in 1888. Since Reed objected to the use of his name for the city, the directors of the Pacific Improvement Company added the suffix.

Reeds Springs. *See* Santa Catarina Springs.

Reefer [Kern Co.] is railroad slang, referring to refrigerator cars, which were used here for housing (Darling).

Reflection Lake. There are several lakes so named, the best known of which is the one in Lassen National Park, reflecting Lassen Peak. Reflection Lake in Kings Canyon NP, one of the most beautiful lakes in the High Sierra, was named by Howard Longley and his party in 1894 (Farquhar).

Refugio. The Spanish word for 'refuge' was repeatedly used as a place name; it appears in the names of land grants in Santa Barbara, Santa Clara, and Santa Cruz Cos. **Refugio** (rə fyōō′ jē ō) **Creek** and **Valley** [Contra Costa Co.] are shown on the chart of the Coast Survey. **Cañada del Refugio** [Santa Barbara Co.] preserves the name of the land grant Nuestra Señora del Refugio, awarded provisionally in Nov. 1794 and definitively in July 1834. The name is also preserved in Refugio Pass.

Regulation Peak [Yosemite NP]. In 1895, Lt. Harry C. Benson, accompanied by a trumpeter named McBride, was placing copies of the park regulations on trees in the park; McBride suggested the name Regulation Peak for a peak between Smedberg Lake and Rodgers Lake, and Benson put it on his map of 1897. On the 1901 edition of the Mount Lyell atlas sheet, the name was put on the wrong peak; to the original Regulation Peak, thus left vacant, the name of Volunteer Peak was assigned, and it still goes by that name (Farquhar).

Reiff (rēf′) [Lake Co.]. Named for John Reiff, on whose ranch a post office was established on May 18, 1881.

Reina de los Angeles. *See* Los Angeles.

Reinhardt Redwoods [Contra Costa Co.]. Named in 1941 by the East Bay Regional Park Board, in honor of Aurelia Henry Reinhardt, for many years president of Mills College (R. E. Walpole).

Reinstein (rīn′stīn), **Mount** [Kings Canyon NP]. Named by R. B. Marshall for Jacob B. Reinstein, a charter member of the Sierra Club, and from 1897 until his death in 1911 a regent of the University of California (Farquhar).

Reister (rē′ stər): **Knoll, Canyon, Rock** [Lake Co.]. The knoll was named for George Reister, an old settler.

Relief: Valley, Creek, Peak, Reservoir, Camp [Tuolumne Co.]. The valley was named in the early 1850s when an immigrant train, having abandoned its wagons, found shelter here, sent a party ahead, and got help (Rensch-Hoover, p. 502). The broad valley marked Relief Valley on the Dardanelles atlas sheet is not the original Relief Valley. This was the narrow valley 5 miles north, mentioned in Hutchings' *Illustrated California Magazine* in May 1858 (2:494). On Hoffmann's map of 1873 the present Relief Valley is called Green Flat. The Wheeler Survey (atlas sheet 56–B) is responsible for the naming of the creek and the peak, and for the transfer of the name from the former to the present Relief Valley. The name has been applied to several other geographic features. There is a small settlement, Relief, in Nevada Co., east of North Bloomfield, doubtless named after Relief Hill. According to an item in *CHSQ* 10:351, this hill was so named because James F. Reed here came to the relief of nineteen members of the Donner Party in 1847.

Reliz. Mexican Spanish *reliz* 'landslide' was repeatedly used in place naming and has survived in at least two places. **Reliez** (rə lēs′) **Valley** [Contra Costa Co.]: This is the local spelling for the valley west of Walnut Creek; the name is elsewhere spelled Raliez. **Reliz Creek** [Monterey Co.] has sometimes been spelled Release Creek by popular etymology.

Represa (ri pres′ ə) [Sacramento Co.]. The post office for Folsom prison was established about 1895. It is not certain whether the Spanish name, which means both 'dam' and 'restriction', was chosen because of its sound, or because there was a dam nearby in the American River, or because someone thought it was appropriate for a prison.

Requa (rek′ wä) [Del Norte Co.]. From Yurok *rek'woy* 'river mouth' (Waterman, p. 231; Robins).

Rescue [El Dorado Co.]. Among the names submitted by townspeople about 1895, the Post Office Dept. chose the one suggested by Andrew Hare, who owned a mine called Rescue (State Library).

Reseda (rə sē' də) [Los Angeles Co.]. The name Reseda, the botanical name for the garden plant mignonette (*Reseda odorata*) was originally applied to a station on the Southern Pacific branch line that was built from Burbank to Chatsworth about 1895. After 1920 the name was transferred to the station of the Pacific Electric previously known as Marion.

Reservation Point [Los Angeles Co.]. So named in 1915 when the U.S. Treasury Dept. acquired the site east of the entrance to the inner harbor of San Pedro and established a quarantine station there.

Respini (res pē´ nē) **Creek** [Santa Cruz Co.]. Jeremiah Respini, from Switzerland, developed a dairy near here in the late 19th century (Clark 1986).

Resting Spring [Inyo Co.]. On Apr. 29, 1844, Frémont had named the place Hernandez' Spring in commemoration of a Mexican traveler murdered here by Indians. In the 1850s the spring was given the present name because Mormon immigrants bound for San Bernardino stopped there to rest and recuperate.

Return Creek [Yosemite NP]. The name appears on Wheeler atlas sheet 56-D. Its application was doubtless prompted by some incident, as were a number of other names given by the Wheeler Survey.

Reversed: Creek, Peak [Mono Co.]. "The ancient drainage has been reversed by the deposition of morainal debris; we have therefore called the stream draining June and Gull lakes, Reversed Creek" (USGS *Report*, 1887: 343). The peak is not at all reversed but takes its name from the creek.

Revolon (rev´ ə lon) **Slough** [Ventura Co.]. Jean-Marie Revolon, born in France, was farming here in 1875 (Ricard).

Rey. The term *del Rey*, found in geographical nomenclature in Spanish times, designated a place belonging to the king, i.e. national or public property. After Mexico became independent, the term was officially changed to *de la Nación* or *nacional;* the old term survived, however, and is even today often used by realtors in naming new subdivisions.

Reyes (rāz, rā´ əz, rā´ əs), **Point** [Marin Co.]. The cape is probably the one discovered by Cabrillo on Nov. 14, 1542, and named Cabo de Pinos. The Vizcaíno expedition passed the point on Jan. 6, 1603, the day of *los reyes magos*, the "three holy kings." Finding shelter in the present Drakes Bay, they named it Puerto de los Reyes, a name that did not stick. The point was probably named at the same time and appears quite regularly as Punta de los Reyes on the maps of the following centuries. On Mar. 17, 1836, and again on Jan. 4, 1842, the name was given to land grants. Hood's map (1838) misspells the name as Point Keys. The present hybrid form is shown on Wilkes's map of 1841. The Spanish version is shown as late as 1860 (Goddard), but the Coast Survey and most maps have used the modern form since 1855. **Point Reyes**: The name Point Reyes Station was applied to the former Olema station of the Wells Fargo Company on Apr. 1, 1883. The post office, Point Reyes, is listed in 1887; when a new post office was established on the headland and named Point Reyes, the old station south of Tomales Bay was renamed Point Reyes Station. Both post offices have been listed since 1892. *For* Pico y Cerro de Reyes, *see* Tamalpais; *for* Río de Reyes, *see* Kings River.

Reyes Peak [Ventura Co.]. The peak may have been named for Jacinto D. Reyes, a resident at Cuyama in 1892.

Reynolds Peak [Alpine Co.]. Named in memory of G. Elmer Reynolds, for many years editor of the *Stockton Record* and long active in forest conservation (BGN, 1929).

Rheem [Contra Costa Co.]. Donald L. Rheem, owner of the property on the east slope of Mulholland Hill, started the development in the fall of 1944 and called it Rheem Center.

Rhett Lake. *See* Tule: Tule Lake [Siskiyou Co.].

Rhodonite Creek [Siskiyou Co.]. Named for a mineral, manganese metasilicate, sometimes used as an ornamental stone.

Rialto [San Bernardino Co.]. The colony was founded and named in 1887 by a group of Methodists from Halstead, Kansas. The Rialto in Venice, Italy, corresponds to Latin *rivus altus* 'deep canal'; it gradually became synonymous with a business center, and this accounts for its popularity as a place name in various parts of the world. Bolinas Bay and Bolinas Lagoon are shown as Rialto Cove on Ringgold's General Chart of 1850.

Ribbon: Fall, Fall Creek [Yosemite NP]. The Whitney Survey maps record the high but trickling fall as Virgin Tears Creek and Virgin Tears Fall. According to Clarence King, p. 136, James M. Hutchings is responsible for this sentimental name: "We [camped] at the Head-waters of a small brook, named by emotional Mr. Hutchings, I believe, the Virgin's Tears, because from time to time from under the brow of a cliff just south of El Capitan there may be seen a feeble waterfall. I suspect this sentimental pleasantry is intended to bear some relation to the Bridal Veil Fall opposite." However, it was apparently not Mr. Hutchings who was responsible for the Virgin Tears; Hutchings gave the falls the more commonplace name, Ribbon Fall, which is first recorded on Wheeler's Yosemite atlas sheet, 1870.

Ricardo [Kern Co.]. So named for the innkeeper, an Indian, when this was a stop on the old stage route from Bishop to Los Angeles.

Rice: station, **Air Base** [San Bernardino Co.]. The station of the Parker branch of the Santa Fe was named before 1919, for Guy R. Rice, chief engineer of the California Southern Railroad (Santa Fe).

Riceville. *See* Corning.

Rich. Combinations with this adjective were very popular in mining days and have retained their popularity, although they are applied no longer to gold deposits but to fertile soil. **Rich Bar** [Plumas Co.], on the North Fork of the Feather River, is the only survivor of several mining camps so named. The gold deposits that three Germans found in Aug. 1850, at the time of the Gold Lake excitement, turned out to be one of the most spectacular discoveries of the Gold Rush. Bruff records the news on Aug. 18: "Brittle, from Myer's diggings, gives us a rich account of the mining there. . . . He says that the spot where Lassen and comrades prospected, and found only a trace of gold, was yielding not less than 50 dolls. worth pr day to each miner, and some had taken out 60 lbs. of gold in a day. That 3 men there had made in three days an average of $2,000 each, and then sold out for several thousand dollars more, each, and left for home. They have named this place 'Rich Bar'" (2:1066). **Richvale** [Butte Co.] was developed and named in 1909 by Samuel J. Nunn. When the post office was established in 1912, the original name, Richland, had to be changed because there already was a post office by that name in the state. **Richgrove** [Tulare Co.] was developed and named in 1909 by S. R. Shoup and W. H. Wise of the Richgrove Development Company. The post office is listed in 1912. *For* Rich Dry Diggings, *see* Auburn.

Richardson Bay [San Francisco Bay]. The name commemorates one of the foremost pioneers of San Francisco, William A. Richardson, a native of England. He arrived in San Francisco in 1822, put up the first house, a tent-like structure, in San Francisco in 1835, and in the same year was made captain of the port of San Francisco. In 1838 he was grantee of Rancho Sausalito, adjoining the bay that bears his name. Ringgold's maps of 1850 show Richardson's Bay, and the Coast Survey had a station called Richardson in the early 1850s. The Spanish name of the bay had been Ensenada de la Carmelita, named by Cañizares on Aug. 6, 1775, "because in it was a rock resembling a friar of that order" (Eldredge, *Portolá*, p. 56). It is shown as Ensenada del Carmelita on Ayala's 1775 map. Another name was Puerto de los Balleneros, which was translated later into Whalers Harbor (*see* Ballena).

Richardson Grove [Humboldt Co.]. Created in 1922 and named for Friend W. Richardson (1865–1927), publisher of the *Berkeley Gazette* for many years, governor of California in 1923–27.

Richardson Peak [Yosemite NP]. Named in 1879 by Lt. M. M. Macomb of the Wheeler Survey, for Thomas Richardson, an early sheepman (Farquhar).

Richardson Springs [Butte Co.]. The springs became known by this name because they were on the property of the four Richardson brothers, cattle and sheep raisers. The first hotel was built by J. V. Richardson in 1889 (Co. Hist.: Butte 1918: 1098–99).

Richfield [Orange Co.]. *See* Atwood.

Richgrove. *See* Rich.

Richland. *See* Orange; Rich: Richvale.

Richmond, Point; Richmond [Contra Costa Co.]. The point was named Point Stevens by the U.S. Coast Survey in 1851. In 1852 the name was changed to Richmond Point, probably after one of the many Richmonds in various sections of the United States. The geological map of the Williamson Reports of 1853 places the name on the point now called Shoal Point. In 1897 the Santa Fe secured the site just north of the point for its terminal and called it Point Richmond; the settlement that developed there was called Santa Fe. The town that was laid out on the Barrett ranch by A. S. Macdonald and his associates was named Richmond. The post office, established at "old town" in Aug. 1900, had also been called Richmond. The city was incorporated on Aug. 7, 1905.

Richter Creek [Inyo Co.]. The name should

probably be Rittgers Creek, for Israel P. and John A. Rittgers, who settled in the district in the 1860s.

Richvale. *See* Rich.

Rickeyville. *See* Ione.

Ridgecrest [Kern Co.]. The name describes the city's topographic location.

Ridge Point. *See* Table: Table Bluff.

Riggs [San Bernardino Co.]. The station on the Tonopah and Tidewater Railroad was named for the owner of a silver mine (Myrick).

Rincon (ring′ kon); **Rinconada** (ring kə nä′ də). In modern Spanish geographical nomenclature *rincón* 'inside corner, nook' often designates a small portion of land, while the derivative *rinconada* is an outside corner formed by hills, woods, roads, or slopes. In California the two terms were apparently used interchangeably—frequently for a projection extending into the sea, into the plains, or into the neighboring property, and sometimes for just a corner or piece of land. Rincon is found in the names of more than twenty land grants and has survived in the names of a number of hills, points, and valleys. Rinconada appears in three land grant names. The English word "rincon" occurs in the southwest for 'nook, secluded place, bend in the river'; but it does not seem to have been used actively in California place naming after the American conquest. In 1857 it was proposed for a newly formed county in the northwest "corner" of the state, but was rejected in favor of Del Norte. In the Kettleman Hills, where many names are Spanish (on the map), we find without surprise El Rincon, south of Avenal Gap. **Rincon: Point, Creek, Mountain** [Ventura, Santa Barbara Cos.]: Crespí called the Indian village near this point Santa Clara de Monte Falco, but Portolá's soldiers called it El Bailarín 'the dancer' because of the dancing ability of one of the chiefs (Crespí, p. 162). On Feb. 24, 1776, the Anza expedition disregarded these names: Anza mentions Rancherías del Rincón, and Font (*Compl. Diary*) speaks of La Rinconada. The land grant El Rincon (or Matilija) is dated Jan. 8, 1834, and June 22, 1835. Rincon Point is recorded on the Land Office map of 1862. **Rincon Point** [San Francisco Co.]: The elevation at the San Francisco terminus of the Bay Bridge was a true *rincón* until part of the waterfront was filled in. The hill was a prominent landmark often mentioned in early reports; it was used by the Coast Survey as a secondary triangulation point as early as 1850.

Ringgold Creek [El Dorado Co.]. The tributary to Weber Creek is a remembrance of the once-prosperous town of Ringgold, the first settlement on the road from Carson Pass in the early Gold Rush days. It might have been named for Cadwalader Ringgold of the Wilkes Expedition, who explored the region in 1841. The name of the creek is also spelled Ringold on the Placerville atlas sheet. *See* Ringgold in the Glossary.

Rio (rē′ ō). Spanish *río* 'river' was naturally a common generic term and is found in the names of at least seven land grants, which are listed under their specific names. It is still found in a few names as a true generic term: Rio Hondo [Los Angeles Co.], Rio Bravo [Kern Co.]. More frequently it is found in names of post offices and communities, sometimes without regard to meaning or Spanish usage. **Rio Vista** [Solano Co.] was founded in 1857 by Col. N. H. Davis and was called Brazos del Rio because it was near the three arms of the Sacramento. In 1860 the name was changed to Rio Vista. When the town was wiped out by a flood on Jan. 9, 1862, it was rebuilt at the present site and at first called New Rio Vista. **Rio Dell** [Humboldt Co.]: The place was first called Eagle Prairie. When the post office was established about 1890, the name River Dell was proposed as an appropriate name; but it was rejected because of its similarity to Riverdale, in Fresno Co. The Spanish word for river was then substituted, and the name was accepted. **Rio Bravo** [Kern Co.] was the name given to the station when the Southern Pacific built the line from Bakersfield to the asphaltum beds in the early 1890s. It preserves the old name of the Kern River, which was called Río Bravo 'wild river' because it was difficult to cross. This is confirmed by the fact that Edward Kern, for whom the river was then named by Frémont, almost drowned in 1845 while swimming it. Rio Bravo had a post office from 1912 to 1919. **Rio Oso** [Sutter Co.], a station on the Sacramento Northern, was named about 1907 because of its location on the Bear River (*see* Oso). **Rio Linda** [Sacramento Co.] was founded and named in 1913 by a member of the firm of Sears Roebuck. It is supposed to mean 'pretty river' (Spanish *río lindo*). **Rio Nido** [Sonoma Co.]: The resort was developed about 1910 and was apparently so named because this combination of Spanish words has a pleasing sound. The post office is listed in 1912. **Rio del Mar** [Santa Cruz Co.] is Spanish for 'river of the sea'; the name was given in the 1920s. *See* El Rio; Nido; Dos Rios; River. *For* Rio Grande de Buena Esperanza, *see* Colorado River.

Ripgut Creek [Shasta Co.]. This tributary of the Pit River was so named by William Bowersok, a cattleman, because his clothes were torn to shreds on the haw shrub that abounds here (Steger).

Ripley [Riverside Co.]. The terminus of the Ripley branch of the Santa Fe was named in 1921 for E. P. Ripley, former president of the railroad.

Ripon (rip′ on) [San Joaquin Co.]. In 1876 the first postmaster of the town, Applias Crooks, chose the name of his former home in Wisconsin to replace the earlier name, Stanislaus City.

Rising River [Shasta Co.]. This tributary of Hat Creek was so named because it has no distinct source but apparently rises out of a marshy meadow (Steger). Hat Creek is perhaps the only creek that has the distinction of having a "river" as a tributary.

Ritter, Mount; Ritter Range [Madera Co.]. Named by the Whitney Survey in 1864, in memory of Karl Ritter (1779–1859), progenitor of scientific geography. Ritter was one of the luminaries of the University of Berlin when Whitney was a student there in the 1840s.

Rittgers Creek. *See* Richter Creek.

River. The term is often used in the names of settlements at or near a river. California has sixteen or more communities with names in which River is combined with -bank, -bend, -dale, -glen, -side, -view, etc. A town in El Dorado is simply called River. **Riverside**: city, **County, Mountains**: A settlement was started on the Rancho Jurupa in 1869 by Louis Prevost, who had learned silk culture in France and intended to develop a silk-producing colony. After Prevost's death, John W. North of the Southern California Colony Association acquired the property and retained the name Jurupa. In Sept. 1870, the company started building the upper canal of the Santa Ana River, and when the canal reached the settlement in June 1871, the name was changed to Riverside. Riverside post office is listed on June 12, 1871. It is, however, possible that both Jurupa and Riverside existed side by side: the von Leicht–Craven map (1874) shows both places on two different channels of the Santa Ana, about 3 miles apart. The county was created by act of the legislature on Mar. 11, 1893, and named after the city. **Riverdale** [Fresno Co.] was originally known as Liberty Settlement. When the post office was established in 1875, the new name was chosen because of its proximity to Kings River. *For* Riverdale [Los Angeles Co.], *see* Glen: Glendale. **Riverton** [El Dorado Co.] was long known as Moore's Station because it was on the toll road between Sacramento and Virginia City, built and operated by John M. Moore, a former member of the San Francisco Vigilance Committee (J. W. Winkley, in Knave, Sept. 22, 1946). When a post office was established about 1895, another name had to be chosen in order to avoid confusion with Moore's Station in Butte Co. The location at the bank of the South Fork of American River suggested the new name. **Riverbank** [Stanislaus Co.] was founded in 1911 when the Santa Fe Railroad established a new terminal and division point south of Burneyville and named it Riverbank for its location on the Stanislaus River. The older settlement, now on the edge of Riverbank, had been named for Maj. James Burney, an early settler and the first sheriff of Mariposa Co. *For* Riverbank [Yolo Co.], *see* Bryte. *For* River Dell, *see* Rio: Rio Dell; *for* River of the Lake, *see* Kings River.

Rivera (ri vâr′ ə) [Los Angeles Co.]. The district was named Maizeland in 1866 because the chief crop was corn. In 1886, when the Santa Fe reached the town, the name was changed to Rivera, a name considered appropriate because the place is between two rivers: Rio Hondo and San Gabriel River (Ada Moss). Rivera is also a common Spanish family name.

Rixford, Mount [Kings Canyon NP]. Named in 1899 by Vernon L. Kellogg and a party from Stanford University, for Emmet Rixford, professor of surgery at Stanford University, who had previously climbed the peak (Farquhar).

Riz (riz) [Glenn Co.]. French for 'rice', a major crop of the area.

Roach Creek [Humboldt Co.]. The stream may have been named for M. Roach, a dairyman at Alliance in the 1890s.

Roads End [Tulare Co.]. The post office was named Roads End because it was at the end of the road from Kernville. Now the road goes on, but the old name remains.

Roaring River. Of the several "roaring" rivers in the state, the best known are in Shasta Co. and in Kings Canyon National Park. Why the lazy tidal channel in Suisun Bay should ever have been called Roaring River (Hoffmann's map, 1873), and in modern times Roaring River Slough, is no longer determinable.

Robbers. The word is found in about ten place names, probably all applied to places where bandits had their hideout or where a robbery had been committed. **Robbers Roost** [Los Angeles Co.]: The rocks on the high ridge between Soledad and Mint canyons were once the hiding place of Tiburcio Vásquez, a famous bandit of the early 1870s (Doyle); *see* Vasquez. **Robbers Roost** [Monterey Co.] was likewise named because Vasquez and his band robbed a stage here (Co. Hist.: Monterey 1881:152). **Robbers Ravine** [Sacramento Co.] was so named because notorious highwaymen of stagecoach days had their meeting place here. **Robbers Creek** [Lassen, Plumas Cos.] was named after James Doyle was robbed here by armed men in 1865 or 1866 (Co. Hist.: Lassen 1916:406).

Robbins [Sutter Co.]. Named in 1925 by the Sutter Basin Company for its president, George B. Robbins. The former name had been Maddox, for a manager of the same company (W. J. Duffy).

Robbs Peak [El Dorado Co.]. Named for Hamilton D. Robb, an early-day stockman (W. T. Russell, in Knave, July 23, 1944); however, USFS records say that it was for a Lieutenant Robb, who, while leading a detachment of cavalry on an exploration trip, climbed the mountain and left a tin cylinder on it (E. F. Smith).

Robert [Alameda Co.]. The Southern Pacific station was doubtless named after Robert's Landing when the line from Alameda Point to Newark was built in 1876–78. Capt. William Robert (or Roberts), a native of England, was one of the earliest settlers of Eden township; he had hay and grain warehouses at the landing in the 1870s and 1880s.

Roberts, Camp [Monterey, San Luis Obispo Cos.]. For Cpl. Harold W. Roberts, a native of San Francisco, who died in 1918 serving as a tank crewman at Montrebeau, France (Clark 1991).

Roberts Canyon [Los Angeles Co.]. Named for H. C. Roberts, a settler of 1856 (USFS).

Robertson Creek [Mendocino Co.]. The stream was named for William E. Robertson, who was in charge of the Indian agency in Ukiah in the 1870s (Asbill).

Robertsville [Santa Clara Co.]. Named for J. G. Roberts, owner of a ranch in the district (Wyatt and Arbuckle).

Robinson: Creek, Peak [Mono Co.]. In the 1860s Moses Robinson operated a sawmill on the creek that bears his name.

Robinson: Point, Reef [Mendocino Co.]. The names were applied by the Coast Survey, doubtless for Cyrus D. Robinson, a native of Pennsylvania who came to California in 1849 and settled at Gualala in 1858, where he engaged in hotel keeping, shipping, and farming.

Robinson's Ferry [Calaveras Co.]. The ferry and trading post were established by John W. Robinson and Stephen Mead on the Stanislaus River in 1848; *see* Melones.

Roblar (rō blär′) [Sonoma Co.]. Spanish for 'the place where deciduous oaks grow' (*see* Roble). A land grant, Roblar de la Miseria 'oak grove of misery', was registered here on Nov. 15, 1845.

Roble. The Spanish word for 'deciduous oak' (*Quercus* spp.) was used for place naming less often than *encina* 'live oak.' It is found in the name of a land grant, Paso de Robles [San Luis Obispo Co.], May 12, 1844. The Rincón de San Francisco grant [Santa Clara Co.] was known as the Robles Rancho, not because *robles* were present but because the grantees were Teodoro and Secundino Robles. In modern times the name is found in several counties; e.g., **Robles del Rio** (rō′ blās del rē′ ō) [Monterey Co.] means 'river oaks'. The names of the two railroad stations called Robla [Sacramento, Tulare Cos.] are probably derived from the same word. *See* Paso: Paso Robles; Encina; Oak.

Rock, Rocky. Besides the wide use of Rock as a generic term, California has more than two hundred names for natural features in which the noun or the adjective is used as a specific term. The oldest is doubtless **Rocky Point**, for the point at the north entrance of Trinidad Bay, given by Vancouver in 1792. There is a Rockpile in Sonoma Co., and a Rockpile Peak and Creek in Mendocino Co. Rock, or Rocky, in some form or combination, appears in the names of about ten inhabited places; a few of these may be for a person. **Rocklin** [Placer Co.] was so named when the Central Pacific built the line from Sacramento to Newcastle in 1863–64. The extensive quarries doubtless suggested the name; *lin* is a Celtic generic term for 'spring, pool', as well as for ravine or precipice. **Rockport**: post office, **Bay, Creek** [Mendocino Co.]: A chute and a wharf were erected on the rocky coast in 1876; in 1880 Rockport was given as an alternate name for Cottaneva (Co. Hist.: Mendocino 1880:471). Rockport post office is listed in 1892. **Rock Creek** [Shasta Co.], a tributary of Pit River, was named for the "Rock Indians," who buried their dead in an upright position among the rocks (Steger). **Rock Island: Lake, Pass; Rock Canyon** [Yosemite NP]: "I named the stream Rock Creek, and the lake Rock Island Lake, from a large granite island that was visible near the northern end" (N. F. McClure, *SCB* 1:178). *For* Rocky Island, *see* Brooks Island.

Rockaway Beach [San Mateo Co.]. Doubtless named after Rockaway, Long Island, a name that was derived from the Delaware (Algonquian) *regawihaki* 'sandy land' (Leland, *CFQ* 4:405). Folk etymology derives the name from the large quarry where they take "rock away." The post office is listed in 1910. In 1957 the community was incorporated into the city of Pacifica.

Rockefeller Forest [Humboldt Co.]. The large area of redwood trees in the Humboldt Redwoods State Park was donated by the philanthropist John D. Rockefeller Jr. (1874–1960).

Rockwood [Imperial Co.]. Named for Charles R. Rockwood, engineer of the California Development Company and the Imperial irrigation district; he had been actively engaged in promoting the irrigation of the Colorado Desert since 1892.

Rodeo (rō dā′ ō) is used in Spanish for an enclosure at a fair or market where cattle were exhibited for sale. In Mexico it developed as a designation for the periodic rounding up of cattle for counting and marketing. It appears in the names of three land grants in Los Angeles, San Francisco and San Mateo, and Santa Cruz Cos. Since "rodeo" has become the common American term for a cowboy contest, it is often pronounced (rō′ dē ō). Some of the Rodeo Canyons, Creeks, Lagoons, etc., may have been named in American times. **Rodeo Creek** and town [Contra Costa Co.]: A place called Rodeo is shown on a *diseño* of the Pinole grant near the center of the rancho. Rodeo Valley appears on a plat of the rancho in 1860, and Rodeo Creek on another in 1865. The post office is listed in 1898.

Rodeo (rō′ dē ō): **Beach, Lagoon, Lake** [Marin Co.]. From Rodier, the name of a local family; Rodier Lagoon is recorded in 1856 (Teather 1986).

Rodgers: Peak, Lake [Yosemite NP]. The peak was named in 1895 by Lt. N. F. McClure, for Capt. Alexander Rodgers, acting superintendent of the national park at that time. Independently, Lt. H. C. Benson christened Rodgers Lake in the same year and gave the same name to the peak south of it. To avoid the duplication, the USGS substituted for the latter the name Regulation Peak, which had previously been intended for another peak. On LeConte's map of 1896 Rodgers Peak is called Mount Kellogg, a name probably given by John Muir for the botanist Albert Kellogg (Farquhar).

Rodriguez, Rancho de. *See* Jacinto.

Roeding Park [Fresno Co.]. The name commemorates Frederick Roeding, a native of Hamburg, Germany, who came to California in 1849 and settled in Fresno Co. in 1868. He was one of the leaders of the "German Syndicate," which subdivided an area of 80,000 acres in 1868, on which the modern city of Fresno was developed. The first of his famous nurseries was established in 1883.

Rogers Peak. *See* Manly.

Rogue Canyon. *See* Roque Canyon.

Rohnert Park [Sonoma Co.]. The town was developed by Paul Golis and incorporated in Aug. 1962. It was named after the Waldo Rohnert Seed Farm.

Rohnerville [Humboldt Co.]. Named for Henry Rohner, a native of Switzerland who, with Joseph Feigenbaum, opened a store here in 1859. The post office existed from 1874 to 1959 (Turner 1993).

Rojo Grande (rō′ hō grän′ dā) [Imperial Co.] is a hill, named 'big red' because of its color (BGN, list 7601).

Rollin [Siskiyou Co.]. Named for Rollin Fagrendes, who discovered a mine here. There was a post office from 1889 to 1927 (Luecke).

Rolling Hills and **Rolling Hills Estates** [Los Angeles Co.] are cities named because of their topography.

Rolph [Humboldt Co.]. Named for James

("Sunny Jim") Rolph (1869–1934), mayor of San Francisco in 1911–30, governor of California in 1931–34 (Laura Prescott). The Governor James Rolph Grove in the Humboldt Redwoods State Park was dedicated in his memory in 1934.

Romanzov, Port. *See* Bodega.

Romero Creek [Merced Co.]. Named for José Romero, who led an exploring party to the creek and, according to F. F. Latta (p. 28), was killed by the Indians on this trip.

Romoland [Riverside Co.]. Founded as Romola Farms in 1925, perhaps inspired by the heroine of George Eliot's novel *Romola* (1863), a story of medieval Italy (Gunther). The Post Office Dept. requested another name, to avoid confusion with Ramona, and the present name was substituted. *See* Ethanac.

Romualdo. *See* Cerro: Cerro Romualdo.

Roops Fort [Lassen Co.]. The blockhouse was built in 1854 by Isaac N. Roop, and there the settlers of Honey Lake Valley met in 1856 to form the "Territory of Nataqua." It earned the title "fort" when it served as fortified headquarters during the "Sagebrush War" in 1863. The settlers of Honey Lake Valley, with Roop as provisional governor, were in authority for some time.

Roosevelt Lake [Yosemite NP] commemorates a visit in 1934 by the first lady, Eleanor Roosevelt (1884–1962), according to Browning 1988.

Root. *See* Alpaugh.

Root Creek [Shasta Co.]. Named for Orin T. Root, who diverted the waters of a spring near Castle Dome to form the stream. Root settled at Castle Rock Spring in 1864 and kept an inn there for fifteen years (Steger).

Rootville. *See* Millerton Lake.

Roque (rō´ kā) **Canyon** [Santa Barbara Co.]. A Spanish given name (*see* San Roque). Made official by BGN, list 6603, supplanting Rogue Canyon.

Rosamond: town, **Dry Lake** [Kern Co.]. The station was named about 1888 for the daughter of an official of the Southern Pacific.

Rosario, Arroyo del. *See* Tres Pinos Creek. *For* Bahía del Rosario, *see* Marin.

Rosasco (rō zas´ kō) **Lake** [Tuolumne Co.]. Perhaps for John Rosasco, an early settler from Italy (de Ferrari).

Rosaville. *See* Cambria.

Roscoe. *See* Sun: Sun Valley.

Rose. The common wild rose, *Rosa californica*, especially noteworthy in the hot summer when few flowers bloom, has given rise to many Rose, Rosebush, and Wildrose Canyons and Valleys. Las Pulgas ('the fleas') Canyon [San Diego Co.] was named Cañada de Santa Praxedis de los Rosales by the Portolá expedition on July 21, 1769, because of the many rosebushes. Ten or more California communities and railroad stations include the word "rose" in their names, but none of these need have any connection with the flower; the word is very commonly used throughout the United States to form pleasant-sounding place names.

Rose Canyon [San Diego Co.]. The canyon and the former settlement, Roseville, were named for Louis Rose, who built a wharf there in 1870.

Rosecrans (rō´ zə kranz), **Fort** [San Diego Co.]. The site of the old Spanish Fort Guijarros was reserved by the War Dept. in 1852, and the building of fortifications was begun in 1897. In 1899 the fort was named in memory of Gen. William S. Rosecrans (1819–98). Rosecrans was one of the ablest strategists among the Union generals and was commander of the Army of the Cumberland, 1862–63. After the war he settled in the Los Angeles district (where Rosecrans Avenue is named for him) and was a congressman from 1880 to 1884. His Confederate opponent in the Tennessee campaign is honored in the name of Fort Bragg.

Rose Lake [Fresno Co.]. Named by R. B. Marshall for the painter of miniatures, Rosa Hooper, sister of Selden S. Hooper, a USGS assistant (Farquhar).

Rosemead [Los Angeles Co.]. The name was originally (in the 1870s) applied to the famous horse farm on Leonard J. Rose's Sunny Slope estate. The post office was established on Sept. 1, 1927.

Rosena. *See* Fontana.

Roseville [Placer Co.]. The name was applied to the station when the Central Pacific reached the place in the spring of 1864. It was chosen by the residents, at a picnic, for the most popular girl present (*Sacramento Bee*, Oct. 20, 1931). Other similar stories are told about the origin of the name, but it is just as possible that it was chosen because of its pleasant sound, like the many other Rosevilles then in existence.

Ross, Fort. *See* Fort Ross.

Ross: town, **Valley** [Marin Co.]. Named for James Ross, who acquired the Rancho Punta de Quintín in 1859. Ross Landing is shown on Hoffmann's map of the Bay region.

Rossmoor. The first of the Rossmoor retirement communities in California was established in 1961 and named for Ross W. Cortese, the originator of the idea. In 1967 there were three Rossmoor Leisure World communities in the state, situated in or near Laguna Hills [San Diego Co.], Seal Beach [Orange Co.], and Walnut Creek [Contra Costa Co.].

Roth Spur. *See* Strathmore.

Rough and Ready. The phrase was popular as a place name after the Mexican War because it was the nickname of General (later President) Zachary Taylor (1784–1850). California had three towns so named, of which the one in Nevada Co. survives. It was founded in 1849 by the Rough and Ready Company, led by Capt. A. A. Townsend, who had served under Taylor. The post office was reestablished in 1948. The name of the town in Siskiyou Co. was changed to Etna by statute in 1874. The name in San Joaquin Co. is preserved in Rough and Ready Island, later a naval supply depot. **Rough and Ready Creek** [Tuolumne Co.], a tributary of the North Fork of the Tuolumne River, was named for the Rough and Ready Company, which started working its rich claim in 1854 (Browne, pp. 41–42).

Roughs, The [Sonoma Co.]. A topographical name applied to the steep slope of Red Oat Ridge (Cazadero atlas sheet).

Round. The descriptive adjective, applied to about a hundred features, is used mainly for Mountains and Valleys. Half of all the elevations designated as "tops" are Round Tops. The best-known feature named Round is **Round Valley** [Mendocino Co.], where an Indian reservation is situated. Big Valley [Modoc Co.] was named Round Valley by Frémont in 1846: "the next day [Apr. 30] again encamped . . . at the upper end of a valley, to which, from its marked form, I gave the name Round Valley" (*Memoirs* 1:480). There is a Round Corral Meadow in Fresno Co., a Round Potrero in San Diego Co., a Round Corral Canyon in Santa Barbara Co., and a Roundtop Hill in Riverside Co. The post office at **Round Mountain** [Shasta Co.] was named after the mountain northwest of the town.

Roussillon Bay. *See* Avalon Bay.

Routier (ro͞o′ tē âr) [Sacramento Co.]. A railroad station was established here in 1866 and named for Joseph Routier, a native of France who had settled here in June 1853 as agent of Capt. Folsom (*see* Folsom) and who later became an assemblyman, a state senator, and a fish commissioner.

Rovana (rō van′ ə) [Inyo Co.]. For Round Valley Vanadium, referring to mining operations here (Sowaal 1985).

Rowdy Creek [Del Norte Co.]. The name was probably applied to distinguish this creek from the commonplace Rough, Roaring, and Wild Creeks.

Rowell Meadow [Tulare Co.]. The name became attached to the meadow because Chester and George Rowell, uncles of the late well-known newspaperman Chester Rowell, used to run sheep there and had a sort of "shotgun" title to the meadow (Farquhar).

Roxie Peconom [pi kon′ əm] **Creek**. *See* Peconom Creek.

Royal Arches, The; Royal Arch: Cascade, Creek [Yosemite NP]. The name Royal Arch, for the seventh degree of tbe Masonic fraternity, was applied to the rock formation by a member of the Mariposa Battalion in 1851 (Bunnell, p. 212). The plural form is shown on the 1865 "Map of the Yosemite Valley" by Clarence King and J. T. Gardner (Washington, D.C.: Geological Survey).

Royce Peak [Fresno Co.]. The name Mount Royce was proposed in 1929 by the State Board on Geographic Names and was accepted by the federal BGN in memory of Josiah Royce (1855–1916), philosopher and educator, a native of Grass Valley, and the author of *California: From the conquest in 1846 to the Second Vigilance Committee in San Francisco* (Boston: Houghton Mifflin, 1886). The present form of the name was made official by the BGN in 1982 (list 8203).

Rube Creek [Tulare Co.]. Formerly called Nigger Rube Creek (BGN, list 8801), probably from a nickname for Reuben. Another Rube Creek [Humboldt Co.] carries the surname of a Yurok Indian family.

Rubicon: River, Point, Peak, Lodge [El Dorado Co.]. The upper course of the Rubicon River is called "The Rubicon" on the von Leicht–Hoffmann Lake Tahoe map (copyrighted 1873), probably in jesting analogy to Julius Caesar's crossing of the Rubicon River—the ancient boundary between Cisalpine Gaul and Italy, which Caesar crossed to attack Pompey in 49 B.C. The name for the point on Lake Tahoe appears on the same map. The less romantic Amos Bowman has Rubicon River on his map of the same year, and Wheeler's atlas sheets 47-B and 47-D have Rubicon Creek. In 1889 the USGS applied the name to the entire river and also named the peak. Wells Drury related that a resort owner thereabouts, a German, called his place Rubicund.

Rubidoux (ro͞o′ bi dō), **Mount** [Riverside Co.]. Named for Louis Robidoux (as he spelled his name), a French pioneer who acquired large land holdings, including a part of Rancho Jurupa, on which the hill is situated.

Rubio (ro͞o′ bē ō) **Canyon** [Los Angeles Co.]. Named for Jesús Rubio, a native Californian who became an American citizen by the peace treaty of 1847 and settled as a squatter at the mouth of the canyon in 1867 (Reid, p. 379).

Ruch Ranch. *See* Carrville.

Rucker Lake [Nevada Co.]. The reservoir was created by the Meager Mining Company before 1871 and was probably named for one of its officers.

Ruda (ro͞o′ də) **Canyon** [San Luis Obispo Co.]. For the Ruda family, pioneers in the area (Hall-Patton).

Rudolph Hagen Canyon [Kern Co.]. Named for Rudolph Hagen, a gold hunter who settled in the Mojave Desert (Doyle).

Ruffy Lake [Siskiyou Co.]. Named for an Indian called Ruffy, who seined some trout out of Etna Creek and planted them in the lake (Ronald Baker).

Ruins Creek [Santa Cruz Co.]. For a sandstone formation called The Ruins: "When they were 'discovered' in the 1850s visions of an ancient city were conjured up" (Clark 1986).

Rumsey [Yolo Co.]. The town was laid out in 1892 and named for Capt. D. C. Rumsey, the owner of the land (Co. Hist.: Yolo 1940: 228–29). The post office is listed in 1892.

Run. The use of the word as a generic term for a small creek originated in Virginia (Stewart, p. 60). It has never achieved general currency, but the two battles of Bull Run in the Civil War made this particular combination known everywhere. **Bull Run Peak** [Alpine Co.]: The presence of a strong Southern element in the district during the Civil War probably explains the application of the name, which commemorates the two Confederate victories. Bull Run Pass [Tulare Co.], a name used for a stock driveway from Lynns Valley to Bull Run Basin by Joel Carver in the early 1880s (Mitchell)—as well as the tautological Bull Run Creek [Plumas, Tulare Cos.], Mud Run Creek [Fresno Co.], and Oak Run Creek [Shasta Co.]—indicates that "run" may sometimes mean trail and not creek. There are, however, a number of "runs" where the term actually designated a small stream: Bloody Run [Nevada Co.], Honey Run, Sucker Run [Butte Co.], Negro Run [Plumas Co.], Whiskey Run [Placer Co.].

Runyon [Sacramento Co.]. The Western Pacific station was named for the journalist and short-story writer Damon Runyon (1884–1946). The place was earlier called Sims, for an old settler (*Headlight*, July 1942).

Ruppert (ro͞o′ pərt), **Point** [Lake Co.]. Named for a member of a pioneer family who settled here in the 1880s (Annie Lovelady).

Ruskin, Mount [Kings Canyon NP]. Named in 1895 by Bolton C. Brown in honor of John Ruskin (1819–1900), the English critic and author (Farquhar).

Ruso, Río. *See* Russian: River.

Russell, Mount [Sequoia NP]. For Israel C. Russell (1852–1906), assistant geologist with the Wheeler Survey in 1878; professor of geology at the University of Michigan, 1892–1906 (Farquhar). Mount Russell in Riverside Co. was apparently also named for this geologist around 1898 (Gunther).

Russelsville. *See* Parkfield.

Russ Grove [Humboldt Co.]. Named in memory of Joseph Russ, a native of Maine, who settled in Humboldt Co. in 1852 and became a large landowner and prominent politician (Co. Hist.: Humboldt 1881:168).

Russian. Besides the names in Sonoma Co. that commemorate the Russian establishment at Fort Ross, there are about fifteen other names that include the word. Most of them are in northern counties (including a Yellow Russian Gulch in Siskiyou Co.) and were probably all given for settlers from Russia or other Slavic countries. **Russian: River, Gulch** [Sonoma Co.]: An early Spanish name for the river was San Ygnacio, mentioned on Nov. 8, 1821, in the diary of Padre Blas Ordaz (Arch. MSB 4:169 ff.). The Russian colonists themselves called it Slavianka 'Slav woman' (Kotzebue, *New Voyage*, 1830:2.119). In Mexican records and in land grant papers, the river (or part of it) appears with various names. The Spanish version of the modern name is mentioned, perhaps for the first time, in the petition for the Bodega grant, dated July 19, 1843: *la boca del Río Ruso* 'the mouth of the Russian River'. On the *diseños* of the Bodega and Muniz grants the name is recorded as "Río Ruso o San Ignacio." The American form, Russian River, is shown on Gibbes's map of 1852 but was doubtless current before the publication of this map. *For* Russian River as the name of a town, *see* Healdsburg. **Russian Spring** [San Diego Co.], rediscovered in 1952, owes its name to a rather fantastic story, according to Dallas Wood in the *Palo Alto Times*. About 1830 a Russian whaler was wrecked on the strand south of the present Hotel Coronado, and the sailors dug a well with their bare hands. Later a girl led sailors from a Mexican boat to the spring, where the bodies of seven Russians were found. **Russian Gulch** [Mendocino Co.], according to local tradition, was so named because a deserter from Fort Ross had settled there (Stewart). **Russian Hill** [San Francisco Co.], according to an undocumented story, was named for the Russian soldiers or sailors who were buried near the top of the hill in the city's early days (Loewenstein).

Rust. *See* Cerrito: El Cerrito.

Ruth [Trinity Co.]. The post office was established in 1904 and named for Ruth McKnight, the daughter of a pioneer family (Albert Burgess).

Rutherford [Napa Co.]. Railroad station and post office were named in the 1880s for Thomas L. Rutherford, who had married a granddaughter of the pioneer George C. Yount of nearby Yountville (Doyle).

Rutherford Lake [Madera Co.]. Named for Lt. Samuel M. Rutherford of the 4th Cavalry, who was stationed in Yosemite NP in 1896 (Farquhar).

Ruwau (rōō′ wô) **Lake** [Inyo Co.]. A combination of the names of two power-company engineers, Clarence H. Rhudy and E. J. Waugh (Browning 1986).

Ryan: town, **Wash** [Death Valley NP]. The old town at the site of the Lila C. Mine, as well as the present town on the site formerly called Devair, was named for John Ryan, for many years manager of the Pacific Coast Borax Company. In 1936 the Park Service gave this name to the southern fork of Furnace Creek Wash (Death Valley Survey).

Ryde [Sacramento Co.]. William A. Kesner bought the land from Judge Williams in 1892 and laid out the town. When the post office was established in 1893, Kesner became the first postmaster but refused to have it named for himself. Judge Williams then suggested the name Ryde, after a town on the Isle of Wight (England), the site of which is similar to that of the California town on Grand Island (Patricia Kesner).

Ryer Island [Solano Co.] commemorates the former owner of the island, Dr. W. M. Ryer, a pioneer physician of Stockton, who vaccinated some two thousand Indians in the San Joaquin Valley in the summer of 1852. The name appears on Ringgold's chart of 1850.

S

Sablon (sä blōn′) [San Bernardino Co.]. When the Santa Fe line from Cadiz was built in 1909, the station was named Randolph. In 1912 the present name (from Spanish *sablón* 'gravel') was applied.

Sabrina, Lake [Inyo Co.]. Named about 1908 for Sabrina Hobbs, wife of C. M. Hobbs, general manager of the Nevada-California Power Company.

Saca. *See* Zaca.

Sacalanes. *See* Acalanes.

Sacatara (sä kə tâ′ rə) **Creek** [Kern Co.]. Probably from the Mexican Spanish adjective *zacatero, zacatera* 'having to do with *zacate*, hay'; or one of the nouns *zacatero* 'man who makes hay; a species of bird' or *zacatera* 'hayfield'. The present form replaces the apparent folk etymologies Canyon del Secretario, del Sectario 'canyon of the secretary, of the sectarian' (BGN, list 6702).

Sacate (sä kä′ tā) [Santa Barbara Co.]. The name is shown for the station on the Southern Pacific map of 1905. Usually spelled *zacate*, the word is Mexican Spanish for 'grass, hay', from Aztec *zacatl*.

Sacaton (sä kə tōn′) **Flat** [Riverside Co.]. For Mexican Spanish *zacatón*, a type of bunchgrass usually found in alkaline soil. This is the augmentative form of *zacate* 'grass, hay' (*see* Hay). The same term is reflected in Sacatone Spring [San Diego Co.], and probably also in Sacraton Flat [Tulare Co.].

Saco (sā′ kō) [Kern Co.]. The Spanish word for 'sack' was applied to the Southern Pacific siding after 1900. Since the word is used as a place name in Maine and other states, it may be a transfer name.

Sacramento (sak rə men′ tō): **River**, city, **County**. The Spanish name for '[the Holy] Sacrament' is often found as a place name in Latin countries. On Oct. 8, 1808, Gabriel Moraga gave the name Sacramento to the Feather River (*Diario de la tercera expedición*). The lower Sacramento was still being called Río de San Francisco by Abella in Oct. 1811 (Bancroft 2:322), but the name given by Moraga soon came to refer to the lower Sacramento as well as the Feather. In May 1817 Durán records going up the Río del Sacramento with Luis Argüello; they probably got almost as far as the mouth of the Feather River (*APCH:P* 2:332 ff.). On the *Plano topográfico de la Misión de San José* (about 1824) the lower course of the river is shown as R. del Sacram[en]to. The trappers of the Hudson's Bay Company called the stream "the big river" or Bonaventura, and as Buenaventura River it appears on Burr's map of 1839. Hood's map of 1838 shows Sacramento River, and the Wilkes expedition in 1841 definitely established the name Sacramento for the river south of the confluence of the Pit and the Little or Upper Sacramento Rivers. The latter is labeled Destruction River on Wilkes's map. In the fall of 1848, John A. Sutter Jr. and Sam Brannan, against the wishes of the elder Sutter, laid out a town at the embarcadero of Sutters Fort and named it Sacramento, after the river. This name had been mentioned as Fort Sacramento for Sutter's settlement (*San Francisco Californian*, July 31, 1847: Fort Sacramento). The county, one of the original twenty-seven, was established and named on Feb. 18, 1850 (*see* Pit River; Buenaventura). Sacramento Buttes [Sutter Co.] was a former name of Sutter Buttes. The name Sacramento is also given to the mountain range west of Needles [San Bernardino Co.].

Sacratone Flat. *See* Sacaton.

Saddle, Saddleback. These descriptive terms are found in the names of more than twenty-five Buttes, Peaks, and Mountains. Sometimes they are even used in a generic sense, as in Salt Tree Saddle [Sonoma Co.]. A number of creeks and lakes have been named after the orographic features. **Old Saddleback** [Orange Co.] has two peaks, Santiago and Modjeska

(Stephenson 1932:117). *For* Saddleback Mountain [Shasta Co.], *see* Bass Mountain.

Sadler: Peak, Lake [Madera Co.]. The name was applied in 1895 by Lt. N. F. McClure, for a corporal in his detachment (Farquhar).

Sage. The presence of sage (genus *Salvia*) or of the similarly aromatic sagebrush (genus *Artemisia*) has been the reason for a number of names in the state. Until 1942 there was a post office named Sage in Riverside Co. A mining camp in 1860 in Kern Co. was called Sageland. A railroad station in Lassen Co. and several physical features were named for the sage hen—the popular name of the sage grouse, *Centrocercus urophasianus*.

Sailor. The characteristics peculiar to seafaring men made sailors readily recognizable among miners, and a number of Bars, Flats, and Ravines were named for them. Some of these names have survived in Humboldt, Nevada, Placer, and Sierra Cos. One may commemorate "Sailor Jack," the Finn who turned the joke on the miners of Pinchemtight by finding a $50,000 pocket in the "worthless" claims that they had induced him to stake out (Goethe, pp. 143–44).

Saint George, Point; Saint George: Reef, Channel [Del Norte Co.]. The name was given to the point by Vancouver on Apr. 23, 1792, the day of Saint George, patron of England. It is shown on most American maps and on the first sketch of the western coast by the Coast Survey in 1850. Vancouver named the rocks below the point Dragon Rocks, recalling Saint George's dragon; this name is preserved in Dragon Channel, south of the Reef. Vizcaíno had named the point Cabo Blanco de San Sebastián on Jan. 19, 1603, "the eve of that glorious martyr" (*CHSQ* 7:367).

Saint Helena (hə lē′ nə), **Mount** [Sonoma Co.]; **Saint Helena Creek** [Lake, Napa Cos.]. The first known mention of the name Mount Saint Helena is in a report on the geysers in the *American Journal of Science and Arts*, Nov. 1851, p. 154. No evidence has been found to show when and why the peak was so named. On *diseños* of the Caymus (1836) and the Mallacomes (1841) land grants, the mountain is apparently designated as Serro [Cerro] de los Mallacomes, for the Indians living on the west slope. Mayacmas Mountains is still the official name of what is commonly known as the Saint Helena Range. In 1853 a plate was discovered on the peak which showed that the Russian scientist and traveler J. G. Woznesenski had climbed the mountain in June 1841; thereafter it was generally assumed that the Russians had named the peak. This belief received the sanction of Whitney: "named in honor of the Empress of Russia" (*Geology*, p. 86). But the name of the empress in 1841 was Alexandra, not Helena. The story became more romantic as the years passed: no less than Princess Helena de Gagarin, "a niece of the Czar," braved the chaparral and the rattlesnakes—and, while the Russian flag fluttered above in the breeze, christened the mountain in honor of her patron saint, Saint Helena, the mother of Constantine the Great (of Constantinople). Erman's map of 1849, based on Russian information, shows the mountain range but leaves it nameless. It is nevertheless possible that the Russians did give the name, though no record of it has been found. One of the Russian vessels bore the name *Saint Helena;* and since the peak is the highest point in this latitude and visible as far as 75 miles offshore, the naming from aboard ship would be more likely than the christening by a princess. After the American conquest the mountain was known as Devil's Mount. On Jan. 29, 1848, the *California Star* printed this item: "It is related of a Russian botanist, who eight or ten years ago, climbed the 'Devil's Mount,' and on its summit affixed a plate of brass commemorative of the feat." Bartlett (*Personal narrative* 2.28 [1854]) wrote in Mar. 1852 of "Mount Helena or Moyacino of the Russians." Until the time of the Whitney Survey, the name was often given as Helen or Hellen, with and without the Saint. **Saint Helena** [Napa Co.]: Henry Still, who settled here in 1853, founded the town in 1855 and named it "from the name given to the Division of the Sons of Temperance established there about that time. . . . On account of the fine view obtained of St. Helena mountain, the Division was named St. Helena, and the Division gave the name to the town" (Co. Hist.: Napa 1873:186, 119). The post office is listed in 1858.

Saint James, Islands of [San Francisco Co.]. This name was given by Sir Francis Drake in 1579 to the islands that later became known as the Farallons. In 1985 the BGN (list 8504) restored the name Island of Saint James to one of the Farallon group.

Saint John, Mount [Glenn Co.]. According to the Colusa County History of 1880, peak and settlement were named for A. C. St. John, a settler in the early 1850s. Mt. St. John is shown on George Baker's map of 1856 and is mentioned in Hutchings' *Illustrated California Magazine* (1857:484).

Saint John Mountain [Napa Co.]. The peak was named Mount Henry by Brewer in 1861, for Prof. Joseph Henry of Princeton University. This name, however, did not appear on maps; and when the region was surveyed between 1896 and 1899, the USGS applied the present, probably local name.

Saint Johns River [Tulare Co.]. Named for Loomis St. John, who had a cabin near Kaweah River in 1850. St. John was a member of the Court of Sessions in 1852 and county supervisor in 1853 (Mitchell).

Saint Jose or **Saint Joseph, Mount** [Lassen Co.]. *See* Lassen Peak.

Saint Louis Mountain [Sonoma Co.]. The mountain was named after the vanished town of Saint Louis, mentioned in the *California Star* of Dec. 25, 1847. In 1859 a Saint Louis post office [Sierra Co.] is listed. Both settlements were probably named by settlers who came from the city in Missouri.

Saint Marys College [Contra Costa Co.]. The Catholic institution was established in Oakland in 1863 and moved to the present site in 1927. The post office is listed in 1928. The name of the saint occurs elsewhere in California: Saint Marys Creek [Lake Co.], Saint Marys Peak [Merced Co.].

Saint Orres (ôrz) **Gulch** [Mendocino Co.]. According to a report received by the BGN, 1962, the gulch was named for the St. Orres family, early settlers of the region.

Sal (sal), **Point** [Santa Barbara Co.]. Named by Vancouver in 1792 for Hermenegildo Sal, at that time *comandante* at San Francisco, in recognition of favors received from him. It appears on the Lahainaluna Carta of 1839 (Wheat, no. 12), and was used by the Coast Survey in 1852.

Salada. The Spanish word for 'salty', from *sal* 'salt', was used for surface waters with a strong saline content and has survived in a number of place names. **Laguna Salada** (sə lä′ də) [San Mateo Co.], lit., 'salty lagoon', is shown on a *diseño* of the San Pedro grant, 1838; **Salada Beach** is nearby. The masculine form occurs in **Arroyo Salado** (sə lä′ dō) 'salty creek' [Orange Co., Imperial Co.]. *See* Salinas; Salt; Sharp Park.

Salal (sə lal′) **Gulch** [Siskiyou Co.]. Named for a shrub, *Gaultheria shallon*. The term is derived from the Chinook Jargon language of the Pacific Northwest and entered English in 1805 (Cutler, p. 80).

Salida (sə lī′ də) [Stanislaus Co.]. When the Southern Pacific reached the place in 1870, it named the station Salida, Spanish for 'departure', as an appropriate name for a railway station.

Salinas (sə lē′ nəs): **River, Valley**, city [Monterey Co.]. The origin of the name is found in the *salinas* 'salt marshes' near the river mouth, which were commercially important in Spanish times. The word is included in the names of several land grants in Monterey Co., the oldest of which was granted before 1795 (PSP 13:269). The river itself had various names in Spanish times: Santa Delfina, San Antonio, Río de Monterey. In the 1820s a legend arose that a mythical river, the Buenaventura, came from the Rocky Mountains and entered the ocean somewhere south of San Francisco; the Salinas was at times confused with this river and bore its name, often with a San in front of it, long after the myth of the great river was exploded (*see* Buenaventura). Salines River is mentioned in a letter by Larkin on Mar. 6, 1846 (Castro Docs. 2:31), and Rio San Buenaventura or Rio Salinas was recorded by Frémont-Preuss in 1848. Both names were used until about 1860—except by those cartographers who could not decide one way or the other and left the important waterway without a name (Wilkes 1849, Derby 1850). The post office is listed in 1858; the town, named after the river, is shown on Goddard's map of 1860. **Rincón de las Salinas y Potrero Viejo** [San Francisco, San Mateo Cos.] was the name of a land grant dated Oct. 10, 1839, and May 30, 1840. The name means 'corner of the salt marshes and old horse-pasture'.

Saline (sā′ lēn) **Valley** [Inyo Co.]. So named because of its salt deposits. This seems to be the only important geographic feature in the state which bears this name. *See* Salt.

Salmon. The name of the common and valuable fish (genus *Salmo*) is found in the names of almost fifty features, mainly Creeks. Since the fish seldom appears south of 38° latitude, the names are confined to the northern counties, except Salmon Creek and Falls [Tulare Co.], Salmon Cove and Creek [Monterey Co.]. In some cases the term may refer to the steelhead (genus *Onchorhynchus*). **Salmon Mountains, Salmon Alps**, and the town **Forks of Salmon** are named after the Salmon River in Siskiyou Co.; the river is mentioned in the *Statutes* of 1852 (p. 233), but the name had probably been in use since 1849. The Truckee River was called Salmon Trout River by Frémont in 1844 (*Expl. Exp.*, p. 323). A post office was established at Salmon Creek [Humboldt Co.] on Feb. 19, 1884, and named Beatrice (*q.v.*). *For* Salmon Bend, *see* Colusa.

Salsberry (sälz′ bər ē): **Spring, Pass** [Inyo Co.]. Named for John Salsberry (or Salisbury), a prospector in the early 1900s.

Salsipuedes. The Spanish expression, meaning 'get out if you can,' was repeatedly used in geographical nomenclature, usually for narrow enclosures or for canyons where the terrain is very rough. One is in Santa Cruz and Santa Clara Cos.: a *sitio* in *el plan de los corralitos llamados* ('the plain of the little corrals called') *Sal si puedes* is mentioned on May 25, 1806 (Estudillo Docs. 1:160). In 1817 the padres of Santa Cruz Mission were granted land called Bolsa de Salsipuedes (or *sitio de Salsipuedes*) on which to put their cattle (Arch. Arz. SF, 3:1.134–35, 3:2.79). In June 1823 the name was applied to a private land grant. The stream shown on a *diseño* (1836) of the grant as Arroyo de Salsipuedes seems to be Corralitos Creek on modern maps. **Salsipuedes** (sal sē po͞o ä′ dəs, säl sē pwä′ dəs) **Creek** [Santa Barbara Co.]. A *sitio de Sal si puedes*, belonging to the Misión de la Purísima, is mentioned in 1817 in *Documentos . . . Historia de California* (4:321 ff.), and on May 18, 1844, the name was given to a grant, Cañada de Salsipuedes. The stream is shown on a *diseño* of the grant as Arrollo [arroyo] de Salsipuedes. The term has also survived as a name for creeks in Monterey and San Luis Obispo Cos.

Salt. Since earliest times this name has been used for places in all parts of the world. California has more than a hundred Salt names, mostly for Creeks, Sloughs, Wells, Springs, and Lakes. There is a Salt Marsh in Ventura Co. and a Salt Pool in Death Valley NP. Only a few names indicate the presence of salt in sizable quantities; most of them, especially on the coast and in the Great Basin, simply describe the peculiar taste of the water, which may come from the presence of common salt or medicinal salts. Two settlements are named Saltdale [Kern Co.] and Salt Works [San Diego Co.]. A number of names are doubtless translations from the Spanish (*see* Salada; Salinas; Saline).

Saltmarsh [San Bernardino Co.]. Named in 1921 by the Santa Fe, not for salt beds but for S. M. Saltmarsh, car accountant for the Santa Fe.

Saltmarsh Canyon [Ventura Co.] was named for John B. Saltmarsh, an oilman who worked here in 1865 (Ricard).

Salton: station, **Sea, Creek** [Riverside Co.]. The ancient lake bed was discovered and explored by William Blake of the Pacific Railroad Survey in 1853–54; it was called Lake Cahuilla for the Indians of the region (*see* Coachella). This name, although still used by geologists, has never become current. The present name, apparently coined from "salt" and first given to the Southern Pacific station, was transferred by Frank Stevens to the lakebed in 1892 and was retained for the sheet of water (much smaller than the old bed) that formed in 1907 after the overflow of the Colorado.

Saltus [San Bernardino Co.]. The name, probably coined from "salt," was given to the Santa Fe station in 1915.

Salud, Cañada de la. *See* Waddell Creek.

Salvador (sal′ və dôr) **Canyon** [San Diego Co.]. Named for Salvador Ignacio Linares, who was born in Coyote Canyon during the Anza Expedition in 1775; formerly called Thousand Palms Canyon (Brigandi).

Salvia (sal′ vē ə) [Riverside Co.]. The name is Latin and Spanish for sage, a common desert plant.

Salyer (sal′ yər) [Trinity Co.]. The post office was established on Apr. 16, 1918, and named for Charles Marshall Salyer, a prominent mining man (Bess Moore).

Samagatuma (sä mə gə to͞o′ mə) **Valley** [San Diego Co.]. The name is derived from that of a Diegueño rancheria Jamatayune shown on a *diseño* (about 1845) of the Sierra de Cuyamaca grant. The origin may be *'ehaa 'emat aayum* 'water place spread-in-the-sun' (Langdon).

Sambo Gulch [Siskiyou Co.]. Officially named in 1981 (BGN, list 8101) for a Shasta Indian family that settled here in the 1870s. Sargent Sambo, who died in the 1950s, was one of the last speakers of the Shasta language.

Samoa [Humboldt Co.]. The place was originally known as Brownsville, for James D. H.

Brown, who established a dairy ranch in 1859. In 1889 a group of Eureka businessmen formed the Samoa Land and Improvement Company, so named because the crisis in the Samoan Islands was emphasized in the newspapers and because Humboldt Bay was assumed to be similar to the harbor of Pago Pago. The lumber town, which developed in the 1890s, was called Samoa.

Sampsons Flats [Inyo Co.]. Named for an Indian chief called Sampson, who piloted a party across Kearsarge Pass in 1858 (Chalfant, *Inyo*, p. 76).

Sam's Neck [Siskiyou Co.] was named after "a tricky white stallion named Sam"; there was homesteading here in the 1880s (Luecke).

Samuel P. Taylor State Park [Marin Co.]. The park was created in 1955 and named for Samuel P. Taylor, who had come to San Francisco in 1850 and established the first paper mill here in 1856 (*see* Taylorville).

Samuel Soda Springs. *See* Grigsby Soda Springs.

San Agustin Creek [Santa Cruz Co.]. The name San Gustrín—i.e. Saint Augustine (354–430 C.E.), bishop of Hippo—was given to a land grant, dated Nov. 23, 1833, and Apr. 21, 1841. The stream, which may have been named before the grant, is on the rancho and was mentioned as Rio San Augustine in 1854 (Trask, *Report* 1854:24). The Southern Pacific station San Agustin in Santa Barbara Co. (also spelled San Augustine) is probably another survivor from Spanish times.

San Alejo (or **Alexos**). *See* San Elijo.

San Ambrosio, Isla de. *See* Santa Rosa: Santa Rosa Island.

San Amidio. *See* San Emigdio.

San Andrés. The Spanish name of St. Andrew, one of the apostles and patron saint of Scotland, is found in several living place names and in several obsolete names. San Clemente Island, San Pedro Bay, and Pinole Point once bore the name San Andrés. In the names of California places, we often find the alternative form **San Andreas** (an drā′ əs): **Valley, Lake** [San Mateo Co.]. The valley was named Cañada de San Andrés by Palou on Nov. 30, 1774, the feast day of the saint (Anza 2: 418–19). Hoffmann's map of the Bay region (1873) has "Canada San Andres"; the name of the reservoir that was created in 1875 was spelled San Andreas from the beginning (county map, 1877). **San Andreas Fault** was named after this valley by geologist Andrew Lawson in 1893 (Teather 1986). **San Andres** [Santa Cruz Co.] was the name of a land grant, dated May 21, 1823, and Nov. 21, 1833. **San Andreas** [Calaveras Co.], on San Andreas Gulch, was settled by Mexicans in 1848– 49. The town is mentioned in the *Statutes* of 1854 (p. 222), and the post office is listed in 1858.

San Anselmo (an sel′ mō) [Marin Co.]. Cañada de Anselmo appears in the papers of the Punta de Quintin grant of 1840 and was applied in the present form to the North Pacific Coast Railroad station in the 1890s. It is very likely that the *cañada* (valley) was named for a baptized Indian, and that the San was added to the name later.

San Antonio. The name was extremely popular as a place name, in mission days especially, because Saint Anthony of Padua (1195–1231) was a patron of the Franciscan Order. The name appears in the titles of many land grants and claims; it has survived in a number of places. **San Antonio River, Mission, Creek**, and **Valley** [Monterey Co.]: The mission was founded by Serra on July 14, 1771; the river had been named by him previously (Engelhardt 2:87–88). A post office San Antonio is listed from 1867 to 1887. **San Antonio Creek, Canyon, Mountains, Peak**, and **Heights** [San Bernardino Co.]: The arroyo called San Antonio is mentioned by Garcés on Mar. 21, 1774 (Anza 2:346). A place named San Antonio is recorded on Mar. 24, 1821 (PSP 20:287), and Arroyo de San Antonio is shown on a *diseño* of the Cucamonga grant (1839). Von Leicht–Craven (1874) have San Antonio Mountain for the highest peak of San Gabriel Mountains. San Antonio Heights district was known by this name before 1888 (Minnie Goodrich); San Antonio post office is listed after 1892. Locally San Antonio Peak is now called Old Baldy because winds have denuded its top of vegetation (Doyle); it is labeled Old Baldy Peak on some maps. **San Antonio Creek** [Alameda Co.] was the name applied to the channel between Alameda and Oakland, popularly known as "The Estuary," because it is on the territory of the San Antonio grant, first granted to Luis María Peralta on Aug. 3, 1820. Río San Antonio and the rancho San Antonio are shown on Narváez's Plano of 1830. The Coast Survey charts show the name in 1850. San Antonio Creek, a tributary of Alameda Creek (south of Sunol), is not on the territory of the grant (*see also* Brooklyn Basin). **San An-**

tonio Creek [Ventura Co.]: A place called San Antonio is shown on a *diseño* of the Ojai grant (1837) and is mentioned on Apr. 4, 1841 (DSP Ang. 6:2). **San Antonio Creek** [Sonoma, Marin Cos.] is on the Laguna de San Antonio grant, dated May 6, 1839, and Nov. 25, 1845. A station of the Northwestern Pacific was also called San Antonio after the grant; but the Estero de San Antonio, flowing into Bodega Bay, is outside the territory of the grant. **San Antonio Creek** [Santa Barbara Co.] traverses Rancho Todos Santos y San Antonio, dated Aug. 28, 1841. A *rancho San Antonio* is mentioned on Mar. 11, 1813 (Arch. MSB 6:173), and the place is shown on Narváez's Plano of 1830. The grants called San Antonio in Los Angeles, Santa Clara, and Sonoma Cos. seem to have left no traces in present-day place names; there is, however, another San Antonio Creek in Santa Clara Co., northeast of Mount Hamilton. *For* Cerrito de San Antonio, *see* Cerrito: El Cerrito; *for* Rio de San Antonio, *see* Salinas River.

San Apollinares. *See* Cristianitos Canyon.

San Ardo [Monterey Co.]. The town was laid out in 1886 when the Southern Pacific reached the place; it was named San Bernardo by M. J. Brandenstein, who had bought the San Bernardo Rancho, originally granted June 16, 1841. When the Post Office Dept. objected to the name because of possible confusion with San Bernardino, Brandenstein created a new saint by lopping off "Bern."

San Augustine. *See* San Agustin.

San Benancio (bi nän′ sē ō) **Gulch** [Monterey Co.]. A *diseño* (1834) of Rancho El Toro shows the Cañada de San Benancio running north and south parallel to Arroyo del Toro; another shows it in approximately its present location. The name probably honors one of the four saints named Venantius; *b* and *v* are often interchanged in non-standard Spanish spelling.

San Benito: River, Valley, Mountain, County, town. The Spanish name for St. Benedict (480–543), founder of the Benedictine Order, was applied to what is now San Juan Creek by Crespí on the saint's feast day, Mar. 21, 1772. On Nov. 24, 1774, Palou remarked that the valley had been named Cañada de San Benito by the expedition of 1772. The name became well known and is often mentioned in mission and provincial records. What is now the San Benito River was formerly the San Juan River, and it is still so labeled on the Land Office map of 1861. However, Goddard's map of 1860 calls the stream San Benito down to its confluence with Pacheco Creek, where it becomes the Pajaro River, an alignment still in use. The legislature created the county from a portion of Monterey Co. and named it on Feb. 12, 1874. The name was also used for a land grant in the same county, dated Mar. 11, 1842.

San Benvenuto. *See* Oso: Cañada de los Osos.

San Bernabé [Monterey Co.]. The land grant, dated Mar. 10, 1841, and Apr. 6, 1842, derived its name from the *Cañada de San Bernabé*, named for St. Barnabas (or Barnaby), a companion of St. Paul; the place is mentioned by Font on Apr. 15, 1776.

San Bernardino (bûr nə dē′ nō): **Valley, Mountains, Peak**, city, **County, National Forest**. St. Bernardino of Siena (1380–1444), the great Franciscan preacher, was honored during mission days in a number of place names. An *asistencia* of Mission San Gabriel was established southeast of the site of the modern city, in 1819, at a place called *guachama* (or *guachinga*) by the Indians and named San Bernardino by the padres (Arch. MSB 3:268). It is mentioned as a rancho of the mission in the early 1820s (Arch. Arz. SF 4:2129). The name was supposedly applied first to a temporary chapel by Padre Dumetz and a party from Mission San Gabriel on May 20, 1810, the feast day of the saint. On June 21, 1842, the name was applied to a land grant. In 1851 a group of Mormons bought the rancho and founded the colony that developed into the modern city. The county, created from part of Los Angeles Co., was named on Apr. 26, 1853. The peak is shown as Mt. Bernardino on Wilkes's map of 1841. Locally this peak and San Gorgonio Peak, which rise from a common base, are known as Old Grayback (Doyle). For the mountain range W. P. Blake had suggested the name Bernardino Sierra because he considered it a prolongation of the Sierra Nevada (Pac. R.R. *Reports* 5:1.8), but the name San Bernardino Mountains has been generally used since the 1850s. The name is now restricted to the chain between Cajon and San Gorgonio Passes. The National Forest was established and named in 1893 by order of President Harrison; from 1908 to 1925 it was part of Angeles National Forest. *See* Berdoo Canyon.

San Bernardo refers to St. Bernard of Clairvaux

(1091–1153), a French theologian and leader of the Second Crusade. **San Bernardo** (bûr när′ dō) **Valley** [San Diego Co.]: The name of Rancho San Bernardo appears from 1800; in 1961 the name was given to a modern subdivision on the site. The name San Bernardo Valley became official in 1976 (BGN, list 7601). **San Bernardo Creek** [San Luis Obispo Co.] was named for the San Bernardo land grant dated Feb. 11, 1840. *For* Río de San Bernardo, *see* Santa Ynez River; *for* San Bernardo Island, *see* San Miguel: San Miguel Island. *See also* San Ardo; Bernardo.

San Bruno: Mountain, Point, Canal, town, **Creek** [San Mateo Co.]. The name, apparently applied by Palou in Nov. or Dec. 1774 (Font, *Compl. diary*, p. 348), was given to the creek that rose on the north slope of San Bruno Mountain, paralleled the highway between Colma Schoolhouse and Baden Station, and emptied into the sloughs south of San Bruno Point. This creek is mentioned by Beechey in 1826 and is shown on *diseños* in the 1830s. What is left of it no longer bears the name, but an intermittent stream flowing east from the Buriburi Ridge is now called San Bruno Creek. St. Bruno of Cologne (1030–1101) was a German saint, founder of the Carthusian Order. A place called San Bruno, apparently a cattle range of the San Francisco presidio, is repeatedly mentioned in early provincial and mission records. The mountain and its foothills are mentioned as Sierra de San Bruno by Beechey (1826:2.42) and as Montes San Bruno by Eld on Oct. 29, 1841; they are shown as San Bruno Mountains on Robert Greenhow's "Map of the western and middle portions of North America" (in his *The history of Oregon and California* [Boston: Little Brown, 1844]). Pta. San Bruno appears on Duflot de Mofras's plan 16, and Point San Bruno on the Coast Survey chart of 1851. The town developed around Richard Cunningham's San Bruno House (1862) on the San Bruno toll road; the name of the station of the San Francisco and San Jose Railroad was announced in the itinerary, dated Oct. 16, 1863. There is a San Bruno Canyon in Santa Clara Co.; it is shown as Cañada de San Bruno on a map of the Rancho de la Laguna Seca (1847).

San Buenaventura (bwā nə ven to͝or′ ə) [Ventura Co.]. The establishment of a mission at Santa Barbara Channel named for St. Bonaventure (1221–74), an Italian scholastic, was one of the projects with which José de Gálvez, the *visitador general* of New Spain, had charged the Portolá expedition. When the expedition reached the site on Aug. 14, 1769, Crespí considered it well suited for a mission, and he named the populous Indian village there La Asunción de Nuestra Señora 'The Assumption of Our Lady'. The mission, however, was not established and named until Easter Sunday, Mar. 31, 1782. *For* Río de San Buenaventura, *see* Salinas River. *See also* Buenaventura; Ventura.

San Campistrano. *See* San Juan Capistrano.

San Carlos. The mission San Carlos Borromeo [Monterey Co.] was established on June 3, 1770, near the presidio of Monterey but was transferred to the present site in 1771; it was named for St. Charles Borromeo (1538–84), archbishop of Milan. The mission is now generally known as Mission Carmel. **San Carlos Pass** [Riverside Co.]: On Mar. 16, 1774, the Anza expedition crossed the mountains out of the desert by this pass. The Puerto Real de San Carlos was named by Anza on Mar. 15, 1774 (Anza 2:199), and is mentioned repeatedly in Anza's and Font's diaries. The old name was revived in 1924 when Bolton reconstructed the route (Brigandi). **Potrero de San Carlos** [Monterey Co.]: The name 'pasture of San Carlos' was given to a land grant on Oct. 9, 1837. **San Carlos de Jonata** [Santa Barbara Co.] was the name of a land grant dated Sept. 24, 1845. Jonata was the name of a Chumash rancheria, mentioned in the mission records in 1791 and later. **San Carlos** [San Mateo Co.] does not go back to Spanish times; the name was applied to the town in 1887 when "Captain" N. T. Smith and his associates founded it. The name San Carlos was chosen because it was believed that the Portolá expedition first saw San Francisco Bay on Nov. 4, 1769, the feast day of St. Charles, from the hills behind the present town (Stanger, pp. 128–29). **Punta de San Carlos**, on the Marin shore of the Golden Gate (probably Lime Point), appears on Ayala's map of 1775 and on some later maps; it was apparently named for the first ship to enter the Golden Gate, as well as for the saint. There is a San Carlos Peak in San Benito Co., and a San Carlos Canyon in Monterey Co.

San Carpoforo (kär pō′ fə rō) **Creek** [Monterey and San Luis Obispo Cos.]. In Sept. 1769 the

Portolá expedition entered the Santa Lucia Range through the canyon of this creek, called Santa Humiliana by Crespí (p. 190). The present name, for one of several saints called Carpophorus, was the name of a rancho of Mission San Antonio in 1836 (SP Mis. 5:51). Camino de S. Carpoforo is shown on a *diseño* (1841) of the San Miguelito grant. In Wood's *Gazetteer* the name is spelled San Carvoforo. "This name suffers many variations. I have read it 'San Carpoco,' 'San Carpojo,' 'Zanjapoco,' and 'Zanjapojo,' while in speech the changes are rung on 'Sankypoco,' and 'Sankypoky'" (Chase, *Coast trails*, p. 167). These pronunciations doubtless came about by association with the more familiar Spanish *zanja* 'ditch', often pronounced (sang′ kē) by Anglos in California.

San Cayetano is the name of the 16th-century St. Cajetan (or Gaetano) of Lombardy; it was given to a grant called Bolsa de San Cayetano [Monterey Co.], dated Oct. 12, 1822; to a rejected grant, Huerta de San Cayetano [Alameda Co.]; and as an alternate name of the Moro y Cojo grant [San Luis Obispo Co.]. The Moraga expedition of 1806 named a stream along its route *Arroyo que llamamos S. Cayetano* ('creek which we call St. Cajetan'; Muñoz, p. 35), but the name has not survived there. **San Cayetano** (kä yə tä′ nō) **Mountain** [Ventura Co.] was so named because it is on the territory of the Sespe or San Cayetano land grant, dated Nov. 22, 1833. *For* Arroyo de San Cayetan [Alameda Co.], *see* Cayetano Creek.

Sanchez, Paraje de [Monterey Co.]. *El parage de Sánchez* 'Sánchez's place' is mentioned on Oct. 16, 1806 (Arch. Arz. SF 2:11), and was made a private land grant on June 8, 1839. Out of many people named Sánchez, José Antonio, a well-known Indian fighter and enemy of the clergy, seems to be the one most likely connected with the *paraje*.

San Clemente refers to St. Clement, the third pope and bishop of Rome (Wagner, *Saints' names*, p. 408), whose feast day is Nov. 23. **San Clemente** (klə men′ tē) **Island** [Los Angeles Co.] was named by Vizcaíno about Nov. 25, 1602. Among the many saints' names applied on or near feast days, this one seems to be fitting for the place: St. Clement discovered, through a miracle, a clear spring of water on a barren island. The name was definitely affixed to the island when Costansó used it on his map of 1770. **San Clemente** town and **Beach State Park** [Orange Co.]: The seaside resort was developed in 1925 by Ole Hanson, former mayor of Seattle, and was probably named after the island 60 miles offshore. The state park was created and named in 1931. The name San Clemente was used in other parts of the state; it is still found in San Clemente Canyon [San Diego Co.] and San Clemente Creek and Ridge [Monterey Co.]. *For* Río de San Clemente, *see* Alameda.

Sand (or **Sandy**). About fifty Sand or Sandy Hills, Creeks, Canyons, Flats, Points, etc., are shown on the maps, and many more similar names are used locally (*see* Arena). There is a Sand Spit Park in Santa Barbara Co., and a Sand Rock Peak in Los Angeles Co. Several orographic features are named for Sandstone. **Sandy Mush Country** (Merced Co.) may refer to acorn mush, a California Indian staple, which has been leached through sand, of which some particles remain. **Sand City** [Monterey Co.] is named for its beach location. *For* Sand Cut, *see* Aromas.

San Daniel, Laguna Grande de. *See* Guadalupe: Guadalupe Lake; Laguna: Punta de la Laguna.

Sandberg [Los Angeles Co.]. The post office was established Apr. 22, 1918, and named for the first postmaster, Harold Sandberg, who operated an inn on the old Ridge Route (G. Hamilton).

Sanders [Fresno Co.]. Probably named for C. E. Sanders, who was the first postmaster when the post office was established on Oct. 15, 1879. *For* Sander's Bend [Tehama Co.], *see* Bend.

Sandia (san dē′ ə) [Imperial Co.]. The Spanish word for 'watermelon' was applied to the station when the Southern Pacific line from Calipatria was built in 1924. Like Casaba, the name advertises an important product of the Imperial Valley.

San Diego (dē ā′ gō): **Bay, Mission**, city, **River, County**. The bay was discovered by Cabrillo on Sept. 28, 1542, and named San Miguel for St. Michael the Archangel, whose feast was the following day. From Nov. 10 to 14, 1602, Vizcaíno anchored in the port and renamed it in honor of San Diego de Alcalá de Henares (St. Didacus), a Franciscan saint of the 15th century, whose feast fell on Nov. 12 and whose name had also been given to Vizcaíno's flagship. The first published map to show the modern name San Diego is that of

Abraham Goos (Wagner, no. 292), in the *West-indische Spieghel* (Amsterdam, 1624). When Serra founded the first mission in Alta California on July 16, 1769, he dedicated it to the same saint and called it La Misión de San Diego de Alcalá. The original Mexican pueblo at Presidio Hill is now Old Town. New San Diego was laid out in 1850 but did not flourish until the coming of A. E. Horton in 1867. The county, one of the original twenty-seven, was created and named by act of the legislature on Feb. 18, 1850.

San Dieguillo. *See* San Dieguito.

San Dieguito (dē ä gē′ tō): **Valley, River** [San Diego Co.]. The stream and a place spelled San Dieguillo are mentioned by Font on Jan. 10, 1776. The diminutive was doubtless used to distinguish the place from San Diego. An Indian rancheria, San Dieguito, under the jurisdiction of San Diego Mission, is mentioned in 1778 (Arch. MSB 1:158, PSP Ben. Mil. 1:41); and an Indian pueblo, San Dieguito, was established before 1834 (Engelhardt 3:531). In 1840 or 1841 the name was applied to a rancho, a grant for which was finally issued on Aug. 11, 1845.

San Dimas (dē′ məs): town, **Canyon, Wash, Junction** [Los Angeles Co.]. In 1886–87 the San Jose Land Company laid out and named the town at the site formerly known as Mud Springs (Santa Fe). St. Dismas was the penitent thief crucified at the side of Christ. The name may go back to Spanish times, but there is no evidence to support the story that Ygnacio Palomares, grantee of Rancho San Jose, chose the name in allusion to the unrepenting Indian robbers who stole his cattle.

San Domingo. The name of St. Dominic (1170–1221), a Spanish saint and founder of the Dominican Order, was occasionally used in place naming in Spanish times; the standard Spanish form would be Santo Domingo. Creeks in Calaveras and Sacramento Cos. still bear the name of this saint.

San Elijo (ə lē′ hō): **Valley, Lagoon** [San Diego Co.]. The Portolá expedition camped not far from Batequitos Lagoon on July 16, 1769, and called the place where they camped San Alejo (Crespí, pp. 127–28) in honor of St. Alexius, whose feast day is July 17. Costansó mentions the *cañada de San Alexos* on July 16, 1769. The name of the valley was preserved in the alternate name of the Encinitos grant, Cañada de San Alejo, on July 13, 1842. The present San Elijo Lagoon is the southern boundary of this grant. At some point the name was evidently confused with that of San Eligio, St. Eligius or Eloi (d. 659), the patron of goldsmiths. The name does not appear on the older charts of the Coast Survey, but San Elejo Creek is shown on the Land Office map of 1891. Cardiff-by-the-Sea (*q.v.*) was formerly called San Elijo.

Sanel (sə nel′) **Mountain** [Mendocino Co.]. A populous Pomo Indian village called Se-nel or Shanel was situated southeast of Hopland; its site is still shown on the Hopland atlas sheet as Rancho del Sanel. The origin of the name is Central Pomo from *šane* 'ceremonial house' plus locative *-l* (Oswalt). On Nov. 9, 1844, the name was used for the land grant Sanel; the *diseños* show Ranchería de Sanel and Río permanente de Sanel. In 1859 the name was applied to the settlement west of the river, now Hopland. The name for town and mountain are shown on the von Leicht–Craven map of 1874.

San Emigdio (ə mēd′ ē ō): **Canyon, Creek, Mountain** [Kern Co.]. The names are derived from the name of a rancho of Mission Santa Barbara, mentioned in 1823–24 (Arch. Arz. SF 4:2.91, DSP 1:42). On July 13, 1842, the property became a private land grant, usually spelled "San Emidio." The saint for whom the name was given was apparently Emygdius (or Emidius), a German martyr (d. 304); he is supposed to provide protection from earthquakes, which are frequent in this region. The name is mentioned repeatedly as San Amidio and Emidio in Brewer's Notes from the spring of 1863.

San Felipe. The name has survived as a place name in San Diego, San Benito, and Santa Clara Cos. There were at least four well-known holy Philips; Wagner (*Saints' names*) ascribes the name to the 1st-century saint who was crucified in Asia Minor and whose feast day is May 1 (or Aug. 1). **San Felipe** (fə lē′ pē, fə lē′ pā): **Valley, Creek** [San Diego Co.]: On Apr. 18, 1782, Pedro Fages and his party gave the name San Phelipe to the place now called Vallecito. The name soon spread to the valley and to an Indian village in it: the *gentiles de San Felipe* are mentioned on Apr. 27, 1785 (PSP 5:210). The name of the creek, which is also in Imperial Co., probably comes from this source. The ranchera, the valley, and the Sierra de San Felipe are repeatedly mentioned in early records. On Feb. 21, 1846, the name Valle de San Felipe was applied to a land grant. The place, San Felipe, is recorded on Thomas Coulter's and Wilkes's maps and is mentioned as San Felippe by Emory in 1846–47 (Mil. Rec. 104). However, the valley and creek in San Diego Co. that are now called San Felipe were so named around the 1820s, when Mexican authorities began to explore the area (Brigandi). **San Felipe** (fə lip′ ē, fə lip′ ā) **Creek, Hills, Lake**, town, and **Valley** [San Benito, Santa Clara Cos.]: The names were preserved through three land grants: San Felipe, Apr. 1, 1836; Bolsa de San Felipe, Nov. 13, 1837; and Cañada de San Felipe y las Animas, Aug. 27, 1839, and Aug. 1, 1844. The stream is recorded as San Felipe River on Wilkes's map of 1841. Creek and Lake are mentioned in the *Statutes* of 1850 (p. 59). Padre Garcés had given the name to the Kern River on May 1, 1776, and it is shown as Río de San Philipe on Font's map of 1777 (Coues, *Trail*, 282–83).

San Fernando (fûr nan′ dō): **Mission, Valley, Pass**, city, **Reservoir** [Los Angeles Co.]. The valley was mentioned under various names by the Portolá and Anza expeditions. The present name comes from the Misión San Fernando Rey de España, founded on Sept. 8, 1797, and named in honor of Ferdinand III, king of Castile and Leon (1200–1252). Many other features named after the mission by the Pacific Railroad Survey (peak, range, hills, pass) received different names when the USGS mapped the region in the 1890s. The city was laid out in 1874 on part of the lands in San Fernando Valley formerly owned by Eulogio de Celis.

San Francisco. The number of saints named Francis, and especially the fame of St. Francis of Assisi (1182–1226), has made this name one of the most popular in Roman Catholic countries. Franciscan friars accompanied Portolá to California, and in 1776 the Franciscans were entrusted with the spiritual work in both Californias. The name was used for the land grants San Francisco de las Llagas 'of the wounds' [Santa Clara Co.] on Feb. 3, 1834, and San Francisco [Los Angeles, Ventura Cos.] on Jan. 22, 1839, and as an alternate name for several other grants. **San Francisco** (frən sis′ kō): **Bay**, city, **County**: The name San Francisco appears on Petrus Plancius's map of 1590; and on his map of 1592 (Wagner, pl. 20), it is shown for a cape in a location that—judging from the distances to Cape Mendocino and to Cabo San Lucas, on the peninsula of Lower California—appears almost exactly in the latitude of present-day San Francisco. Wagner (p. 500) is doubtless right in assigning Plancius's Cabo de San Francisco, which is shown on some later maps, to the realm of imaginary geography; however, it is a remarkable coincidence that a Dutch cartographer in 1592 should attach this name to a spot on the map where, centuries later, it would actually occur as such a famous place name. On Nov. 6, 1595, Rodríguez Cermeño entered the harbor now known as Drakes Bay, and the following day he named it Bahía or Puerto de San Francisco (Wagner, p. 499). The date is not close to any of the feast days of the several saints named Francis; but since the christening was performed by a member of the Franciscan Order, it may be assumed that it was in honor of the saint of Assisi. Bolaños's *derrotero* (1603) records this Puerto de San Francisco (Wagner, p. 118) about one degree north of the actual latitude of present San Francisco Bay. Until 1769 the name remained a vague geographical conception; however, by another coincidence, George Anson's map of 1748 shows the outline of the bay, the Farallones, the strait now known as the Golden Gate, the peninsula, and Point Reyes—all in fairly proper position (Wagner, pl. 30). British and other non-Spanish maps after 1625 usually label the uncertain bay with the name of Francis Drake in accordance with the English claim to this part of the coast (*see* New Albion). On Oct. 31, 1769, a detachment of the Portolá expedition beheld from the hills near Pedro Point [San Mateo Co.] the large body of water between them and Point Reyes (now known as the Gulf of the Farallones), and they believed they had rediscovered the original Bahía de San Francisco. A few days later, when they saw the bay now called San Francisco, they called it merely an estuary (*estero*) of the port. Costansó's map of 1770 records both the Puerto (Drakes Bay) and the Estero de San Francisco. On Ayala's map of 1775 the name Puerto de San Francisco indicates San Francisco Bay. The presidio was dedicated and formally named on the day of the Wounds of St. Francis, Sept. 17, 1776; La Misión de Nuestro Seráfico Padre San Francisco de Asís a la Laguna de los Dolores 'the mission of our seraphic father Saint Francis of Assisi at the Lake of [Our Lady of] the Sorrows' was founded on June 29 and dedicated on Oct. 8 or 9, 1776. The pueblo at the presidio was established on Nov. 3, 1834, and the pueblo on Yerba Buena Cove was founded by order of Gov. Figueroa in 1835. William A. Richardson and Jacob P. Leese were the first Anglo settlers. This village became the most important of the three settlements, eventually absorbing the villages at the presidio and the mission. It was known as Yerba Buena; but when W. A. Bartlett, its chief magistrate, recorded the agreement to found the town of Francisca (*see* Benicia), he feared that the name of the new port at Carquinez Strait would be confusing and also detrimental to the interests of Yerba Buena; hence he decided to establish the name as San Francisco. He published the following proclamation in the issues of Jan. 23 and 30, 1847, of the *California Star:* "Whereas the local name of Yerba Buena, as applied to the settlement or town of San Francisco, is unknown beyond the immediate district; and has been applied from the local name of the Cove on which the town is built—therefore to prevent confusion and mistakes in public documents, and that the town may have the advantage of the name given on the published maps, it is hereby ordered that the name of San Francisco shall hereafter be used in all official communications, or records appertaining to the town." Bartlett, however, had no authority to confirm the name, and it was not made official until Mar. 10, 1847, when Gen. Stephen W. Kearny, the military governor, decided for San Francisco (Bowman). The nickname "Frisco" has been traditionally disfavored by local people, but many people in northern California refer to San Francisco as "The City." The county, one of the original twenty-seven, was named on Feb. 18, 1850. The name is now sometimes pronounced (san ran sis′ kō). *For* Rio de San Francisco and Llano de San Francisco, *see* San Joaquin: San Joaquin River; Sacramento River; Tule: Tule River.

San Francisco Solano Mission. *See* Solano.

San Francisquito. The diminutive suffix *-ito* or *-ita* was frequently used in Spanish times to designate a geographic feature smaller than another one previously named. When applied to saints' names, this way of naming one feature after another often results in ambiguous combinations and sometimes produces a comic effect. Arroyo de San Francisquito might properly be translated as 'Little San Francisco Creek'—but it is not the Holy Francis who is little, but the creek. The name San Francisquito is found as the principal or alternate name of eight land grants, and at least three such names have survived. **San Francisquito** (fran sis kē′ tō) **Creek** [San Mateo, Santa Clara Cos.] was the site of Palou's camp on Nov. 28, 1774, and he selected the spot as a suitable place for a mission to be dedicated to St. Francis of Assisi (*Diary*, pp. 411–12). Anza's diary of Mar. 26, 1776, mentions the Arroyo de San Francisco; and he adds that the spot had been found unsuitable for a mission because of lack of water there in the dry season. After the San Francisco Mission (popularly called Mission Dolores) was established, the stream came to be known as Arroyo de San Francisquito. This name was given to three land grants: San Francisquito, May 1, 1839, on which Stanford University is now situated; Rinconada del Arroyo de San Francisquito, Feb. 16, 1841; and Rincón de San Francisquito, Mar. 29, 1841. The modern hybrid name for the stream is mentioned in the *Statutes* of 1850 (pp. 59–60). **San Francisquito Flat** and **Sierra de San Francisquito** [Monterey Co.] are both on the San Francisquito land grant of Nov. 7, 1835. The flat is probably *el llanito de San Francisquito*, mentioned as early as 1822 (Arch. MSB 3:297). **San Francisquito Canyon** [Los Angeles Co.] is on the San Francisco land grant. The name is associated with various gold discoveries in the San Fernando Valley, six years before the discovery at Sutters Mill. Francisco López discovered the gold deposits in Mar. 1842 and was for some time in partnership with a Frenchman, Charles Baric. Duflot de Mofras's map of 1844 records the *Mine d'or de Mr. Baric*. Rancho San Francisquito and San Francisquito Pass are shown on the Parke-Custer map of 1855 and are repeatedly mentioned in the Pac. R.R. *Reports*.

San Gabriel (gā′ brē əl): **Valley, Mission, River, Mountains, Peak**, city, **Dams; Gabriel Canyon** [Los Angeles Co.]. The valley was discovered by the Portolá expedition and named San Miguel Arcángel on July 30, 1769 (Crespí, pp. 144–45). The Misión del Santo Arcángel San Gabriel de los Temblores 'Mission of the Holy Archangel St. Gabriel of the Earthquakes' was established and named on Sept. 8, 1771, and was moved to the present location in 1776. The word *temblores* was added to the name because earthquakes had been felt at this place by the Portolá expedition (Arch. MSB 1:101, PSP 1:118), or because the intended site for the mission had been on the Río de los Temblores, now the Santa Ana River (Palou 2:323). The Arroyo de San Gabriel is mentioned by Font on Jan. 4, 1776, but this was probably not the present San Gabriel River. However, on Pantoja's map of 1782 the name is already identified with that river. The stream is shown but not labeled on the Frémont-Preuss map of 1848. Williamson called the upper part Johnson's River, "after the soldier who found it for us" (Pac. R.R. *Reports* 5:1.30). The mountain chain of which the San Gabriel Mountains are an important part was vaguely called the Sierra Madre by the missionaries, but a Cierra [Sierra] de San Gabriel is mentioned in Aug. 1806 (Arch. MSB 4:67) and later. Blake, in 1853, mentions as local names Qui-Quai-mungo Range and San Gabriel Range (Pac. R.R. *Reports* 5:2.137). The Whitney Survey adopted the name San Gabriel Range, but the BGN in 1927 decided for San Gabriel Mountains. The San Gabriel post office was established on July 26, 1854. *For* Río de San Gabriel, *see* Hondo; Kaweah. *For* Caxon de San Gabriel de Amuscopiabit, *see* Cajon.

Sanger [Fresno Co.]. The station was named for Joseph Sanger Jr., an official of the Pacific Improvement Company (an affiliate of the Southern Pacific) when the branch from Fresno to Porterville was built in 1888.

Sanger Meadow [Inyo Co.]. Named for the Sanger family, pioneers of Inyo and Kern Cos., who ran horses here (Brierly).

San Geronimo (jə ron′ i mō): **Creek**, town [Marin Co.]. The name Cañada de San Gerónimo was applied to a land grant of Feb. 12, 1844. Cañada or Arroyo de San Gerónimo is shown on the *diseños* and maps of the grant. It is not known whether the place was named for one of several St. Jeromes, or whether it was named for an Indian who had received the name Gerónimo at his baptism. On Hoffmann's map of the Bay region, present Lagunitas Creek is called Arroyo San Geronimo, and present San Geronimo Creek is labeled Arroyo Nicasio. The name was applied to the station of the North Pacific Coast Railroad in 1875. The post office is listed in 1898. **San Geronimo** [San Luis Obispo Co.], a land grant dated July 24, 1842, seems to have left no traces in present-day place names.

San Gorgonio (gôr gō′ nē ō): **Pass, River, Creek, Mountain** [San Bernardino, Riverside Cos.]. San Gorgonio was a cattle ranch of Mission San Gabriel in 1824 (DSP 1:27), named for Gorgonius, a 3d-century martyr, whose feast day is Sept. 9 (Wagner, *Saints' names*). The name appears in the land grant San Jacinto y San Gorgonio, dated Mar. 22, 1843. Arroyo and Valle de San Gorgonio are shown on the *diseño*. The pass, now traversed by the Southern Pacific, is shown as Pass Saint Gorgonie on Gibbes's map of 1852. The Pacific Railroad Survey gave this name to a peak and mountain, but these appear on modern maps as San Jacinto Peak and Mountains. The present-day San Gorgonio Mountain has also been called Grayback. The town Beaumont (*q.v.*) [Riverside Co.] was once named San Gorgonio after the mountain.

Sangre, Isla de la. *See* Bloody: Bloody Island [Tehama Co.].

San Gregorio (gri gôr′ ē ō): **Creek, Valley**, post office, **Mountain** [San Mateo Co.]. The name was applied to this area in a land grant petition of 1831 (Alan K. Brown). The name honors the pope St. Gregory the Great (540–604). The post office was established in the 1870s. A place in Los Angeles Co. was called San Gregorio by Crespí on Aug. 4, 1769, and Anza gave the name to a place in Borrego Valley [San Diego Co.] on Mar. 12, 1774; but these names did not survive.

San Guillermo (gē yâr′ mō) **Mountain** [Ventura Co.]. Guillermo is Spanish for William, the name of several saints.

Sanhedrin (san′ hē drin) **Mountain; Sanhedrin Creek** [Mendocino Co.]. The peak is given as Sanhidrim on the von Leicht–Craven map of 1874. A few years later the Coast Survey established a triangulation station there and used the present spelling. The region was first settled by Missourians; it is possible that they named the mountain after Mount Sanhedrin in their home state, which in turn may have been named after the supreme council of the ancient Hebrews.

San Isidro [San Benito, Santa Clara Cos.]. The name of one of several St. Isidores was given to a land grant conveyed to Ignacio Ortega in

1808–9 and regranted in three parts to his heirs in 1833. The name is shown on Narváez's Plano of 1830 and is often mentioned in mission and land grant records, usually spelled San Ysidro (*q.v.*).

Sanitarium [Napa Co.]. In 1878 the Seventh-Day Adventists established the St. Helena Sanitarium and called the place Crystal Springs. The community developed around the institution, and the name was soon changed by popular usage to Sanitarium (E. L. Place).

San Jacinto (jə sin′ tō, hə sin′ tō): town, **Lake, Indian Reservation** [Riverside Co.]. San Jacinto (Viejo) was a cattle ranch of Mission San Luis Rey as early as 1821 (Arch. MSB 4:223). Between 1842 and 1846 the name of the saint appears in the names of four different land grants. The name honors a Dominican, St. Hyacinth of Silesia (d. 1257). The mountain is mentioned in Hutchings' *Illustrated California Magazine* of Feb. 1859. Anglo settlers began to arrive in the 1860s, and San Jacinto School district was formed in 1868. A settlement developed around the store built around 1869 by a Russian exile, Procco Akimo, and a formal townsite was laid out in 1883, a couple of miles away (Brigandi); the post office, however, had already been in existence since July 27, 1870. *For* San Jacinto Hot Springs, *see* Gilman Hot Springs; *for* San Jacinto Plains, *see* Perris Valley.

San Jacome de la Marca. *See* Jamacha.

San Joaquin (wä kēn′). St. Joachim, honored by Roman Catholics as the father of the Virgin Mary, was often commemorated in place names; his name has survived in one of the most important geographical names of the state and in the names of several lesser features. **San Joaquin: River, County, Valley, Bridge;** post office [Fresno Co.]. Crespí saw the river on Mar. 30, 1772, when he and Fages were attempting to reach Point Reyes, and he named it San Francisco for St. Francis of Assisi. Gabriel Moraga gave the name San Joaquín to the river when he reached its southern part in 1805–6, according to Padre Muñoz's diary (Arch. MSB 4:1 ff.). Before and after Moraga's visit, various sections of the river had different names. However, in the records after 1810, San Joaquin is mentioned as if it were a well-known name. The name appears for the upper course of the river on Estudillo's map of 1819, and three main channels of the lower course are shown with this name on the *Plano topográfico de la Misión de San José* (about 1824). The accounts of Kotzebue, Beechey, and Wilkes allow no doubt about the river's identity. On Narváez's Plano of 1830, to be sure, the San Joaquin Valley is shown covered by an enormous swamp, Ciénegas o Tulares. The maps of Wilkes (1841) and Frémont-Preuss (1845) definitely identify the name with the major part of the river. The former has San Joachim, but Frémont uses the Spanish version. The county is one of the original twenty-seven, created and named on Feb. 18, 1850. A San Joaquin City existed in 1850 (*San Francisco Alta California*, Feb. 7, 1850), but it soon vanished. The name San Joaquin Valley seems to have come into general use at the time of the Pacific Railroad Survey, 1853–54. The bridge across the river was built by the Central Pacific in 1869, and the station was named San Joaquin Bridge (Mary Seamonds). A post office in San Joaquin Co. is listed as San Joaquin in the 1880s; the post office in Fresno Co. was established in 1915. **San Joaquin Peak** [San Benito Co.] was so named because it is on the San Joaquin or Rosa Morada grant, dated Apr. 1, 1836. **San Joaquin Hills** [Orange Co.] were so named because they are on the Bolsa de San Joaquín grant, dated May 13, 1842. **Castillo de San Joaquin** [San Francisco Co.] preserves the name of the old Spanish fort that stood at the site of Fort Winfield Scott. **San Joaquin** [Contra Costa Co.] was the name given by W. H. Davis to his mother-in-law's ranch, a part of Rancho Pinole. To the middle name of his father-in-law, José Joaquín Estudillo, Davis "added San . . . then it became the name of a Saint" (William Heath Davis, p. 531). (*For* San Joaquín de la Laguna, *see* Santa Barbara.)

Sanjon (sän hōn′) **Barranca** [Ventura Co.]. For Spanish *zanjón* 'big ditch', augmentative of *zanja* (*q.v.*). The name is attested from 1877 (Ricard).

San Jose. Of all saints' names, that of San José—St. Joseph, husband of the Virgin Mary—is probably the most popular for place names in Spanish-speaking countries. In California, from the beginnings of colonization, the name has been intimately connected with geographical nomenclature. José de Gálvez, *visitador* of all New Spain, in a solemn proclamation of Nov. 21, 1768, named St. Joseph the patron saint of the first expedition to settle

Alta California; and the new ship built at San Blas in Mexico, which was to bring part of the Portolá expedition to Monterey, was named in his honor. The vessel was disabled, and after being repaired was lost at sea—an ill omen, but it did not deter the Spaniards from their venture. The name appears as the principal or alternate name for about fifteen land grants and claims and has survived for more than ten geographic features. **San Jose** (san hō zā′, san ō zā′) [Santa Clara Co.] was founded by José Joaquín Moraga on Nov. 29, 1777, under instructions from Gov. Felipe de Neve, who named it Pueblo de San José de Guadalupe, for St. Joseph and for the river on which the town was situated. It was the first Spanish pueblo in what is now the state of California; thus the modern city has the distinction of being the oldest civic municipality of the state. In 1849 it became the state's first capital when the legislature convened there on Dec. 15. On older maps Río de San José is found for Guadalupe River, Port San Jose for Alviso, and Estrecho de San José for the southern arm of the bay. The mission named San Jose [Alameda Co.] was established on June 11, 1797, at a place known as San Francisco Solano—at some distance from the pueblo, in accordance with Spanish methods of colonization (*see* Mission: Mission San Jose). **San Jose Valley** [San Diego Co.], now partly covered by the Henshaw Reservoir, was named San Josef (or San José) by Fr. Juan Mariner, of San Diego Mission, in 1795 (Brigandi); in 1821 it appears as *el Valle de San José o Guadalupe*. The name is recorded for two land grants called Valle de San José, dated Apr. 16, 1836, and Nov. 27, 1844. The grantee of the latter was Juan J. Warner, who developed his famous ranch on this grant. **San Jose: Creek, Hills, Peak, Wash** [Los Angeles Co.] were so named because they are partly on the territory of the San José grant, dated Apr. 15, 1837, and Mar. 14, 1840. **Arroyo San Jose** [Marin Co.], mentioned in 1828 (Registro, p. 147), had its name preserved through the San José grant, dated Mar. 24, 1838, and Oct. 3, 1840. **San Jose Creek** [Monterey Co.] is on the territory of the San José y Sur Chiquito grant, dated Apr. 16, 1839. *For* Punta de San José, *see* Fort Point. *For* Arroyo de San José Cupertino, *see* Stevens Creek; Cupertino. *For* San José de Alvarado, *see* Alvarado.

San Josef o Cantil Blanco, Punta de. *See* Fort Point.

San Juan. Besides the important place names derived from the missions San Juan Bautista and San Juan Capistrano, the name was used in several land grants and has survived in a number of geographical names. Wagner (*Saints' names*) lists six St. Johns honored in California nomenclature.

San Juan Bautista (wän bô tēs′ tə): **Mission**, post office; **San Juan: Creek**, town [San Benito Co.]. The mission was founded, and named in honor of St. John the Baptist, by Padre Lasuén on June 24, 1797, the saint's feast day. The present San Benito Valley was called Llano de San Juan by Beechey (1831:2.49); it appears as San Juan Valley on Wilkes's map of 1849 and is so called by Trask in his *Report* (1854:67). In the 1850s the Gabilan Range was called the San Juan Range. The post office was established under the name San Juan in 1852; this name was changed to the present form about 1905. The name Río de San Juan Baptista is given to the San Joaquin River on Ayala's map of 1775. On Mar. 30, 1844, the name San Juan Bautista was applied to a land grant in Santa Clara Co.

San Juan Capistrano (kap i stran′ ō, kap i strä′ nō): **Mission**, post office; **San Juan: Creek, Canyon, Rock, Hot Springs, Seamount** [Orange Co.]. The name San Juan Capistrano was first given by Crespí, on July 18, 1769, to the valley in San Diego Co. now known as San Luis Rey Valley. In 1775 Viceroy Bucareli gave instructions to name the next mission in honor of the fighting priest, St. John Capistran (1385–1456), who took a heroic part in the first defense of Vienna against the Turks. The site of the mission was first dedicated on Oct. 30, 1775; but because of the Indian revolt at San Diego in Nov. 1775, the mission was not formally founded until Nov. 1, 1776. Serra gave the new mission the name San Juan Capistrano de Quanis-savit, the last name doubtless referring to the Indian rancheria at the site (Palou, *Vida . . . Serra*, pp. 174–75; Engelhardt). On early American maps the name is usually shortened to San Juan, and on a military map of 1847 it is "San Campistrano." Goddard 1860 records the name San Juan Capistrano. The short form Capistrano was used for the post office from the time it was established on June 5, 1867, until Oct. 23, 1905, when Zoeth Eldredge prevailed upon the Post Office Dept. to restore the original name (*see* Capistrano). Point San Juan appears on Duflot de Mofras's map of 1844. Von Leicht–Craven 1874 give Point Capistrano; the Coast Survey gives Dana Point; and the USGS calls the cape San Juan Capistrano Point. Dana Cove was known in Spanish times as Bahía de San Juan Capistrano. The creek is shown as San Juan River on the Coast Survey charts of the 1880s. The name San Juan Capistrano was apparently applied to the San Joaquin River at some time, and Wilkes's map of 1841 still has a "R. St. Juan" for a tributary to the San Joaquin. A land grant, San Juan Capistrano del Camote ('of the sweet potato') [San Luis Obispo Co.], is dated July 11, 1846. *For* Rio San Juan, *see also* Calaveras: Calaveras River; *for* Mount San Juan Bautista [Contra Costa Co.], *see* Diablo, Mount. *For* San Juan by the Sea, *see* Serra; *for* San Juan Capistrano el Viejo [San Diego Co.], *see* San Luis Rey.

San Juan Hill [Nevada Co.]. *See* North San Juan.

San Julian [Santa Barbara Co.]. San Julián, the name of one of several St. Julians, was applied to a rancho in the jurisdiction of the presidio of Santa Barbara before 1827 (Docs. Hist. Cal. 4:811 ff.). It is shown on Narváez's Plano of 1830 and was made a land grant on Apr. 7, 1837.

San Justo [San Benito Co.]. It is uncertain which St. Justus was honored in the names of two land grants, dated Feb. 18, 1836 (not confirmed), and Apr. 15, 1839. In modern times the name is preserved only in the name of a school (*see* Hollister).

Sankey [Sutter Co.]. The station of the Sacramento Northern was probably named for Calvin Sankey, a native of Missouri, who was a farmer in the Vernon district in 1866.

Sankypoco. *See* San Carpoforo Creek.

San Ladislao. *See* Buchon Point.

San Leandro (lē an′ drō): **Creek**, city, **Bay, Hills, Reservoir** [Alameda Co.]. An Arroyo de San Leandro is mentioned in 1828 (Registro, p. 6) and Río San Leandro is shown on Narváez's Plano of 1830. The saint honored was probably St. Leander, archbishop of Seville, "Apostle of the Goths" in the 6th century. A land grant, San Leandro, is dated Aug. 16, 1839, and Oct. 16, 1842; the name also appears in the titles of several other grants. The settlement of the grantee, José J. Estudillo, is shown north of the creek on Duflot de Mofras's map of 1844. The town south of the creek was laid out in 1855; the post office, however, was already established on May 3, 1853.

San Lorenzo (lə ren′ zō) **River** [Santa Cruz Co.] was named by the Portolá expedition on Oct. 17, 1769 (Crespí, p. 215). The saint honored was probably St. Laurence, a martyr of the 3d century. The name appears in the title of a grant, Cañada del Rincón ('canyon of the corner') en el Río San Lorenzo (July 10, 1843), and the Americanized version is mentioned by Trask (1854, p. 8). **San Lorenzo: Creek**, town, **Village** [Alameda Co.]: The creek was named Arroyo de San Salvador de Horta by Crespí on Mar. 25, 1772 (p. 287). "It is known also as the Arroyo de la Harina, for so the soldiers called it during the journey of Señor Fages, because in it a load of flour (*harina*) got wet" (Font, *Compl. diary*, p. 355). The name San Lorenzo was applied to two land grants dated Feb. 23, 1841, and Oct. 10, 1842; however, the name had been similarly used much earlier. Joaquín Castro testified on Feb. 18, 1854, in the San Lorenzo (1842) land grant case: "In 1812, it was then called by the same name." Rancho de San Lorenzo is also shown on the *Plano topográfico de la Misión de San José* (about 1824). The creek forms part of the boundary of the two grants, but it is not known whether the creek or one of the ranchos was named first. The town was laid out in 1854 by Guillermo Castro, grantee of the 1841 rancho (*see* Castro: Castro Valley). The town is mentioned in the *Statutes* of 1854 (p. 223). **San Lorenzo Creek** [Monterey Co.]: A *sitio de San Lorenzo* is mentioned on July 20, 1823 (SP Sac. 10:13), and Arroyo de San Lorenzo is shown on a *diseño* of San Bernabé (1841). The name was applied to three grants, dated Aug. 9, 1841, Nov. 16, 1842, and Feb. 18, 1846.

San Lucas [Monterey Co.]. The Spanish name of St. Luke the Evangelist was applied to a land grant dated May 9, 1842. When the Southern Pacific built the section from King City to Paso Robles in the fall of 1886, the station was named after the grant, although it is outside of its boundaries. The post office is listed in 1892. St. Luke's name was used for other place names in Spanish times. Crespí's Arroyo de San Lucas for Coja Creek in Santa Cruz Co. (p. 216) did not survive; but there is a San Lucas Creek, a tributary of the Santa Ynez River, in Santa Barbara Co. *For* Islas de San Lucas, *see* Channel Islands; San Miguel Island.

San Luis (lo͞o′ is): **Creek, Hill** [Merced Co.]. An

expedition sent out from the presidio of San Francisco, probably in 1805, named this place San Luis Gonzaga because it was discovered on June 21, the day of St. Aloysius Gonzaga (1568–91), an Italian Jesuit. Padre Muñoz mentions the place on Sept. 22, 1806, and Padre Viader on Aug. 24, 1810. On Oct. 3, 1843, the name San Luis Gonzaga was applied to a land grant. *For* San Luis [Santa Barbara Co.], *see* Gaviota.

San Luis Beltrán, Cañada de. *See* Waddell Creek.

San Luisito (lo͞o i sē′ tō) **Creek** [San Luis Obispo Co.] was named after the San Luisito grant, dated Aug. 6, 1841. There is no "little Saint Louis"; the name is simply a diminutive of nearby San Luis Obispo (*see* San Francisquito).

San Luis Obispo (lo͞o′ is ə bis′ pō): **Mission, Bay**, city, **County, Creek, Peak; San Luis: Canyon, Hills, Range; Port San Luis**. The mission was founded by Serra on Sept. 1, 1772, and named in honor of St. Louis, bishop of Toulouse, son of the King of Naples and Sicily (13th century). The Indian name of the site was *tixlini*, and in early records the terms *de tixlimi* and *de los tichos* are found attached to the name of the mission (Arch. MSB 1:125, Docs. Hist. Cal. 4:12). The bay (*ensenada*) is recorded as San Luis on Narváez's Plano of 1830. Potrero de San Luis Obispo is the name of a grant dated Nov. 8, 1842. The county, one of the original twenty-seven, was established and named on Feb. 18, 1850. The town was laid out and named by William R. Hutton in Aug. 1850; the post office is listed in 1851. Port San Luis, the city's harbor, was formerly called Port Harford after John Harford, who had built a wharf in 1872–73.

San Luis Rey (lo͞o′ is rā′): **Mission, River**, post office [San Diego Co.]. The valley through which the river flows was named San Juan Capistrano by the Portolá expedition on July 18, 1769, and was intended as the site of a mission (Crespí, p. 131). In Oct. 1797 the site was again recommended and chosen, but was referred to as San Juan Capistrano el Viejo 'Old Capistrano' because in the meantime the name had been used for a mission elsewhere (Engelhardt 2:496). Padre Lasuén founded the new mission on June 13, 1798 (Prov. Recs. 5:277) and on the request of the viceroy named it in honor of St. Louis (1215–70), king of France. The site had been called Jacayme by the natives (Prov. Recs. 6:98). Boca del Río ('mouth of the river') San Luis is shown on a *diseño* (1845) of Santa Margarita y Las Flores grant. The post office was established before 1867.

San Manuel (man′ yo͞o əl) **Indian Reservation** [San Bernardino Co.]. St. Manuel, supposedly of Persian origin, lived in the 4th century.

San Marcos (mär′ kəs, mär′ kōs). St. Mark the Evangelist was often honored in place names in Spanish times. The Tehachapi Range and the south end of the Sierra Nevada had been named Sierra de San Marcos by Garcés in 1776 (Coues, *Trail*, pp. 270–71). The name is shown on Font's map of 1777 (Davidson copy), and on Humboldt's map (1811) it designates the central part of the Sierra Nevada. **San Marcos Valley, Creek, Mountains**, and town [San Diego Co.]: El Valle San Marcos is mentioned in a letter in 1797. In 1835 San Marcos was the name of one of the ranchos of Mission San Luis Rey (SP Mis. 6:10–11). The name appears in the land grant Los Vallecitos ('little valleys') de San Marcos, dated Apr. 22, 1840; the *vallecitos* are shown on a *diseño*. The patent of the grant was not issued until 1883; the town developed around the Santa Fe station after the "boom year" of 1887 (Santa Fe). **San Marcos Pass** [Santa Barbara Co.] is named for a rancho of Mission Santa Barbara mentioned in 1817 (PSP 20:178). The place is shown on Narváez's Plano of 1830. On June 8, 1846, the name was given to the land grant. "San Marcus Pass" across the Santa Ynez Mountains is mentioned by John G. Parke in 1855 (Pac. R.R. *Reports* 7:1.2). **San Marcos Creek** [San Luis Obispo Co.] is mentioned as El Arroyo de San Marcos as early as 1795 (Arch. MSB 2:18 ff.). *For* Sierra de San Marcos, *see* Sierra Nevada.

San Marino (mə rē′ nō) [Los Angeles Co.]. In 1878, James de Barth Shorb built a home on the land that he and his wife had received from his father-in-law, "Don Benito" Wilson. He named the estate San Marino after his birthplace in Emmitsburg, Maryland—which in turn had probably been named after the tiny European republic, and this in turn after a saintly Italian stonemason and hermit of the 4th century. In 1903 Henry E. Huntington bought the property and retained the name for his estate, the site of the Huntington Library and Art Gallery (Edith Shorb Steele).

San Martin (mär tēn′) [Santa Clara Co.]. Martin Murphy, a native of Ireland, came to California in 1844 and settled on the San Francisco de las Llagas grant, which was later patented to James Murphy, one of his sons. Martin Murphy followed the Spanish custom and named his settlement in honor of his patron saint—probably St. Martin of Tours, who lived in 4th-century France. *For* Sierras de San Martín, *see* Santa Lucia Range.

San Martin, Cape [Monterey Co.]. The name was originally applied by Cabrillo in 1542, but there has been much confusion as to exactly what cape was designated by the term at various periods (Clark 1991).

San Mateo. The name of St. Matthew, the evangelist and apostle, was repeatedly used in Spanish times and has survived in two large cluster names. **San Mateo** (mə tā′ ō): **Creek, Point**, city, **County, Slough, Bridge**. The Arroyo de San Matheo is mentioned in the diaries of Anza and Font on Mar. 26, 1776; and in the vicinity of the creek, in the 1790s, Mission Dolores had a sheep ranch called San Mateo, which developed into an unofficial "Mission" San Mateo (*CHSQ* 23:247 ff.). Esteros de San Mateo and Punta de San Mateo are mentioned by Argüello in 1810 (PSP 19:280). The name appears in a petition for a land grant, Dec. 22, 1836; and on May 5 or 6, 1846, it was applied to another grant. Arroyo, point, and settlement are shown on most maps, and until 1850 their names are usually spelled San Matheo. The county, carved out of San Francisco Co., was created and named on Apr. 19, 1856. The modern city was laid out by C. B. Polhemus in 1863 when the San Francisco–San Jose Railroad was built; the first train reached the station on Oct. 18, 1863. **San Mateo Canyon, Creek, Point**, and **Rocks** [San Diego Co.] take their names from the Arroyo de San Mateo and a place, probably a rancheria, mentioned in Mar. 1778 (PSP 2:1). A petition for a land grant, San Mateo, was filed in 1839, but no grant is recorded. A Rancho San Mateo existed in 1828 (Registro, p. 41) and is shown on Duflot de Mofras's map of 1844. The creek is shown as Río San Mateo on a *diseño* of 1845, and San Mateo Creek is mentioned in the *Statutes* of 1850 (pp. 58–59).

San Miguel. The Spanish form of the name St. Michael was a favorite in Spanish California nomenclature. Among places that once bore the name are San Diego Bay, San Gabriel Valley, San Joaquin River, and a stream of the Kaweah River system. All the San Miguel names still in existence were probably bestowed, according to Wagner (*Saints' names*), in honor of the Archangel Michael. **San Miguel** (mi gel′) **Island, Passage** [Santa Barbara Co.]: Juan Rodríguez Cabrillo discovered this island and Santa Rosa Island, probably on Oct. 18, 1542, and named them Islas de San Lucas. He renamed one island Posesión after having taken possession of it; and this was afterward named Juan Rodríguez by Ferrer, in his memory. The island appears on the maps of the following centuries with various names until about 1790, when the name San Miguel became attached to it, a name that Miguel Costansó had in 1770 applied to what is now Santa Rosa Island (Wagner, p. 411). This new nomenclature of the Channel Islands became fixed when Vancouver used it for his maps—although, until Wilkes adopted Vancouver's alignment in 1841, older names persisted: Juan Rodriguez (Bonnycastle, Tanner); San Bernardo (Humboldt, Narváez). **San Miguel Mission, Canyon**, and town [San Luis Obispo Co.]: The mission was founded and named San Miguel Arcángel on July 25, 1797. When the Southern Pacific reached the place in the fall of 1886, the name of the mission was applied to the station. The post office is listed in 1887. **San Miguel Canyon** [Monterey Co.]: San Miguel was an alternate name of the Bolsa de los Escarpines grant, dated Oct. 7, 1837; Cañada de San Miguel is shown on a *diseño* of the grant. **San Miguel Hills** [San Francisco Co.]: The elevation, now generally known as Mount Davidson, has been labeled San Miguel Hills on maps because it is on the San Miguel land grant, dated Dec. 23, 1845. The San Miguel grants in Sonoma Co. (Nov. 2, 1840) and Ventura Co. (July 6, 1841) have apparently left no trace in modern nomenclature; but San Diego Co. has a San Miguel Mountain, shown as Sierra de San Miguel on a *diseño* of Rancho de la Nación, 1843. The name has often been pronounced in English as (mə gil′), as if referring to a "St. McGill."

San Miguelito. The name was applied to three land grants dated Apr. 8, 1839 [San Luis Obispo Co.]; Sept. 25, 1839 [Monterey Co.]; and May 30, 1845 [Ventura Co.]. There is no "little Saint Michael"; San Miguelito is a diminutive of the place name San Miguel. A place called San Miguelito in San Luis Obispo Co. is shown on Narváez's 1830 Plano. (*See*

San Francisquito; Santa Anita.) The name is preserved in modern geography in San Miguelito (mig ə lē′ tō) Creek, an intermittent stream in Santa Barbara Co.

San Nicolas (nik′ ə ləs) **Island** [Ventura Co.]. The island was evidently given this name by the crew of Vizcaíno's launch *Tres Reyes* on Dec. 6, 1602, the feast day of St. Nicholas of Myra, the prototype of Santa Claus (Wagner, p. 412). It is also known as Passing Island because the sand is gradually being blown away and the island will eventually be reclaimed by the sea (Doyle).

San Onofre (ō nō′ frē): **Creek, Canyon, Mountain, Hill, Bluff**, post office [San Diego Co.]. The name of St. Onuphrius, an Egyptian, is mentioned as the name of a rancho of San Juan Capistrano Mission in 1828 (Registro, p. 41). It appears in the name of the Santa Margarita y San Onofre grant, dated Feb. 23, 1836, and May 10, 1841, but it does not appear on early American maps or on the charts of the Coast Survey. It was given to the station when the Santa Fe built the coast line from Los Angeles to Oceanside in the late 1880s (*see* Flores; Santa Margarita). **Cañada San Onofre** [Santa Barbara Co.], shown as Cañada de San Onofre on the Lompoc atlas sheet, is probably the San Onofre mentioned in 1795 (PSP 14:62) and was doubtless named for the same saint.

San Pablo (pab′ lō): **Point, Strait, Bay, Creek**, town, **Reservoir** [Contra Costa Co.]. The names San Pedro and San Pablo (Peter and Paul) for the points on opposite shores of San Pablo Strait probably originated at the same time. They are mentioned in Abella's diary in Oct. 1811, and Punta de San Pablo is recorded by Durán on May 13, 1817. The name San Pablo was used for a land grant, dated Apr. 15 and 23, 1823. On the *Plano topográfico de la Misión de San José* (about 1824) the narrows are called Estrecho de San Pablo, and Rancho San Pablo is shown. The name of the bay itself had been called Bahía Redonda by Crespí in 1772 and by Cañizares in 1776; on the *Plano topográfico* it is called Bahía de Sonoma. The two points are mentioned by Beechey (2:425); on Duflot de Mofras's plan 16, the name San Pablo appears also for the bay. The town of San Pablo is shown on Butler's map of 1851; the post office was established on Nov. 15, 1854. *For* San Pablo in Imperial Co., *see* Pilot: Pilot Knob; in Kern Co., *see* Castac: Castaic.

San Pascual (pas kwäl′) [Los Angeles Co.]. Rancho San Pascual is mentioned on Dec. 27, 1833 (Arch. LA 4:74). The land, which had belonged to Mission San Gabriel, became public property with secularization of the mission; one part was made into a land grant dated July 27, 1838, and another, Sept. 24, 1840, and July 10, 1843. This grant and the places called San Pasqual in San Diego Co. were probably all named for St. Pascal Baylon (1540–92), a Spanish Franciscan. After the American occupation the spelling San Pasqual was ordinarily used in official reports and in land grant records. **San Pasqual**: town, **Valley** [San Diego Co.]: The name was evidently applied to an Indian village under the jurisdiction of San Diego Mission. It appears as S. Pascual on Bancroft's map of the San Diego district, 1800–1830 (2:105), and was applied in 1835 to the pueblo composed of former neophytes of Mission San Luis Rey (San Diego Public Library). The present spelling was used in army reports after the "battle of San Pasqual" between Kearny's and Andrés Pico's forces on Dec. 6, 1846.

San Pedro. The Spanish form of the name St. Peter was very popular for place names in Spanish times. It was given to five land grants, and about twelve geographic features still bear it. Not all the names honor the apostle; some honor other saints named Peter. **San Pedro** (pē′ drō, pā′ drō): **Bay, Channel, Hill, Hills**, post office [Los Angeles Co.]: The bay had been named Bahía de los Fumos or Fuegos by Cabrillo on Oct. 8, 1542, because smoke from numerous fires was visible. According to Wagner (p. 412), the name San Pedro was applied to the bay by one of Vizcaíno's men, probably because it was sighted on Nov. 26, 1602, the feast day of the St. Peter who was martyred in Constantinople. It appears in the Bolaños Ascensión *derrotero* and on the maps based upon Vizcaíno's discoveries. Bahía de San Pedro is mentioned in Vila's diary on Apr. 26, 1769 (*APCH:P* 2:86), and repeatedly in the *Reports* of the Anza expedition. Later the name San Pedro was applied to one of the land grants bordering the bay on the north, granted before Nov. 20, 1784, to Juan José Domínguez and regranted at various times to other members of the Domínguez family. Narváez's Plano of 1830 shows an Ensenada and a Punta San Pedro. On Duflot de Mofras's map (1844) San Pedro is shown with an anchorage; Los Angeles River and Long Point are also labeled San Pedro. The town of San Pedro is mentioned in the *Statutes* of 1854 (p. 223); the post office is listed in 1858. **San Pedro Point** [Marin Co.] recalls an Indian rancheria in the region called San Pedro by Abella on Feb. 28, 1807 (Arch. Arz. SF 2:54 ff.); Punta de San Pedro is mentioned in 1811 (Arch. MSB 4:321). San Pedro is shown near the point on the *Plano topográfico de la Misión de San José*, about 1824. The name is also found in the name of the land grant "San Pedro, Santa Margarita y las Gallinas," dated Nov. 18, 1840, and Feb. 12, 1844. Beechey mentions Puntas San Pablo and San Pedro (2:425), and these are shown on Duflot de Mofras's plan 16, and on sketch J (1850) of the Coast Survey. Point San Pedro is now generally called McNears Point (*see* McNears Point; San Pablo). *For* San Pedro in San Mateo Co., *see* Pedro; *for* Bahía de San Pedro, *see* Monterey Bay; *for* Río de San Pedro, *see* Tule: Tule River.

San Philipe, Río de. *See* San Felipe: San Felipe Creek.

San Quentin [kwen´ tin, kwin´ tin], **Point; San Quentin** [Marin Co.]. The name was not given in honor of the Roman officer who resigned his commission to become a Christian missionary, but for his namesake, an Indian named Quintín—a notorious thief, according to Bancroft; a subchief and daring warrior, according to Vallejo. The point was called Punta de Quintín because the chief and/or thief was captured there (1824). On Sept. 24, 1840, the name was applied to a land grant. Punta Quintín is shown on the *diseños* of the Corte de Madera del Presidio grant (1834) and of the Punta de Quintín grant (1840). In the meantime a point on the bay shore of San Francisco, just north of Potrero Point, had been named for the saint (Forbes's map of 1839). When the Coast Survey charted the bay in 1850, it Americanized both points to Quentin and added a "San" to the point in Marin Co. For many years both names were on the maps, until the old Mission Bay was filled in and the confusion ended. "San" has often been added gratuitously by Americans to Spanish names; one can even hear Mount Diablo [Contra Costa Co.] referred to as "Mount San Diablo."

San Rafael. The archangel Raphael, the guardian angel of humanity, was frequently honored in place names, and a number of these have survived in Los Angeles, Marin, Santa Barbara, and Ventura Cos. **San Rafael** (rə fel´) **Hills** [Los Angeles Co.] are on the territory of the Rancho San Rafael, one of the oldest land grants in California, dated Oct. 20, 1784, and Jan. 12, 1798. The grant, conveyed to José María Verdugo, was known as Hahaonuput, or Arroyo Hondo, or Zanja, and later as San Rafael. It is one of two known grants made to soldiers marrying Indian girls, in accordance with a decree of Aug. 12, 1768 (Bowman Index). **San Rafael Mission, Bay**, city, **Creek**, and **Rock** [Marin Co.]: The mission was founded as an *asistencia* of Mission Dolores on Dec. 14, 1817 (Engelhardt, *The Franciscans in California* [Harbor Springs, Mich.: Holy Childhood Indian School, 1897], p. 440), and was named San Rafael Arcángel. The city had its beginnings when Timoteo Murphy was granted a lot near the mission and built a house on it sometime before 1841. Both the mission and Murphy's residence are shown on Duflot de Mofras's plan 16. The post office was established on Nov. 6, 1851. **San Rafael Mountains** [Santa Barbara Co.] are shown as Sierra de San Rafael on a *diseño* of the San Marcos grant, 1846. The American version is mentioned in the Pac. R.R. *Reports* 7:1.6.

San Ramon (rə mōn´): **Valley, Creek**, town [Contra Costa Co.]. The name is used for several land grants, the oldest dated June 5, 1833. This name did not originally honor a saint; the "San" was simply added to make the whole name conform to the usual name of this type. In the San Ramon land grant case (1855), José María Amador testified that "the name was given it [the creek] by a mayordomo [of Mission San Jose] by the name of Ramón who had the care of some sheep there a long time ago. It was also called Arroyo del Injerto ['graft'] from the fact that there is a singular tree growing there, that is an oak with a willow ingrafted on it" (*WF* 6:373). The town came into existence in the 1850s and is shown on Goddard's maps. The post office is listed in 1859. The early Bancroft maps designate Pacheco Creek as San Ramon Creek. On Hoffmann's map of the Bay region, the name is applied to the present creek. In Santa Barbara Co., Crespí had honored St. Raymond Nonnatus on Aug. 31, 1769, by calling a lake or lagoon San Ramón Nonato; but the soldiers' name, La Graciosa, prevailed (*see* Graciosa).

San Remo (rā′ mō) [Monterey Co.]. This name was given by Joseph Victorine to his ranch in the late 19th century, possibly from San Remo on the Italian Riviera. The Italian name was originally based on the name of San Romulo, a 5th-century bishop of Genoa—presumably from the association with Romulus and Remus, founders of Rome (Clark 1991).

San Roque (rō′ kē): **Canyon, Creek** [Santa Barbara Co.]. The names are derived from a *paraje llamado* ('place called') *San Roque*, one and a half leagues from Mission Santa Barbara, mentioned on June 22, 1824 (DSP 1:49). The name of St. Roch (d. 1327), a French Carmelite, was repeatedly used for place names. The rancheria at the site of Carpinteria was called San Roque by Crespí on Aug. 17, 1769; the lower course of the Sacramento is called Río de San Roque on Cañizares' map of 1776.

San Salvador de Horta, Arroyo de. *See* San Lorenzo: San Lorenzo Creek.

San Salvador Island. *See* Santa Catalina.

San Sebastián refers to St. Sebastian, a Roman martyr of the 3rd century. **San Sebastian Marsh** [Imperial Co.] was an important desert watering place at the junction of San Felipe and Carrizo Creeks, named by Anza in 1774 in honor of his native guide, Sebastián Tarabal (Brigandi). *For* Cabo Blanco de San Sebastián, *see* Saint George, Point; *for* Isla de San Sebastián, *see* Santa Cruz: Island; *for* Río Grande de San Sebastián, *see* Tomales Bay.

San Sevaine (sə vān′) **Flats** [San Bernardino Co.]. This place in the wilderness area above Etiwanda was named for the well-known French pioneer of 1839, Pierre ("Don Pedro") Sainsevain, obviously spelled to reflect the way his Spanish neighbors pronounced the name. In the 1870s Sainsevain had a well dug here to irrigate his vineyard in Cucamonga (Buie); *see* Don Pedro Reservoir.

San Simeon (sim′ ē ən): **Bay, Creek, Point**, town [San Luis Obispo Co.]. A *rancho de San Simeón*, belonging to Mission San Miguel, is mentioned on Jan. 31, 1819; on Nov. 26, 1827, it is recorded that the mission kept cattle and horses there (Engelhardt, *San Miguel*, p. 28). The name honors St. Simeon of Jerusalem, the cousin of Jesus. The Arroyo de San Simeón is shown on a *diseño* of the Santa Rosa grant, 1841. The rancho was made a private land grant, dated Oct. 1, 1842; the settlement and anchorage appear on Duflot de Mofras's map of 1844. The name of the bay is shown on the Coast Survey charts since 1852. The post office was established three times, in 1864, 1867, 1873, and each time it was soon discontinued. In 1874–75 Leopold Frankl, an Austrian emigrant of 1849, opened the first store here and became the postmaster of the definitive post office, established on Aug. 9, 1878. When Sen. George Hearst began to develop his vast estate here, Frankl sold most of his land to him. The Hearst Castle became a state park after the heirs of William Randolph Hearst donated it to the state.

Santa Ana. The name of St. Anne was repeatedly used in Spanish times for place names; probably all honor the mother of the Virgin Mary. These names have survived in various sections of the state. **Santa Ana** (san′ tə an′ ə, san′ tē an′ ə) **River** [San Bernardino, Riverside, Orange Cos.], **Mountains**, and city [Orange Co.]: The Portolá expedition camped on the stream on July 28, 1769, and the padres bestowed on it *el dulcísimo nombre de Jesús de los Temblores* 'the most sweet name of Jesus of the Earthquakes'—"on account of the earthquakes that we felt when on it" (Crespí, pp. 32, 142). To the soldiers the stream was known as Santa Ana (ibid., p. 142), and thereafter Río de Santa Ana is mentioned repeatedly in the early records. This seems to be the only time that the soldiers of the expedition applied a holy name to a place. Río de Santa Ana is shown on Narváez's Plano (1830). Five land grants in Orange, Riverside, and San Bernardino Cos. were named after the river. The mountain chain along the Orange-Riverside county line, southwest of Corona, is recorded as Sierra de Santa Ana on the maps of Parke-Custer 1855 and von Leicht–Craven 1874, and as Santa Ana Mountains on the Land Office map of 1879. A settlement Santa Anna is mentioned by Emory (1846–47:118), and it is shown on the maps of Butler 1851 and Bancroft 1858; but the modern city, at the present site on the Rancho Santiago de Santa Ana, was not founded and named until 1869 (Stephenson). **Santa Ana: Creek, Valley, Peak** [San Benito Co.]: The name may go back to the Portolá expedition when Crespí, on Oct. 8, 1769, added the name of La Señora Santa Ana to the name of Río del Pájaro (Crespí, p. 211). Santa Ana Creek is connected with the Pajaro River through Tequesquite Slough. On Apr. 16, 1836, the name was used in the Santa Ana y Santa Anita grant, unconfirmed by the U.S. government; and on Apr. 8, 1839, in the Santa Ana y Quien Sabe grant. The creek, valley, and peak are on this grant. Two mountains are called Picacho de Santa Ana on a *diseño* of the grant. The present Santa Ana Peak is mentioned in Hoffmann's notes of July 10, 1865, and it is still commonly so known. A pass north of Pacheco Pass is called Pass of Santa Aña [*sic*] on Derby's map (1850). **Santa Ana Creek** and **Valley** [Ventura Co.] got their names from the Santa Ana land grant, dated Apr. 14, 1837.

Santa Anita (ə nē′ tə): **Canyon, Creek**, station, **Park** [Los Angeles Co.]. These places are on the Santa Anita land grant, dated Apr. 16, 1841, and Mar. 31, 1845, which was developed in the 1880s by "Lucky" Baldwin into one of the spectacular ranchos of southern California. The name is found in the titles of five other land grants or claims. There is no "Saint Annie"; the name Santa Anita is simply a diminutive of the place name Santa Ana. According to the USFS, the canyon was not named after the land grant, but for Baldwin's daughter Anita. Place names formed by the addition of "San" or "Santa" to a personal name are not unusual (*see* San Quentin, San Ramon).

Santa Barbara (bär′brə): **Channel, Island, Mission, Harbor, County**, city, **Point, Canyon**. The name Canal de Santa Bárbara was applied to the passage between the mainland and what are now the Channel Islands by Vizcaíno on Dec. 4, 1602, the feast day of the Roman maiden who was beheaded by her father because she had become a Christian. Vizcaíno also named an island for St. Barbara, probably the easternmost of the Anacapa Islands; however, Palacios's chart (Wagner, no. 236) shows it in almost the same relative position as the present Santa Barbara Island (farther southeast, in Los Angeles Co.). The name "Canal de St. Barbaria" is shown on a Briggs type map of 1625 (Wagner, pl. 23). The Presidio de Santa Bárbara, Virgen y Mártir, was established on Apr. 21, 1782, on the land called Yamnonalit by the Indians (Engelhardt 2:369) and San Joaquín de la Laguna by the Spaniards (Prov. Recs. 2:61). On Dec. 4, 1786, St. Barbara's day, a great cross was raised at a spot called Pedregoso 'stony', which had been selected as the site for the mission (Arch. MSB 5:3), and the mission was formally declared founded on Dec. 16 (ibid. 2:434). The county, one of the original twenty-seven, was named on Feb. 18, 1850; the city was incorporated on Apr. 9, 1850. What are now the Channel Islands were once named the Santa Barbara Islands by the Coast Survey; the mountains opposite the town were called the Santa Barbara Mountains in the Pac. R.R. *Reports* 7:1.7. *See* Los Padres National Forest.

Santa Catalina (kat ə lē′ nə): **Island; Gulf of Santa Catalina; Catalina: Harbor, Head** [Los Angeles Co.]. Cabrillo named the island San Salvador in Oct. 1542, after one of his ships. Vizcaíno renamed it Santa Catalina on Nov. 25, 1602, the feast day of St. Catherine, the royal virgin and martyr of Alexandria. (The usual Spanish form of the name Catherine is Catalina; a rarer variant is Catarina—*see* Santa Catarina.) All older maps consulted show this name except Diogo Homem's of 1559 (Wagner, no. 42). Yet the name given by Cabrillo lasted until modern times. Bonnycastle (1818) has San Salvador for Santa Catalina Island, Duflot de Mofras (1844) has it as an alternate name for San Clemente, and Tanner (1846) has a San Salvador Island at the latitude of San Diego. Ysla de Santa Catalina was one of the last private land grants made under Mexican rule (July 4, 1846). Catalina harbor has been recorded on the Coast Survey charts since 1852; the name is used for the gulf between San Pedro and San Diego by Davidson in the 1889 edition of his Coast Pilot. *For* Santa Catalina de Bononia de los Encinos, *see* Encina: Encino Park.

Santa Catarina (kat ə rē′ nə) **Springs** [Anza-Borrego Desert State Park]. Anza's expedition camped here on Mar. 14, 1774: "We arrived at a spring or fountain of the finest water . . . and to the place I gave the name Santa Catarina" (Anza 2:86–87). The springs are known as Willow or Reeds Springs, but in recent years the Spanish name has come into use again.

Santa Clara (klâr′ ə) **River** [Los Angeles, Ventura Cos.]. The Portolá expedition rested at the river on Aug. 9, 1769, and Crespí named the valley Santa Clara in honor of St. Clare of Assisi (d. 1253), co-founder of the Franciscan Order of Poor Clares. The river soon came to be known as Río de Santa Clara; this name appears on Pantoja y Arriaga's map of 1782 and many others. The name was used in the titles of two land grants on the river: Santa Clara on May 18, 1837, and Río de Santa Clara on May 22, 1837. Gibbes's map of 1852 shows Santa Clara Creek, but the Pacific Railroad Survey used the name Santa Clara River. **Santa Clara Mission**, city, **County**, and **Valley**: The mission was founded by Padre Tomás de la Peña on Jan. 12, 1777; according to instructions from Mexico, it was named Misión de Santa Clara de Asís. On the *Plano topográfico de la Misión de San José* (about 1824) the lower part of San Francisco Bay is called Estero de Santa Clara. The highest peak of Montara Mountain is designated as Mont Santa Clara on Duflot de Mofras's plan 16. On Feb. 29, 1844, the name was used for a land grant, Potrero de Santa Clara. The county, one of the original twenty-seven, was named on Feb. 18, 1850. The name Santa Clara Valley seems to have come into general use in the 1850s and is repeatedly found in the Pac. R.R. *Reports*. *For* Santa Clara de Monte Falco [Ventura Co.], *see* Rincon: Rincon Point.

Santa Clarita (klə rē′ tə) [Los Angeles Co.]: **River, Valley**, city. The "little Saint Clare" is so named as a tributary of the Santa Clara River.

Santa Conefundis. *See* Pitas Point.

Santa Cora (or **Santa Cota**) **Creek**. *See* Zanja: Zanja Cota Creek.

Santa Cruz. The words for 'holy cross' are frequently found as a place name in Spanish-speaking countries. Santa Cruz might easily have become the name of the state, for when Cortés came to the peninsula of what was later called California, he named his landing place Santa Cruz. One of the first names bestowed by the Portolá expedition in what is now California was Triunfo de la Santa Cruz, a place near San Elijo Lagoon [San Diego Co.] (Crespí, p. 126). **Santa Cruz** (kro͞oz): **Island, Channel** [Santa Barbara Co.]. The name, given to one of the Channel Islands in Santa Barbara Channel early in Apr. 1769, had no immediate religious significance but arose from an incident. A party from the *San Antonio*, a ship commanded by Juan Pérez, landed on the island. A friar in the party lost a staff with a cross on it; this was found by an Indian and returned to him the next day (Wagner, p. 414). From the maps and reports of the expedition, it is not clear which of the Channel Islands was so named. The name was not definitely fixed upon the present Santa Cruz Island until the visit of Vancouver (*Voyage* [1798 ed.], 2.448 and atlas). Ferrer in 1543 had named it San Sebastián; Vizcaíno in 1602 called it Isla de Gente Barbuda because one of his men said that he had seen bearded natives there. **Santa Cruz: Mission**, city, **County, Mountains, Harbor, Point**: An arroyo of running water (probably Majors Mill Creek) was named Santa Cruz by the Portolá expedition on Oct. 18, 1769 (Crespí, p. 216). The mission was founded and named on Aug. 28, 1791. On Jan. 27, 1797, the viceroy, the Marqués de Branciforte, ordered the establishment of a pueblo near the mission, although this was contrary to the royal decrees that forbade white settlements within a league of an Indian mission. The pueblo was founded by Governor Borica on July 24, 1797, and named Villa de Branciforte in honor of the viceroy (Engelhardt 2:454, 519). The town was established in 1849, and from the beginning bore the name Santa Cruz (*San Francisco Alta California*, Sept. 27, 1849). The post office is listed as Santa Cruz in 1850. The county, one of the original twenty-seven, was created on Feb. 18, 1850, and was named Branciforte; but the name was changed to Santa Cruz by act of the legislature on Apr. 5, 1850. The Santa Cruz Mountains are mentioned in the *Statutes* of 1850 (p. 59). The mountains are shown as Cierra [Sierra] Madre de Santa Cruz on a *diseño* of the San Antonio grant (1838). Frémont called the mountains Cuesta de los Gatos (*see* Gatos). **Santa Cruz Creek** and **Peak** [Santa Barbara Co.] take their name from the Cañada de Santa Cruz, shown on *diseños* of the Tequepis (1845) and San Marcos (1846) grants. There is also a Santa Cruz Mountain in Mariposa Co.

Santa Delfina, Río de. *See* Salinas River.

Santa Fe. The Spanish term for 'Holy Faith', which has given to the Southwest one of its best-known place names of Spanish origin, seems to have been little used in California. The alternate name of a land grant in San Luis Obispo Co., usually designated in documents as "1,000 *Varas*," was Ranchito de Santa Fe (Sept. 18, 1842); and the Gabilan Range [Monterey, San Benito Cos.] is labeled Sierra de Santa Fe on several *diseños*. The Coast Survey gazetteer gives the channel between Potrero Point and Brooks Island [San Francisco Bay] as Santa Fe Channel—a name derived either from that of the railroad settlement nearby or from the name of the railroad itself. **Santa Fe** (fā) **Springs** [Los Angeles Co.]: In 1886 the Santa Fe Railroad purchased and renamed the mineral springs where J. E. Fulton had established a sanitarium in 1873 under the name of Fulton Sulphur Springs and Health Resort (*see* Rancho: Rancho Santa Fe).

Santa Gertrudis Creek [Riverside Co.]. An *ojo de agua* (spring) called Santa Gertrudis is mentioned on Sept. 24, 1821 (Arch. MSB 4:223), and is shown on a *diseño* of the Temecula grant. The saint honored is perhaps St. Gertrude of Saxony (1256–1302), according to Raymund F. Wood (1987:56). **Santa Gertrudis** [Los Angeles, Orange Cos.]: One of the five parts into which the large Nieto grant (dated Nov. 20, 1784) was divided after the death of Manuel Nieto in 1804 was called Santa Gertrudis; this was regranted on July 27, 1833. Another Santa Gertrudis grant, dated Nov. 22, 1845, was apparently in Lower California.

Santa Humiliana, Cañada de. *See* San Carpoforo Creek.

Santa Inez. *See* Santa Ynez.

Santa Isabel. *See* Santa Ysabel.

Santa Lucia (lo͞o sē′ ə) **Range** [San Luis Obispo, Monterey Cos.]. In Nov. 1542 Cabrillo named this important link in the Coast Ranges Sierras de San Martín, and called its northern part Sierras Nevadas because there was snow on it. Around Dec. 14, 1602, Vizcaíno named it Sierra de Santa Lucía in honor of St. Lucy of Syracuse (d. 303?), whose feast day is Dec. 13. The Sierra de Santa Lucía is repeatedly mentioned in the diaries of the Portolá expedition, apparently for the entire range. Font's *Compl. diary*, Mar. 1, 1776, states: "It is the very high, rough and long *Sierra de Santa Lucía*, which begins here [near San Luis Obispo] and ends at the mission of Carmelo near Monterey." The Parke-Custer map (1855) extends the name south to the Santa Clara River, but the Coast Survey established the divide southeast of San Luis Obispo as the southern limit. **Santa Lucia Canyon** [Santa Barbara Co.] is shown on a *diseño* of the Jesús María grant (1837).

Santa Manuela [San Luis Obispo Co.]. The name was applied to the land grant on Apr. 6, 1837. It honors the St. Manuela whose feast day is June 24 (Wagner, *Saints' names*).

Santa Margarita (mär gə rē′ tə): **River, Canyon, Mountains** [San Diego, Riverside Cos.]. The Portolá expedition camped near the river on July 20, 1769: "Because we arrived at this place on the day of St. Margaret [of Antioch] we christened it with the name of this holy virgin and martyr" (Crespí, p. 133). A *ranchería Santa Margarita* is mentioned in Aug. 1795 (Arch. MSB 4:200 ff.). The name was given to the land grant Santa Margarita y San Onofre, dated Feb. 23, 1836, and May 10, 1841. **Santa Margarita Creek** and town [San Luis Obispo Co.]: The site and river of Santa Margarita are mentioned in Anza's diary on Mar. 4, 1776. The name had doubtless been applied by the padres of San Luis Obispo Mission in honor of St. Margaret of Cortona, who in 1769 had been honored in a place name near Santa Barbara (Crespí, p. 168). Before 1790 the mission used the place near San Luis Obispo for a hog farm (Engelhardt 3:643–44); it is mentioned on July 15, 1790 (DSP San Jose 1:37). Later Santa Margarita became an *asistencia*. *El rancho de Santa Margarita de Cortona* is mentioned on Dec. 16, 1833 (Guerra Docs. 1:246), and the name Santa Margarita was given to a land grant on Sept. 27, 1841. The present town developed in the 1870s. It is shown on the Land Office map of 1879; the post office is listed in 1880. **Santa Margarita Lake** replaced the name Salinas Reservoir in BGN, list 6802. **Santa Margarita Valley** [Marin Co.] is named from the land grant called San Pedro, Santa Margarita y las Gallinas 'St. Peter, St. Margaret, and the hens', dated Feb. 12, 1844. *For* Santa Margarita de Cortona, *see also* Mescalitan Island; *for* Santa Margarita de los Peñasquitos, *see* Peñasquitos.

Santa Maria was a favorite place name and has survived in San Diego, Santa Barbara, and Kern Cos. These names were probably given in honor of a saint other than the Virgin, who was usually referred to as Nuestra Señora. **Santa Maria** (mə rē′ ə): **Valley, Creek** [San Diego Co.]: The name was preserved through a land grant, Valle de Pamo or Santa María, dated Nov. 21, 1843. The rancho is mentioned by Emory in 1846 (*Notes of a military reconnoissance*, p. 107), and the valley by Audubon in Oct. 1849 (p. 170). **Santa Maria River** [San Luis Obispo, Santa Barbara Cos.], city, and **Valley** [Santa Barbara Co.]: The name was applied in Spanish times and was preserved through a land grant, Tepusquet or Santa María, dated Apr. 6, 1837. Rancho and river (the latter called Creek) are recorded in the

Statutes of 1850 (p. 59). At least three different versions of the founding of the modern city have been published. Y. A. Storke's account—that it was laid out in 1874 by R. D. Cook, the owner of the land (Co. Hist.: Santa Barbara 1891:515)—seems the most plausible. The place is shown on the Land Office map of 1879. The post office is listed in 1880. The river is called the Cuyama River above its junction with the Sisquoc, its principal tributary.

Santa Monica (mon′ i kə): **Mountains, Bay, Canyon**, city [Los Angeles Co.]. Wagner (p. 415) thinks that the mountains were named by Portolá's party on May 4, 1770—the feast day of St. Monica, mother of St. Augustine—while on their way to found the mission and presidio at Monterey. In June 1822, *parage de Santa Mónica* and *sierra de Santa Mónica* are recorded in the Guerra Documentos (7:107). The name was used for two grants: Boca de Santa Monica on June 19, 1839, and San Vicente y Santa Monica on Dec. 20, 1839, and June 8, 1846. The all-Spanish name for the mountain range was retained for several decades: Sierra de la Monica (Parke-Custer), Sierra Monica (Goddard), Sierra de Santa Monica (Land Office, 1879). The popular designation Santa Monica Mountains appears on the county map of 1881. The town was founded in 1875 on Rancho San Vicente y Santa Monica by Nevada senator John P. Jones and Col. Robert S. Baker. The post office is listed in 1880. **Santa Monica** [San Diego Co.] was the alternate name of the Cajón de San Diego grant, dated Sept. 23, 1845. A mission rancho Santa Mónica at this place is mentioned in 1821 (Arch. MSB 4:2). There is also a Santa Monica Creek west of Carpinteria in Santa Barbara Co.

Santa Nella (nel′ ə) [Merced Co.]. There is no 'St. Nella'; the name probably comes from the Centinela Adobe (Spanish *sentinela* 'sentinel'). There is also a Santa Nella in Sonoma Co.

Santa Paula (pôl′ ə): city, **Canyon, Creek, Peak, Ridge** [Ventura Co.]. A stock ranch called Santa Paula, belonging to Mission San Buenaventura, is mentioned in July 1834 (DSP Ben. Mil. 79:93). The place was probably named for St. Paula, a noble Roman matron who became a disciple of St. Jerome. The name was applied to a land grant, Santa Paula y Saticoy, dated July 31, 1834, and Apr. 28, 1840. The modern town was founded on the grant by Nathan W. Blanchard and E. L. Bradley in 1872.

Santa Praxedis de los Rosales. *See* Flores; Rose.

Santa Rita. The name appears in the names of four land grants and has survived in a number of places. The saint honored is the Italian St. Rita of Cascia (1381–1457), according to Raymund F. Wood (1987). **Santa Rita** (rē′ tə): **Slough, Park** [Merced Co.]: Muñoz mentions the place Santa Rita in his diary on Sept. 22, 1806; he says that it had been discovered by Gabriel Moraga and named previously (Arch. MSB 4:4–5). Arroyo de Santa Rita is repeatedly mentioned in provincial and mission records and is shown on Estudillo's map of 1819 for Fresno Slough. On Sept. 7, 1845, the name Sanjón ('ditch') de Santa Rita was applied to a land grant. The post office Santa Rita Park is listed in 1941. **Santa Rita** [Monterey Co.] was laid out in 1867 and was so named because it is on the Los Gatos or Santa Rita grant, dated Sept. 30, 1837. **Santa Rita Hills** and settlement [Santa Barbara Co.] were so named because they are on the Santa Rita land grant, dated Apr. 10, 1839. **Santa Rita** [Alameda Co.] preserves the name of the Santa Rita grant, dated Apr. 10, 1839. There is a Santa Rita Peak southeast of New Idria [San Benito Co.], named or confirmed by the BGN on Mar. 6, 1912.

Santa Rosa. The name was extremely popular for naming places and appears as part of the name of at least ten land grants. Most places were doubtless named for the Dominican St. Rose of Lima (1586–1617), until recent times the only female saint of the Americas, but some may have been named for the Italian Franciscan St. Rose of Viterbo (1235–53). **Santa Rosa** (rō′ zə): **Creek, Hills** [Santa Barbara Co.]: "We named this river, . . . the largest that we have encountered, for San Bernardo and his companion, but, because we arrived on this day [Aug. 30, 1769, the feast day of St. Rose of Lima], it is also called Santa Rosa" (Crespí, p. 178). Although the name of the river was changed to Santa Ynez sometime after 1801, the name Santa Rosa remained for a *ranchería* and a *cañada*. These and the *cuchilla* ('ridge') *de Santa Rosa* are repeatedly mentioned in the records, and Cañada and Lomas de Santa Rosa are shown on various *diseños*. On July 30, 1839, the name was given to a land grant. **Santa Rosa Island** [Channel Islands NM]: Cabrillo in

1542 gave the name Islas de San Lucas to the islands now called San Miguel and Santa Rosa. Vizcaíno named the latter San Ambrosio in 1602. The name Santa Rosa was used, apparently for what is now San Miguel Island, in the journal of Juan Pérez of his expedition of 1774 (Wagner, p. 415). On a map of about 1794 (Wagner, no. 825) Santa Rosa is the middle one of the three chief channel islands, and this name became fixed when it was adopted by Vancouver. The island was made a private land grant on Oct. 4, 1843. **Santa Rosa Creek** and city [Sonoma Co.]: According to a clipping dated June 5, 1876, in Bancroft's Scrapbooks (5:263), "the stream was named by a missionary priest who, before the settlement of the country, captured and baptized in its waters an Indian girl, and gave her the name of Santa Rosa, in honor of the saint on whose day in the calendar this interesting ceremony was performed." Although this story has never been substantiated, it has found its way into literature complete with the name of the priest and the exact date. The name was doubtless applied by the padres of Sonoma Mission, who claimed the site. The name appears in the *expediente* of a land grant dated Jan. 8, 1831 (PSP Ben. Mil. 71:7), a grant that was not confirmed because of the mission's opposition. Arroyo de Santa Rosa is mentioned by Vallejo in May 1833 (SP Mis. and C. 2:102). In 1841 there was a grant Cabeza de Santa Rosa, and in 1844 another grant called Llano de Santa Rosa, southwest of the present city. Camp Santa Rosa is mentioned by John McKee in the Indian Report on Aug. 11, 1851; the post office was established and named Apr. 23, 1852. **Santa Rosa Mountain** and **Mountains** [Riverside, San Diego Cos.]: There was a Santa Rosa land grant, dated Apr. 28, 1845, and Jan. 30, 1846, in the western part of what is now Riverside Co.; however, the peak and range are not near that grant. The name is preserved in the Santa Rosa Plateau Ecological Reserve and the Santa Rosa Indian Reservation. *For* Santa Rosa [San Luis Obispo Co.], *see* Cambria; *for* Santa Rosa de las Lajas, *see* Yuha [Imperial Co.].

Santa Rosalia (rō sə lē′ ə) **Mountain** [Santa Cruz Co.]. For St. Rosalia (1130–60), patroness of Palermo, Sicily (Clark 1986).

Santa Susana (so͞o zan′ ə): **Mountains, Pass, Tunnel**, station [Los Angeles, Ventura Cos.]. *El camino* ('the road') *de Santa Susana y Simí* is mentioned as early as Apr. 27, 1804 (Arch. Arz. SF 2:36). *Una gran cuesta conocida por Santa Susana* 'a large ridge known as Santa Susana', probably Chatsworth Peak between Simi and San Fernando Valleys, is recorded in July 1834 (DSP Ben. Mil. 79:89). Sierra de Santa Susana is shown on a *diseño* of Las Virgenes (1837). Santa Susana Mountains are mentioned in the *Statutes* of 1850 (p. 59). Station and tunnel were named when the Somis branch of the Southern Pacific was built in 1902. The original name apparently honored St. Susanna, the Roman virgin and martyr of the 3d century.

Santa Teresa (tə rē′ sə) **Hills** [Santa Clara Co.] were so named because they are on the Santa Teresa or Laguna de Santa Teresa land grant, dated July 11, 1834. The name probably has no connection with the Laguna de Santa Teresa named by Crespí on Oct. 15, 1769, the feast day of the Carmelite St. Theresa of Avila; this lagoon was several miles south of the land grant. *For* Rancheria de Santa Teresa [Santa Barbara Co.], *see* Cojo.

Santa Venetia (və nē′ shə) [Marin Co.]. The name of the non-existent "St. Venice" was given to a real estate development of 1914 that was intended to resemble Venice, Italy (Teather 1986).

Santa Ynez (ē nez′): **Mission, River, Valley, Mountains, Peak**, town [Santa Barbara Co.]. The mission was founded on Sept. 17, 1804, and named in honor of St. Agnes (modern Spanish Inés), one of the four great virgin martyrs of the early Roman church (about the year 300). The river had been called Santa Rosa and San Bernardo by the Portolá expedition on Aug. 30, 1769. In 1817, the site of the mission had the Indian name Calahuasa and the stream is mentioned as Río de Calaguasa (PSP 20:176). After the founding of Mission La Purisima, it was sometimes called Río de la Purísima. On a *diseño* (1835) of the Santa Rosa grant, the name Río de Santa Ynés o la Purísima appears, and on an 1865 map of the Lompoc grant the stream is labeled Santa Ynez or La Purisima River. The mountain range is mentioned as Santa Inez range by W. P. Blake in the Pac. R.R. *Reports* (5:2.137), but the all-Spanish name Sierra de Santa Inés persisted on many maps until the Whitney Survey fixed the commonly used name, Santa Inez Mountains. The spelling varied until the BGN (6th Report) decided for Santa Ynez. The town was founded in 1882 as a trading center for the large College Ranch.

Santa Ysabel (iz′ ə bel): **Creek, Indian Reservation**, town [San Diego Co.]. The name of St. Elizabeth (modern Spanish Isabel) appears in the records in 1818 for a place where the missionaries planned to build a chapel (Prov. Recs. 12:165), and there was some kind of mission establishment there in 1822 (Arch. MSB 3:261). The saint honored is probably Elizabeth or Isabella of Portugal (1271–1336), daughter of the king of Aragon. The name was applied to a land grant dated Nov. 8, 1844. The Indian Reservation was set aside by executive order of President Grant on Dec. 27, 1875. The town is shown on the Land Office map of 1879; the post office is listed in 1892. **Santa Ysabel** [San Luis Obispo Co.]: Santa Ysabel and Arroyo de Santa Ysabel are mentioned on Aug. 27, 1795 (SP Mis. 2:56–57), and the name of a rancho, Santa Ysabel, belonging to Mission San Antonio, is recorded on May 2, 1801 (Estudillo Docs. 1:58) and later. On May 12, 1844, the rancho was made a private land grant. It is not known for which of several holy Elizabeths the place was named (*see* Isabel). In Spanish times the name was much more popular than the surviving names indicate. One of the first names applied by the Portolá expedition (July 15, 1769) was Valle de Santa Isabel, for Soledad Canyon in San Diego Co. Font, on Feb. 22, 1776, gave the name Santa Ysabel to what is now San Fernando Valley.

Santee (san tē′) [San Diego Co.]. The post office, named for Milton Santee, the first postmaster, is listed in 1892 (Stein). The community was first known as Fanita, for Mrs. Fanita McCoon, then also as Cowles, until by popular vote the people accepted the name of the post office in 1902 (Dorothea Hoffmann).

Santiago. The name of St. James the Apostle was frequently used in Spanish times for place names (*Santiago* from Latin *Sanctus Jacobus*); but except for the names in Orange Co., it seems to have survived only in Santiago Creek [Kern Co.]. Point Bonita appears as Punta de Santiago on Ayala's map (1775); and Poso Creek [Kern Co.] was named Río de Santiago by Garcés in 1776. **Santiago** (san tē ä′ gō) **Creek, Hills, Peak**, and **Reservoir** [Orange Co.]: The creek was named by the Portolá expedition on July 27, 1769, two days after the feast day of St. James (Crespí, p. 140). Arroyo de Santiago is mentioned in a petition for a grant, Dec. 8, 1801 (Bowman). Santiago de Santa Ana was the name of a land grant dated July 1, 1810; and the name Lomas de Santiago or Lomerías de Santiago was given to another grant on May 26, 1846. The mountain known locally as Old Saddleback was labeled Santiago Peak when the USGS mapped the Corona quadrangle in 1894.

San Timoteo (tim ō tä′ ō): **Canyon, Creek** [Riverside, San Bernardino Cos.]. A place called San Timoteo is mentioned on Dec. 31, 1830 (SP Mis. 4:37–38). The name was used as an alternate name of the Yucaipa grant, dated Mar. 22, 1843. The saint honored may be Timothy, a disciple of St. Paul and bishop of Ephesus.

San Tomas (tō mäs′) **Creek** [Santa Clara Co.]. In the 1850s the stream was shown on the plats of several land grants as Arroyo de San Tomás Aquinas (standard Spanish would be Santo Tomás de Aquino).

Santos. *See* Strathmore. *For* Río de los Santos Reyes, *see* Kings: River.

San Vicente was very popular as a place name; it is found in the names of eight land grants and claims, and has survived in at least five localities. There were a number of holy Vincents, three of whom Wagner (*Saints' names*) has identified in California place names. **San Vicente** (vi sen′ tē) **Mountain** [Los Angeles Co.]: A *parage* ('place') *de San Vicente* is mentioned as early as Aug. 31, 1802 (Docs. Hist. Cal. 4:121). A provisional land grant was so named in 1828, and later the name was incorporated in the San Vicente y Santa Monica land grant, dated Dec. 20, 1839, and June 8, 1846. The mountain was named San Vicente because it is on the grant. **San Vicente Creek** [Santa Cruz Co.] is on the land grant San Vicente, dated Apr. 16, 1839. Arroyo de S. Vicente is shown on a *diseño* (about 1836) of the Arroyo de la Laguna grant. **San Vicente Valley** and **Creek** [San Diego Co.] are on the Cañada de San Vicente y Mesa del Padre Barona land grant, dated Jan. 25, 1846. The valley is shown as Cañada de S. Vicente on a *diseño* of the grant. The San Vicente grants in Monterey Co. (Sept. 20, 1836) and Santa Clara Co. (Aug. 1, 1842) have apparently left no traces in geographical nomenclature, but there is another San Vicente Creek in San Mateo Co.

San Ygnacio, Río de. *See* Russian: River.

San Ysidro (i sē′ drō): **Mountains**, town [San Diego Co.]. A *rancho San Ysidro*, probably a rancho of San Diego Mission, is mentioned in 1836 (Hayes Docs., p. 110). The name probably honors San Isidro, the Spanish St. Isidore the Plowman (d. 1130). In the 1870s and for many years thereafter, town and post office were called Tia Juana, from the Tia Juana or Tijuana River that adjoins the international border; but when the Little Landers Colony was established in 1909, William E. Smythe, the founder, considered the name too "sporty" and applied the present name. **San Ysidro Creek** [San Benito, Santa Clara Cos.], also spelled Isidro, bears the name of a land grant conveyed to Ignacio Ortega in 1808–9, and in 1833 regranted in three parts to his heirs. The name is shown on Narváez's Plano of 1830.

Sapague (sə pä′ wē) **Creek** [Monterey Co.]. Apparently a Rumsen Costanoan name, perhaps from *shapewesh* "to put out fire" (Harrington in Clark 1991). Also spelled Sapaque.

Sapphire Lake [Trinity Co.]. Formerly called Upper Stuarts Forks Lake; renamed by Anton Weber in the 1920s, perhaps because of its color (Stang). *See* Emerald: Emerald Lake.

Saranap (sâr′ ə nap) [Contra Costa Co.]. The name was coined in 1913 from Sara Naphthaly, the name of the mother of Samuel Naphthaly, vice president of the Oakland and Antioch Railway, and was applied to the station previously known as Dewing Park (Jessie Lea).

Saratoga [Santa Clara Co.]. The town was founded in 1851 and called McCarthysville, for the miller, Martin McCarthy. When the post office was established in 1867, it received the present name, chosen because the waters of nearby Pacific Congress Spring resemble those of Congress Spring at Saratoga, New York, from Mohawk (Iroquoian) *sharató:ken*, etymologized by present-day Indians as 'where you get a blister on your heel' (Mithun). The town was known by both names until the 1870s. (*For* Saratoga Creek [Santa Clara Co.], *see* Campbell: Creek.) A now-vanished Saratoga in El Dorado Co. is mentioned in the *Statutes* of 1854 (p. 222). Saratoga Springs in Lake Co. was settled in 1874 by J. W. Pearson; it was formerly known as Pearson Springs (Mauldin). **Saratoga Springs** [Death Valley NP]: The name was apparently applied to the pool of water at the southern boundary of the monument by the USGS in 1900.

Sarco Creek [Napa Co.] is shown as Arroyo Sarco on a *diseño* of Tulucay (about 1836–41), and as Sarco Creek on a plat (1858) of Rancho Yajome. The name is from Spanish *zarco* 'clear blue'.

Sardella Lake [Tuolumne Co.]. The little lake in the Sierra Nevada was named in memory of Giovanni Domenico Sardella, a local resident, who died in 1955 (BGN, 1965).

Sardine Lake [Mono Co.]. According to a local story (Chase, *Yosemite trails*, p. 299), a mule, carrying a load of sardines, rolled off the trail and was drowned in the lake. There is a Sardine Lake and Creek in Sierra Co.

Sargent [Santa Clara Co.]. When the Southern Pacific reached the place in 1869, the station was named for James P. Sargent, a farmer and stock breeder.

Sarvorum (sär vôr′ əm) **Mountain** [Humboldt Co.]. From *sahvúrum*, a Karuk village south of the confluence of Boyce Creek and Klamath River; another English spelling is Savorum. The Karuk name apparently contains *sah-* 'downhill' and *vur-* 'to flow'.

Sastise (or **Sasty**) **River**. *See* Shasta River.

Satcher. *See* Sotcher.

Saticoy (sat′ i koi) [Ventura Co.] represents the name of a Chumash rancheria; its meaning is not known. A *sitio de Saticoy* is mentioned on May 20, 1826 (Dep. Recs. 4:47), and the name appears repeatedly in land grant papers. The Santa Paula y Saticoy grant is dated July 31, 1834, and Apr. 28, 1840. The modern town started when J. L. Crane settled there in 1861; the post office is listed in the 1870s. The Indian village, Saticoy, existed as late as 1863 (Alexander B. Taylor, July 24, 1863).

Sattley [Sierra Co.]. Named in 1884 for Mrs. Harriet Sattley Church, the oldest woman in the town at that time.

Sauce; Sauz. This word for 'willow', pronounced in Spanish (sou′ sā, sous′), occurs in place names in its plural *sauces* and its derivatives: *saucito* (also spelled *sauzito, sausito*) 'little willow', *sauzal* (or *sausal*) 'willow grove', and *sauzalito* (*sausalito*) 'little willow grove'. Several species of willow (genus *Salix*) grow in the state. **Sausal Redondo** [Los Angeles Co.]: The name of the land grant, dated Mar. 15, 1822, and May 20, 1837, means 'round willow grove'; its specific part has survived in the name of the city Redondo Beach. **Sausalito** (sôs ə lē′ tō) town, **Point**, and **Cove** [Marin Co.]: The cove was a convenient place to get water and wood, and served generally as anchorage for foreign whalers. On the charts of the Coast Survey it is called Horseshoe Bay. Beechey mentions the name Sausalito in Nov. 1826, and on a *diseño* of the Corte Madera del Presidio grant (1834) a sort of building, probably John Reed's shanty, is labeled Sausalito. On Feb. 11, 1838, a land grant called Sausalito was conveyed to William A. Richardson, captain of the port of San Francisco. After the American occupation the spelling grew confused: Saucilito, Sausolita, Sausolito, Sousoleto, Sausaulito, Sauselito, etc., until the form Saucelito, used by the Pacific Railroad Survey, came to be generally accepted. The present town was started when the Saucelito Land and Ferry Company subdivided the land in 1868 and established ferry service to San Francisco. The post office was established and named Saucelito on Dec. 12, 1870. The spelling Sausalito was restored by the Post Office Dept. on Nov. 12, 1887. Nevertheless, the other spelling persisted until after 1900. In Monterey Co., Saucito, mentioned as a rancho on June 12, 1822 (PSP 53:32–33), is the name of a grant dated May 22, 1833; Sausal is another, dated Aug. 2, 1834, and Aug. 10, 1845. **Los Sauces Creek** [Ventura Co.] is shown as Arroyo de los Sauces on a surveyor's map of the El Rincón grant. **Sausal** (sou sal′) **Creek** [San Mateo Co.] is mentioned in 1853 (Alan K. Brown).

Saucos [Tehama Co.]. The Rancho Los Saucos was granted Dec. 20, 1844, to Robert H. Thomes. This seems to be the only recorded place name using the Spanish word *saúco* for the elder or elderberry bush, genus *Sambucus* (*see* Elder; Tehama).

Sauerkraut Gulch. *See* Beer Creek.

Saugep (sô′ gep) **Creek** [Del Norte Co.]. Perhaps from Yurok *segep* 'coyote' (Turner; Robins).

Saugus (sô′ gəs) [Los Angeles Co.]. The name of the railroad station was originally Newhall (*q.v.*), for Henry M. Newhall. When this name was transferred on Feb. 15, 1878, to the station 2 miles south, the old station was renamed Saugus after Newhall's birthplace in Massachusetts. The name is an Algonquian word for 'outlet' and by chance fits the locality near the outlets of San Francisquito, Bouquet, Mint, and Soledad Canyons (*CFQ* 4:404–5).

Sauquil. *See* Soquel.

Saurian (sôr′ ē ən) **Crest** [Tuolumne Co.]. Named in 1911 by William E. Colby because the crest resembles the sawtoothed back of a dinosaur (Farquhar).

Sausal. *See* Sauce.

Sausalito. *See* Sauce.

Sauz; Sauzal. *See* Sauce.

Savage Dam [San Diego Co.]. The building of the dam, first called Lower Otay, was begun in 1917; in 1934 the name was changed, in honor of H. N. Savage, the engineer who had planned and directed the construction.

Saver Peak. *See* Shaver Lake.

Savorum Mountain. *See* Sarvorum.

Sawmill. About twenty features in California bear this name; the best known are Sawmill Creek, Pass, and Point in Inyo Co., and Sawmill Flat on Woods Creek in Tuolumne Co.

Sawpit Flat [Plumas Co.]. A pit was built on the flat in 1850 for the use of a whipsaw; hence the name (Co. Hist.: Plumas 1882:290).

Sawtooth. About fifteen orographic features in the state are so named because of their appearance. Among these are the well-known peak in Sequoia NP and the imposing ridge on the northern boundary of Yosemite NP. Since 1960, Glacier Pass on the boundary of Kings and Tulare Cos. has been called Sawtooth Pass (BGN, 1960).

Sawyer Ridge [San Mateo Co.]. Named after Sawyer's Camp, the headquarters of a trainer of circus horses in San Andreas Canyon (Wyatt).

Sawyers Bar [Siskiyou Co.]. The bar was named in the 1850s for Dan Sawyer, who lived and mined near the site of the present town. Later the name was applied to the settlement recorded on the von Leicht–Craven map of 1874.

Sayante, Río de. *See* Zayante Creek.

Scaffold Meadow [Kings Canyon NP] was so named because sheepmen had a scaffold there to protect their supplies from bears and other animals (Farquhar).

Scarface [Modoc Co.]. The name of the Great Northern station preserves the name of Scarface Charlie, who fought with Captain Jack in the Modoc War of 1873 (Beth Bowden).

Scarper Peak [San Mateo Co.]. The name should have been spelled Scarpa because the peak was named for a family by that name residing in the district (Wyatt).

Schallenberger (shä′ lən bûr gər) **Ridge** [Placer Co.]. The name for the mountains south of Donner Lake was applied by the USGS in honor of young Moses Schallenberger, who wintered at the lake in 1844–45. He was a member of the Stevens-Murphy-Townsend party.

Scheelite (shē′ līt) [Inyo Co.]. For the mineral scheelite, calcium tungstate, an important ore of tungsten. The mineral in turn was named for the German chemist K. W. Scheele.

Schellville [Sonoma Co.]. The Northern Pacific Railroad named the station before 1888 for Theodore L. Schell, who owned 160 acres of land there in the 1850s and was active as a merchant in Sonoma and San Francisco. He died in San Francisco in 1878 (*San Jose Pioneer and Register*, Jan. 12, 1878). On some maps the name is spelled Shellville. There is a prominent Schell Mountain in Shasta Co. and a Schell Creek in Sierra Co., but they have not been identified with the family of Theodore Schell.

Schilling [Shasta Co.]. Settled in 1849, the place was first known as Franklin City, then as Whiskey Creek, and finally as Whiskeytown (*see* Whiskey). The post office was established in 1881; it successively bore the names Blair, Stella, and in 1917 Schilling, for John Frederick Schilling, a resident and storekeeper in the early 1900s (Steger). The post office was named Whiskeytown again on July 1, 1952, following the modern trend of reverting to bawdy but "historical" names.

Schofield (skō' fēld) **Peak** [Yosemite NP]. Named by Maj. William W. Forsyth in honor of Gen. John M. Schofield, secretary of war, 1868–69, commander-in-chief of the Army, 1888–95 (Farquhar).

Schonchin (skon′shin) **Butte** [Lava Beds NM]. The name commemorates the old chief of the Modocs who signed a treaty with representatives of the government in 1864. Looking for something to give emphasis to his pledge, so the story goes, he pointed to the dominant lone butte and declared that that mountain would fall before Schonchin ever would raise his hand against the whites (William S. Brown). He kept his pledge and lived peaceably with his people in the Klamath Reservation; but his younger brother, Schonchin John, became a leader in the Modoc War eight years later.

Schoolcraft Island. *See* Sutter: Sutter Island.

School House Station. *See* Colma.

Schramsberg [Napa Co.]. Named for Jacob Schram, a pioneer vineyardist.

Schultheis (shoolt′ hīs): **Lagoon, Pass** [Santa Clara Co.]. Named for John Martin Schultheis (or Schultheiss), a native of Germany, who homesteaded here in 1852 (Hoover, p. 533).

Schumann Canyon. *See* Shuman Canyon.

Schwan Lake [Santa Cruz Co.]. Often pronounced "Swan Lake." Jacob Schwan, a native of Germany, was a farmer in the area in the 1860s (Clark 1986).

Schwaub (shwôb) [Death Valley NP]. According to the Death Valley Survey, the mining camp was promoted in 1906 by Charles M. Schwab of the Bethlehem Steel Works, who had a large interest in the Greenwater copper mining district. But the Death Valley literature often mentions a Charles Schwab as an active prospector and promoter of mineral projects after 1900, when the great industrialist was already the president of the Carnegie Steel Corporation.

Sciad. *See* Seiad.

Scodie: Canyon, Spring [Kern Co.]. Named for William Scodie, who established a store at the mouth of the canyon in the 1860s (*see* Onyx). *For* Scodie Mountain, *see also* Kiavah Mountain.

Scorpion Hills. *See* Escorpion.

Scotia (skō′ shə) [Humboldt Co.]. When the Pacific Lumber Company built its mill in 1885, the town was named Forestville. When the post office was established on July 9, 1888, the present name was chosen because many of the mill workers were natives of Nova Scotia in Canada (Borden).

Scott. About forty geographic features in California are named Scott; several more are called Scotty and Scotchman. Most of the Scott names are from the family name; some were meant to be Scot but were spelled like the surname. In Tehama Co. the name is combined with a rare generic term: Scotts Glade. **Scotts Valley** [Santa Cruz Co.] was named for Hiram Daniel Scott, a sailor who came to Monterey Bay in 1846; in 1852 he purchased Rancho San Agustin (Hoover, p. 574). **Scott Bar, Valley, River**, and **Mountain** [Siskiyou Co.]: John W. Scott discovered gold here in 1850, and his name has been perpetuated in a number of geographic features. The bar and the valley are mentioned in 1851 (Indian Report, pp. 170–71). The stream is shown as Scotts Fork on Gibbes's map of 1852 and is mentioned as Scott's River in the *Statutes* of 1852 (p. 233). The river had been called Beaver River by the Hudson's Bay Company trappers.

Scottys: Castle, Canyon [Death Valley NP]. In

1923 Walter Scott (alias "Death Valley Scotty"), a former champion roughrider in Buffalo Bill's show, bought the old Staininger Ranch along with A. M. Johnson of Chicago; they then began building the fabulous "castle" for which the millionaire Johnson, who had come to Death Valley for his health, apparently provided the money.

Screwdriver Creek [Shasta Co.]. The tributary of Pit River was so named because of its "screwy movement through the canyon and flat" (Steger).

Scylla (sil′ ə) [Kings Canyon NP]. In 1895, when T. S. Solomons and E. C. Bonner came down from Mount Goddard to Simpson Meadow by way of Disappearing Creek and the Enchanted Gorge, they passed between two peaks that they named Scylla and Charybdis, from Homer's *Odyssey* (Farquhar): "Sheer, ice-smoothed walls arose on either side, up and up, seemingly into the very sky, their crowns two sharp black peaks of most majestic form. A Scylla and a Charybdis they seemed to us, as we stood at the margin of the lake and wondered how we might pass the dangerous portal" (*Appalachia* 1896:55).

Seal. This name is given to the sea lion (*Zalophus californianus*), common along our coast. The term has been used to name a number of places, including the well-known Seal Rocks below the Cliff House in San Francisco. The charts of the Coast Survey show six other Seal Rock or Rocks, two Coves, one Island, and one Creek. The sea lion was called *lobo del mar* 'sea wolf' in Spanish (*see* Lobo). **Seal Beach** [Orange Co.] was first called Bay City; it was given the present name by Philip A. Stanton and other interested persons (Stephenson). The post office is listed in 1915.

Seal Creek [San Mateo Co.] may have been named for Henry W. Seale, who at one time owned part of Rancho Rinconada del Arroyo de San Francisquito.

Sea Lion. The name is much less frequently used than the popular form "seal," but there are three barren Sea Lion Rocks and one Sea Lion Gulch on the California coast.

Seamans Gulch [Shasta Co.]. Named for George Seaman, who located a claim here in the 1860s (Steger).

Sea Ranch, The [Mendocino Co.] was named for its location on the seashore when it was built in the 1960s.

Searles (sûrlz) **Lake** [San Bernardino Co.]; **Searles** [Kern Co.]. The dry lake was named for John and Dennis Searles, brothers, who discovered borax there in 1863; in 1873 they began to exploit the rich deposits of minerals. A post office at the shipping point across the line in Kern Co., established on Aug. 20, 1898, was named Searles.

Sears Point [Sonoma Co.]. The name was applied to the station of the Northwestern Pacific before 1888, probably for Franklin Sears, who came to California in 1845, took part in the Mexican War, and settled in Sonoma Valley in 1851.

Searsville Lake [San Mateo Co.]. The name recalls the old town of Searsville, named for John H. Sears, who came from New York in 1849 and built the Sears Hotel in 1854. In 1890 the lake was created as a reservoir of the Spring Valley Water Company and was named after the town, now vanished.

Seaside [Monterey Co.]. The town was laid out in 1888 by Dr. J. L. D. Roberts and named East Monterey. When the Post Office Dept. declined this name in 1890, Roberts chose the present name.

Seavey Pass [Yosemite NP]. Named by R. B. Marshall for Clyde L. Seavey, president of the State Railroad Commission in 1923; he later was a member of the Federal Power Commission (Farquhar).

Sebastopol (sə bas′ tə pol) [Sonoma Co.] is the sole survivor of five Sebastopols—all named in or soon after 1854, when the siege of the Russian seaport by the British and French, during the Crimean War, excited the whole world. The town in Napa Co. is now Yountville; those in Tulare, Sacramento, and Nevada Cos. no longer exist. The town in Sonoma Co. was founded by H. P. Morris in 1853 as Pine Grove; according to tradition, the name was changed at the time of the Crimean War because of a local fight in which one party found its "Sebastopol" in the general store. The post office is listed in 1859.

Seco (sā′ kō); **Seca** (sā′ kə). Since many watercourses in California dry out in summer and many others have become permanently dry, the Spanish word for 'dry' was a common specific name in Spanish and Mexican times, especially combined with Arroyo and Laguna. It was also used with Campo 'field' and Llano 'plain' to indicate the absence of water. The adjective is still found in more than ten place names, including the settlements Arroyo Seco [Monterey Co.] and Campo Seco [Calaveras Co.].

Second Garrote (gə rō´ tē) [Tuolumne Co.]. Modern Groveland (*q.v.*) was formerly called Garrote because a Mexican was hanged from a tree there in July 1850 for stealing two hundred dollars from two miners (Bancroft Scrapbooks 16:127). A similar incident in a camp 2 miles southeast led to naming it Second Garrote, thereby distinguishing it from the original Garrote. *Garrote* is the Spanish designation for capital punishment by strangulation or for the scaffold on which the punishment is inflicted. The name of the post office was spelled Garrote from 1854 until it became Groveland in the 1870s, but the maps never got the name straight. Trask's map of the mineral districts has two Garots. Lapham and Taylor (1856) show a 2nd Garota but no first; Goddard (1860) has a Garota and a 2nd Garota. Hoffmann and Gardner (1867) use the Spanish spelling Garrote, but the other maps of the Whitney Survey have Garrota.

Secret. The word was frequently used for Camps, Gulches, Springs, Ravines, etc., during the Gold Rush. The places were named "secret" either because they were hard to find or because some successful miners did not want to make their finds known. Still recorded on modern maps are two small localities: Secret Ravine [Placer Co.] near Newcastle, and Secret [Lassen Co.], an old railroad station. Both formerly had post offices.

Secretario or **Sectario, Arroyo del**. *See* Sacatara Creek.

Secuan. *See* Sycuan.

Sedco (sed´ kō) **Hills** [Riverside Co.]. Established about 1915 and called Sedco, coined from South Elsinore Development Company (L. E. Burnham).

Seeley [Imperial Co.]. After the destruction of Blue Lake and the surrounding district by the overflow of the New River, 1905–7, a new town was started several miles to the north and was named for Henry Seeley, one of the pioneers in the development of Imperial Valley.

Segunda [Monterey Co.]. La Cañada llamada de la Segunda is mentioned in 1822 (Arch. MSB 3:296), but it is not known why the canyon should be called 'of the second'. On Apr. 4, 1839, the name Cañada de la Segunda was applied to a land grant. On the Monterey atlas sheet the *cañada* is called Canyon Secundo.

Seiad (sī´ ad): town, **Creek** [Siskiyou Co.]. The place is in the former territory of the Shasta Indians, and the name, originally spelled Sciad, may be derived from their language. On Nov. 15, 1863, Brewer of the Whitney Survey (p. 480) mentions Sciad Creek and Sciad Ranch, a thriving place settled in 1854 by a farmer from New York named Reeves. The post office is listed in 1867 with the spelling Seiad. This spelling is now generally accepted, although Wood's *Gazetteer* lists the stream as Sciad Creek.

Seigler (sēg´ lər): **Springs, Valley, Creek, Mountain** [Lake, Napa Cos.]. The springs were named for the discoverer and original owner, Thomas Sigler, a native of New York. They are listed as Hot Sigler Springs [*sic*] in the Pacific Coast Business Directory, 1876–78. The name is spelled Siegler in the County History (Napa 1881); but the spelling Seigler, used by the Whitney Survey, finally won out.

Selby [Contra Costa Co.]. The post office was established about 1887 and named for Prentiss Selby, superintendent of the Selby Smelting and Lead Works and first postmaster.

Selden Pass [Fresno Co.]. Named by R. B. Marshall of the USGS, for Selden S. Hooper, who was a member of the survey from 1891 to 1898 (Farquhar). In a decision of Feb. 3, 1915, the BGN misspelled the name Seldon, and this version is found on some maps.

Selma [Fresno Co.]. Established by the Southern Pacific in 1880 and named for the daughter of Max Gruenberg, assertedly at the request of Leland Stanford, to whom Gruenberg had shown a picture of his baby daughter. Another version was published in the *Centennial Almanac* of 1956 (p. 170): according to Ernest J. Neilsen of Selma, the station was named for Selma Michelson Kingsbury, the wife of an official of the Central Pacific Railroad.

Semas (sē´ məs) **Mountain** [Monterey Co.]. Probably for José Simas, of Portuguese descent, who acquired land near here in 1891 (Clark 1991).

Semig (sem´ ig) **Basin** [Siskiyou Co.]. For Bernard G. Semig, civilian doctor with the U.S. Army, wounded near here in a battle with the Modoc Indians, 1873 (BGN, list 7503).

Semi Pass. *See* Simi Hills.

Sempervirens (sem pər vī´ rənz) **Creek** [Santa Cruz Co.]. Latin for 'evergreen'; named for the coast redwood tree, *Sequoia sempervirens* (Clark 1986).

Seneca (sen´i kə) [Plumas Co.]. Named by R. K.

Dunn, a pioneer settler, after the township, which had been named about 1854, doubtless after one of the many places of this name in the East and Middle West. The name refers to the Seneca (Iroquoian) tribe of New York State. The Iroquoian name originally referred to another tribe, the Oneida, but was applied by whites to various Iroquoian groups other than the Mohawk (Chafe). By folk etymology, the native name came to be spelled like the name of the Roman philosopher. **Seneca Creek** [Monterey Co.] is perhaps a transfer name or may be a corruption of Spanish *ciénega, ciénaga* 'marsh' (Clark 1991).

Senel. *See* Sanel Mountain.

Senger, Mount; Senger Creek [Fresno Co.]. Named in 1894 by T. S. Solomons, for Joachim H. Senger (1848–1926), professor of German at the University of California and one of the four founders of the Sierra Club (Farquhar).

Sentenac (sen′ tə nak) **Canyon** [San Diego Co.]. Named for the brothers Paul and Pierre Sentenac, from France, who settled here in the 1880s (Brigandi).

Sentinel. About twenty orographic features are so named because of their real or fancied resemblance to a watchtower, including the Imposing Grand Sentinel in Kings Canyon NP. **Sentinel Dome** and **Rock** [Yosemite NP]: In 1851 the Mariposa Battalion applied the name South Dome to the mountain. At the time of the Whitney Survey the present name was current: "A prominent point, which . . . from its fancied likeness to a gigantic watch-tower, is called 'Sentinel Rock'" (Whitney, *Geology*, p. 412). **Sentinel Cave** [Lava Beds NM] was so named because nine pillar-like formations stand like sentinels before the entrance. *For* Sentinel Rock [Alpine Co.], *see* Jeff Davis Peak.

Sepulveda (sə pul′ və də) [Los Angeles Co.]. The name was applied to the station east of the present town, when the railroad from Los Angeles to San Fernando was built in 1873, probably for Fernando Sepúlveda, whose adobe was near the base of the Verdugo Mountains.

Sepulveda Canyon [Los Angeles Co.]: The name commemorates Francisco Sepúlveda, grantee of the San Vicente y Santa Monica grant in 1839.

Sequoia (sə kwoi′ ə). In 1847 the Austrian scholar, Stephan L. Endlicher, published his monumental *Synopsis coniferarum*, in which he classified the California giant trees as a genus *Sequoia*. (Current taxonomy distinguishes the coast redwood, *Sequoia sempervirens*, and the 'big tree' of the Sierra, *Sequoiadendron giganteum*.) In applying the name, Endlicher doubtless wished to honor Sequoya (George Gist), the creator of the Cherokee writing system. Among the few places bearing the name are Sequoia Ridge [Fresno, Tulare Cos.], and two railroad stations, Sequoia [Humboldt Co.] and Sequoya [Contra Costa Co.]. **Sequoia National Park** was established by act of Congress on Sept. 25, 1890, and christened by John W. Noble, secretary of the interior, at the suggestion of George W. Stewart. **Sequoia National Forest** was established and named by executive order of President Theodore Roosevelt on July 2, 1908 (Farquhar).

Serra (sâr′ ə) [Orange Co.]. In 1950 the Santa Fe named the station after the newly formed school district, which was presumably named for Padre Junípero Serra. The former name of the station was San Juan by the Sea. This locality has been absorbed into the City of Dana Point (Brigandi).

Serrano (sə rä′ nō) [San Luis Obispo Co.]. The station was named in 1893 by the Southern Pacific for the Serrano family, from whom the right-of-way was obtained. Miguel Serrano, a native of Mexico, had settled in the county in 1828. The name Serrano, Spanish for 'mountaineer' (from *sierra*), is also applied to a Takic (Uto-Aztecan) Indian tribe of Southern California.

Serrano Boulder [Riverside Co.]. This historical monument was named for Leandro Serrano, majordomo of Mission San Juan Capistrano and, after 1818, ranchero in Temescal Valley.

Serret (sâr′ ət) **Peak** [Kern Co.]. The high mountain north of Kernville was named for a Basque sheep owner of the region in "the early days."

Sespe (ses′ pē): town, **Creek, Gorge, Hot Springs, Oilfields** [Ventura Co.]. The name can be traced to a Chumash rancheria, called Cepsey in 1791 (AGN, California, p. 46). It was often recorded, with various spellings, in the archives of Mission San Buenaventura. The meaning of the word is probably 'knee cap,' according to Harrington. The present spelling is used in a letter of May 5, 1824 (Arch. Arz. SF 4:2.102), and for the land grant Sespe or San Cayetano, dated Nov. 22, 1833.

The Parke-Custer map of 1854 has Cespai River; Goddard's map of 1860, Sespe River. The name was applied to the Southern Pacific station in 1887 and to the post office in 1894. Henshaw, however, about the same time gives the spelling Cespe Ranch and calls the Indian village Sek-Pe.

Setimo (set′ i mō) **Creek** [Kern Co.]. Perhaps for Spanish *Séptimo*, a man's given name (literally 'seventh'). By BGN, list 7501, this name supplants Set Creek and Setino Creek.

Seven Gables [Fresno Co.]. Named by T. S. Solomons and Leigh Bierce on Sept. 20, 1894: "The south wall of the gap we found to be the side of a peak, the eccentric shape of which is suggested in the name Seven Gables, which we hastened to fasten upon it" (*SCB* 1:230).

Seville (sə vil′) [Tulare Co.]. In 1913, when the Santa Fe built the line from Minkler to Exeter through the newly developed citrus region, the location engineers chose Spanish-sounding names as appropriate station names: Primero, Orosi, Seville, Rayo, Sur, Fruta, etc.

Seward Creek [Mendocino Co.]. The tributary of Forsythe Creek was named, about 1880, for Anson J. Seward (H. McCowen).

Shackleford Creek [Siskiyou Co.]. The stream was named for John M. Shackleford, who built an eight-stamp quartz mill on the creek in 1852 (Co. Hist.: Siskiyou 1881:215).

Shadow: Mountains, Mountain [San Bernardino Co.]. The range was so named because a dark-colored peak appears to be the shadow of a neighboring lighter-colored peak. Shadow Mountain is not a part of the range but stands 7 miles east. **Shadow Lake** [Madera Co.] is known as one of the most beautiful lakes in the High Sierra; the same name was once applied by John Muir to Merced Lake in Yosemite.

Shafter [Kern Co.]. Named in memory of Gen. William ("Pecos Bill") Shafter, 1835–1906, commander of U.S. forces in Cuba during the Spanish-American War. General Shafter lived on his ranch of 10,520 acres near Bakersfield from the time of his retirement in 1901 until his death. In 1914 the land was subdivided, and the town of Shafter was established.

Shagoopah. *See* Chagoopa.

Shake City. *See* Shingle.

Shamrock Lake [Yosemite NP]. The outline of the little lake north of Benson Pass suggested the name; it was probably applied when the Bridgeport quadrangle was mapped in 1905–7.

Shandon [San Luis Obispo Co.]. In 1891 the Post Office Dept. accepted the name, which was unique at that time among post offices in the United States. It was suggested by Dr. John Hughes, supposedly from a story published in *Harper's Magazine*.

Shanty Creek [Trinity Co.]. So named because of the bark-covered Indian dwellings called shanties by the white people (L. P. Duncan).

Sharktooth Peak [Fresno Co.]. The peak was so named by T. S. Solomons in 1892 because of its shape (Farquhar).

Sharp Park [San Mateo Co.]. When the district was developed in 1905, the Ocean Shore Land Company named the place Salada Beach after nearby Laguna Salada. When the Post Office Dept. in 1936 requested another name to avoid confusion with a similar one, the name was changed to Sharp Park, after the park and recreation area given to the city of San Francisco by Honora Sharp. The county map of 1877 shows that two sections of Rancho San Pedro were owned by G. F. Sharp.

Shasta, Mount; Shasta: River, Valley, Springs, Retreat; Mount Shasta (town) [Siskiyou Co.]; **Shasta: County**, town, **Dam, Lake** [Shasta Co.]; **Shasta National Forest**. The origin of the place name is found in the name of the Indians who inhabited the region of the upper Rogue, middle Klamath, Scott, and Shasta Rivers in the early part of the 19th century. The name of the tribe is mentioned in an entry under the date of Jan. 31, 1814, in the journals of Alexander Henry and David Thompson: "They [the Indians] said they were of the Wallawalla, Shatasla, and Halthwypum nations" (Coues, *New light*, p. 827). On Feb. 10, 1827, Peter Ogden records: "Here we are among the Sastise," and four days later he gave the name of the tribe to a river and a mountain: "I have named this river Sastise River. There is a mountain equal in height to Mount Hood or Vancouver, I have named Mt. Sastise. I have given these names from the tribes of Indians." In later entries, Mar. 9 and 13, he spells the name Sasty (*OHQ* 11:213–14, 216 [1910]). Ogden's sketchy diary does not make clear which peak and river he named. It is, however, fairly certain that it was Mount Shasta, and equally certain that it was not Shasta River. Merriam (*Journal of the Washington* [D.C.] *Academy of Sciences* 16:522–29) believes that Ogden referred to Mount McLoughlin and Rogue River in Oregon, and the maps of the 1830s support this belief. Abert's map of 1838, a U.S. government publication, shows Shasty River (and a tributary, Nasty River!) flowing into the Klamath River from the north, as well as Mount Shasty, all in Oregon. In the summer of 1841 Wilkes expressly mentioned the mountain in his instructions to Emmons and Ringgold, who were to explore the interior: "particularly the Shaste Peak, the most southern one in the territory of Oregon" (Wilkes 5:521). However, the Emmons detachment definitely applied the name to the present Mount Shasta; and Eld's sketches have Sasty Peak and River in California, with Sasty Country and the Sasty or Cascade Mountains straddling the line. On Duflot de Mofras's map of 1844 the names for both mountain and river (Mont Sasté and R. des Sastés) are in their new location. There are numerous spelling variants on early maps (Frémont-Preuss 1848: Mt. Tsashtl), and as late as 1852 Horn's Overland Guide uses the name Shaste for the Rogue River. Brewer, on Oct. 5, 1862, states that an Indian pronounced the name "tschasta." Since the spelling Chasty or Chasta occurs on some early maps, it may be assumed that this was the approximate native pronunciation and that there was a relation to the Chastas in Oregon. Other early American maps called the peak Mount Jackson, and Gibbs says the Indian name was Wy-e-kah (Schoolcraft 3:165); *see* Yreka. The sponsor of the modern version was apparently Madison Walthall, assemblyman from the Sacramento district, who proposed the name Shasta Co. for the county in the northeast corner of the state—instead of the name Reading Co., proposed by others. With the creation of the twenty-seven original counties, on Feb. 18, 1850, the name became official. The spelling Shasta is used on Scholfield's map of 1851. Some of the other maps of the 1850s have Mt. Shasté and Shasta Butte. The town known as Reading's Springs or Diggings was changed to Shasta City in 1850. The name was abbreviated to Shasta with the establishment of the post office in 1851. The town in Siskiyou Co., first known as Strawberry Valley—and from 1886 to 1922 as Sisson (for H. J. Sisson, an early settler)—was changed to Mount Shasta in 1922. The national forest was created and named by presidential proclamation in 1905, and Shasta Dam was officially named by John C. Page, commissioner, Bureau of Reclamation, on Sept. 12, 1937. **Shasta Bally Mountain** [Shasta Co.] contains Wintu *boli* 'mountain' (*see* Bally).

Shastina (shas tē′ nə) [Siskiyou Co.]. This playful diminutive form of the name Shasta has been applied to the west peak of Mount Shasta.

Shauls Lake [Lake Co.]. Named for Ben F. Shaul, a settler of 1865, on whose ranch the lake was situated (Mauldin).

Shaver Lake, reservoir, post office; **Shaver Crossing** [Fresno Co.]. According to L. A. Winchell, the place was named for C. B. Shaver, of the Fresno Flume and Irrigation Company, which built a reservoir and sawmill here in the 1890s. The Southern California Edison Company retained the name for the enlarged reservoir (Farquhar). However, a settlement called Saver, as well as Saver's Peak, are shown in this locality on the maps of the Whitney Survey as early as 1873. It is quite possible that this original name was changed when C. B. Shaver became associated with the region (*see* McCloud River for a similar shifting of names).

Shaws Flat [Tuolumne Co.]. Named by Mandeville Shaw, who planted an orchard here in 1849 (Kazmarek).

Sheep. Among the fifty or more features so named on the maps of the state are four peaks that rise more than ten thousand feet, several Sheep Hollows and Sheep Corrals, a Sheephead Mountain and Pass [Death Valley NP], a Sheep Thief Creek [Stanislaus Co.], a Sheep Crossing [Madera Co.], and a Sheep Repose Ridge [Sonoma Co.]. The majority of the names doubtless refer to herds of domestic sheep, but some were applied because of the presence of the mountain or desert bighorn sheep (*Ovis canadensis*); *see also* Carneros. **Sheep Rock** [Siskiyou Co.], the oldest Sheep name in California, was given by Hudson's Bay Company trappers, probably in the 1820s: "It is said to be one of only three places, where the bighorn, or mountain sheep, is at present found, west of the Sierra Nevada" (Gibbs 1851, in Schoolcraft 3:168). **Sheepranch** [Calaveras Co.] was named after a gold mine, which in turn had been named after a sheep corral. *For* Sheep Island, *see* Brooks Island. *For* Sheep Mountain, *see* Langley, Mount; Tucki Mountain.

Sheetiron: Springs, Mountain [Lake Co.]. Some

time between 1860 and 1880 a man camped by the springs for his health. His shelter was a sheet-iron hut. After he abandoned it, the hut fell apart, and for years hunters and campers used the pieces for fire plates and camp stoves (Mauldin).

Sheldon [Sacramento Co.]. The settlement that developed on the Omochumnes (or Cosumnes) land grant was named for Jared Sheldon, grantee in 1844.

Shell. About twenty places in California, chiefly along the ocean shore, are named for the presence of sea shells; included are one post office, Shell Beach [San Luis Obispo Co.], and several communities. Along the coast are numerous places designated as shell mounds, some of which are now protected in state parks; these were refuse heaps and burial mounds that had developed near Indian villages and were so named because they consist mainly of fragments of shells. *For* Shellville, *see* Schellville.

Shelter Cove [Humboldt Co.]. The name was applied to the small natural harbor by the Coast Survey when making a hydrographic survey of it in 1854. It afforded "an anchorage from northwest winds, and may, perhaps, be regarded as a harbor of refuge for small coasters which have experienced heavy weather off Cape Mendocino, and are short of wood and water, both of which may be obtained here from one or two gulches opening upon the sea" (*Coast Pilot* 1858:66).

Shenandoah (shen ən dō´ ə) **Valley** [Amador Co.]. Blood Gulch running through the valley seems to indicate that some miniature battle took place here that reminded the name giver of the many bloody engagements in the valley in Virginia during the Civil War. According to Doyle, however, John Jameson, who settled in the valley in the early 1850s, named it after his birthplace; it originated in an Algonquian language, with a meaning something like 'spruce stream'.

Shepherd. More than ten geographic features are so named because some shepherd had his cabin there. The spelling Shepard occurs in Alameda, Glenn, and Inyo Cos. The creek and pass in Inyo Co., however, owe their name to a pioneer family named Shepherd (Farquhar).

Sheridan [Placer Co.]. Since the Roseville-Marysville branch of the Central Pacific was built during and immediately following the Civil War, the station was doubtless named in honor of Gen. Philip Sheridan (1831–88).

Sheridan Creek [Shasta Co.]. The tributary of Bear Creek was named for George Sheridan, who settled here in the 1850s (Steger).

Sheriffs Springs [Orange Co.]. So named because Sheriff James Barton and his posse were murdered here by Juan Flores and his band in 1857 (*see* Flores Peak).

Sherlock Gulch [Mariposa Co.]. The gulch northeast of Bullion Mountain was named after Sherlock's Diggings, discovered in the summer of 1849 "by a man named Sherlock, who with a company of seventy Mexicans worked these deposits on shares. . . . Sherlock has gone into the mountains . . . and we have elected a new Alcalde, who . . . issued an order for all Mexicans to decamp" (*San Francisco Alta California*, Oct. 25, 1849).

Sherman Indian High School (Riverside Co.). Earlier Sherman Institute, named in 1901 after James Schoolcraft Sherman, congressman from New York and chair of the House Committee on Indian Affairs at the time; he later became U.S. vice president (Gunther).

Sherman Island [Sacramento Co.]. Named for Sherman Day, state senator, 1855–56, and U.S. surveyor general for California, 1868–71 (*see* Day, Mount). The island, on which Day owned a ranch, is shown on Ringgold's General Chart of 1850.

Sherman Oaks [Los Angeles Co.]. Developed around 1910 by Moses H. Sherman (Atkinson).

Sherwin: Creek, Lakes, Hill [Mono Co.]. Named for James L. C. Sherwin, a native of Kentucky, who came to California in 1850 and settled in Round Valley in the early 1860s. Sherwin took an active interest in Indian affairs and was a member of the California state assembly in 1858.

Sherwood, Lake [Ventura Co.]. An artificial body of water originally called Canterbury Lake, later Mathiessen Lake. It was renamed in 1922 for Robin Hood's Sherwood Forest, after the movie *Robin Hood* was made here in 1922 (Ricard).

Sherwood Valley [Mendocino Co.]. Named for Alfred Sherwood, who settled in the valley in 1853 and lived there until his death in 1900.

Shiloah (shī´ lō) [Shasta Co.]. The mineral spring was named by Charles Dougherty after the ancient spring near Jerusalem, noted for its healing qualities (Steger).

Shingle [El Dorado Co.]. The first house in the settlement known as Shingle Springs was built in 1850 by a man named Bartlett. The name came "from a shingle machine used for the manufacture of shingles at a cluster of springs, situated on the western extremity of the village" (Co. Hist.: El Dorado 1883:199–200). Shingle Spring post office was established on Feb. 3, 1853. It was changed to Shingle on May 11, 1895, and back to Shingle Spring on Jan. 13, 1955. There is a **Shingletown** [Shasta Co.], and Mendocino Co. has a settlement named Shake City.

Shinn, Mount [Fresno Co.]. Named in 1925 by members of the Forest Service in memory of Charles H. Shinn (1852–1924), who was for many years connected with the USFS and was a charter member of the Sierra Club (Farquhar).

Ship. The several Ship Mountains and Rocks in the state apparently were given this name because of their shape. The mountain in Siskiyou Co. was originally known as Four Brothers and was given the present name in 1909 by A. W. Lewis (USFS).

Shirttail. This favorite name in the colorful nomenclature of mining days has survived in several places. **Shirttail: Gulch, Peak** [Shasta Co.]: In the late 1850s a group of miners working on the East Fork of Clear Creek were swept down the stream by a slide. All were killed except Wilbur Dodge, whose wife was able to rescue him because his shirttail had caught on a snag at the mouth of the creek (Steger). **Shirttail Canyon** [Placer Co.]: The explanation that it was so named in 1849 because a prospector was found working there clad only in a shirt is the most plausible of the half dozen theories about the origin of the name.

Shivelly (shīv′ lē) **Gulch** [Amador Co.]. Named for Charles Shivelly, who came to California in 1849 and apparently found a fortune in the gulch in 1852 (Knave, Sept. 22, 1946).

Shively [Humboldt Co.]. Probably named for William R. Shiveley, a pioneer of the area, who came to California from Ohio in 1852.

Shoal Point. *See* Richmond.

Shoemaker Bally [Shasta Co.] was named for Simon Shoemaker, with Wintu *buli* 'mountain'; *see* Bally.

Shoquel. *See* Soquel.

Shorb Junction [Los Angeles Co.]. A station of the Southern Pacific was named before 1900 for James de Barth Shorb (*see* San Marino).

Shore Acres [Contra Costa Co.]. Named for its location on Suisun Bay. *See also* Manhattan Beach.

Shoshone (shō shō′ nē) [Inyo Co.]. The name was applied to the station when the Tonopah and Tidewater Railroad was built shortly after 1900. The post office is listed in 1915. It is apparently the only place in California named for the widespread family of Shoshonean Indians. The term was first applied in the early 19th century to the Eastern Shoshone of Wyoming; its origin is not known (*HNAI* 11:334).

Shot Gun. A number of places are so named because of some incident in which a shotgun played a role, such as Shotgun Pass [Sequoia NP]. **Shot Gun Creek** [Shasta Co.], a tributary of the Sacramento River, was supposedly named because a miner named Brand, instead of paying his helper, ran him out with a shotgun (Steger). A bend in the road to Angels Camp was once called Shot Gun Bend because highwaymen could easily hold up a stage at this point.

Showalter (shō′ wôl tər) **Hill** [Calaveras Co.]. Named for John Showalter, who had a cabin near the top of the hill on which extensive coyote mining was conducted (*Las Calaveras*, Oct. 1957).

Showers Pass [Humboldt Co.]. Probably named for Jacob O. Showers, a farmer at Rohnerville, who summered his cattle here in the 1860s. According to local tradition, the pass was so called because Showers once rode across it at breakneck speed to escape a band of Indians (Arthur Renfroe).

Shuman Canyon [Santa Barbara Co.]. The canyon north of Casmalia was long known by the name of an early settler, who spelled it Schumann. The BGN decided for the present spelling in June 1949.

Shuteye: Peak, Pass [Madera Co.]. Named for Old Shuteye, an Indian who was blind in one eye. One of the main trails across the Sierra passed through his rancheria. The peak has a namesake called Little Shuteye Peak (*see* Little).

Siberia [San Bernardino Co.]. Named by the Southern Pacific Railroad when the original line (now Santa Fe) was built in 1883. The choice of the name was probably inspired by the striking contrast between the temperature of the desert and that of Siberia (*see* Klondike).

Siberian: Outpost, Pass, Pass Creek [Sequoia

NP]. The region was named Siberian Outpost by Harvey Corbett in 1895 because of its bleak appearance. The pass had been called Rampart Pass by Wallace, Wales, and Wright in 1881 (Farquhar). In 1907 the USGS used the present name for the pass and the creek.

Sibleyville. *See* Dinuba.

Sicard (si kärd′): **Bar, Flat** [Yuba Co.]. Named for Theodore Sicard, a French sailor who came to Sutters Fort in 1842; in 1848 he mined and traded at the places that became known by his name and that now exist in name only. In 1849 he was associated with Covillaud in the establishment of Marysville, and some people favored the name Sicardoro for the new place.

Sidewinder: Mountain, Road [San Bernardino Co.]. The peak was named after the Sidewinder Mine northwest of it, which in turn was probably named because of the presence of the rattlesnake called a sidewinder (*Crotalus cerastes*), so called from its lateral looping movement.

Siegler Springs. *See* Seigler Springs.

Sieroty (si rō′ tē) **Beach** [Marin Co.]. Named in 1982 after former state senator Alan Sieroty of Los Angeles, in recognition for his work in helping to protect the coast (Teather 1986).

Sierra (sē âr′ ə). The Spanish word for 'saw blade', or by extension for 'mountain range', was frequently used in Spanish times: any two or more peaks in a row might be called a *sierra*. The word appears also in the chief or secondary names of six land grants. It is occasionally used as a generic term in modern times, as in Sierra Azul 'blue range' [Santa Cruz Co.], Sierra Morena 'brown range' [San Mateo Co.], Sierra de Salinas [Monterey Co.], and Sierra Pelona 'bald range' [Los Angeles Co.]. However, in modern usage, "Sierra" is usually an abbreviation for the Sierra Nevada mountain range. **Sierra Peak** and **Canyon** and **La Sierra** [Riverside, Orange Cos.] take their names from two land grants—Sierra, and Sierra de Santa Ana—both dated June 15, 1846, and granted to Bernardo Yorba and Vicente Sepúlveda respectively. **Sierra City, County, Valley,** and **Buttes** and **Sierraville** were named directly or indirectly for their location in the northern part of the Sierra Nevada. The county, formed from a portion of Yuba Co., was established and named on Apr. 16, 1852. Sierraville post office is listed in 1867. **Sierra Point** [Yosemite NP], with its view of Yosemite Valley, was discovered by Charles A. Bailey, Warren Cheney, W. E. Denison, and Walter E. Magee, who named it on June 14, 1897, as a compliment to the Sierra Club (Farquhar). **Sierra National Forest** is one of five national forests created in 1908 by division of the area that included the original Sierra Forest Reserve, established and named by executive order of President Harrison in 1893.

Sierra Madre (mä′ drē, mä′ drā) [Los Angeles Co.]. The name means 'mother range' and was applied in Spanish times to a number of ranges in different sections of California. The various ranges in Kern, Los Angeles, San Bernardino, and Riverside Cos. were called Sierra Madre de California by Font in 1775 (*Compl. diary*, p. 122). "The whole chain . . . was formerly called Sierra Madre by the Padres—probably from the fact that the Sierra Nevada, the Coast Mountains, and other ranges seem to spring from it" (Blake, 1853, Pac. R.R. *Reports* 5:2.137). There is today no collective name for these ranges. The name was given to the town in Los Angeles Co. when Nathaniel C. Carter subdivided part of Rancho Santa Anita in 1881; Sierra Madre Villa is recorded in that year. By decision of the BGN, 1965, the range between the Cuyama and Sisquoc Rivers in Santa Barbara Co. is now tautologically called Sierra Madre Mountains. *For* Sierra Azul, *see* Blue; Azul.

Sierra Nevada (nə vad′ ə). The name existed in California more than two centuries before it was identified with the lofty mountain range. It is a common descriptive Spanish name for a range covered with snow. In California, the name was first applied by Cabrillo in Nov. 1542, as Sierras Nevadas, to the Santa Lucia Mountains south of Carmel. On a map of 1556 (Wagner, no. 10) a large section of the present state is covered by a glacial mass, called *Sierra Neuados;* and on Bolognino Zaltieri's map of 1566 (Geographical Report, opp. p. 504) a well-defined Sierra Nevada runs east-west in the latitude of Alaska. For more than a century the cartographers used the name for mountains or points along the coast, variously spelled and in various latitudes, until California had become an island on the maps and names were dropped for lack of space. The present-day Sierra Nevada was seen but not named by Crespí in 1772. It was seen again by Font on Apr. 3, 1776, from a hill near the juncture of the Sacramento and San Joaquin rivers: "If we looked to the east we saw on the other side of the plain at a distance of some thirty leagues a great *sierra nevada,* white from the summit to the skirts, and running about from south-southeast to north-northwest." On Font's map of 1776 the name appears for the northern part of the Sierra Nevada, and also for the Coast Ranges in the southern counties. The latter were then already known by this name, as Padre Garcés records on Apr. 25, 1776, "I came upon another large sierra which makes off from the Sierra Nevada and extends north-eastward, to which I gave the name of San Marcos" (Coues, *Trail,* pp. 270–71). On Font's map of 1777 (Davidson's copy), which incorporated both his and Garcés's discoveries, the entire range first appears. The San Gabriel Mountains are called Sierra Nevada; the Tehachapi and Piute Mountains and the south end of the modern Sierra Nevada are left nameless; between latitude 36° and 38° north it is Sierra de San Marcos; and between about 38° and 39° north again Sierra Nevada. This extraordinary map presents, on the whole, the same orographic picture of the southern half of the state that we see on modern maps. Unfortunately, Font's map of 1777 was unknown to later cartographers, and Humboldt's map of 1811 has only a vague Sierra de San Marcos between latitude 36° and 39° north. Langsdorff in 1806 repeatedly mentions the name Sierra Nevada, but he did not publish a map. American and European cartography remained vague about the location of the range, and it was sometimes confused with the Rocky Mountains. Narváez's Plano of 1830, finally, gave a fairly accurate location but left the range nameless. Wilkes's map of 1841 calls it the California Range, and the Frémont-Preuss map of 1845 records Sierra Nevada of California. Several American cartographers attempted in vain to substitute Snowy Mountains or California Range. Muir suggested the poetic name Range of Light. The present northern limitation of the range is indicated at 40°20' north on Erman's map of 1849, where it is labeled Sierra Nevada or *Blaue Berge* (German for 'blue mountains'). By decision of the BGN (6th Report) the range is limited on the north by the gap south of Lassen Peak, on the south by Tehachapi Pass. The range is often spoken of colloquially as "the Sierra Nevadas," "the Sierras," or the tautological "Sierra Nevada Mountains."

Sierra Verde de Pinos. *See* Pinos; Santa Cruz: Santa Cruz Mountains.

Sifford Lakes [Lassen Volcanic NP] were named for the Sifford family, which acquired Drakesbad in 1900; the name was approved by the BGN in 1937. *For* Siffords, *see* Drakesbad.

Sigler Springs. *See* Seigler Springs.

Signal. It is common usage in place naming to attach this word to an eminence that serves as a signal station in warfare, explorations, and surveys. California has many such names, mainly Buttes and Hills, but also a number of Creeks. **Signal Lake** and **Mount Signal** [Imperial Co.] were both named after Signal Mountain, which is just across the border in Mexico. This mountain was named during the Boundary Survey: "The prominent mountain lying about four miles S. 10° E. from camp, and apparently 2,000 feet in height, must serve as a beacon to travelers crossing from the Colorado, and may probably be found a convenient point from which to flash gunpowder for the determination of the difference of longitude between San Diego and the mouth of the Rio Gila. Hence it may be called 'Signal mountain,' and this lake so near its foot 'Signal Lake'" (Whipple, *Extract,* 1849:9). **Signal Hill** [Los Angeles Co.] had been known since Spanish times as Los Cerritos. When the Coast Survey established the Los Angeles Base Line in 1889–90, John Rockwell erected a signal on the hill, and it became known as Signal Hill: "I recognize a hill near the town on which I erected a signal and which bears the name of Signal Hill to this day" (Rockwell to Davidson, Feb. 5, 1905). The Coast Survey charts and other maps retained the old Spanish name for some time, but the popular name became official after the discovery of oil in 1921.

Silaxo. *See* Oxalis.

Silicon Valley refers to the area of the computer industry around San Jose, in Santa Clara and San Mateo Cos., so called because of the silicon chips used in semiconductors.

Sill, Mount [Kings Canyon NP]. Joseph N. LeConte named the peak in 1896, in memory of Edward R. Sill (1841–87), poet and professor of English at the University of California from 1874 to 1882.

Silliman, Mount; Silliman: Creek, Crest, Lake, Pass [Sequoia NP]. Named by the Whitney

Survey in June 1864 in honor of Benjamin Silliman Jr. (1816–85), professor of chemistry at Yale (Farquhar).

Silsbee [Imperial Co.]. The town was laid out in 1902 and named for Thomas Silsbee, a cattleman of San Diego, who had pastured his stock here.

Silurian (si lo͞o′ rē ən): **Valley, Lake** [San Bernardino Co.]. The valley southwest of Death Valley was named because of the Silurian (Paleozoic) rocks.

Silva: Flat, Reservoir [Lassen Co.]. The flat became known as Silva's after a herder who drove his sheep from the Sacramento Valley to pasture here in summer. The reservoir was constructed in the 1920s and named after the flat (R. B. Sherman).

Silver. This word is found in the names of more than seventy-five physical features in California. A number of these—including the Peak, Creek, Valley, and the old mining towns Silver Mountain and Silverking in Alpine Co.—were named for the presence of silver ore. Most of them, however, especially the many Creeks and Lakes, were given the name because of their silvery appearance. **Silver Lake** [San Bernardino Co.]: The station of the old Tonopah and Tidewater Railroad was named in 1906 because it is at the edge of a large dry lake known as Silver Lake (W. W. Cahill). **Silver Creek, Peak, Pass**, and **Divide** [Fresno Co.]: The cluster name had its origin in Silver Creek, so named by T. S. Solomons in 1892 because of its silvery appearance (Farquhar). The peak was named after the creek, by Solomons also, but the name for the pass and divide were applied by the USGS when the Mount Goddard quadrangle was surveyed in 1907–9. **Silver Creek** and **Mountain** [Alpine Co.]: Rich silver deposits were discovered here about 1860; a mining town, now disappeared, is said to have been established by Scandinavian miners and called Kongsberg after the mining town in Norway. The post office established on May 12, 1863, was "Germanized" to Konigsberg; in 1865 it was changed to Silver Mountain, and in 1879 to Silver Creek. The "Silver" in some names refers to the presence of quicksilver (mercury), often called "silver." *For* Silver Lakes [Alpine Co.], *see* Kinney Lakes.

Silverado (sil və rä′ dō): **Canyon**, town [Orange Co.]. The name was coined in analogy to "Eldorado," suggesting silver instead of gold. It was given to the canyon when silver was discovered there in the 1870s (Stephenson). The post office is listed in 1932. **Silverado** [Napa Co.], a now-vanished mining community, flourished in the early 1870s. In the summer of 1880 Robert Louis Stevenson lived here with his bride and wrote *The Silverado squatters*. There is also a Silverado Canyon in Mono Co.

Silveyville [Solano Co.]. Named for Elijah S. Silvey, who built a tavern there in 1852. A post office was established on July 25, 1864.

Simi (si mē′, sē′ mē): **Hills, Peak, Valley** [Ventura Co.]. According to Kroeber, the name is from the "Ventura dialect Chumash *shimiyi* or *shimii*, a place or village." The name is recorded in its present form in 1795 as the name of a land grant, regranted on Mar. 8, 1821, and Apr. 25, 1842. *Un valle que se llama Simí* 'a valley called Simi' is mentioned on Feb. 3, 1796 (Arch. MSB 2:10), and the name appears repeatedly in mission and provincial records. It is shown on Narváez's Plano of 1830. The Parke-Custer map of 1855 shows Rancho Semi, Semi Pass, and Rio Semi (for Calleguas Creek). The spelling Simi is restored on Goddard's map. When the post office was established on Jan. 19, 1889, it bore the impressive name Simiopolis; six months later, on July 15, the name Simi was adopted. The name Simi Valley was made official for the city by BGN, list 7304.

Simmler [San Luis Obispo Co.]. Named for John J. Simmler, a native of Alsace, who came to California in 1853, opened the first hotel at San Luis Obispo, and was for many years the postmaster there.

Simmons Peak [Yosemite NP]. Named in 1909 by R. B. Marshall for Dr. Samuel E. Simmons, of Sacramento (Farquhar).

Simons [Los Angeles Co.]. The name developed by common usage after Walter Simons of Pasadena established a brickyard here in 1905.

Simpson Meadow [Kings Canyon NP]. Probably named for S. M. Simpson, a member of the family that ran sheep here in the 1880s. It was previously known as Dougherty Meadow, for Bill and Bob Dougherty, who pastured horses there; but that name has since been applied to a meadow on the Monarch Divide (Farquhar).

Sims [Shasta Co.]. Brewer relates in an entry of Sept. 14, 1862, that his party had camped at "Mr. Sim Southern's (long be his name remembered)," whose marvelous tales earned him the reputation that "with him 'truth is stranger than fiction'" (pp. 306, 310). Simeon Fisher Southern had settled there in 1859; the railroad station, previously known as Welch Station, was named Sims in 1887. The Post Office Dept. used the name Hazel Creek, given May 15, 1877, after the creek where hazelnuts grow (*see* Hazel: Hazel Creek). *For* Sims in Sacramento Co., *see* Runyon.

Sing Peak [Yosemite NP]. R. B. Marshall named this peak in 1899 for Tie Sing, the Chinese cook for the USGS from 1888 to 1918 (Farquhar).

Sink. A common generic term designating the spot where a stream disappears into the ground. The most notable of California's "sinks" is Soda Sink, where the Mojave River is absorbed by the sand.

Sinkyone (sing′kē ōn) **Wilderness State Park** [Humboldt Co.] is named for the Sinkyone (Athabaskan) tribe. The term is derived from the name applied to the Sinkyone by the neighboring Kato tribe, namely *sin-kiyahan* 'coast tribe' (William Anderson, p.c.).

Siquico, Arroyo de. *See* Sisquoc River.

Sir Francis Drake Bay. *See* Drakes Bay.

Sirretta (sə ret′ ə) **Peak** [Tulare Co.]. For Hippolyte Sarret, a French sheepman who settled in Poso Flat (Darling 1988).

Sisar (si sär′): **Canyon, Creek, Peak** [Ventura Co.]. The name is derived from that of a Chumash village, *sisá*, repeatedly mentioned in the archives of Mission San Buenaventura in and after 1807. Ranchería de Sisá is shown on the *diseño* of the El Rincón grant (1834).

Siskiyou (sis′ ki yo͞o): **Mountains, County**. Siskiyou was the Chinook Jargon word for 'bobtailed horse', apparently borrowed from Cree (Algonquian) *kîskâyowêw* 'bobtailed' (from *kîsk-* 'cleanly cut off', *-âyow-* 'tail', according to John Nichols): "This name, ludicrously enough, has been bestowed on the range of mountains separating Oregon and California, and also on a county in the latter State. The origin of this designation, as related to me by Mr. [Alexander C.] Anderson, was as follows. Mr. Archibald [Alexander] R. McLeod, a chief factor of the Hudson's Bay Company, in the year 1828, while crossing the mountains with a pack train, was over taken by a snow storm, in which he lost most of his animals, including a noted bob-tailed race-horse. His Canadian followers, in compliment to their chief, or 'bourgeois,' named the place the Pass of the Siskiyou, an appellation subsequently adopted as the veritable Indian name of the locality, and which thence extended to the whole range, and the adjoining district" (Gibbs, *Chinook Jargon*). Another interpretation derives the name from the French *six cailloux* 'six stones': In an argument delivered in the State Senate on Apr. 14, 1852, Sen. Jacob R. Snyder of San Francisco, who had been instrumental in creating the new county on Mar. 22, 1852, maintained that the name "Six Callieux" [*sic*] was given to a ford on the Umpqua River in Oregon, where Michel La Framboise had led a party of Hudson's Bay Company trappers over six stones in the river in 1832.

Sisquoc (sis kwok′): **River**, station, **Condor Sanctuary** [Santa Barbara Co.]. The name is Chumash and means 'quail', according to local tradition. It was given to a land grant, dated Apr. 17, 1845. The river is shown as Arroyo de Siquico on a *diseño* (1846) of the Cuyama grant. The station was named when the Pacific Coast Railroad was built in 1887. The Condor Sanctuary was set aside by the USFS in 1937 in an effort to preserve the California condor from extinction.

Sissel Gulch [Siskiyou Co.]. The gulch near the head of Moffett Creek was named for John Baker Sissel (*see* Cecilville).

Sisson. *See* Shasta.

Sister Elsie Peak [Los Angeles Co.]. The mountain was named for a nun who was in charge of a nearby Indian orphanage. It is more commonly known as Mount Lukens—named for Theodore P. Lukens, Mayor of Pasadena, who encouraged the building of a fire tower on the summit (Wheelock); *see* Lukens Lake.

Sisters. *See* Brothers.

Sites [Colusa Co.]. Named in 1887 by C. E. Grunsky, for John H. Sites, landholder (E. Weyand).

Sitio. The Spanish word as used in land grant cases designated one square league; but like *paraje*, it was sometimes used in a general sense for 'place, location'.

Six Rivers National Forest. The name for the 18th national forest of the state was suggested by Peter B. Kyne, San Francisco author, and was accepted by the USFS in Dec. 1946. The new forest embraces the watersheds of the Smith, Klamath, Trinity, Mad, Van Duzen, and Eel Rivers.

Skaggs Springs [Sonoma Co.]. Named for

William and Alexander Skaggs, who acquired the land on which the springs are situated in 1856 and built a resort in 1864. The name is shown on the von Leicht–Craven map.

Skeggs Observation Point [San Mateo Co.]. Named for Col. Ino H. Skeggs, later assistant state highway engineer, when that section of Skyline Boulevard was built in 1922.

Skeleton Peak [Death Valley NP]. The high peak of the Funeral Mountains was so named because its slopes are marked like ribs (*Chuckwalla*, Feb. 15, 1907).

Skidoo [Death Valley NP]. Established as a mining camp by Harry Ramsey and Bob Montgomery in 1906, when the slang phrase of leave-taking "Twenty-three, skidoo!" was at the height of its popularity.

Ski Heil Peak [Lassen Volcanic NP]. Winter sports enthusiasts chose this name for the peak, which is the center of their activities in the park. The BGN approved the name in 1937. *Ski Heil!* is a German and Swiss salutation among winter sportsmen.

Skinner Grove. *See* Atwell Grove.

Skookum (sko͞o′ kəm): **Rock, Gulch** [Siskiyou Co.]. The name preserves a word of Chinook Jargon, the trade language of the Pacific Northwest. It is derived from the word *skukum* of the Chehalis Indians, meaning 'strong' and designating also a 'ghost' or an 'evil spirit' (Gibbs, *Chinook Jargon*). This language has left other traces in the geographical nomenclature of Oregon, Washington, and California (*see* Siskiyou).

Skull Cave [Lava Beds NM]. According to J. D. Howard, the cave was named by E. L. Hopkins in 1892, when he and three companions discovered two human skeletons in the cave, as well as numerous skulls of bighorn, antelope, and goats.

Sky High Valley [Siskiyou Co.]. The unusual depression on the south end of Marble Mountains was named Sky High Valley because it is at an altitude of over 5,500 feet (USFS). **Sky High** [Butte Co.] is the name of the ridge between the Middle and Little North forks of Feather River. **Sky Forest** [San Bernardino Co.] was founded in 1928 and was so named because it is on top of a ridge more than a mile high, surrounded by a forest of pine, fir, cedar, and oak (Mary Henck). **Sky Parlor** [Sequoia NP] is a rocky formation on the Chagoopah Plateau.

Sky Londa [San Mateo Co.]. Earlier Sky L'Onda, named by a realtor in the 1920s, from the intersection of La Honda Road and Skyline Boulevard (Alan K. Brown).

Slabtown. *See* Cambria.

Slate. Good slate suitable for building purposes is found in various sections of the state, and many geographic features are named for outcrops of this valuable stone. Most of these are Slate Creeks because the stone is easily recognized when laid bare by the action of running water. **Slate Castle Canyon** and **Creek** [Sierra Co.]: Near the head of this canyon is a sharp outcrop of slate-gray rock, in shape and apparent size resembling a castle (Stewart).

Slate Hot Springs [Monterey Co.]. Named for Thomas B. Slate, who settled nearby in 1868 (WSP 338:56).

Slater Canyon [Riverside Co.]. The canyon in the Cleveland National Forest, named in 1960 in memory of Durward F. Slater, a USFS employee who lost his life in the Decker fire in Aug. 1959.

Slaughters Bar. *See* Goodyears Bar.

Slavianka. *See* Russian: River.

Sleepers Bend [Imperial Co.]. The bend in the Colorado River was named in Jan. 1858: "While turning a bend . . . we suddenly noticed upon the summit of a little hill on the left bank a ludicrous resemblance to a sleeping figure. The outlines and proportions were startlingly faithful" (Ives, p. 50).

Slide. *See* Fortuna.

Slinkards Valley [Alpine Co.]. A. James Slinkard, who was road supervisor of Douglas Co., Nevada, from 1862 to 1865, later built a road up Slinkards Creek into what became known as Slinkards Valley (Maule).

Slippery Ford [El Dorado Co.]. "The gold days' immigrants found here near the foot of the Lover's Leap peak a shallow stream crossing where the waters spread out over a large flat rock." This ford "was also a dangerous one because of the precarious footing on the water-polished rock and the swift-flowing stream clashing across the smooth surface" (John W. Winkley). A post office called Slippery Ford existed from Nov. 21, 1861, to Jan. 13, 1911.

Sloat [Plumas Co.]. The Western Pacific named the station in 1910 in honor of Commodore John Drake Sloat (1781–1867), who sailed into Monterey in July 1846 and peacefully took possession of California for the United States.

Slough. In England the generic term was used

for a muddy or miry place. In America, in the Middle West, it came to designate any good-sized backwater, and in California it took on the meaning of tidal creek, estuary, river channel (Stewart, pp. 265–66). The tidal channels of the San Francisco Bay district are often called sloughs, although on maps they may appear as creeks. The term "creek" is usually applied where the channel is really the lower end of a running stream. The generic term "slough" is also used for the channels and branches of the lower Sacramento and San Joaquin River systems, as well as for channels in the Tulare Lake section (*see* Lucerne; Mussel: Mussel Slough), and in other regions where similar hydrographic conditions prevail. In Los Angeles Co. the name was given to two swampy lakes that have no connection with the ocean or with a river: Bixby Slough and Nigger Slough, north of Wilmington. The name of Nigger Slough was changed to Laguna Dominguez by order of the Board of Supervisors on Mar. 30, 1938 (Co. Surveyor). The term is also found in the name of a settlement: **Sloughhouse** [Sacramento Co.], named after the hotel called the Slough House, which was built in 1850 by Jared Sheldon on Deer Creek, a branch of the Cosumnes River (Rensch-Hoover, p. 317). The post office is listed in 1916.

Slover Mountain [San Bernardino Co.]. The isolated hill at Colton is a fitting monument to Isaac Slover, a Kentuckian trapper, who first came to California with Pattie's party in 1828. He came back with a group from New Mexico about 1841–43 and settled in what is now San Bernardino Co. (Bancroft 5:722).

Slug. The word was a favorite miners' term for coarse gold nuggets and was also used for the large gold coins of the 1850s; it is preserved in Slug Canyon [Sierra Co.] and Slug Gulch [Placer Co.].

Slumgullion. *See* Carson Creek; Melones.

Sly Park; Sly Park: Creek, House [El Dorado Co.]. The flat, 12 miles east of Placerville, was named for James Sly, one of the Mormons who discovered the valley on July 5, 1848, on their trek from Sutters Fort to Salt Lake (Bigler).

Smartsville [Yuba Co.]. Named for James Smart, who built a hotel here in 1856. The post office is listed after 1867 as Smartville.

Smedberg Lake [Yosemite NP]. Named in 1895 by Lt. Harry Benson, for Lt. William R. Smedberg Jr., who was on duty in the park at that time (Farquhar).

Smeltzer [Orange Co.]. Named for D. E. Smeltzer, who discovered that the drained peat lands in this district were adapted for growing celery (Co. Hist.: Orange 1911:65). The area is now part of the City of Huntington Beach (Brigandi).

Smith: Meadow, Peak [Yosemite NP]. Named for a sheep owner who claimed the Hetch Hetchy Valley and drove his stock into it every summer (BGN, list no. 30).

Smith, Mount [Los Angeles Co.]. Probably named for "Bogus" Smith, a miner in San Gabriel Canyon about 1860 (S. B. Show).

Smithflat [El Dorado Co.]. Named for Jeb Smith, a pioneer rancher. The name appears on maps since the 1880s.

Smith Mountain [Death Valley NP]. Named by the Merriam expedition in 1891 for Francis M. ("Borax") Smith (1846–1931), president of the Pacific Coast Borax Company, "who aided the expedition in Death Valley in every possible way" (Palmer).

Smith Mountain [Riverside, San Diego Cos.]. Said by some to have been named for Thomas L. ("Pegleg") Smith (1801–66), a native of Kentucky, and a well-known trapper and horsethief, who was in California probably as early as 1829. Smith is supposed to have discovered gold to the east of the mountain that now bears his name while leading a party of trappers in 1836, but the mine was never found again. For another story, however, *see* Palomar Mountain.

Smith River [Del Norte Co.]. Jedediah Smith (1799–1831) traversed this region in May–June 1828; and either he or the cartographer, David H. Burr, applied the name to what is now the lower Klamath River. He believed this was a separate stream and that the upper Klamath and the Rogue River were one and the same. This error appears as late as 1851 in Findlay's Directory (p. 347, where Findlay in the same breath accuses Wilkes of inaccuracies). George Gibbs comments in Sept. 1851 on the confusion over the identity of the stream: "The name of 'Smith's river,' which, as a matter of tradition, has been bandied from pillar to post, shifting from Eel to Rogue's river, has recently vibrated between a stream running into Pelican bay, and another, called by some Illinois river, and supposed to be the south fork of Rogue's river. Of the former, called by the

Klamath Indians the Eenag'h-paha, or river of the Eenagh's, we received, at different times, information from those who had visited it" (Schoolcraft 3:137). Gibbs applied Smith's name to the hitherto nameless river north of the Klamath. The river appears in this location on Goddard's map of 1857. The town of Smith River on Highway 101 received a post office on Aug. 12, 1863. (*See* Jedediah Smith State Park.)

Smiths Landing. *See* Antioch.

Smiths Neck. *See* Loyalton.

Smithson Springs. *See* Piñon: Piñon Hills.

Smithville. *See* Stone: Stonyford.

Smokehouse: Canyon, Creek [Lake Co.]. This affluent of Salmon Creek was noted for its heavy run of salmon; in the early days both Indians and whites had camps along the stream for smoking and drying fish. It received its present name about 1870, when someone rigged up a crude smokehouse (Mauldin).

Smugglers Cove [Santa Cruz Island]. The name was applied by the Coast Survey when the east shore of the island was surveyed in 1855–56. It is reminiscent of Spanish and Mexican days, when stringent laws against free trade made smuggling a dangerous but profitable enterprise along the coast.

Snag Lake [Lassen Co.]. A snag is a dead tree, and the lake received its name from the snags that stand in it. About two hundred years ago a lava flow from Cinder Cone formed a dam across a preexisting lake, flooding the timbered area now forming the lake and killing the trees. There are several other Snag Lakes, probably all named for similar reasons.

Snake. California's twenty-odd Snake Rivers, Creeks, Lakes, Gulches, and Sloughs, as compared with the many more Rattlesnake place names, indicate that prospectors and settlers paid little attention to any but the venomous serpent. Some of the streams were not named because of the presence of snakes but because of the meandering course of the stream. Mendocino Co. has a Snakehouse Creek.

Snelling [Merced Co.]. A hotel was built here in the summer of 1851 by John M. Montgomery, Samuel Scott, and David W. Lewis. When the Charles V. Snelling family bought the hotel in the fall of the same year, the place became known as Snelling's Ranch, and it is mentioned by this name in the *Statutes* of 1854 (p. 222). From 1855 to 1872 it was the county seat; the post office is listed as Snelling's Ranch in 1867.

Sniktaw (snik′ tô): **Meadow, Creek** [Siskiyou Co.]. Years ago a man by the name of Watkins resided on this small tributary of Scott River. When his neighbors decided to name it for him, he objected, saying he would not have such a "snick" of a creek named for him. Thereupon the stream was called Sniktaw, Watkins' name in reverse (Schrader). This was probably William F. Watkins, assemblyman from Siskiyou Co. in 1859, who wrote for the *Golden Era* and the *Sacramento Union* under the pseudonym Sniktaw.

Snow; Snowy. These words are found in the names of a number of orographic features, including a Snowball Mountain [Santa Barbara Co.]. They have not, however, attained the same prominence in California nomenclature as the corresponding Spanish word, *nevada*. Mount Lassen was formerly known as Snow Butte, and on early American maps the name Sierra Nevada was translated as 'Snowy Mountains'.

Snow Peak [San Bernardino Co.]. Named by the BGN (list 8803) for Charles Alden Snow (1904–83), founder of Snow Peak Communications.

Soberanes (sō bə rä′ nəs): **Creek, Point** [Monterey Co.]. Named for a member of the large Soberanes family, prominent in the political life of the Monterey district during the Mexican regime. Their ancestor was probably José María Soberanes, a soldier of the Portolá expedition.

Soboba (sə bō′ bə): **Hot Springs, Indian Reservation** [Riverside Co.]. Kroeber gives the Luiseño form as *sovovo* (AAE 8:39); the Gabrielino form is *shovóvanga* (Munro, after Harrington). The underlying stem is perhaps to be found in Luiseño *ṣuvo-ya* 'be cold', *ṣuvóo-wut* 'winter'.

Sobrante. This Spanish generic term, used in connection with land grants, refers to surplus land in an area after granted land has been measured and separated from the public domain. This surplus land was often made a new grant, and the name Sobrante was attached to the specific name. The most celebrated Sobrante land grant case in California annals is that of the New Helvetia Sobrante, granted to Sutter by Micheltorena on Feb. 5, 1845, but rejected by the U.S. Supreme Court on Feb. 14, 1859 (Bowman). **El Sobrante** (sə brän′ tē); **Sobrante Ridge** [Contra Costa Co.]: When the Central Pacific was built from

Berkeley to Tracy in 1878, the name Sobrante was applied to the station because it was on El Sobrante de San Ramón land grant, dated Apr. 23, 1841. When the post office was established in 1945, the Spanish article *El* was added to the name. A local joke is that the name means 'the leftovers'.

Socouan. *See* Sycuan.

Soctish (sok′ tish): **Creek, Mountain** [Humboldt Co.]. On the Hupa Indian Reservation (BGN, list 8002). Members of the local Socktish family (note spelling), who took their name from the creek, believe that the word is from Hupa *sawhjich* 'I put (seeds, granular substance) into my mouth' (Golla).

Soda. The collective name for natural mineral waters, characteristically effervescent, is applied to more than a hundred hydrographic features, half of which are springs. In Lake, Nevada, Placer, San Bernardino, and Siskiyou Cos., the word appears in the names of communities. **Upper Soda Springs** [Siskiyou Co.]: The name Soda Springs was probably already used by Hudson's Bay Company trappers who crossed the Sacramento at this spot. It is mentioned in the Pac. R.R. *Reports* and became widely known when Ross McCloud built the famous inn and developed the springs in 1857. About 1889 the name was changed to Upper Soda Springs to distinguish it from Lower Soda Springs 5 miles south (Marcelle Masson). **Soda Springs** [Nevada Co.] was developed by Mark Hopkins and Leland Stanford about 1870 and was known as Hopkins Springs until the post office was established on Mar. 8, 1875 (Morley). From 1867 to 1873 the station of the Central Pacific was known as Tinkers Station (Doyle); *see* Tinkers.

Solana (sō lä′ nə) **Beach** [San Diego Co.]. The town was platted in 1923 by Col. Ed Fletcher and given the Spanish name meaning 'sunny place' (Santa Fe).

Solano (sō lä′ nō, sə lä′ nō) **County**. The county, one of the original twenty-seven, was created on Feb. 18, 1850, and was named, at the request of Vallejo, in honor of both the apostle of South America in the 16th century, St. Francis Solanus (1549–1610), and his namesake Francisco Solano in Alameda Co., the chief of the Suisun (Patwin) Indians. The latter had probably received the name of the patron of Mission San Francisco Solano when he was baptized there. Of all great Franciscan friars, St. Francis Solanus probably stood next to St. Francis of Assisi in the esteem of the missionaries of California. As many as three places were named for him by members of the Portolá expedition (Crespí, pp. 9, 92, 109, 138). On Nov. 25, 1795, in a place called San Francisco Solano, Padre Antonio Danti raised a cross where Mission San Jose was later set up (Arch. MSB 4:192 ff.). In 1823 Padre José Altimira, a young, ambitious friar at San Francisco (Dolores) Mission, took up a previously discussed idea of giving up the mission in San Francisco and founding a new mission in what is now Sonoma Co., which was to absorb the missions of San Francisco and San Rafael. He had the support of Governor Argüello, and on July 4, 1823, he raised the cross at the site of the new mission, calling it Nuevo San Francisco—after the old San Francisco Mission that was to be discontinued (Engelhardt 3:178). The ecclesiastic authorities naturally did not approve of this illegal step, but after many bitter words a compromise was effected: old San Francisco as well as San Rafael remained as missions, but Altimira was permitted to found the new mission. On the day of dedication, Apr. 4, 1824, it was named for St. Francis Solanus (Bancroft 2:504–5), the last mission established on California soil. It is now usually referred to as Sonoma Mission, and the name of the county alone preserves the memory of the South American missionary.

Soldier. The names of more than ten geographic features in California include the word "soldier"; most of them were probably given by early settlers to places where soldiers were stationed for their protection. It is significant that most of these names are found in the northeastern counties, where complete pacification of the Indians was not achieved until about 1870. A few old mining camps bore the name because of the presence of a soldier, most likely a deserter. Soldier Meadow and Detachment Meadow at the headwaters of the San Joaquin River [Madera Co.] date back to the Army's administration of nearby Yosemite National Park (David Brower). *For* Soldiers Gulch, *see* Volcano.

Soledad. The Spanish word for 'solitude' was repeatedly used for place naming in Spanish times. Not all Soledad names referred to an isolated place; some are a shortening of Nuestra Señora de la Soledad 'Our Lady of Solitude', referring to the loneliness of the Blessed Virgin after the Crucifixion. **Soledad** (sol′ ə dad): **Valley, Canyon, Mountain** [San Diego Co.]: The Portolá expedition crossed the valley on July 15, 1769, and named it Santa Isabel in honor of St. Elizabeth, Queen of Portugal in the 14th century (Crespí, p. 124). Anza records in his diary on Jan. 10, 1776, that the Indian rancheria in the valley had been Christianized by the San Diego mission and was called La Soledad. On Apr. 13, 1838, the name was applied to a land grant. Emory gives Solidad (creek) in 1846; Williamson, in 1853, gives Soledad for a settlement at the site of present Sorrento; the Coast Survey, 1874, gives Soledad Valley. **Soledad** [Monterey Co.]: The name is a folk-etymological rendering of an Indian name. "They told me that they gave it this name because in the first expedition of Portolá they asked an Indian his name and he replied, 'Soledad,' or so it sounded to them" (Font, *Compl. diary*, pp. 287–88). When the mission was founded on Oct. 9, 1791, Padre Lasuén included the old term in the name: La Misión de Nuestra Señora de la Soledad. The post office was established and named after the mission in 1870. **Soledad: Canyon, Pass** [Los Angeles Co.]; **Mountain** [Kern Co.]: The pass was discovered by Williamson in 1853: "As the existence of this pass was supposed previously to be unknown, I named it New Pass" (Pac. R.R. *Reports* 5:1.30). Blake renamed it Williamson's Pass, in honor of the discoverer. Bancroft's map of 1859 again has New Pass; but the Land Office map of 1859 relabels it La Soledad Pass, after an Indian village so named, shown on the *diseño* of Rancho San Francisco, 1838. The von Leicht–Craven map of 1874 has the present version.

Solid Comfort. *See* Lokoya.

Solimar (sō′ li mär) [Ventura Co.]. Subdivided in the 1920s and named from Spanish *sol y mar* 'sun and sea' (Ricard).

Solis (sō lēs′) **Canyon** [Monterey Co.]. The name of the canyon commemorates Joaquín Solís, a convict ranchero, one of the ringleaders of the revolt of 1828 at Monterey and *comandante general* after its temporary success. It is not known whether the Solís land grant in Santa Clara Co., dated Feb. 27, 1831, has any connection with the family name.

Solitaire Lake [Siskiyou Co.]. For a bird, Townsend's solitaire (*Myadestes townsendi*), by order of the BGN (list 7803).

Solomon Canyon; Mount Solomon [Santa Barbara Co.]. The name arose in the 1850s when the bandit Salomón Pico used the canyon as a base for his operations against the stagecoaches plying between Santa Barbara and San Luis Obispo. Salomón, a cousin of Governor Pío Pico, is said to have taken to highway robbery as a protest against American occupation. By decision of the BGN, 1965, Mount Solomon and the group of hills between Santa Maria and Los Alamos Valleys are now called Solomon Hills.

Solomons, Mount [Kings Canyon NP]. Theodore Seixas Solomons (1870–1947) explored and mapped the area in the 1890s (Browning 1986). *For* Solomons Hole, *see* Moaning Cave.

Solromar (sōl rō mär′) [Ventura Co.]. The name was selected by the Post Office Dept. from a list of twenty names submitted by residents. It was coined from the Spanish words *sol, oro,* and *mar* to suggest a "golden sunset on the sea" (G. West). The post office is listed in 1945.

Solstice: Canyon, Creek [Los Angeles Co.]. This astronomical term may have been used in the applied sense of 'highest or farthest' for these places high in the Santa Monica Mountains, or it may have been given at the time of a solstice.

Solvang (sol′ vang) [Santa Barbara Co.]. The colony was founded in 1911 by the Danish-American Corporation, an organization headed by professors of the Danish college in Des Moines, Iowa; the name is Danish for 'sun meadow'.

Somerset. *See* Bellflower.

Somersville [Contra Costa Co.]. This community, thriving at the time of the Mount Diablo coal boom, was named for Francis Somers, one of the men who located the Black Diamond vein in 1859.

Somes [Ventura Co.]. *See* Somis.

Somes Bar (sōmz bär′); **Somes Mountain** [Siskiyou Co.]. Various accounts say that this mining camp, dating from the 1850s, was named for a settler Abraham Somes or George Somes. The name is shown on the Land Office map of 1879 as Some's Bar, and on the Mining Bureau map of 1891 as Sumner's Bar. The post office is listed as Somesbar after 1892.

Somis (sō′ mis) [Ventura Co.]. A rancheria named Somes is mentioned in the records as early as 1795 and 1796 (PSP 14:35, Prov. Recs. 3:47, 56). The name is spelled Somes and Somo by Taylor (July 24, 1863), and it was transferred with the new spelling to a Southern Pacific station when the section of the Ventura-Burbank line was built in 1899–1900.

Sonoma (sə nō′ mə): **Mission, Creek, Valley, County, Mountain**, town. The name of the Indian tribe is mentioned in baptismal records of 1815 as "Chucuines o Sonomas," by Chamisso in 1816 as Sonomi, and repeatedly in Mission records of the following years. The name is doubtless derived from a Patwin word for 'nose', which Padre Arroyo (*Vocabularies*, p. 22) gives as *sonom* (Suisun). Bowman (*CFQ* 5:300–302 [1946]) plausibly theorizes that Spaniards found an Indian chief with a prominent protuberance and applied the nickname of "Chief Nose" to the village and the territory (cf. Kroeber, AAE 29:354 [1932]). Beeler believes that the name applied originally to a nose-shaped orographic feature (*WF* 13:268–72 [1954]). The interpretation 'valley of the moon' is more poetic but less authentic. Sonoma Mission, as it is commonly called, was established in 1824 and named San Francisco Solano. The town was founded by M. G. Vallejo in 1835; it is recorded as Zanoma on Wilkes's map of 1841, and with the present spelling by Duflot de Mofras (1844). The post office was established on Nov. 8, 1849. Town, creek, and valley are mentioned in the *Statutes* of 1850. The county, one of the original twenty-seven, was created and named on Feb. 18, 1850. *For* Bahía de Sonoma [Contra Costa Co.], *see* San Pablo Bay.

Sonora (sə nôr′ ə): town, **Creek, Pass** [Tuolumne Co.]; **Peak** [Alpine Co.]. The camp was established in the summer of 1848 by miners from the state of Sonora in northwest Mexico. It was first called Sonorian Camp to distinguish it from American Camp, as Jamestown was then called. When it was made the county seat on Feb. 18, 1850, the name was changed to Stewart, for Maj. William E. Stewart (*Statutes* 1850:63). On Apr. 18, 1850, the name was changed to the present form (*Statutes* 1850:263), probably as the result of local pressure.

Sopiago (sō pē ä′ gō) **Creek** [El Dorado Co.]. Named for a Basque family (Witcher).

Soquel (sō kel′): **Creek**, town, **Canyon, Cove, Point, Valley** [Santa Cruz Co.]. Río de Zoquel, apparently named after an Indian village, is mentioned as early as 1807 (Arch. Arz. SF 2:60–61). The name, spelled Soquel and Shoquel, was given to two land grants, dated Nov. 23, 1833, and Jan. 7, 1844. Town and creek are spelled Sauquil on Eddy's map (1854), and Shoquel on Hoffmann's map (1873); but the name of the post office, established July 5, 1857, is spelled Soquel. The term is supposedly a Costanoan Indian name meaning 'place of the willows', but Costanoan vocabularies show no comparable words for 'willow'. However, Pinart's Santa Cruz vocabulary shows *sokkoce, sokoci, sokkotc* for 'laurel' (Clark 1986).

Sorass Lake. *See* Dry: Dry Lake.

Sorenson Hill [Butte Co.]. The hill northwest of Oroville was named for Neils Sorenson, who homesteaded here about 1880 (BGN, 1965).

Sorrento [San Diego Co.]. The townsite was laid out during the boom of 1886–88 and named Sorrento after the Italian city—to which, however, it had no resemblance. The town did not develop, but the name was preserved in the name of the post office (Grace Diffendorf).

Soscol. *See* Suscol.

Sotcher (soch′ ər) **Lake** [Madera Co.]. The lake, as well as Reds Meadow and Creek, was named for "Red" Satcher or Sotcher, who lived at the meadow and is said to have raised vegetables there for sale at Mammoth during the Mammoth mining boom in the years after 1878 (Robinson).

Sotoyome [Sonoma Co.]. Río de Satiyome is mentioned by Vallejo on Mar. 28, 1836 (Vallejo Docs. 3:347), and Mark West Creek is labeled Sotoyome on a *diseño* of the Molino grant. The name was given to a land grant, dated Sept. 28, 1841, and Nov. 12, 1844. The term contains the Lake Miwok and Bodega Miwok element *yomi* 'village' (Oswalt). According to Barrett (*Pomo*, p. 218), the name means 'the home of Soto', referring to a chief; according to Josefa Carrillo Fitch, patentee of the grant, the name consists of *sati* 'brave' and *yomi* 'rancheria'.

Soulajulle [Marin Co.]. The rancho of this name was established in 1844; it is probably from Coast Miwok *sówlas* 'laurel' and *húyye* 'promontory' (Callaghan). **Soulajoule** (so͞o lə ho͞o′ lē) **Reservoir**, with a variant spelling, was built in 1979 (Teather 1986).

Soulsbyville [Tuolumne Co.]. The gold deposits at this place were discovered in 1856 while the

Platt brothers were hunting for their cattle. In 1858 Benjamin Soulsby and his sons struck even richer deposits. The place was named for them and became one of the richest gold-producing localities in the county. A post office was established on July 16, 1877.

South America, Lake [Sequoia NP]. The name was applied in 1896 by Bolton C. Brown because the outline of the lake resembles the map of the southern continent (Farquhar).

Southampton: Bay, Shoal [San Francisco Bay]. This bay on the north shore of Carquinez Strait and the shoal southwest of Richmond were named by the Coast Survey in the 1850s for the USS *Southampton*, which led Commodore Jones's fleet to its anchorage at Benicia in the spring of 1849. The name is shown on Ringgold's General Chart of 1850.

South Butte. *See* Sutter.

South Clearwater. *See* Hynes.

South Cucamonga. *See* Guasti.

South Dome. *See* Half Dome; Sentinel: Sentinel Dome; Starr King, Mount.

South Dos Palos [Merced Co.]. The post office was established about 1907 and named after nearby Dos Palos (*q.v.*).

Southeast Palisade. *See* Split Mountain.

Southerns [Shasta Co.]. Named after Southern's Station, established on the Sacramento before 1879 by Sim F. Southern (*see* Sims).

South Fork [Humboldt Co.]. A post office of this name was established on June 19, 1861, and so named because of its situation on the South Fork of the Eel River. It lasted only one year. When the Northwestern Pacific Railroad reached the place in 1910, the station was called Dyerville, after the (now-vanished) town and post office across Eel River (Borden). On Oct. 17, 1933, a new post office was established with the old name. *For* South Fork [Modoc Co.], *see* Likely.

South Fork Peak [Lassen Co.] was so named because it is near the South Fork of the Pit River.

South Gable Promontory. *See* Harvard, Mount.

South Gate [Los Angeles Co.]. Named in 1918 after the South Gate Gardens on the Cudahy Ranch, which had been opened to the public in 1911. In 1923, when the city was incorporated, the shortened form was adopted. The name indicates the location south of Los Angeles.

South Guard. *See* North Guard.

South Laguna [Orange Co.]. The resort was founded in 1926 by Lewis H. Lasley. When the post office was established in 1933, it was named Three Arches because of a picturesque rock formation. In 1934 the Post Office Dept., on petition of residents, changed the name to South Laguna because it is situated just south of the well-known city of Laguna Beach (J. C. Lasley). It is now part of Laguna Beach (Brigandi).

South Mountain [Ventura Co.]. Known earlier as Lomas de Santa Paula 'Santa Paula hills'. This early site of the California oil boom is so called because it is to the south of the city of Santa Paula (Ricard).

South Pasadena [Los Angeles Co.]. The city was laid out on the land of the Indiana Colony in 1885, by O. R. Dougherty of the South Pasadena Land Office.

South Riverside. *See* Corona.

South San Francisco [San Mateo Co.]. The name was given by G. F. Swift, the original developer, around 1891. Abbreviated colloquially to South City (Alan K. Brown).

Spadra (spad′ rə) [Los Angeles Co.]. Established as a stage station in the 1850s and named after Spadra Bluffs, Arkansas, the former home of William Rubottom, first American settler in the valley and owner of the local tavern. In 1874 the name was given to the Southern Pacific station; the post office is listed in 1880.

Spanish. This adjective of nationality is comparatively rare in the toponymy of the state. It appears in probably not more than thirty place names, most of which are survivors of the mining days when the camps of the Mexicans, South Americans, and Portuguese were usually designated as "Spanish." It is preserved in the place called Spanish Ranch and the post office Spanish Creek [Plumas Co.]; in two former post offices, Spanish Flat and Spanish Diggings [El Dorado Co.]; in Spanish Peak [Fresno Co.]; and in a number of Hills and Creeks. Half Moon Bay [San Mateo Co.] was formerly called Spanishtown. *For* Spanish Corral, *see* Ophir.

Spaulding, Lake [Nevada Co.]. The reservoir was created by the South Yuba Water and Mining Company in 1892; it was named for "Uncle" John Spaulding, a stagecoach driver who was one of its organizers. In 1912–13 the Pacific Gas and Electric Company built a new dam half a mile below but kept the old name.

Speckerman Mountain [Madera Co.]. Named in the 1850s for a settler who lived at the foot of the mountain (Blanche Galloway).

Spence [Monterey Co.]. The name was applied to the Southern Pacific station when the section from Salinas to Soledad was built in 1872–73. It commemorates David Spence, grantee of the Encinal y Buena Esperanza grant, on which the station was established. Spence, a native of Scotland, came from Peru to Monterey in 1824 and became one of the most influential foreigners in California.

Spencer, Mount [Kings Canyon NP]. For Herbert Spencer, 1820–1903, English philosopher and evolutionist (Browning 1986). *See* Evolution Peaks.

Spenceville [Nevada Co.]. A post office was established in the 1870s and named after the Spenceville school district; this had been formed in 1868 and named for Edward Spence of Nevada City, who donated the lumber for the school building (Knave, Mar. 8, 1942).

Spiller: Creek, Lake [Yosemite NP]. Named for J. Calvert Spiller, a topographical assistant with the Wheeler Survey, 1878–79 (Farquhar).

Split Mountain [Kings Canyon NP]. The peak that the Wheeler Survey had called Southeast Palisade, which was known for a while as South Palisade, was renamed by Bolton C. Brown in 1895: "To the north . . . the crest rises into a huge mountain with a double summit . . . which I called Split Mountain" (*SCB* 1:309). In Kern Co. there is another Split Mountain, and there are about ten Split Rocks in various sections of the state.

Spottiswood. *See* Famoso.

Spreckels [Monterey Co.]. Named for Claus Spreckels, a native of Germany and one of the leaders in the industrial development of California, who established a sugar refinery here in 1899.

Spring. Most of the place names that include the word were applied because of the presence of an active source of water; some may be transfer names, since the combination with Spring is a favorite in U.S. place naming. The word is included in the names of several post offices and communities, such as Springdale [Los Angeles Co.], Springfield [Tuolumne Co.], Spring Garden [Plumas Co.], Springs [San Luis Obispo Co.], Springside [Contra Costa Co.], and Springville [Tulare Co.]. It is also found frequently as the specific name for physical features; the best known of these is **Spring Valley** [San Mateo Co.], the site of the reservoirs of the San Francisco water supply. The Spring Valley Water Company was not named after the valley, but vice versa. The original Spring Valley was somewhere between Mason and Taylor Streets, between Washington and Broadway, in what is now downtown San Francisco. When the supply from the "spring" in this valley became inadequate for the fast-growing population, the company turned to a new source and took the name along. *For* Spring Valley in Calaveras Co., *see* Valley: Valley Springs.

Springville. *See* Fortuna.

Sproul (sprou′ əl) **Grove** [Humboldt Co.]. The grove in Prairie Creek Redwoods State Park was established in 1959 in honor of Robert G. and Ada W. Sproul. Robert Sproul (1891–1975) was president of the University of California, 1930–58, and treasurer of the Save-the-Redwoods League for many years.

Sprowell Creek [Mendocino Co.]. Named for brothers who settled here in the 1850s and were killed by Indians (Asbill). *See* Bell Spring Mountain.

Spy Mountain [San Bernardino Co.]. According to O'Neal (p. 77), it was named for a citizen of Giant Rock, suspected of being a spy in World War II.

Spyrock: Peak, post office [Mendocino Co.]. According to local tradition, the high rock, which commands a good view of the country, was used by Indians to spy on white settlers. The name was applied to the station when the Northwestern Pacific was completed in 1915.

Squash Ann Creek [Humboldt Co.]. Also spelled Squashan; a folk etymology from Yurok *kwosan* (Waterman, map 29, item 8).

Squaw. This Algonquian word for 'woman' is found, variously spelled and pronounced, in many related languages of the eastern United States; as a loan from Massachusett *squà*, it is first documented in English in 1622 (Cutler, p. 34). In its Anglicized spelling it has spread throughout North America, although it is now considered offensive by Indians. In California the word has been loosely applied since Gold Rush days and has been a favorite term for place naming. There are very few counties in the state that do not have at least one Squaw Valley, Creek, Canyon, Hill, or Hollow. There is also Squaw Dome [Madera Co.], Squaw Leap [Fresno Co.], Squaw Queen Creek [Plumas Co.], Squaw Tank [Joshua Tree NM], and two Squaw Tits [Humboldt, San

Bernardino Cos.]. **Squaw Rock** [Mendocino Co.]: Supposedly a jilted Indian girl took revenge by springing with a large rock upon her lover and his bride, who were sleeping below; all three were killed. **Squaw Valley, Creek**, and **Peak** [Placer Co.]: The name goes back to the early mining and lumbering activities and is already recorded on Goddard's map of 1857. It has become a popular winter sports area, where the Winter Olympic Games were held in 1960 (*see* Olympic Valley).

Stacy [Lassen Co.]. The post office was established in 1914 and named for Mrs. Stacy Spoon (Mamie Dicting).

Stacy Creek [Shasta Co.]. This tributary of Clear Creek was probably named for Thomas Stacy, a native of Kentucky, who was a resident of the county as early as 1871.

Stag. The use of this word, referring to an adult male deer, is much less frequent in California place naming than "buck" and "deer." Places named Stag include Stag Dome [Kings Canyon NP] and Stags Leap [Napa Co.].

Stagg. *See* Ludlow.

Staininger Ranch. *See* Scottys.

Stalder. *See* Mira: Mira Loma.

Standard [Tuolumne Co.]. The post office was established in 1912 and named for the Standard Lumber Company of Sonora (J. C. Rassenfoss). **Standard City** [Siskiyou Co.] was named for the same logging firm (Luecke).

Standard Canyon [Kern Co.]. The canyon near Inyokern was named for an old settler named Standard, who had his ranch here and who died about 1960 (Wheelock).

Standish [Lassen Co.]. The town was laid out in 1897. In 1899 H. R. T. Coffin of New York settled at the place and named it in honor of Miles Standish of *Mayflower* fame (F. J. Winchell).

Standish-Hickey State Park [Mendocino Co.]. The Edward Hickey Memorial State Park, established on July 21, 1950, was enlarged and renamed in 1953 when land (including the "Miles Standish Tree") was given for the park by Mr. and Mrs. S. M. Standish.

Standley State Recreation Area [Mendocino Co.]. Created in 1944 in honor of Admiral William T. Standley, a native of the area.

Stanfield Hill [Yuba Co.]. Named for William Stanfield, who opened the Stanfield House in Long Bar Township in 1856, just above the Galena House—another hotel on the old Foster Bar turnpike (Co. Hist.: Yuba 1879:87).

Stanford, Mount. *See* Crocker, Mount; Stanford University.

Stanford Lakes [Yosemite NP]. A misspelling, referring to the Kenneth J. Staniford family of Fresno (Browning 1988).

Stanford University [Santa Clara Co.]. The university was established in 1885 by Leland Stanford (1824–93), railroad builder, governor of California, and U.S. senator; it was named Leland Stanford Junior University in memory of Stanford's son, who had died the preceding year. **Stanford, Mount** [Kings Canyon NP]: Prof. Bolton C. Brown, who made the first ascent in Aug. 1896, named the peak Mount Stanford for the university. He suggested as an alternate name Stanford University Peak, if the name Mount Stanford should be declared ineligible because of another peak so named in Placer Co. However, the original Mount Stanford, which had been named by the Whitney Survey, was renamed as Castle Peak. Later, when the USGS mapped the Pioneer Basin region [Fresno, Mono Cos.] in 1907–9, its chief geographer, R. B. Marshall, named the four peaks of the divide in memory of the "Big Four" magnates of the Central Pacific Railroad: Charles Crocker, Mark Hopkins, Collis P. Huntington, and Leland Stanford. So now there are, after all, two Mount Stanfords in the state (*see* Crocker, Mount).

Stanislaus (stan′ is lô, stan′ is lôs): **River, County**, post office [Tuolumne Co.]; **Peak** [Alpine Co.]; **National Forest**. The river was discovered and named Río de Nuestra Señora de Guadalupe by an expedition under Gabriel Moraga in Oct. 1806 (Arch. MSB 4:1–47). In 1827–28, a neophyte named Estanislao, probably in honor of one of two Polish saints called St. Stanislaus, ran away from Mission San Jose and became the leader of a band of Indians in the San Joaquin Valley. Because it was feared that he was instigating a general uprising, two expeditions were sent against him in 1829. The first accomplished nothing; but the second, under Mariano Vallejo, then *alférez* at Monterey, succeeded in breaking up the band in a bloody engagement at the Río de los Laquisimes (SP Mis. and C. 2:15 ff.). The river later came to be known by the name of the Indian leader. It is mentioned as Río Estanislao in 1839 (Arch. SJ 1:43). On Dec. 29, 1843, the name was used for a land grant, Ranchería del Río Estanislao. Frémont used the Americanized version, Stanislaus River, in Mar. 1844 (*Expl. exp.* 1853:359), and it appears on the Frémont-Preuss map of 1845. A Stanislaus City was founded at the site of Brannan's New Hope in 1849 (*San Francisco Alta California*, Sept. 27, 1849). That has now vanished, but it had a post office in 1875, and it was still shown on the Land Office map of 1879 at the site of Ripon. The later post office Stanislaus is in Tuolumne Co. and was established on Aug. 17, 1911. Stanislaus Co. was created from part of Tuolumne Co. on Apr. 1, 1854. The national forest was created and named in 1897.

Stanton [Orange Co.]. Named for Philip A. Stanton, Republican assemblyman from Los Angeles, 1903–9, who helped the area incorporate in 1911 (Brigandi). The post office is listed in 1912.

Star City Creek [Shasta Co.]. This tributary of McCloud River preserves the name of the once-prosperous Star City mining district, which was probably named for John B. Star, said to have operated a trading post there in the late 1850s (Steger).

Starr, Mount [Fresno, Inyo Cos.]. Named by the Sierra Club in memory of Walter A. Starr Jr., a mountain climber of renown and author of the *Guide to the John Muir Trail and the High Sierra region*. The BGN approved the name in 1938. The Walter and Carmen Starr Grove in Humboldt Redwoods State Park was established in 1956 through the efforts of the Save-the-Redwoods League. The Walter A. Starr Grove in Calaveras Big Trees State Park was established in 1959.

Starr King, Mount; Starr King: Lake, Meadow [Yosemite NP]. The dome was named during the Civil War for Thomas Starr King (1824–64), a Unitarian minister of San Francisco, who was influential in keeping California in the Union. The mountain was known as South Dome before Starr King's name became attached to it. The northeast peak of Mount Diablo [Contra Costa Co.], now labeled North Peak, had been named Mount King by Whitney when he, accompanied by King and others, ascended the mountain on May 7, 1862 (Brewer, pp. 263, 267).

Starvation. The term was popular in early mining and exploring days, when there was often danger that food might give out. More than ten Canyons, Creeks, and Flats are still so named. Siskiyou Co. has a place called Starveout (*see* Grenada).

Starwein (stär′ win, stär′ wīn): **Ridge, Flat** [Humboldt, Del Norte Cos.]. From the Yurok village name *stowen* (Waterman, p. 235).

Stateline [El Dorado Co.]. The post office was established about 1900 and so named because it is just this side of the Nevada boundary.

Stauffer (stô′ fər) [Ventura Co.]. Named for the Stauffer Chemical Company, which worked a borax deposit here before the commercial development of the extensive borax beds in Death Valley began in the 1880s.

Steamboat. The word is found in the names of no fewer than ten physical features in the state. It is interesting to observe that there were at least three different reasons for giving the name. **Steamboat Spring** [Sonoma Co.]: "An opening in the rocks . . . through which is constantly ejected, with the noise of a number of steamers, a body of steam sufficient, could it be controlled, to propel a large amount of machinery" (Cronise pp. 172–73). **Steamboat Slough** [Yolo Co.]: "When the Sacramento was first navigated fewer obstructions to navigation were encountered in Steamboat Slough than in old Sacramento River, as the other branch is called. For many years the slough was therefore the channel preferred by navigators" (Wood's *Gazetteer*). **Steamboat Rock** [Humboldt Co.] was named by the Coast Survey: "The upper part is white and the lower black, somewhat resembling a steamer with a low black hull and white upper works" (*Coast Pilot* 1903:95).

Steelys Fork [El Dorado Co.]. This fork of the Cosumnes River was named for Dr. J. W. Steely, who discovered and located a gold ledge here in Mar. 1852 and built two mills (*Placerville Democrat*, July 1, 1876). The name is misspelled Steeley Fork by the USGS.

Stege (stēj) [Contra Costa Co.]. The railroad station and the school district preserve the name of Richard Stege, a native of Germany and a veteran of both the California gold rush and the British Columbia gold rush. About 1890 Stege settled in what is now part of Richmond on a large farm. Stege post office is listed in 1892.

Steinberger Slough [San Mateo Co.]. The slough is the only reminder of the once-promising town named for Baron Steinberger (*CHSQ* 23:176). It had a post office from 1853 to 1856.

Stella. *See* Whiskey: Whiskeytown.

Stephens Grove [Humboldt Co.]. The grove in

the Humboldt Redwoods State Park was established in 1922 in honor of Gov. William D. Stephens (1917–23). It is one of the earliest properties acquired by the state for state park purposes.

Stevens, Point. *See* Richmond, Point.

Stevens Creek [Santa Clara Co.]. Named for Capt. Elisha Stevens, of the Stevens-Murphy-Townsend party of 1844, who once owned a ranch here. The Spanish name was Arroyo de San José Cupertino; Hoffmann's map of the Bay region (1873) shows Cupertino or Stephens Creek (*see* Cupertino). Stephens was apparently the original spelling of the name.

Stevens Mountain [Del Norte Co.]. Named by the USFS for Phil Stevens, who drove the stage between Crescent City and Grants Pass for many years (Co. Hist.: Del Norte 1953).

Stevenson Memorial State Park [Napa Co.]. Created in 1949 and named for Robert Louis Stevenson (1850–94), who spent his honeymoon at the site of the old town of Silverado in 1880.

Stevens Peak [Alpine Co.]. The mountain was named for J. M. Stevens, a county supervisor, who operated the stage station in nearby Hope Valley in the 1860s. The name was placed on the map by the USGS when the Markleeville quadrangle was surveyed in 1889.

Stevinson [Merced Co.]. Named for James J. Stevinson, who acquired a large parcel of land on the lower Merced River in Aug. 1852. The post office is listed in 1908.

Stewart. *See* Sonora.

Stewart, Mount [Sequoia NP]. The mountain was named in Apr. 1929 at the suggestion of the Kiwanis Club of Visalia, for George W. Stewart, one of the men responsible for the creation of Sequoia National Park in 1890 (State Library).

Stewart Peak [Nevada Co.]. This peak was officially named George R. Stewart Peak by the BGN in 1984 (list 8404), in honor of the Berkeley scholar (1885–1980) who wrote *American place names*.

Stewarts Point, Creek [Sonoma Co.]. When members of the Coast Survey charted this section of the coast in 1875, they gave the name Point Stewart to a secondary triangulation point. It is possible that the point was named for Lt. Col. C. S. Stewart of the Corps of Engineers, who in that year was engaged in the removal of Noonday Rock near the Farallones.

Stillwater: Creek, Plains [Shasta Co.]. In 1853 De Witt Clinton Johnson settled at the creek and named it after his home town, Stillwater, New York. There was a post office Stillwater between 1870 and 1890 (Steger).

Stingleys Hot Springs [Ventura Co.]. Named for the owner, S. G. Stingley (WSP 338:63).

Stink Creek [Siskiyou Co.]; **Stinking Canyon Creek** [Shasta Co.]. Several names of this type refer to places with springs containing hydrogen sulfide (*see* Hedionda).

Stinson Beach [Marin Co.]. Named for Nathan H. Stinson, who settled in 1866 at Point Reyes and acquired the land that included the beach in 1871 (Co. Hist.: Marin 1880:426).

Stirling City [Butte Co.]. When the Diamond Match Company erected a sawmill here in 1903, J. F. Nash, the superintendent, applied the name because he had seen the firm name Stirling Boiler Works on the boilers ordered for the mill (M. W. Hedge).

Stockton [San Joaquin Co.]. In 1845, Charles M. Weber, next to Sutter the most noted pioneer of the interior valley, purchased the Rancho del Campo de los Franceses from William Gulnac. The settlement that developed there was appropriately called Tuleburg; but Weber applied the present name shortly after Commodore Robert F. Stockton (1795–1866) had taken possession of California for the United States. It is recorded in the *New Helvetia diary* on Oct. 14, 1847: "The two Nemshau boys that arrived yesterday . . . brought a passport from Weber in Stockton." **Fort Stockton** [San Diego Co.]: The old Spanish earthworks built in 1840 were improved by Commodore Stockton in Nov. 1846 and occupied by U.S. military forces until Sept. 1848.

Stoil [Tulare Co.]. The name of the station was coined from Standard Oil Company, which once had a pumping station here (Santa Fe).

Stone; Stony. The names of more than a hundred physical features contain these descriptive terms. Fully half of these are Creeks, the oldest of which are probably Stony Creek and its tributary Grindstone Creek [Glenn Co.]. The former is shown as Stone Creek on a *diseño* of about 1846. In the last hundred years the terms "rock" and "rocky" have generally been preferred in the American language in the naming of orographic features; only an occasional Stony Butte, Hill, Peak, Point, or

Top is found on the map of California. About ten Stone Corrals designate natural rock enclosures. Stonewall Canyon [Monterey Co.] was named because of its steep rocky slopes. Some of the Stone toponyms may represent family names. **Stony Creek** and **Valley** [Monterey Co.] were translated from Spanish: Arroyo de las Piedras and Cañon de las Piedras are shown on a *diseño* of the San Miguelito grant, 1841 (*see* Piedra). *For* Stony or Stoney Peak [Shasta Co.], *see* Freaner Peak; *for* Stony Creek [Mendocino Co.], *see* Mendocino. **Stonyford** [Colusa Co.]: The town of Smithville on Stony Creek was established and named by John L. Smith in 1863. In 1890 the Stony Creek Improvement Company bought the place and moved the town one-half mile to a new site called Stony Ford (Co. Hist.: Colusa 1891:284). **Stony Gorge Reservoir** [Glenn Co.] was created in 1963, when Stony Creek was impounded by the Black Butte Dam.

Stonemans Mountain [Los Angeles Co.]. Named in 1853 by R. S. Williamson of the Pacific Railroad Survey, for Lt. George Stoneman, a member of the party. Stoneman (1822–94), a native of Busti, New York, had been quartermaster of the Mormon Battalion in 1846–47; he became a famous cavalry leader in the Civil War, and later governor of California (1883–87). **Camp Stoneman** [Contra Costa Co.]: The San Francisco army port of embarkation in World War II was named in honor of George Stoneman by order of the War Dept. on May 28, 1942.

Stone Valley [Contra Costa Co.]. Named for Silas Stone, an early settler in the valley (Co. Hist.: Contra Costa 1926:471).

Storrie [Plumas Co.]. The post office was established on Apr. 26, 1926, and named for R. C. Storrie, the first postmaster and builder of the Bucks Creek Power House.

Stovepipe Wells [Death Valley NP]. There are two wells here with an abundance of good water. One or both were once protected by a stovepipe; hence the name.

Stover Mountain [Riverside Co.]. Named for Cristobal Stover, an early cattleman and bear hunter, who came to this region with Louis Robidoux (*Desert Magazine*, Feb. 1939).

Straight Spring Gulch [Calaveras Co.]. So named to distinguish it from nearby Crooked Spring Gulch.

Strancos Creek. *See* Tranca.

Stratford [Kings Co.]. The town was laid out in the spring of 1907 on the old Empire Ranch and was named after William Stratton, manager of the ranch. When the Post Office Dept. rejected this name, the present name was substituted (Co. Hist.: Kings 1940:177–78).

Strathmore [Tulare Co.]. The Balfour Guthrie Company, a Scottish corporation, laid out the town about 1908. At the suggestion of Mrs. Hector Burness, wife of one of the officials, the Scottish name Strathmore was given, meaning 'broad valley'. The place had formerly been called Roth Spur and Santos.

Stratton. *See* Stratford.

Strawberry is the most popular of all folk names derived from an edible wild fruit (*Fragaria californica*). There are more than fifty Strawberry names on the map, and many more are used locally—mainly for Creeks, Valleys, and Flats, but occasionally also for Hills, Peaks, and Points. The word is also found in the names of communities in El Dorado, Los Angeles, San Bernardino, Tuolumne, and Yuba Cos. **Strawberry** [Tuolumne Co.] is the oldest of the still surviving names. Although the post office was established as early as June 16, 1849, it was apparently not a mining camp. **Strawberry Valley** [Yuba Co.] was settled in 1850 and is shown on the maps as early as 1851. The post office was established on July 6, 1855. According to Joseph Booth's diary, June 6, 1852, it was actually so called for the abundance of wild strawberries. The County History (1879:97) agrees and even mentions the name giver, Capt. William Mock. But other sources maintain that the name was combined from those of two settlers; a deputy sheriff of the county described the Straw Berry Valley in 1856 (De Long, Jan. 26, 1856). **Strawberry Valley** [El Dorado Co.]: Two different stories agree that the place was not named for strawberries but for Mr. Berry, owner of the old roadhouse there. The details of the naming vary. Since Berry's straw mattresses were insufficiently stuffed, his customers were wont to cry: "More straw, Berry" (*Out West*, Nov. 1904). The teamsters called him "Straw" Berry because he kept their oats and barley and fed straw to the horses (Wells Drury, *To Old Hangtown or bust* 1912:12–13). The premises of later Strawberry Valley Station had actually been preempted by I. F. Berry in May 1858, so these stories may have their merits (Ruth Teiser, *CFQ* 5:302). *For*

Strawberry Valley in Siskiyou Co., *see* Shasta. **Strawberry Peak** [Los Angeles Co.] was named "by some wags at Switzer's camp in 1886, from its fancied resemblance to a strawberry standing with its blossom end up; but one of them said, 'We called it Strawberry Peak because there weren't any strawberries on it.' The joke took; and that burlesque name has been commonly used by the old settlers; but the peak is waiting some worthy occasion for a worthy name" (Reid, p. 370). **Strawberry Valley** [Shasta Co.] and the settlement of Berryvale (now Mount Shasta) were so named because wild strawberries were actually found there in abundance. **Strawberry Flat** [San Bernardino Co.] was also named for the abundance of the fruit; but since July 29, 1916, it has been known as Twin Peaks because the Post Office Dept. refused to add another Strawberry name to those already in existence.

Stringfellow Acid Pits [Riverside Co.]. A notorious series of seventeen quarry pits in the Jurupa Mountains, named for the owner, James B. Stringfellow. The site was used for the dumping of toxic wastes without approval of authorities for seventeen years; it was closed down in 1972 and investigated by the Environmental Protection Agency in the 1980s (Gunther).

Striped Mountain [Kings Canyon NP]. "That nearest the pass is strikingly barred across its steep craggy summit with light streaks. As this is an unusually marked case of this peculiarity . . . I called it Striped Mountain" (B.C. Brown, *SCB* 1:309). There is a Striped (formerly Curious) Butte in Inyo Co., and a Striped Rock in Mariposa Co.

Strong Creek [Humboldt Co.]. Named for Samuel Strong, who came to the area in 1853.

Stronghold [Modoc Co.]. When the Great Northern built its extension into California to meet the Western Pacific in 1930, the station was named after Capt. Jack's Stronghold—locally so known because, in the Modoc War, the Indian leader held out here against U.S. troops from Dec. 1872 to Apr. 1873. *See* Lava: Lava Beds NM; Modoc.

Stuarts Forks Lakes. *See* Emerald: Emerald Lake.

Stubbs Island [Clear Lake Co.]. The island preserves the name of Charles ("Uncle Jack") Stubbs, an English sailor who purchased a large estate on the shore of East Lake. A post office was established on June 14, 1926, as Stubbs; but the name was changed to Clearlake Oaks on June 1, 1935 (Mauldin).

Studio City [Los Angeles Co.]. The Republic Studios were opened here by Mack Sennett in the late 1920s (Atkinson).

Styx [Riverside Co.]. When the Santa Fe branch from Rice to Ripley was built in 1921, this desert station was given the name of the river that, in Greek mythology, flows through Hades.

Subaco (sōō bā′ kō) [Sutter Co.]. An acronym for Sutter Basin Corporation.

Subeet (sōō′ bēt) [Solano Co.]. The name of the Southern Pacific siding in the sugar-beet region was coined from the words "sugar" and "beet."

Success Lake [Tulare Co.]. The reservoir was formed by "successfully" damming Tule River east of Porterville (BGN, 1961).

Sucker. Of a number of places so named in mining days, only Sucker Run [Butte Co.] and Sucker Flat [Yuba Co.] remain on modern maps. The latter was named for an early settler named Gates, who came from Illinois, the "Sucker State" (Co. Hist.: Yuba 1879:84). It has not been ascertained whether the other names were applied for the same reason, or because the well-known carp-like sucker fish was caught there, or because a camp was inhabited by "suckers" in the sense of gullible people. Respectable Californians were doubtless relieved when the BGN (6th Report) changed the spelling of Sucker Flat [Placer Co.] to **Succor** (suk′ ər) **Flat**. To be sure, Oregonians had to submit at the same time to the board's decision for Sucker Creek ("not Succor") in Malheur Co.

Suey (sōō ā′) **Creek**, station [Santa Barbara Co.]. Suey was the name of a land grant dated Apr. 6, 1837. The *diseño* of the grant is labeled Sitio de Sue, and an Arroyo of Suey is shown on an 1850 map of the rancho. It is probably derived from a Chumash word, perhaps referring to tarweed (Hall-Patton).

Sugarbowl Dome [Sequoia NP]. So named because a depression, filled with snow most of the year, has the aspect of a bowl filled with sugar. There is another Sugar Bowl near Donner Summit, named for the same reason.

Sugar Hill [Modoc Co.]. A wagon, partly loaded with sugar on its way from Yreka to the military garrison at Fort Bidwell in the early 1870s, broke down, scattering soldiers' sugar ration all over the hillside; hence the name (William S. Brown). It is not known whether Sugar Lake and Creek in Siskiyou Co. were named for a similar reason.

Sugarloaf. In former centuries sugar was not put up in bags or boxes, but was delivered in the form of a "loaf" to the grocer, who would break off pieces and sell it by weight. So familiar was the sight of the conoidal sugar loaf that the word was applied to any object of similar form. It finally developed into a geographical generic term, applied to a mountain so shaped, and the map of the United States became dotted with thousands of "sugar loafs." California has its share of about a hundred orographic features so named. When used as a descriptive generic term, the name is usually spelled in two words, Sugar Loaf; but when the term is used with Hill, Mountain, Peak, and Butte, it is usually spelled as one word. There is, naturally, an assortment of Sugarloaf Creeks, Lakes, and Meadows, named after a nearby Sugar Loaf. Placer Co. has a group of peaks called Sugarloaves (*Names* 4:241–43 [1956]). **Sugarloaf** [San Bernardino Co.]: When the old Big Bear Park post office was reopened on Jan. 1, 1947, Mary E. Hebert proposed the present name because the post office nestles (at 7,200 feet elevation) on Sugarloaf Mountain (Carmel Botkin). *For* Sugarloaf in Orange Co., *see* Pleasants Peak.

Sugar Pine. This valuable tree (*Pinus lambertiana*), which grows to an imposing height and is seldom found in solid stands, has given its name to more than thirty geographical features (*see* Pine). **Sugar Pine** [Madera Co.] owes its origin to a 200-foot tree near the sawmill (Blanche Galloway).

Suisun (sə so͞on′): **Bay, Creek, City, Valley, Hill, Point, Slough, Cutoff** [Contra Costa, Solano Cos.]. The bay was explored by Cañizares in Aug. 1775 and is labeled Junta de los quatro Evangelistas on Ayala's map. Since Cañizares reported that it contained fresh water (Eldredge, *Portolá*, p. 67), it became known as Puerto Dulce and is mentioned as Freshwater Bay as late as 1842 (Simpson 1:405). Abella's diary (Oct. 28, 1811) mentions *estero de los Suisunes*, named after the Patwin band or village on the north shore, whose name appears in records from 1807 on, with a great variety of spellings. L. A. Argüello speaks of Bahilla de Suysun on May 13, 1817, and Bahía de los Suysunes is shown on the *Plano topográfico de la Misión de San José* (about 1824). The modern spelling appears in the name of Suisun land grant, dated Jan. 28, 1842. Suisun Valley, River (Creek), and Embarcadero (City) are mentioned in the *Statutes* of 1850 (pp. 61, 100). The original meaning of the name is not known.

Sullivan Creek [Tuolumne Co.]. The name of the creek recalls one of the richest gold deposits in the Sonora district, which was discovered by John Sullivan in 1848 and became known as Sullivan's Diggings (Buffum, p. 126). Sullivan's Creek is shown on Gibbes's map of 1852.

Sulphur. The frequent occurrence of sulphurous springs has given the state more than 150 place names containing the word; most of these are Springs, but there are also Sulphur Creeks, Canyons, Gulches, Mountains, and one Sulphur Gap [Colusa Co.].

Sultana (sul tan′ ə) [Tulare Co.]. The station was called Atla when the San Francisco and San Joaquin Valley Railroad reached the place in 1898. Because of the possible confusion with Alta [Placer Co.], the name was changed when the post office was established about 1900. The Sultana grapes grown there at that time were later largely supplanted by Thompson Seedless.

Summer Home. *See* Balch Park.

Summerland [Santa Barbara Co.]. The town was laid out in 1888 by H. L. Williams, shortly after the section of the Southern Pacific was built from Santa Barbara to Ventura. Although the United States was then dotted with place names containing the word "summer," the combination Summerland was unusual.

Summers: Creek, Back Pasture [Mono Co.]. In the 1860s Jesse Summers, a butcher, grazed cattle in the natural meadows that became known as Summers Back Pasture (Maule).

Summersville [Tuolumne Co.]. Franklin and Elizabeth Summers settled at this place on Turnback Creek in 1854. The settlement developed after the discovery of gold in the spring of 1856.

Summit. When the highest point of a mountain pass is crossed by a trail, a highway, or a railroad, the name Summit is almost invariably applied. Six railroad stops in California have been called Summit. The term also appears in the names of more than fifty Creeks, Lakes, Meadows, Rocks, Springs, etc., at or near a summit. *For* Summit in Plumas Co., *see* Vinton; in Riverside Co., *see* Beaumont. **Summit Range** [Kern, San Bernardino Cos.] was so named because the Owenyo branch of the Southern Pacific reaches its highest point here. **Summit City** [Shasta Co.] was founded when construction of Shasta Dam was started in 1938 and was so named because it is near the summit where Highway 209 begins its descent to the dam (Steger). *For* Summit City in Alpine Co., *see* Jeff Davis. *For* Summit Peak, *see* Ina Coolbrith, Mount.

Sumners Bar. *See* Somes Bar.

Sun; Sunny. These specific terms are included in the names of at least ten California communities. **Sunnyslope** [Los Angeles Co.]: In 1861 Leonard J. Rose, a Pennsylvania German, acquired a portion of the Santa Anita grant and developed his famous Sunny Slope Farm, after which the Southern Pacific station was named. **Sunnyside** [San Diego Co.] was founded and named by J. C. Frisbie in 1876. **Sunland** [Los Angeles Co.] was established as a post office in 1887 (Atkinson). **Sunnyvale** [Santa Clara Co.]: The name was applied by W. E. Crossman about 1900 to the subdivision of a portion of the Pastoría de las Borregas grant (Thelma Miller). **Sunnymead** [Riverside Co.]: The Sunnymead Orchard Tract was laid out and named in 1913; it is now part of the City of Moreno Valley (Brigandi). **Sunny Hills** [Orange Co.]: The Sunny Hills Ranch Company bought the old Bastanchury citrus ranch and subdivided it in 1940 (Santa Fe). **Sun Valley** [Los Angeles Co.]: The name of the post office, Roscoe, established on Jan. 15, 1924, was changed on Jan. 1, 1949, to the present name by vote of the people in the area. **Sun City** [Riverside Co.]: The post office was established on Aug. 3, 1963, for the retirement community developed by the Del Webb corporation. *For* Sunshine Summit, *see* Deadman: Deadman's Hole.

Sunflower Valley [Shasta Co.]. The valley was once covered with sunflowers (*Helianthus* spp.), the seeds of which were highly valued by the Indians. Pierson B. Reading, Indian agent after the American occupation, had to divide the area between the Yana and the Wintu so that each tribe would receive its share of the harvest (Steger).

Sunkist [San Bernardino Co.]. From a brand name of California oranges, supposedly "sun-kissed."

Sunol (sə nōl′): town, **Valley, Hills** [Alameda Co.]. Named for Antonio Suñol (1800–1865), a native of Spain, who deserted the French ship *Bordelais* in 1818 and was part owner in 1839 of the Rancho El Valle de San José, on which town and valley are situated. Suñol's hospitality and his rich garden are repeatedly mentioned by early travelers. His place is shown on Duflot de Mofras's map, 1844.

Sunset. The term is found in the names of more than ten Cliffs, Peaks, Rocks, and Valleys and of five communities, the best known of which is **Sunset Beach** [Orange Co.]. There are also a few Sunrise Hills, Peaks, and Valleys in the state.

Superstition Mountain [Imperial Co.]. "While [the hill] appears to be composed entirely of sand, there is a rocky mass below, and over this the sand plays, constantly shifting to and fro in the desert winds, and because of this instability the Indians of the region speak ill of it, hence its name" (James 1:13).

Sur (sûr, sŏŏr) [Monterey Co.]. The name El Sur 'The South' (i.e. the area south of Monterey) was applied to Juan B. Alvarado's land grant, dated July 30, 1834. In 1851 the Coast Survey applied the name Point Sur to the cape that was called Morro de la Trompa and Punta que Parece Isla in Spanish times, from the big rock extending into the ocean that looks like an island in the shape of a trumpet. The Coast Survey name has given rise to a large name cluster—Sur Breakers, Canyon, River, and Rock; Big Sur River, Little Sur River, Little Sur Creek, and False Point Sur. The post office Sur was established on Oct. 30, 1889; on Mar. 6, 1915, the name of another post office, Arbolado, was replaced by the name Big Sur.

Surf [Santa Barbara Co.]. The post office was established on June 22, 1897, and was so named because of its location near the ocean shore. **Surfside** [Orange Co.]: The post office was established on Apr. 5, 1953.

Surgone (sûr′ gôn) [Humboldt Co.]. From the Yurok village name *sregon* (Waterman, p. 244; Harrington).

Surprise. This term was often used by early explorers, surveyors, and prospectors for a place that they came upon unexpectedly. It is still found in the names of more than twenty Canyons, Creeks, Springs, Lakes, and Valleys. **Surprise Valley** [Modoc Co.]: Of a number of explanations given for the origin of the name, the one given by William S. Brown seems the most plausible: "Emigrants who had just traversed the Black Rock Desert were 'surprised' as they came out of the arid sagebrush hills into the smiling valley." The name is shown on maps since about 1860. The lakes labeled Alkali Lakes on the maps are locally known as Surprise Valley Lakes. **Surprise Canyon** [Death Valley NP]: "The prospector was surprised when he found the lead, the mining sharks were surprised to hear of his luck, the growth of the camp was more surprising still" (J. R. Spears, *New York Sun*, Feb. 21, 1892).

Surpur (sûr′ pûr) **Creek** [Humboldt Co.]. From the Yurok village name *srpr* (Waterman, p. 238; Harrington).

Susan River; Susanville [Lassen Co.]. Both names were applied by Isaac Roop, pioneer of the Honey Lake district, for his daughter, Susan. The town developed around Roops Fort (*q.v.*) and was known as Rooptown until 1857. The river is frequently mentioned in the Pac. R.R. *Reports* in the early 1850s.

Suscol (sus′ kəl): **Creek**, station [Napa Co.]. The name of a Patwin Indian village recorded with various spellings in records and *diseños* since 1835. A Corral de Soscol is shown on a *diseño* (1838) of the Napa grant; a Sierra de Soscol (lower Howell Mountains), on a *diseño* of the Suisun grant. On June 19, 1844, the name Suscol or Soscol was used for a land grant. The name appears as Susqual in the Indian Report of 1853 (p. 405). The creek is mentioned with the present spelling in the *Statutes* of 1850 (p. 61).

Sutil (sōō til′) **Island** [Los Angeles Co.]. The island southwest of Santa Barbara Island, formerly known as Gull Island, was named Sutil Island by the BGN in 1939. The *Sutil* ('subtle') was a ship of the Galiano expedition of 1792.

Sutro (sōō′ trō): **Forest, Heights; Mount Sutro** [San Francisco Co.]. Named for Adolph Sutro (1830–98), a native of Germany, builder of the famous tunnel in the Comstock mines in Nevada in the 1870s, and mayor of San Francisco in 1894–98. The forest was planted in the 1880s at his instigation; his estate above the Cliff House was willed to the city for a public park. The official name of the hill is Sutro Crest (BGN, Apr. 6, 1910).

Sutter. The name of the pioneer John A. Sutter (*see* Glossary) is properly honored in a number of place names in the state. **Sutters Fort** [Sacramento Co.]: In Aug. 1839, Sutter settled at the site of what is now the city of Sacramento; in 1841 he began the construction of his famous fort, which soon became the terminus of the emigrant trail from Missouri. Sutter himself called the settlement Nueva Helvetia (shown on maps also in the English, German, French, and Latin forms), using the ancient name of Switzerland; but popular usage preferred Sutter's, Fort Sutter, Sutters Fort, or simply "the Fort." **Sutterville** [Sacramento Co.], now vanished, was laid out in Jan. 1846 by Bidwell and Hastings, and was envisioned by Sutter as the coming metropolis of the valley. The name is preserved in Sutterville Bend (of the Sacramento River), Sutterville Lake, and Sutterville Road. **Sutter Buttes** [Sutter Co.]: The dominant orographic features of the Sacramento Valley were called Los Tres Picos in Sutter's grant and were known to Hudson's Bay Company trappers in the early 1830s as the Bute or Buttes. On the maps of the 1840s and 1850s they appear as Three Buttes, Sutter's Buttes, Los Picos de Sutter, Prairie Buttes, Sacramento Buttes, or simply Butte or Bute Mountains. The Whitney Survey applied the name Marysville Buttes, and most maps used this name until the BGN (Oct. 1949) chose the present name. **Sutter County** was one of the twenty-seven original counties; it was created and named on Feb. 18, 1850. **Sutter Island** in the Sacramento River is shown on Ringgold's General Chart of 1850 for part of Ryer Island. The Coast Survey transferred the name to what Ringgold called Schoolcraft Island, and named the channel west of it Sutter Slough. **Sutter Creek** [Amador Co.]: The city was named after the creek, which had been locally known by that name since Sutter operated a mining camp there in 1849. The name of the village, township, and post office is mentioned in the *Sacramento Daily Democratic State Journal* of Dec. 2, 1853; the town is mentioned in the *Statutes* of 1854 (p. 222). **Sutter** [Sutter Co.]: The name of the town known as South Butte was changed to Sutter City during the boom of the 1880s. To avoid confusion with Sutter Creek, the post office dropped the "City."

Sutter [Napa Co.]. Named after the Sutter Home Winery by Mrs. E. C. Leuenberger, whose father, John Sutter, became manager of the winery in 1888 (Margaret Klausner).

Suwanee River [Sequoia NP]. This tributary of the Kaweah River shows no resemblance to the stream in Florida and Georgia, after which it apparently was named; the original name is probably from a Muskogean Indian language (Stewart).

Suysunes, Bahía de los. *See* Suisun Bay.

Swager. *See* Swauger.

Swan Lake [Shasta Co.] is visited by the whistling swan, *Olor colombianus* (Stang). *See also* Schwan Lake.

Swanton [Santa Cruz Co.]. Named for Fred W. Swanton, a leader in the development of public utilities in Santa Cruz Co.

Swauger: Canyon, Creek [Mono Co.]. A parcel of land here was patented to Samuel A. Swauger in 1880 (Maule). This name is sometimes misspelled as Swager. *For* Swauger in Humboldt Co., *see* Loleta.

Swede Creek [Shasta Co.]. The tributary of Cow Creek was named for H. O. Akerstrom, a native of Sweden, who settled near the mouth of the creek in Jan. 1854 (Steger). More than fifteen other Swede Canyons, Creeks, and Gulches, mainly in northern counties, were named for Swedish or other Scandinavian settlers. Swede was the rural California designation for a Scandinavian: only two names with the adjective Danish, and none with Norwegian, have been found on available maps (*see* Dutch).

Sweet Brier Creek [Shasta Co.]. So named because of the occurrence of the sweetbrier rose, *Rosa eglanteria* (Steger).

Sweetland [Nevada Co.]. Named for Henry P. Sweetland, who had a trading post here in 1852 (Doyle) and was an assemblyman in the fifth session (1854). The place is mentioned in the *Statutes* of 1854 (p. 222), and a post office is listed in 1858.

Sweet Pizzlewig Creek. *See* Pizzlewig Creek.

Sweetwater. The name is used chiefly for surface waters in regions where many springs and streams have "bitter" water. In the names of coastal features, the term "fresh water," as contrasted to "ocean water," is ordinarily used instead. **Sweetwater River** [San Diego Co.] was known as Agua Dulce in Spanish times, and Suisun Bay was once called Puerto Dulce.

Swifts Point [Glenn Co.]. The name commemorates a colorful early character, Granville P. Swift, a native of Kentucky who came with the Kelsey party from Oregon in 1844. He took part in the Micheltorena campaign, was a leader in the Bear Flag uprising, and settled in Colusa Co. in 1847. In 1848–49 he made a fortune in the mines.

Swiss. More than twenty physical features in various sections of the state bear witness to the presence of Swiss settlers, who have played an important role in California since Mexican times.

Switchback Peak [Sequoia NP]. "So named because of the striking zigzag trail (now road) over its eastern slope" (BGN, 6th Report).

Switzer Canyon [San Diego Co.]. The canyon was named for E. D. Switzer, a San Diego jeweler in the 1870s, whose home was here (C. Gunn).

Switzerland [San Bernardino Co.]. Formerly known as Valley of the Moon because of the proximity of Moon Lake. About 1935 a Los Angeles investment company bought the tract, established the St. Moritz Club, and called the place Switzerland because the mountains are somewhat suggestive of the Alps.

Sword Meadow [Alpine Co.]. Apparently named for the crossed swords carved in a tree in the meadow, commemorating a duel between two Frenchmen, in which one was killed. A letter from the fiancée of one contestant was found in the possession of the other (J. H. Hall).

Sycamore. The habitat of the western sycamore or plane tree (*Platanus racemosa*) is clearly indicated by the locations of the geographic features that bear the name. The tree thrives best in the moderate moisture and the average temperature of the southern Coast Ranges, and to a lesser degree in the Sierra foothills along the interior valleys to about latitude 40° north. The name (and the tree) appear most frequently in Ventura, San Diego, Kern, and San Luis Obispo Cos. No Sycamore names are found east or west of Butte and Colusa Cos., or in the High Sierra, or in the Great Basin. The oldest and northernmost name found was Sycamore River (now Battle Creek in Tehama Co.), mentioned by John Work on Nov. 28, 1832, and shown on Abert's map of 1838. Since the sycamore prefers streambeds that are dry in summer, more than half the names refer to canyons. Only one Sycamore Hill [Butte Co.] and no Sycamore Mountain can be found on maps, since the tree rarely grows on slopes. In several places the name was transferred to settlements [Colusa, Los Angeles, San Bernardino Cos.]. In Spanish times the name *aliso* 'alder' (*q.v.*) was often used for sycamore: the term Cañada de los Alisos is on the *diseño* of the Conejo grant, 1835. **Sycamore Canyon: Big** and **Little** [Ventura Co.] are named for the size of the trees, not the canyons, since Little Sycamore Canyon is larger than Big Sycamore Canyon (Ricard). **Sycamore Grove** [San Bernardino Co.] was one of the Mormon settlements in Cajon Canyon. *For* Sycamore in Fresno Co., *see* Herndon.

Sycuan (sə kwän′): **Creek, Peak, Indian Reservation** [San Diego Co.]. According to Kroeber, the name is derived from Diegueño *sekwan*, a kind of bush. An application for a grant called Secuan or Sequan is recorded in 1835 (San Diego Arch., Hayes, C and R, p. 58), and an Indian *rancheria* (or *rancho*) called Socouan is mentioned on Feb. 1, 1836 (Hayes, Mis. 1:292). On May 2, 1839, the grant was made, but no claim was ever filed. The current spelling was ratified by the BGN in 1977 (list 7702).

Sylmar (sil′ mär) [Los Angeles Co.]. Once the site of the world's largest olive groves; the name is a fanciful creation supposed to mean 'sea of trees' (Atkinson). Presumably the elements are Latin *silva* 'forest' and Spanish *mar* 'sea'.

Symbol Bridge [Lava Beds NM]. "I named this natural bridge in 1917 because of the red, white, and black writings on both walls, and the pictures of an Indian head and a Spanish priest with a ruffled collar on the south walls" (Howard). Symbol Cave and Painted Cave in the monument were so named because Indian pictographs were seen on the walls.

Symmes Creek [Inyo Co.]. The stream probably owes its name to J. W. Symmes, pioneer settler, who was county superintendent of schools, 1870–73 and 1876–82 (Farquhar).

Syncline (sin′ klīn) **Hill** [San Luis Obispo Co.]. The prominent elevation at the western edge of the Carrizo Plain was designated as Syncline Hill by a decision of the BGN in 1909. "Syncline" is a geological term applied to a fold formed by strata dipping toward a common line.

T

Tabaseca (tä bə sā′ kə) **Tank** [Riverside Co.]. A desert water hole, from Cahuilla *távish héki'* 'home of the redshafted flicker', according to Cahuilla leader Katherine Siva Saubel (Gunther).

Table. There are about fifty Table Mountains and Hills in the state, as well as a number of Table Bluffs and Rocks. There is also a Table Lake [Yosemite NP] near an unnamed flat-topped mountain. In southern California, the corresponding Spanish term *mesa* is frequently used instead. Disregarding Beechey's name, Table Hill, for Mount Tamalpais, the oldest feature still so named is probably **Table Bluff** [Humboldt Co.], called Punta Gorda in 1793 (*CHSQ* 10:330), Ridge Point by the *Laura Virginia* party, and Brannan Bluff by Sam Brannan, but known by the present name as early as Sept. 1851. **Table Mountain** [Sequoia NP] and **Creek** [Kings Canyon NP]: The mountain is first mentioned by the Whitney Survey and recorded as "Table" on Hoffmann's map of 1873. **Table Mountain** [Calaveras, Tuolumne Cos.]: "One of the most striking features in the topography and geology of Tuolumne County is the so-called 'Table Mountain,' a name given, throughout the state, to the flat table-like masses of basaltic lava which have been rendered so conspicuous by the erosion of the softer strata on each side, and which now exist as elevated ridges, dominating over the surrounding country, and remarkable for their picturesque beauty, but still more so on account of the important deposits of auriferous detritus which lie beneath them" (Whitney, *Geology*, p. 243). The USGS made four Table Mountains out of Whitney's one in 1914—but forgot to number them, although they are within a few miles of one another. The word is sometimes used as a generic term: Big Table and Kennedy Table northeast of Millerton Lake [Madera Co.].

Taboose (tə bo͞os′): **Creek, Pass** [Inyo Co.]. An edible groundnut found in Owens Valley was called "taboose" in the Paiute language (Farquhar). The name was also applied formerly to Division Creek, which rises not far from the pass, and to the pioneer stage station at the creek (Robinson).

Tache (tä′ chē) also spelled Tachi, Tadji, Dachi, etc.) refers to a subtribe of the Yokuts, dwelling north of Tulare Lake; however, mission records mention only a village and a lake of that name. Padre Juan Martín, of Mission San Miguel, mentions *una de las rancherías tulareñas llamada Tache* 'one of the tulare villages called Tache' in 1815 (Arch. MSB 6:85–86). Estudillo labels Tulare Lake as Laguna de Tache o Bubal on his map of 1819. Laguna de Tache was the name of two land grants, dated Dec. 4, 1843, and Dec. 12, 1843. On American maps Tulare Lake appears as Lake Chintache, Tontache, etc., as well as Tula Lake, until the late 1850s. The Yokuts form has been recorded as *t'ac^{h}c^{h}iy'* (Gamble).

Tachevah (tä chē′ vä) **Canyon, Creek** [Riverside Co.]. Said to mean 'a plain view' in Cahuilla (Gunther).

Tadpole Creek [Placer Co.]. The larval form of the frog figures in this place name, and Shasta Co. also has a Tadpole Creek. The synonym "polliwog" is not attested in California place names.

Taft [Kern Co.]. The post office was established in 1909 and named for the newly elected president, William H. Taft (1857–1930). The branch railroad from Pentland reached the place in the same year but the station was called Moron, a name that persisted until about 1918. **Taft Point** [Yosemite NP]: The name was given to the lookout station on the Pohono Trail by Robert B. Marshall when President Taft was in office.

Tagus (tā′ gəs) **Ranch** [Tulare Co.]. The name was applied to the station when the San Joaquin Valley line reached the place in the fall of 1872. Tagus is the English name of a river in Spain and Portugal (called Tajo in Spanish).

Tahana (tə han′ ə) **Gulch** [San Mateo Co.]. "Said to have been named for a rancher here over 60 years ago. . . . Probably the name was Tehaney" (Alan K. Brown).

Tahichipa, also **Tahichipi**. *See* Tehachapi.

Tahoe (tä′ hō), **Lake; Tahoe City, Pines, Vista, National Forest** [Placer Co.]. The name is clearly from Washo *dá'aw* 'lake' (William Jacobsen). However, there has been much speculation about the meaning and the application of the name. When Frémont discovered the lake on Feb. 14, 1844, he wrote: "We had a beautiful view of a mountain lake at our feet, about fifteen miles in length, and so entirely surrounded by mountains that we could not discover an outlet" (*Expl. exp.*, p. 334). On some of the maps in earlier editions of the report the lake is shown as Mountain Lake. Later Frémont named it Lake Bonpland in honor of Aimé Bonpland, the French botanist who had accompanied Humboldt on his great journey in South America, and it is so labeled on Preuss's map of 1848. This name would probably have stuck had not the friends of John Bigler, governor of California from 1852 to 1856, succeeded in imposing the name Lake Bigler (Eddy's official map of 1854). During the Civil War, the Union sentiment objected strenuously to this name because Bigler was an outspoken secessionist, and a movement was started to restore to the lake its Washo Indian name, understood to be Tahoe and to mean 'big water'. Various persons claim the credit for having brought about the change. It seems likely, however, that Henry de Groot, John S. Hittell, and William H. Knight selected the name. At any rate, it was de Groot who explored the mountains in 1859 and suggested the Indian name of the lake, and it was Knight who placed the name Lake Tahoe on Bancroft's map of the Pacific States in 1862. This, however, did not end the controversy. An act of the Democratic legislature, approved Feb. 10, 1870, declared that the lake "shall be known as Lake Bigler, and the same is hereby declared to be the official name of said lake, and the only name to be regarded as legal in official documents, deeds, conveyances, leases and other instruments of writing to be placed on State or county records, or used in reports made by State, county or municipal officers" (*Statutes* 1869–70:64). The Whitney Survey (von Leicht–Craven) naturally has the name Lake Bigler, since the survey was dependent on the legislature; but the maps of the Land Office and the Wheeler Survey, as well as most private maps, use the name Tahoe exclusively. The "official" name was so completely forgotten that most Californians heard it for the first time when, in 1945, the legislature solemnly enacted that the lake "designated as Lake Bigler by Chapter 58 of the Statutes of 1869–70, is hereby designated and shall be known as Lake Tahoe" (*Statutes* 1945:2777). A post office Taho in El Dorado Co. is listed in 1868; Tahoe in Placer Co. is listed in 1880 but is now called Tahoe City. Lake Tahoe National Forest was created in 1899 and consolidated with Yuba Forest Reserve under the present name in 1906. **South Lake Tahoe** [El Dorado Co.] has been the official name since 1967 (BGN, list 6701) of what was earlier called Tahoe Valley or Al Tahoe.

Tahoma (tə hō′ mə) [Placer Co.]. Built as a resort in 1916; coined from "Tahoe" and "home" (Lekisch).

Tahquitz (tä′ kwits, tä′ kēts): **Peak, Canyon, Creek, Valley** [Riverside Co.]. The name is that of a supernatural being—Luiseño *táakwish*, Cahuilla *tákush*—that was said to dwell in the San Jacinto Mountains and to manifest itself as a fireball with a train of sparks. The name does not appear on older maps and was apparently applied when the San Jacinto quadrangle was surveyed in 1897–98.

Tail Canyon. *See* Jail Canyon.

Tailholt. *See* White: White River.

Tailings Gulch [Mariposa Co.]. The name of the ravine through which Sand Creek runs is left over from Gold Rush days. "Tailings" was the term used for the remnants of the ore that had been washed for gold, especially in hydraulic mining.

Tajanta [Los Angeles Co.]. The name of the land grant, dated July 5, 1843, is probably derived from a Gabrielino word, but its meaning is unknown. On a *diseño* of the grant a group of trees is shown as Monte de Tujunta. On Land Office maps the name is spelled Tajauta.

Tajea (tä hā′ ə): **Flat, Spring** [San Luis Obispo Co.]. The Spanish word means 'a channel, a culvert'.

Tajiguas (tə hig′ wəs, tə hē′ wəs): **Creek**, station [Santa Barbara Co.]. According to Alexander

Taylor, the name is derived from that of a Chumash village. El rancho del Refugio en Tajiguas (or Jajiguas) is mentioned on Oct. 22, 1841 (Arch. SB, Juzgado, p. 33). The Southern Pacific station was named after the creek, probably when the last link of the coast line was completed between 1894 and 1901. The Indian village received its name perhaps from *tayiyas*, another Chumash word for the islay or holly-leaved cherry. The padres of Mission Santa Barbara wrote in 1800: *el Yslay, o Tayiyas, que es una fruta algo parecida a la guinda* 'a fruit that somewhat resembles the cherry' (Arch. MSB 2:113–14).

Talawa (tä´ lə wə): **Lake, Slough** [Del Norte Co.]. The western arm of Lake Earl and the channel preserve the name of the Athabaskan tribe that occupied the extreme northwest corner of the state; the name is also spelled Tolowa. The English term comes from its Yurok name, *tolowel* (Waterman, p. 266; Robins).

Talbert [Orange Co.]. The post office was established about 1900 and named for James T. Talbert, a native of Kentucky and a veteran of the Civil War, who bought land in the district known as Gospel Swamp in 1898. His sons, Samuel and Thomas, built the Talbert Drainage System by which the swampland was reclaimed (Co. Hist.: Orange 1921:1186–87, 1560–61). The community is now part of the City of Moreno Valley (Brigandi).

Talc City Hills [Inyo Co.]. This "city" of low mountains was so named because of the deposits of the mineral talc, commonly known as soapstone (BGN, 1960).

Talega (tə lā´ gə) **Canyon** [Orange, San Diego Cos.]. Probably from Spanish *talega* 'bag, sack', applied because of the shape of the canyon.

Taliaferro Ridge [Mendocino Co.]. For Nicholas L. Taliaferro (1890–1961), biologist of the University of California, Berkeley; cf. BGN, list 6804.

Tallac (tə lak´): **Mount, Creek, Lake** [El Dorado Co.]. The mountain is shown as Crystal Peak on the maps of the Whitney Survey, but the Wheeler Survey applied the Washo Indian term *dalá'ak* 'mountain' in 1877 (William Jacobsen).

Tallulah (tə lo͞o´ lə) **Lake** [Yosemite NP]. The name appears for the first time on the Bridgeport atlas sheet in 1911 (Farquhar). It may be a transfer name from Georgia, where it exists as a place name, or the lake may have been named for Tallulah LeConte (David Brower).

Talmage [Mendocino Co.]. The post office is listed in 1892, but the origin of the name is not known. One Junius Talmadge, from Big River, was registered as a voter in the county in 1872, but he cannot be identified with the place.

Talus (tā´ ləs, tal´ əs) [Inyo Co.]. The geological term refers to a sloping mass of rocky fragments at the base of a cliff.

Tamales. *See* Tomales.

Tamalpais (tam əl pī´ əs), **Mount; Tamalpais Valley** [Marin Co.]. From Coast Miwok *tamal páyiṣ*, lit. 'coast mountain'; *tamal* means 'west, west coast' (Callaghan). The name has nothing to do with Spanish *país* 'land', or with the Mexican food *tamales*, or even less with tamale pies (*see* Tomales). A familiar nickname is "Mount Tam" (Teather 1986). The first Spanish name given to the mountain appears as Pico y Cerro de Reyes on Cañizares's Plano of 1776. Another Spanish name, Picacho Prieto 'dark peak', is mentioned as late as 1849. In 1826 Beechey applied the name Table Hill to the mountain because it looks like a flat table from the ocean off Point Reyes. For half a century it was so known to cartographers, or as Table Mountain, or Table Butte. On a *diseño* (about 1842) of the Baulenes grant the name Tamal paiz appears near the mountain, and on Nov. 25, 1845, the name is used for a land grant, Tamalpais or San Clemente, not confirmed by the U.S. government. In American publications of the 1850s there appear such absurd typographical errors as Tamel Pisc Mountain, Tama el Paris, and Tannel Bume. The old name was used again by the Whitney Survey. Hoffmann, observing from Black Mountain [Santa Cruz Co.], records on Sept. 3, 1861: "Highest Blue Mt. (Tamal Pais) back of San Francisco." After 1866 the popular Bancroft maps used the name, and Beechey's appellation thereafter disappeared from the maps, although the Coast Survey did not drop it until 1883. Another alleged Indian name of the mountain, as given to Barrett (*Pomo*, p. 308) by a speaker of the Southern Moquelumnan (Coast Miwok) dialect, was Pa´-lemus. It is possible that this was the Indian version of the name Palmas, given to the mountain in the 1830s by Francisco de Haro, onetime alcalde at San Francisco, because it reminded him of a hill in Mexico. This "Indian" name in turn may have become Mount Palermo, the name which Wilkes's expedition applied to the mountain in 1841. **Tamalpais Valley** [Marin Co.]. The settlement at the foot of the mountain was known as Big Coyote until the post office was named Tamalpais on Feb. 15, 1906. In 1908 Valley was added to the name.

Tamarack. There is no true tamarack (eastern larch) native in California, but the *Pinus murrayana*, or lodgepole pine, is often called tamarack, hence the confusion. A mountain and four lakes [Shasta Co.], a peak [Fresno Co.], a flat [Yosemite NP], a lake [Mono Co.], and a number of creeks are named Tamarack because lodgepole pines grow on or near them. The word is probably of Algonquian origin.

Tamarisk, Lake [Riverside Co.]. The tamarisk (*Tamarix* sp.) is a desert tree imported to California from the Near East.

Tambo [Yuba Co.]. This Spanish-American word for 'hotel, inn' (borrowed from Quechua, the language of the Incas) was applied to the Western Pacific station when the line was built in 1907.

Tan Bark Creek [Monterey Co.]. For the bark of the tan oak, *Lithocarpus densiflorus*, used in tanning leather (Clark 1991). There is also a Tanbark Creek in Los Angeles Co.

Tanforan (tan fə ran′, tan′ fə ran) [San Mateo Co.]. Toribio Tanfarán was a farmer here from the late 1850s. The misspelling was established by the Land Commission in the 1850s (Alan K. Brown).

Tanganyika (tan gən yē′ kə) **Lake** [San Bernardino Co.]. Named for a major lake in Tanzania, Africa.

Tanners Peak [Siskiyou Co.]. Probably named for John Pryor Tanner, a native of Kentucky, who was registered as a miner from Sawyers Bar in 1878.

Tanquary (tang′kwə rē) **Gulch** [Shasta Co.]. The gulch near Millville was named for O. H. P. Tanquary, who operated the Buscombe Mill here with James R. Keene in the late 1850s (Steger).

Tantrum Glade [Mendocino Co.]. So named because a crazy mule once upset a camp there (USFS).

Tapia (tä pē′ ə) **Canyon** [Los Angeles Co.]. Probably named for a member of the Tapia family, some of whom settled in the Los Angeles district before 1800.

Tapie (tap′ ē) **Lake** [Trinity Co.]. The lake was named for Raymond E. Tapie, who stocked it with trout (BGN, May 1954).

Tapo (tap′ ō): **Canyon, Creek** [Ventura Co.]. In 1821 it was recorded that Rancho San José de Gracias y Simi extended as far north as *el Volcán de Azufre y Tapi en lengua de gentiles* 'the sulphur volcano and Tapi in the tongue of the heathen' (Registro, p. 33). A rancho Tapo is mentioned on June 16, 1829 (SP Mis. and C. 2:6), and two "cañadas . . . de Tapo" in July 1834 (DSP Ben. Mil. 79:91).

Tar. *See* Asphalt.

Tara (tä′ rə): **Brook, Hills** [Contra Costa Co.]. Named in 1948 by the owner of the land, Abe Doty, for his daughter Taralin ("Tara"), whose name had been taken from the novel *Gone with the wind*.

Tarolkes (tə rōl′ kəs) [Trinity Co.] is a name once used for what is now called Long Ridge or South Fork Mountain. It is from Wintu *toror kelas*, lit. 'long ridge' (Shepherd).

Tarro, El. *See* Jarro: El Jarro Point.

Tartarus Lake [Lassen NP]. This lake in Hot Spring Valley was given the name of the infernal regions in classical mythology; other places in the region have similar hell-born names, given because of the hot springs.

Tarup Creek. *See* Turup Creek.

Tarzana (tär zan′ ə) [Los Angeles Co.]. When Edgar Rice Burroughs bought the Otis estate in 1917, he bestowed upon it the name derived from his famous fictional character Tarzan (AGS: Los Angeles, p. 381). The post office is listed in 1931.

Tassajara (tas ə hâr′ ə). *Tasajero* is a Spanish-American word designating a place where meat is cut in strips and hung in the sun to cure. The name is preserved in **Tassajara: Creek, Valley**, settlement [Alameda, Contra Costa Cos.], and **Tassajara: Creek, Hot Springs** [Monterey Co.]. The creek in Alameda Co. is shown as Arroyo de la Tasajera on a *diseño* (1841) of Valle de San José.

Taurusa. *See* Dinuba.

Taylor Canyon [Mono Co.]. Named for "Black" Taylor, one of the discoverers of the Bodie mines, who wintered some cattle in Hot Springs Valley and was killed there by Indians (Robinson).

Taylor Meadow [Kern Co.]. Named for Charlie Taylor, who was manager of the A. Brown interests in Kernville for many years (Crites, p. 269).

Taylorsville [Plumas Co.]. Named for J. T. Taylor, who built the first barn, mill, and hotel there in 1852.

Taylorville [Marin Co.]. The place, now commonly called Camp Taylor, was named for Samuel P. Taylor, who built the first paper mill on the Pacific Coast (*see* Paper Mill Creek).

Tea: Bar, Creek [Siskiyou Co.]. The name of the creek and settlement on the east side of the Klamath, once a mining community, is a folk-etymological rendering of *tíih*, a Karuk Indian village. The spellings Tee Bar and Ti Bar also occur.

Tea Canyon [Lake Co.] is so called because of the growths of a wild herb used for making tea (Mauldin).

Teal Lake [Lassen Volcanic NP] is named for the wild ducks called blue-winged teal (*Anas discors*) and green-winged teal (*Anas carolinensis*).

Tebo. *See* Thibau.

Tecabala. *See* Bally.

Tecate (ti kä′ tē): post office, **Mountain** [San Diego Co.]. A place called Tecate, probably a rancheria, is mentioned in the archives in 1830 (PSP Ben. Mil. 52:59) and a *cañada de Tecate* on Sept. 7, 1831 (Dep. Recs. 9:112). On Dec. 14, 1833, the name was used for a land grant, the territory of which was outside of the United States when the international boundary was established in 1847. Tecate Mountain is shown on Williamson's map of 1855. The post office was named after the mountain in 1913; and when the San Diego–Arizona Railroad was built in 1915, the name was applied to the station. Aztec origins have been proposed, but none has been confirmed. A possible Diegueño derivation is from *tuukatt* 'to cut with an ax', referring to a place where a tree was felled (Langdon).

Tecolote. This Mexican Spanish word for 'owl' is borrowed from Aztec *tecolotl*. It was repeatedly used for geographical names and has survived in **Tecolote** (tek ə lō′ tē) **Valley** [San Diego Co.] and **Tecolote Canyon** [Santa Barbara Co.]. The name of an *aguage corriente* 'running spring' mentioned in 1817 (PSP 20:177) is probably the origin of the latter name. The Real del Tecolote, so called by the Moraga expedition of 1806 and mentioned by Muñoz *por la mucha abundancia de estos animales* ('because these animals are so abundant') has not survived. It was probably in what is now Merced or Mariposa Co.

Tecolotito (tek ə lō tē′ tō) **Creek** [Santa Barbara Co.]. 'Little owl', from the Mexican Spanish diminutive of *tecolote* 'owl'; established by BGN, list 7802 (1978).

Tecopa (tə kō′ pə) [Inyo Co.]. The old mining camp was named before 1892 by J. B. Osbourne for a Southern Paiute elder, *tuku-pïda* 'wildcat-arm' (Fowler), who later demanded $200 for the use of his name. It is not known whether Osbourne paid, but it is a fact that Jim Slauson (of Resting Springs) paid tribute to Tecopa by sending him a plug hat every year (Gill). Tecopa once saved the people of Pahrump Valley from being killed by Indians (O. J. Fisk). When the Tonopah and Tidewater Railroad was built in 1908, the station was named for the camp, and mining operations were resumed. A picture of the Indian, plug hat and all, was published in *Desert Magazine*, Sept. 1943.

Tectah (tek′ tä) **Creek** [Humboldt Co.]. From the Yurok village *tektoh*, meaning literally 'log' (Waterman, p. 239; Robins).

Tecuya (ti ko͞o′ yə): **Creek, Mountain, Ridge** [Kern Co.]. The name is said by Kroeber to be derived from *tokya*, the name applied by the Yokuts to the Chumash Indians, a division of whom occupied the region. The origin is probably Yokuts *tʰoxil* 'west' (Gamble).

Tedoc. *See* Tidoc.

Tee Bar. *See* Tea Bar.

Tehachapi (tə hach′ ə pē): **Creek, Pass, Mountains, Valley**, town [Kern Co.]. From Kawaiisu *tïhachïpía*, supposedly meaning something like 'hard climbing', from *tïhaa* 'difficult' plus *chipii-* 'to climb' (Munro). The name was placed on a map as Tah-ee-chay-pah Pass by the Pacific Railroad Survey in 1853: "From the Indians we learned that their name for the creek was Tah-ee-chay-pah" (Pac. R.R. *Reports* 5:1.19). Frémont had crossed the pass in 1844 and called the creek Pass Creek. The name used by the survey remained on the maps with many variants in spelling. The post office is listed as Tehichipa in 1870; the von Leicht–Craven map of 1874 has Tahichipi Pass, Creek, Valley, town. The spelling Tahichipi is used by Kroeber for the name given by the Yokuts to the region or part of it. In 1876 the Southern Pacific built the railroad through the canyon of Cache Creek, transferred the name from the old wagon route, and fixed the present spelling, which Brewer had used in his journal on June 7, 1863. The "new town" developed at the railroad station; the post office was at first about a mile away at Greenwich (named for the postmaster, P. D. Green). It is listed as Greenwich in the Postal Guides between 1880 and 1892, but appears as Tehachapi in 1898.

Tehama (tə hā′ mə): city, **County**. Kroeber records Tehama or Tehemet as the name of a Nomlaki (Wintun) village, but gives no further etymology. The Indian origin is discussed by Clara Hisken, *Tehama: Little city of the big trees* (New York: Exposition Press, 1948), pp. 3–4; by James F. Bryant, "Derivation of the name 'Tehama'," *Tehama County memories* (1955), p. 57; and by Eugene F. Serr, "Solving the 'Tehama' puzzle," *Tehama County memories* (1997). Derivations proposed from Arabic or Spanish are less convincing. The town was laid out on R. H. Thomes's Saucos Rancho, and lots were offered for sale in the *San Francisco Alta California* on Aug. 5, 1850. *See* Thomes Creek; Toomes Creek. Tehama Co. was created from portions of Shasta, Colusa, and Butte Cos. on Apr. 9, 1856.

Tehipite (tə hip′ i tē): **Valley, Dome** [Kings Canyon NP]. The name is shown on the Tehipite atlas sheet of 1905. Nothing is known about its origin beyond the fact that it is Indian, probably Mono. Winchell's interpretation of the meaning as 'high rock' (Farquhar) sounds plausible.

Tejon (tə hōn′): **Canyon, Creek, Hills** [Kern Co.]. The name Cañada del Tejón 'badger valley' was given to the canyon by the expedition of Lt. Francisco Ruiz in 1806 because a dead badger was found at its entrance (*San Francisco Evening Bulletin*, June 5, 1863). On Nov. 24, 1843, the name was applied to the land grant El Tejon. Sutter in his *Reminiscences* (p. 125), dictated in 1876, states that his men found Tejon Pass on the return march from the Micheltorena campaign in Mar. 1844 and implies that the name Tejon was in use then. The pass is shown on Gibbes's map of 1852. **Tejon Pass** [Los Angeles Co.] and **Old Fort Tejon** [Kern Co.]: In Sept. 1853, R. S. Williamson of the Pacific Railroad Survey found the Tejon Pass too difficult and established a new route through the Cañada de las Uvas, over the pass that had been called Buena Vista by Fages in 1772. In the same month Edward F. Beale, superintendent of Indian affairs, selected a site in the canyon for a fort and an Indian reservation (Pac. R.R. *Reports* 5:1.20 ff.). The new pass became known as Fort Tejon Pass; the name was abbreviated to the present form when the old route was abandoned.

Telegraph. Several elevations were so named because they were used for telegraphing by signal before the invention of the electrical apparatus. **Telegraph Hill** [San Francisco Co.]: On July 11, 1846, a midshipman from the *Portsmouth* put up a signal pole here and "telegraphed," as the first message, that the British frigate *Juno* had entered the harbor (Rogers, *Montgomery*, p. 67). **Telegraph Peak** [San Bernardino Co.]: The peak was originally called Heliograph Hill because the USGS had a heliograph station there (Wheelock). The two Telegraph Canyons in Orange and San Diego Cos. were so named because a telegraph line passed through them.

Telephone Spring [Death Valley NP]. In 1906, a telephone line passed through the canyon in which the spring is situated (Death Valley Survey). **Telephone Flat** [Modoc Co.] was so named in 1917, when the new telephone line of the USFS was built across the flat (William S. Brown).

Telescope: Peak, Mountains [Death Valley NP]. The peak was named in Apr. 1861 by W. T. Henderson because of the wide, clear view that it afforded. Henderson was one of the California Rangers who are credited with having killed the bandit Joaquín Murieta. Telescope District is shown on Farley's map of 1861. The mountains are mentioned in the Bendire Report (1867) and are shown on the von Leicht–Craven map (1874) west of Emigrant Canyon, on both sides of Wildrose Canyon. Now the term is applied to the section of the Panamints south of Wildrose Canyon.

Tells: Peak, Creek [El Dorado Co.]. The places were not named for the Swiss hero Wilhelm Tell, but for a homesteader named Tell, who had settled with several other Swiss a few miles to the west of the peak about 1875. Tell's house, as well as Tells Creek and Peak, are shown on Wheeler atlas sheet 56-B. However, W. T. Russell states that the peak was named for Ciperano Pedrini, better known as Bill Tell, early storekeeper of Garden Valley (Knave, July 23, 1944).

Temascal. *See* Temescal.

Tembladera (tem blə dâr′ ə) [Monterey Co.].

The name given to an area of several hundred acres of swampy land south of Castroville, covered by a dense overgrowth of vegetation; Spanish *tembladero* means 'marsh, quagmire'.

Temblor. In Spanish times, *temblor* 'earthquake' was frequently applied to a place where an earthquake had been felt. The Portolá expedition named the Santa Ana River Río de los Temblores on July 28, 1769, because while encamped there they were shaken by a violent quake, which was repeated four times that day (Costansó, p. 17). Crespí called the place Jesús de los Temblores; but the soldiers, contrary to their custom, gave it a saint's name, Santa Ana, and this name the Anza expedition adopted seven years later. Among the surviving Temblor names is **Temblor** (tem blôr′) **Range**, a long mountain chain in Kern and San Luis Obispo Cos.

Temecula (tə mek′ yo͞o lə, tem ə kyo͞o′ lə): town, **Canyon, Valley, Indian Reservation** [Riverside, San Diego Cos.]. An Indian rancheria Temeca is mentioned as early as 1785 (Brigandi), and the present spelling is found as early as 1820 (Arch. MSB 3:179). According to Sparkman (p. 191), the Luiseño name is really *temeko;* this version too occurs in early records: Temeco, 1802 (Arch. MSB 8:160). Temecula was the name of a rancho of Mission San Luis Rey in 1828 or before (Registro, p. 39). Later the name was used in the titles of several land grants, two of which, dated Aug. 25, 1844, and May 7, 1845, were recognized by the United States. The name is spelled Temecola on Wilkes's map of 1841, and Temecula on Ord's sketch of 1849. The post office opened in 1859. The meaning of the name is unknown; Kroeber leaves open the possibility that it may contain Luiseño *temét* 'sun'.

Temescal. This Mexican Spanish word, also spelled *temascal,* means 'sweathouse', referring to the small house used by American Indians, in California as well as Mexico, for sweating and bathing. The word is from Aztec *temaxcalli,* containing *tema* 'to bathe' and *calli* 'house'. The word survives, now invariably spelled Temescal, in several California place names. **Temescal** (tem ə skal′, tem′ ə skal) **Creek** and **Wash** [Riverside Co.]: The name was preserved through the Temascal land grant of 1818–19. Wilkes's map of 1841 shows a place called Temascal east of Santa Ana Mountains. Brewer, in Feb. 1861, mentions Temescal Indian village, sulphur springs, Overland station, and hills (pp. 39, 34, 44). **Temescal, Lake** [Alameda Co.]: A *diseño* of the San Antonio grant (dated Aug. 3, 1820) shows Arrollo (arroyo) de Temescal o Los Juciyunes flowing toward Loma de Temescal, apparently the elevation later known as Emeryville Shellmound. Southwest of the arroyo, on the point from which the Oakland mole was later extended, the Encinal ('oak grove') del Temascal is also shown. The name became well known, and in 1844 the landing place for the eastern bay shore was called Embarcadero Temescal (William Heath Davis, p. 98). The lake was created when William F. Boardman dammed the creek for the reservoir of the Contra Costa Water Company in 1870. **Temescal: Canyon, Creek** [Ventura, Los Angeles Cos.]: Rancho del Temascal is mentioned on Dec. 16, 1834 (DSP 3:205). On Mar. 17, 1848, the name was given to a land grant.

Temetate (tem ə tä′ tē) **Creek** [San Luis Obispo Co.]. The name apparently reflects Mexican Spanish *temetate,* designating a stone slab used for grinding corn, etc., or its Aztec prototype *temetatl,* from *tetl* 'stone' and *metlatl* 'grinding slab'. However, this is probably a popular etymology, distorting the Obispeño Chumash place name *stemeqtatimi,* of unknown etymology (John Johnson). Monte y Arroyo de Temetatl are shown on a *diseño* of the Bolsa de Chamisal grant (1837). The alternative spelling Temettati is found.

Temple City [Los Angeles Co.]. Named in 1923 for Walter P. Temple, founder of the town and president of the Temple Townsite Company (*Los Angeles Times,* Dec. 4, 1931, p. 16).

Temple Crag [Inyo Co.]. The beautiful crag was earlier named Mount Alice for Mrs. Alice Ober of Big Pine, who chaperoned a party of young people on a trip to this region (Brierly). The BGN (6th Report) decided in favor of the present descriptive name.

Templeton [San Luis Obispo Co.] was laid out by the West Coast Land Company with the coming of the railroad in 1886 and named Crocker. Because the name was changed shortly afterward to Templeton, it has been assumed that the town was named for Templeton Crocker of San Francisco—a grandson of Charles Crocker, one of the founders of the Central Pacific Railroad.

Templeton: Meadows, Mountain [Tulare Co.].

Named for Benjamin S. Templeton, a sheep man (Farquhar).

Tenaja (tə nä´ hä) **Canyon; El Potrero de Tenaja** [Riverside, San Diego Cos.]. *La poza llamada Tinaja* 'the waterhole called Tinaja' is mentioned in the Registro (pp. 44–45), and **La Tinaja** is shown on a *diseño* of the Santa Margarita y Las Flores grant (1841). *Tinaja* is the Spanish word for a 'large earthen jar', often used in the Southwest for a natural waterhole. The name Las Tinajas de los Indios [Kern Co.] is doubtless from the same source.

Tenaya (tə nī´ ə): **Lake, Creek, Canyon, Peak** [Yosemite NP]. The Southern Sierra Miwok name is *ṭïyenna* 'sleeping place', from *ṭïye-* 'to sleep' (Broadbent; Callaghan). Another story, however, is that the lake was named on May 22, 1851, by the Mariposa Battalion, upon the suggestion of L. H. Bunnell: "Looking back to the lovely little lake, where we had been encamped during the night, and watching [the old chief] Ten-ie-ya as he ascended to our group, I suggested to the Capt. [Boling] that we name the lake after the old chief, and call it 'Lake Ten-ie-ya.' The Capt. . . . readily consented to the name. . . . He said, 'Gentlemen, I think the name an appropriate one, and shall use it in my report of the expedition. Beside this, it is rendering a kind of justice to perpetuate the name of the old chief.' . . . I called [Ten-ie-ya] up to us, and told him that we had given his name to the lake and river. At first, he seemed unable to comprehend our purpose, and pointing to the group of glistening peaks, near the head of the lake, said: 'It already has a name, we call it Py-we-ack.' Upon my telling him that we had named it Ten-ie-ya, because it was upon the shores of the lake that we had found his people, who would never return to it to live, his countenance fell and he at once left our group and joined his own family circle. His countenance as he left us indicated that he thought the naming of the lake no equivalent for the loss of his territory" (Bunnell, pp. 236–37). The spelling Tenaya is used on the Hoffmann-Gardner map of 1867. The peak had been named Coliseum Peak by Joseph LeConte and his party, a name that did not last in spite of the name-giver's exhortation: "We called this Coliseum Peak. So let it be called hereafter to the end of time" (*Journal*, pp. 76–77).

Ten Lakes [Yosemite NP] were mentioned by Muir in 1872: "On the north side of the Hoffmann spur . . . there are ten lovely lakelets lying near together in one general hollow, like eggs in a nest" (Muir, *The mountains of California* [New York: Century, 1911], p. 100).

Ten Mile River. *See* Mile: Ten Mile River.

Tennant [Siskiyou Co.]. The post office was established about 1922 and named for Graner Tennant, an official of the Long Bell Lumber Company (Luecke).

Tennessee: Point, Cove, Valley [Marin Co.]. So named because the steamer *Tennessee* was wrecked offshore in 1853. The Coast Survey named the point and the cove; the county map of 1873 shows the name for the valley.

Tent Meadow [Kings Canyon NP]. A large block of granite, seen from a distance, resembles a white tent; hence the name (Farquhar).

Teofulio (tā ō fo͞o´ lē ō) **Summit** [San Diego Co.]. Named for Teofulio Helm (1874–1967), prominent member of the Cupeño Indian tribe, who homesteaded here; cf. BGN, list 7502.

Tepona (ti pō´ nə) **Point** [Humboldt Co.]. From Yurok *tepoona*, derived from *tepoo* 'tree' (Waterman, p. 272; Robins).

Tepo (ti pō´) **Ridge** [Del Norte Co.]. From a form of Yurok *tepon-* 'to stand, be vertical' (Waterman, p. 269; Robins).

Tepusquet (tep´ əs kē): **Creek, Peak** [Santa Barbara Co.]. Arroyo de Tepusque and Sitio de Tepusque are shown on *diseños*. On one map the name is misspelled Tepusquet, and in this form it was applied to the land grant dated Apr. 6, 1837. In colonial times, Mexican Spanish *tepusque* designated a copper coin of low value; it is from Aztec *tepuztli* 'copper' (Santamaría).

Tequepis (tek´ ə pis) **Canyon** [Santa Barbara Co.]. A Ranchería de Tequeps is mentioned in Oct. 1798 (Arch. MSB 8:164) and in later documents. In 1837 the name appears as Tequepis for a land grant, and the canyon is shown on a *diseño* of the grant as Cañada de Tequepis. The name may originally have been a Chumash place name.

Tequesquite (tek əs kē´ tē) **Slough** [San Benito Co.]. The name was preserved through the Llano de Tequesquite grant, dated Oct. 10, 1835. Sanjón del Tequesquite is shown on a *diseño*, and in 1861 John Gilroy testified in the Juristac case that this "ditch" is now called the Pajaro (River). *Tequesquite* is Mexican Spanish for 'saltpeter', from Aztec *tequixquitl*, containing *tetl* 'rock' and *quixquitl* 'efflorescent' (Santamaría). Muñoz wrote on Sept. 25, 1806 (Arch. MSB 4:1 ff.) that the pastures near San Joaquin River have patches of *tequesquite*. **Tequesquite Arroyo** [Riverside Co.]: The arroyo was doubtless so named because, in hot weather, "white alkali" appeared here like patches of snow (Loye Miller).

Terminal Island [Los Angeles Co.]. This was probably the Isla Raza de Buena Gente 'island [of the] tribe of good people' named on Palacios's chart of the bay (Wagner, p. 418). In American times it became known as Rattlesnake Island and was so designated by the Coast Survey until 1951. After the Los Angeles Terminal Railway had built a line from the city to the island, it became known as Terminal Island and is so shown on the Official Railway Map of 1900.

Termination Valley. *See* Eureka: Eureka Valley.

Terminous [San Joaquin Co.]. The place was established about 1900 by John Dougherty; and because it was at the end of the road that ran into the delta region, he wished to call it Terminus. In the application to the Post Office Dept., the name was erroneously spelled with an *o*, and this version was adopted (Amy Boynton).

Termo (tûr′ mō) [Lassen Co.]. From "terminal." When the Nevada-California-Oregon Railroad reached the place in 1900, the general manager applied the name to the station just because he had a liking for station names that ended in *o* (Myrick).

Terra Bella (ter′ ə bel′ ə) [Tulare Co.]. The Latin (or Italian) phrase meaning 'beautiful land' (cf. Spanish *tierra bella*) was applied to the station when the Southern Pacific branch from Exeter to Famoso was built in 1889. The name was applied in the spring, when the country is covered by a carpet of wildflowers (W. W. Hastings).

Terra Buena. *See* Tierra: Tierra Buena.

Terra Linda [Marin Co.], meaning 'beautiful land', is said to have been named by Portuguese landowners: hence *terra* instead of Spanish *tierra*.

Terwah (tûr′ wä) **Creek** [Del Norte Co.]. From the Yurok site called *trwrr* (Waterman, p. 235; Harrington), perhaps from the word for 'locust'. On Oct. 12, 1857, Lt. George Crook established a fort here and named it Fort Ter-Waw (F. B. Rogers, *CHSQ* 26:1). It appears on the von Leicht–Craven map as Fort Terwah. The spelling used by the USGS on the Preston Peak atlas sheet is Turwar.

Terwilliger Valley [Riverside Co.]. *See* Anza: Anza Valley.

Tesla (tes′ lə) [Alameda Co.]. The name still appears on maps and recalls the thriving town named in 1898 in honor of Nicola Tesla (1856–1943), the Austrian-American inventor. Plans were projected at that time to build an electric power plant at the local coal mines to supply power to Oakland (Still).

Teutonia (to͞o tō′ nē ə) **Peak** [San Bernardino Co.]. Named after the Teutonia (now known as Dutch Silver) Mine on the northeast slope of the mountain. The mine had been named by a prospector, "Charlie" Toegel, a native of Germany (Gill).

Texas Spring [Death Valley NP] was perhaps named for a Charles Bennett, nicknamed "Texas" or "Bellerin Teck," who started Furnace Creek Ranch in 1870 (Death Valley Survey). **Texas Springs** [Shasta Co.] was also named for a settler nicknamed "Texas." A number of physical features bearing the name were, however, named by miners or settlers after their home state.

Thalheim. *See* Valley: Valley Home.

Tharps Rock [Sequoia NP]. Named for Hale D. Tharp (1828–1912), a native of Michigan who settled in the Three Rivers region in 1856 and was the first white man to explore this part of the High Sierra (Farquhar). *For* Tharp's Peak, *see* Alta: Alta Peak.

Thatcher: Butte, Creek [Mendocino Co.]. Named for three brothers who camped in the vicinity while hunting deer in the winter of 1855–56. In the spring, two of the brothers were killed by Indians; the third had gone to the Sacramento Valley for supplies (USFS).

Thermal [Riverside Co.]. The name was applied to the station before 1888, because of the extreme heat in the Salton Sea basin (Southern Pacific).

Thermalito (thûr mə lē′ tō): town, **Forebay, Afterbay** [Butte Co.]. The name is apparently coined from "thermal," referring to hot springs, plus the Spanish diminutive suffix *-ito*. The town had a post office from 1895 to 1920. The two sheets of water were created when the Oroville dam project was constructed in the 1960s and were given the euphonious name of the town.

Thibau (tē′ bō) **Creek** [Inyo Co.]. The stream was named for a French family that in the

1890s lived where the old country road crossed the creek. The name is sometimes spelled Tebo or Thibault (Sowaal 1985).

Thimble. There are several peaks and hills so named because of their shape. The two highest and best-known Thimble Peaks are in Alpine and Inyo Cos.

Thing Valley [San Diego Co.]. In the early days the valley was called Hollister Valley for the man who raised sheep there. When Damon Thing bought the valley about 1870 and developed a cattle ranch there, it became known as Thing Valley (C. F. Emery). The family tradition says the name had originally been "Hogg," and they petitioned to change it to "anything" (Stein).

Thomes (tom′ əs, tōmz) **Creek** [Tehama Co.]. The stream, spelled Toms Creek on Eddy's map of 1854 and Thoms and Thomes on later maps, emerged as Thomas Creek when the USGS mapped the Vina quadrangle. The name commemorates Robert H. Thomes, a native of Maine, who came to California in 1847; he worked in partnership with A. G. Toomes as a carpenter in San Francisco and Monterey and settled in 1847 on his Rancho Los Saucos, through which the creek flows. His partner was grantee of the rancho on the opposite side of the Sacramento (*see* Toomes Creek; Tehama).

Thompson, Mount [Kings Canyon NP]; **Thompson Ridge** [Inyo Co.]. The peak was named by R. B. Marshall in honor of Almon H. Thompson (1839–1906), who was associated with J. W. Powell in the exploration of the Colorado River, 1870–78, and was geographer of the USGS, 1882–1906 (Farquhar).

Thompson Peak [Alpine Co.]. Named for John A. "Snowshoe" Thompson (1827–76, originally Tostensen). From 1853 to 1876 he carried mail across the High Sierra from Placerville to Carson Valley; in the winter he used skis, then often called Norwegian snowshoes.

Thompson Peak [Trinity Co.]. In the 1870s Packer Thompson climbed the peak on a dare and chiseled his name on the summit (J. D. Beebe). When the peak was used as a triangulation point, the USGS accepted the name.

Thompson's Dry Diggings. *See* Yreka.

Thom's Creek. *See* Thomes Creek.

Thorn: post office, **Valley** [Humboldt Co.]. The valley was named because of the abundance of native white-flowering thornbush (Laura Mahan). The post office at the settlement was in existence between 1888 and 1923, and was reestablished on Feb. 16, 1951. There is a Thorn railroad station near Oro Grande in San Bernardino Co.

Thornton [San Joaquin Co.]. Arthur Thornton, a native of Scotland, established his New Hope Ranch here about 1855. The place was known as New Hope until the Western Pacific built across the property in 1907 and named the station for the owner of the land (*see* Newhope Landing).

Thousand. This number is used in geographical nomenclature as a convenient expression for "many." Among the best-known features in California designated in this manner are Thousand Lakes Primitive Area [Shasta Co.], Thousand Oaks [Ventura Co.], and Thousand Springs [Shasta Co.]. **Thousand Island Lake** [Madera Co.] in fact has around a hundred islets (Browning 1986). *For* Thousand Palms [Riverside Co.], *see* Palms; for Thousand Palms Canyon [San Diego Co.], *see* Salvador Canyon.

Three Arch Bay [Orange Co.]. The name was suggested by the three natural rock arches; *see* South Laguna.

Three Brothers [Yosemite NP]. A group of three rocky peaks, named by members of the Mariposa Battalion because Chief Tenaya's three sons were captured nearby.

Three Buttes. *See* Sutter: Sutter Buttes.

Three Rivers [Tulare Co.]. The post office was established in 1878 and so named because the town is near the junction of three forks of Kaweah River.

Three Sirens [Kings Canyon NP]. The mountain with three peaks was so named because of its location near Scylla and Charybdis, from Homer's *Odyssey*.

Throop (thro͞op) **Peak** [Los Angeles Co.]. Named for Amos G. Throop, who founded Throop University at Pasadena, now the California Institute of Technology, in 1891. The mountain is also known as North Baldy Peak (USFS).

Thumb, The [Inyo Co.]. Windsor B. Putnam named the peak when he made the first ascent on Dec. 12, 1921. The shape of the mountain somewhat resembles the end of a thumb. It is sometimes called East Palisade.

Thunder Mountain [Calaveras Co.]. The name of the peak near Railroad Flat is a translation of an Indian name, according to a letter of June 29, 1872, by Phil Schuhmacher of the

Whitney Survey (Davidson Papers). A Thunder Mountain in San Bernardino Co. is listed in the BGN, Decisions, 1965. *See also* Dunderberg. **Thunder Mountain** [Sequoia NP]: Named by George R. Davis, of the USGS, when he made the first ascent in Aug. 1905 to establish a benchmark on the summit (Farquhar).

Thurston Lake [Lake Co.]. Apparently named for Charles Edwin Thurston, a native of New York, a sheepherder at Lower Lake in 1867.

Tia Juana River. *See* Tijuana River.

Ti Bar. *See* Tea Bar.

Tiburon (tib′ ə ron): **Point, Peninsula**, town [Marin Co.]. Punta de Tiburón 'shark's point' is mentioned in the diary of José Sánchez on July 6, 1823 (SP Sac. 11:47). The name appears again on Beechey's map of 1826 and on a *diseño* of Corte de Madera del Presidio land grant (1834). The early charts of the Coast Survey leave the point nameless, but a Tiburn [*sic*] Point is shown on Hoffmann's Bay map of 1873. The name, properly spelled, was applied to the post office about 1885.

Tice Valley [Contra Costa Co.]. Named for James and Andrew Tice, who owned Rancho El Sobrante de San Ramón, now the site of Saranap and Rossmoor.

Tick. A canyon in Los Angeles Co. and several other features in the state were probably so named because the surveyors were annoyed by this parasite (*see* Garrapata).

Tico (tē′ kō) [Ventura Co.]. Probably named for Fernando Tico, grantee of Ojai rancho in 1837.

Tidoc (tē′ dok): **Mountain, Gap** [Tehama Co.]. Also spelled Tedoc; from Wintu *t'idooq* 'red ant' (Shepherd).

Tie Canyon [Death Valley NP]. About 1925, some 120,000 old railroad ties were purchased from the abandoned Tonopah and Tidewater Railroad to provide fuel for the eighteen fireplaces in Scotty's Castle. They were stacked in a nearby gorge, which became known as Tie Canyon (Randall Henderson, *Desert Magazine*, Sept. 1952, p. 8).

Tiefort (tē′ fôrt) **Mountains** [San Bernardino Co.]. Tiefort is an old German family name as well as a German place name, but no connection with the name of the mountain range has been found. The theory of C. B. McCoy that the name goes back to Tie Foot—a Ute Indian chief who operated in the region with his band, molesting travelers on the old Mormon Trail—sounds like folk etymology but is not impossible (Arda Haenszel). Locally pronounced (tē′ fôrd).

Tierra. The Spanish word for 'land, earth' is repeatedly found in California place names. Corral de Tierra 'earthen enclosure' was the name of a land grant in Monterey Co., dated Apr. 10, 1836, and of two grants in San Mateo Co., dated Oct. 5, 1839, and Oct. 16, 1839; it was also the alternate name of the San Pedro grant [San Mateo Co.], dated Jan. 26, 1839. The grant that included part of the lands of Mission Santa Clara (Nov. 28, 1845) was called Tierra Alta 'high land'. In modern geography the term is found in Tierra Buena 'good land' [Colusa Co.], Tierra Blanca ('white') Mountains [San Diego Co.], and Tierra Redonda ('round') Mountain [San Luis Obispo Co.]. **Tierra Buena** (tē âr′ ə bo͞o ā′ nə) [Sutter Co.]: The name of the settlement was changed from Terra Buena to the more correct Spanish for 'good earth' by the BGN in May 1954. **Tierra del Sol** (del sōl′) [San Diego Co.]: The name of the old post office Hipass was changed to the Spanish name, meaning 'Sunland', on Aug. 20, 1956. **Tierra Rejada** (rā hä′ də) [Ventura Co.] is local Spanish for 'plowed land' (from *reja* 'plow').

Tijera (tē hâr′ ə) [Los Angeles Co.]. Ciénega o Paso de la Tijera 'marsh or ford of the drainage channel' is the name of a land grant dated Feb. 23, 1823, and May 12, 1843. The name is preserved in La Tijera, a Los Angeles branch post office.

Tijuana (tē wä′ nə) **River** [San Diego Co.]. The name of a *parage*, of unidentifiable Diegueño Indian origin, is spelled Tiajuan in 1829 (Dep. Recs. 7:62) and appears with similar spellings in later records. Even in Spanish times it was changed by folk-etymological process to Tía Juana 'Aunt Jane' (Dep. Recs. 13:52). The river is shown as Arroyo de Tijuan on a *diseño* of the unconfirmed Milijo grant (1833). The name on the Mexican side of the border is now Tijuana, but BGN, list 6801, also established this spelling for the river on the U.S. side of the border.

Tilden: Lake, Canyon, Creek [Yosemite NP]. On Hoffmann's map of 1873 the lake appears as Lake Nina, named for Hoffmann's sister-in-law, Nina F. Browne. The present name appears on McClure's map of 1895 (Farquhar). No explanation for the origin of the name has been found.

Tilden Park [Alameda, Contra Costa Cos.]. Named in 1937 by the East Bay Regional Park Board in honor of Maj. Charles Lee Tilden (1857–1950), leader of the movement to create the regional parks (R. E. Walpole).

Tiltill: Creek, Mountain, Valley [Yosemite NP]. From Southern Sierra Miwok *tiltilna* 'tarweed', genus *Madia* (Broadbent; Callaghan). The name does not appear on older maps and was probably not applied until the Dardanelles and Yosemite quadrangles were surveyed between 1891 and 1896.

Timalula (ti mä′ lo͞o lə) **Falls** [Yosemite NP]. A name no longer current, of Indian origin, probably referring to the falls of Avalanche Creek into the Merced River (Browning 1988:210). Likely to be from Southern Sierra Miwok *temaalï-la* 'trading place', from *temaalï* 'to trade' (Callaghan).

Timber. The word is found in the names of some twenty physical features, including Timbered Mountain in Modoc Co. This term and its companion term "wooded" are applied particularly when some bald features are nearby. **Timber Cove** and **Gulch** [Sonoma Co.] were so named in the 1850s when the cove was used as a lumber-shipping point (Co. Hist.: Sonoma 1877:100). *See* Madera.

Timbuctoo [Yuba Co.]. The name of the prosperous mining town of the 1850s was perhaps inspired by an African-American miner nicknamed "Timbuctoo," or by a popular song ("for he was a man from Timbuctoo"). According to Drury (p. 497), the first white miners "found a blackamoor busy with pick and pan, smiling a golden smile, and their ready fancy fixed on Timbuctoo as the name for the new camp."

Timms: Landing, Point [Los Angeles Co.]. The names commemorate A. W. Timm, an early pioneer of San Pedro, who died in 1888. Avalon Bay on Santa Catalina Island was once known as Timms Cove.

Tin. Although tin ore has been found in varying amounts in several localities in the state, the name has been applied sparingly. Tin Mine Canyon and Creek are found in Riverside Co., and Tin Mountain in Death Valley NP.

Tinaja, La. *See* Tenaja Canyon.

Tinaquaic [Santa Barbara Co.]. The name of the land grant dated May 16, 1837, is of unidentified Chumash origin: Ranchería de Tinoqui is mentioned in 1790 (PSP Ben. Mil. 9:6). Arroyo de Tinaquaic is shown on a *diseño* of the grant.

Tin Cup Gulch [Shasta Co.]. The miners who worked here once claimed that their daily find was as much gold as would fill a tin cup (Steger), hence the name.

Tinemaha (tin′ ə mə hä), **Mount; Tinemaha: Creek, Reservoir** [Inyo Co.]. The peak was known by this name to the early prospectors and cattlemen of Owens Valley; it was named for a legendary Paiute chief, Tinemaha or Tinemakar, brother of Winnedumah (*q.v.*).

Tinkers: Defeat, Knob [Placer Co.]. J. A. Tinker was a "rough, hard-driving, hard-drinking teamster" who hauled freight between Soda Springs and the mines on Forest Hill Divide. From 1867 to 1873 Soda Springs Station was known as Tinkers Station. The name "Tinkers Defeat" was applied to a hairpin curve of the Soda Springs Road—because one day, so the story goes, the teamster's load, drawn by eight horses, came to grief at this place (Knave, May 10, 1953). The knob was named for him with humorous reference to his nose (Doyle). According to a directory of 1867, James A. Tinker was the owner of the Tinker and Fenton Hotel on Donner Lake road.

Tinoqui, Rancheria de. *See* Tinaquaic.

Tioga (tī ō′ gə): **Pass, Road** [Yosemite NP], **Peak, Crest, Lake** [Mono Co.]. The name is applied to counties and rivers in Pennsylvania and New York, where it also occurs in a number of town names. The origin is probably Mohawk (Iroquoian) *teió:ken* 'where it branches in two' (Mithun). The name was transferred to California when the Tioga Consolidated Mine was registered at Bodie on Mar. 14, 1878. In 1880 there was another Tioga Mine near Mount Dana. The famous road was built in 1882–83 but was abandoned in 1884 because of financial failure. In 1915 it was bought by private subscription and donated to the government (Farquhar).

Tionesta (tī ə nes′ tə) [Modoc Co.]. Named in 1931 by J. R. Shaw, of the Shaw Lumber Company, after Tionesta Forest in Pennsylvania; the origin is apparently Iroquoian.

Tippecanoe [San Bernardino Co.]. The sobriquet of President William H. Harrison (1773–1841), which refers to his victory over the Indians at Tippecanoe River, Indiana, on Nov. 7, 1811, was a popular place name, especially after the campaign of 1840 when the "country resounded with the Whig battle cry: 'Tippecanoe and Tyler too!'" The name was applied to the Pacific Electric station about

1908. The Indiana name is said to be a folk etymology from Potawatomi (Algonquian) *ki-tap-i-kon* 'buffalo-fish' (Stewart).

Tipton [Tulare Co.]. The name was applied by the Southern Pacific when the Tulare-Delano section was built between July 1872 and July 1873. The name of the English town in the county of Stafford had already become a popular place name in the eastern United States.

Tip Top. The name is found for several orographic features. **Tip Top Mountain** [San Bernardino Co.] is appropriately applied to the highest peak of a ridge; but **Tip Top Ridge**, east of Crannell [Humboldt Co.], seems to be a misnomer in view of the higher elevations east and north of it.

Tish-Tang-a-Tang (tish täng′ ä ting [*sic*]) **Creek; Tish Tang Point** [Humboldt Co.]. From the Hoopa village name *diysh-taang'aading*, probably meaning 'grouse-promontory' (Golla).

Tit. It is a common custom in the western states to designate an elevation shaped like a woman's breast as Tit or Teat. The name is ordinarily used only in local speech and is seldom found on maps. *See* Squaw; Two Teats; Pecho.

Titanothere (tī tan′ ō thēr) **Canyon** [Death Valley NP]. Applied in 1936 by the National Park Service because the fossil bones of a titanothere, an extinct mammal, were found there (Death Valley Survey).

Titus Canyon [Death Valley NP] was named for Morris Titus, a mining engineer, who perished in this canyon while on a prospecting tour in 1905.

Toadtown. *See* Johnstonville.

Tobias Peak [Tulare Co.]. Named in 1884 by John and Tobe Minter, in memory of their father, Tobias Minter, who homesteaded a meadow at the base of the mountain (R. J. Beard).

Tocaloma (tō kə lō′ mə) [Marin Co.]. According to testimony in the Los Baulenes land grant case in 1862, the stream called San Geronimo (now Lagunitas Creek) was also known by its Indian name, Tokelalume. This may contain Coast Miwok *lúme* 'willow' (Callaghan). Tocaloma post office existed from Apr. 27, 1891, to Sept. 30, 1919.

Todd: Valley, Creek [Placer Co.]. Named for Dr. F. Walton Todd, a cousin of Abraham Lincoln's wife according to Doyle. Dr. Todd opened a store there in June 1849. Todd's Valley Diggings are mentioned in the *Sacramento Daily Transcript*, Apr. 19, 1851, and Todd's is shown on Gibbes's map of 1852.

Todos Santos. *See* Concord. *For* Puerto de Todos Santos, *see* Cojo.

Todos Santos y San Antonio [Santa Barbara Co.] 'All Saints and St. Anthony' is the name of a land grant dated Aug. 28, 1841. A *cañada nombrada Todos Santos* 'valley named All Saints' is mentioned on July 25, 1834 (DSP Ben. Mil. 79:106), and Cajón de Todos Santos (probably Harris Canyon) is shown on a *diseño* of 1840. Todos Santos was used elsewhere as a place name but does not seem to have survived. *See* Concord; San Antonio.

Toiyabe (toi yä′ bē) **National Forest** is based in Nevada but includes land in Mono and Alpine Cos. The name is from Shoshone *toyapin* 'mountain' (McLaughlin), or from a similar form in some other Numic language. Toiyabe Mountains are recorded in 1867; in 1908 the forest reserve in Nevada was named after the mountains. The name became a California place name in 1946 when the northern part of Mono National Forest was incorporated into Toiyabe National Forest. The BGN decided for the spelling Toyabe in 1908 but later reversed its decision to conform with local usage.

Tokopah (tō′ kə pä): **Valley, Falls** [Sequoia NP]. The upper valley of the Marble Fork of Kaweah River was named by former Superintendent John R. White and Col. George W. Stewart. The word may be Yokuts for 'high mountain valley' (Browning 1986).

Tolay (tō′ lā) **Creek** [Sonoma Co.]. The tidal channel was named after Tolay Lake, now drained. Padre José Altimira records in his diary on June 27, 1823: *Laguna de Tolay así llamada del capitán de los Indios* 'Tolay Lake so called for the chief of the Indians' (Arch. Arz. SF 4:2.28).

Tolenas (tə lē′ nəs): **Creek**, town [Solano Co.]. The Tolenas and Tolenos are mentioned in the records of baptisms of Indians at Mission San Francisco Solano in 1824 and later. On Mar. 4, 1840, the name appears in the Tolenas grant. In 1850 José Berreyesa testified in the land grant case that the creek was named after the Indian (River Patwin) village on its banks (*WF* 6:373).

Tollhouse [Fresno Co.]. In stagecoach days, toll was collected at this point on the road that led to the Saver and Thorne mines. When the county took over the road, the toll was abolished but the name was kept (M. H. Yancey). Toll House [Napa Co.], Toll Gate Creek [Plumas Co.], and Tollgate Canyon [Fresno Co.] also recall the days when building toll roads was a profitable business. In Nevada Co. there is a Tollhouse Lake northwest of Donner Pass.

Tollones, Arroyo de los. *See* Toyon.

Tolowa. *See* Talawa.

Toluca (tō lo͞o′ kə, tə lo͞o′ kə) **Lake** [Los Angeles Co.]. North Hollywood was called Toluca when the town was founded in 1888. Toluca post office is listed from May 26, 1893, to Oct. 17, 1906. This is probably a transfer name from the city of Toluca, southwest of Mexico City; the name is originally *Tolocan* in Aztec. *For* Toluca [Napa Co.], *see* Tulucay.

Tom, Mount [Inyo Co.]. Thomas Clark, a resident of the now-vanished town of Owensville, is credited with having made, in the 1860s, the first ascent of the peak, which was subsequently named for him (Farquhar).

Tomales (tə mä′ ləs): **Bay, Bluff, Creek, Point**, town [Marin Co.]. These geographical names (as well as Mount Tamalpais, *q.v.*) were given because the features were in the territory of the Tamal Indians, the name of which appears in the baptismal records of Mission Dolores as early as 1801 (Arch. Mis. 1:80 ff.). The tribal name is from Coast Miwok *támal* 'west, west coast' (Callaghan); there is no connection with the Mexican food known as *tamales*. The bay was discovered by Vizcaíno's expedition in Jan. 1603 and based on the assumption that it was a river, was named Río Grande de [San] Sebastián (Wagner, p. 419). It was later known to navigators and cartographers by a variety of names. Tamales as a place name is mentioned on Sept. 26, 1819; Padre Amorós of Mission San Rafael reports that he had baptized about a hundred natives, *todos de un rumbo llamado los Tamales* [these words are underscored in the MS], *que son unos esteros que comunican con el mar de la Bodega* 'all from the region called the Tamales, which are several estuaries that communicate with the ocean at Bodega' (Arch. Arz. SF 3:2.552). The name, spelled both Tamales and Tomales, appears in the titles of three land grants: Punta de los Reyes or Cañada de Tamales, Mar. 17, 1836, and June 8, 1839; Tomales y Baulenes, Mar. 18, 1836; Bolsa de Tomales, June 12, 1846. The tip of the peninsula formed by the bay is called Point Tomales by Tyson (1850:18), and Tomales Point on Gibbes's map of 1852. The town was settled by John Keys and Alexander Noble in 1850, and Tomales post office was established on Apr. 12, 1854 (Co. Hist.: Marin 1880:406, 408). Tomales Bay, Creek, Point, and the settlement are shown on Eddy's map.

Tombstone, The [Fresno Co.]. The shape of the mountain on the South Fork of San Joaquin River suggested the name, which was probably applied by the USGS when the Mount Goddard quadrangle was mapped in 1907–9. The name occurs in Tombstone Mountain [Shasta Co.], where marble is found; in Tombstone Ridge and Creek [Fresno Co.]; and probably elsewhere in local usage.

Tom Creek [Modoc Co.]. Named for Tom Cantrell, who settled here with his brothers in the 1870s (Irma Laird).

Tom Dye Rock [Lake Co.]. Tom Dye, a resident of Middletown, who had killed a man in 1878, used this rock as his hideout before giving himself up (Mauldin).

Tomka (tom′ kə) **Valley** [San Diego Co.]. Probably of Luiseño origin, but not clear.

Tomki (tôm′ kī) **Creek** [Mendocino Co.]. From Northern Pomo *miṭhóm kháy*, lit. 'splash valley', originally the name of Little Lake Valley; but it has been applied by whites to an entirely different creek and valley (Oswalt).

Tom Martin: Creek, Peak [Siskiyou Co.]. Probably named for a member of an Indian family that lived in the district (G. D. Gleason).

Tom's Creek [Tehama Co.]. *See* Thomes Creek.

Tontache, Lake. *See* Tule: Tulare Lake; Tache.

Toolwass. *See* Toowa.

Toomes Creek [Tehama Co.]. Named for Albert G. Toomes, a native of Missouri, who came to California in 1841 and was grantee of the Río de los Molinos grant, through which the stream flows. The rancho is shown as Rancho de Toomes on Bidwell's 1844 map (*see* Thomes Creek).

Toorup, Mount. *See* Turup Creek.

Toowa (to͞o′ wä) **Range** [Tulare Co.]. The name, of unknown origin, is shown on the Olancha atlas sheet of 1907, but apparently not on older maps. A similar name, Toolwass [Kern Co.], is listed for a post office in 1898.

Top. This generic term is frequently used for round or broad-topped orographic features that stand out in a range of mountains or hills. Most elevations called "Top" are found in the Sierra Nevada and the Coast Ranges of central California, between Sierra and San Luis Obispo Cos. About a third of them are Round Tops; another favorite combination is Bald Top. Monterey Co. has a Quail Top and a San Martin Top; Plumas Co., a Lava Top; Fresno Co., a Long Top; Madera Co., a Redtop; and Calaveras Co., a plain Mountain Top.

Topanga (tō pang′ gə, tə pang′ gə) **Canyon**, post office, **Beach, Park** [Los Angeles Co.]. The name of the canyon was preserved through the Topanga Malibu land grant, dated July 12, 1805. The origin is the Gabrielino place name *topa'nga*, containing the place suffix *nga* (Munro). Serro (Cerro) de Topango is mentioned in 1833 (Carrillo Docs., p. 70), and Punta de Topanga is shown on a *diseño* of the Boca de Santa Monica grant (1839).

Topatopa (tō pə tō′ pə): **Bluff, Mountains** [Ventura Co.]. Named after a rancheria of Chumash Indians near Ojai, also given as *topotopow* and *si-top-topo* (Kroeber). The inscription Top Top is shown on a *diseño* (1840 or 1841) of the Ojai grant. M. S. Beeler proposed an etymology in terms of Barbareño Chumash *tïp* '*monte*, brush'; thus *tïp tïp* would be 'brushy place'.

Topaz: Lake, post office [Mono Co.]. The post office was established on the Kirman and Rickey Ranch about 1885 and was named after the artificial lake, probably so called to indicate the clearness of the water.

Tophet (tō′ fit) **Springs** [Lassen NP]. The Hebrew word of uncertain meaning, often used to designate 'hell' or a place likened to it, was applied to the hot springs 3 miles southwest of Mount Lassen (*see* Bumpass Hell).

Topo is Spanish for 'mole', in California also applied to the gopher. **Topo** (tō′ pō) **Creek** [Monterey, San Benito Cos.] was so named about 1865, presumably because of the many burrowing animals (E. W. Palmtag).

Topock (tō′ pok) [San Bernardino Co.]. From Mojave *tuupák*, derived from the verb *tapák-* 'to drive piles' (Munro). The place was formerly called Red Rock, or Mellen for Jack Mellen, captain of a river steamboat.

Top Top. *See* Topatopa.

Tormey [Contra Costa Co.]. Named for John Tormey and his brother Patrick, who bought a part of the Pinole grant in 1867 (Co. Hist.: Contra Costa 1882:682–83).

Toro. The Spanish word for 'bull' was frequently applied to places and has survived in several counties in the southern half of the state. **Toro** (tôr′ ō) **Creek; Mount Toro** [Monterey Co.]: A place called El Toro Rabón 'the bob-tailed bull' is mentioned by Font on Mar. 10, 1776 (*Compl. diary*, p. 289), and appears repeatedly as El Toro in documents of the following decade. The *diseño* (1834) of the land grant El Toro shows an Arroyo del Toro, which is now Toro Creek. The place El Toro is shown on Wilkes's map of 1841. The mountain was apparently named by the Whitney Survey (Hoffmann's map, 1873). **El Toro** [Orange Co.]: A *parage del Toro* is mentioned in a decree, dated July 12, 1838 (DSP Ang. 2:18). It was so called "because there was a nice tame bull there" (Boscana, p. 217). The old name was revived for the post office in 1888, when the Post Office Dept. declined to accept the proposed name, Aliso, because of its similarity to Alviso. It was spelled Eltoro until 1905, when it was officially changed to the present form at the insistence of Z. S. Eldredge. The name is also found in Santa Clara, San Luis Obispo, Riverside, and Santa Barbara Cos. but in some instances may have been applied in American times. **Toro Indian Village, Canyon, Creek**, and **Peak** [Riverside Co.]: The Cahuilla Indian village was referred to as "Toro's" in 1856, also spelled 'Torros"; it was subsequently said to be named after an Indian chief called Toro (Spanish 'bull') and also referred to as El Toro, El Torro, Los Toros, Los Torros, and Torres. (*Torres* is of course a common Spanish surname.) When the present Torres-Martinez Reservation was established in 1876, for Indians from Toro Village, it was called Torres (Gunther).

Torote (tō rō′ tā) **Canyon** [San Diego Co.]. The term is Mexican Spanish for a desert plant, the elephant tree (*Bursera microphylla*).

Torquay (tôr′ kā) [San Mateo Co.]. Probably named after the famous watering place in Devonshire, England (Wyatt). Torqua Springs on Santa Catalina Island may have been named for the same reason.

Torquemada (tôr ki mä′ də), **Mount** [Los Angeles Co.]. The peak on Santa Catalina Island was probably named for Juan de Torquemada, whose *Monarquía Indiana* (Madrid, 1615) contained Father Antonio de la Ascensión's account of the Vizcaíno expedition, which named Santa Catalina Island (S. K. Gally).

Torrance [Los Angeles Co.]. Planned as a model

city by Frederick Law Olmsted, the great landscape architect, and named in 1911 by the owner, Jared S. Torrance, a financier and philanthropist (Santa Fe).

Torre (tôr′ ē) **Canyon** [Monterey Co.]. For José de la Torre, who claimed land here in 1891 (Clark 1991).

Torres-Martinez Indian Reservation [Riverside Co.]. *See* Toro: Toro Indian Village.

Torrey Pines Park [San Diego Co.]. The grove was set aside as a reserve to protect the rare pine, identified in 1850 and named *Pinus torreyana* for Prof. John Torrey (1796–1873) of Columbia University (Drury).

Tortuga (tôr too̅′ gə) [Imperial Co.]. The Spanish name for 'turtle, tortoise' was applied shortly before 1900 to the Southern Pacific station on the Yuma division. There is a Cañada Tortuga in Santa Barbara Co.

Towalumnes, Río de los. *See* Tuolumne River.

Tower Peak [Yosemite NP]. The Whitney Survey named that peak after the first ascent was made by Charles F. Hoffmann and his party in 1870. The name Castle Peak had been given to it more than ten years before by George H. Goddard but was transferred by mistake to a rounded peak about 18 miles away.

Towle [Placer Co.]. The name was applied to the Southern Pacific station in the 1880s. It is shown as Towles on the Mining Bureau map of 1892. Allen and George Towle were early lumbermen on the Dutch Flat divide (Co. Hist.: Placer 1924:194).

Towne Pass [Inyo Co.]. Named in 1860 by Darwin French for Capt. Towne, a member of a party that crossed Death Valley in 1849. It is mentioned in the *Sacramento Daily Union*, July 31, 1861, as Town's Pass, and this form is shown in Wheeler's atlas (sheet 65-D). The pass is sometimes called Townes or Townsend Pass.

Townsend Flat [Shasta Co.]. Named for Nathan A. Townsend, who developed mining claims near Briggsville; he built the Townsend Dam and Ditch in the 1850s (Steger).

Townsend Pass. *See* Towne Pass.

Township. Although the word is no longer a current geographical term in California, it is used in the U.S. surveys of public lands as the name of a territorial unit comprising thirty-six sections of one square mile each. In California there are three township base lines and meridians: Mount Diablo, San Bernardino, and Humboldt. Each township has a double number: the township number (T) north and south of the base line, and the range number (R) east and west of the meridian.

Toyabe. *See* Toiyabe.

Toyon (toi′ on). The name of this beautiful shrub, also known as Christmas berry and California holly (*Heteromeles arbutifolia*), is used for the names of a number of places, including two communities [Calaveras Co., Shasta Co.]. The California Spanish word *toyón* is from Costanoan *totčon* (Harrington 1944:37). An Arroyo de los Tollones is shown on a *diseño* of Rancho Cañada de los Osos y Pecho y Yslai [San Luis Obispo Co.].

Trabuco (trə boo̅′ kō): **Canyon, Creek, Highlands, Peak** [Orange Co.]. The name, meaning 'blunderbuss' (an archaic type of gun), was given to the place by soldiers of the Portolá expedition in July 1769, "because at this place, where there is a small arroyo, they lost a blunderbuss" (Font, *Compl. diary*, p. 188). Sierra del Trabuco, the Santa Ana Mountains southeast of Santa Ana, is mentioned by Font on Dec. 30, 1775. The name El Trabuco for a *paraje*, the creek, and the peak is found repeatedly in Spanish and Mexican documents. On July 31, 1841, the name was applied to a land grant. Arroyo del Trabuco is shown on a *diseño* of Rancho de la Nación, 1843.

Tracy [San Joaquin Co.]. When the Southern Pacific line from Berkeley reached the place on Sept. 8, 1878, the station was named for Lathrop J. Tracy, an official of the railroad.

Traer Agua (trä âr′ ä′ gwä) **Canyon** [San Bernardino Co.]. The name indicates the aridity of the box canyon, since the Spanish phrase means 'to carry water'.

Trafton Mountain [Mono Co.]. For George G. Trafton (1907–76), a longtime resident of the area (BGN, 7804).

Tragedy Springs, Creek [El Dorado, Amador Cos.]. Bigler records on July 20, 1848, that they had buried three of their scouts murdered by Indians: Daniel Browett, Ezrah Allen, Henderson Cox. The men had gone in advance of the party of Mormons in search for a pass and were killed probably on June 27: "We called this place Tragedy Springs."

Tranca. The Spanish word for 'crossbar' was often used in the plural to designate a barrier put up for protection. Las Trancas, erected about 1841 by the Vallejos at the head of the tidewater on Napa River to prevent cattle from crossing at low tide, was a well-known landmark before the Gold Rush. The name Trancas y Jalapa was given to a part of Salvador Vallejo's grant, dated Sept. 21, 1838. Paso de las Trancas was the alternate name of the Yajome grant [Napa Co.], dated Mar. 16, 1841. Agua Puerca y las Trancas grant [Santa Cruz Co.] is dated Oct. 31, 1843. **Trancas** (trang′ kəs): **Creek, Canyon** are in the Santa Monica Mountains [Los Angeles Co.]. **Los Trancos Creek** [Santa Clara Co.]: On a map of the Corte de Madera grant, this creek is shown as Strancos Creek, apparently a corruption of the original name.

Tranquility [Fresno Co.]. This soothing name, which was used in several places in the East, was applied to the station by the Southern Pacific when the Ingle-Hardwick branch was built in 1912. The name of the post office, listed in 1912, is spelled with a double *l*.

Tranquillon (trang kwil′ yən) **Mountain** [Santa Barbara Co.]. The prominent landfall for making Point Arguello was used by the Coast Survey as a triangulation station and was simply called Arguello after the point. In 1873 the present local name was applied by William Eimbeck of the Coast Survey: "Arguello station is a misnomer; it is not at Arguello Point, and is on a mountain the name of which (Indian) is Tranquillon (Tran-quel-yon), and the pronunciation of which would have one think it was Spanish, but it is not" (Lawson, Mar. 28, 1883). The possible Chumash origin has not been confirmed.

Trap Creek [Shasta Co.]. The tributary of the Pit River was probably so named because the Indians had fish traps in it (Steger). A connection with the "traps" that gave the name to the Pit River is possible.

Traver (trā′ vər) [Tulare Co.]. Founded in 1884 and named for Charles Traver, of Sacramento, who was interested in a land and canal development project there.

Travertine (trav′ ər tēn): **Rock** [Imperial Co.], **Spring** [Death Valley NP]. These were so named because of the incrustation of travertine, a crystalline calcium carbonate formed by deposition from the waters of the spring. Other features in the desert regions that are covered with travertine are sometimes called "Coral," e.g. Coral Reef south of Indio.

Travis Air Force Base [Solano Co.] was named for Brig. Gen. Robert F. Travis, who died in Korea in 1950 (Teresa Russell).

Treasure Island [San Francisco Co.]. The name was applied to the 400-acre artificial island built by U.S. Army engineers in San Francisco Bay, adjacent to Yerba Buena Island and used by the 1939–40 Golden Gate International Exposition. "The name was selected in the autumn of 1936 because it perfectly expressed a glamorous, beautiful, almost fabulous island that would present the treasures of the world during the 1939 World's Fair. It was no direct attempt to capitalize upon Robert Louis Stevenson, although the fact that he had made 'Treasure Island' a household word was a factor in their choice" (W. L. Wright).

Treasure Lakes [Inyo Co.]. The "treasure" consists of golden trout that were planted in the lakes. The name was approved by the BGN in 1938.

Trench Canyon [Mono Co.]. A narrow canyon named in 1884 by I. C. Russell of the USGS (Maule).

Tres Ojos de Agua [Santa Cruz Co.]. The name, meaning 'three springs,' was given to a land grant before Mar. 18, 1844.

Tres Pinos (trās pē′ nōs): town, **Creek** [San Benito Co.]. The name of the town was not applied because of the presence of 'three pines'; it is a transfer name from a nearby community. It was originally given to the settlement on the Ciénega de los Paicines grant, probably because there were three pines at the site. Lapham and Taylor's map of 1856 shows, doubtless from a mistake of the mapmaker, two places called Tres Pinos, neither in the right location. Goddard's map of 1860 shows the name at the site of modern Paicines (*q.v.*). It is mentioned by Brewer on July 21, 1861 (p. 135), as a ranch 15 miles from San Juan, exactly the distance from San Juan to Paicines. When the plan for a railroad from Gilroy via Pacheco Pass to San Joaquin Valley was projected, the name Tres Pinos was appropriated by the Southern Pacific for the station, which was reached on Aug. 12, 1873, and remained the terminus of the uncompleted line. Tres Pinos post office is shown on Hoffmann's map, and the locality formerly called Tres Pinos appears as Los Paicines Rancho House. Both Paicines and Tres Pinos are listed as post offices in 1880. Tres Pinos Creek is shown as Arroyo del puerto del Rosario on a *diseño*, and as Arroyo del Rosario on American maps until the Mining Bureau map of 1891 applied the new name. The pronunciations (tres pē′ nəs, trās . . .) have been noted; *see* Pinos (cf. Shafer, pp. 243, 246).

Trevarno [Alameda Co.]. Named after the home of George Bickford in Cornwall, England, by the Coast Manufacturing and Supply Company, which made the safety fuses invented by William Bickford, George's father.

Tri-Dam [Stanislaus Co.]. The $52 million irrigation project on the Stanislaus River was constructed by the Oakdale irrigation district and the South San Joaquin irrigation district in conjunction with the Pacific Gas and Electric Company, and was dedicated on June 15, 1957. It was named for the three principal dams: the Donnells, the Beardsley, and the Goodwin Dams (*see* Donnells Reservoir; Goodwin Dam).

Trigo (trē′ gō) [Madera Co.]. The Spanish term means 'wheat'.

Trimmer Spring [Fresno Co.]. Named in memory of Morris Trimmer, the first settler and owner of the place (Co. Hist.: Fresno 1956).

Trinidad (trin′ i dad): **Bay, Head**, town, **Harbor** [Humboldt Co.]. The Spanish term refers to the Holy Trinity. The Bruno de Hezeta expedition entered the bay on June 10, 1775; they took possession on June 11, and named it Puerto de la Trinidad because it was Trinity Sunday (Wagner, p. 419). The name appears on the maps of Vancouver, Humboldt, Wilkes, and Duflot de Mofras. The town was founded in 1850. Both Trinidad Bay and City are shown on the Coast Survey chart of 1850. (Another Trinidad City near Red Bluff is recorded on Gibbes's map.) The name was repeatedly used in place naming in Spanish times.

Trinity: River, County, Center, Mountains, Alps, National Forest. The name Trinity River owes its origin to an error. In late Mexican and early American times, it was assumed by many people that Trinidad Bay ranked next to San Diego and San Francisco Bays as the third great California port. When Pierson B. Reading came upon the river in 1845, he gave it the name Trinity, the English version of Trinidad, in the mistaken belief that the stream entered Trinidad Bay (*q.v.*). The name gained wide popularity after Reading discovered gold at Reading's Bar [Trinity Co.] in July 1848. Trinity Mountains are mentioned in 1850. The county, one of the original twenty-seven, was named on Feb. 18, 1850; Trinity Center was settled by miners from San Francisco in 1850; the post office is listed in 1858. Trinity National Forest, comprising areas in Humboldt, Shasta, Tehama, and Trinity Cos., was created and named in 1905. *See* Humboldt Bay. **Trinity Alps** is the name of a post office in Trinity Co.; it is also the generally accepted name for the mountains north of Weaverville. **Trinity Bally** [Trinity, Tehama Cos.] contains Wintu *buli* 'mountain'; *see* Bally.

Tripas (trē′ pəs) **Canyon** [Ventura Co.]. The Spanish term means 'guts, intestines'.

Triunfo (trī un′ fō): town, **Canyon, Pass** [Ventura Co.]. The Portolá expedition camped probably in what is now Potrero Valley on Jan. 13, 1770, and Crespí called the valley El triunfo del Dulcísimo Nombre de Jesús 'The triumph of the sweet name of Jesus'. Next day Crespí called an Indian rancheria in what is now Russell Valley El triunfo de Jesús. The abbreviated form El Triunfo appears in mission and land grant papers. The settlement is shown on maps since the early 1850s. A post office with the name Triumfo was established on Aug. 27, 1915; the name was changed to the present spelling on Apr. 21, 1917.

Trocha [Tulare Co.]. The Spanish term is translated as 'a narrow path across a high road'.

Trojan Peak [Inyo Co.]. The peak southeast of Mount Tyndall was named in honor of the athletic teams of the University of Southern California (BGN, May 1954).

Trona (trō′ nə) [San Bernardino Co.]. The name, referring to a mineral consisting of sodium carbonate and bicarbonate, was applied to the post office on Mar. 27, 1914, and to the railroad terminal from Searles to the lakebed in 1916. In 1908 Alfred de Ropp, a German engineer from the Russian Baltic provinces, succeeded in making soda and potash from the brine of Searles Lake.

Trout. Among California place names given for species of fish, Trout stands next to Salmon in popularity. There are more than twenty-five Trout Creeks, and a few Trout Buttes, Camps, and Meadows. *See* Golden Trout Creek.

Troy. Since 1789, when Troy in the state of New York was named, the name of the ancient city in Asia Minor has always been a favorite American place name; it is represented in California by a Southern Pacific station in Placer Co., and a Santa Fe station in San Bernardino Co.

Truckee (truk′ ē): **River, Pass**, town [Placer, Nevada Cos.]. The name is that of a Northern

Paiute leader. His granddaughter, Princess Winnemucca, reported that John C. Frémont had given him that name (Lekisch). An account of how the name was given to the river is based on the recollections of Moses Schallenberger, a member of the Stevens-Murphy-Townsend party (*San Jose Pioneer*, Mar. 15, Apr. 15, 1893): "Finally [near Humboldt Sink], an old Indian was found, called Truckee, with whom . . . Greenwood talked by means of signs diagrams [*sic*] drawn on the ground. From him it was learned that fifty or sixty miles to the west there was a river that flowed easterly from the mountains. [Three of them explored ahead with Truckee and found the river. Early in Oct. 1844] they reached the river which they named the Truckee, in honor of the old Indian chief, who had piloted them to it." An article in the *Sacramento Bee* of Aug. 26, 1880, adds a touch of piquancy to Schallenberger's account: "In traversing this region, the Indian told them of a rapid river that flowed from one great lake to another. The party did not reach this river as soon as they expected, and they began to look upon 'Truckee's river' as a river of the mind . . . When at last they reached the stream . . . they had already named it. From 'Truckee's river' to 'the Truckee' was a transition natural and easy." The names seem to have been generally used after 1846 for the river and the pass, sometimes spelled Truckey or Truchy; but Frémont's name Salmon Trout River appears on Bancroft's map as late as 1858. Donner Lake was known as Truckee Lake until the Whitney Survey changed the name. The town came into existence when the Central Pacific surveyed the route across the pass in 1863–64; it was first called Coburn Station for the owner of the saloon there. After the fire of 1868, the station was renamed after the river.

Trujillo (trōō hē′ yō) **Creek** [San Diego Co.]. For Gregory Trujillo, an Indian who raised sheep near here (Stein).

Trumble Lake [Mono Co.]. The land adjacent to the lake was patented in 1880 to John S. Trumble. The name is misspelled Trumbull on the Bridgeport atlas sheet (Maule).

Trumbull Peak [Mariposa Co.]. Probably for Lyman Trumbull (1813–96), U.S. senator from Illinois, who may have visited the park (Browning 1988).

Tuba Canyon [Death Valley NP]. From Panamint *tïpa* 'pine nut' (McLaughlin).

Tub Spring [Trinity Co.]. So named because of a barrel sawed in two and set in the spring (C. D. Willburn).

Tucalota (tōō kə lō′ tə) **Creek** [Riverside Co.]. According to Gunther, a distortion of Mexican Spanish *tecolote* 'owl' (*q.v.*), from Aztec *tecolotl*.

Tucho [Monterey Co.]. A piece of land called *El Tucho* is mentioned on Feb. 21, 1833 (Arch. Mont. 7:56), and a rancho El Tucho on Mar. 19, 1840 (Docs. Hist. Cal. 1:413), and later. The name appears to have been quite important: it was given to two confirmed land grants, dated June 21, 1841, and Dec. 4, 1843, and appears in eight other land grant cases. The name is spelled El Tuche in at least one source (Docs. Hist. Cal. 3:94).

Tucker Flat [El Dorado Co.]. Named for a stockman who took up land at the Upper Truckee River and built a road to the pass through which U.S. Highway 50 now crosses the range (E. F. Smith).

Tucker Mountain [Tulare Co.]. Named for a homesteader on the east side of the mountain (A. L. Dickey).

Tucki (tuk′ ī, tuk′ ē) **Mountain** [Death Valley NP] is not labeled on older maps but was called Tucki or Sheep Mountain by Mendenhall in 1909. The source may be Shoshone *tukku* 'mountain sheep', or a similar word in some other Numic language (McLaughlin).

Tueeulala (tōō ē lä′ lə) **Falls** [Yosemite NP]. The Indian name was given by John Muir in 1873 as Tu-ee-u-lá-la (Browning 1988). Perhaps from Miwok *tï"ele-la* 'shallow place' (Callaghan).

Tufa Falls [Tulare Co.]. The geological term refers to porous limestone deposited by springs or the like.

Tujunga (tə hung′ gə): **Canyon, Creek, Valley, Wash**, town [Los Angeles Co.]. An Indian rancheria, Tuyunga, is mentioned in Aug. 1795 (Arch. MSB 2:12 ff.). *La Sierra llamada* ('called') *Tujunga* is recorded in 1822 (ibid., 3:237) and *un arroyo . . . conocido con el nombre de* ('known by the name of') *Tujunga* on July 25, 1834 (DSP Ben. Mil. 79:88). The name was given to a land grant dated Dec. 5, 1840. The post office, first spelled phonetically as Tuhunga, is listed in 1887. The name is probably Gabrielino (with place suffix *-nga*), but its original meaning is not known. BGN, list 6802, differentiated Big and Little Tujunga Creeks and Canyons.

Tujunta, Monte de. *See* Tajanta.

Tulainyo (tōō lə in′ yō) Lake [Sequoia NP]. The

name was coined in 1917 by R. B. Marshall because the lake almost touches the boundary of Tulare and Inyo Cos. (David Brower).

Tule (to͞o′ lē); **Tular**. The common California word for the cattail or bulrush (*Scirpus* sp.) and similar aquatic plants is derived from Mexican Spanish *tule*, from Aztec *tollin;* the place where they grow is called a *tular*. Today the names Tule, Tulare, and the diminutive Tularcitos are used as specific terms in more than fifty place names—chiefly with Lake, Lagoon, Slough, and Creek, but also with Peak, Mountain, and Hill. Tuleburg was a former and very appropriate name for Stockton. **Tule River** [Tulare Co.], and **Tulare** (to͞o lâr′ ē, to͞o lâr′) **Basin, County**, city [Tulare Co.], and **Lake** [Kings Co.] are major exemplars. Before the reclamation of the San Joaquin Valley, much of this area was marshland covered by a luxurious growth of reeds. The first man to see the valley in the wet season was probably Fages in 1772. He called it the Llano de San Francisco and described it as a great plain, "a labyrinth of lakes and *tulares*" (*CHSQ* 10:218–19). On Mar. 4, 1776, while crossing the Cuesta Pass north of San Luis Obispo, Font speaks of *los tulares* as though it were an established name: "another range which we kept on our right [Mount Diablo Range], and behind which are the tulares" (*Compl. diary*, p. 274). San Joaquin Valley or parts of it are repeatedly mentioned in later records as Valle de los Tulares and Llano del Tular, and Narváez's map of 1830 shows "Ciénegas ['marshes'] o Tulares" in the valley. After Smith's and Wilkes's explorations, the name became restricted to the lower part of the valley. Most American maps record one or two Tule Lakes, the identity of which is often uncertain because of the peculiar hydrographic conditions in the basin. The Indian name of what is now Tulare Lake was *chintache* (Wilkes 5:157; map, 1841); *see* Tache. The Frémont-Preuss map of 1848 calls it Laguna de los Tulares, and other early American maps use this name, or the old Indian name with various spellings. The name Tulare Lake is recorded in the *Statutes* of 1855 (p. 162) but was current before that date. Tule River had been called Río de San Pedro by an expedition in Apr. 1806 (Arch. MSB 4:1 ff.) and appears as Tule River or San Pedro on Derby's map of 1850. Tulare Co. was created and named on Apr. 20, 1852. The disease tularemia was so named because it was first identified in Tulare Co.; it has nothing to do with *tulares* as such. When the Indian reservation was set aside, it was named after Tule River, on which it was first established. The city came into existence when the Southern Pacific reached the place in 1872 and named the station after the lake. **Tularcitos Creek** and **Ridge** [Monterey Co.]: Cañada de los Tularcitos is recorded in 1822 (Arch. MSB 3:296), and a place called Tularcitos is mentioned, together with Laureles and Chupines, in a report of 1828 (Registro, 11 ff.). On Jan. 8, 1831, the name was given to a land grant. "This name is derived from a chain of small ponds, which are infested with tules and which are the source of Tularcitos Creek, main tributary of Carmel River" (W. I. Wilson). Another Tularcitos grant [Santa Clara Co.], dated Oct. 4, 1821, does not seem to have left further traces in modern geographical nomenclature. **Tule Lake** [Siskiyou Co.] was apparently the lake named in 1846 by Frémont for his friend Barnwell Rhett. At any rate, it appears on most maps as Rhett Lake until about 1900. Von Leicht–Craven (1874) have Tule or Rhett Lake, and the former is now the generally accepted name. **Tulelake** post office was named after the lake about 1900.

Tuledad (to͞o′ lə dad): **Canyon, Creek** [Modoc, Lassen Cos.]. Named for Tuledad Matney, early frontiersman, and later farmer and stockman; he ran cattle and horses here (W. S. Brown to O. M. Evans).

Tulucay (to͝ol′ ə kā) **Creek** [Napa Co.]. The name was preserved through the Tulucay land grant, dated Oct. 26, 1841. According to Barrett (*Pomo*, p. 293), the name is derived from Patwin *tu'luka* 'red,' and the names Tulkays and Ulucas were applied to inhabitants of an Indian village. The name Toluca (or Joluca) is mentioned in the records of Mission San Francisco Solano, and Tular 'place where tules grow' is shown on a diseño of the grant; these may have some connection with the name.

Tumco (tum′ kō) [Imperial Co.]. The name was coined in 1910 from the name of The United Mines Company. It is now a ghost town, although it once claimed a thousand inhabitants. The original name was Hedges (*Desert Magazine*, Sept. 1940).

Tumey (to͞o′ mē) **Gulch** [Fresno, San Benito Cos.]. Since the name was formerly spelled

Toomey, it is possible that the gulch was named for William Maloy Toomey of Kentucky, listed in the Great Register of 1878.

Tuna Canyon [Los Angeles Co.] does not refer to the fish, but rather is from Spanish *tuna*, the fruit of the prickly pear cactus. The word originally comes from the Taino Indian language of the West Indies.

Tunawee (tun′ ə wē) **Canyon** [Inyo Co.]. From Panamint *tïnapi* 'mountain mahogany'—pronounced something like (tun′ ə vē)—a species of *Cercocarpus* (McLaughlin).

Tune (to͞o′ nē, to͞o′ nə) **Creek** [Shasta Co.]. From Wintu *tune* 'forward, front' (Shepherd).

Tunemah (to͞o′ nə mä): **Trail, Lake, Pass, Peak** [Kings Canyon NP]: "The sheepherders frequenting that part of the country employed Chinese cooks. Owing to the roughness of the path they gave vent to their disgust by numerous Chinese imprecations. Gradually the most prominent settled itself on to the trail and it became known as 'Tunemah!'" (Elesa M. Gremke, *Sunset Magazine*, 6:139). The other features were named when the Tehipite and Mount Goddard quadrangles were surveyed in the 1900s. Chinese linguist colleagues in Hong Kong have identified "Tunemah" as probably reflecting Cantonese *diu2 nei5 aa3 maa1* (the numbers indicate tones), meaning 'Fuck your mother!'

Tungsten Hills [Inyo Co.]. In 1913 valuable tungsten deposits were discovered west of Bishop. Adolph Knopf, professor at Yale University, suggested the name in his report in USGS *Bulletin* 640-L.

Tunitas (to͞o nē′ təs) **Creek** [San Mateo Co.]. *Arroyo de las Tunitas* is shown on the *diseños* (about 1839) of the San Gregorio and Cañada Verde grants. *Tunita* is the diminutive of Spanish *tuna* (*see* Tuna Canyon)—the fruit of the *nopal*, or prickly pear cactus (genus *Opuntia*).

Tunnabora (to͞o nə bôr′ ə) **Peak** [Sequoia NP]. The first ascent was made by George R. Davis of the Geological Survey in Aug. 1905 (Farquhar). The name is probably from Panamint *tuu* 'black' plus *napatïn* 'canyon' (McLaughlin).

Tuolumne (to͞o ol′ ə mē): **River, County**, town, **Canyon, Meadows, Falls, Peak, Pass**. The name is that of Indians who once lived on the banks of the Stanislaus River, probably somewhere west of modern Knights Ferry. The origin is probably Central Sierra Miwok *ṭaawalïmni* 'squirrel place', from *ṭaawalï* 'squirrel' (Callaghan). This location would indicate that they were a branch of the Central Miwok. They were mentioned as Taulámne (also spelled Tahualamne) in Padre Muñoz' diary (1806) and as Taualames by Padre Viader on Oct. 23, 1810 (Arch. MSB 4:85 ff.). The river is shown by Frémont and Preuss on their map of 1845, labeled Río de los Merced by mistake; on their map of 1848 it appears as Río de los Towalumnes. The modern spelling was used on Derby's 1849 map, and for the county, one of the original twenty-seven, when it was created and named, Feb. 18, 1850; until 1860 various spelling variants are found on maps. Tuolumne Canyon and Meadows were named by the Whitney Survey; the falls are shown on Hoffmann's map, and the peak on Wheeler's atlas sheet 56-D. A Tuolumne City, founded in 1850 about 3 miles from the mouth of the river, was short-lived. The present town had its nucleus in Summerville, named for Franklin Summers, who settled there in the fall of 1854. To avoid confusion with Somersville in Contra Costa Co., the Post Office Dept. changed the name to Carters, for C. H. Carter, a pioneer merchant. In 1899 the station of the Sierra Railroad was named Tuolumne; the post office is listed in 1904.

Tupman [Kern Co.]. Named at a public meeting in 1920, for H. V. Tupman, from whom the Standard Oil Company purchased the land. Naval Petroleum Reserve No. 1 is situated nearby in the Elk Hills (Louise Stine). The post office is listed in 1922.

Turlock [Stanislaus Co.]. When the railroad reached the place in 1871, John W. Mitchell modestly declined to have the station on his property named for him and suggested the present name, after Turlough in the county of Mayo, Ireland.

Turn [Imperial Co.]. When the Southern Pacific line from Calipatria to Sandia was built in 1923–24, the name was given to the station because the railroad makes a sharp turn at this point.

Turnback Creek is a common western name for streams that, because of topography, seem to reverse themselves. The best-known California creek so called is the one in Tuolumne Co., the scene of a gold rush in 1856. However, according to A. H. Dexter, *Early days in California* (Denver, 1886), p. 134, this creek was named because a group of white and Cherokee miners were forced to "turn back" here after an undecisive fight with local Indians.

Turner Lake [Yosemite NP] was named for Henry Ward Turner, 1857–1937, a geologist of the U.S. Geological Survey, who pioneered some of the mapping of the park area (BGN, 1963).

Turntable Creek [Shasta Co.] was so named because there was a place to turn on the old road just north of the Pit River bridge (Schrader).

Turpentinom (tûr pən tī´ nəm) **Creek** [Shasta Co.]. From Wintu *čur-pantinom* (Pitkin, p. 103); the word may contain *čur-* 'spawn', *pan-* 'top', and *nom* 'west' (Shepherd), presumably with folk-etymological influence of English "turpentine."

Turret. There are several mountains in the state so named because of their turret-like formation; the best known are Turret Peak [Fresno Co.] and Turret Mountain [Modoc Co.].

Turup (to͞o´ rəp) **Creek** [Del Norte Co.]. For the Yurok village *turip* (Waterman, p. 235; Harrington); also spelled Tarup. A Mount Toorup is mentioned by Cleveland Rockwell of the Coast Survey in 1878.

Turwar (or **Turwer**) **Creek**. *See* Terwah Creek.

Tuscan Springs [Tehama Co.]. Borax was first discovered in California in the waters of these springs, Jan. 8, 1856, by Dr. John A. Veatch, who named them after Tuscany in Italy, an important source of borax. The springs had formerly been called Lick Springs.

Tustin [Orange Co.] was laid out in 1870 by Columbus Tustin, who named it for himself. A post office was authorized in 1872 under the name Tustin City (Brigandi).

Tuttle [Merced Co.]. The Santa Fe named the station for R. H. Tuttle, superintendent at Fresno, shortly after it purchased the line in 1900 (Santa Fe).

Tuttle Creek [Inyo Co.]. Named for Lyman Tuttle; he was the recorder of Russ mining district in 1862, one of the organizers of Inyo Co. in 1866, and county surveyor, 1860–72 (Robinson).

Tuttletown [Tuolumne Co.]. The town bears the name of Judge Anson H. Tuttle, who built the first log cabin in 1848 and became the first judge of the county. Some called the town Tuttleville; others, Mormon Gulch, for the Mormons who began mining there in 1848.

Tuxedo, The. *See* Mount: Mount Hermon.

Twain [Plumas Co.]. According to local tradition, the Western Pacific station was named in 1907 in memory of the author Mark Twain, the pen name of Samuel Clemens (R. Batha).

Twain-Harte [Tuolumne Co.]. The mountain resort was named in 1924 by Katurah F. Wood for two great writers of the California scene, Mark Twain (1835–1910) and Bret Harte (1839–1902) (A. L. Nevins).

Tweedy Lake [Los Angeles Co.]. The lake west of Elizabeth Lake was named for Robert Tweedy, a homesteader and the original owner (J. Hindman).

Twenty-Mile House. *See* Cromberg.

Twenty Mule Team Canyon [Inyo Co.]. The name was applied by the Pacific Coast Borax Company to commemorate the famous twenty-mule teams, which had been used for hauling borax before the railroad was built in 1915.

Twentynine Palms. *See* Palm.

Twin. It has long been customary to designate paired geographic features as Twins. California has more than two hundred places so named, chiefly Twin Peaks and Twin Lakes, but also numerous Twin Rocks, Meadows, Buttes, Sisters, Springs, and Sloughs. Sometimes each Twin has been given a special designation for the purpose of distinguishing it from its companion: Upper and Lower Twin Lake [Lassen NP], Big and Small Twin Lake [San Luis Obispo Co.], Big and Little Twin Gulch [Tuolumne Co.]. Occasionally Twins is used as a generic for orographic features: The Twins, Liebre Twins [Kern Co.]. The modifying adjective is also included in the names of a number of communities: Twin Bridges [El Dorado Co.], Twin Buttes [Tulare Co.], Twin Cities [Sacramento Co.], Twin Lakes [Santa Cruz Co.], Twin Oaks [San Diego Co.], Twin Peaks [San Bernardino Co.], Twin Rocks [Mendocino Co.]. **Twin Peaks** [San Francisco Co.] were not commonly so called until about 1890. In Spanish times the name was Las Papas 'the potatoes'; on Beechey's map (1828) and on some other non-Spanish maps, the peaks are labeled Paps, but the charts of the Coast Survey have Las Papas. Another Spanish name, Los Pechos de la Chola 'the breasts of the Indian girl', probably owes its origin to American romanticism. Kirchhoff (p. 14) in 1885 gives Twin Sisters as the commonly used name. For Twin Peaks [San Bernardino Co.], *see* Strawberry: Strawberry Flat.

Two Rock [Sonoma Co.]. The name can be traced to Mexican times, when the two big rocks were a prominent landmark called Dos

Piedras. They are shown on several *diseños* of the 1840s. When John Schwobeda settled there before 1854, he kept the Spanish name, which is shown on Eddy's map and as late as 1873 on Hoffmann's map of the Bay region. When a post office was established on Schwobeda's farm on July 17, 1857, the name was translated to Two Rocks. In 1874 it is listed as Two Rock, and this version was adopted when the post office was reestablished in 1915. Both Dos Piedras and Two Rock appear on recent maps.

Two Teats [Mono, Madera Cos.]. On older maps the two peaks San Joaquin Mountain and Two Teats appear as two breast-shaped elevations. However, the mountain now called Two Teats has actually two protrusions, one pointing east, the other west (Genny Schumacher).

Tyee (tī′ ē) **Lakes** [Kings Canyon NP] were named in 1936 for Tyee, a famed brand of salmon eggs (Browning 1986). The word originates in the Nootka language of Vancouver Island; it means 'chief' in Chinook Jargon, once used as a trade language among Indians and Whites in the Pacific Northwest.

Tyndall (tin′ dəl), **Mount** [Sequoia NP]. The name honors John Tyndall (1820–93), well-known British scientist and Alpinist. It was given by Clarence King and Richard Cotter of the Whitney Survey, who made the first ascent in July 1864: "I rang my hammer upon the topmost rock; we grasped hands, and I reverently named the grand peak Mount Tyndall" (King, *Mountaineering* 1872:75).

Typhoon Mesa [Siskiyou Co.]. The reason for the name is not known; it was established by BGN list 6702.

Tzabaco [Sonoma Co.]. A land grant dated Oct. 14, 1843, was given this name. It is probably the name of a Coast Miwok group, perhaps meaning 'sweat-people' (Callaghan).

U

Ubehebe (yo͞o bē hē′ bē): **Peak, Crater** [Death Valley NP]: The name is mentioned in the 1870s as the name of a mine, and it appears again as the name of a camp east of the peak during the boom of 1906–7. The Panamint name for the peak is in fact *wosa* 'burden-basket'; but local Indians suggest that the English name is from Owens Valley Paiute *hïbíbici* 'woman's breasts'—originally applied to the Wahguyhe Peaks (*q.v.*) in the Grapevine Range, as seen from above Ubehebe Crater (Fowler).

U-Fish (yo͞o′ fish) **Creek** [Siskiyou Co.] represents Karuk *yufísh-thuuf* 'salt creek', from Karuk *yúfish* 'salt'.

Uhlen (yo͞o′ lən) **Valley** [Nevada Co.] became well known when the new route of Highway 40 was built across the Sierra in the 1960s. It was apparently named for an early pioneer, whose identity could not be established.

Ukiah (yo͞o kī′ ə): **Valley**, city [Mendocino Co.]. On May 24, 1845, the name appears in the *expediente* of a land grant as Yokaya and Llokaya. (The term Jukiusme on Duflot de Mofras's map of 1844, evidently applying to the Indians north of Clear Lake, may be a variant.) The name is derived from Central Pomo *yó-qhaaya* 'south valley' (Oswalt 1980). The spelling Yokayo has also been used for the Indian rancheria near the town. The first white settler was Samuel Lowry in 1856, and in the same year Judge J. B. Lamar applied the present name to the settlement. The area was first known as Parker Valley but appears in 1858 under the name Ukiah in Hutchings' *Illustrated California Magazine* (3:148).

Ukonom (yo͞o′ kə nom): **Creek, Mountain, Lake** [Siskiyou Co.]. From the Karuk village *yuhnaam*, which may be from *yu-* 'downriver' plus *naam* 'a flat'.

Ulatis (yo͞o lat′ is) **Creek** [Solano Co.]. The Ululato Indians are mentioned by Chamisso in 1816 (Rurik, p. 89) as living north of the Suisun. They formed a division of the Patwin and are repeatedly mentioned, with various spellings, in mission records. The stream is shown as Arrollo de Ululatos on a *diseño* of the Capay grant. The present spelling was determined by a decision of the Geographic Board of Apr. 1, 1908.

Ulistac [Santa Clara Co.]. The name of a land grant dated May 19, 1845. It is from a Costanoan place name, with the locative case ending *-tak* added to *uli, uri* 'head, hair' (Callaghan).

Ullathorne (yo͞o′ lə thôrn) **Creek** [Humboldt Co.]. For T. A. Ullathorne, an early settler (Turner 1993).

Ulpinos [Solano Co.]. Los Ulpinos was the name of the land grant conveyed to John Bidwell, provisionally on July 26, 1844, and definitively on Nov. 20, 1844. The name is derived from that of the Indian tribe on the west side of the Sacramento just above its junction with the San Joaquin, mentioned as *nación Julpones* on Jan. 31, 1796 (PSP 14:15), and with various spellings in later years.

Umalibo. *See* Malibu.

Umpa (o͝om′ pə) **Lake** [El Dorado Co.]. Said to be an Indian name, of unidentified origin.

Umunhum (yo͞o′ mə num), **Mount** [Santa Clara Co.]. Perhaps from a Costanoan word for 'hummingbird', as suggested by Beeler 1954; the Mutsun form is *humuunya*, and the Rumsen is *ummun* (Callaghan). The peak is shown as Picacho de Umenhum and Umurhum on C. S. Lyman's maps of the New Almaden Mine (1848). Hoffmann in his notes spells the name Unuhum on Aug. 10, 1861, corrects it to Umunhum on Aug. 20, and records the pronunciation on Aug. 26: "oomoonoom." Brewer gives a similar version in his Notes on Sept. 6, 1861: "oomoonoon."

Uncle Sam. The colloquial name for the United States was a favorite name for mines; it is preserved in Uncle Sam Mountain in the Santa Lucia Range [Monterey Co.]. Mount Konocti [Lake Co.] was once called Uncle Sam Mountain; the nearby town of Uncle

Sam, named after the mountain, was later renamed Kelseyville (Feltman, p. 23). *See also* Kelseyville; Konocti.

Unicorn Peak [Yosemite NP] was named by the Whitney Survey in 1863 because the peak resembles the horn of the mythical animal: "A very prominent peak, with a peculiar horn-shaped outline, was called 'Unicorn Peak'" (Whitney, *Geology*, p. 427).

Union. Of the many idealistic and patriotic names for places in the United States, Union is the most popular. As the struggle for upholding the Union in the middle of the 19th century increased in intensity, the number of Unions, Uniontowns, and Unionvilles multiplied, especially in states where the Union was in danger, like Kansas and Missouri. In California it was a favorite name for mining towns; only one of these, however, has survived: Union Hill [Nevada Co.]. Today the name is found for communities or railroad stations in Napa, San Luis Obispo, San Mateo, and Santa Barbara Cos., as well as for several topographical features. **Union City** [Alameda Co.] has been recorded since 1851 and is probably the oldest Union name. *For* Uniontown [El Dorado Co.], *see* Lotus.

Universal City [Los Angeles Co.]. The post office was established on Apr. 24, 1915, and named after the Universal Pictures Company, which was organized the same year (R. W. Gracey).

University Park. *See* Palo: Palo Alto.

University Peak [Kings Canyon NP]. In 1890 Joseph N. LeConte named the peak north of Kearsarge Pass in honor of the University of California. On July 12, 1896, he transferred the name to the present University Peak when he and his party made the first ascent (Farquhar). *See* Gould, Mount.

Unnamed Wash [Imperial Co.]. Nameless features are often so designated by surveyors in their reports, but the army engineers who edited the Picacho Peak atlas sheet in 1943 did something unique when they created the name "Unnamed Wash."

Upland [San Bernardino Co.]. Named in 1902 by the county board of supervisors at the request of the citizens. The original tract, opened in 1887 by the Bedford brothers, had been called Magnolia Villa, and later became identified with North Ontario. The elevation is slightly higher than that of Ontario.

Upper Lake [Lake Co.]. The post office was established before 1867 and named Upper Clear Lake because of its proximity to the northern shore of Clear Lake. The name was changed to the present form in 1880 (*see* Lower Lake).

Urbita (ûr bē′ tə) **Springs** [San Bernardino Co.]. The name, coined from Latin *urbs* (city) plus Spanish diminutive *-ita*, to suggest the meaning 'little city', was given to a nearby settlement before 1899. When the springs were developed, Dr. S. C. Bogart named the place Urbita Springs Park. From 1926 to 1930 the name was Pickering Mineral Hot Springs (for the owner, Ernest Pickering), and since 1930 it has been Urbita Springs (Helen Luce).

Usal (yo͞o′ sôl): **Creek, Rock, Mountain**, town [Mendocino Co.]. From Northern Pomo *yoo-sal*, perhaps containing *yoo* 'south' (Harrington). Usal Mountain is mentioned by John Rockwell of the Coast Survey in 1878. The community came into existence when Robert Dollar formed the Usal Redwood Company and built one of his early mills there (Borden).

Usona (yo͞o sō′ nə) [Mariposa Co.]. When the post office was to be established in 1913, the residents met at a public meeting and decided to name their community with an acronym for the United States of North America. This rather ingenious way of creating a new euphonious name has remained unique, as far as gazetteers and postal guide tell the story. According to Stewart (p. 173), it was once proposed as a name for the United States.

Utopian colonies. Since the times of the Gold Rush, repeated economic experiments with cooperative, religious, and socialist colonies were made in the state. Some of these undertakings, like Anaheim and Fresno, were very successful after their initial cooperative phase had passed; all the utopian colonies failed after a short existence. Robert V. Hine has treated the stories of the colonies in *California's utopian colonies*, 1953. For the names of such colonies that remained as place names, *see* Colony, Kaweah, Llano.

Uva (yo͞o′ və) [Fresno Co.] is Spanish for 'grape'; the name was given in modern times. **Uvas** (yo͞o′ vəs) **Creek** [Santa Clara Co.] takes its name from the land grant Cañada de las Uvas, dated June 14, 1842. The word *uvas* 'grapes' was frequently used in Spanish times for places where wild grapes (genus *Vitis*) were found; Grapevine Canyon [Kern Co.] was once Cañada de las Uvas. Harter (*q.v.*) in Sutter Co. was formerly called Las Uvas.

V

Vaca (vak′ ə): **Valley, Mountains; Vacaville** (vak′ ə vil) [Napa, Solano Cos.]. These places perpetuate the name of the Vaca family, which came to California in 1841 from New Mexico, where Vaca is a well-known surname. Juan Manuel Vaca was co-grantee of the Los Putos or Lihuaytos grant, dated Aug. 30, 1845, on which the town is situated. The plat of the town was filed in Dec. 1851, and the township was created and named on Nov. 1, 1852. Vaca Valley and Mountains are mentioned in the *Statutes* of 1855. *For* Vaca in Solano Co., *see* Elmira.

Vaca (vä′ kə) **Canyon** [Contra Costa Co.]. Vaca ('cow') Creek is shown on a plat of Rancho Pinole in 1865.

Vade (vād) [El Dorado Co.]. The post office was established on Sept. 3, 1912, and named for a resident of Phillips whose name was Nevada, nicknamed "Vade" (Burt Perkins); *see* Little: Little Norway.

Valale, Arroyo. *See* Gualala.

Valinda [Los Angeles Co.]. Probably from the Spanish masculine noun *valle* 'valley' plus the feminine adjective *linda* 'pretty'.

Valle. The generic term *valle* 'valley' was frequently used in Spanish times and appears in the names of more than fifteen land grants (*see* Posita). Except for Arroyo Valle [Alameda Co.] and the names of a few subdivisions, the term is no longer found in California toponymy. Its diminutive, however, is still actively used. **Vallecito** (val ə sē′ tō): station, **Wash, Mountains** [San Diego Co.]: The little valley with several cold and warm springs was a convenient stopping place on the trail from San Diego to the Colorado. The Indian rancheria of Hawi was in the valley, and it is likely that San Diego Mission maintained some sort of station there. In 1846 Kearny's "Army of the West" camped here, and the place was known to the soldiers as "Vallo Citron" and "Bayou Cita" (*CHSQ* 1:145). The proper spelling is used by Whipple on his map of 1849 and by the men of the Pacific Railroad Survey. In 1858 the name was given to a station of the Butterfield Overland Mail. *See also* San Felipe. **Vallecitos de San Marcos** [San Diego Co.] is the name of a land grant dated Apr. 22, 1840; it is 45 miles west of the Vallecito station. **Vallecito** [Calaveras Co.]: Daniel and John Murphy discovered gold here in 1848; later the place became known as Murphy's Old Diggings, to distinguish it from Murphy's New Diggings, now the town of Murphys (J. A. Smith). The new name, spelled Vallicito, is mentioned in the *Statutes* of 1854 (p. 222), and the post office was established as Vallicita on Aug. 17, 1854. The place was known under this name for almost a century, until the Post Office Dept. restored the original Spanish spelling on Oct. 1, 1940. **Vallecitos Creek**, valley, and settlement [San Benito Co.]: Vallecitos Canyon is mentioned in Hoffmann's notes of July 1861. By decision of Mar. 6, 1912, the valley is officially called Vallecitos without the generic, although common usage continues to call it the equivalent of "Little Valley Valley."

Valle de San Jose [Alameda Co.] was the name of a land grant dated Feb. 23, 1839. The generic term is preserved in Arroyo Valle. *See* Arroyo: Arroyo Sanitorium. *For* Valle de San Jose [San Diego Co.], *see* San Jose.

Vallejo (və lā′ hō, və lā′ ō): city, **Heights, Hill** [Solano Co.]. The city was laid out in 1850 by Gen. Mariano G. Vallejo (*see* Glossary) on his land and was the capital of the state in 1851–52. It is said that Vallejo had wanted to bestow the name Eureka upon the prospective capital; but the maps of 1850 show the name Vallejo, and on Mar. 13, 1851, this name was made official. *For* Vallejo Mills, *see* Niles.

Valle Vista [Riverside Co.]. Intended to mean 'valley view', for Spanish *vista del valle*.

Valley. The word is often used as a specific term in naming towns, and California has seven places so named. **Valley Ford** [Sonoma Co.]

dates back to the 1860s and was so named because the old trail crossed the Estero Americano at this point. **Valley Center** [San Diego Co.] has been thus known since the 1870s because the settlement is situated in the center of the valley. **Valley Springs** town and **Peak** [Calaveras Co.]: The old mining town was known as Spring Valley because of the mineral springs in the valley of Cosgrove Creek. When the post office was established on Dec. 3, 1872, the name was changed to Valley Springs. **Valley Home** [Stanislaus Co.]: The name of the German settlement Thalheim became a victim of war hysteria in 1917–18 and was replaced by the English version. *For* Valley of California, *see* Great: Great Central Valley. *For* Valley of the Moon, *see* Sonoma Valley; Switzerland.

Vallo Citron. *See* Valle: Vallecito.

Valona (və lō′ nə) [Contra Costa Co.]. This was an Italian community, and the name may be of Italian origin.

Valpe (val′ pē) **Ridge** [Alameda Co.]. Apparently named for Capt. Calvin Valpey, a pioneer of 1851, or for one or more of his three sons (Still).

Val Verde. *See* Verde.

Valyermo (väl yâr′ mō) [Los Angeles Co.]. Named in 1909 by W. C. Petchner, owner of the Valyermo Ranch. The name is a combination of two Spanish elements: *val* for *valle* 'valley', plus *yermo* 'barren'. The post office is listed in 1912.

Van Arsdale Lake [Mendocino Co.]. The power reservoir was named in 1905 for W. W. Van Arsdale of San Francisco, president of the Eel River Power and Irrigation Company.

Vanauken Creek [Humboldt Co.]. For Jackson Vanauken, who patented land here in the 1890s; established by BGN, list 7304.

Vance Canyon [Santa Barbara Co.]. Named for J. N. Vance, who settled on Willow Creek, "otherwise called Vance Cañon," in 1868 (Co. Hist.: Santa Barbara 1883:268).

Van Damme Beach State Park [Mendocino Co.]. Named for Charles Van Damme, whose gift of beach and other land was included in the park, formed in 1934 (E. P. Hanson).

Vandenberg: Air Force Base, Village [Santa Barbara Co.]. For Gen. Hoyt S. Vandenberg (1899–1954), Air Force chief of staff from 1948 to 1953.

Vanderbilt [San Bernardino Co.]. The mining district was so named in the hope that it would prove as rich as the fortune created by Cornelius Vanderbilt (1794–1877). A post office was established on Feb. 1, 1893.

Vandever (van dē′ vər), **Mount** [Tulare Co.] commemorates the man who introduced the bills in Congress in 1890 establishing Yosemite, Sequoia, and General Grant National Parks. William Vandever (1857–93) was a brevet major general in the Civil War and represented the sixth district of California in Congress, 1887–91 (Farquhar).

Van Duzen (van do͞o′ zən) **River** [Humboldt, Trinity Cos.]. The name Van Dusen's Fork (of the Eel River) was given to the stream in Jan. 1850 by the Gregg party, for James Van Duzen, one of its members. It appears on Blake's map of 1853 as Vandusen's Fork. The "Fork" was changed to "River" by popular usage and was adopted by the USGS (Turner 1993).

Van Norden. *See* Norden.

Van Nuys (van nīz′) [Los Angeles Co.]. The post office was established in 1912 and named for Isaac N. Van Nuys, son-in-law of Isaac Lankershim, with whom he was associated in 1876 in the first successful cultivation of wheat on a large scale in southern California.

Vaquero (və kâr′ ō). The Spanish word for 'cowboy' (from *vaca* 'cow') is commonly used throughout the Southwest, particularly in referring to Mexican cowboys. It is found in California nomenclature as the name for several places. **Cañada de los Vaqueros** [Contra Costa, Alameda Cos.] was a land grant dated Feb. 29, 1844. The rancho subsequently became known as "The Vasco" (Still), perhaps reflecting Spanish *vasco* 'Basque'.

Vasquez (vas kwez′, vas′ kəz): **Canyon, Rocks** [Los Angeles Co.]. So named because the notorious bandit Tiburcio Vásquez (1835–75), who was hanged in San Jose, had his hideout among the rocks of "Robbers Roost" (Drury). **Vasquez Monolith** [Pinnacles NM] constitutes the roof of the largest "cave" in the covered canyon. It was named in 1934, at the time of the first ascent, in the belief that this cave was another of the hideouts (David Brower). There is a Vasquez Creek north of Nipomo [San Luis Obispo Co.], but it is not known if it was also named for the outlaw.

Veeder, Mount [Napa, Sonoma Cos.]. Named for the Rev. Peter V. Veeder, who was minister of the Presbyterian church in Napa from about 1858 to 1861, and president of San Francisco City College, 1861–71.

Vega (vā′ gə) [Monterey Co.]. The Southern Pacific station was named when the line from Gilroy to Watsonville was opened on Nov. 27, 1871. The name is derived from that of the land grant Vega del Rio del Pájaro 'plain of the Pajaro River', dated Apr. 17, 1820, and June 14, 1833.

Venado (və nä′ dō) [Sonoma Co.]. The post office was established in 1921 and named with this Spanish word for 'deer'. In the 1890s there was a post office by the same name in Colusa Co.; this appears on some maps as Venada. The name is found in other parts of the state. There was a post office named El Venado in Los Angeles Co. between May 8, 1914, and Dec. 15, 1917. *For* Arroyo de los Venados, *see* Deer: Deer Creek.

Venice [Los Angeles Co.]. Designed and built in 1904 by Abbot Kinney on a part of his Ocean Park Tract, with a system of canals for thoroughfares, in imitation of Venice, Italy.

Ventana (ven tan′ ə): **Cone, Double Cone, Creek** [Monterey Co.]. The Spanish name for 'window' was applied because of a window-like opening in one of the hills.

Ventucopa (ven tə kō′ pə) [Santa Barbara Co.]. The post office was established in 1926 at a point between Ventura and Maricopa. It was named at the suggestion of Dean ("Dinty") Parady, a resident, who coined the name from the names of these two places (L. L. Curyea).

Ventura (ven to͞or′ ə): city, **River, County**. In his instructions of Sept. 15, 1768, José de Gálvez, inspector general, directed Padre Serra to name a mission between San Diego and Monterey in honor of San Buenaventura (Engelhardt 2:7). The saint so honored was a learned doctor and prelate of the Franciscan Order in the 13th century. The mission was not founded until Mar. 31, 1782, at the site called La Asunción de Nuestra Señora by Crespí on Aug. 14, 1769. The river appears as Río S. Buenaventura on Narváez's Plano of 1830. An attempt to found a town near the mission in 1848 was not successful, but a post office was established and named after the mission in 1861. The county was created from part of Santa Barbara Co. on Mar. 22, 1872, and was given the abbreviated name Ventura. In 1891, on petition of the residents, the Post Office Dept. changed the post office name to Ventura: "Much mail and express matter designed for this office found its way to San Bernardino, and vice versa. Then the name was too long to write and too difficult for strangers to pronounce" (Co. Hist.: Santa Barbara 1891:230). The new name was generally accepted, although the Southern Pacific did not change the name of the station until 1900. In 1905 Z. S. Eldredge wrote the following obituary to the old name in his campaign to restore Spanish names: "And now comes the Post Office Dept., which is the most potent destroyer of all. I have spoken before of the injury done the people of San Buenaventura. They cling to that name and use it among themselves. But they are doomed. Mapmakers, from the Director of the Geological Survey to the publisher of a pocket guide following the lead of the post office, call the place Ventura, and the historic name will be lost" (*San Francisco Chronicle*, Apr. 10, 1905). The name of the river appears in its abbreviated form on most maps after 1895. **Ventura Rocks** [Monterey Co.]: The name of the rocks off Point Sur may be reminiscent of Buenaventura River (*see* Buenaventura), as the Salinas was once called.

Vera, Lake [Nevada Co.]. The lake is a power reservoir, built in 1909 and named for the daughter of Eugene de Sabla, the promoter of the project.

Verde. The Spanish adjective for 'green' was commonly used in place naming and occurs in the names of two land grants—Cañada Verde y Arroyo de la Purísima [San Mateo Co.], dated Mar. 25, 1838, and June 10, 1839; and Cañada Larga o Verde [Ventura Co.], dated Jan. 30, 1841. The word is still found in modern place names, e.g. Loma Verde (vûr′ dē) [Los Angeles Co.], Val Verde [Riverside Co.], Verde Canyon [Marin Co.]—and sometimes in combinations coined by Americans, such as Anaverde Valley [Los Angeles Co.]. **Verdemont** [San Bernardino Co.], suggesting 'green mountain', was applied to the Santa Fe station after the line from Barstow to San Bernardino was built in 1883.

Verdi (vûr′ dī) **Peak** [Sierra Co.]. The mountain is labeled Crystal Peak on the Mining Bureau map of 1891. The present name was probably applied by the USGS, after the station across the Nevada line, which had been named for the Italian composer Giuseppe Verdi (1813–1901).

Verdugo (vûr do͞o′ gō): **Canyon, Mountains, City, Wash** [Los Angeles Co.]. The name commemorates the Verdugo family, established in California by Mariano de la Luz Verdugo (1746–1822). José María Verdugo, a corporal of the San Diego company, who had served in the mission guard at San Gabriel, received one of the first land grants, dated Oct. 20, 1784, and Jan. 12, 1798. The name Rancho de los Verdugos is repeatedly mentioned in documents and appears on Narváez's Plano of 1830 as Berdugo. The modern town was laid out by Harry Fowler in 1925, but Verdugo Park is shown on the maps of the 1910s as the terminal of a local railroad. The mountains are shown as Sierra de los Berdugos on a *diseño* of La Cañada grant (1843), and Sierra or Cañada de los Verdugos was the name of an unconfirmed grant of 1846.

Vergeles [Monterey, San Benito Cos.]. When José J. Gómez was placed in possession of his two grants, which were designated by various names and dated Aug. 2, 1834, and Aug. 28, 1835, he requested that the name of the combined grants be changed to Los Vergeles (Bowman Index). *Vergel* is Spanish for 'flower and fruit garden'.

Vermilion Valley [Fresno Co.]. The name was applied by T. S. Solomons and Leigh Bierce in Sept. 1894 (Farquhar). The word is quite commonly used for naming features that by nature have a bright orange-red hue or appear so in the evening sun. The lake in the valley is now called Feather Lake, and the name Vermilion Lake was transferred to a small lake nearby.

Vernal Fall [Yosemite NP]. Dr. L. H. Bunnell of the Mariposa Battalion applied the name, derived from the Latin word for 'springtime', because "the cool, moist air and the newly-springing Kentucky bluegrass at the Vernal, with the sun shining through the spray as in an April shower, suggested the sensation of spring" (p. 205).

Vernalis (vər nal′ is) [San Joaquin Co.]. The Latin word meaning 'pertaining to springtime' was applied to the Southern Pacific station when the line from Tracy to Newman was built in 1887.

Vernon [Los Angeles Co.]. Post office and station were known as Vernondale, for Capt. George R. Vernon, who rose from the ranks during the Civil War and became a settler of the district after 1871. The abbreviated form was used when the city was incorporated in 1905. *For* Vernon in Sutter Co., *see* Verona.

Verona [Sutter Co.]. The town was laid out in 1849 on property purchased from Sutter and given the familiar American place name Vernon. The post office was discontinued in the 1880s. When it was reestablished about 1906, a new name had to be chosen, because in the meantime a Vernon post office had been opened in Stanislaus Co.

Verruga. *See* Cañada: Cañada Verruga.

Versteeg (vûr′ stēg), **Mount** [Sequoia NP]. The mountain southeast of Mount Tyndall was named in 1964 in memory of Chester Versteeg —businessman, author, and public-spirited citizen, who died in 1963.

Vestal [Tulare Co.]. The California Edison Company built the substation here in 1919 and named it for the virgin priestesses who tended the sacred fire of Vesta, the Roman goddess of the hearth (Mary Sharp).

Vesuvius, Mount [Los Angeles Co.]. In 1893 fireworks were displayed on the summit every Saturday at 9:00 P.M., ending with a miniature representation of Mount Vesuvius in Italy (Reid, p. 369).

Vetter, Mount [Los Angeles Co.]. Named for Victor P. Vetter, district ranger, Angeles National Forest (USFS).

Viboras (vē′ bə rəs) **Creek** [San Benito Co.]. From Arroyo de las Víboras 'creek of the rattlesnakes', the name by which it was known in Mexican times and later (J. P. Davis).

Vicente (vi sen′ tē): **Point, Creek** [Los Angeles Co.]. The point was named on Nov. 24, 1793, by Vancouver, for Padre Vicente Santa María of Mission San Buenaventura. The name appears on Duflot de Mofras's map of 1844; it was adopted by the Coast Survey in 1852 but was spelled Vincente for years.

Vichy (vish′ ē) **Springs** [Mendocino Co.]. The springs were named before 1890 because their waters recall those of the spa in Vichy, France.

Victor [San Joaquin Co.]. Named in 1908 by the Southern Pacific for Victor Morden, whose father, A. E. Morden, was instrumental in securing the station. The post office is listed in 1924.

Victorville; Victor Valley [San Bernardino Co.]. The station was named Victor in 1885 for J. N. Victor, construction superintendent of the California Southern Railroad, 1888–89. At the request of the Post Office Dept., the name was changed in 1901 to Victorville, in order to avoid confusion with Victor, Colorado.

Vidal (vi däl′) [San Bernardino Co.]. Founded in

1907 by Hansen Brownell and named for his son-in-law, whose name was Vidal (Santa Fe).

Vidette, East, West; Vidette Creek [Kings Canyon NP]. "Two of these promontories, standing guard, as it were, the one at the entrance to the valley and the other just within it, form a striking pair, and we named them Videttes" (C. B. Bradley, *SCB* 2:272). The usual spelling of the word is "vedette," and the meaning is 'a mounted sentry placed in advance of an outpost'.

Vieja, Mission [Los Angeles Co.]. After Mission San Gabriel was built on its present site, the original mission building, 5 miles away, was called Misión Vieja 'old mission'. The adjectives *viejo* and *vieja* were often used in Spanish times in the same way that Americans apply the word "old" in geographical nomenclature.

Viejas (vē′ əs): **Valley, Creek, Indian Reservation, Mountain** [San Diego Co.]. Valle de las Viejas is shown on a *diseño* of the Cuyamaca grant in 1845. The name was given to a land grant of May 1, 1846, but was rejected by the United States on Dec. 26, 1854. A Spanish expedition supposedly called an Indian village Valle de las Viejas 'valley of the old women', because at their approach the inhabitants fled, leaving behind only the old women.

Vigia, Cerro de la. *See* Lavigia Hill.

Villa. *See* Pueblo.

Villa (vil′ ə) **Park** [Orange Co.]. The original name of the town, Mountain View, was changed to Villa Park when the post office was established in 1888, because there was already a Mountain View post office in Santa Clara Co. (Brigandi). *For* Villa de Branciforte, *see* Branciforte Creek; Santa Cruz.

Viña. The Spanish word for 'vineyard' was often used in Spanish times, especially in connection with the missions, most of which had at least one vineyard. La Viña was the name of an unconfirmed grant in San Luis Obispo Co., dated Jan. 4, 1842. The word has also been used for place naming in American times, albeit without the tilde and with a variety of pronunciations: **Vina** (vī′ nə) [Tehama Co.], listed in 1880 as a post office on Leland Stanford's once famous vineyard; note also La Viña (vēn′yə) [Los Angeles Co.] and Las Vinas (vē′ nəs) [San Joaquin Co.]. **Cañada de la Viña** [Santa Barbara Co.] was mentioned as Arroyo de la Viña in 1839 (Arch. SB, Juzgado, p. 5).

Vinagre (vin′ ə gər) **Wash** [Imperial Co.]. From the Spanish for 'vinegar'.

Vincente. *See* Vicente.

Vincent Gulch [Los Angeles Co.]. Named by the USFS for Charles (Tom) Vincent, a pioneer miner and hunter (S. B. Show).

Vine. Because of the extensive grape culture in California, the word "vine" has been repeatedly used in place names. The best known are Vineburg [Sonoma Co.], Vineyard [Los Angeles Co.], and Vine Hill [Contra Costa Co.].

Vino [Fresno Co.]. Spanish for 'wine'; *see* Viña.

Vinton [Plumas Co.]. The name of the old post office Summit was changed on Feb. 16, 1897, to the present name, for Vinton Bowen, daughter of Henry Bowen of the Sierra Valleys Railway (Myrick).

Viola (vī ō′ lə) [Shasta Co.]. Named by B. F. Loomis for his mother, Viola Loomis. Loomis filed a homestead here in 1888 and later built the Viola Hotel (Steger).

Violin Canyon (or **City**). *See* Fiddletown.

Virgilia [Plumas Co.]. Named by the Western Pacific for Virgilia Bogue, daughter of Virgil G. Bogue and queen of the Portolá celebration in 1909 (Drury). The post office is listed in 1929 (*see* Bogue).

Virgin. Several creeks in various parts of the state and a spring in Death Valley were probably so named because of their pure water (*see also* Las Virgenes; Ribbon Fall).

Virginia. During the Gold Rush a number of mines, towns, and physical features were named by miners from Virginia. The name of **Virginia Creek** [Mono Co.] goes back to "early days" and is shown on Hoffmann's map of 1873. It is not known whether Virginia Lake in Yosemite belongs to the same group, or whether it was named, like many other mountain lakes, for a lady friend of a surveyor. Red Peak in Yosemite was changed to Virginia Peak by the BGN in 1932 upon recommendation of the Park Service, because there is another, better-known Red Peak in the park. *For* Virginia Mills, *see* Berry Creek.

Visalia (vī sāl′ yə) [Tulare Co.]. "On the 1st of November, [Nathaniel] Vise and O'Neil located and surveyed a new town called Visalia, in the finest section of that county" (*San Francisco Alta California*, Dec. 11, 1852). The town was probably named after Visalia, Kentucky, which in turn had been named about 1820 for a relative of "Nat." After the county seat was moved from Woodsville to Visalia, the supervisors, on Dec. 21, 1853, changed the name to Buena Vista. In Feb. 1854 the name was changed back to Visalia at the insistence of the residents (Mitchell).

Visitacion (viz i tā′ shən) **Valley, Point** [San Francisco, San Mateo Cos.]. A cattle ranch of the San Francisco Presidio, la Visitación, is mentioned in 1798 (SP Mis. and C. 1:74). Jacob P. Leese was authorized in 1839 to build houses in the place named La Visitacion (Prov. Recs. 10:12), and on July 31, 1841, he was granted the Cañada de Guadalupe, la Visitación y Rodeo Viejo. The place names, originally given for the Visitation of the Virgin Mary to Saint Elizabeth, were spelled Visitacion but were gradually Americanized, and the BGN decided for Visitation (6th Report). In Mar. 1949 the board revised the earlier decision and decided for the Spanish spelling (*see* Brisbane).

Vista. The Spanish name for 'view' was occasionally used for place naming in Spanish times (*see* Buena Vista), but most names that include the word are modern applications. Besides the popular combination Buena Vista, the word is found as the generic or specific part of many other combinations, often without regard to Spanish usage: Vista del Mar, Vista del Valle [Los Angeles Co.], Vista Grande [San Mateo Co.], Vista Robles [Butte Co.], Alta Vista [Sonoma Co.], Sierra Vista [Merced Co.], Valle Vista [Contra Costa, Riverside Cos.]. The name is especially popular in San Diego Co., which has a Vista de Malpais (a view of the Borrego Badlands), Chula Vista, Monte Vista, Sunny Vista, and plain Vista.

Vitzthum (vits′ təm) **Gulch** [Trinity Co.]. Named for the Vitzthum family, early settlers.

Víboras, Las. *See* Oso: Oso Flaco.

Vizcaino (vis kä ē′ nō), **Cape; Vizcaino: Dome, Seavalley** [Mendocino Co.]. The name of the explorer (*see* Glossary) appears first on Malaspina's chart for what Wagner (p. 523) believes was Punta Gorda. On Humboldt's map (1811), "C. Vizcayno" is shown at latitude 40° north, about halfway between Punta Gorda and the present location of Cape Vizcaino. Duflot de Mofras (1844) identifies the name with the present cape. The charts of the Coast Survey do not show the name until about 1900.

Vodega. *See* Bodega.

Vogelsang (vō′ gəl sang): **Peak, Lake, Pass, Camp** [Yosemite NP]: The name was given to the peak in 1907 by H. C. Benson, in honor of Charles A. Vogelsang, for many years executive officer of the State Fish and Game Commission (C. A. Vogelsang to F. P. Farquhar). Although named for a man, the name is singularly fitting to the beautiful place, since in older German it means 'a meadow in which birds sing'.

Volcano, Volcanic. About twelve peaks are so named, including an imposing Knob [Madera Co.] and a Ridge [Fresno Co.]. A group of Volcanic Lakes [Fresno Co.], as well as Volcano Lake [Sierra Co.], Volcano Canyon [Placer Co.], and Volcano Falls [Tulare Co.], are named after nearby craters. **Volcano** [Amador Co.]: The miners called the place "The Volcano." "Bayard Taylor [1850], in his travels, speaks of several craters near, but no one else has ever seen them—they probably exist only in the poet's imagination" (Brewer, p. 435). "This marble or limestone is much broken up, and contains many caves, and chimney like openings, from which the first explorers presumed that it had been the seat of an active volcano, at some time, and hence the name" (Bancroft Scrapbooks 1:89). The name appears in Bowen's Post-Office Guide of 1851. An earlier name was Soldiers Gulch, because some discharged soldiers of Stevenson's New York Volunteers mined there in 1848. **Volcanoville** [El Dorado Co.] was once an important gold-mining town. It was so named because a nearby mountain seemed to be an extinct volcano, and the miners had to work through lava cement. The place had a post office until 1953. *For* Volcano Creek [Tulare Co.], *see* Golden Trout Creek.

Vollmer (vōl′ mər) **Peak** [Alameda Co.]. The peak formerly known as Baldy was named in 1940, by the East Bay Regional Park Board, for August Vollmer (1876–1955), of Berkeley, well-known authority on criminology, who had served as a member of the park board since its organization (R. E. Walpole).

Volta [Merced Co.]. The town was laid out and named by the Volta Improvement Company in 1890; the post office is listed in 1892. The name probably was derived from the name of the Italian physicist Alessandro Volta (1745–1827).

Volunteer Peak [Yosemite NP]. Replaces the former name, Regulation Peak (*q.v.*), which was given by Lt. H. C. Benson in 1895 and transferred by mistake to another peak by the USGS (Farquhar).

Von Schmidt Line. Some maps record this old California-Nevada boundary line, established in 1872–73 (*see* Glossary).

Vontrigger (von′ trig ər) **Spring** [San Bernardino Co.]. The spring at the eastern end of the New York Mountains was named for Erick Vontrigger, who made some mining locations and camped at the spring in the 1870s (O. J. Fisk).

Vultee Field [Los Angeles Co.]. Established in 1942 and named for the Vultee Aircraft Corporation.

W

Wabena (wä bē´ nə) **Creek** [Placer Co.]. Margaret Fitzmorris of Tahoe City reports that this was the family name of her great-grandparents, who had summer pasturage on this site in the 1850s. The family name, also spelled Wubbena, was originally from the Netherlands.

Waddell (wä del´) **Creek** [San Mateo, Santa Cruz Cos.]. Named for William Waddell, of Kentucky, who came to California in 1851 and built a sawmill at the creek in 1862. He was killed by a bear in 1875 (Hoover, p. 588). The stream had been named Cañada de San Luis Beltrán by Crespí of the Portolá expedition on Oct. 20, 1769, and the valley was called Cañada de la Salud 'valley of health' when the sick members of the party miraculously recovered after resting there two days.

Waddington [Humboldt Co.]. The post office was established about 1890 and named for Alexander Waddington, a native of England, who came to California in 1867.

Wages (wā´ jəz) **Creek** [Mendocino Co.]. The stream north of Westport was named for Alfred Weges, who settled here in 1864 (Co. Hist.: Mendocino 1914:117). Maps show both spellings, Weges and Wages.

Wahguyhe (wä´ gī) **Peaks** [Death Valley NP]. The name was mentioned by Mendenhall in 1909 (WSP 224:88–89) and is still in use. The term may be from Panamint *waa-kko'i* 'pinyon-pine peak', referring to the whole Grapevine Range (McLaughlin); or from *waha-ko'i* 'two-peaks'.

Wahtoke (wä´ tōk): **Creek**, town, **Park** [Fresno Co.]. A Yokuts or Shoshonean "tribe" is mentioned in 1857 as Wattokes, high up on Kings River, and in 1861 as Wartokes, in the Fresno Reservation (Hodge, 922). The stem appears to be the Yokuts *watak* 'pine nut' (Kroeber). The stream is shown on the Land Office map of 1859 as Wahtohe Creek. The present spelling is used by J. N. LeConte on the Kings-Kern sheet of his Sierra map, 1899–1904. The name was applied to the station when the Santa Fe built the branch line in 1910–11.

Waitisaw (wē´ ti sô) [Shasta Co.]. From Wintu *wayti sawal* (Pitkin, p. 685). This may be 'north pond', containing *way* 'north' and *saaw-* 'pond' (Shepherd).

Walalla. *See* Gualala.

Waldo [Marin Co.]. A Waldo Street appears on the map of the Sausalito Land and Ferry Company, filed Apr. 26, 1869. The station was named Waldo in 1891 (G. H. Harlan). Since many of the streets were named after towns and one was named Eugene Street, it is quite possible that Waldo Street was named after Waldo, Oregon. The Oregon town was named for William Waldo, who had been of service to the people of the community and was Whig candidate for governor of California in 1853 (McArthur).

Walerga (wä lâr´ gə, wä lûr´ gə) [Sacramento Co.]. The origin of this name has not been identified.

Wales Lake [Sequoia NP]: The name was proposed by the Sierra Club in 1925, in memory of Frederick H. Wales, a native of Massachusetts, veteran of the Civil War, and for many years a minister, editor, and farmer in Tulare Co. (Farquhar).

Wal Hollow. *See* Gualala.

Walker: Lake, Creek [Mono Co.]. The lake northeast of Mono Pass was named for William J. Walker, who settled on the land adjoining the lake in 1880 (Maule). The name is often erroneously connected with Joseph R. Walker (*see* Walker: Pass). **Walker Creek** [Glenn Co.]: Named for Jeff Walker, a sheepherder in the early 1850s (Co. Hist.: Colusa 1918:202).

Walker: Pass [Kern Co.]; **East Walker River** [Mono, Nevada Cos.]. Joseph R. Walker (1798–1876), of Tennessee, was one of the great pathfinders of the American West. The first time he crossed the pass that bears his name was on his return from the famous expedition of 1833 from Great Salt Lake to Monterey, on which he saw Yosemite Valley. Frémont suggested the name in his journal on Mar. 24, 1844: "This pass, reported to be good, was discovered by Mr. Joseph Walker whose name it might, therefore, appropriately bear" (*Memoirs* 1:354). Walker was Frémont's guide on the latter's third expedition (1845–46), and Frémont then gave his name to the Walker River in Nevada. By BGN, list 8902, the tributary of this river that lies within California is called the East Walker River.

Walker Creek [Inyo Co.]. Named for Gus Walker, a native of Germany, who came to the San Joaquin Valley in 1859, helped in the survey of the valley, and settled in the Olancha district as a farmer in 1864 (Robinson).

Wallace [Calaveras Co.] was named for an unidentified man by David S. Burton, the founder of the town (Frances Bishop). The post office was established on Jan. 1, 1883.

Wallace: Lake, Creek [Sequoia NP]. The lake northwest of Mount Whitney was named in 1925 at the suggestion of the Sierra Club, for Judge William B. Wallace, well known as an early mountaineer in the Kings, Kern, and Kaweah regions. Wallace came to Tulare Co. in 1876, where he later served for many years as judge of the superior court (Farquhar).

Wallace, Mount. *See* Evolution.

Walloupa. *See* Guadalupe [Nevada Co.].

Walnut. About ten features in the state are named after the most widely cultivated species of the nut family. Some of the names are translations of the Spanish name Nogales 'walnut trees', given to places where the native walnut was found (*Juglans californica*). **Walnut Creek** [Contra Costa Co.] is a translation from the Spanish name of the creek recorded as Arroyo de los Nogales 'creek of the walnut trees' by Padre Viader on Aug. 16, 1810 (Arch. MSB, vol. 4). The stream is called Arroyo de las Nueces 'creek of the walnuts' on July 22, 1834 (Legis. Recs. 2:162), and this is the name of a land grant dated July 11, 1834. The town was named after the creek when the post office was established in the 1860s. **Walnut Grove** [Sacramento Co.] was settled by "woodchoppers" as early as 1850, but the present name was not applied until the early 1860s when the post office was established. **Walnut** [Los Angeles Co.]: The Southern Pacific station was called Lemon until 1912, when the present name was applied. The change was probably made because, shortly before 1912, "lemon" came to be used in the slang sense and made the name humorous. There is a Walnut Creek about 5 miles north of the town.

Walong (wā′ long) [Kern Co.]. The name of W. A. Long, a Southern Pacific trainmaster, contracted to form the name Walong, was applied to a siding on the Tehachapi slope in 1876 (Santa Fe).

Walsh Station [Sacramento Co.]. Named by the Post Office Dept. in 1875 for J. M. Walsh, the town's first storekeeper and postmaster (Co. Hist.: Sacramento 1890:202, 252).

Walteria (wäl târ′ ē ə) [Los Angeles Co.]. The community was named for Captain Walters, who built the Walters Hotel early in the 20th century (R. S. Sleeth). The post office was established on Sept. 18, 1926.

Walters. *See* Mecca.

Wamelo (wä′ mə lō) **Rock** [Madera Co.]. The name appears on Hoffmann's map of 1873 and was probably applied by the Whitney Survey. In 1875 Muir stated that the granite dome was called Wamello by the Indians (Farquhar). It is otherwise known as Fresno Dome.

Wanda Lake [Kings Canyon NP] was named for Annie Wanda Muir Hanna, daughter of John Muir. Mount Wanda [Contra Costa Co.] is also named for her (BGN, list 1995).

Wapama (wə pä′ mə) **Falls** [Tuolumne Co.]. Perhaps from a Southern or Central Sierra Miwok form such as *wepaama*, from *weepa* 'uphill' (Callaghan).

Wapaunsie (wə pôn′ sē) **Creek** [Plumas Co.]. Formerly also spelled Wahponsey; perhaps a Maidu Indian name.

Ward: Mountain, Tunnel [Fresno Co.]. Named for Dr. George C. Ward, who directed construction of the hydroelectric plant in the Sierra Nevada. Approved by the BGN in 1935.

War Eagle Field [Los Angeles Co.]. On Mar. 3, 1942, Maj. C. C. Moseley bestowed the symbolic name on the airfield, which was established as a training center for American and British pilots.

Warm Springs [Alameda Co.]. The post office at the site is listed as Harrisburgh from 1867 to 1880. The new name was taken from the warm springs that are 2 miles northeast on Agua Caliente ('warm water') Creek; the "new" Warm Springs are on Agua Fria ('cold water') Creek. The community became part of the city of Fremont in 1956. There are a number of other Warm Springs in the State, including two settlements [Lassen and Los Angeles Cos.].

Warner Mountains [Modoc Co.]. Named in memory of Capt. William H. Warner, who was killed by Indians at the foot of this range, just south of the Oregon line, on Sept. 26, 1849. Warner was in command of a reconnaissance party in search of a route across the mountains. Warner Valley in Oregon is named for the same man (cf. McArthur).

Warner Springs; Warners: Ranch, Pass [San Diego Co.]. Named for Jonathan Trumbull (Juan José) Warner (1807–95), who arrived in California in 1831, became a Mexican citizen, and was granted the Rancho Agua Caliente or Valle de San José in 1844. The rancheria at the springs was called *kupa* by the local Indians, who received the Spanish (and English) name Cupeño. To the early immigrants who came via the southern route, Warners played a role like that of Sutters Fort for the immigrants who entered California via the central route.

Warren, Mount [Mono Co.]. The name was applied by the Whitney Survey before 1867 for Gouverneur K. Warren, a member of the Pacific Railroad Survey and a distinguished Union general in the Civil War (Farquhar).

Warren Grove [Humboldt Co.]. The grove in the Prairie Creek Redwoods State Park was established in 1954 in honor of Earl Warren (1891–1974), governor of California (1943–53) and later chief justice of the U.S. Supreme Court.

Wasco (wäs′ kō) [Kern Co.]. The Santa Fe station was established in 1898 and named Dewey in honor of the admiral's victory in Manila Bay on May 1, 1898. When it was learned that a place named Dewey already existed in the state, William Bonham, one of the oldest settlers, was asked to select a new name. He chose Wasco after his home county in Oregon, which had been named for a Chinookan Indian tribe (A. D. Jackman).

Waseck (wä′ sek) [Humboldt Co.]. From the Yurok village name *wahsek* (Waterman, p. 254).

Wash. In England a "wash" designates a piece of ground washed by the action of the sea or a river (Knox). In western America the term is generally used for the dry bed of an intermittent stream, but in the desert regions of California it is applied to a creek bed that is dry most of the time. *For* Wash [Plumas Co.], *see* Clio.

Washapie (wä′ shə pī) **Mountains** [Tulare Co.]. Named after one of the Indian villages in adjacent Drum Valley (A. L. Dickey).

Washboard, The [Fresno Co.]. So named because the contours make the northeast slope of Kettleman Hills look like a washboard.

Washburn Lake [Yosemite NP] was named by Lt. N. F. McClure in 1895 in honor of Albert H. Washburn, of Wawona (Farquhar).

Washington. The state naturally has a number of places named directly or indirectly in honor of the Father of the Country, including the old mining town of Washington [Nevada Co.] and Washington Column [Yosemite NP]. In former years there were many more: Washington Corners [Alameda Co.] is now Irvington, and Washingtonville [Sierra Co.] is Downieville. Washington Peak [Del Norte Co.] is named after the Washington Ranch situated at its foot; Washington Lake (euphemistic for Slough), in the Yolo Basin, was named after the town now called Broderick.

Washingtons Flat [Del Norte Co.] is named for a pioneer hunter and trapper, George Washington, who was killed there by a landslide.

Washoe (wä′ shō) **Creek** [Sonoma Co.]. Probably for the Washoe Mine, developed after 1859; the name was transferred from Washoe Co., Nevada, of which Reno is the county seat. The term refers to the Washo (or Washoe) Indians, who originally lived in the Lake Tahoe region of California and Nevada; it is from *wáašiw*, the tribe's name for itself (Jacobsen).

Wasioja (wä sē ō′ ə) [Santa Barbara Co.]. The post office (discontinued in 1933) was established in 1894 and named after Wasioja, Minnesota, the former home of a resident (Margaret B. Richardson). The origin is in a Siouan language, supposedly meaning 'pine-clad'.

Wasna, Río. *See* Huasna River.

Wassuma (wä so͞o′ mə) **Creek** [Madera Co.]. Named after the Southern Miwok village on its bank (Merriam, *Mewan stock*, p. 346). The name also appears as Wassaiuma. It may be related to Southern Sierra Miwok *wasaama*, the location of an Indian round-house at Ahwanee (*q.v.*), perhaps containing *wassa* 'ponderosa pine'; or it may be from *wasayya-* 'mush-stirrer, coarse acorn flour' (Broadbent; Callaghan).

Watchorn Basin [Los Angeles Co.]. The basin in Los Angeles harbor was named about 1912 for Robert Watchorn, an official of the Outer Harbor Dock and Wharf Company operating in that area (Co. Surveyor).

Water. *See* Newberry Springs.

Water Dog is a term used for several large species of salamanders; Siskiyou Co. has a Water Dog Lake.

Waterford [Stanislaus Co.]. When the Southern Pacific Railroad built the Stockton-Merced branch in 1890, this name was given to the Southern Pacific station because it was near a much-used ford on the Tuolumne River. Waterford is a famous old city of Ireland, and the name is quite common as an American place name.

Waterhouse Peak [El Dorado Co.]. The peak was named by the USFS in memory of Clark Waterhouse, who had been in charge of the Angora Lookout and lost his life in World War I (E. Smith).

Waterloo. The British and Prussian victory over Napoleon in 1815 at Waterloo made the name a favorite for places in America before the Civil War. The place in San Joaquin Co., however, owes its name to an actual fight over a land title in the early 1860s, although it is not clear from the reports which party met its Waterloo. The post office is listed in 1867.

Waterman, Mount [Los Angeles Co.]. According to tradition in the USFS, the mountain was named for Robert Waterman (1826–91), from 1904 to 1908 a ranger in the old San Gabriel Timberland Reserve. The original name was Lady Waterman Mount, because Mrs. Robert Waterman was the first woman to climb it (Wheelock).

Waterman Canyon [San Bernardino Co.]. When the road was built in the 1880s, Dr. Ben Barton and his associates named the canyon after the famous Waterman silver mine. The mine was owned by Robert W. Waterman, of San Bernardino, who was governor of California, 1887–91. It is not known whether Waterman post office [Amador Co.] was named for the same man. *For* Waterman Junction, *see* Barstow.

Waterman Park [Solano Co.]. The post office, listed in 1945, preserves the name of Robert H. Waterman, a famous clipper ship captain (*see* Fairfield).

Waterwheel Falls [Yosemite NP]. "The water here encounters shelves of rocks projecting from the river bottom which cause enormous arcs of water to be thrown into the air, some rising 20 feet" (Doyle).

Watkins, Mount [Yosemite NP]: The name was affixed, probably in the 1860s, for Carleton E. Watkins (1829–1916), an early photographer of Yosemite, whose picture of Mirror Lake reflecting the mountain was especially popular (Farquhar).

Watson [Los Angeles Co.]. Established as a station of the San Pedro line of the Pacific Electric Railway in 1904 and so named because it was on that part of Rancho San Pedro that was allotted in 1885 to María Domínguez de Watson (Co. Library).

Watson Gulch [Shasta Co.]. Named after Watson Diggings, which had been named for a miner from Oregon, credited with having taken the first wagon and ox team over Trinity Divide in 1849 (Steger).

Watsonville [Santa Cruz Co.]. The town was laid out in 1852 on a part of Rancho Bolsa del Pájaro by D. S. Gregory and Judge John H. Watson, owners of the land, and was named for the latter. The post office is listed in 1854. **Watsonville Junction** [Monterey Co.]: The earlier name Pajaro (*q.v.*), which had been given to the station when the railroad was built in 1871, was changed by the Southern Pacific in 1912.

Watterson Canyon [Inyo Co.]. Named for George Watterson, a native of the Isle of Man, who operated a small mine in the canyon on the north side of the road (Robinson).

Watts [Los Angeles Co.]. The district, which gained widespread notoriety because of race riots in 1965, was named for C. H. Watts, a Pasadena realtor around 1900. He had a large ranch at this place. According to the *Los Angeles Times*, Oct. 10, 1965, it was originally called Mud Town, and it was annexed to Los Angeles in 1926 (Brother Henry).

Watts Valley [Fresno Co.]. Named for C. B. Watts, who settled in the valley formerly known as Popes Valley (Co. Hist.: Fresno 1933:48).

Waucoba (wə kō´bə): **Canyon, Mountain** [Inyo Co.]. From Owens Valley Paiute *wokóbï* or Northern Paiute *wogópi* 'bull pine' (Fowler); the English name is used locally for various species of pine. C. D. Walcott, former director of the USGS, gave this name, spelled Waucobi, to the ancient lake, the remnant of which is Owens Lake in Owens Valley (Chalfant, *Inyo*, p. 3).

Waugh Lake [Mono Co.]. Created by the construction of a dam in 1918 and named for E. J. Waugh, chief construction engineer of the Nevada-California and Southern Sierras Power Companies (W. L. Huber to P. Farquhar).

Waukell (wô kel′): **Creek, Flat** [Del Norte, Humboldt Cos.]. From the Yurok village name *wo'ke'l* (Waterman, p. 234).

Waukena (wô kē′ nə) [Tulare Co.]. The town was laid out by a development company in 1886. It is not certain whether the name is of Indian origin or whether it is Joaquin, given an English spelling and a feminine ending.

Wawona (wä wō′ nə) [Yosemite NP] is from Southern Sierra Miwok *wohwohna* 'sequoia' (Broadbent; Callaghan). The cabin that Galen Clark built there in 1857 was first known as Clark's Station—and, when Edwin Moore acquired half-interest in 1869, as Clark and Moore's. In 1875 the Washburn brothers bought the place and changed the name to Wawona.

Weaver Bally. *See* Bally.

Weaverville; Weaver Creek [Trinity Co.]. The place was called Weavertown or Weaverville for George Weaver, a prospector, who built the first cabin in 1850. It is mentioned as Weaverville in the *San Francisco Alta California*, May 1, 1852.

Webber: Lake, Peak [Sierra Co.]. The former Little Truckee Lake received its present name from David Gould Webber, who bought the surrounding land for a stock range in 1852. His place of residence is mentioned as Webber's Station in Brewer's Notes, Nov. 1861.

Weber Creek [El Dorado Co.] is the only place name that commemorates one of the most notable pioneers of central California. Charles M. Weber, a native of Homburg, Germany, and the founder of the city of Stockton, came to California with the Bartleson party in 1841. In 1848 he mined successfully at the creek, calling his camp Weberville. Weber Creek is mentioned by Col. Richard B. Mason on Aug. 17, 1848. On some maps the name of the creek is misspelled Webber.

Wedding Cake [Trinity Co.]. The elevation was so named by James King and his bride on their wedding trip about 1870 because, through their rose-colored glasses, it looked like a huge old-fashioned wedding cake (J. D. Beebe).

Wedertz (wed′ ərts) **Flat** [Mono Co.]. Named for Louis Wedertz, a pioneer, who owned the land adjacent to the flats (Maule).

Weed [Siskiyou Co.]. Named in 1900 for Abner Weed, founder of the Weed Lumber Company. Weed, a native of Maine, was present at Lee's surrender at Appomattox and came to California in 1869; he was county supervisor for many years and from 1907 to 1909 was a state senator.

Weed Patch [Kern Co.]. The place was so named because of a luxuriant growth of weeds caused by subirrigation (*see* Algoso; Patch).

Weeks Poultry Community. *See* Winnetka.

Weges. *See* Wages.

Weimar (wē′ mär) [Placer Co.]. When the post office was established in 1886, the Post Office Dept. rejected the name of the railroad station, New England Mills, because it was too long. In the 1950s the residents decided to call the place in memory of "old Weimah," a colorful chief of the "Oleepas." His mark X and the name "Weima" appear in a document published in the *Placer Times* on May 27, 1850. Since this spelling was apparently unknown in the 1880s, the name of the post office was spelled, though not pronounced, like that of the German city of Goethe and Schiller. **Weimar Hill** [Nevada Co.], mentioned by Browne in 1869 (p. 115), may also have been named for the Indian.

Weisel (wī′ səl) [Riverside Co.]. Named by the Santa Fe for Peter J. Weisel, owner of the silica plant to which a spur was built in 1927 (G. A. Van Valin).

Weitchpec (wich′ pek) [Humboldt Co.]. This town where the Klamath and Trinity Rivers meet takes its name from Yurok *wecpek* or *wecpus* 'confluence' (Waterman, p. 257; Harrington). In 1859 a post office was established at the white settlement opposite the Indian village and was called Weitchpec.

Welch's Store. *See* Plainsburg.

Welch Station. *See* Sims.

Weldon [Kern Co.]. Named for William Weldon, a stockman of the 1850s (R. Palmer).

Wells Peak [Mono Co.]. The name was applied by the BGN at the suggestion of the USFS, in memory of John C. Wells, supervisor of Mono National Forest, 1901–22.

Wells Peak [Yosemite NP] was named by R. B. Marshall in the 1890s in honor of Lt. Rush S. Wells, a native of New Mexico (Farquhar).

Wendel [Lassen Co.]. The station was named about 1905 by Thomas Moran, president of the Nevada-California-Oregon Railroad, for a friend whose first name was Wendel (A. C. Riesenman). The post office is listed in 1915.

Wendling. *See* Navarro.

Wengler [Shasta Co.]. The post office was established about 1900 and named for E. M. Wengler, part owner of the Wengler and Buick mill and first postmaster (Steger).

Weott (wē′ ot) [Humboldt Co.]. An alternative spelling of Wiyot, the name given by anthropologists to the Indians who inhabited the Eel River and Humboldt Bay regions. The term is from *wíyo't*, the Wiyot name for the Eel River (Teeter 1958). Weott post office is listed in 1926.

West Butte [Sutter Co.]. When the post office was established in 1867, it was so named because of the location of the settlement on the west side of the Sutter Buttes.

Westend [San Bernardino Co.]. The name was changed in 1919 from Hanksite by M. ("Borax") Smith, president of the West End Consolidated Mining Company. The post office was established on Oct. 11, 1919.

Westfalls Meadow [Yosemite NP] was named for a German family who had a homestead near the junction of the Mariposa and old Mono trails; it is shown on the Hoffmann-Gardner map of 1867.

Westgard Pass [Inyo Co.]. A toll road was built across the pass by Scott Broder in 1873, and the route is still known as Toll Road. About 1913 A. L. Westgard of the American Automobile Association crossed the pass in search of the best transcontinental route, and the Good Roads Club of Inyo Co. named the pass for him (Brierly).

Westlake Village [Ventura Co.]. The city was named for its location on an artificial lake, to the west of the Los Angeles metropolitan area.

Westley [Stanislaus Co.]. Named by the citizens in 1888, in memory of John Westley Van Benschoten—a butcher who came with Frémont in 1846, served in the Mexican War, and settled on the San Joaquin in 1850 (W. W. Coz).

West Lodi. *See* Kingdon.

Westminster [Orange Co.]. Named in the 1870s by the Rev. L. P. Webber, who founded a colony here for people sympathetic with the ideals of the Presbyterian Church as laid down in the Westminster Assembly (1643–49). It is one of the older settlements in the county.

Westmorland [Imperial Co.]. Local sources maintain that the place was named by the developers in 1910 to call attention to more land to the west in Irrigation District no. 8—and not after a Westmorland "back East," or after the county in England.

Weston Beach [Monterey Co.]. Named for the photographer Edward Weston (1886–1958), who lived and worked in this area.

Weston Meadow [Tulare Co.]. Named for Austin Weston, of Visalia, who made this mountain meadow his headquarters for summer stock grazing (Farquhar).

West Point [Calaveras Co.]. An old mining town, called Indian Gulch in 1852 and West Point in 1854. It is mentioned in Hutchings' *Illustrated California Magazine* of 1859 (p. 490).

Westport [Mendocino Co.]. The place was called Beal's Landing for some time, for the first settler, Samuel Beal: "In 1877 James T. Rodgers began the construction of a chute, and to him belongs the honor of giving to the place its present name, he being from Eastport, Maine, naturally called the new town Westport" (Co. Hist.: Mendocino 1880:470).

Westwood [Lassen Co.]. In 1913, when the Red River Lumber Company of Minnesota began extensive operations in northern California, its officials called the headquarters Westwood, probably because the company already had eastern holdings. The Southern Pacific reached the place the following year and applied the name to the station. Several other places in the state bear this name, including Westwood in the city of Los Angeles.

Wether Ridge [Sonoma Co.]. No other feature in the state seems to have been named for the wether, a castrated male sheep. The highest peak of the ridge is Buck Mountain.

Whale, Whaler. The once-important industry of whaling has left its imprint on names of coastal features in all parts of the world (*see* Ballena). Along the California coast are three Whale Rocks, one Whaleboat Rock, one Whalers Rock, two Whaler Islands, and one Whalers Knoll. **Whaleman Harbor** [San Mateo Co.] was probably named for someone called Whaleman. **Whaleback** [Kings Canyon NP]: The shape of the mountain, which extends for 3 miles along Cloud Canyon, suggested the name; it was probably applied when the USGS mapped the Tehipite quadrangle in 1903. *For* Whaler's Harbor, *see* Ballena; Richardson Bay.

Wheatland [Yuba Co.]. When the railroad was built from Sacramento to Marysville in 1867, the name was applied to the station because it was in a wheat-growing district. The post office is listed in 1867.

Wheaton Dam [Stanislaus Co.]. Named for M. A. Wheaton, who purchased the dam and water rights on the Tuolumne River in 1855 (S. P. Elias, *Stories of Stanislaus* 1924:20).

Wheatville. *See* Kings: Kingsburg.

Wheeler. The state has four orographic features so named: Crest [Inyo, Mono Cos.], Peak [Mono Co.], Ridge [Kern Co.], and Peak [Tuolumne Co.]. The peak was probably named for some unidentified army officer (Farquhar); the other three may have been named for the army engineer George M. Wheeler (*see* Glossary). By decision of June 7, 1911, the BGN designated the chain west of Round Valley as Wheeler Ridge, but it continued to appear on some maps as Wheeler Crest. None of the features are shown in the Wheeler atlas. **Wheeler Springs** [Ventura Co.] was probably named for some other Wheeler.

Wheelock. *See* Fort Jones.

Whipple: Mountains, Well, Wash [San Bernardino Co.]. In 1858 Lt. Joseph C. Ives named a mountain for Lt. Amiel W. Whipple, a member of the Mexican Boundary Commission and of the Pacific Railroad Survey: "The Needles and a high peak of the Monument range, which I have called Mount Whipple, are the most conspicuous landmarks, and designate points where the river enters and leaves Chemehuevis valley" (Ives 1:60). The von Leicht–Craven map shows the highest peak of this range as Mount Whipple. When the USGS mapped the Parker quadrangle in 1902–3, it applied the name to the great mountain mass in the bend of the Colorado—including the Monument Range, a fitting monument to the man who played an important role in the scientific delineation and recording of the geography of California, and who later was one of the Union generals killed at Chancellorsville. The old name is preserved in Monument Peak.

Whiskey. In the nomenclature of the state, this liquor is represented more often than all other alcoholic beverages combined. There are more than fifty Whiskey names, applied to all types of features, from a Butte [Siskiyou Co.]. to a Run [Placer Co.]. Various reasons have been given for the application of the name: a mule stumbled and dropped a barrel of the amber liquid in a creek; a moonshine still was operated in a secluded gulch; the miners at a certain place could consume a barrel a day; the drinking of the water of a spring made a person "rattle-weeded," etc. The many mining towns named Whiskey naturally changed their names when the country became self-conscious and respectable. **Whiskeytown** post office and **Reservoir** [Shasta Co.]: The place developed as a gold-mining camp in the early 1850s and was named Whiskey Creek, after the stream on which it was located. Whiskey Creek post office was established on Feb. 18, 1856, and lasted until 1864. When it was reestablished on Jan. 6, 1881, it bore more "respectable" names—Blair, Stella, Schilling (*q.v.*)—until July 1, 1952, when the alcoholic name was restored. Whiskeytown Reservoir was built in 1963. *For* Whiskey Creek [Shasta Co.], *see* Kern: Kernville; *for* Whiskey Hill [Santa Cruz Co.], *see* Freedom.

White. Next to "black" and "red," this is the adjective of color used most frequently in place naming. More than a hundred places in California are called White. Two-thirds of these are orographic features; among them are more than twenty White Rocks and about fifteen White Mountains. Although there are many Black Buttes, there is apparently not a single White Butte. The chief reasons for naming elevations White include a pale-colored rock formation, a covering of snow, or a deposit of guano. A number of White Rivers, Creeks, Lakes, etc., owe their name to white rock formations, or to the milky appearance of the water. The adjective often refers not to the feature or place itself, but rather to the specific term: White Wolf, White Oaks, White Pines, White House Landing, White Cabin Creek, White Mans Ravine, White Chief Mountain. (*For* White Horse in Siskiyou Co., *see* Kinyon.) About ten communities have White in some form or combination. Whites Bridge [Fresno Co.], White Hill [Marin Co.], and doubtless other places and features were named for a person or persons named White. (*For* White Mountain in Lassen NP, *see* Reading Peak.) **White Mountains** [Mono Co.] were named for what was once called White Mountain Peak: "White Mountain peak is named from the appearance of its summit which seems to be composed of this rock [dolomite], often mistaken for snow, and which is found in great quantities at its base" (Hanks, Report, p. 178). Whitney believed in 1864 that snow actually had given the name to the mountain: "There is a high peak called the 'White Mountain,' which is doubtless 14,000 feet or more in elevation, as it has . . . snow on its southern side, . . . [which] never entirely disappears." In 1917 the USGS transferred the name to Montgomery Peak. **Whitehouse Creek** [San Mateo, Santa Cruz Cos.] received its name from a white house, built in 1857 or before by Isaac Graham. It is said that this house served as a landmark to navigators until the growing trees obscured it (Hoover, pp. 411–12). **White River** [Tulare Co.] was originally called Tailholt because (according to the most convincing of several stories) a miner had nailed a cow's tail on his cabin door to serve as a handle. When a post office was about to be established in the 1860s, Levi Mitchell, who thought that Tailholt would not make a very good name, substituted the name White River, after the river on which the mining camp was situated (Mitchell). **White Horse Valley, Reservoir**, and post office [Modoc Co.]: According to local tradition, surveyors named the valley about 1870 because they saw wild white horses there (Co. Library). The post office is listed in 1930. **White Water** [Riverside Co.]: The post office was established in 1926 and named after the river, which in turn received its name because of its milky appearance, caused by deep fine sand. The name was earlier applied to a ranch east of the spot where the river disappears in the sand (Co. Hist.: Riverside 1912:215). The name was formerly given to the Palm Springs railroad station. **White Pines** [Calaveras Co.]: The post office, listed in 1941, was named because of the presence of a species of pine (*Pinus monticola*). **White Plains** [Mendocino Co.] were so named because the topsoil is highly charged with alkaline salts and appears white in many places. This peculiar soil condition causes the dwarf cypress to develop in miniature and other conifers to grow in dwarf form.

White Island [San Pedro Bay, Los Angeles Co.]. For astronaut Edward H. White, 1930–67 (BGN, list 7902).

Whitesboro [Mendocino Co.]. Named in 1876 for Lorenzo E. White, the principal owner of the Salmon Creek Mill Company (Borden). A post office was established on Oct. 11, 1881.

Whites Bridge [Fresno Co.]. The name was applied to the railroad station for James R. White, a Fresno pioneer, who built the bridge there (Co. Hist.: Fresno 1919:279). The station is on the Southern Pacific section between Newman and Fresno, completed in 1892.

White Wolf [Yosemite NP]. According to Paden and Schlichtmann (p. 153), two cattlemen, Diedrich and Heinrich Meyer, were pursuing horse thieves in the 1850s when they came upon a temporary Indian encampment and named the place for the chief of the Indian band, "White Wolf."

Whitlow [Humboldt Co.]. The post office was established on May 21, 1927, and named for Al Whitlow, a licensed surveyor (Virginia Wolf).

Whitmore [Shasta Co.]. Named for Simon H. Whitmore, a blacksmith, at whose instigation a post office was established in 1883 (Steger).

Whitmore's Ferry [King Co.]. *See* Kings: Kingston.

Whitney, Mount [San Diego Co.]. This mountain took its name from an eccentric old character who had a homestead on the west side of the mountain (Charles Kelley).

Whitney, Mount; Whitney: Pass, Creek, Meadow, Portal [Sequoia NP; Inyo Co.]. Josiah Dwight Whitney (1819–96), chief of the State Geological Survey, 1860–74, had forbidden his subordinates to name what is now Mount Hamilton for him. In July 1864 Whitney's four chief assistants—William Brewer, Charles Hoffmann, James Gardiner, and Clarence King—beheld from Mount Brewer what they rightly assumed to be the culminating peak of the Sierra Nevada. On this occasion they stood upon their privilege as discoverers and named it in honor of their chief. In 1871 Clarence King climbed the peak now known as Mount Langley, a few miles south, supposing it to be Mount Whitney. His error was discovered by others two years later. He hastened to the scene; but before he could get there, three fishermen—John Lucas, Charles D. Begole, and A. H. Johnson, all of Inyo Co.—made the first ascent on Aug. 18, 1873. There was an attempt to name the peak Fishermans Peak in their honor; but the name Mount Whitney was firmly established by 1881, when the summit was occupied by Prof. S. P. Langley for observations on solar heat.

(For a more detailed account, consult Farquhar, *Place names*.)

Whitney Creek. *See* Golden Trout Creek.

Whittemore Grove [Humboldt Co.]. Dedicated on Sept. 21, 1929, and named for Harris Whittemore, a philanthropist and pioneer in reforestation.

Whittier [Los Angeles Co.]. Founded in 1887 by the Pickering Land and Water Company, an organization of Quakers, and named at the suggestion of Micajah D. Johnson, in honor of a fellow Quaker, John Greenleaf Whittier (1807–1892), poet and reformer.

Whitton. *See* Planada.

Wible (wī′ bəl) **Orchard** [Kern Co.]. The station was named after S. W. Wible's orchard when the Southern Pacific line from Bakersfield to McKittrick was built in 1895. Simon W. Wible, a native of Pennsylvania, came to California in 1852. In 1872 he took up a homestead, and later he became superintendent for Miller and Lux. He was one of the notable pioneers of the county (Emily Easton).

Wickiup (wik′ ē up), from Fox (Algonquian) *wiikiyaapi* 'house' (Cutler, p. 111), designates a hut used by Indian tribes. It is found as a place name for a tributary of Tujunga Creek [Los Angeles Co.] and for a place near Viola [Shasta Co.].

Widow Reed Creek [Marin Co.]. The cumbersome name Arroyo Corte Madera del Presidio (Hoffmann's map of the Bay region, 1873) was replaced by the USGS when it named the stream in memory of the wife of John Reed (Read, or Reid), who built here the first sawmill in the area in 1834.

Wilbur Springs [Colusa Co.]. Named about 1868 for a Mr. Wilbur, who owned the resort.

Wilcox: Springs, Peak [Shasta Co.]. Named for the Wilcox family, which had a homestead here from 1872. Charles W. Wilcox came to San Francisco in 1851 (Steger).

Wildcat. More than fifty features in the state, mainly Canyons, Creeks, and Mountains, are named Wildcat (*see* Gatos). Most of these were named because wildcats (*Lynx rufus*) were seen, encountered, or killed there. Some names, however, may have been derived from the applied sense of the word, as a designation for an unsound scheme. *See* Red Dog.

Wildman Meadow [Kings Canyon NP]. "About 1881 Brother Jeff and I camped there with a band of sheep. After dark we were startled by a lot of unearthly yells like someone in distress. After spending a large part of the night we were unable to locate anyone and finally concluded that it must have been a wild man, and so named the meadow. Later we found the noise was caused by a peculiar looking owl" (Frank Lewis to P. Farquhar).

Wildomar (wil′ də mär) [Riverside Co.]. In 1883, when the old Rancho Laguna was subdivided, Margaret Collier Graham coined the name from the given names of the new owners: her brother William Collier, her husband Donald Graham, and herself (information from Ronald Baker). *See* Elsinore.

Wild River. *See* American River.

Wildrose: Spring, Canyon [Inyo Co.]. The spring was named in 1860 by Dr. S. G. George and the party with him, who were searching for the "lost" Gunsight Lode (Hanks, Report, p. 34); *see* Rose.

Wildwood [Shasta Co.] is a descriptive term, also found in Trinity and Sonoma Cos. **Wildwood Canyon** [San Bernardino Co.] was originally called Hog Canyon.

Wilfred Canyon [Inyo Co.]. Named for Wilfred Watterson, who in his younger days ran sheep successfully in that locality, and in his later days, until 1927, was an unsuccessful banker at Bishop (Robinson).

Willard Valley. *See* Cuddeback Lake.

Williams [Alpine Co.]. The name became attached to the place because Billy Williams settled at Red Lake "in the early days" (L. T. Price).

Williams [Colusa Co.]. Named for W. H. Williams, who laid out the town in 1876. The post office is listed in 1880.

Williams Butte [Mono Co.]. Named for Thomas Williams, who was granted a patent to the adjacent land in 1882 (Maule).

Williams Grove [Humboldt Co.]. The State Forestry Board, in 1922, named the grove for Solon H. Williams, a former chairman of the board (Drury).

Williamson, Mount [Inyo Co.]. Named in 1864 by Clarence King of the Whitney Survey, for Lt. Robert S. Williamson (*see* Glossary) of the Pacific Railroad Survey.

Williamson's Pass. *See* Soledad: Soledad Pass.

Williams Reservoir [Modoc Co.]. Constructed and named about 1910 by Curtis J. Williams, a local rancher (Co. Library).

Willis Hole [Del Norte, Siskiyou Cos.]. According to a 1965 decision of the BGN, this bog, 19 miles southeast of Gasquet, is named for the renowned botanist Willis Jepson (1867–1946)—but it is not Willis Jepson Hole, as the board's Decision List no. 6503 expressly states.

Willits: town, **Creek, Valley** [Mendocino Co.]. The post office was established in the late 1870s and named for Hiram Willits, who had settled in the district in 1857. The name is shown on the Land Office map of 1879 and is listed in the Postal Guide of 1880.

Willow. The willow (genus *Salix*) has always been extremely popular for place naming because its presence usually denotes running water. In California there are about twenty species, widely distributed in the state (*see also* Sauce). The maps show more than two hundred Willow Creeks, and more than a hundred Willow Springs, Sloughs, Lakes, and Valleys. Willow is also repeatedly found with the rare generic Glen—and once, in Humboldt Co., with the even rarer generic Brook. There are also a number of Lone Willow Creeks, and a Five Willows Springs [San Luis Obispo Co.]. The name is a favorite for communities and post offices, e.g. Willow Ranch [Modoc Co.], listed as a post office in 1880; Willowbrook [Los Angeles Co.]; Willow Creek [Humboldt Co.]; and Willow Springs [Kern Co.]. **Willows** [Glenn Co.]: "There was but one watering place in the plains south of Stony Creek. That was a willow pond, from which the present town of Willows took its name" (John B. De Jarnatt and Ellis T. Crane, *Colusa County* 1887:11). The town was laid out in 1876; the post office is listed as Willow from 1880 to 1917. *For* Willow Springs [Borrego State Park], *see* Santa Catarina Springs.

Will Rogers Beach State Park [Los Angeles Co.]. Established in 1930 and named in memory of the famous actor (1879–1935). His home was made a state park in 1944.

Will Thrall Peak [Los Angeles Co.]. The peak was named in memory of William H. Thrall, an early advocate of conservation, who had died in 1963 (BGN, 1963).

Wilmar Lake [Yosemite NP]. Named by R. B. Marshall, for Wilmar Seavey, daughter of Clyde L. Seavey (Farquhar). By decision of the BGN (1964), the name was changed from Wilmer.

Wilmington [Los Angeles Co.]. Named by Phineas Banning (*see* Banning), a leader in the early development of Los Angeles, after his birthplace in Delaware. It had been the terminal of the U.S. mail stage line since 1858 and was known as New San Pedro or Newtown until 1863.

Wilseyville [Calaveras Co.]. The post office was established on Sept. 16, 1947, and named for Lawrence A. Wilsey, an official of the American Forest Products Company, which has a subsidiary here (Marjorie Dietz).

Wilsie [Imperial Co.]. Named for W. E. Wilsie, a farmer, who came to the Imperial Valley in 1901 (*Desert Magazine*, July 1939). The name was applied to the station about 1917, when the San Diego and Arizona Eastern Railroad was under construction.

Wilson [Sutter Co.]. Probably named for George W. Wilson, a merchant in Marysville, who owned a large grain farm nearby about 1900 (Co. Library).

Wilson, Mount [Los Angeles Co.]. Named for Benjamin D. ("Don Benito") Wilson (1811–78), who built a burro trail up the mountain in 1864. Wilson, one of the best known of the American settlers in southern California after 1841, was the first mayor of Los Angeles under U.S. rule.

Wilson Creek [Yosemite NP]. Named by Lt. H. C. Benson for his friend Mountford Wilson, of San Francisco (Farquhar).

Wilton [Sacramento Co.]. The station of the Central California Traction line was named for the owner of the land, Seth A. Wilton, dairy and poultry rancher, who had lived in Sacramento Co. since 1887 (Co. Hist.: Sacramento 1923:311). The post office is listed in 1915.

Winchell, Mount [Kings Canyon NP]. The name honors Alexander Winchell (1824–91), professor of geology for many years at the University of Michigan. His cousin, Elisha C. Winchell, had given the name in 1868 to the present Lookout Point, overlooking Kings River Canyon. In 1879 Elisha's son, Lilbourne A. Winchell, bestowed the same name upon a peak south of the Palisades. The USGS transferred the name to a peak north of North Palisade, where it now rests (Farquhar).

Winchester [Riverside Co.]. Named for Mrs. Amy Winchester, an owner of the land when it was subdivided in 1886 (Gunther).

Windfield. *See* Winton.

Window Cliff [Tulare Co.]. "These cliffs are perforated with a window-like opening at the head of a gorge dropping into Kern Canyon, through which inspiring views of the [Sequoia National] park may be obtained" (BGN, 6th Report). *See* Ventana.

Windsor: town, **Creek** [Sonoma Co.]. The post office was established on Aug. 31, 1855, and was named at the suggestion of Hiram Lewis, a native of England, presumably after Windsor Castle. The place had previously been known as Poor Man's Flat.

Windy Gap. *See* Wingate Pass.

Winemas (wī nē′ məz) **Chimneys** [Lava Beds NM]. The rock formation was named for Winema, daughter of the chief of a small tribe at the outlet of Upper Klamath Lake. She married a white man, Frank Riddle, before the Modoc War and acted as interpreter for Gen. Edward Canby (Howard).

Wineville. *See* Mira: Mira Loma.

Winfield Scott, Fort [San Francisco Co.]. The fort was under construction from 1854 to 1876 and named in 1882 in memory of Gen. Winfield Scott (1786–1866), commander-in-chief of the U.S. Army, 1841–61 (*see* Fort Point).

Wingate: Pass, Wash [San Bernardino Co.]. When Lt. Bendire crossed the Panamint Mountains through this pass in 1867, he may have named it in memory of Maj. Benjamin Wingate, who died in 1862 of wounds received in the battle of Valverde, New Mexico. Wheeler's atlas sheet 65 records a Windy Gap at the west side of the pass, possibly because he misread Bendire's notes. This name was used more frequently, but the USGS decided for Wingate in 1915.

Winnedumah (win ə do͞o′ mə) [Inyo Co.]. The name of the remarkable 80-foot granite monolith commemorates Winnedumah, a Paiute medicine man—who, according to an often-repeated legend, was turned into this pillar while he was invoking the help of the "great spirit" during a battle with an enemy tribe. His name may contain Owens Valley Paiute *wïnï* 'to stand' or *wïnïdï* 'tree' (Fowler). The monolith is also known as Paiute Monument. *See* Tinemaha.

Winnemucca (win ə muk′ ə) **Lake** [Alpine Co.]. The local name is Roundtop Lake, after Roundtop Peak, under which the lake nestles. The map name was perhaps given to the lake by a fisherman from Nevada, where Winnemucca Lake is named for an eminent Paiute chief, and was recorded on the Markleeville atlas sheet of 1936 (Maule). The Paiute name is of unclear etymology, perhaps containing *moko* 'shoe' (Fowler).

Winnetka [Los Angeles Co.]. The post office, established on Apr. 1, 1947, was probably named after Winnetka, Illinois; that name is said to have been coined by whites on the basis of Algonquian *winne* 'beautiful'. The place was formerly known as Weeks Poultry Community.

Winnibulli (win′ ē bo͝ol ē) [Shasta Co.] is Wintu *wenem buli* 'middle mountain' (Shepherd); the English spelling Winnibully also occurs (*see* Bally).

Winston Ridge [Los Angeles Co.]. Named for L. C. Winston, a Pasadena banker, who died in a blizzard while hunting in the mountains in 1900 (S. B. Show).

Winterhaven [Imperial Co.]. The town that developed on the homestead of "Don Diego" Yeager, a pioneer of 1857, is surrounded by the Yuma Indian Reservation. When a post office was to be established in 1916, a group of women, enjoying a card party when the temperature reportedly was 120° in the shade, chose the name Winterhaven (T. J. Worthington).

Winters [Yolo Co.]. The town developed when the Southern Pacific built the line from Elmira to Madison in 1875. It was named for Theodore W. Winters, who donated half the land for the town.

Wintersburg [Orange Co.]. When the Southern Pacific extension was built to Huntington Beach in 1897, the station was named for Henry Winters, specialist in celery culture, who donated the right-of-way and part of the townsite (Co. Hist.: Orange 1921:873–74). The area is now part of the City of Huntington Beach (Brigandi).

Winters Peak [Inyo Co.]. The peak in the Funeral Mountains was named for Aaron Winters, who discovered borax deposits in Death Valley in 1881 (BGN, 1961).

Winton [Merced Co.]. The old Winn (or Wynn) Ranch was subdivided by the Coöperative Land and Trust Company in 1910, and the new town was called Windfield. When the railroad station was established in 1911, the Santa Fe requested a change; N. D. Chamberlain suggested the new name, apparently for Edgar Winton, one of the surveyors who laid out the town.

Wintun (win′ to͞on): **Glacier, Butte** [Siskiyou Co.]. The southeast glacier of Mount Shasta and the butte southwest of the mountain were doubtless named for the Wintu or Wintun Indian tribe; the spellings Wyntoon and Wintoon are also found. The Wintu form is *winthuun* 'Indian', from *win-* 'people' (Shepherd). The butte, however, is generally known as Black Butte, a name which was made official by a BGN decision of June 6, 1934.

Wishon (wish′ on): **Dam, Lake,** post office [Madera Co.]. The post office was established on Nov. 12, 1923, and named for A. Emory Wishon, of the San Joaquin Light and Power Corporation, later vice president and general manager of the Pacific Gas and Electric Company. The dam was built in 1955–58.

Witch Creek [San Diego Co.]. The tributary of Santa Ysabel Creek was so named because its Indian name sounded to early settlers like the Spanish *hechicera* 'witch'.

Wit-So-Nah-Pah (wit sō nä′ pə), **Lake** [Mono Co.]. Probably from a Numic language, but no etymology has been confirmed.

Wittawaket (wī′ ti wä kət) **Creek** [Shasta Co.]. From Wintu *witee waqat* 'turn creek', containing *wit-* 'to turn' and *waqat* 'creek' (Shepherd).

Wittenberg, Mount [Marin Co.]. The mountain was not named after Prince Hamlet's alma mater, but for a man named Wittenberg who had a ranch there (J. C. Oglesby).

Witter Springs [Lake Co.]. Named in 1871 for Dr. Dexter Witter, who bought the land in partnership with W. P. Radcliff from Benjamin Burke, the discoverer of the springs in 1870. The post office was established as Witter's Springs on Mar. 7, 1873; reestablished as Witter on June 7, 1901; and changed to the present name on May 5, 1913.

Wofford Heights [Kern Co.]. The post office was established on May 16, 1953, and named after the community founded by I. L. Wofford (Helen Hight).

Wohlford, Lake [San Diego Co.]. Named in 1924 by the Escondido Mutual Water Company for one of its officers, A. W. Wohlford, an Escondido banker and citrus grower. The lake was formerly known as Lake Escondido (Co. Hist.: San Diego 1936:145, 344, 399).

Wolfskill: Camp, Canyon, Creek, Falls [Los Angeles Co.]. The names commemorate one of the most notable pioneers of southern California. William Wolfskill (1798–1866), a native of Kentucky, came to California over the Old Spanish Trail in 1831, settled in Los Angeles in 1836, and became a pioneer of California's greatest industry, fruit growing.

Wolfskill [Solano Co.]: In 1842 Wolfskill was granted the Río de los Putos grant. His brothers John and Sarchel managed the rancho, and later two other brothers, Malthus and Milton, also settled on the vast estate. The place is shown as Wolfskills on Gibbes's map (1852), but as Wolfs Hill on Eddy's map (1854).

Wolverton Creek [Sequoia NP]. Named for James Wolverton, of Three Rivers, hunter, trapper, and veteran of the Union Army, who named the General Sherman Tree in 1879 (Farquhar).

Wonoga (wō nō′ gə) **Peak** [Inyo Co.]. Perhaps contains Mono *wono'* 'burden basket' (McLaughlin).

Wood. The combination of the word "wood" with a generic name is a favorite for American place names. California has at least twelve communities so named, including Woodacre [Marin Co.], Woodcrest [Riverside Co.], Woodlake [Tulare Co.], Woodville [Tulare Co.], and Woodside [San Mateo Co.]. **Woodland** [Yolo Co.] was settled before 1855 by Henry Wyckoff, who opened a trading post there. When the post office was established in 1859, it was named Woodland, at the suggestion of Maj. S. Freeman, who had bought Wyckoff's store.

Wood, Mount [Mono Co.]. Named in 1894 by Lt. N. F. McClure, for Capt. Abram E. Wood, veteran of the Civil War and acting superintendent of Yosemite National Park, 1891–93 (Farquhar).

Woodall Creek [Shasta Co.]. George Woodall was one of the owners of the Dry Mill at Shingletown; his sons later gardened at the stream that now bears the pioneer family's name (Steger).

Woodbridge [San Joaquin Co.]. In 1852 Jeremiah H. Woods began operating a ferry across the river, and the settlement that developed there became known as Woods' Ferry. In 1859 a bridge was built, and the name of the place was changed to Woodbridge. The post office is listed in 1857 as Wood's Ferry, with J. Woods as postmaster.

Wooden Valley [Napa Co.]. The valley, a part of Rancho Chimiles, was formerly called Corral Valley but became known by the present name when John Wooden purchased the land in 1850 (D. T. Davis).

Woodfords [Alpine Co.]. Known first as Careys Mill, for John Carey, who established a sawmill here in 1853 or 1854. When Daniel Woodford became owner of the mill in 1869, the name was changed to Woodfords, and this name was adopted by the Post Office Dept. (Maule).

Woodland. *See* Wood.

Woodleaf [Yuba Co.]. James Wood purchased the Barker Ranch in 1858, and the settlement that developed there became known as Woodville. When the post office was established about 1900, the name was changed to Woodleaf to avoid confusion with Woodville in Tulare Co.

Woodman: Creek, Hill [Shasta Co.]. Named for the brothers L. C. and George Woodman, who settled on Little Cow Creek in 1852 (Steger).

Woods: Creek, Crossing [Tuolumne Co.]. Named for an early miner whose claim was on the North Fork of the Tuolumne, some 8 miles southeast of Jamestown. The name Woods is recorded on Derby's map of 1849.

Woods: Creek, Lake [Kings Canyon NP] were named by J. N. LeConte, for Robert M. Woods, a sheepman of the Kings River region, who spent many summers from 1871 to 1900 in the Sierra (Farquhar).

Wood's Dry Diggings. *See* Auburn.

Wood's Ferry. *See* Woodbridge.

Woodside [San Mateo Co.]. According to the County History, 1883, the place was named after a lumber camp as early as 1849. The post office was established on Apr. 18, 1854.

Woodson Bridge Recreation Area [Tehama Co.]. The property was purchased by the state in 1959 and was named after the bridge across the Sacramento River, called after Warren Woodson.

Woodson Mountain [San Diego Co.]. Named for Dr. Marshall Clay Woodson, who had 320 acres of land at the base of the mountain in the 1890s (J. Davidson).

Woodville. *See* Woodleaf.

Woodworth, Mount [Kings Canyon NP]. Named about 1888 for Benjamin R. Woodworth, who lived for a time in Fresno and was with a camping party in Simpson Meadow when the peak was named (Farquhar).

Woody [Kern Co.]. Named for Sparrell W. Woody, a pioneer rancher of the 1860s.

Woosterville. *See* Ione.

Workman Hill [Los Angeles Co.]. The name of the elevation commemorates William ("Julian") Workman, a native of England who came to California in 1841 and was co-grantee in 1845 of La Puente grant, on which the hill is situated.

Worthla (wûrth′ lə) **Creek** [Del Norte Co.]. From Yurok *wrɬri* (Waterman, map 9, item 57).

Wouk Grove [Humboldt Co.]. The Abe Wouk Memorial Grove in Grizzly Creek State Park was named in 1966 by the Save-the-Redwoods League in memory of the son of Herman Wouk, well-known American novelist.

Wounded Knee Mountain. *See* Broken Rib Mountain.

Wragg: Canyon, Creek [Napa Co.]. The canyon opening into the lower end of Berryessa Valley was named for the first white settler in the vicinity, whose name was Wragg (D. T. Davis).

Wright: Lakes, Creek [Sequoia NP]. The name was proposed by the Sierra Club in 1925 in honor of James W. A. Wright, who had accompanied W. B. Wallace and H. Wales on a trip to the Kern River and Mount Whitney in 1881. A mountain on the Great Western Divide was named for him in 1881, but the name never became current (Farquhar).

Wright Chimneys [Lava Beds NM] were named in memory of Ben Wright, whose party was massacred by Indians on Apr. 26, 1873, in the Modoc War.

Wrights [Santa Clara Co.]. A station on the Los Gatos–Santa Cruz railroad, named in 1880 for John V. Wright, son of James R. Wright, a pioneer of the region (Hoover, p. 532).

Wrightwood [San Bernardino Co.]. The old Circle C Ranch was made a residential community about 1924 and was named for the subdivider, a Mr. Wright. The post office is listed in 1929.

Wunpost (wun′ pōst) [Monterey Co.]. A railroad point named after a Chinese worker named Wun; the "post" refers to his grave marker (S. Stevens).

Wutchumna (wə chum′ nə) **Hill** [Tulare Co.]. Probably reflects an early misspelling of Wukchumne, the name of a Yokuts Indian group (Verstegge); the name of the Indian group is also found as Wikchamni.

Wyandotte (wī′ ən dot): **Creek**, town [Butte Co.]. Named for a party of Wyandot (Iroquoian) Indians from Kansas who prospected in the vicinity in 1850 (Co. Hist.: Butte 1882:266). The Wyandott Mining Company had been organized in 1849 by William Walker, a Wyandot, who was later (1853) provisional governor of Nebraska (Bruff, p. 608). The name is found also in Los Angeles and

San Joaquin Cos. The tribe's name in its own languge is *wendat* (Chafe).

Wyeth (wī′ əth) [Tulare Co.]. An interesting and unique name, derived from the term "wye," the joining of two railroad tracks to form the letter *Y*. The name was applied to the junction in 1913–14, when the Santa Fe built the section from Cutler to connect the Minkler-Exeter line with the Reedley-Visalia line (Santa Fe). In the decades of great railroad expansion, stations near this type of junction often were given the word "Wye" as the generic part of the name. **Napa Wye** [Napa Co.] seems to be the only survivor.

Wyman Creek [Inyo Co.]. Named for Dan Wyman, who came to the White Mountains in 1861 seeking placer mines (Chalfant, *Inyo*, p. 143).

Wynema. *See* Hueneme.

Wynne, Mount [Kings Canyon NP] was named in memory of Sedman W. Wynne, former supervisor of Sequoia National Forest, who lost his life while on duty (BGN, 6th Report).

Wyntoon. *See* Wintun.

XYZ

Xamacha. *See* Jamacha.

Ximeno (hi mā′ nō) [Los Angeles Co.]. Spanish *j* often appeared as *x* in earlier Spanish spelling, and still in a few examples like *México* = *Méjico*. The name may preserve the memory of Manuel Jimeno Casarín; *see* Jimeno.

Yager (yä′ gər): **Creek**, settlement [Humboldt Co.]. The name for the stream is mentioned in Oct. 1858 in connection with Indian attacks and was doubtless given for one of the enterprising settlers who found the rich grazing lands suitable for stock raising.

Yahi (yä′ hē) **Indian Camp** [Tehama Co.]. For the local Yahi tribe, a branch of the Yana; *see* Ishi.

Yajome [Napa Co.]. The name of a land grant dated Mar. 16, 1841, also spelled Llajome; the origin is probably Indian but cannot be traced. Yajome Creek is mentioned by Pablo de la Guerra on Jan. 4, 1850 (California Senate *Journal*, 1st sess., p. 415), but the name has not survived.

Yalloballey. *See* Bally; Linn, Mount.

Yaney Canyon [Mono Co.]. The canyon is called Huntoon Valley on the maps of the Wheeler Survey (1877), but it was later named for the owner of the sawmill, a Mr. Yaney (Maule, p. 9).

Yankee. Next to Germans, who were invariably called Dutch, the Americans from east of the Mississippi and north of the Ohio seemed to be the only people who became known by a nickname. Numerous topographic features as well as some settlements still bear the name Yankee. Most, if not all, of these names go back to the periods of the Gold Rush when California had numerous mines and camps called Yankee Jack, John, Jim, Maid, Girl, Doodle, etc.; there was even a Wild Yankee Hill. **Yankee Jims** [Placer Co.] is probably the best-known surviving settlement bearing the appellation. It had a post office from 1852 to 1940. According to the Directory of Placer County (1861:12–13), the character known as "Yankee Jim" was a Sydneyite whose name was Robinson. He built a corral for stolen horses at this place and was hanged in 1852. (*See also* Divide, The.) **Yankee Hill** [Tuolumne Co.] was another famed mining camp of the 1850s and seems to have been named for a Mr. Hill nicknamed "Yankee." **Yankee Hill** [Butte Co.]: The post office established with this name on Oct. 19, 1856, lasted almost a century.

Yaqui (yä′ kē) **Camp** [Calaveras Co.]. The place on Willow Creek was named for the Yaqui Indians who came from Mexico after the discovery of gold (*Las Calaveras*, July 1957). The Yaqui, who now live in Sonora and Arizona, call themselve *hiaki;* this term was recorded by the Spanish missionary Andrés Pérez de Ribas in 1645 (*HNAI*, 10:262). There is also a Yaqui Pass and Well in San Diego Co. (Brigandi).

Yarro, El. *See* Jarro.

Ybarra (ē bär′ ə) **Canyon** [Los Angeles Co.]. Probably named for a member of the Ybarra (or Ibarra) family, some of whom settled in the Los Angeles district as early as 1814.

Ycatapom (wī kä′ tə pōm, wī kat′ ə pōm) **Peak** [Shasta and Trinity Cos.]. From Wintu *wayk'odipom* 'north step place', containing *way* 'north', *k'od-* 'to step', and *-pom* 'place' (Shepherd). **Ycotti** (wī kot′ ē) **Creek** [Shasta Co.] may represent a contracted form of the same name.

Ydalpom (wī dal′ pom) [Shasta Co.]. From Wintu *waydalpom* 'place to the north', from *way* 'north' + *-dal* 'direction' + *-pom* 'place' (Shepherd).

Yegua. The Spanish word for 'mare' was used more often for place names than *caballo* 'horse' (*see* Cavallo; Mare Island). It has survived in Yeguas (yā′ gwəs) Mountain [Kern, San Luis Obispo Cos.] and Las Yeguas Canyon [Santa Barbara Co.].

Yellow Jacket Creek [Trinity Co.]. Named for an Indian family by that name, owners of a claim of 160 acres in the Trinity National Forest. A nearby claim is listed in the name of Sally Jacket (K. Smith).

Yeomet (yō′ mət) **Creek** [Amador Co.]. The mining town at the forks of Cosumnes River and Big Indian Creek was probably named for an Indian rancheria. When it was still in El Dorado Co., it had a post office, spelled Yornet, from 1854 to 1861.

Yerba Buena 'good herb' is Spanish for the sweet-scented creeper *Micromeria chamissonis*, named for Adelhert von Chamisso, the botanist of the first Kotzebue expedition. The herb was found by Font near Mountain Lake in San Francisco in Mar. 1776 (*Compl. diary*, p. 333). By 1792 it had become a place name: on Nov. 14, 1792, the *comandante del puerto* (at San Francisco Bay) reported the arrival of Vancouver, who had anchored at a *parage que llamamos la Yerba buena* 'place which we call Yerba Buena' (SP Sac. 1:116). The cove that was used as an anchorage by later navigators and near which Richardson and Leese built their houses in 1835 and 1836, as well as the settlement that developed around these houses, was called Yerba Buena until Mar. 10, 1847 (*see* San Francisco). **Yerba Buena** (yûr′ bə bwā′ nə) **Island** in San Francisco Bay, which now alone preserves the old name, was called Isla de Alcatraces 'island of pelicans' on Ayala's map of 1775, but it is mentioned as la Ysla de la Yerba Buena as early as 1795 (Arch. MSB 2:33). In American times it became known as Goat Island because it was populated with hundreds of goats in the 1840s (*see* Goat). By act of the legislature in June 1931, the old Spanish name was restored. The name Yerba Buena was also applied to a land grant in Santa Clara Co., dated Nov. 25, 1833, and Feb. 24, 1840. There is a Yerba Buena Creek in San Luis Obispo Co.

Yermo (yûr′ mō): town, **Valley** [San Bernardino Co.]. When the San Pedro, Los Angeles, and Salt Lake Railroad (now the Union Pacific) was built in 1904–5, the station was called Otis. When the post office was established about 1908, the name was changed to Yermo (Spanish for 'barren'), a very appropriate name.

Yettem (yet′ əm) [Tulare Co.]. In 1902 the Rev. Mr. Jenanyan applied the Armenian word for 'paradise' to the Armenian settlement. The railroad station now called Calgro was called Yettem from 1910 to 1936 (*see* Cal-).

Ygnacio (ig nä′ sē ō, ig nā′ sē ō) **Valley** [Contra Costa Co.]. The name for the region north of Walnut Creek appears as Ignasio Valley on Hoffmann's map of the Bay region (1873) and is mentioned with the present spelling in the County History of 1878. The proximity to Martinez (*q.v.*) seems to indicate that the valley was named for Ygnacio Martínez. An old-time pronunciation was (nash′ ō).

Yokaya; Yokayo. *See* Ukiah.

Yokohl (yō′ kôl): **Creek, Valley** [Tulare Co.]. The names in the Blue Ridge country were derived from *yokol* or *yokod*, the name of a Yokuts tribe. The meaning is not known (Kroeber).

Yolano (yō lä′ nō) [Solano Co.]. The name of the Sacramento Northern station was coined from Yolo and Solano.

Yolla Bolly (yō′ lə bō lē, bō lə) **Mountains** [Trinity, Tehama Cos.] represents Wintu *yoola buli* 'snow mountain', from *yoola* 'snow' and *buli* 'mountain' (Shepherd); *see* Bally. BGN, list 8202, distinguishes the North and South Yolla Bolly Mountains.

Yolo (yō′ lō): **County**, town, **Basin**. According to Vallejo's Report, the name is a "corruption of the Indian word 'Yoloy,' signifying a place abounding with rushes (*tular*)." Kroeber states that Yodoi was the name of a Patwin village at the site of Knights Landing (AAE 29:261). Barrett was informed that there had been a chief named Yodo (*Pomo*, p. 294). A Rancho del Ioleo or Rancho de Dioleo is shown on the south side of Cache Creek on Bidwell's map of 1844. It is coincidental that a captain of a rancheria at Tomales Bay in 1810 bore the name Yolo (PSP 19:278). The county, one of the original twenty-seven, was created and named on Feb. 18, 1850. Yolo post office was established on Feb. 3, 1853. The place was formerly called Cacheville (*see* Cache).

Yontockett (yon′ tok ət) [Del Norte Co.]. From the Tolowa (Athabaskan) village *yan'-dagəd*, lit. 'southward uphill' (Golla).

Yorba Linda (yôr′ bə lin′ də) [Orange Co.]. This place preserves the name of one of the oldest pioneer families of southern California, whose ancestor, José Antonio Yorba (1743–1825), was one of Fages's Catalonian volunteers in 1769. The name Yorba is shown on Narváez's map of 1830 and on Duflot de Mofras's map of 1844. The community of Yorba, now gone, was on the Cañón de Santa Ana grant of Bernardo Yorba, the son of Antonio; a post office is listed in 1880. About 1913 real estate promoters coined the name of the new community from Yorba plus nearby Olinda (*q.v.*), not from Spanish *linda* 'pretty'.

Yorkville [Mendocino Co.]. Named around 1870 for an early settler, Richard H. York of Tennessee.

Yornet Creek. *See* Yeomet Creek.

Yosemite (yō sem′ i tē): **Valley, Falls, Creek, National Park**. From Southern Sierra Miwok *yohhe'meti* or *yoṣṣe'meti* 'they are killers', derived from *yoohu-* 'to kill', evidently a name given to the Indians of the valley by those outside it (Broadbent; cf. Beeler 1955). The Indians who lived in the valley called it *awooni*, modern Ahwahnee (*q.v.*). A derivation from Southern Sierra Miwok *ìhììmaṭi* or *ìṣììmaṭi* 'grizzly bear', which has often been proposed, cannot be supported. The valley was first seen but not named by the Walker party in 1833 (*see* Walker Pass). Edwin Sherman (*CHSQ* 23:368) claimed discovery of the valley in the spring of 1850, naming it "The Devil's Cellar." In Mar. 1851 it was entered by the Mariposa Battalion and named at the suggestion of L. H. Bunnell. "I then proposed 'that we give the valley the name of Yo-sem-i-ty, as it was suggestive, euphonious, and certainly *American;* that by so doing, the name of the tribe of Indians which we met leaving their homes in this valley, perhaps never to return, would be perpetuated'; upon a viva voce vote being taken, it was almost unanimously adopted" (Bunnell, pp. 61–62). In 1852 Lt. Tredwell Moore replaced the final *y* with an *e*, and this version was adopted for the publications and maps of the Whitney Survey. The name "Yo Semite Valley" was officially used when Congress by act of June 30, 1864, granted the valley to the state of California as a recreational reserve. The national park was created by act of Congress on Oct. 1, 1890, and the name was given by John W. Noble, secretary of the interior. On Mar. 3, 1905, the legislature receded Yosemite Valley to the United States, and on June 11, 1906, it became a part of Yosemite National Park. The name is found elsewhere in the state: Yosemite Incline [Nevada Co.], Yosemite Lake [Merced Co.].

You Bet [Nevada Co.]. According to the County History (1880:71), augmented by local tradition, the name originated in 1857 when Lazarus Beard, the saloonkeeper, was discussing a name for the settlement with two of his customers. Since Beard was providing free whiskey, one customer, hoping to prolong the discussion, suggested jokingly that "You bet," the saloonkeeper's favorite phrase, would make a good name. To the customer's chagrin Mr. Beard immediately accepted the suggestion, the discussion was ended, and the place was called You Bet.

Young, Mount [Sequoia NP]. The Rev. F. H. Wales climbed the mountain on Sept. 7, 1881, and named it for Charles A. Young (1834–1908), professor of astronomy at Princeton.

Young Lake [Yosemite NP] was named for Gen. Samuel B. M. Young (1840–1924), who was acting superintendent of the national park in 1896. General Young was a veteran of the Civil and Spanish-American Wars and of many Indian campaigns; in 1903–4 he was chief of staff of the U.S. Army (Farquhar).

Youngstown [San Joaquin Co.]. A group of settlers, led by their pastor, the Rev. J. O. Boden, named the place in 1903 after Youngstown, Pennsylvania, their former home (Wallace Smith, p. 433).

Yountville (yount′ vil) [Napa Co.]. Named for George C. Yount (1794–1865), of North Carolina, who came to California with the Wolfskill party in 1831; in 1836 he was grantee of Rancho Caymus, on which the town is situated. The name Yountville is shown on a map of the grant in 1860. The name of the settlement was formerly Sebastopol.

Yparraguirre (pâr ə gâr′ ē) **Canyon** [Mono Co.]. A Basque surname.

Yreka (wī rē′ kə): city, **Creek** [Siskiyou Co.]. The city started as a mining camp in the summer of 1851; it was first called Thompson's Dry Diggings, and later Shasta Butte City. The name was changed by the legislature on Mar. 22, 1852, when Siskiyou Co. was established: "The County Seat of said County shall be located at Shasta Butte City, and shall be known by the name of Yreka" (*Statutes* 1852:233). The name is derived from *wáik'a'*, the Shasta name for Mount Shasta (as recognized by Gibbs, in Schoolcraft 3:165). The name is recorded with various spellings: Gibbes's map of 1852 has Wyreka. "It was intended the county seat should bear the name of I-eka, the Indian name of Mount Shasta, but by mistake the name of Wyreka was substituted and the error continued, with the exception of dropping the letter W, thought to be superfluous" (*Yreka Journal*, July 12, 1876).

Yridisis (i ri dē′ sis) **Creek** [Santa Barbara Co.]. The name may be from California Spanish *iriris* 'dogwood tree, Cornus sp.' (John Johnson).

Ysidora (iz ə dôr′ ə) [San Diego Co.]. When the railroad to Fallbrook was built in 1882, this name was given to a station on the Santa Margarita rancho. It was doubtless named for the sister of Andrés and Pío Pico, Ysidora Forster, whose husband, John Forster, had purchased the rancho in 1864.

Yslais, Arroyo de los. *See* Islay.

Yuba (yo͞o′ bə): **River, City, County**. The river was discovered by Jedediah Smith on Mar. 14, 1828, and given the Indian name Henneet. When Sutter came to the valley he named the stream Yubu after the Maidu village (spelled Yubu, Yupu, Jubu by early settlers) near the confluence of Yuba and Feather Rivers: "The tribe I found at, and which still remains at the old rancheria at Yuba City, informed me that the name of their tribe was Yubu (pronounced Yuboo). As this tribe lived opposite the mouth of the river from which your county takes its name, I gave that river the name Yubu, which it has ever since borne" (*Marysville Herald*, Aug. 13, 1850). The name may be related to Nisenan (Southern Maidu) *yubuy* 'shade, shadow' (Shipley). Río de los Yubas is shown on several *diseños*. The often-repeated statement that the river was originally called Río de las Uvas 'river of the grapes' is not supported by evidence. On Dec. 22, 1844, the name Yuba was applied to a land grant, which was later rejected by the Supreme Court. Yuba City was laid out and named after the river in Aug. 1849. The county, one of the original twenty-seven, was created and named on Feb. 18, 1850. In 1950 the BGN abolished the confusing names of the numerous branches of the Yuba River (Little North Fork of the Middle Fork of the North Fork!) by naming the three main streams North, Middle, and South Yuba River, and by giving individual names to the minor branches.

Yucaipa (yo͞o kī′ pə): **Creek, Valley**, town [San Bernardino Co.]. The name is mentioned on Oct. 15, 1841 (Arch. LA 2:118 ff.). It is supposedly derived from the Guachama (Serrano) dialect and means 'wet or marshy land' (O. J. Fisk).

Yucca (yuk′ ə): **Creek, Mountain, Ridge** [Sequoia NP]. These were doubtless named for the yucca that grows there in abundance (David Brower). **Yucca Valley** [San Bernardino Co.]: The post office was established on Nov. 15, 1945, and was so named because of the abundance of *Yucca brevifolia*, the Joshua tree. Other places in the southern counties are named after various species of Yucca (*see* Joshua Tree; Palmdale).

Yuha (yo͞o′ hä): **Well, Plain, Basin, Desert** [Imperial Co.]. The name dates from around 1900; its meaning and origin are unknown. Since it is in Yuman territory, the same root may be indicated. The well is probably the one called Santa Rosa de las Lajas by Anzón on Mar. 8, 1774.

Yuhwamai (yo͞o wä′ mī) [San Diego Co.]. Probably from Luiseño *yuxwáamay* 'little mud', from *yuxwáala* 'mud'.

Yulupa (yo͞o lo͞o′ pə): **Creek**, station [Sonoma Co.]. The name Yulupa is found on various *diseños* and probably designated an Indian village, possibly the one spelled Jalapi in the records of Mission San Francisco Solano in 1824. Jalapa is the name of a part of the Napa grant. In the *expedientes* of an unconfirmed land grant the spellings Yulupa, Ulupa, and Julupa are found (Bowman Index).

Yuma (yo͞o′ mə). The name is applied by anthropologists to an Indian tribe that occupied the region on both sides of the Colorado River, with Fort Yuma as the approximate center; they are now officially called by their own name, Quechan—in English (kwə chän′). The name Yuma was first recorded by Kino in 1699, and it has been claimed that this was the name applied to the Quechan by the O'odham (Pima-Papago) of Arizona (*HNAI* 10:97); however, the O'odham language has no initial *y* except in loanwords (Munro), so the original source probably lies elsewhere. For spelling variants, cf. Hodge. Fort Yuma was established and named by Heintzelman in Nov. 1850. Fort Yuma Indian Reservation was established in 1863 and named after the Fort.

Zabriskie Point [Death Valley NP]. Named for Christian B. Zabriskie, who came to Death Valley in 1889 as a representative of the Pacific Coast Borax Company and later became its executive head. The name Zabriskie was also applied to a station of the Tonopah and Tidewater Railroad.

Zaca (zak′ ə): **Lake, Peak**, town; **La Zaca Creek** [Santa Barbara Co.]. The name Saca, mentioned by Padre Zalvidea on July 20, 1806 (Arch. MSB 4:49–68), refers to an Indian village or its chief. In 1838 the name, spelled Saca and La Zaca, was applied to a rancho, granted to an Indian named Antonio. According to Isobel Field, its meaning is 'peace' or 'quiet place'. The name does not seem to appear on early maps but was apparently rescued by members of the Whitney Survey. The von Leicht–Craven map of 1874 shows Zaca Creek.

Zamora (zə môr′ ə): station, **Creek**, post office [Yolo Co.]. The Southern Pacific station was called Blacks until 1910, when the name was changed to Zamora; this is the name of places in Spain and Mexico and is a Spanish family name. The post office is listed in 1916.

Zanita (zə nē′ tə) **Point** [Yosemite NP] was so named because the heroine of Maria Teresa Yelverton's novel *Zanita* (1872) hurled herself from this point into the Merced River (cf. Farquhar, *Place names*, p. 119).

Zanja; Zanjon. The Spanish words for 'ditch, drain, channel' were often used in place naming in Spanish times and have survived in several places. On Estudillo's map of 1819, the two channels connecting Tulare and Buenavista Lakes are labeled Zanjón; the straight lines outlining the channels seem to indicate that they were man-made. The name appears in the titles of four land grants, usually spelled Sanjon, in Monterey, Merced, Fresno, and Sacramento Cos. The San Rafael grant [Los Angeles Co.] was first called La Zanja. **Zanja Cota** (zän′ hä kō′ tə) **Creek** [Santa Barbara Co.]: *La zanja que llaman de Cota* 'the channel that is called Cota's' is mentioned on June 17, 1795 (PSP 13:28), and repeatedly in later years. The specific name probably refers to a member of the Cota family. The stream has also been labeled Santa Cota and Santa Cora. **Zanja** (zang′ kē, zän′ hä): **Creek, Peak** [San Bernardino Co.]: The stream that runs through Redlands was dug by Mission Indians in 1819 to divert water from Mill Creek. The peak was doubtless named after the stream in American times (Beattie). From 1917 to 1920 there was a Zanja post office. **Sanjon de los Alisos** [Alameda Co.] is one of the channels through which Alameda Creek discharges into San Francisco Bay, as shown on *diseños*. *For* Zanja Onda, *see* Honda [Los Angeles Co.]; *for* Zanjapoco or Zanjapojo, *see* San Carpoforo.

Zanoma. *See* Sonoma.

Zanone [zə nō′ nē] **Creek** [Humboldt Co.]. Probably named for Domingo Anthony Zanone, a prominent rancher who came here in 1865 (Turner 1993).

Zante (zan′ tē) [Tulare Co.]. Perhaps for a type of currant, named after the town of Zante in Greece.

Zapato. The name is found for several creeks in southern counties. *Zapato* is the Spanish word for 'shoe'; some of the names, however, may be derived from Spanish *zapote*, referring to the sapote tree, which was introduced into California about 1810. (Both Spanish *zapote* and English "sapote" are from Aztec *tzapotl*.) **Zapato Chino Creek** [Fresno Co.] contains the Spanish phrase for 'Chinese shoe'.

Zayante (zī an′ tē) **Creek** [Santa Cruz Co.]. Probably from Rumsen (Costanoan) *sayyan-ta* 'at the heel' (Callaghan). A *parage llamado Sayanta* 'place called Sayanta' is mentioned in July 1834 (Legis. Recs. 2:151), and the stream appears as Río or Arrollo de Sayante on several *diseños*. On July 21, 1834, the name was applied to land grants.

Zelzah. *See* Northridge.

Zem-Zem Creek [Napa Co.]. The stream is named after the Zem-Zem Ranch. In 1867 J. C. Owen acquired the property, developed the sulphur spring, and gave his ranch the mythical name Zem-Zem, said to mean 'healing waters' (D. T. Davis). The name is also spelled Zim-Zim. **Zem Zem Springs** [Los Angeles Co.] were also named for the 'healing waters'.

Zenia (zēn′ yə) [Trinity Co.]. The name was applied in 1900 by George Croyden, the first postmaster, for a girl of that name (Albert Burgess). **Zenia Ridge** [Mendocino Co.] appears on some maps as Zeni.

Zim-Zim Creek. *See* Zem-Zem Creek.

Zinfandel (zin′ fan dəl) [Napa Co.] was named before 1900, when the Zinfandel grape was the most important product of the region. The origin of the grape and its name have been much discussed; the topic is summarized by David L. Gold, *Names* 44:59–77 (1996), who finds the most likely immediate origin in a Czech grape name, *cinifádl*, derived in turn from one of several variant German terms such as *Zierfahndler*.

Zmudowski (zum dou′ skē) **Beach State Park** [Monterey Co.]. On Oct. 10, 1952, the park commission changed the name of the old Pajaro River Park in recognition of the generous contributions of the Zmudowski family.

Zorrilla is Mexican Spanish for 'skunk'. It occurs

only rarely in California place names, as in Cañada de Las Zorrillas (zō rē´ yəs) [Santa Barbara Co.], lit. 'valley of the skunks'.

Zumwalt Meadow [Kings Canyon NP]: Named for its former owner, Daniel K. Zumwalt (1845–1904), who was active in the movement to preserve the Big Trees (Farquhar). Zumwalt was the attorney for the Southern Pacific Railroad and was instrumental in creating the Sequoia and Grant National Parks (Oscar Berland).

Zuniga (zo͞o´ ni gə): **Shoal, Point** [San Diego Co.]. The name Barros [Bajos] de Zuñiga for the shoal in the outer San Diego harbor is shown on Pantoja's Plano of 1782 (Wagner, no. 687). The name was probably given for José Zúñiga, a lieutenant in the San Diego company, 1781–93. The name for the point appears in Vancouver's atlas (1798).

Zurich (zo͞o´ rik) [Inyo Co.]. The station of the narrow-gauge railroad was first given the name Alvord, but about 1913 this was changed to Zurich to avoid confusion with another station. The Alpine scenery a few miles to the west suggested the name of the Swiss city.

Zzyzx (ziz´ iks, zī´ ziks): locale, **Spring** [San Bernardino Co.]. This was the site of a local radio station run by an eccentric desert dweller, with the supposed call letters ZZYZX; cf. Anne Q. Duffield-Stoll, *Zzyzx: History of an oasis* (Northridge: Santa Susana Press, 1994). The name was established by BGN, list 8402.

GLOSSARY AND BIBLIOGRAPHY

A combined bibliography and glossary was considered more practical and more convenient for the user of this book than separate lists of manuscripts, maps, books, contributors, and technical expressions.

The bibliographical material in this section contains items that are repeatedly cited in the text, as well as special items which are not ordinarily found in the reference department of libraries. Well-known general reference books like *Encyclopedia Americana, Dictionary of American Biography*, the Army and Navy registers, Lippincott's, Rand McNally's, and Thomas's gazetteers and general dictionaries are not listed here, although they were used in preparing this work. Manuscripts or books that are cited only a few times are also omitted, but enough of the titles of such works is given in the main text to identify the source easily.

Each source is listed in the Glossary and Bibliography with a key word (or with some other convenient abbreviation), and this key word is given in the text as a reference. It was not considered necessary to give the page numbers of references in works that have an alphabetical arrangement or a comprehensive index, or when (in diaries) the passage can be easily identified by the date cited. For maps, the key word is ordinarily the name of the man who made the map and (or) was responsible for the names on it. No consistency could be maintained, because frequently only the publisher of the map is known.

The name of a person given in parentheses in the text indicates that the information, or most of it, was received from this source. It does not mean that he or she is responsible for the wording of the entry, unless quotation marks are used. Individuals who contributed repeatedly are listed and identified in this section.

Many manuscripts, maps, and books exist in different versions or editions. An attempt has been made to refer only to the version listed in this bibliography; but since the work on this book extended over a period of several years, it is not impossible that there will be some discrepancies.

Any attempt to give all the sources of all the items in this book would clutter it with professional paraphernalia that would be of little value to the scholar and would be a nuisance to the general reader. Hence, references to authorities in the text have been limited; they are given, first, where all or most of the information was taken from a single source; second, where it was hard to decide on a controversial question; third, where it seemed preferable to leave the responsibility for a statement to the source; fourth, where the first mentioning of the name or its variant is cited; fifth, where a direct quotation from a source is given.

AAE. University of California Publications in American Archaeology and Ethnology (Berkeley and Los Angeles: University of California Press, 1903–).

Abbot, Henry Larcom, a lieutenant of topographical engineers, assisted Lt. R. S. Williamson in making the railroad survey from the Sacramento Valley to the Columbia River in 1855. His report, which is in Pac. R.R. *Reports*, vol. 6, pt. 1, includes "Route from Shasta Valley . . . to Fort Reading, explored by Williamson . . . in 1851" (pp. 127–29). *See* Pac. R.R. *Reports;* Williamson; and (in text) Abbot, Mount.

Abella, Padre Ramón (1764–1842), served at Mission Dolores from 1798 to 1819; he participated in explorations of the lower Sacramento and San Joaquin rivers in 1811 and 1817. His *diario* of the 1811 expedition is in Arch. MSB 4:101–34.

Abert's map. *See* Hood's map.

Adams, Ramon F., *Western words: A dictionary of the range, cow camp, and trail* (Norman: University of Oklahoma Press, 1944).

AGN. Archivo General de la Nación (Mexico City).

AGS. The American Guide Series, prepared by the U.S. Work Projects Administration.

Alameda County. *See* Mosier and Mosier.

Alegría, F., "Nombres españoles en California," *Atenea* (1951) 28:217–27. This is the critical review mentioned by Gudde in his 1960 preface.

Algonquian (pronounced and sometimes spelled Algonkian). This widespread American Indian family of eastern and central North America has been the source of hundreds of place names, some of which have been transferred to California (e.g. Michigan, Saugus).

Alviso Docs. "Documentos para la historia de California: Archivo de la familia Alviso, 1817–1850" (original MSS, Bancroft Library).

Anza, Juan Bautista de (1735–88), opened up the route from the province of Sonora, Mexico, to San Gabriel in 1774, and in 1775 he brought the first colonists to Nueva California. **Anzas's diaries**: Those from Jan. 8 to May 27, 1774, are in Herbert E. Bolton, *Anza's California expeditions* (5 vols.; Berkeley: University of California Press, 1930), 2:1–243; those from Oct. 23, 1775, to June 1, 1776, are in 3:1–200. *See* Garcés; Palou; Font.

APCH:P. Academy of Pacific Coast History: Publications. The four volumes of the series were published by the University of California Press between 1909 and 1919.

Applegate, Richard B., is the author of "Chumash place names," *Journal of California Anthropology* 1:186–205 (1974), and of "An index of Chumash placenames," *Papers on the Chumash* (San Luis Obispo County Archaeological Society, Occasional papers 9:19–46 [1975]). These publications use data from the unpublished field notes of J. P. Harrington.

Arbuckle, Clyde, secretary of the San Jose Historic Landmarks Commission and co-author (with Roscoe D. Wyatt) of *Historic names, persons, and places in Santa Clara County* (San Jose Chamber of Commerce, 1948).

Arch. Arz. SF. "Archivo del Arzobispado de San Francisco: Cartas de los Misioneros de California, 1772–1849" (5 vols. in 3, transcripts and extracts in the Bancroft Library).

Archives of California, Miscellany. *See* Archivo de California.

Archivo de California. Documents formerly in the U.S. Surveyor General's Office, San Francisco, destroyed in the fire of 1906; transcripts are in the Bancroft Library. This includes "Archives of California, Miscellany," as well as the following collections listed separately below: Dep. Recs.; DSP; DSP Ang.; DSP Ben. C and T; DSP Ben. CH; DSP Ben. Mil.; DSP Ben. P and J; DSP Mont.; Legis. Recs.; Prov. Recs.; PSP; PSP Ben. Mil.; PSP Pres.; SP Mis. and C; SP Sac.

Arch. LA. "Archives of Los Angeles, Miscellaneous Papers, 1821–50" (5 vols.; transcripts and extracts, Bancroft Library).

Arch. Mis. "Archivo de las Misiones, 1769–1825" (1 vol.; original MSS, Bancroft Library).

Arch. Mis. S.Buen. "Archivo de la Misión de San Buenaventura" (2 vols.; transcripts, Bancroft Library).

Arch. Mont. "Archives of Monterey County, Miscellaneous Documents" (16 vols. in 1; transcripts and extracts, Bancroft Library).

Arch. MPC. "Archivo de la Misión de la Purísima Concepción" (transcripts, Bancroft Library).

Arch. MSB. "Archivo de la Misión de Santa Bárbara, 1768–1836" (12 vols.; transcripts, Bancroft Library).

Arch. SB. "Archivo de la Misión de Santa Bárbara, Libros de Misión, 1782–1853" (abstracts, Bancroft Library).

Arch. SB, Juzgado. "Archivo de Santa Bárbara, Oficios del Juzgado de Primera Instancia, 1839–1849, 1852–1853" (abstracts and copies, Bancroft Library).

Arch. SJ. "San Jose Archives, 1796–1849" (6 vols.

and loose papers in the office of the City Clerk, San Jose; in 1 vol. of transcripts and extracts, Bancroft Library).

Argüello, Luis Antonio (1784–1830), commander of an expedition into the Sacramento Valley in Oct.–Nov. 1821, left a diary (original MSS, Bancroft Library). Padre Blas Ordaz chronicled the same expedition; in his diary (transcript, Arch. MSB 4:169–90) the spelling of some place names differs from Argüello's.

Armendariz, Frank, of Santa Barbara, provided valuable information on his area.

Arrowsmith, Aaron, "Chart of the world . . . ," 1790 (Wagner, no. 744).

Arroyo de la Cuesta, Padre Felipe, "Collected vocabularies of several central California Indian dialects" (MSS, Lecciones de Indios, Bancroft Library).

Asbill, Frank, "Place naming in the Wailaki country," *Western Folklore* (1949) 8:252–55.

Ashley, Leonard R. N., "The Spanish place-names of California: Proposition 1994," *Names* 44:3–40 (1996).

Ashton, William E., "Presidential place name covers," *Weekly Philatelic Gossip*, Jan. 12, 1952; June 25, 1955.

Athabaskan. Several groups of this most widespread linguistic family of North American Indians dwelt in the northwest corner of the state. The Tolowa, the Hupa, and the Mattole have left traces in geographical names. The name is also spelled Athapascan. *See* Golla.

Atkinson, Janet I., *Los Angeles County historical directory* (Jefferson, N.C.: McFarland, 1988).

Atlas sheet. An atlas sheet is a section of the topographical atlas of the United States covering one of the many thousands of quadrangles into which the Geological Survey has divided the country. Unless otherwise stated, the reference is to a sheet published by the USGS. *See* Helm Index.

Audubon, John Woodhouse, son of the great American ornithologist John James Audubon, was in California from Sept. 1849 to May 1850. His diary is of great interest for early mining towns. The references are to Audubon's *Western journal: 1849–1850* (Cleveland: Clark, 1906).

Ayala, Juan Manuel de. "Plano del puerto de San Francisco 1775," based on Ayala's survey and drawn by José de Cañizares; reproduced in Bolton, *Anza's California expeditions*, vol. 1, opp. p. 385, and in Palou, vol. 4, opp. p. 40 (Wagner, no. 640). This is the first chart of San Francisco Bay; some of the place names on it are still in use. Ayala was commander of the first official survey of the bay, but most of the responsibility fell to Cañizares.

Aztec is a conventional term used for the Nahua tribes, who lived in Mexico at the time of the Spanish conquest. Many words, including geographical terms, were taken from their language into Mexican Spanish and thus found their way into California geography.

Badianus. *The Badianus manuscript: An Aztec herbal of 1552*, ed. Emily Walcott Emmart (Baltimore: Johns Hopkins Press, 1940).

Bailey, G. E. A number of articles published as "History and origin of California names and places," in the *Overland Monthly* (San Francisco) from July to Dec. 1904.

Bailey, Richard C., *Kern County place names* (Bakersfield: Kern County Historical Society, 1967).

Baker, George H., *Map of the mining region of California* (Sacramento: Barber, 1856). Listed as Wheat, no. 289.

Baker, Ronald, of the Riverside Public Library, provided valuable information on his county.

Ballard, Diana, of the Napa County Historical Society, kindly provided information on her area.

Bancroft, Hubert Howe (1832–1918), *History of California* (7 vols. [vols. 18–24 of his *Works*]; San Francisco, 1884–90). Contains rich factual material. The "Pioneer register and index" in vols. 2–5 includes biographical sketches of many persons mentioned in the present work. **Bancroft scrapbooks** refers to materials collected in the preparation of Bancroft's Pacific States handbooks (1860–64), vols. 1–17, California Counties (collection of clippings, vols. 1–13, Bancroft Library).

Bancroft's maps. For several decades after 1858, H. H. Bancroft and Co., San Francisco, periodically published nicely engraved maps of the state, which were very popular and to some extent influential in fixing place names.

Barnes, Will C., "Arizona place names," *University of Arizona Bulletin* (1935) 6:1; revised as *Will C. Barnes' Arizona place names*, ed. Byrd H. Granger (Tucson: University of Arizona Press, 1960).

Barrett, Samuel A., wrote several important works of the Miwok tribes, including *The ethno-geography of the Pomo and neighboring Indians*, AAE 6:1–332 (1908); *The geography* *and dialects of the Miwok Indians*, AAE 6: 333–68 (1908); and *Myths of the southern Sierra Miwok*, AAE 16:1–28 (1919).

Bauman, James, *The Harrington collection of Indian placenames in North Central California* (Redding: Shasta Trinity National Forest, n.d.), compiles Wintu place names from the manuscripts of J. P. Harrington.

Beattie, George W., *Heritage of the Valley: San Bernardino's first century* (Pasadena: San Pasqual Press, 1939).

Beckwith, Lt. E. G., 3d Artillery, "Report of explorations for a route for the Pacific Railroad . . . forty-first parallel . . . 1854," Pac. R.R. *Reports*, vol. 2 [pt. 2].

Beechey, Captain F. W., *Narrative of a voyage to the Pacific . . . in the years 1825, 26, 27, 28*. The references are to the original octavo edition in two volumes (London, 1831). Chaps. 1–2 of the second volume give an account of Beechey's stay in California. A number of important names in the San Francisco Bay district (some of them misspelled) were fixed by Beechey. His map of "The harbour of San Francisco, Nueva California . . . 1827 and 8," the first chart of the bay to be fairly accurate, is reproduced in *An account of a visit to California 1826–27*, ed. Edith M. Coulter (San Francisco: Grabhorn Press, 1941).

Beeler, Madison S., late professor of German and linguistics, University of California, Berkeley; authority on the Chumash languages and on California names of Indian origin. His book review of Couro and Hutcheson (*q.v.*), published in *Names* 22:137–41 (1974), has especially valuable comments on American Indian etymologies of California place names. Other relevant articles by him are "Sonoma, Carquinez, Umunhum, Colma: Some disputed California names," *Western Folklore* 13:268–77 (1954); "Yosemite and Tamalpais," *Names* 3:185–88 (1955); "On etymologizing Indian place-names," *Names* 5:236–40 (1957); "The California oronym and toponym Montara," *Romance Philology* 20:35–39 (1966); "Hueneme," *Names* 14:36–40 (1966); "Inyo," *Names* 20:56–59 (1972); "Inyo once again," *Names* 26:208 (1978).

Belcher, Sir Edward, *Narrative of a voyage round the world . . . during the years 1836–1842* (2 vols., London, 1843).

Belden, L. Burr, historian and contributor to the *San Bernardino Sun-Telegram*.

Bendire, Lt. Charles, made a report of his expedition of 1867, but it has been misplaced in the archives of the War Dept.

BGN. The U.S. Board on Geographic Names was established on Dec. 23, 1891, as a bureau of the U.S. Dept. of the Interior, for the purpose of making uniform the usage and spelling of geographical names. In 1906 its powers were extended to include the approval of new names submitted to the board. For a number of years the bureau was known as the Geographic Board. Since the method of publishing decisions has changed from time to time, references will be found to reports and to decisions, with or without the date of publication.

Bidwell, John (1819–1900), a native of New York, arrived in California in 1841 and was Sutter's most faithful assistant before the Gold Rush. In 1849 he acquired the Arroyo Chico rancho and became the leader in the development of Butte Co. In later years he was a congressman, a general of the California militia, and in 1892 the presidential candidate on the Prohibition ticket. The references are to his *Echoes of the past . . .*, published in Chico shortly after his death and later reprinted. **Bidwell's map** of 1844, "Mapa del Valle del Sacramento" (MS in California State Library; Wheat, no. 15), is the first known map of the Sacramento Valley showing the Mexican grants and rivers in detail: "On my return from Red Bluff in March, 1843, I made a map of the upper Sacramento Valley on which most of the streams were laid down, and they have since borne the names then given them" (Co. Hist.: Butte 1877).

Bieber, Ralph P., ed., *Southern trails to California in 1849* (Glendale: Clark, 1937).

Bigler, Henry William, *Diary of a Mormon in California*. The Bancroft version is published in Bigler's *Chronicle of the West*, ed. Erwin G. Gudde (Berkeley and Los Angeles: University of California Press, 1962).

Blake, William P., was the geologist of the Pacific Railroad Survey, 1853–54. His report and map are published in the Pac. R.R. *Reports* 5:2. In it are several maps including the "Geological map of a part of . . . California explored in 1853"; on this a number of place names were recorded for the first time.

Blanco S., Antonio, *La lengua española en la historia de California* (Madrid: Cultura Hispánica, 1971). The "Vocabulario de californianismos," pp. 163–249, is useful for Spanish terms characteristic of California usage.

Blue Book. Reference is to the California *Blue Book and State Roster,* published (usually biennially) by the California secretary of state from 1891.

Bodega y Quadra, Juan Francisco de la, commander of the schooner *Sonora* of the Hezeta expedition of 1775, made an important voyage up the coast as far as Alaska; he named a number of places, including Bodega Bay. For an account of his discoveries and for three of his maps, *see* Wagner (pls. 37, 39).

Boggs, Mae Hélène Bacon, *My playhouse was a Concord coach* . . . (Oakland, 1942). An "anthology of newspaper clippings and documents" from 1839 to 1888; contains excellent reproductions of a number of valuable maps, and several lists of post offices of the 1850s.

Bolton, Herbert E. (1870–1933). The indispensable critical editions of the diaries and reports of the Anza and Portolá expeditions by this eminent historian are listed under Anza, Crespí, Font, and Palou. His edition of Kino's *Memoir* is listed under Kino. **Bolton,** *Span. expl.* refers to Bolton's *Spanish exploration in the Southwest, 1542–1706* (New York: Scribners, 1916).

Bonnycastle, R. H. *Spanish America* . . . (London, 1818). References are to the map.

Borden, Stanley T., supplied many names in the northern counties, especially those derived from the lumber industry.

Borthwick, J. D., *The gold hunters* (New York, 1917). This popular account was first printed in 1857 under the title *Three years in California;* it has appeared in several other editions. Borthwick was in California in 1851.

Boscana, Padre Gerónimo (1775–1831), served at Mission San Juan Capistrano from 1814 to 1826. His account of Juaneño Indian life and religion was translated into English and first published in Alfred Robinson, *Life in California* (1846); it has frequently been reprinted. The edition used here is *Chinigchinich*, with notes by J. P. Harrington (Santa Ana: Fine Arts Press, 1933; reprinted with a preface by William Bright, Banning: Malki Museum Press, 1978).

Boston ships refers to the sailing ships from New England that traveled around Cape Horn to trade for hides in California during the early 19th century (cf. Dana, *Two years before the mast*).

Boundary Commission. *See* Nevada Boundary Survey.

Bowen's Guide. *See* Postal Guide.

Bowman, Amos, *Map of Georgetown Divide* . . . (San Francisco, 1873). *See* von Leicht-Hoffmann.

Bowman, Jacob N. The reference indicates that the item was contributed by J. N. Bowman of Berkeley, who died in 1968. **Bowman Index** refers to his unpublished "Index of California private landgrants and private landgrant papers and cases," now in the Bancroft Library. This monumental work on one of the most important phases of California's political, legal, and economic history is compiled from the land-grant documents in the U.S. District Court, San Francisco; the National Archives, Washington, D.C.; the State Archives, Sacramento; the U.S. Public Survey Office, Glendale; published government documents; and various county archives. The names and dates of land grants given in this book are taken from it. For researchers who do not have access to the Bowman Index, Ogden Hoffman's *Index in report of land cases* (San Francisco, 1862), and the lists of land grants given in the reports and on the maps of the U.S. Land Office, must serve for general reference. Bowman's article "The names of the California missions " was published in *HSSC:Q* 39:351 ff. *See* Land grant.

Brannan, Samuel, brought a party of Mormons from New York in 1846 to settle in California. He played an important role in California history during and after the Gold Rush. The quotation concerning the naming of Calistoga is taken from Reva Scott's *Samuel Brannan and the golden fleece* (New York: Macmillan, 1944).

Brewer, William H. (1828–1910), *Up and down California in 1860–1864,* ed. Francis P. Farquhar (Yale University Press, 1930). **Brewer's notes**: Brewer's field notes, 1861–64, consist of sixteen MS notebooks in the Bancroft Library. His diaries and correspondence in that library were also checked. *See* Brewer, Mount, in the text.

Brierly, A. A., Inyo Co. surveyor, supplied information on a large number of place names in his county.

Brigandi, Phil, museum curator of the Ramona Pageant, in Hemet, provided especially valuable information about place names in Orange, Riverside, and San Diego Counties.

Briggs, Henry, *The north part of America,* 1625 (Wagner, no. 295, pl. 23).

Bright, William, the editor of this volume, is professor emeritus of linguistics and anthropology at UCLA and the author of works on several California Indian languages, including Karuk and Luiseño. He has also written "Some place names on the Klamath River," *Western Folklore* (1952) 1:121–22; a review of *California place names*, 2d ed., by E. G. Gudde, *Journal of American Folklore* (1962) 75:78–82; and "Place name dictionaries and American Indian place names," in his *American Indian linguistics and literature*, 63–75 (Berlin: Mouton, 1984).

Broadbent, Sylvia M., professor of anthropology, University of California, Riverside, is the author of *Language of the Sierra Miwok* (Berkeley and Los Angeles: University of California Press, 1964), and co-author with L. S. Freeland of *Central Sierra Miwok dictionary with texts* (Berkeley and Los Angeles: University of California Press, 1960). Both works contain vocabulary relevant to place-name research.

Brown, Alan K., *Place names of San Mateo County* (San Mateo: County Historical Society, 1975).

Brown, Thomas P., "What's in a name," a column published from July 1942 to Oct. 1943 in *The Headlight*, monthly publication of the Western Pacific Railroad.

Brown, William S., formerly of the USFS, provided valuable information from rangers and settlers on place names in the national forests.

Browne, J. Ross (1821–75), *Report on the mineral resources of the states and territories west of the Rocky Mountains* (Washington, D.C.: Government Printing Office, 1868). This was issued also as 40th Cong., 2d sess. (1867), House Ex. Doc. no. 202.

Browning, Peter, is the author of *Place names of the Sierra Nevada* (Berkeley: Wilderness Books, 1986), covering the area from Alpine Co. to Walker Pass; and *Yosemite place names* (Lafayette: Great West Books, 1988).

Bruff, J. Goldsborough (1804–89), *Gold Rush*, ed. Georgia Willis Read and Ruth Gaines (2 vols.; Columbia University Press, 1944). The work is an important source for the period of California history from Apr. 1849 to July 1851.

Bryant, Edwin (1805–69), *What I saw in California* . . . (New York, 1849).

Buffum, Edward Gould (1820–67), *Six months in the gold mines* . . . (London, 1850).

Buie, Earl E., columns in the *San Bernardino Daily Sun*.

Bunnell, Lafayette Houghton (1824–1903), *Discovery of the Yosemite and the Indian War of 1851*. . . . Quotations refer to the Chicago edition of 1880. Bunnell was a member of the Mariposa Battalion, which discovered Yosemite Valley. He is responsible for retaining some of the Indian names in the valley, including "Yosemite" itself. For a biographical sketch, consult Farquhar under "Bunnell Point."

Burr, David H., created a map in 1839, based in part on Jedediah Smith's notes, or on a map made by Smith but not now extant. Burr's map is reproduced in Maurice S. Sullivan, *The travels of Jedediah Smith*.

Butler, B. F., *Map of the state of California* (San Francisco, 1851); listed as Wheat, no. 185.

Cabrillo, Juan Rodríguez, a Portuguese navigator in Spanish service, was commander of the first expedition to sail along the coast of California (1542) and was the first to apply Spanish names to coastal features. None of the names survived. *See*, in the text, Cabrillo, Mugu, San Martin, Sierra Nevada; and, in the Glossary, Ferrer. Although his surname was Rodríguez, we follow common practice in listing him under "Cabrillo."

Cahuilla. A tribe and language of the Takic (Uto-Aztecan) family, still spoken in Riverside Co. *See* Cahuilla and Coachella in text.

California Blue Book. *See* Blue Book.

Callaghan, Catherine A., professor of linguistics at Ohio State University, is the author of *Lake Miwok dictionary* (1965); *Bodega Miwok dictionary* (1970); *Plains Miwok dictionary* (1984); and *Northern Sierra Miwok dictionary* (1987); all these, published by the University of California Press, contain materials useful in place-name research. Dr. Callaghan has also, in personal correspondence, supplied abundant information on place names of Miwok and Costanoan origin.

Cañizares, José de (fl. 1769–90), *Plano del puerto de San Francisco*, 1776 (Wagner, no. 653). Cañizares, as first pilot under Ayala, made the first official survey of San Francisco Bay in 1775 (*see* Ayala), and made further explorations in 1776.

Carranco, Lynwood, and Andrew **Genzoli**, "California Redwood Empire place names," *Journal of the West* (1968) 7:363–80.

Carrillo, Domingo, "Documentos para la historia de California" (transcripts, Bancroft Library).

Cary, John, created a map of North America, which was published in London in 1806.

Castro, Manuel, "Documentos para la historia de California, 1821–1850" (2 vols.; originals, Bancroft Library).

Cermeño. *See* Rodríguez Cermeño.

CFQ. *California Folklore Quarterly*, published by the University of California Press for the California Folklore Society beginning in 1942. In 1947 the name was changed to *Western Folklore*.

Chafe, Wallace, professor of linguistics, University of California, Santa Barbara, is a specialist in the Iroquoian languages of the northeast United States and has provided etymologies for Iroquoian names transferred to California.

Chalfant, Willie Arthur, for many years editor and publisher of the *Inyo Register*, is the best authority on the old names of Inyo Co. and Death Valley. His books include *The story of Inyo* (Chicago, 1922); *Death Valley: The facts* (Stanford: Stanford University Press, 1930, 1936); and *Tales of the pioneers* (Stanford: Stanford University Press, 1942).

Chamisso, Adelbert von (1781–1838), a German botanist and poet, of French extraction, was the botanist of the first Kotzebue expedition, 1815–18. In 1816 he collected and described what is now California's state flower, commonly known as the golden poppy, and named it *Eschscholtzia californica* for Johann F. Eschscholtz, the zoologist of the expedition. *See* Kotzebue; Rurik.

Chapman, Charles E., *A history of California: The Spanish period* (New York: Macmillan, 1939).

Chase, J. Smeaton (1864–1923), was an observant traveler in California who related his experiences in three interesting books published by the Houghton Mifflin Company: *Yosemite trails . . .* (1911); *California coast trails . . .* (1913); *California desert trails . . .* (1959).

Chemehuevi (chem′ ə wā vē). A southern California tribe and language, closely related to Southern Paiute, in the Numic branch of the Uto-Aztecan language family; *see* Chemehuevi in text.

Chinook Jargon. A trade language of American Indian origin, used on the Northwest coast as far south as the Klamath River; some place names of Chinook Jargon origin were transferred to California. References are to George Gibbs, *A dictionary of the Chinook Jargon* (New York, 1863).

CHSQ. *California Historical Society Quarterly*, published since 1922 in San Francisco.

Chuckwalla refers to the *Death Valley Chuckwalla*, Greenwater, Calif., Jan. 1–June, 1907.

Chumash (chōō′ mash). The term refers to a language family, comprising related tribes in San Luis Obispo, Santa Barbara, and Ventura Counties. They are known in modern times by names derived from the missions with which they became associated, i.e. Ventureño, Barbareño, Purisimeño, Ynezeño, and Obispeño. *See* Applegate; Harrington.

Clark, Donald Thomas, prepared two definitive volumes on place names of the central California coast: *Santa Cruz County place names* (Santa Cruz: Santa Cruz Historical Trust, 1986); and *Monterey County place names: A geographical dictionary* (Carmel Valley: Kestrel Press, 1991).

Cleland, Robert Glass, *A history of California: The American period* (New York: Macmillan, 1927).

Clemson, Beverly, of the Contra Costa History Center, kindly provided information on place names in Contra Costa Co.

Coast Pilot. This invaluable book for navigation on the Pacific Coast was first published by George Davidson under the titles *Directory of the Pacific Coast* (1858 and 1862) and *Coast pilot of California, Oregon, and Washington* (1869 and 1889). Since 1903 it has been published at irregular intervals by the Coast Survey. The older editions are a valuable source of information concerning names along the coast.

Coast Survey. The U.S. Coast Survey, since 1878 officially designated the Coast and Geodetic Survey, commenced the charting of the Pacific Coast in 1850; it is in general responsible for many names along the coast, including the retention, translation, or hybridization of previously existing names. The names applied to certain points sometimes differ in the various publications of the survey. A reference with a date but no page number refers to the charts at the end of the annual reports; with a page number, it refers to the report itself. **Coast Survey gazetteer** refers to *Geographic names in the coastal areas of California, Oregon, and Washington*, compiled by personnel of the Work Projects Administration in Philadelphia, 1939–40. *See* Coast Pilot; Davidson.

Co. Hist. The annals of most, if not all, California counties have been recorded in a county history at some time or other. Some of these volumes not only contain valuable information about local history but also are important sources for the origin of place names. The references cited in the text may be found in the following list, in which the year, the publisher, and (when known) the author or editor are indicated.

Alameda. 1876, William Halley, *The centennial year book* (Oakland: Halley). 1878, "Official atlas map" (Oakland: Thompson and West). 1883, *History of Alameda County* (Oakland: M. W. Wood). 1928, Frank C. Merritt, *History of Alameda County* (Chicago: Clarke). 1932, Roy C. Beckman, *The romance of Oakland* . . . (Oakland: Landis and Kelsey).

Amador. 1881, Jesse D. Mason, *History of Amador County* (Oakland: Thompson and West). 1917, *Amador County history*, ed. Mrs. J. L. Sargent (Jackson: Amador County Federation of Women's Clubs).

Butte. 1877, *Butte County . . . illustrations* . . . (Oakland: Smith and Elliott). 1882, Harry L. Wells and W. L. Chambers, *History of Butte County* (San Francisco: Wells). 1918, George C. Mansfield, *History of Butte County* (Los Angeles: Historic Record Co.).

Calaveras. 1892, *see* Co. Hist.: Merced.

Colusa. 1880, W. S. Green, *Illustrations* . . . (San Francisco: Elliott and Moore). 1891, Justus H. Rogers, *Colusa County* (Orland). 1918, Charles Davis McComish and Mrs. Rebecca T. Lambert, *History of Colusa and Glenn Counties* (Los Angeles: Historic Record Co.).

Contra Costa. 1878, *Illustrations* . . . (Oakland: Smith and Elliott). 1882, *History of Contra Costa County*, Preface signed by J. P. Munro-Fraser (San Francisco: Slocum). 1917, *The history of Contra Costa County*, ed. F. J. Hulaniski (Berkeley: Elms). 1926, *History of Contra Costa County* (Los Angeles: Historic Record Co.). 1940, Mae Fisher Purcell, *History of Contra Costa County* (Berkeley: Gillick).

Del Norte. 1881, A. J. Bledsoe, *History of Del Norte County* (Eureka: Wyman). 1953, Esther Ruth Smith, *The history of Del Norte County* (Oakland: Holmes).

El Dorado. 1883, *Historical souvenir* . . . (Oakland: Paolo Sioli). 1915, Herman Daniel Jerrett, *California's El Dorado* . . . (Sacramento: Anderson).

Fresno. 1882, *History of Fresno County* (San Francisco: Elliott). 1892, *A memorial and biographical history of the Counties of Fresno, Tulare, and Kern* (Chicago: Lewis). 1919, Paul E. Vandor, *History of Fresno County* (Los Angeles: Historic Record Co.). 1933, Lilbourne Alsip Winchell, *History of Fresno County and the San Joaquin Valley* (Fresno: Cawston). 1941, Ben R. Walker, *The Fresno County blue book* (Fresno: Cawston). 1946, *Fresno community book*, ed. Ben R. Walker (Fresno: Cawston). 1956, *Fresno County centennial almanac* (Fresno: Centennial Committee).

Glenn. 1918, *see* Co. Hist.: Colusa.

Humboldt. 1881, *History of Humboldt County* (San Francisco: Elliott). 1890, Lillie E. Hamm, *History and business directory of Humboldt County* (Eureka: Daily Humboldt Standard). 1915, Leigh H. Irvine, *History of Humboldt County* (Los Angeles: Historic Record Co.).

Imperial. 1918, *The history of Imperial County*, ed. F. C. Farr (Berkeley: Elms and Franks). 1931, *see* Tout.

Inyo. *See* Chalfant.

Kern. 1883, *History of Kern County* (San Francisco: Elliott). 1892, *see* Co. Hist.: Fresno. 1914, Wallace M. Morgan, *History of Kern County* (Los Angeles: Historic Record Co.). 1929, Thelma B. Miller (Chicago: Clarke). 1934, *see* Comfort.

Kings. 1913, *see* Co. Hist.: Tulare. 1940, Robert R. Brown and J. E. Richmond, *History of Kings County* (Hanford: Cawston).

Lake. 1873, *see* Co. Hist.: Napa. 1881, *see* Co. Hist.: Napa. 1914, *see* Co. Hist.: Mendocino.

Lassen. 1882, *see* Co. Hist.: Plumas. 1916, Asa M. Fairfield, *Fairfield's pioneer history* . . . (San Francisco: Crocker).

Los Angeles. 1876, J. J. Warner, Benjamin Hayes, and J. P. Widney, *An historical sketch* . . . (Los Angeles: Lewin). 1880, John Albert Wilson, *History of Los Angeles County* (Oakland: Thompson and West). 1889, *An illustrated history* . . . (Chicago: Lewis). 1890, *see* Co. Hist.: San Diego. 1908, Luther A. Ingersoll, *Ingersoll's century history, Santa Monica Bay cities* (Los Angeles). 1920, *History of Pomona Valley* . . . (Los Angeles: Historic Record Co.; also contains history of part of San Bernardino Co.). 1923, *History of Los Angeles County*, ed. John Steven McGroarty (Chicago and New York: American Historical Society). 1939, William J. Dunkerley, *Know Los Angeles County* (Los Angeles County Board of Supervisors). 1962–65, *The historical volume and reference works*, ed. Robert P. Studer (5 vols.; Los Angeles: Historical Publishers).

Marin. 1880, J. P. Munro-Fraser, *History of Marin County* (San Francisco: Alley, Bowen).

Mariposa. 1891, *see* Co. Hist.: Merced.

Mendocino. 1873, *see* Co. Hist.: Napa. 1880, Lyman L. Palmer, *History of Mendocino County* (San Francisco: Alley, Bowen). 1914, Aurelius O. Carpenter and Percy H. Millberry, *Mendocino and Lake Counties* (Los Angeles: Historic Record Co.).

Merced. 1881, *History of Merced County* (San Francisco: Elliott and Moore). 1892, *A memorial and biographical history of the counties of Merced, Stanislaus, Calaveras, Tuolumne and Mariposa* (Chicago: Lewis). 1925, John Outcalt, *A history of Merced County* (Los Angeles: Historic Record Co.).

Monterey. 1881, *History of Monterey County* (San Francisco: Elliott and Moore); also contains history of San Benito Co. 1903, *see* Co. Hist.: Santa Cruz. 1910, J. M. Guinn, *History and biographical record of Monterey and San Benito Counties* (Los Angeles: Historic Record Co.).

Napa. 1873, C. A. Menefee, *Historical . . . sketch book of Napa, Sonoma, Lake and Mendocino* . . . (Napa City: Reporter Pub. House). 1881, *History of Napa and Lake Counties*, Preface signed by Lyman L. Palmer (San Francisco: Slocum, Bowen). 1901, *History of Napa County*, ed. W. F. Wallace (Oakland: Enquirer). 1912, *see* Co. Hist.: Solano.

Nevada. 1867, *Bean's history* . . . , comp. Edwin F. Bean (Nevada: Daily Gazette). 1880, Harry Laurenz Wells, *History of Nevada County* (Oakland: Thompson and West). 1924, *see* Co. Hist.: Placer.

Orange. 1890, *see* Co. Hist.: San Diego. 1911, *History of Orange County*, ed. Samuel Armor (Los Angeles: Historic Record Co.). 1921, revised edition of the preceding.

Placer. 1882, *History of Placer County*, ed. Myron Angel (Oakland: Thompson and West). 1924, W. B. Lardner and M. J. Brock, *History of Placer and Nevada Counties* (Los Angeles: Historic Record Co.).

Plumas. 1882, *Illustrated history of Plumas, Lassen and Sierra Counties* (San Francisco: Fariss and Smith).

Riverside. 1912, Elmer Wallace Holmes, *History of Riverside County* (Los Angeles: Historic Record Co.). 1922, *see* Co. Hist.: San Bernardino. 1935, John Raymond Gabbert, *History of Riverside City and County* (Riverside: Record Pub. Co.).

Sacramento. 1880, *History of Sacramento County*, ed. George F. Wright (Oakland: Thompson and West). 1890, Hon. Win. J. Davis, *An illustrated history* . . . (Chicago: Lewis). 1913, William L. Willis, *History of Sacramento County* (Los Angeles: Historic Record Co.). 1923, ed. G. Walter Reed (Los Angeles: Historic Record Co.). 1931, J. W. Wooldridge, *History of the . . . Valley* (Chicago: Pioneer Historical Pub. Co.).

San Benito. 1881, *History of San Benito County* (San Francisco: Elliott and Moore). 1903, *see* Co. Hist.: Santa Cruz. 1910, *see* Co. Hist.: Monterey.

San Bernardino. 1883, *History of San Bernardino County* (San Francisco: Elliott). 1890, *see* Co. Hist.: San Diego. 1904, Luther A. Ingersoll, *Ingersoll's century annals* . . . (Los Angeles). 1920, *History of Pomona Valley* (Los Angeles: Historic Record Co.). 1922, *History of San Bernardino and Riverside Counties*, ed. John Brown (Chicago: Western Historical Association).

San Diego. 1888, Theodore Strong Van Dyke, *The city and county* . . . (San Diego: Leberthon and Taylor). 1890, *An illustrated history of southern California* . . . (Chicago: Lewis; also contains history of San Bernardino, Los Angeles, and Orange Cos.). 1922, Clarence Alan McGrew, *City of San Diego and San Diego County* (Chicago and New York: American Historical Society). 1936, *History of San Diego County*, ed. Carl H. Heilbron (San Diego: San Diego Press Club).

San Joaquin. 1879, Frank T. Gilbert, *History of San Joaquin County* (Oakland: Thompson and West). 1890, *An illustrated history* (Chicago: Lewis). 1923, George H. Tinkham, *History of San Joaquin County* (Los Angeles: Historic Record Co.).

San Luis Obispo. 1883, Myron Angel, *History of San Luis Obispo County* (Oakland: Thompson and West). 1891, *see* Co. Hist.: Santa Barbara. 1903, *see* Co. Hist.: Santa Cruz. 1917, Mrs. Annie L. Morrison and John H. Haydon, *History of San Luis Obispo*

County and environs (Los Angeles: Historic Record Co.).

San Mateo. 1878, *Moore and De Pue's illustrated history* . . . (San Francisco: Brown). 1883, *History of San Mateo County* (San Francisco: Alley). 1916, Philip W. Alexander and Charles P. Hamm, *History of San Mateo County from the earliest times* (Burlingame: Burlingame Pub. Co.). 1946, *see* Stanger.

Santa Barbara. 1883, Jesse D. Mason, *History of Santa Barbara County* (Oakland: Thompson and West); also contains history of Ventura Co. 1891, Mrs. Yda Addis Storke, *A memorial . . . history of the counties of Santa Barbara, San Luis Obispo and Ventura . . .* (Chicago: Lewis). 1939, *History of Santa Barbara County*, ed. Owen H. O'Neill (Santa Barbara: Meier).

Santa Clara. 1876, "Historical atlas map . . ." (San Francisco: Thompson and West). 1881, *History of Santa Clara County*, Preface signed by J. P. Munro-Fraser (San Francisco: Alley, Bowen). 1888, *see* Foote. 1922, Eugene T. Sawyer, *History of Santa Clara County* (Los Angeles: Historic Record Co.).

Santa Cruz. 1879, *Santa Cruz County . . . with historical sketch* (San Francisco: Elliott). 1892, Edward S. Harrison, *History of Santa Cruz County* (San Francisco: Pacific Press). 1903, J. M. Guinn, *History of the state . . . and biographical record of Santa Cruz, San Benito, Monterey and San Luis Obispo Counties* (Chicago: Chapman). 1911, Edward Martin, *History of Santa Cruz County* (Los Angeles: Historic Record Co.).

Shasta. 1949, Rosena A. Giles, *Shasta County California, a History* (Oakland: Biobooks).

Sierra. 1882, *see* Co. Hist.: Plumas.

Siskiyou. 1881, Harry Laurenz Wells, *History of Siskiyou County* (Oakland: Stewart).

Solano. 1878, "Historical atlas map . . ." (San Francisco: Thompson and West). 1879, *History of Solano County* (San Francisco: Wood, Alley). 1912, Thomas Gregory, *History of Solano and Napa Counties* (Los Angeles: Historic Record Co.).

Sonoma. 1873, *see* Co. Hist.: Napa. 1877, "Historical atlas map . . ." (Oakland: Thompson). 1879, Robert A. Thompson, *Historical sketch* . . . (San Francisco: Bancroft). 1880, *History of Sonoma County* (San Francisco: Alley, Bowen). 1911, Thomas Gregory, *History of Sonoma County* (Los Angeles: Historic Record Co.). 1937, Ernest Latimer Finley, *History of Sonoma County* (Santa Rosa: Press Democrat).

Stanislaus. 1881, *History of Stanislaus County* (San Francisco: Elliott and Moore). 1892, *see* Co. Hist.: Merced. 1921, George H. Tinkham, *History of Stanislaus County* (Los Angeles: Historic Record Co.).

Sutter. 1879, *History of Sutter County* (Oakland: Thompson and West). 1924, *see* Co. Hist.: Yuba.

Tehama. 1880, *Tehama County . . . illustrations . . . with historical sketch* (San Francisco: Elliott and Moore).

Trinity. 1858, Isaac Cox, *The annals* . . . (San Francisco: Commercial Book; reprinted in Eugene, Ore., 1940).

Tulare. 1883, *History of Tulare County* (San Francisco: Elliott). 1888, *Business directory and historical . . . hand book* . . . (Tulare City: Pillsbury and Ellsworth). 1892, *see* Co. Hist.: Fresno. 1913, Eugene L. Menefee and Fred A. Dodge, *History of Tulare and Kings Counties* (Los Angeles: Historic Record Co.).

Tuolumne. 1882, *A history of Tuolumne County*, comp. Herbert O. Lang (San Francisco: Alley). 1892, *see* Co. Hist.: Merced.

Ventura. 1883, *see* Co. Hist.: Santa Barbara. 1891, *see* Co. Hist.: Santa Barbara.

Yolo. 1879, *The illustrated atlas and history* (San Francisco: De Pue). 1913, Thomas Gregory, *History of Yolo County* (Los Angeles: Historic Record Co.). 1940, *History of Yolo County*, ed. William O. Russell (Woodland).

Yuba. 1879, William H. Chamberlain and Harry L. Wells, *History of Yuba County* (Oakland: Thompson and West). 1924, Peter J. Delay, *History of Yuba and Sutter Counties* (Los Angeles: Historic Record Co.).

Co. Library; Co. Surveyor. The reference indicates that the information was received from the county librarian or the county surveyor.

Comfort, Herbert G., *Where rolls the Kern: A history of Kern County, California* (Moorpark: Enterprise, 1934).

Cordua, Theodor, settled at the site of present-day Marysville in 1841 and called his farm Neu Mecklenburg, after his home province in Germany. References are to his memoirs (*CHSQ* 12:4).

Costanoan. This Native American language

family included tribes in the coastal area between Suisun Bay and Monterey Bay. A relationship to the Miwokan family is possible.

Costansó, Miguel, was an engineer attached to the Portolá expedition. References are to "The Portolá Expedition of 1769–1770: Diary of Miguel Costansó," ed. Frederick J. Teggart, published (in both Spanish and English) in *APCH:P* 2:161–327. **Costansó's map**: *Carta reducida del Oceano Asiático o Mar del Sur*, 1771 (reproduced in Wagner, pl. 33); the earliest version is dated Oct. 30, 1770.

Coues, Elliott. Two works of this historian are cited here. **Coues, *New light***: *New light on the early history of the greater North West . . .* , ed. E. Coues (New York: Harper, 1897), includes the journals of Alexander Henry and David Thompson of the Northwest Company. **Coues, *Trail***: *On the trail of a Spanish pioneer*, 2 vols. (New York: Harper, 1900), concerns the travels of Francisco Garcés (*q.v.*).

Coughlin, Kari, of Death Valley National Park, kindly provided information on names in that area.

Coulter, John, *Adventures on the western coast of South America and the interior of California . . .* (2 vols.; London, 1847).

Coulter, Thomas. "Upper California" is the title of the map that accompanies Dr. Coulter's "Notes on Upper California," *Journal of the Royal Geographical Society of London* 5:59 ff. (1835; Wheat, no. 8).

Couro, Ted, and Christina **Hutcheson**, *Dictionary of Mesa Grande Diegueño* (Banning: Malki Museum Press, 1973); includes information on Diegueño place names.

Coy, Owen C. Two works by this historian are cited here. **Coy, Counties**: *California county boundaries . . .* (Berkeley: California Historical Survey Commission, 1923). **Coy, Humboldt**: *The Humboldt Bay region, 1850–1875* (Los Angeles: California State Historical Association, 1929).

Craigie, William Alexander, ed., *A dictionary of American English on historical principles* (4 vols.; Chicago, 1936–44).

Crapsey, Malinee, of the Public Information Office, Sequoia National Park, kindly provided information on names in Sequoia and Kings Canyon National Parks.

Crespí, Padre Juan (1721–82), was the chronicler of the Portolá expedition of 1769–70. He probably placed more names on the map of California than any other person in Spanish times. The references are to Herbert E. Bolton's *Fray Juan Crespí . . .* (Berkeley and Los Angeles: University of California Press, 1927), which contains Crespí's diary of the Portolá expedition. (This diary is also in Palou 4:109–260.) Crespí's diary of the Fages expedition of 1772, from Monterey to the San Francisco Bay region, is in Bolton's *Fray Juan Crespí*, pp. 277–303.

Crites, Arthur S., *Pioneer days in Kern County* (Los Angeles, 1951).

Cronise, Titus Fey, *The natural wealth of California . . .* (San Francisco, 1868).

Crosby, Jeff, of the Kings County Library, provided valuable information on his area.

Cutler, Charles, *O brave new words! Native American loanwords in current English* (Norman: University of Oklahoma Press, 1994).

Cutter, Donald C., translated and edited the accounts of the Moraga expedition of 1808 and the Malaspina expedition of 1791.

Dakin, Susanna Bryant, *A Scotch paisano: Hugo Reid's life in California, 1832–1852* (Berkeley and Los Angeles: University of California Press, 1939). Appendix B reprints Reid's "Letters on the Los Angeles County Indians," first published in the *Los Angeles Star* in 1852.

Dalrymple, Alexander. *Plan of Port St. Francisco* (1789); *Plan of Port San Francisco in New Albion* (1790); *Chart of the west-coast of California* (1790); published from Spanish sources. These maps were reproduced in the *Pacific Historical Review* (1947) 16:4; however, references in this work are to Davidson's copies (*q.v.*).

Dana, Richard Henry Jr. (1815–82), *Two years before the mast: A personal narrative* (New York, 1840). References are to the Houghton Mifflin edition of 1911.

Darling, Curtis, *Kern County place names* (Bakersfield: Kern County Historical Society, 1988). Mr. Darling has also kindly provided further information in personal communication.

Dart, Louisiana C., *What's in a name?* (San Luis Obispo: Mission Federal Savings, 1979), deals with place names of San Luis Obispo Co.

Davidson, George (1825–1911), a native of Nottingham, England, was for many years California's leading scientist in the fields of astronomy, geodesy, and geography. In 1850 he was a member of the first party sent by the U.S. Coast Survey to chart the Pacific coast, and from 1868 to 1895 he was in charge of the Coast Survey on the west coast. His work was naturally of great importance for the nomenclature of our coastal features. Unless otherwise stated, his name refers to his papers and correspondence, now in the possession of the University of California.

Davis, D. T., of the faculty of Napa Junior College, provided information to E. G. Gudde.

Davis, H. P., of Nevada City, provided information to E. G. Gudde.

Davis, William Heath (1822–1909), *Sixty years in California . . .* (San Francisco, 1889).

Dayley, John P., *Tümpisa (Panamint) Shoshone dictionary* (UCPL, 116).

Death Valley Guide. Federal Writers Project, *California, Death Valley, a guide* (American Guide Series; Boston: Houghton Mifflin, 1939).

Death Valley Survey. "Death Valley National Monument: Place name survey" (typewritten MS). An excellent regional study of geographical names by officials of the national monument: former superintendents J. R. White and T. R. Goodwin, former ranger H. D. Curry, and former naturalist E. C. Alberts. Much of the material is from the work of T. S. Palmer (*q.v.*).

de Ferrari, Carlo M., Tuolumne County Historian, in Sonora, provided valuable information on his area.

Delano, Alonzo (1802–78), wrote, under the pen name "Old Block," numerous satires, letters, travelogues, and human interest stories depicting the times of the Gold Rush. Unless otherwise stated, the reference is to his *Penknife sketches, or Chips of the old block* (Sacramento, 1853), as republished by the Grabhorn Press (San Francisco, 1934).

De Long, Charles E., "'California's bantam cock': The journals of Charles E. De Long, 1854–1863," ed. Carl I. Wheat, *CHSQ*, vols. 8–10.

de Massey, Ernest. *See* Gibbs.

Dep. Recs. "Departmental records, 1822–45" (14 vols. in 4). *See* Archivo de California.

Derby, Lt. George H., of the Topographical Engineers (known in literature as John Phoenix) was engaged in California exploration in 1849–50. His map of 1849 cited in this work is entitled "Sketch of General Riley's route through the mining districts" (Wheat, no. 79); that of 1850 is "Reconnaissance of the Tulares Valley." Derby's maps are reproduced in *CHSQ*, vol. 11, opp. pp. 99, 102, 247.

Desert Magazine. A well-edited periodical, published monthly from 1937 in El Centro.

Diament, Henri, "L'emploi systématique et inconscient du pseudo-espagnol dans les noms donnés à la propriété foncière et aux odonymes californiens contemporains," *Onomastica Canadiana* (1991) 73:1.9–26, describes the "pseudo-Spanish" names often given to California place names and street names.

Dictionary etymologists. *See* Etymology.

Diegueño. A tribe and language of the Yuman family, named for San Diego Mission; now sometimes called Kumeyaay. *See* Couro and Hutcheson.

Diseños were the maps or plats of Spanish and Mexican land grants, which usually accompanied petitions for such grants. For many grants there are a number of *diseños*. Many *diseños* in the land-grant records of the U.S. District Court in San Francisco, some in the National Archives in Washington, D.C., and some in the state archives in Sacramento have been examined. These maps are of the greatest importance as sources for Indian and Spanish names. *See* Bowman Index; *Expediente;* Land grant.

Disturnell, John, "Mapa de los Estados Unidos de Mejico . . ." (New York, 1847). This was the reference map used in the negotiations of the Treaty of Guadalupe Hidalgo of 1848 (Wheat, no. 33).

Docs. Hist. Cal. "Documentos para la historia de California, 1770–1875" (4 vols.; originals in Bancroft Library).

Donaldson, Karen, director of the Kentucky Mine Museum, Sierra City, kindly provided information on Sierra Co.

Douglas, Edward M., "Gazetteer of the mountains of the state of California, preliminary (incomplete) edition" (mimeographed; Washington, 1929). A list of the orographic features as far as they had been put on the topographical atlas of the USGS in 1929.

Doyle, Thomas B., of San Francisco, compiled an extensive file of California place names collected over many years.

Drury, Aubrey. When the name is followed by a page number, the reference is to Drury's *California: An intimate guide* (New York: Harpers, 1939); when the name alone is given, the information was received directly from the author.

DSP. "Departmental State Papers, 1821–46" (20 vols. in 7). *See* Archivo de California.

DSP Ang. "Departmental State Papers, Angeles, 1825–47" (12 vols. in 4). *See* Archivo de California.

DSP Ben. "Departmental State Papers, Benicia, 1821–46" (5 vols. in 2). *See* Archivo de California.

DSP Ben. C and T. "Departmental State Papers, Benicia, Commissary and Treasury, 1825–42" (5 vols. in 1). *See* Archivo de California.

DSP Ben. CH. "Departmental State Papers, Benicia, Custom House, 1816–48" (8 vols. in 1). *See* Archivo de California.

DSP Ben. Mil. "Departmental State Papers, Benicia, Military, 1772–1846" (36 vols., numbered 53 to 88, in 3 vols., continued from PSP Ben. Mil.). *See* Archivo de California.

DSP Ben. P and J. "Departmental State Papers, Benicia, Prefecturas y juzgados, 1828–46" (6 vols. in 1). *See* Archivo de California.

DSP Mont. "Departmental State Papers, Monterey, 1777–1845" (8 vols. in 1). *See* Archivo de California.

DSP San Jose. "Departmental State Papers, San Jose" (7 vols. in 2; transcripts, Bancroft Library).

Duflot de Mofras, Eugène (1810–85), attaché of the French legation in Mexico, visited California and Oregon in 1841–42, charged by the French government to report on the conditions of these territories. His reports—based on Humboldt's studies, on his own research in Mexico, and on direct observation—were published in 1844 in Paris under the title *Exploration du Territoire de l'Orégon, des Californies et de la Mer Vermeille* . . . (2 vols. and atlas). His name and a page number refer to the book. **Duflot de Mofras's map**: His great map, faulty in many details, is nevertheless a milestone in western American cartography. **Duflot de Mofras's plan** plus a number refers to the maps of smaller areas included in the atlas, plan 16 being "Port de San Francisco dans la Haute Californie" and "Entrée du port de San Francisco."

Dumke, Glenn S., *The boom of the Eighties in southern California* (San Marino: Huntington Library, 1944).

Durán, Padre Narciso (1776–1846), served at Mission San Jose, 1806–33, and with Padre Abella accompanied Luis Argüello up the lower Sacramento and San Joaquin Rivers in May 1817. His diary of this expedition (in both Spanish and English), "Expedition on the Sacramento and San Joaquin Rivers in 1817," ed. Charles E. Chapman, is in *APCH:P* 2:329–49.

Durham, David L., *Geographic names of Madera County* (Fresno: Pioneer, 1987).

Eade, Linda, of the National Park Service, Sequoia National Park, provided valuable information on the park area.

Eddy, W. M. "Approved and declared to be the official map of the state of California by an act of the legislature passed Mar. 25th 1853" (New York: J. H. Colton, 1854; Wheat, no. 257). This map has been much maligned, and Eddy has been accused of improvising his geography and toponymy (see the scathing criticism in Rep. Sur. Gen. 45:236–37 [1856]). The adverse criticism, however, was directed at the map probably because of its somewhat bombastic title. It is a good map, certainly not worse than most other maps before Goddard's. An excellent German map, reproduced in *Zeitschrift für allgemeine Erdkunde, neue Folge,* vol. 1, p. 208 (Berlin, 1856), is called Eddy's map but differs considerably from the original because it has used additional sources.

Egli, J. J., *Nomina geographica* (2d ed.; Leipzig, 1893). This work deals with the origin and etymology of more than 42,000 geographical names in all parts of the world.

Eld, Henry. As a midshipman, Eld was a member of Lt. G. F. Emmons's detachment of the Wilkes expedition on its return from the Willamette Valley to San Francisco in 1841. His diary and sketches were acquired by Yale University. *See* Wilkes.

Eldredge, Zoeth Skinner (1846–1918), was the author of several books on California history. In 1905 he carried on a somewhat hysterical campaign to restore "correct" Spanish place names. Supported by George Davidson, he prevailed on the Post Office Dept. to change the spelling of a number of names, including a few that were not Spanish at all. He also published a *History of California* in five volumes (New York: Century History Co., 1915). **Eldredge, Portolá** refers to his *The march of Portolá and the discovery of the Bay of San Francisco* . . . (San Francisco, 1909).

Ellis, Ruth, of the Sacramento Public Library, kindly provided information on place names in her area.

Emmert, Lou, of the Madera County Historical Society, provided valuable material on her county.

Emmons, Lt. George Falconer, was in command of a detachment of Charles Wilkes's expedition, exploring the interior from the Columbia River to San Francisco in 1841. *See* Eld; Wilkes.

Emory, Maj. William H., *Notes of a military reconnoissance from Fort Leavenworth . . . to San Diego . . . 1846–7* (Washington, 1848). **Emory, Report** refers to his *Report on the survey of the boundary between the United States and Mexico* (2 vols.; Washington, 1857); vol. 1 contains the reports by Major Emory and his subordinate officers of their expeditions in 1849.

Engelhardt, Zephyrin (1851–1934), *The missions and missionaries of California* (4 vols., San Francisco: Barry, 1908–16). The standard work of its kind. Readers desiring more information on the missions are referred to Engelhardt's comprehensive index. Where the name of a mission is added to the name Engelhardt, the reference is to one of his monographs on missions published between 1920 and 1934.

Erman's map. "Karte von Californien," a map based on Russian sources, is published in Janus Hoppe, *Californiens Gegenwart und Zukunft* (Berlin, 1849); Wheat, no. 96.

Estudillo, José María. This refers to a sketch map of exploration in the San Joaquin Valley in 1819, from Mission San Juan Bautista to the Merced River and southward to Buena Vista Lake, reproduced in Priestley's *Fages*. Estudillo's diary of his expedition from Monterey to the *rancherías situadas en los tulares* has been translated by Anna H. Gayton as *Estudillo among the Yokuts* (Berkeley: University of California Press, 1936).

Estudillo Docs. "Documentos para la historia de California, 1776–1850: Archivo particular de la familia Estudillo" (2 vols.; originals, Bancroft Library).

Etymology. The branch of philology devoted to the study of the origin, derivation, and evolution of words. **Folk etymology** (or popular etymology) is the process of changing strange terms to make them resemble well-known words; thus the Yurok Indian place name *kwosan* has become Squash Ann Creek (*q.v.*) in popular speech. **Dictionary etymologist** is a term applied to people who look up the meaning of an unfamiliar name in a dictionary of the language from which the name seems to have been taken. If they find the word, they give the translation and the foreign pronunciation. If they do not find the word, they declare the name to be a "corruption" of a similar word in the dictionary. Through the methods of such a dictionary etymologist, Chiles (chīlz) Valley, named for the well-known pioneer Joseph R. Chiles, is interpreted as 'Red Pepper Valley' and is pronounced (chē´ lās).

Expediente. In Spanish and Mexican times, the file of legal papers dealing with the separation of a parcel of land from the public domain, and the granting of it to an individual or institution, was known as the *expediente*.

Fages, Pedro (1730–94), a lieutenant of Catalonian volunteers and one of the leaders of the Portolá expedition of 1769, was an outstanding figure in the early years of Spanish occupation and from 1782 to 1791 was governor of the Californias. The name "Fages" alone refers to his *Historical, political, and natural description of California*, trans. Herbert I. Priestley (Berkeley: University of California Press, 1937), from a manuscript dated 1775. The book is of great value for California ethnology, and in it are mentioned numerous place names. Fages does not seem to have bestowed any names himself: his Article 6 is headed "From the Real Blanco to a place without a name in 36°44'." **Fages's diary** refers to "The Colorado River campaign, 1781–1782: Diary of Pedro Fages" (in both Spanish and English), ed. Herbert I. Priestley, in *APCH:P* 3:133–233. *See* Crespí.

Farley, Minard H., "Map of the newly discovered tramontane silver mines in southern California . . ." (San Francisco, 1861; Wheat, no. 318).

Farquhar, Francis P. (1887–1974). The name indicates that the information was received directly from Farquhar, or that it was taken from the field copy of his *Place names of the High Sierra* (San Francisco: Sierra Club, 1926).

Fay, Albert H., *A glossary of the mining and mineral industry* (Washington: U.S. Bureau of Mines, Bulletin 95, 1920).

Feltman, Erving R., *California's Lake County: Places and postal history* (Lake Grove, Ore.: The Depot, 1993).

Ferrer (or Ferrelo), Bartolomé, was second in command of the first known exploring expedition along the coast of California and became commander of the expedition after Cabrillo's death on Jan. 3, 1543. *See* Cabrillo.

Ferrier, William W., *Berkeley, California . . .* (Berkeley, 1933).

Findlay, Alexander G., *A directory for the navigation of the Pacific Ocean . . .* (London, 1851).

Fisk, O. J., of San Bernardino, provided information to E. G. Gudde.

Fitch, Josefa Carrillo de (Mrs. Henry D. Fitch), "Narración" (MS, Bancroft Library).

Folk etymology. *See* Etymology.

Font, Padre Pedro (d. 1781), was the chaplain of the second Anza expedition (1775–76). His records and maps are sources of great importance for early Spanish place names. His name followed by a page number refers to his "Short diary," written soon after his return from the expedition, and translated and printed in Bolton's edition of *Anza's California expeditions*, vol. 3. **Font, *Compl. diary*** refers to his expanded diary, vol. 4 of Bolton's work. The references to his map are to Davidson's copy of the large map dated Tubutama, 1777. This map includes the result of the explorations of Garcés; it is more nearly complete than either the one published in the *Complete diary* or the one in Coues, *On the trail of a Spanish pioneer*.

Foote, Horace S., ed., *Pen pictures from the garden of the world or Santa Clara County, California* . . . (Chicago, 1888).

Forbes, Alexander (1778–1864). Reference is to the map and the sketches of California harbors in his *California: A history of Upper and Lower California* . . . (London, 1839). The map, drawn by John Arrowsmith, is "The coasts of Guatemala and Mexico, from Panama to Cape Mendocino" (London, 1839). Inserted on it are a map by Captain Beechey of the harbor of San Francisco, with "Sketches" by Captain John Hall of Port Bodega, the port of Diego, Monterey harbor, Santa Barbara harbor, and "Port S. Gabriel, or S. Pedro."

Fowler, Catherine S., professor of anthropology, University of Nevada, Reno, supplied valuable linguistic materials from the Numic tribes of the Great Basin.

Fox, Georgia, of the Amador County Museum, provided valuable information on her county.

Fox, Pheron, of San Jose, provided valuable information on Santa Clara Co.

Freeland, Lucy S., provided information to E. G. Gudde about several important Miwok names. Material credited to "Freeland," with no reference to a printed source, was received directly from her. The results of her research are published in two monographs (*International Journal of American Linguistics*, vol. 13, and its Memoir, no. 6), and in work by Broadbent and Freeland (*see* Broadbent).

Frémont, John Charles (1813–90), was in California on two of his major explorations (1843–44 and 1845–46) and took a prominent part in the acquisition of the province. He is intimately connected with California place naming; not only was he influential in perpetuating many older terms, but he also bestowed a number of major names such as Golden Gate, Great Basin, Kern River, Mojave River, and Owens Lake. **Frémont, *Expl. exp*.** or ***Expedition*** refers to his *Report of the exploring expedition to the Rocky Mountains . . . 1842, and to Oregon and North California . . . 1843–1844*, first printed as a Senate document in 1845. Unless otherwise stated, the reference is to the edition of 1853. **Frémont, *Geog. memoir***: *Geographical memoir upon Upper California* . . . 30th Cong., 1st sess. (1848), Senate Misc. Doc. No. 148. **Frémont, *Memoirs***: *Memoirs of my life* . . . (Chicago and New York, 1887), vol. 1 (only one volume was published). **Frémont-Preuss maps**: Of these two maps, that of 1845 is the one accompanying the *Report* mentioned above. That of 1848, although entitled "Map of Oregon and Upper California," includes all U.S. territory west of El Paso (Texas), Pikes Peak, and Fort Laramie. It was made by Charles Preuss from Frémont's surveys and from other authorities.

French, Dr. Darwin, led a party of fifteen into Death Valley in May 1860, in search of the "lost" Gunsight Mine. They did not find the mine but left a number of place names.

Frey, Sandi, of the Lake County Library, provided valuable information on her county.

Frickstad, Walter N., *A century of California post offices, 1848 to 1954* (Oakland: Philatelic Research Society, 1955). A valuable contribution to the history and nomenclature of the state.

Frye, Thomas, of the Oakland Museum, provided valuable information on names in Alameda Co.

Gabrielino. A tribe and language of the Takic (Uto-Aztecan) family. The language was once spoken in Los Angeles Co., named after San Gabriel Mission. Now sometimes called Tongva.

Gamble, Geoffrey, professor of anthropology, Washington State University, kindly provided information on names of Yokuts origin.

Gampp, Ute, of the San Joaquin County Historical Society and Museum, provided valuable information on her area.

Gannett, Henry, *The origin of certain place names in the United States* (2d ed.; Washington:

USGS, Bulletin 258, 1905). *American names: A guide to the origin of place names in the United States* (Washington: Public Affairs Press, 1947) is a literal reprint of this book, which contains too many inaccuracies (together with some good information) to be called a "guide."

Garber, Elaine K., "Hueneme: Origin of the name," *Ventura County Historical Society Quarterly* (1967), vol. 12, no. 3, pp. 11–15.

Garcés, Padre Francisco (1738–81), came to California with the Anza expedition in 1774 and made important explorations in southern California. **Garcés, 1774**: His accounts of his 1774 explorations are in Bolton's *Anza's California expeditions* (Berkeley: University of California Press, 1930), 2:307–92. Garcés was in California again in 1775–76, starting with the Anza expedition but taking a different route after reaching Yuma. His account of this expedition is in Coues, *On the trail of a Spanish pioneer*.

Generic name. The names of most physical features consist of a specific element and a generic element. The specific element is the actual name given: Diablo, Pit, Yosemite, etc. The generic element designates the feature: Mountain, River, National Park, etc. In the names of communities, the generic parts (e.g. burg, town, ville) are no longer considered separate elements.

Geographic; Geographical. An attempt has been made in this book to distinguish these two terms by using the shorter form when referring to a natural feature, hence "geographic feature" and "geographic landmark," and the term "geographical" when referring to the science of geography, hence "geographical name" and "in a geographical sense."

Geographical Report; Geographical Survey. *See* Wheeler Survey.

Geographic Board. *See* BGN.

Gibbes, Charles Dayton, "New map of California" and "Map of the southern mines" are in J. H. Carson, *Early recollection of the mines* (Stockton, 1852). Both maps are important sources for the names of mining settlements.

Gibbs, George, accompanied Col. Redick McKee, U.S. Indian agent, on his expedition through northwestern California in the summer and fall of 1851. His "Journal" of this expedition and his "Observations on Indian dialects of northern California" were published in Schoolcraft 3:99 ff., 420 ff. Both sources are important for place names in the territory traversed. Gibbs's map is in the Indian Office in Washington, D.C. A part of it was reproduced in Ernest de Massey, *A Frenchman in the Gold Rush* (San Francisco: California Historical Society, 1927). *See* Chinook Jargon; Schoolcraft.

Gifford, Edward W., professor of anthropology, University of California, supplied information to E. G. Gudde.

Gill, R. Bayley, of Cima, supplied information about names in the Ivanpah area.

GNIS. Geographic Names Information System, a computerized file, organized by state, available on CD-ROM from the Geographical Names Branch, U.S. Geological Survey, Reston, Virginia. This is the most complete and authoritative listing of U.S. place names with their geographical locations. However, historical and etymological information is not included. *See also* Omni Gazetteer.

Goddard, George H., "Britton and Rey's map of the state of California" (3d ed.; San Francisco, 1860). This is sometimes referred to as "Goddard's map" or "Goddard 1860." Goddard, a native of England, was one of the great engineers of the 1850s who played an important role in the geographical delineation of the state. His map, first published in 1857, is the first reliable map of California that made use of all the official and private surveys executed in the first decade of American occupation.

Goddard, Pliny Earle, "Life and culture of the Hupa," AAE (1903) 1:1–88; "Kato texts," AAE (1909) 5:65–238.

Goethe, C. M., *Sierran cabin . . . from skyscraper . . .* (Sacramento, 1943).

Golla, Victor, professor of Native American studies, Humboldt State University, Arcata, is a specialist in the Athabaskan languages; he has provided information on several names in northwestern California.

González. Cabrera Bueno, José, compiled a *Navegación especulativa y práctica* (Manila, 1734), principally for the use of the Philippine galleons. The part relating to the California coast is based on the *derrotero* of the Vizcaíno expedition of 1602–3, made by Francisco de Bolaños and Padre Antonio de la Ascensión. Bolaños had sailed along the coast with Rodríguez Cermeño in 1595 (*see* San Francisco in text). The Portolá expedition had with them a copy of the *Navegación*.

Great Register. A list of the names of registered voters of a county, compiled by the county clerk. The older registers often give not only the place of residence of the voter, but also his age, country or state of birth, date of naturalization, occupation, and date of registration.

Gregg party. A party of prospectors led by Dr. Josiah Gregg (1806–50) named many of the rivers between Humboldt and San Francisco Bays in the winter of 1849–50. Gregg, an outstanding explorer of western America, died from exposure at the end of the journey (*see* L. K. Wood).

Grizzle, Donna, of the Imperial County Historical Society, provided useful information on her area.

Grizzly Bear. A monthly magazine, published in Los Angeles from 1907 as the official organ of the Native Sons and Native Daughters of the Golden West.

Gudde (go͞od′ ē), Erwin G. (1889–1969), authored the first three editions of this book. He was professor of German at Berkeley; for his personal history, see *Names* 7:1–11 (1959), 19:33 (1971). He was the author of several books on California history, and his work on the place names of the state was translated into Chinese as *Chia chou ti ming tzu tien* (Taipei: Chu pan, 1989).

Guerra y Noriega, José de la, "Documentos para la historia de California" (7 vols.; transcripts, Bancroft Library).

Guilford-Kardell, Margaret, of Blaine, Washington, is a former resident of Shasta Co. and provided valuable information on names in the upper McCloud River area.

Guinn, J. M., "Some California place names," *HSSC:Q* 7:39 ff.

Gunther, Jane Davies, *Riverside County, California, place names* (Riverside: The author, 1984).

Hall, Joyce, of the Fresno County Library, provided valuable information on her county.

Hall-Patton, Mark P., *Memories of the land: Placenames of San Luis Obispo County* (San Luis Obispo: EZ Nature Books, 1994).

Hanks, Henry G., "Report on the borax deposits of California and Nevada," published in *Third Annual Report of the State Mineralogist* (1883); and "Catalogue and description of the minerals of California . . . ," published in *Fourth Annual Report* . . . (1884).

Hanna, Phil Townsend, *The dictionary of California land names* (Los Angeles: Automobile Club of Southern California, 1946).

Harder, Kelsie B., *Illustrated dictionary of place names, United States and Canada* (New York: Van Nostrand Reinhold, 1976), is the most up-to-date reference book of its kind.

Harrington, John P., senior ethnologist of the Smithsonian Institution and legendary field worker on the linguistic anthropology of Native California, was the author of "A tentative list of the Hispanized Chumashan placenames of San Luis Obispo, Santa Barbara, and Ventura Counties, California," *American Anthropologist*, n.s. (1911) 13:725–26; he provided information through personal communication to E. G. Gudde for earlier editions of this volume. The present edition also draws on Harrington's "Indian words in southwest Spanish," *Plateau* 17:2.27–40 (1944), and on his voluminous unpublished notes, now in the Smithsonian Institution, but made available on microfilm by Kraus International Publications (distributed by Norman Ross Publications, New York).

Harris, Lola, "Place names of eastern Shasta County" (Redding: Shasta College, 1967), is a manuscript made available through the library at California State University, Chico.

Hart, James D., *A companion to California* (New York: Oxford University Press, 1978), is an especially convenient and valuable source of facts on California history and geography; a revised and expanded edition was published by the University of California Press in 1987.

Hayes, Benjamin (1815–77), "Documentos para la historia de California, 1826–1850." Originals, copies, and extracts from the Archives of San Diego Co. collected by Benjamin Hayes. **Hayes, C and R** refers to the section called "Commerce and Revenue." **Hayes, Docs.**: Benjamin Hayes, "Documents for the history of California" (transcripts, Bancroft Library). **Hayes, Mis.**: Benjamin Hayes, "Missions of Alta California" (2 vols.; MSS, clippings, etc., Bancroft Library).

Headlight. *See* Brown, Thomas P.

Heceta. *See* Hezeta.

Heintzelman, Gen. Samuel P., commander of the 3d Corps of the U.S. Army in the Peninsula campaign in 1862, had been (as a major) a military commander in southern California from 1850 to 1855, then for several years commander and sub-Indian agent on the Klamath River. His name with the date 1853 refers to his report on the Yuma Indians (34th Cong., 3d sess., House Ex. Doc. no. 76); with the date 1858, it refers to his report on the Klamath River Indians (35th Cong., 2d sess., House Ex. Doc. no. 2).

Helm Index refers to Mary H. Helm, "Index to topographic quadrangles of California," with an introduction by Olaf P. Jenkins, *California Journal of Mines and Geology* (1945) 41:251–360.

Hendryx, Michael, Director Curator, Siskiyou County Museum, provided valuable data from his area.

Henshaw, Henry W., a member of the Bureau of American Ethnology, collected much California Indian language material intermittently between 1884 and 1893. His extensive word and phrase lists of the Chumash and the Costanoan languages, edited by Robert F. Heizer, are published in *Anthropological Records* 15:2 (Berkeley and Los Angeles: University of California Press, 1955).

Hezeta, Bruno de, Spanish navigator, fl. 1775 (also spelled Heceta). Reference is to the journey of the *Santiago*, commanded by Hezeta in 1775 (Wagner, chap. 25).

Hine, Robert V., *California's utopian colonies* (San Marino: Huntington Library, 1953).

Historic landmarks. The movement to preserve landmarks of historical importance or interest was inaugurated originally by the Native Sons of the Golden West; *see* the *Reports on registered landmarks* issued at irregular intervals by the State Dept. of Natural Resources and the State Park Commission in cooperation with the California State Chamber of Commerce.

Hittell, John S. (1825–1901), *Yosemite: Its wonders and its beauties* (San Francisco, 1868). The same author has a chapter on place names in his excellent book *The resources of California* (4th ed.; San Francisco, 1868).

Hittell, Theodore H.(1830–1917), *History of California* (4 vols.; San Francisco, 1897–98).

HNAI. The *Handbook of North American Indians*, ed. William Sturtevant, has been appearing from the Smithsonian Institution since 1978; twenty volumes are planned. Vol. 8, on California, was edited by Robert F. Heizer.

Hodge, Frederick W., ed., *Handbook of American Indians* (Washington: Smithsonian Institution, Bureau of American Ethnology, Bulletin 30, 1907–10; citations in text are to the two-volume edition, Washington, 1912). A monumental reference on Indians north of Mexico.

Hoffmann, Charles F., a native of Frankfurt am Main, Germany, was topographer and cartographer of the State Geological (Whitney) Survey throughout its existence (1860–74). Hoffmann was one of the pioneers of modern topography and is responsible for the adoption of the contour line for the topographical atlas of the United States made by the USGS. **Hoffmann's map**: "Topographical map of Central California together with a part of Nevada," 1873, was the most detailed and most reliable map of this part of the state until the issuance of the atlas sheets by the USGS. Unfortunately, the Whitney Survey was discontinued before the map was completed; the northwest section exists only as a sketch. **Hoffman's map of the Bay region**: "Map of the region adjacent to the Bay of San Francisco," 1873. **Hoffmann-Gardner map**: "Map of a portion of the Sierra Nevada adjacent to the Yosemite Valley from surveys made by Ch. F. Hoffmann and J. T. Gardner," 1863–67. The three maps were publications of the Whitney Survey. **Hoffmann's notes**: Hoffmann's field notes of 1861 remained in the possession of his son, Ross Hoffmann. Unfortunately, the books containing notes of subsequent years have been misplaced.

Hood, Washington, "Map of the United States territory of Oregon west of the Rocky Mountains . . . ," compiled in the Bureau of Topographical Engineers under the direction of Col. J. J. Abert (1838; Wheat, no. 11).

Hoover, Mildred Brooke, *Historic spots in California*, vol. 3: *Counties of the Coast Range* (Stanford: Stanford University Press, 1937).

Howard, J. D., of Klamath Falls, Oregon, made a thorough examination of the lava beds (now a national monument) from 1917 to 1927 and supplied E. G. Gudde with valuable information about the names in this region.

HSSC:P. *Historical Society of Southern California: Annual publication* (Los Angeles, 1884–1934). ***HSSC:Q***. *Historical Society of Southern California: Quarterly*. This publication, which began with vol. 17 in 1935, is a continuation of ***HSSC:P***. It is now called *Southern California Quarterly*.

Humboldt, Alexander von, German geographer and pioneer of modern scientific exploration (1769–1859). He has probably been honored in more place names in the western United States than any other man of science. Many explorers of the American West considered as a standard the methods of research employed by Humboldt in his epoch-making journey through South America (1797–1802). His essay on the kingdom of New Spain, which included the first reliable account of

Upper California published in Europe, was first issued in Stuttgart in 1808. **Humboldt's map**: His map of Mexico was published in Paris in 1811.

Humboldt Bay. "The discovery of Humboldt Bay," *Geographical Society of the Pacific: Proceedings* (1891) 2:2. *See* Coast Pilot; Coast Survey.

Humboldt County. *See* Turner.

Hunzicker, Lena B., compiled an index of San Diego Co. place names, the larger part of which was incorporated in the file of the State Geographic Board. Additional names from this file were received by E. G. Gudde from the San Diego Public Library.

Hupa. An Athabaskan tribe and language of Humboldt Co.; also spelled Hoopa.

Hutchings, James M. (1818–1902), published several works of importance as sources of California place names. **Hutchings, Scenes**: *Scenes of wonder and curiosity in California* (San Francisco, 1870). **Hutchings, Sierras**: *In the heart of the Sierras* (Oakland, 1886). His often-mentioned *Illustrated California Magazine* (or *California Magazine*) was published from July 1856 to June 1861. *See* Hutchings, Mount, in text.

Hydrography; Hydrographic. The phase of physical geography referring to surface waters.

Indian Report. 38th Congress, special session (1853), Senate Executive Document no. 4, a lengthy report of the Secretary of the Interior including correspondence from Indian agents and commissioners in California. It is valuable for many Indian and some other early names. **Indian Report** with the year added refers to the annual publication of the Commissioner of Indian Affairs.

Iroquoian. This language family of the eastern United States includes such languages as Mohawk, Oneida, Seneca, and Cherokee. Many local place names were derived from these languages, and some were transferred to California, e.g. Alleghany and Tioga. Information on Iroquoian linguistics has been kindly provided by Wallace Chafe and Marianne Mithun.

Iventosch, Herman, "Orinda," *Names* 12:103–7 (1964).

Ives, Joseph C., "Report upon the Colorado River of the West, explored in 1857 and 1858 . . ." (Washington, D.C., 1861), 36th Cong., 1st sess., House Ex. Doc. no. 90.

James, George Wharton (1858–1923), *The wonders of the Colorado Desert (southern California)* . . . , 2 vols. (Boston: Little Brown, 1906).

Jayhawker party. One of the parties of gold seekers who crossed Death Valley in 1849 and who were directly or indirectly responsible for a number of names in the region. The party had been organized in Illinois. See John G. Ellenbecker, *The Jayhawkers of Death Valley* (1938); and Carl Wheat in *HSSC:Q* 22:102 ff.

Jepson, Willis Linn, *The silva of California* (Berkeley and Los Angeles: University of California Press, 1911).

Johnson, Charles, of the Ventura County Library, kindly provided information on place names of his area.

Johnson, John, anthropologist of the Santa Barbara Museum of Natural History, provided information on place names of the Chumash area. His dissertation, "Chumash social organization" (University of California, Santa Barbara, 1988), contains useful material on Chumash place names.

Johnston, Col. Adam, "Report of the Secretary of the Interior, communicating . . . a report of the Commissioner of Indian Affairs," 32d Cong., 1st sess., Senate Ex. Doc. no. 61. It includes communications to and from Col. Johnston, U.S. Indian agent, and others, in which names of Indian tribes and of streams on the east side of the San Joaquin Valley are mentioned.

Jones, Alice Goen (ed.), *Trinity County historical sites* (Weaverville: Trinity County Historical Society, 1981).

Jones, William A., of Meriam Library, California State University, Chico, kindly provided information on Butte Co.

Juaneño. A tribe and language of the Takic (Uto-Aztecan) family, closely related to Luiseño. The name derives from San Juan Capistrano Mission. Now also called by the native name Ajachemem.

Karuk (kä′ ro͞ok). A tribe and language of the upper Klamath River, in Humboldt and Siskiyou Cos., sometimes classified in the Hokan stock. Also spelled Karok. *See* Bright.

Kazmarek, F. A., *Ghost towns and relics of '49* (Stockton: Chamber of Commerce, 1936).

Keffer, Frank M., *History of San Fernando Valley* . . . (Glendale: Stillman, 1934).

Kern County. *See* Richard C. Bailey; Darling.

King, Clarence (1842–1901), *Mountaineering in the Sierra Nevada* (Boston, 1872). *See* Clarence King, Mount, in text.

Kino, Eusebius, a German Jesuit of Italian ancestry, an outstanding missionary and explorer from 1683 to 1711, established the first mission in Lower California. The references are to Herbert E. Bolton's edition of *Kino's historical memoir of Pimería Alta* (2 vols.; Cleveland: Clark, 1919).

Kirchhoff, Theodor, *Californische Kulturbilder* (Cassel, 1886). A very informative account of California in the 1880s by a German resident of San Francisco.

Klar, Kathryn, "Pismo and Nipomo," *Names* 23:26–30 (1975); "An addendum to Applegate's 'Chumash place names'" (San Luis Obispo County Archaeological Society, Occasional papers 11:52–54 [1977]).

Knave. Reference is to "The Knave," a page in the Sunday edition of the *Oakland Tribune*, ably conducted for many years by the late Ad Schuster, and later by Leonard Verbarg. Interesting place-name material was contributed by John Winkley, Henry Mauldin, Rockwell Hunt, Alex Rosborough, and others.

Kneiss, Gilbert H., *Bonanza railroads* (Stanford: Stanford University Press, 1944).

Knowland, Joseph R., *California: A landmark history* . . . (Oakland: Tribune, 1941).

Knox, Alexander, *Glossary of geographical and topographical terms* (London, 1904).

Kobayashi, Deanna, of the Merced County Library, kindly provided information on names of Merced Co.

Kohl, Johann G. (1808–78), a German geographer, was the pioneer of North American cartographical history. Unfortunately, his great cartographical work remained unfinished. Part of the text was published in the *Reports* of the Coast Survey for 1857 and 1885. His great collection is in the Library of Congress: Justin Winsor, *The Kohl collection of maps relating to America* (Cambridge, 1886); *see* Coast Survey *Report* 1855:374. In official reports after 1855, the Coast Survey used the orthography of the names of coastal features established by Kohl. Some of Kohl's misspellings were not corrected until about 1900.

Kotzebue, Otto von (1786–1846), Russian navigator, a son of the once-famous German playwright August von Kotzebue. He commanded two Russian exploring expeditions around the world in 1815–18 and 1823–26. On both journeys he stayed in San Francisco several weeks, and his accounts, as well as those of his staffs, are sources of some value for place names. The references to the second expedition are from *A new voyage round the world in the years 1823, 24, 25, and 26* (2 vols.; London, 1830). *See* Chamisso; Rurik.

Kroeber, Alfred Louis (1876–1960), founder of the Department of Anthropology at the University of California, Berkeley, was for many years the foremost authority on California Indians. Unless otherwise stated, the information was either taken from the field copy of "California place names of Indian origin" in AAE 12:2 or received from Kroeber directly by E. G. Gudde. **Kroeber's Handbook**: *Handbook of the Indians of California* (Washington: Bureau of American Ethnology, Bulletin 78 [1925]).

Lahainaluna. *Carta esférica de la costa de la Alta California* . . . (Lahainaluna, Sandwich Islands, 1839; listed as Wheat, no. 12).

Lake County. *See* Feltman.

Land grant. After the landed aristocracy of Europe entered the last period of its struggle for existence against new economic forces, California underwent a phase of unadulterated feudalism. From 1775 to the end of the Mexican regime, a class of large landholders was created by separating 666 land units from the public domain and granting these to citizens who had rendered services or who just promised to cultivate the land. Of these grants, 524 were confirmed by U.S. courts, of which 517 were patented to the grantees or their legal successors. The date given in connection with a land grant is that of the *concedo*, the day on which the governor signed the concession. Often the name that was given to the grant had been in existence long before this date; sometimes, the name was applied later. The name of the grantee is given in the text entry only if he was important historically, or if it is needed to identify the grant. *See* Bowman Index; *Diseño; Expediente*.

Land Office Report and Map. Annual *Report of the* [U.S.] *Commissioner of the General Land Office*. Public Survey maps of various sections of the United States are frequently appended to the reports. These maps have been instrumental in fixing the current form or spelling of many local names. The maps referred to are naturally those of California.

Landrum, Elizabeth A., "Maps of the San Joaquin Valley up to 1860 . . ." (MS, School of Librarianship, University of California, Berkeley), includes photostats of thirty-six of the more important maps of the southern half of the state.

Lang, Robert, of the Library, University of California, Riverside, provided information on English pronunciations of southern California place names.

Langdon, Margaret, professor of linguistics, University of California, San Diego, provided information about Diegueño names.

Langsdorff, Georg Heinrich von (1774–1852), a German surgeon and naturalist, accompanied the Russian "imperial inspector," Nikolai P. Rezanof, to San Francisco in Apr. 1806. An account of his stay there is found in pt. 2 of his *Voyages and travels . . . 1803, 1804, 1805, 1806, and 1807* (2 vols.; London, 1813–14).

La Pérouse, Jean-François de Galaup, comte de (1741–88?), *A voyage round the world, 1785, 1786, 1787, and 1788, under the command of J. F. G. de la Pérouse*, 2 vols. and atlas (London, 1798–99). The original edition, *Voyage autour du monde . . .*, was published in Paris in 1797 in 4 vols. and an atlas. The French explorer was in Monterey in Sept. 1786; in chaps. 11–12 he gives an interesting account of California in the early years of Spanish colonization. His maps are of some value as sources for place names.

Larkin, Thomas Oliver (1802–58). *The Larkin papers: Personal, business, and official correspondence of Thomas Oliver Larkin, merchant and United States Consul in California*, ed. George P. Hammond (10 vols.; Berkeley and Los Angeles: University of California Press, 1951–64).

Las Calaveras, a quarterly published by the Calaveras County Historical Society, includes articles on place names of the county.

Lassen, Peter (1800–1859), a native of Denmark, came to California in 1840 and worked in various places at his trade of blacksmith. In 1844 he was grantee of a land grant in what is now Tehama Co. (*see* Bosquejo in text). He became the outstanding pioneer of the northeastern section of California, and his memory is preserved in the names of more prominent geographic features than that of any of his fellow pioneers.

Latta, Frank F., *El camino viejo a Los Angeles* (Bakersfield: Kern County Historical Society, 1936).

Lawson, James, letters to George Davidson.

Lawson, Scott, of the Plumas County Museum in Quincy, kindly provided information on names in Plumas Co.

LeConte. Both Joseph LeConte (1823–1901), a native of Georgia and professor of geology at the University of California from 1869 to 1901, and his son, Joseph Nisbet LeConte (1870–1950), professor of mechanical engineering at the same university from 1895 to 1937, were intimately connected with place naming in the High Sierra. For a list of their maps, consult Farquhar, pp. 123–24. ***LeConte, Journal***: Joseph LeConte, *A journal of ramblings through the High Sierra of California*, ed. Francis P. Farquhar (San Francisco: Sierra Club, 1930). **J. N. LeConte** given as a reference means that the information was received from Prof. LeConte directly.

Lee, Lila J., director of the Mendocino County Historical Society Library in Ukiah, provided valuable information on her area.

Legis. Recs. "Legislative Records, 1822–46" (4 vols. in 3). *See* Archivo de California.

Lekisch, Barbara, *Tahoe place names* (Lafayette: Great West Books, 1988).

Leland, J. A. C. Information provided to E. G. Gudde.

Loewenstein, Louis K., *Streets of San Francisco: The origins of street and place names* (San Francisco: Lexikos, 1984).

Los Angeles County. *See* Atkinson.

Luddy, William, "Reminiscences by Hardscrabble," *Scott Valley Advance*, Aug. 20, 1900.

Luecke, Mary, is the author of *Post offices of Klamath and Siskiyou Counties, 1948–1954* (Yreka: Klamath National Forest, 1880) and *Two hundred place names in Siskiyou County*, published as *Siskiyou Pioneer*, vol. 5, no. 5 (Yreka: Siskiyou County Historical Society, 1982).

Luiseño. A tribe and language of the Takic (Uto-Aztecan) family, still spoken in Riverside and San Diego Cos.; named after San Luis Rey Mission. *See* Bright.

McArthur, Lewis A., *Oregon geographic names* (6th ed.; Portland: Oregon Historical Society, 1992). One of the best regional place-name studies published so far.

McKenney's Directory. *McKenney's Pacific coast directory . . .* for 1878; 1880; 1882; 1883–84; 1886 (San Francisco and Oakland).

McLaughlin, John E., of the Dept. of English, Utah State University, supplied information on Indian languages of the Numic family, spoken in the Great Basin and along the eastern edge of California.

McNamar, Myrtle, *Way back when* (Cottonwood, 1952), deals with Shasta and Tehama Cos.

Madera County. *See* Durham.

Maidu (mī′ do͞o); **Maiduan**. A linguistic family including Indian tribes east of the Sacramento River from Lassen Co. to El Dorado Co.; sometimes assigned to the Penutian stock. *See* Shipley.

Mariposa Battalion. The unit was formed in the spring of 1851 to punish marauding Indians. The battalion discovered Yosemite Valley in March and made a second expedition in May, bestowing many place names.

Marsh, John (1799–1856), a native of Massachusetts and a graduate of Harvard University, acquired Rancho Los Meganos in 1837 and became the foremost pioneer of the Mount Diablo district. Information that he gave to explorers and surveyors led to the fixing of a number of place names.

Marshall, Robert B. (1867–1949), a native of Virginia, was topographer of the USGS in California from 1891 to 1902 and was geographer for California, Oregon, and Nevada from 1905 to 1907. He is responsible for the application of many names, especially in the High Sierra (*see* Farquhar, p. 114).

Maslin, Prentiss, "Origin and meaning of the names of the counties of California," *California Blue Book* (1907), pp. 275 ff.

Mattole (mə tōl′). An Athabaskan tribe and language of southern Humboldt Co.

Mauldin, Henry K., historian of Lake Co., published the stories of many place names in the *Oakland Tribune*'s "The Knave" (*q.v.*) and supplied others directly to E. G. Gudde.

Maule, William M., *A contribution to the geographic and economic history of the Carson, Walker, and Mono Basins* (San Francisco: USFS, 1938). An excellent local study of geography. Additional information was received by E. G. Gudde directly from Mr. Maule.

Mendenhall. *See* WSP.

Menefee, C. A. *Historical and descriptive sketch book of Napa, Sonoma, Lake, and Mendocino . . .* (Napa City, 1873).

Merriam, Clinton Hart (1855–1942), a well-known biologist and ethnologist, was for many years chairman of the BGN. His extensive files of western U.S. bibliographical references and California place names are in the Museum of Vertebrate Zoölogy of the University of California, Berkeley. **Merriam, Mewan stock**: "Distribution and classification of the Mewan stock of California," *American Anthropologist*, n.s. (1907) 9:338–57. **Merriam, Pit River**: *The classification and distribution of the Pit River Indian tribes of California*, Smithsonian Misc. Coll. (1927) 78:3. **Merriam, Expedition**: "The Death Valley expedition," pt. 2 (*North American fauna*, no. 7; Washington: Dept. of Agriculture, 1893).

Metz, Andrea, of the Merced County Museum, provided valuable information on her county.

Mexican times. This phrase is used for the period from 1822 to 1846 only when it is necessary to distinguish from Spanish times in a political sense.

Miller, Guy, city historian of Palo Alto, supplied E. G. Gudde with information.

Miller, Loye, professor of biology emeritus, University of California, Los Angeles, provided information to E. G. Gudde.

Miller, Miriam, of the Yolo County Historical Museum, kindly provided information on her area.

Miller and Lux. The name of this firm is often mentioned in the text, especially in connection with place names in the San Joaquin Valley—where the firm, in the 1880s, owned nearly all land for fifty miles on both sides of the San Joaquin River. Henry Miller and Charles Lux, German-born butchers, formed their famous partnership in 1856. Neither name is commemorated in a major geographic feature.

Mining Bureau Maps. "Geological map of the State of California," issued by the California State Mining Bureau (1916). For physical geography and toponymy, the most reliable map published between the period of the Whitney Survey (1860–74) and the mapping of California by the USGS. Since 1882 the Mining Bureau has published annual reports under various titles.

Mitchell, Annie, secretary of the Tulare County Historical Society, supplied information on names of her area.

Mithun, Marianne, professor of linguistics, University of California, Santa Barbara, is a specialist in the Iroquoian languages of the northeastern United States and on the Pomo languages of California; she provided information on names originating in those two language families.

Miwok (mē′ wok); **Miwokan**. An Indian language family, sometimes classed in the Penutian stock, found in several locations between the Pacific coast and Yosemite. The languages include Coast Miwok, Lake Miwok, Plains Miwok, and several types of Sierra Miwok; they have been described in the works of L. S. Freeland, Catherine A. Callaghan, and Sylvia Broadbent. *See also* Miwok in text.

Monache (mō′ nə chē); **Mono** (mō′ nō). A tribe and language belonging to the Numic branch of the Uto-Aztecan language family, living in east central California; *see* Monache and Mono in text.

Monterey County. *See* Clark.

Moraga, Gabriel, is repeatedly mentioned as the leader of land expeditions that applied a number of important names. He came to California in 1776, became a soldier, and rose to the rank of lieutenant; he died in 1823. His MS, "Diario de la tercera expedición" (1808), is in the Bancroft Library. Translated and edited by Donald C. Cutter, it was published by Glen Dawson in 1957. For Moraga's 1806 expedition, *see* Muñoz.

Morley, S. Griswold, professor of Spanish, University of California, Berkeley, provided E. G. Gudde with valuable information.

Mormon Battalion. The organization of a battalion of Mormon volunteers to participate in the Mexican War was one phase of the westward movement of the Latter Day Saints. The Mormon Battalion, like Stevenson's New York Volunteer Regiment, is known in history chiefly because many of its members played an important role in the Gold Rush and in the early American period of California.

Morrill, Gil, of the Lassen County Historical Society, in Susanville, provided valuable information on his area.

Morris, Leonard, of the Trinity County Historical Society, Weaverville, kindly provided information on his county.

Mosier, Page, and Dan **Mosier**, *Alameda County place names* (Fremont: Mines Road Books, 1986).

Muir, John (1838–1914), a native of Scotland, was one of the four founders of the Sierra Club and for many years was the best known of California's mountaineers. He has been honored in more California geographical names than any other person, although he himself was very conservative in bestowing names on places. Unless otherwise stated, the reference is to *My first summer in the Sierra* (Boston: Houghton Mifflin, 1911).

Muñoz, Padre Pedro, accompanied Gabriel Moraga on his expedition of 1806. His diary of this expedition is in "Archivo de la Misión de Santa Barbara" 4:1–47.

Munro, Pamela, professor of linguistics at the University of California, Los Angeles, provided information on Yuman and Uto-Aztecan languages of southern California.

Munz, Philip A., and David D. **Keck**, *A California flora* (revised; Berkeley and Los Angeles: University of California Press, 1973), is an authoritative source on California plants.

Myrick, David F., of San Francisco, supplied E. G. Gudde with information about a number of names.

Names: *Journal of the American Name Society*. A quarterly publication, founded by Erwin G. Gudde in 1952.

Narváez, José María (1771–1853?), "Plano del territorio de la alta California . . . 1830," is probably the most valuable of the Mexican maps, especially for the missions and other settlements. It is the first map to show the three mountain ranges, rather correctly but not labeled. The hydrography is mainly imaginary. This MS map is reproduced as Wheat, no. 6.

Nevada Boundary Survey. An unofficial report of the Eastern Boundary Commission, to establish the California-Nevada line, was published in installments in the *Sacramento Daily Union*, June–July 1861. The date in connection with the reference refers to the issue of the *Union*.

New Helvetia diary. The record of events kept at Sutters Fort by Sutter and his clerks from Sept. 9, 1845, to May 25, 1848. Published in 1939 by the Grabhorn Press, San Francisco, for the Society of California Pioneers.

Newmark, Harris (1834–1916), *Sixty years in southern California* (New York: Knickerbocker, 1916; new edition, 1926). The reminiscences of a man who lived through the development of Los Angeles from a sprawling village to a great metropolis.

Nichols, John D., professor of linguistics, University of Manitoba, Winnipeg, provided information on place names of Algonquian origin that have been transferred to California.

Nordby, Gail, of the Calaveras County Library, provided valuable information on her county.

Nordenskiöld, A. E., *Facsimile atlas to the early history of cartography* (Stockholm, 1889). This work by a great Swedish explorer contains several maps valuable for early California nomenclature.

Numic. A branch of the Uto-Aztecan language

family, spoken in the Great Basin. It includes several languages spoken along the eastern side of California, such as Monache, Panamint, and Chemehuevi. *See* Fowler; McLaughlin.

Official Railway Map. *See* Railroad Maps.

OHQ. *Oregon Historical Quarterly*, published by the Oregon Historical Society from 1900.

Old Block. *See* Delano.

Omni Gazetteer *of the United States of America*, ed. Frank R. Abate (Detroit: Omnigraphics, 1991). This is a printed compilation of data from the GNIS (*q.v.*), in 11 vols.; California is in vol. 9.

O'Neal, Lulu Rasmussen, *A peculiar piece of desert: The history of Califonia's Morongo Basin* (Los Angeles: Westernlore Press, 1957).

O'Neill, Owen H., provided information to E. G. Gudde. When followed by a page number, the reference is to the *History of Santa Barbara County* (Santa Barbara, 1939), of which O'Neill was the editor-in-chief.

Ord, Lt. E. O. C. (1818–83). The reports on Lt. Ord's reconnaissance in southern California in the fall of 1849 are found in 31st Cong., 1st sess., Senate Ex. Doc. no. 47, pt. 1, pp. 119 ff. His "Topographical sketch of the Los Angeles plains and vicinity, August, 1849," is attached to the same document. **Ord's Map**: "Topographical sketch of the gold and quicksilver district of California, July 25th, 1848," accompanying the President's Message to Congress, 30th Cong., 2d Sess., House Ex. Doc. no. 1 (Wheat, no. 54.)

Orography; Orographic. The phase of physical geography referring to mountains and hills.

Oswalt, Robert, of Berkeley, Calif., is a specialist in Kashaya Pomo and in the Pomo languages generally, who provided much useful information. He is the author of *Kashaya texts* (UCPL 36, 1964) and two articles on place names of Pomo origin—"Gualala," *Names* 8:57–58 (1960), and "Ukiah," in *American Indian and Indoeuropean studies: Papers in honor of Madison S. Beeler*, ed. Kathryn Klar et al., 183–90 (The Hague: Mouton, 1980).

Pac. R.R. *Reports*. *Reports of explorations and surveys to ascertain the most practicable and economical route for a railroad from the Mississippi River to the Pacific Ocean, made under the direction of the Secretary of War in 1853–4* (12 vols. in 13; Washington, 1855–69). The sections of interest for California's nomenclature are found chiefly in vols. 2, 3, 5, and 11 (pt. 2 of which contains many maps). The survey not only named many features in the territory covered but also rescued a number of local Indian place names. *See* Beckwith; Blake; Parke-Custer map; Whipple; Williamson.

Paden, Irene D., and Margret E. **Schlichtmann**, *The Big Oak Flat Road* (San Francisco: Privately printed, 1955).

Palmer, Theodore Sherman, *Place names of the Death Valley region in California and Nevada* (Morongo Valley: Sage Brush Press, 1980). Palmer was a member of the Death Valley Expedition of 1891; in 1938 he returned to the area and gathered materials on place names. His work was not published until 1948 and was available only in a limited edition until the 1980 reprint.

Palou, Padre Francisco, was one of the outstanding missionaries and the author of several important works relating to early California history. When his name alone is cited, the reference is to *Historical memoirs of New California by Fray Francisco Palou*, ed. Herbert E. Bolton (4 vols.; University of California Press, 1926). **Palou, Vida . . . Serra**: *Relación histórica de la vida del venerable Padre Fray Junípero Serra* (Mexico, 1787). **Palou's diary**: in Bolton, *Anza*, 1:383–405.

Panamint. A tribe and a language, also called Tümpisa, closely related to the Shoshone language of Nevada within the Numic branch of the Uto-Aztecan language family, found in the Death Valley area; *see* Dayley.

Pantoja y Arriaga, Juan, wrote a journal in 1782 which is in the "Archivo General de Mexico, California," vol. 35 (typewritten copy in California Historical Society). Pantoja was pilot of the *Princesa* in the expedition of Estéban José Martínez, which brought supplies from San Blas to the presidios of California in 1782. He made several maps. One of his maps of San Diego harbor is in *CHSQ* 9:1 (Mar. 1930).

Parke, Lt. John G., of the Topographical Engineers, assisted Lt. R. S. Williamson in his surveys in California for a railroad from the Mississippi River to the Pacific Ocean (*see* Williamson.) His reports concerning California are in Pac. R.R. *Reports*, vol. 7. **Parke-Custer map**: *From San Francisco Bay to the plains of Los Angeles, from explorations and surveys made . . . by Lieut. John G. Parke . . . 1854 and 55*, constructed and drawn by H. Custer. It is in Pac. R.R. *Reports* 11:2.

Parker, Horace, *Anza-Borrego Desert guide book* (Palm Desert: Desert Magazine Press, 1957).

Parsons, Mary E., *The wild flowers of California* . . . (San Francisco: Cunningham, Curtis, and Welch, 1909).

Pattie, James O. (1804–50), was in California with a group of trappers in 1828–29; he published his *Personal narrative* in Cincinnati, 1833 (*see* Bancroft 3:162–72).

Payne, Roger L., executive secretary of the U.S. Board on Geographical Names in Reston, Virginia, kindly provided valuable information on California place names.

Peñafiel, Antonio, *Nomenclatura geográfica de México: Etimologías de los nombres de lugar correspondientes a los principales idiomas que se hablan en la república* (3 vols. in 1; Mexico, 1897).

Pico, José Ramón, "Documentos para la historia de California: Papeles originales, 1781–1850" (3 vols.; MSS, Bancroft Library).

Piña, Joaquín, report of an expedition to the "Valle de San José" in May 1829 (original MS, Bancroft Library.)

Pinart, Alphonse, a French scientist, recorded Costanoan, Chumash, and Esselen words and phrases in 1878 from surviving natives at several former missions. The vocabularies, edited by R. F. Heizer, are published in *Anthropological Records* 25:1 (Berkeley and Los Angeles: University of California Press, 1952).

Pitkin, Harvey, *Wintu dictionary* (Berkeley and Los Angeles: University of California Press, 1985).

Plano . . . de San José. "Plano topográfico de la Misión de San José." An extraordinary little map, showing the region east of San Francisco Bay as far as the site of Stockton. It was probably made in the late 1820s, although it is dated 1824, by another hand (MS, Bancroft Library).

Pomo (pō′ mō). A group of Indian tribes and languages, sometimes classified in the Hokan linguistic stock, spoken in Sonoma, Lake, and Mendocino Cos. *See* Oswalt.

Popular etymology. *See* Etymology.

Portolá, Gaspar de (1723–86), a Catalonian and captain of dragoons, in 1769 led the expedition from Old (Baja) California that was to found the first Spanish settlement in what is now the state of California. A detachment of this first land expedition discovered San Francisco Bay and bestowed many place names along the route, several of which are still in use. Unless otherwise indicated, the reference is to the "Diary of Gaspar de Portolá during the California Expedition of 1769–1770" (in both Spanish and English), ed. Donald Eugene Smith and Frederick J. Teggart, *APCH:P* 1:31–89. *See* Costansó; Crespí; Eldredge; Fages; Vila.

Postal, Mitch, of the San Mateo County Historical Association, kindly provided information on his area.

Postal Guide. The dates of the establishment and naming of post offices are taken from the U.S. Official Postal Guide, published from 1878 by the Post Office Dept. Dates of post offices established before 1878 are taken from guides published by commercial firms or from lists published in newspapers and almanacs. For some post offices we can give the exact date of their establishment; these dates were supplied by the Post Office Dept. years ago, when federal offices were less overworked. All the dates can now be found in Frickstad.

Powers, Stephen (1840–1904), *Tribes of California* (Washington: Dept. of the Interior, 1877; reprinted, Berkeley and Los Angeles: University of California Press, 1976). Valuable for Indian nomenclature, and, until the publication of Kroeber's *Handbook of the Indians of California*, the best authority on California Indians.

Preuss, Charles (1803–54), a native of Germany, was Frémont's topographer on the first two expeditions. For his maps, *see* Frémont-Preuss maps.

Prov. Recs. "Provincial Records, 1775–1822" (12 vols. in 5). *See* Archivo de California.

Pseudo-Spanish. The term is used to describe the "Spanish" which is used in place naming by Americans without regard for Spanish grammar or usage (*see* Diament, Henri). The term does not imply censure or criticism. Since there are no standards governing the application of names, a person has a perfect right to apply a name that may be ungrammatical in form but pleasing in sound.

PSP. "Provincial State Papers, 1767–1822" (22 vols. in 14). *See* Archivo de California.

PSP Ben. "Provincial State Papers, Benicia" (2 "volumes"; pt. 1 of Archives of California, Miscellany).

PSP Ben. Mil. "Provincial State Papers, Benicia, Military, 1767–1822" (52 vols. in 3). *See* DSP Ben. Mil.; Archivo de California.

PSP Pres. "Provincial State Papers, Presidios" (2 vols. in 1; transcripts, Bancroft Library). *See* Archivo de California.

Quadrangle. The USGS, in preparing the topographical atlas, divided the surface of the United States into squares, called quadrangles. The atlas sheet covering such a square is sometimes also called a quadrangle. For a detailed listing and discussion of the California quadrangles, consult the Helm Index.

Railroad Maps. The reference is to the maps attached to the annual reports of the principal railroad companies, particularly the Southern Pacific and the Santa Fe. **Official Railway Map** refers to the map formerly published by the state railroad commissioners of California.

Railroad names. Names given by railroads to sidings, loading points, etc., may be found on maps, but the origins are often obscure. These names tend to lack connections with local history. They are arbitrary and sometimes fanciful, but usually easy to spell and to pronounce; an example is Armona.

Ramey, Earl, *The beginnings of Marysville* (San Francisco: California Historical Society, 1936).

Rawlins, Susan, of the Willows Public Library, has kindly given information on names of Glenn Co.

Reading, Pierson B. (1816–68), prepared a manuscript map of the Sacramento Valley in 1849, from actual observations. The original is in the nomenclature file of the California State Library in Sacramento; it is reproduced in Boggs, opposite p. 24. *See* Reading in text.

Rector, Carol, of the San Bernardino County Museum, provided valuable information on her area.

Registro de Sitios, Fierros y Señales . . . (Register of places, branding irons, and [ear] marks [used in Alta California]) is pt. 2 of Archives of California, Miscellany.

Reid, Hiram A., *History of Pasadena* . . . (Pasadena, 1895). One of the few local histories in which place names are properly treated.

Rensch-Hoover, *Historic spots in California*. This important reference work, compiled by Hero E. Rensch, Ethel G. Rensch, and Mildred Brook Hoover, was originally published by Stanford University Press in three separate volumes: *The southern counties* (1932); *Valley and Sierra counties* (1933); *Counties of the Coast Range* (1937). Several revised editions have been published. Citations refer to the original volumes.

Rep. Sur. Gen. *Annual* [sometimes *Biennial*, or *Statistical*] *report of the surveyor-general of the state of California*, published by the state. The early reports, especially that of 1856, are sources of value for names in certain areas.

Reveal, Arlene H., of the Mono County Library, provided valuable information on her area.

Ricard, Herbert F., *Place names of Ventura County* (Ventura County Historical Society, *Quarterly* [1972] 17:2).

Ringgold, Cadwalader, was a lieutenant-commander in command of the *Porpoise* on the Wilkes expedition. In Aug.–Sept. 1841, he was in command of a detachment that explored the Sacramento Valley. Sutter and Marsh supplied him with the names of a number of important places (American, Feather, Cosumnes, and Mokelumne Rivers, Mount Diablo, etc.), which then became definitely established. The charts referred to are in *A series of charts, with sailing directions* . . . , published in Washington in 1851. These charts, based on Ringgold's surveys of 1850, are important sources for geographical names of the San Francisco Bay district. **Ringgold, *General chart*** is *General chart embracing surveys* . . . (of various waters inland from the Golden Gate as far as Sacramento and Stockton).

Ritchie, Robert Welles, *The hell-roarin' forty-niners* (New York: Sears, 1928).

Robertson, Mary, of the Yuba County Library, kindly provided information on her area.

Robins, Robert H., professor of linguistics at the University of London, is the author of *The Yurok language* (Berkeley and Los Angeles: University of California Press, 1958).

Robinson, Douglas. The last name alone refers to Douglas Robinson, who prepared a valuable compilation of names in Inyo National Forest, a typewritten copy of which was made available by the USFS.

Robinson, W. W., *Ranchos become cities* (Pasadena: San Pasqual Press, 1939).

Rockwell, John. The reference is to names used by John Rockwell, of the Coast Survey, in his reports.

Rodríguez, Sebastián. Two diaries, Apr. 2 to May 6, and May 26 to June 22, 1826, record expeditions to capture runaway Indians from San Miguel and San Juan Bautista (MSS, Cowan Collection, Bancroft Library).

Rodríguez Cermeño, Sebastián, a pilot sent out by the viceroy of New Spain, made a reconnaissance along the coast in 1595, as far north as latitude 42°, in the interest of the Philippine galleons, which frequently came to grief in returning from Manila to Acapulco.

Rogers, Fred Blackburn, *Montgomery and the Portsmouth* (San Francisco: Howell, 1958). Where the name is given alone, the information was supplied by Colonel Rogers directly.

Rowntree, Lester, *Flowering shrubs of California* (Stanford: Stanford University Press, 1939).

Rozynsky, Brita, of the Nevada County Local History Library, provided valuable information on her area.

Rumsen was the name of the tribe and language native to the Monterey Bay area, belonging to the Costanoan language family.

Rurik. August C. Mahr, ed., *The visit of the "Rurik" to San Francisco in 1816* (Stanford: Stanford University Press, 1932). The book presents translations of extracts in convenient form from the reports and diaries of the members of the first Kotzebue expedition, which was in San Francisco Bay from Oct. 2 to Nov. 1, 1816 (*see* Chamisso; Kotzebue).

Russell, Carl Parcher, *One hundred years in Yosemite* . . . (Berkeley and Los Angeles: University of California Press, 1947).

Russell, Teresa, of the Fairfield-Suisun Community Library, kindly provided information on Solano Co.

Samson, Karri, of the Placer County Library, provided valuable information on her area.

Sanchez, Nellie van de Grift, *Spanish and Indian place names of California* (San Francisco: Robertson, 1914; 3d ed., 1930). Some good information behind a thick veil of sentimentality.

San Diego Arch. *See* Hayes, C and R.

San Diego County. *See* Stein.

San Luis Obispo County. *See* Hall-Patton.

San Mateo County. *See* Brown, Alan K.

Santa Clara County. *See* Arbuckle.

Santa Cruz County. *See* Clark.

Santa Fe. The reference is to "History of the Santa Fe coast lines," a compilation by the Atchison, Topeka and Santa Fe Railway system, a typewritten copy of which was placed at my disposal through the kindness of Lee Lyles and E. G. Ryder.

Santamaría, Francisco J., *Diccionario de Mejicanismos* (México: Porrúa, 1959). The standard work on the vocabulary of Mexican Spanish, including many terms used in place names.

Sapir, Edward (1884–1930), was a pioneering linguistic anthropologist who studied many Native American languages; his *Southern Paiute, a Shoshonean language* (Boston, 1930) has been useful for tracking etymologies from Numic languages of California.

Saunier, Mary Lou, of the Del Norte County Museum, has kindly provided information on names in Del Norte Co.

Sawyer, Jesse O., *English-Wappo vocabulary* (UCPL 43, 1965).

Scancarelli, Janine, a scholar of the Cherokee language, kindly provided information on transfer names of Cherokee origin.

SCB. The *Sierra Club Bulletin*, published from 1893 in San Francisco.

Schaeffer, L. M., *Sketches of travel in South America, Mexico, California* (New York, 1860).

Schoolcraft, Henry R., *Archives of aboriginal knowledge* (6 vols.; Philadelphia, 1860). Vol. 3 of this imposing work contains George Gibbs's "Journal" and "Observations," plus other articles with valuable information on Indian place names in California (*see* Gibbs).

Schrader, George, of Shasta National Forest, provided information to E. G. Gudde.

Schulz, Paul E., *Stories of Lassen's place names* (Mineral: Loomis Museum Association, 1949). A satisfactory regional study done "in the field."

Scott, Reva. *See* Brannan.

SCP:P. Society of California Pioneers, *Publication*, published annually, in San Francisco, from 1941. **SCP:Q**: *Quarterly* of the Society of California Pioneers, published from 1924 to 1933, in San Francisco.

Serr, Eugene F., of Red Bluff, kindly provided information on names of Tehama Co.

Serra. *See* Junipero Serra Peak in text.

Shafer, Robert, "The pronunciation of Spanish place names in California," *American Speech* (1942) 17:239–46, is unique for its period as a linguistically reliable publication on the topic.

Shasta County. *See* Steger.

Shepherd, Alice (Schlichter), kindly provided information on Wintu and Yuki. She is the author of *Wintu dictionary* (Berkeley: Dept. of Linguistics, Survey of California and Other Indian Languages, 1981) and co-author with Jesse O. Sawyer of *Yuki vocabulary* (Berkeley and Los Angeles: University of California Press, 1984).

Sherer, Lorraine M. "The name Mojave, Mohave," *Southern California Quarterly* 49:1–36, 455–58 (1967).

Sherman, H. L. (comp.), *A history of Newport Beach* (Los Angeles, 1931). An excellent monograph on the development of a modern California community.

Shipley, William, professor of linguistics at the University of California, Santa Cruz, is the leading authority on California Indian languages of the Maidu family; he has published *Maidu texts and dictionary* (Berkeley and Los Angeles: University of California Press, 1963); and, with H. J. Uldall, *Nisenan texts and dictionary* (Berkeley and Los Angeles: University of California Press, 1966).

Sierra Nevada. *See* Browning; Sowaal.

Simpson, Sir George, *Narrative of a journey round the world . . . 1841 and 1842* (2 vols.; London, 1847).

Sitio. *See* entry in the text.

Smith, Jedediah (1799–1831), was the leader in 1826 of the first party of whites to enter California after crossing the deserts west of the Rocky Mountains, and in 1827 was the first to cross the Sierra Nevada, which he did from west to east. He gave names to many features in the state, particularly on his trip in 1827–28, but none of these seems to have survived. However, the name of the great trailblazer is commemorated in Smith River. **Smith, Travels**: *See* Sullivan.

Smith, Wallace. *Garden of the sun* (3d ed.; Los Angeles: Lymanhouse, 1956).

Solomons, Theodore S., born in 1870 in San Francisco, was an explorer of the High Sierra and a man of many accomplishments. In the 1890s he named the peaks of the Evolution Group and numerous other features in the Sierra Nevada.

Southern California Quarterly. *See HSSC:P.*

Southern Pacific. Information about railroad stations was received from the office of the Southern Pacific Company in San Francisco.

Sowaal, Marguerite, *Naming the Eastern Sierra* (Bishop: Chalfant Press, 1985). Ms. Sowaal also gave valuable information on Inyo and Mono Cos. in a personal communication.

Spanish times. The phrase is used in the book to designate the period from 1769 to 1846, when Spanish was the official language of what is now the state of California. When used in a political sense, it designates the period from 1769 to 1822, before California became a Mexican province.

Sparkman, Philip S., *The culture of the Luiseño Indians* (AAE 8:187–234).

SP Ben. Misc. "State Papers, Benicia, Miscellaneous, 1773–1829" (Pt. 4 of Archives of California, Miscellany).

Specific name. *See* Generic name.

SP Mis. "State Papers, Missions, 1785–1846" (11 vols. in 2; transcripts, Bancroft Library).

SP Mis. and C. "State Papers, Missions and Colonization, 1787–1845" (2 vols.). *See* Archivo de California.

SP Sac. "State Papers, Sacramento, 1770–1845" (19 vols. in 3). *See* Archivo de California.

Stang, Janet Lohmeyer, *A study of lake names in northwestern California* (M.A. thesis; San Francisco State University, 1974).

Stanger, Frank M., *Peninsula community book* (San Mateo: Cawston, 1946).

Stark, Julie, of the Sutter County Museum, kindly provided information on her area.

State Library. The reference indicates that the item was taken from the nomenclature file of the California State Library in Sacramento.

State Surveyor General. *See* Surveyor General.

Statutes. The *Statutes* of California, published by the state since 1850, contain many references to place names, including the official naming of places by an act of the legislature.

Steger, Gertrude A., *Place names of Shasta County* (Redding, 1945). One of the better regional studies in California. The reference indicates that the information was taken from her book or supplied by Mrs. Steger directly. A revised edition of her book was edited by Helen H. Jones (Glendale: La Siesta Press, 1966).

Stein, Lou, *San Diego County place-names* (San Diego: Tofua Press, 1975).

Stephens, Stanley D., librarian emeritus, University of California, Santa Cruz, provided valuable information on Santa Cruz and Monterey Cos.

Stephens-Murphy-Townsend party. *See* Stevens-Murphy-Townsend party.

Stephenson, Terry E., "Names of places in Orange County," *Orange County History Series* 1:45–46 (1931), 2:107 ff. (1932). The name Stephenson without a page number indicates that Mr. Stephenson supplied the information to the State Geographic Board.

Stevens-Murphy-Townsend party. The party was led by Elisha Stevens (1804–84) and consisted of about fifty persons. They came to California in 1844 via Fort Hall and the Humboldt River and were the first party to bring wagons across the Sierra Nevada to Sutters Fort.

Stevenson's Volunteers. The regiment of New York Volunteers under the command of Jonathan D. Stevenson came round the Horn to California, arriving in Mar.–Apr. 1847. It was too late to participate in military operations, but many of its members became outstanding pioneers, and their names are frequently mentioned in this book.

Stewart, George R. The name Stewart followed by a page number refers to George R. Stewart, *Names on the land*, 4th ed. (San Francisco: Lexicos, 1982). Citations of Stewart 1970 refer to his *American place-names* (New York: Oxford, 1970). The name alone indicates that the information was received from Prof. Stewart directly. George R. Stewart is not to be confused with George W. Stewart, one of the men responsible for the creation of Sequoia National Park in 1890.

Still, Elmer G., "Livermore history highlights," *Livermore Herald*, Oct.–Nov. 1936. Excellent sketches on place names by the city clerk of Livermore.

Sullivan, Maurice S., *The travels of Jedediah Smith . . .* (Santa Ana: Fine Arts Press, 1934). This contains a transcript of Smith's journal of his second trip to California (1827) and reproduces David H. Burr's map of 1839, which shows Smith's routes.

Surveyor General. The chief geodetic officer of the state, as well as the surveyor sent by the federal government in connection with land-grant cases and the public surveys, bore this title in California after 1851. If preceded by "U.S.," reference is to the federal officer, whose reports were published in the *Report of the Commissioner of the General Land Office;* otherwise the reference is to the *Annual report* of the state officer, published as a document of the California State Assembly. The report of the state officer—particularly for the year 1855, published in 1856—contains valuable information about place names.

Sutter, John A. (1803–80), born in Kandern, Germany, of Swiss lineage, settled in the Sacramento Valley in 1839. In the years before the discovery of gold he was perhaps the most important and most colorful character in California. He applied a number of names and suggested others to Emmons and Ringgold. The references are to the "Reminiscences" dictated by Sutter to Bancroft in 1876 (MS, Bancroft Library). See *New Helvetia diary.*

Tahoe, Lake. *See* Lekisch.

Takic. A subgroup of the Uto-Aztecan family of American Indian languages, comprising several languages of southern California: Luiseño, Juaneño, Cupeño, Cahuilla, Serrano, Kitanemuk, and Gabrielino.

Tanner, H. S., "A map of the United States of Mexico . . ." (3d ed., 1846; Wheat, no. 32). A fair map, based on Wilkes's and Frémont's reports.

Tassin, J. B., "A newly constructed map of the state of California" (San Francisco, 1851; Wheat, no. 208).

Tautology; Tautological. A philological expression for the repetition of the same term in a different form or in a different language. Thus Picacho Peak [Imperial Co.] is literally 'Peak Peak', since *picacho* is a Spanish word for 'peak'.

Taylor, Alexander S. (1817–76), "The Indianology of California," published in four series in the *California Farmer and Journal of Useful Sciences* between Feb. 22, 1860, and Oct. 30, 1863.

Taylor, Bayard (1825–78). Unless otherwise stated, the reference is to the 1855 edition of Taylor's *Eldorado, or Adventures in the path of empire* (New York: Putnam).

Teather, Louise, *Place names of Marin* (San Francisco: Scottwall, 1986). Ms. Teather also kindly provided information by personal communication.

Teeter, Karl, "Notes on Humboldt County, California, place names of Indian origin," *Names* 6:55–56 (1958), 7:126 (1959).

Terstegge, Mary Anne, of the Tulare County Library, kindly provided information on her area.

Tinkham, George H., "From the pages of time—twenty years ago," *Stockton Daily Record*, Jan. 28 to Apr. 6, 1935.

Tognazzini, Wilmar, of Morro Bay, provided much useful information on San Luis Obispo Co.

Tolowa (tol′ ə wə). An Athabaskan tribe and language of Del Norte Co.; also called the Smith River tribe and language.

Topographical atlas. The preparation of a topographical atlas of the United States is one of the major activities of the USGS. *See* Atlas sheet; Helm Index; Quadrangle.

Toponymy. The term designates the geographical names of a certain region or of a certain language.

Tout, Otis B., *The first thirty years . . . history of Imperial Valley . . .* (San Diego, 1931).

Towendolly, Grant, folklorist and etymologist of the Wintu Indians. His information on place names of the Shasta region was communicated through the courtesy of George Schrader and Gertrude Steger.

Township. *See* entry in the text.

Transfer names refers to names transplanted to California from other areas, such as Pittsburg (from the city in Pennsylvania).

Trask, John B. (1824–79), was the author of two important works. **Trask's map** refers to his "Map of the state of California" and "Topographical map of the mineral districts of California . . . ," both published in 1853 and both reproduced in Wheat (nos. 246, 247). ***Trask, Report*** refers to his *Report on the geology . . .* , California State Assembly Document no. 9 (1854) and Senate Document no. 14 (1855). Trask was a member of the Mexican Boundary Survey and the first geologist of the State of California.

Treutlein, Theodore E., "Los Angeles, California: The question of the city's original Spanish name," *Southern California Quarterly* (1973) 55:1–7.

Triangulation. A surveyor's term for the process of determining the triangles into which any portion of the surface of the earth is divided in a trigonometrical survey.

Trinity County. *See* Jones, Alice Goen.

Turner, Dennis W., *Place names of Humboldt County, California* (Orangevale: The author, 1993). Mr. Turner also provided valuable information by personal communication.

Tyson, Philip T., *Geology and industrial resources of California . . . including the reports of Lieuts. Talbot, Ord, Derby, and Williamson, of their explorations in California and Oregon* (Baltimore, 1851). This was the first scientific report after the discovery of gold, based on personal investigations in 1849; it was first published in 31st Cong., 1st sess. (1850), Senate Ex. Doc. no. 47. Among the maps accompanying it are Tyson's *Geological reconnoissances in California* (cited in text as Tyson 1850) and Williamson's map of Warner's route (*see* Williamson).

UCPH. University of California Publications in History (Berkeley and Los Angeles: University of California Press, 1911–).

UCPL. University of California Publications in Linguistics (Berkeley and Los Angeles: University of California Press, 1940–). The many volumes of Indian texts and vocabularies contain a number of Indian place names.

USFS refers to the U.S. Forest Service. The regional office in San Francisco and many of its supervisors and rangers cooperated splendidly.

USGS refers to the U.S. Geological Survey, established as an office under the Dept. of the Interior on Mar. 3, 1879. *See* Atlas Sheet; Quadrangle; Topographical Atlas.

Uto-Aztecan. An American Indian language family of the western United States and Mexico, taking its name from the Ute tribe of the Great Basin and the Aztecs of Mexico. The family also includes the Numic languages, spoken in the Great Basin (including eastern California), and in southern California such languages as Cahuilla, Luiseño, and Gabrielino.

Vallejo, Mariano Guadalupe (1807–90), a native of Monterey, was military commander and director of colonization at the northern frontier after 1835 and was appointed *comandante general* of Alta California in 1836. Gen. Vallejo played an important role in the last decade of Mexican rule and continued his political career after California became part of the Union. **Vallejo Docs.**: "Documentos para la historia de California, 1713–1851" (36 vols., originals, Bancroft Library). **Vallejo's Report**: At the time of the creation of the original counties in the winter of 1849–50, Vallejo was chairman of the "Select Committee [of the state Senate] on the derivation and definition of the names of the several counties of the state of California." His report was first published as an appendix to the Senate Report, 1850, and was republished at irregular intervals in the editions of the *Blue Book*. In the 1958 edition, Vallejo's and Maslin's articles on the names of the counties of California were replaced by a new account written by Erwin G. Gudde.

Vancouver, George (1758–98), visited California three times between 1792 and 1794 while in command of a British exploring expedition. He was the first non-Spanish navigator after Francis Drake to bestow names on features along the coast, most of which have survived. The references are to *A voyage of discovery to the North Pacific Ocean and round the world* . . . and to the atlas accompanying this report. Unless otherwise stated, the six-volume edition published in London in 1801 has been used.

Venegas, Miguel (1680–1764), *A natural and civil history of California*, translated from the original Spanish (2 vols.; London, 1759). This was first published in Madrid in 1757.

Ventura County. *See* Ricard.

Verstegge, Mary Anne, of the Tulare County Library, provided valuable information on her area.

Viader, Padre José, wrote diaries of two expeditions to the San Joaquin River in Aug. and Oct. 1810; transcripts are in Arch. MSB 4: 73–94.

Vila, Vicente, "The Portolá expedition of 1769–1770: Diary of Vicente Vila," ed. Robert S. Rose (in both Spanish and English), *APCH:P* (1911) 2:1–119.

Vischer, Eduard, "A trip to the mining regions in the spring of 1859," in *CHSQ*, vol. 11.

Vizcaíno, Sebastián (1548–1629), was in command of a Spanish expedition (1602–3) sent north by the viceroy of New Spain to find a port where the galleons returning from the Philippines could stop in case of distress. Vizcaíno discovered and named Monterey Bay, and recommended it as a suitable port of refuge. The expedition, however, was the last for a long time, and it was not until 1769 that Portolá was instructed to find the harbor of Monterey. Vizcaíno's voyage is of great interest to the student of place names because, for the sake of prestige, he endowed many important coastal features with names, even those that had been named previously. Since no other navigator followed to replace Vizcaíno's names, most of these have survived to the present day. Father Antonio de la Ascensión's account of Vizcaíno's voyage, translated and edited by H. R. Wagner, is in *CHSQ* 7:295–394 (1928), 8:26–70 (1929).

Von Leicht, Ferdinand, and A. **Craven**, "Map of California and Nevada," 1873. The reference is to the edition of 1874, revised by John D. Hoffmann and A. Craven, and issued by the University of California. It is the official map of the Whitney Survey and was for many years the best map of the entire state.

Von Leicht, Ferdinand, and John D. **Hoffmann**, "Topographical map of Lake Tahoe . . ." This bears the date 1874 but was copyrighted in 1873. Amos Bowman's map of 1873 (*q.v.*) has in general the same nomenclature. The von Leicht–Hoffmann map is obviously based on an original survey, but it is uncertain whether it is responsible for new names that appeared at the same time on the Bowman map.

Von Schmidt, Allexey W., "Map of the eastern boundary of the state of California," based on his survey for the U.S. General Land Office, to establish the California-Nevada boundary in 1872–73. Von Schmidt's line south of Lake Tahoe has since been corrected. Von Schmidt (1821–1906), a German from the Baltic provinces, was one of the most prominent of the pioneer civil engineers after 1850. Besides the boundary survey, his greatest achievements were the initiation of San Francisco's water supply system and the spectacular removal of Blossom Rock.

Von Schmidt, Julius, and C. C. **Gibbs**. "Map of California and Nevada," 1869.

Wagner, Henry Raup (1862–1957), was for many years the leading authority on the geographical history of the northwest coast of America and published the results of his research in numerous monographs and articles. Unless otherwise stated, the reference is to his *Cartography of the northwest coast of America to the year 1800* (Berkeley: University of California Press, 1937). This monumental work was published in two volumes, but its pages are numbered consecutively. **Wagner, no.** refers to the "List of Maps" (pp. 273 ff.); **Wagner, pl.**, to the maps reproduced in vol. 1. **Wagner, Saints' names** refers to "Saints' names in California," *HSSC:Q* (1947) 29:49 ff.

Wahdan, Jo, San Benito County Library, provided valuable information on her area.

Wappo (wä′ pō). An Indian tribe and language around Mount St. Helena, probably belonging to the Yukian family. A vocabulary of the now-extinct language, including a number of Indian place names, was published by Jesse O. Sawyer in *UCPL* (1965) vol. 43.

Warner, Heidi, curator, Stanislaus County Museum, provided valuable information on her county.

Warren, Lt. G. K., "Map of the territory of the United States from the Mississippi to the Pacific Ocean," 1854–57, the general map of Pac. R.R. *Reports*, vol. 9, pt. 2.

Warren, Patricia A., *San Gabriel Valley community names* ([Arcadia:] Dept. of Arboreta and Botanic Gardens, County of Los Angeles, [1969]).

Washo (wä′ shō). A tribe and language, sometimes classified in the Hokan stock, found around Lake Tahoe; the name of the lake was taken from this language.

Waterman, T. T., "Yurok geography," AAE (1920) 16:177–314, is one of the most detailed studies published on place names in California native languages.

Weber, Francis J., *El Pueblo de Nuestra Señora de Los Angeles: An inquiry into early appelation* (Los Angeles: Plantin Press, 1958).

Weight, Harold O., *Lost mines of Death Valley* (Twentynine Palms: Calico Press, 1953).

Westways. The publication of the Automobile Club of Southern California repeatedly published articles on California geographical names, thanks to the interest of its late editor, Phil Townsend Hanna.

WF. *Western Folklore*, the journal of the California Folklore Society. Formerly *California Folklore Quarterly*.

Wheat, Carl I. (1892–1966), *The maps of the California gold region, 1848–1857* (San Francisco: Grabhorn Press, 1942). A beautiful book, containing information and reproductions of maps not easily accessible. Wheat's great historical-cartographical work, *Mapping the Transmississippi West, 1540–1861*, 5 vols. in 6, was published by the San Francisco Institute of Historical Cartography, 1957–63.

Wheeler, Lt. George M., *Report upon United States geographical surveys west of the one hundredth meridian* (Washington, 1889). **Wheeler Survey**: The survey was made between 1869 and 1873 by Lt. Wheeler of the Corps of Engineers, U.S. Army. Vol. 1, *Geographical report*, contains valuable information about names in certain sections of California. Appendix F, "Memoir upon the voyages . . . ," includes reproductions of old maps with notes by J. G. Kohl. **Wheeler atlas**: *Topographical atlas projected to illustrate geographical explorations and surveys west of the one hundredth meridian*.The work is important for east-central and southern California, sections for which the detailed maps of the Whitney Survey were not published.

Wheelock, Walt, owner of La Siesta Press in Glendale, California, supplied information on a large number of names from his intimate knowledge of the geography of the state.

Whipple, Amiel W., who, as a major general, died of wounds received in the battle of Chancellorsville, May 4, 1863, was a lieutenant of the U.S. Topographical Engineers engaged in California exploration in 1849, and again in 1853–54. **Whipple, Extract**: Reference is to the "Extract from a journal of an expedition from San Diego, California to the Rio Colorado . . ." (31st Cong., 2d sess., Senate Ex. Doc. no. 19). **Whipple, Report**: Reference is to Whipple's account in Pac. R.R. *Reports*, vol. 3. **Whipple-Ives map**: Map 2 of Pac. R.R. *Reports*, vol. 11, "From the Rio Grande to the Pacific Ocean." **Whipple's sketch** was published in 31st Cong., 1st sess., Senate Ex. Doc. no. 34 (1850). Cf. Grant Foreman (ed.), *A pathfinder of the Southwest . . . A. W. Whipple* (Norman: University of Oklahoma Press, 1941).

Whitney, Josiah D. (1819–96), was director of the State Geological Survey from 1860 to 1874. He and his assistants were responsible for the naming of many physical features, especially in the mountainous regions of the state (*see* Brewer, Hoffmann, and King, in this glossary; and Whitney, Mount, in the text). **Whitney, Geology**: the first volume of the report of progress of the survey, published by authority of the Legislature of California, 1865. **Whitney, Yosemite book**, with the year, refers to Whitney's *The Yosemite guide book;* cf. Edwin F. Brewster, *Life and letters of Josiah Dwight Whitney* (Boston: Houghton Mifflin, 1909). For the maps of the Whitney Survey, *see* Hoffmann, and von Leicht and Craven. Some of the correspondence and field notes of members of the survey are deposited in the Bancroft Library.

Wilkes, Charles (1798–1877), *Narrative of the United States exploring expedition . . . 1838 . . . 1842* (5 vols.; Philadelphia, 1845), describes the first U.S. expedition to explore the interior of California, in 1841. Its reports and maps are of great value for the history of the geographical delineation of California. The section dealing with California constitutes vol. 5, chaps. 5–6; **Wilkes's map**: this volume also includes Wilkes's "Map of Upper California . . . 1841," which gives a fair picture of the location of rivers of the great valley and the Sierra Nevada.

Williamson, Lt. Robert S., of the U.S. Topographical Engineers, was assistant to Capt. W. H. Warner in 1849, and in 1853 was a member of the Pacific Railroad Survey. His map of 1849 (Wheat, no. 182), sketching Warner's route, is in Tyson, *Geology;* his *Report* is in Pac. R.R. *Reports* 5:1. His map of 1853 (Wheat, no. 250) and the Williamson-Abbot map of 1855 (Wheat, no. 272) are in Pac. R.R. *Reports* 11:2.

Winchell, Lilbourne A., a native of Sacramento, was an outstanding explorer of the High Sierra after 1879 and was responsible for naming a number of features.

Windle, Ernest, *Windle's history of Santa Catalina Island . . .* (2d ed.; Avalon, 1940).

Wintun (win´ to͞on). The Wintun Indians, a large linguistic group sometimes classed in the Penutian stock, had their habitat along the entire length of the Sacramento River, from Siskiyou Co. to Suisun Bay. They are divided into three languages: Wintu, Nomlaki (Central Wintun), and Patwin. *See* Pitkin; Shepherd.

Wistar, Isaac Jones, *Autobiography . . . 1827–1905* (2 vols.; Philadelphia, 1914).

Witcher, Denis, of the El Dorado County Museum, provided valuable information on his county.

Wiyot (wē´ ot). A tribe and language of Humboldt Co. related to the Algonquian linguistic family (*see* Weott in the text).

Wolfe, Linnie Marsh, prepared twenty-six folders of notes and scraps on California history, deposited in the Bancroft Library by R. E. Wolfe in 1947.

Wood, B. D., *Gazetteer of surface waters of California*, USGS, Water-Supply Papers, nos. 295–97 (Washington, 1912–13). A list of all lakes and streams, based on the topographical atlas, the Land Office map, and the official county maps.

Wood, L. K., "Discovery of Humboldt Bay," *SCP:Q*, Mar. 1932 (*see* Gregg party).

Wood, Raymund F., was the author of "Anglo influence on Spanish place names in California," *Southern California Quarterly* (1970) 63:392–413, and *The saints of the California landscape* (Eagle Rock: Prosperity Press, 1987).

Work, John (1792–1861), *Fur brigade to the Bonaventura: John Work's California expedition, 1832–1833, for the Hudson's Bay Company*, ed. Alice B. Maloney (San Francisco: California Historical Society, 1945).

Workman, Boyle, *The city that grew* (Los Angeles: Southland Publishing, 1936).

Wrede, Herman, of Hollister, provided valuable information on San Benito Co.

WSP. Reference is to the Water Supply Papers published by the USGS. Paper no. 224 is Walter C. Mendenhall, *Some desert watering places in southwestern California and Southwestern Nevada* (Washington, 1909); no. 225, Mendenhall, *Ground waters of the Indio region, California desert* (1909); nos. 295–97, see B. D. Wood; no. 338, Gerald A. Waring, *Springs of California* (1915).

Wyatt, Roscoe D., *Names and places . . . in San Mateo County* (Redwood City: San Mateo County Title Co., 1936). With Clyde Arbuckle, Wyatt co-authored *Historical names, persons, and places in Santa Clara County* (San Jose Chamber of Commerce, 1948).

Wynn, Marcia Rittenhouse, *Desert bonanza* (Glendale: Clark, 1963).

Yokuts (yō´ kəts). The Yokuts Indians inhabited the San Joaquin Valley. A number of names in their own territory and in that of their eastern neighbors are of Yokuts origin. The term is not a plural; there is no such thing as a "Yokut." *See* Gamble.

Yosemite National Park. *See* Browning.

Yuki (yo͞o´ kē). A tribe and language of Mendocino Co. The relationship of the Yuki language to others remains uncertain. *See* Shepherd.

Yurok (yo͞or´ ok). A tribe and language of Humboldt and Del Norte Cos.; the language is related to those of the Algonquian family (*see* Robins). Yurok geography and place names have been carefully studied (*see* Waterman).

Zalvidea, Padre José María, wrote a diary of an expedition from Santa Barbara to the interior valley, July 19 to Aug. 14, 1806; it is in Arch. MSB 4:49–68. Anastasio Cabrillo's account of the same expedition is in the *San Francisco Evening Bulletin*, June 5, 1865.

Designer: Barbara Jellow
Compositor: Integrated Composition Systems, Inc.
Printer and Binder: BookCrafters